MODERN ELECTRONIC COMMUNICATION

Fifth Edition

Gary M. Miller

Professor, Division of Mathematics,
Science, and Technology

Monroe Community College
Rochester, New York

Prentice Hall
Englewood Cliffs, New Jersey   Columbus, Ohio

Library of Congress Cataloging-in-Publication Data

Miller, Gary M., 1941-
    Modern electronic communication / Gary M. Miller. — 5th ed.
       p.   cm.
    Includes index.
    ISBN 0-13-217879-6
    1. Telecommunication.   I. Title.
TK5101.M498 1996   95-30184
621.382—dc20   CIP

Cover Art: Michael Agliolo, International Stock
Editor: Charles E. Stewart, Jr.
Developmental Editor: Carol Hinklin Robison
Production, Design, and Photo Coordination: Proof Positive/Farrowlyne Associates, Inc.
Marketing Manager: Debbie Yarnell
Production Manager: Deidra M. Schwartz
Illustrations: Rolin Graphics, Inc.

This book was set in Times Roman by Black Dot Graphics and was printed and bound by Von Hoffmann Press. The cover was printed by Von Hoffman, Inc.

 © 1996 by Prentice-Hall, Inc.
A Simon & Schuster Company
Englewood Cliffs, New Jersey 07632

Printed in the United States of America

10 9 8 7 6 5 4 3 2 1

ISBN: 0-13-217879-6

Prentice-Hall International (UK) Limited, *London*
Prentice-Hall of Australia Pty. Limited, *Sydney*
Prentice-Hall of Canada, Inc., *Toronto*
Prentice-Hall Hispanoamericana, S. A., *Mexico*
Prentice-Hall of India Private Limited, *New Delhi*
Prentice-Hall of Japan, Inc., *Tokyo*
Simon & Schuster Asia Pte. Ltd., *Singapore*
Editora Prentice-Hall do Brasil, Ltda., *Rio de Janeiro*

Dedicated to my parents with thanks for all the
years of caring, love, and support.

Dorothy A. Miller

Reginald M. Miller

C O N T E N T S

PREFACE

Modern Electronic Communication, Fifth Edition, provides a comprehensive introduction to this rapidly expanding body of knowledge. I believe that other books for this market do not come close to the up-to-date coverage included here. Comprehensive and current coverage is not enough, however, if student comprehension is not also possible. This edition continues to make the changes that will allow students to read and learn from the textbook. Certainly instructor assistance and guidance is essential to the overall learning process, but this text will not end up being just the basis for course organization or source of homework assignments.

In preparing the five editions of *Modern Electronic Communication,* I have had the benefit of input from hundreds of users/reviewers. One thing learned from this feedback is the usage pattern. The vast majority of users cover Chapters 1–6, but after that no clear path exists. Most do not have the time to cover the entire book and therefore resort to a pick-and-choose arrangement for the remaining material. To a large extent these choices are influenced by the electronic industries in the geographic area.

A major addition to this edition is significant troubleshooting sections at the end of every chapter. I believe that students will be very interested in applying the knowledge gained in each chapter to some real-world problem solving. Equally important, employers of technology graduates and accrediting agencies have strongly encouraged a greater emphasis on troubleshooting techniques. These all-new sections will greatly facilitate the learning process for these vital skills.

Example 3-3 assumed that the receiver's AGC was operating at some fixed level based on the input signal's strength. As previously explained, the AGC system will attempt to maintain that same output level over some range of input signal. *Dynamic range* is the decibel difference between the largest tolerable receiver input signal (without causing audible distortion in the output) and its sensitivity (usually the minimum discernible signal). Dynamic ranges of up to about 100 dB represent current state-of-the-art receiver performance.

TROUBLESHOOTING

In this section we are going to analyze and troubleshoot the AM converter circuit. The converter, also known as an autodyne circuit, is a combination of the local oscillator and the mixer in a single stage. We're also going to discuss the power supply and audio amplifier problems in this section.

Upon completing this section you should be able to

*   Troubleshoot an AM converter circuit
*   Identify an open input circuit
*   Identify a dead or intermittent local oscillator circuit
*   Identify causes for a dead and intermittent local oscillator
*   Troubleshoot the receiver's power supply
*   Troubleshoot the receiver's audio amplifier

The Converter Circuit

In Sec. 3-3 it was shown that the local oscillator and the mixer play a very important part in AM reception. Figure 3-26 shows the converter stage (autodyne circuit) of an AM radio. The received RF input signal is fed into the base of Q1 from coils L1 and L2. The

NEW AND IMPROVED FEATURES

In addition to the many popular features of the previous edition, several new features, as well as improvements to existing features, have been incorporated into this fifth edition.

Distinctive full-color format used throughout

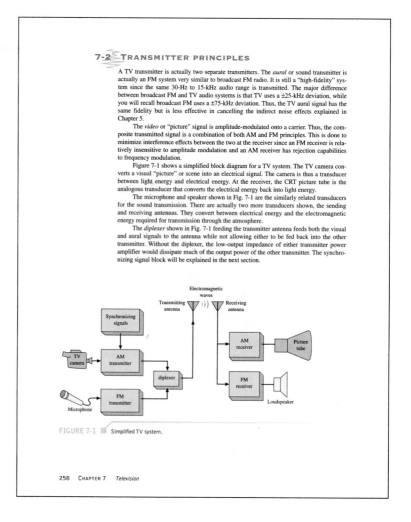

7-2 TRANSMITTER PRINCIPLES

A TV transmitter is actually two separate transmitters. The *aural* or sound transmitter is actually an FM system very similar to broadcast FM radio. It is still a "high-fidelity" system since the same 30-Hz to 15-kHz audio range is transmitted. The major difference between broadcast FM and TV audio systems is that TV uses a ±25-kHz deviation, while you will recall broadcast FM uses a ±75-kHz deviation. Thus, the TV aural signal has the same fidelity but is less effective in cancelling the indirect noise effects explained in Chapter 5.

The *video* or "picture" signal is amplitude-modulated onto a carrier. Thus, the composite transmitted signal is a combination of both AM and FM principles. This is done to minimize interference effects between the two at the receiver since an FM receiver is relatively insensitive to amplitude modulation and an AM receiver has rejection capabilities to frequency modulation.

Figure 7-1 shows a simplified block diagram for a TV system. The TV camera converts a visual "picture" or scene into an electrical signal. The camera is thus a transducer between light energy and electrical energy. At the receiver, the CRT picture tube is the analogous transducer that converts the electrical energy back into light energy.

The microphone and speaker shown in Fig. 7-1 are the similarly related transducers for the sound transmission. There are actually two more transducers shown, the sending and receiving antennas. They convert between electrical energy and the electromagnetic energy required for transmission through the atmosphere.

The *diplexer* shown in Fig. 7-1 feeding the transmitter antenna feeds both the visual and aural signals to the antenna while not allowing either to be fed back into the other transmitter. Without the diplexer, the low-output impedance of either transmitter power amplifier would dissipate much of the output power of the other transmitter. The synchronizing signal block will be explained in the next section.

FIGURE 7-1 ■ Simplified TV system.

Troubleshooting sections added at conclusion of all 17 chapters

Performance-based objectives added at the beginning of each chapter

More end-of-chapter questions and problems

Summaries added to the end of each chapter

Glossary of important terms added at the end of the book for easy reference

Completely revised and expanded
coverage of digital communications
including an additional chapter

CHAPTER

8

COMMUNICATIONS
TECHNIQUES

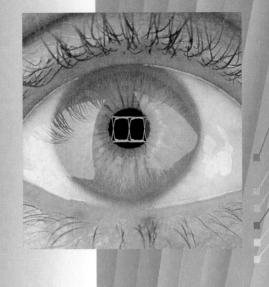

CHAPTER

9

DIGITAL
COMMUNICATIONS:
CODING
TECHNIQUES

CHAPTER

10

DIGITAL
COMMUNICATIONS:
TRANSMISSION

ANCILLARIES

An improved and expanded ancillary package is available to aid in the teaching and learning process. Supplements to the text include:

- Instructor's Solutions with Laboratory Manual Results

- Transparency Masters

- Test Item File

- Prentice Hall Custom Test

- Lab Manual by Mark E. Oliver

ACKNOWLEDGMENTS

I'd like to thank Carol Robison and Charles Stewart of Prentice Hall for all their help and direction. I'd also like to thank the following for their very helpful input to this edition: Jeff Beasley, New Mexico State University; James Fisk, Northern Essex Community College; James Graves, Indian River Community College; Ken Simpson, Stark Technical Institute; Charles Solie, New Mexico State University; Chuck Conner, Capitol College; Kevin Gray, DeVry Institute of Technology; Oleh Kuritza, College of DuPage; Bill Martin, Oklahoma State University; John McCarthy, DeVry Institute of Technology; and Mark E. Oliver, Monroe Community College. Last but not least, thanks to Rosita for her support and patience.

Gary Miller

MODERN ELECTRONIC COMMUNICATION

Fifth Edition

INTRODUCTORY TOPICS

OBJECTIVES

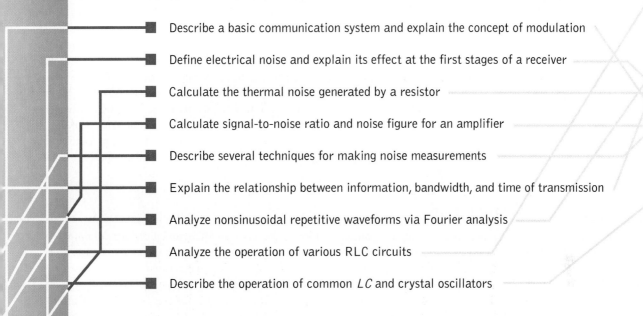

- Describe a basic communication system and explain the concept of modulation

- Define electrical noise and explain its effect at the first stages of a receiver

- Calculate the thermal noise generated by a resistor

- Calculate signal-to-noise ratio and noise figure for an amplifier

- Describe several techniques for making noise measurements

- Explain the relationship between information, bandwidth, and time of transmission

- Analyze nonsinusoidal repetitive waveforms via Fourier analysis

- Analyze the operation of various RLC circuits

- Describe the operation of common *LC* and crystal oscillators

1-1 INTRODUCTION

This book provides an introduction to all relevant aspects of communication systems. These systems had their beginning with the discovery of various electrical, magnetic, and electrostatic phenomena prior to the twentieth century. Starting with Samuel Morse's invention of the telegraph in 1837, a truly remarkable rate of progress has occurred. The telephone, thanks to Alexander Graham Bell, came along in 1876. The first complete system of wireless communication was provided by Guglielmo Marconi in 1894. Lee DeForest's invention of the triode vacuum tube in 1908 allowed the first form of practical electronic amplification and really opened the door to wireless communication. In 1948 another major discovery in the history of electronics occurred with development of the transistor by Shockley, Brattain, and Bardeen. The more recent developments, such as integrated circuits, large-scale integration, and computers on a single silicon chip, are probably familiar to you.

The rapid transfer of these developments into practical communications systems linking the entire globe (and now into outer space) has stimulated a bursting growth of complex social and economic activities. This growth has subsequently had a snowballing

Photo: Crystal oscillators. (Courtesy of MTI-Milliren Technologies, Inc. Photo by Kurt Brown Photography.)

effect on the growth of the communication industry with no end in sight for the foreseeable future. Some people refer to this as the age of communications.

The function of a communications system is to transfer information from one point to another via some communications link. The very first form of "information" electrically transferred was the human voice in the form of a code (i.e., the Morse code), which was then converted back to words at the receiving site. People had a natural desire and need to communicate rapidly between distant points on the earth, and that was the major concern of these developments. As that goal became a reality, and with the evolution of new technology following the invention of the triode vacuum tube, new and less basic applications were also realized, such as entertainment (radio and television), radar, and telemetry. The field of communications is still a highly dynamic one, with advancing technology constantly making new equipment possible or allowing improvement of the old systems. Communications was the basis origin of the electronics field, and no other major branch of electronics developed until the transistor made modern digital computers a reality.

Modulation

Basic to the field of communications is the concept of modulation. *Modulation* is the process of putting information onto a high-frequency carrier for transmission. In essence, then, the transmission takes place at the high frequency (the carrier) which has been modified to "carry" the lower-frequency information. The low-frequency information is often called the *intelligence signal* or, simply, the *intelligence*. It follows that once this information is received the intelligence must be removed from the high-frequency carrier—a process known as *demodulation*. At this point you may be thinking, why bother to go through this modulation/demodulation process? Why not just transmit the information directly? The problem is that the frequency of the human voice ranges from about 20 to 3000 Hz. If everyone transmitted those frequencies directly as radio waves, interference would cause them all to be ineffective. Another limitation of equal importance is the virtual impossibility of transmitting such low frequencies since the required antennas for efficient propagation would be miles in length.

The solution is modulation, which allows propagation of the low-frequency intelligence with a high-frequency carrier. The high-frequency carriers are chosen such that only one transmitter in an area operates at the same frequency to minimize interference, and that frequency is high enough so that efficient antenna sizes are manageable. There are three basic methods of putting low-frequency information onto a higher frequency. Equation (1-1) is the mathematical representation of a sine wave, which we shall assume to be the high-frequency carrier,

$$v = V_P \sin(\omega t + \Phi) \qquad\qquad (1\text{-}1)$$

where v = instantaneous value
    V_P = peak value
    ω = angular velocity = $2\pi f$
    Φ = phase angle

Any one of the last three terms could be varied in accordance with the low-frequency information signal to produce a modulated signal that contains the intelligence. If the amplitude term, V_P, is the parameter varied, it is called amplitude modulation (AM). If the frequency is varied, it is frequency modulation (FM). Varying the phase angle, Φ, results in phase modulation (PM). In subsequent chapters we shall study these systems in detail.

TABLE 1-1 *Radio-Frequency Spectrum*

Frequency	Designation	Abbreviation
30–300 Hz	Extremely low frequency	ELF
300–3000 Hz	Voice frequency	VF
3–30 kHz	Very low frequency	VLF
30–300 kHz	Low frequency	LF
300 kHz–3 MHz	Medium frequency	MF
3–30 MHz	High frequency	HF
30–300 MHz	Very high frequency	VHF
300 MHz–3 GHz	Ultra high frequency	UHF
3–30 GHz	Super high frequency	SHF
30–300 GHz	Extra high frequency	EHF

Communications Systems

Communications systems are often categorized by the frequency of the carrier. Table 1-1 provides the names for various frequency ranges in the radio spectrum. The extra-high-frequency range begins at the starting point of infrared frequencies, but the infrareds extend considerably beyond 300 GHz (300×10^9 Hz). After the infrareds in the electromagnetic spectrum (of which the radio waves are a very small portion) come light waves, ultraviolet rays, X rays, gamma rays, and cosmic rays.

Figure 1-1 represents a simple communication system in block diagram form. Notice that the modulated stage accepts two inputs, the carrier and the information (intelligence) signal. It produces the modulated signal, which is subsequently amplified before transmission. Transmission of the modulated signal can take place by any one of four means: antennas, waveguides, optical fibers, or transmission lines. These four modes of propagation will be studied in subsequent chapters. The receiving unit of the system picks up the transmitted signal but must reamplify it to compensate for attenuation that occurred during transmission. Once suitably amplified, it is fed to the demodulator (often referred to as the detector), where the information signal is extracted from the high-frequency carrier. The demodulated signal (intelligence) is then fed to the amplifier and raised to a level enabling it to drive a speaker or any other output transducer. A *transducer* is a device that converts energy from one form to another.

There are *two basic limitations* on the performance of a communications system: (1) electrical noise and (2) the bandwidth of frequencies allocated for the transmitted signal. Sections 1-2 to 1-5 are devoted to these topics because of their extreme importance.

1-2 NOISE

Electrical noise may be defined as any undesired voltages or currents that ultimately end up appearing in the receiver output. To the listener this electrical noise often manifests itself as *static*. It may only be annoying, such as an occasional burst of static, or continuous and of such amplitude that the desired information is obliterated.

Noise signals at their point of origin are generally very small, for example, at the microvolt level. You may be wondering, therefore, why they create so much trouble. Well, a communications receiver is a very sensitive instrument that is given a very small signal

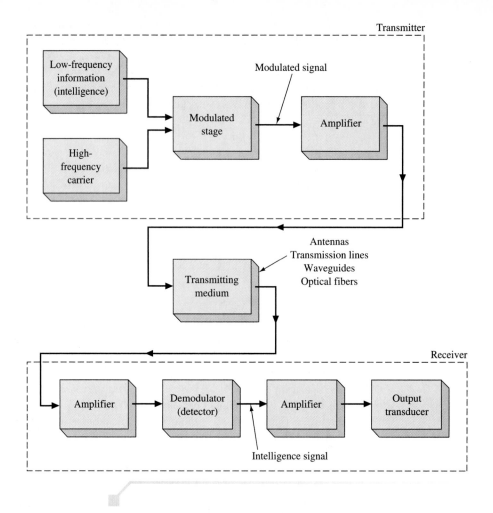

at its input that must be greatly amplified before it can possibly drive a speaker. Consider the receiver block diagram shown in Fig. 1-1 to be representative of a standard FM radio (receiver). The first amplifier block, which forms the "front end" of the radio, is required to amplify a received signal from the radio's antenna that is often less than 10 μV. It does not take a very large dose of undesired signal (noise) to ruin reception. This is true even though the transmitter output may be many thousands of watts, since when received it is severely attenuated. Therefore, if the desired signal received is of the same order of magnitude as the undesired noise signal, it will probably be unintelligible. This situation is made even worse since the receiver itself introduces additional noise.

The noise present in a received radio signal has been introduced in the transmitting medium and is termed *external noise*. The noise introduced by the receiver is termed *internal noise*. The important implications of noise considerations in the study of communications systems cannot be overemphasized.

External Noise

MAN-MADE NOISE   The most troublesome form of external noise is usually the man-made variety. It is often produced by spark-producing mechanisms such as engine ignition systems, fluorescent lights, and commutators in electric motors. This noise is actually "radiated" or transmitted from its generating sources through the atmosphere in the same fashion that a transmitting antenna radiates desirable electrical signals to a receiving antenna. This process is called *wave propagation* and is the subject of Chapter 13. If the man-made noise exists in the vicinity of the transmitted radio signal and is within its frequency range, these two signals will "add" together. This is obviously an undesirable phenomenon. Man-made noise occurs randomly at frequencies up to around 500 MHz.

Another common source of man-made noise is contained in the power lines that supply the energy for most electronic systems. In this context the ac ripple in the dc power supply output of a receiver can be classified as noise (an unwanted electrical signal) and must be minimized in receivers that are accepting extremely small intelligence signals. Additionally, ac power lines contain surges of voltage caused by the switching on and off of highly inductive loads such as electrical motors. It is certainly ill-advised to operate sensitive electrical equipment in close proximity to an elevator! Man-made noise is weakest in sparsely populated areas, which explains the location of extremely sensitive communications equipment, such as satellite tracking stations, in desert-type locations.

ATMOSPHERIC NOISE   *Atmospheric noise* is caused by naturally occurring disturbances in the earth's atmosphere, with lightning discharges being the most prominent contributors. The frequency content is spread over the entire radio spectrum, but its intensity is inversely related to frequency. It is therefore most troublesome at the lower frequencies. It manifests itself in the static noise that you hear on standard AM radio receivers. Its amplitude is greatest from a storm near the receiver but the additive effect of distant disturbances is also a factor. This is often apparent when listening to a distant station at night on an AM receiver. It is not a significant factor for frequencies exceeding about 20 MHz.

SPACE NOISE   The other form of external noise arrives from outer space and is called *space noise.* It is pretty evenly divided in origin between the sun and all the other stars. That originating from our star (the sun) is termed *solar noise.* Solar noise is cyclical and reaches very annoying peaks about every 11 years.

All of the other stars also generate this space noise, and their contribution is termed *cosmic noise.* Since they are much farther away than the sun, their individual effects are small, but they make up for this by their countless numbers and their additive effects. Space noise occurs at frequencies from about 8 MHz up to 1.5 GHz (1.5×10^9 Hz). While it contains energy at less than 8 MHz, these components are absorbed by the earth's ionosphere before they can reach the atmosphere. The ionosphere is a region above the atmosphere where free ions and electrons exist in sufficient quantity to have an appreciable effect on wave travel. It includes the area from about 60 to several hundred miles above the earth (see Chapter 13 for further details).

Internal Noise

As stated previously, internal noise is introduced by the receiver itself. Thus, the noise already present at the receiving antenna (external noise) has another component added to it before it reaches the output. The receiver's major noise contribution occurs in its very first stage of amplification. It is there that the desired signal is at its lowest level, and noise injected at that point will be at its largest value in proportion to the intelligence signal. A glance at Fig. 1-2 should help clarify this point. Even though all following stages

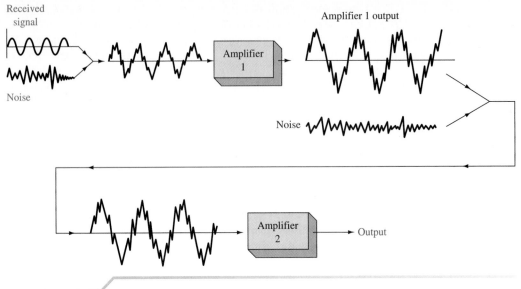

FIGURE 1-2   Noise effect on a receiver's first and second amplifier stages.

also introduce noise, their effect is usually negligible with respect to the very first stage because of their much higher signal level. Note that the noise injected between amplifiers 1 and 2 has not appreciably increased the noise on the desired signal even though it is of the same magnitude as the noise injected into amplifier 1. For this reason, the first receiver stage must be very carefully designed to have low noise characteristics, with the following stages being decreasingly important as the desired signal gets larger and larger.

THERMAL NOISE   There are two basic types of noise generated by electronic circuits. The first one to consider is due to thermal interaction between the free electrons and vibrating ions in a conductor. It causes the rate of arrival of electrons at either end of a resistor to vary randomly, and thereby varies the resistor's potential difference. Resistors and the resistance within all electronic devices are constantly producing a noise voltage. This form of noise was first thoroughly studied by J. B. Johnson in 1928 and is often termed *Johnson noise.* Since it is dependent on temperature, it is also referred to as *thermal noise.* Its frequency content is spread equally throughout the usable spectrum, which leads to a third designator: *white noise* (from optics, where white light contains all frequencies or colors). The terms *Johnson, thermal,* and *white noise* may be used interchangeably. Johnson was able to show that the power of this generated noise is given by

$$P_n = kT \, \Delta f \tag{1-2}$$

where k = Boltzmann's constant (1.38×10^{-23} J/K)
     T = resistor temperature in kelvin (K)
    Δf = frequency bandwidth of the system being considered

Since this noise power is directly proportional to the bandwidth involved, it is advisable to limit a receiver to the smallest bandwidth possible. You may be wondering how the bandwidth figures into this. The noise is an ac voltage that has random instantaneous amplitude but a predictable rms value. The frequency of this noise voltage is just as random as the voltage peaks. The more frequencies allowed into the measurement (i.e.,

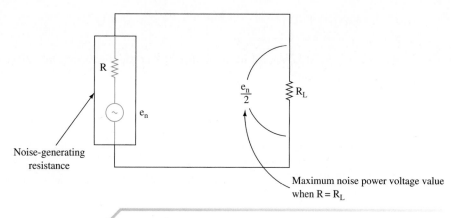

FIGURE 1-3 ▓ Resistance noise generator.

greater bandwidth) the greater the noise voltage. This means that the rms noise voltage measured across a resistor is a function of the bandwidth of frequencies included.

Since $P = E^2/R$, it is possible to rewrite Eq. (1-2) to determine the noise voltage (e_n) generated by a resistor. Assuming maximum power transfer of the noise source, the noise voltage is split between the load and itself, as shown in Fig. 1-3.

$$P_n = \frac{(e_n/2)^2}{R} = kT\ \Delta f$$

Therefore,

$$\frac{e_n^2}{4} = kT\ \Delta f\ R$$

$$e_n = \sqrt{4kT\ \Delta f\ R}$$

(1-3)

where e_n is the rms noise voltage and R is the resistance generating the noise. The instantaneous value of thermal noise is not predictable but has peak values generally less than 10 times the rms value from Eq. (1-3). The thermal noise associated with all nonresistor devices is a direct result of their inherent resistance and to a much lesser extent their composition. This applies to capacitors, inductors, and all electronic devices. Equation (1-3) applies to copper wire-wound resistors, with all other types exhibiting slightly greater noise voltages. Thus, dissimilar resistors of equal value exhibit different noise levels, which gives rise to the term *low-noise resistor,* which you may have heard before but not understood. Standard carbon resistors are the least expensive variety, but unfortunately they also tend to be the noisiest. Metal film resistors offer a good compromise in the cost/performance comparison and can be used in all but the most demanding low-noise designs. The ultimate noise performance (lowest noise generated, that is) is obtained with the most expensive and bulkiest variety: the wire-wound resistor. We use Eq. (1-3) as a reasonable approximation for all calculations in spite of these variations.

EXAMPLE 1-1

Determine the noise voltage produced by a 1-MΩ resistor at room temperature (17°C) over a 1-MHz bandwidth.

It is helpful to know that $4kT$ at room temperature (17°C) is 1.60×10^{-20} joules.

$$
\begin{aligned}
e_n &= \sqrt{4kT\,\Delta f\,R} \\
&= \left[\left(1.6 \times 10^{-20}\right)\left(1 \times 10^6\right)\left(1 \times 10^6\right)\right]^{\frac{1}{2}} \\
&= \left(1.6 \times 10^{-8}\right)^{\frac{1}{2}} \\
&= 126 \ \mu\text{V} \ \ \text{rms}
\end{aligned} \tag{1-3}
$$

From the preceding example we can deduce that an ac voltmeter with an input resistance of 1-MΩ and a 1-MHz bandwidth generates 126 μV of noise (rms). Signals of about 500 μV or less would certainly not be measured with any accuracy. A 50 Ω resistor under the same conditions would generate only about 0.9 μV of noise. This explains why low impedances are desirable in low noise circuits.

EXAMPLE 1-2

An amplifier operating over a 4-MHz bandwidth has a 100-Ω input resistance. It is operating at 27°C, has a voltage gain of 200, and has an input signal of 5 μV rms. Determine the rms output signals (desired and noise), assuming external noise can be disregarded.

SOLUTION

To convert °C to kelvin, simply add 273°, so that K = 27°C + 273°C = 300 K. Therefore,

$$
\begin{aligned}
e_n &= \sqrt{4kT\,\Delta f\,R} \\
&= \sqrt{4 \times 1.38 \times 10^{-23} \ \text{J/K} \times 300 \ \text{K} \times 4 \ \text{MHz} \times 100 \ \Omega} \quad \text{(1-3)*} \\
&= 2.57 \ \mu\text{V rms}
\end{aligned}
$$

After multiplying the input signal e_s (5 μV) and noise signal by the voltage gain of 200, the output signal consists of 1-mV rms signal and 0.514-mV rms noise. This is not normally an acceptable situation. The intelligence would probably be unintelligible!

TRANSISTOR NOISE   In Ex. 1-2, the noise introduced by the transistor, other than its thermal noise, was not considered. The major contributor of transistor noise is called *shot noise*. It is due to the discrete-particle nature of the current carriers in all forms of semi-

---

*When first introduced, all numbered equations appear in a small color box. When used again, these equations are indicated by the number only.

conductors. These current carriers, even under dc conditions, are not moving in an exactly steady continuous flow since the distance they travel varies due to random paths of motion. The name *shot noise* is derived from the fact that when amplified into a speaker, it sounds like a shower of lead shot falling on a metallic surface. Shot noise and thermal noise are additive. The equation for shot noise in a diode is

$$i_n = \sqrt{2qI_{dc}\ \Delta f} \tag{1-4}$$

where i_n = shot noise (rms amperes)
    q = electron charge (1.6×10^{-19} C)
    I_{dc} = dc current (A)
    Δf = bandwidth (Hz)

Unfortunately, there is no valid formula to calculate its value for a complete transistor where the sources of shot noise are the currents within the emitter–base and collector–base diodes. Hence, the device user must refer to the manufacturer's data sheet for an indication of shot noise characteristics. The methods of dealing with these data are covered in Sec. 1-4. Shot noise generally increases proportionally with dc bias currents except in MOSFETS, where shot noise seems to be relatively independent of dc current levels.

EXAMPLE 1-3

(a) *Determine the shot noise current for a diode with a forward bias of 1 mA over a 100-kHz bandwidth.*
(b) *Determine the diode's equivalent noise voltage.*
(c) *If the diode is in a circuit with 500 Ω series resistance, calculate the total output noise voltage at 27°C.*

SOLUTION

(a) $$i_n = \sqrt{2qI_{dc}\ \Delta f}$$

$$= \sqrt{2 \times 1.6 \times 10^{-19}\ C \times 1 \times 10^{-3}\ A \times 10^5\ Hz} \tag{1-4}$$

$$= 0.00566\ \mu A\ rms$$

(b) The dynamic junction resistance of a diode is approximated by the relationship

$$R_j \simeq \frac{26\ mV}{I_{dc}}$$

$$= \frac{26\ mV}{1\ mA} = 26\ \Omega$$

Therefore,

$$e_n = i_n \times 26\ \Omega$$

$$= 0.147\ \mu V\ rms$$

(c) The noise due to the 5-kΩ resistor is

$$e_n = \sqrt{4kT\ \Delta f\ R}$$

$$= \sqrt{4 \times 1.38 \times 10^{-23}\ \text{J/K} \times (27 + 273)\text{K} \times 10^5\ \text{Hz} \times 500\ \Omega}\quad (1\text{-}3)$$

$$= 0.91\ \mu\text{V rms}$$

Therefore,

$$i_n = \frac{0.91\ \mu\text{V rms}}{500\ \Omega} = 0.00182\ \mu\text{A rms}$$

This current generates a noise voltage across the diode of

$$0.00182\ \mu\text{A} \times 26\ \Omega = 0.0473\ \mu\text{V rms}$$

The total noise voltage of the diode is not the sum of the two sources but is given by the square root of the sum of the squares

$$e_{n\,\text{total}} = \sqrt{\left(0.147 \times 10^{-6}\right)^2 + \left(0.0473 \times 10^{-6}\right)^2}$$

$$= 0.154\ \mu\text{V}$$

FREQUENCY NOISE EFFECTS   Two little-understood forms of device noise occur at the opposite extremes of frequency. The low-frequency effect is called *excess noise* and occurs at frequencies below about 1 kHz. It is inversely proportional to frequency and directly proportional to temperature and dc current levels. It is thought to be caused by crystal surface defects in semiconductors that vary at an inverse rate with frequency. Excess noise is often referred to as *flicker noise, pink noise,* or 1/*f* noise. It is present in both bipolar junction transistors (BJT) and field-effect transistors (FET).

At high frequencies device noise starts to increase rapidly in the vicinity of the device's high-frequency cutoff. When the transit time of carriers crossing a junction is comparable to the signal's period (i.e., high frequencies), some of the carriers may diffuse back to the source or emitter. This effect is termed *transit-time noise.* These high- and low-frequency effects are relatively unimportant in the design of receivers since the critical stages (the front end) will usually be working well above 1 kHz and hopefully below the device's high-frequency cutoff area. The low-frequency effects are, however, important to the design of low-level, low-frequency amplifiers encountered in certain instrument and biomedical applications.

The overall noise intensity versus frequency curves for semiconductor devices (and tubes) have a bathtub shape, as represented in Fig. 1-4. At low frequencies the excess noise is dominant, while in the midrange shot noise and thermal noise predominate, and above that the high-frequency effects take over. Of course, tubes are now seldom used and fortunately their semiconductor replacements offer better noise characteristics. Since semiconductors possess inherent resistances, they generate thermal noise in addition to shot noise, as indicated in Fig. 1-4. The noise characteristics provided in manufacturers' data sheets take into account both the shot and thermal effects. At the device's high-frequency cutoff, f_{hc}, the high-frequency effects take over, and the noise increases rapidly.

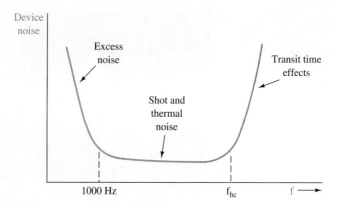

FIGURE 1-4 ▦ Device noise versus frequency.

1-3 NOISE DESIGNATION AND CALCULATION

Signal-to-Noise Ratio

We have thus far dealt with different types of noise without showing how to deal with noise in a practical way. The most fundamental relationship used is known as the *signal-to-noise ratio* (*S/N* ratio), which is a relative measure of the desired signal power to the noise power. The *S/N* ratio is often designated simply as *S/N* and can be expressed mathematically as

$$\frac{S}{N} = \frac{\text{signal power}}{\text{noise power}} = \frac{P_S}{P_N} \qquad (1\text{-}5)$$

at any particular point in an amplifier. It is often expressed in decibel form as

$$\frac{S}{N} = 10 \ \log_{10} \frac{P_S}{P_N} \qquad (1\text{-}6)$$

For example, the output of the amplifier in Ex. 1-2 was 1 mV rms and the noise 0.514 mV rms, and thus (remembering that $P = E^2/R$)

$$\frac{S}{N} = \frac{1^2/R}{0.514^2/R} = 3.79 \ \text{or} \ 10 \ \log_{10} \ 3.79 = 5.78 \ \text{dB}$$

Noise Figure

S/N very well identifies the noise content at a specific point but is not useful in relating how much additional noise a particular transistor has injected into a signal going from input to output. The term *noise figure* (NF) is usually used to specify exactly how noisy a device is. It is defined as follows:

$$\text{NF} = 10 \ \log_{10} \frac{S_i/N_i}{S_o/N_o} = 10 \ \log_{10} \ \text{NR} \qquad (1\text{-}7)$$

Noise figure measurements. (Courtesy of Hewlett-Packard.)

where S_i/N_i is the signal-to-noise power ratio at the device's input and S_o/N_o is the signal-to-noise power ratio at its output. The term $(S_i/N_i)/(S_o/N_o)$ is called the *noise ratio* (NR). If the device under consideration were ideal (injected no additional noise), then S_i/N_i and S_o/N_o would be equal, the NR would equal 1, and NF = 10 log 1 = 10 × 0 = 0 dB. Of course, this result cannot be obtained in practice.

EXAMPLE 1-4

A transistor amplifier has a measured S/N power of 10 at its input and 5 at its output.

(a) *Calculate the NR.*
(b) *Calculate the NF.*
(c) *Using the results of part (a), verify that Eq. (1-7) can be rewritten mathematically as*

$$NF = 10 \, \log_{10} \frac{S_i}{N_i} - 10 \, \log_{10} \frac{S_o}{N_o}$$

SOLUTION

(a)
$$NR = \frac{S_i/N_i}{S_o/N_o} = \frac{10}{5} = 2$$

(b)
$$NF = 10 \ \log_{10} \frac{S_i/N_i}{S_o/N_o} = 10 \ \log_{10} \ NR$$

(1-7)

$$= 10 \ \log_{10} \frac{10}{5} = 10 \ \log_{10} \ 2$$

$$= 3 \ dB$$

(c)
$$10 \ \log \frac{S_i}{N_i} = 10 \ \log_{10} 10 = 10 \times 1 = 10 \ dB$$

$$10 \ \log \frac{S_o}{N_o} = 10 \ \log_{10} 5 = 10 \times 0.7 = 7 \ dB$$

Their difference (10 dB − 7 dB) is equal to the result of 3 dB determined in part (b).

The result of Ex. 1-4 is a typical transistor NF. However, for low-noise requirements devices with NFs down to less than 1 dB are available at a price premium. The graph in Fig. 1-5 shows the manufacturer's NF versus frequency characteristics for the 2N4957 transistor. As can be seen, the curve is flat in the midfrequency range (NF ≃ 2.2 dB) and has a slope of −3 dB/octave at low frequencies (excess noise) and 6 dB/octave in the high-frequency area (transit-time noise). An *octave* is a range of frequency in which the upper frequency is double the lower frequency.

Manufacturers of low-noise devices usually supply a whole host of curves to exhibit their noise characteristics under as many varied conditions as possible. One of the more interesting curves provided for the 2N4957 transistor is shown in Fig. 1-6. It provides a visualization of contours of NF versus source resistance and dc collector current for a 2N4957 transistor at 105 MHz. It indicates that noise operation at 105 MHz will be optimum when a dc (bias) collector current of about 0.7 mA and source resistance of 350 Ω is utilized since the lowest NF of 1.8 dB occurs under these conditions.

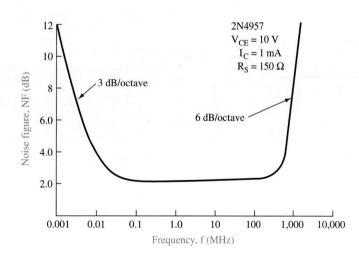

FIGURE 1-5 ▓ NF versus frequency for a 2N4957 transistor. (Courtesy of Motorola Semiconductor Products, Inc.)

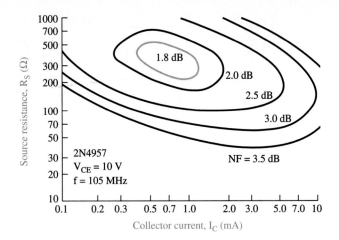

FIGURE 1-6 Noise contours for a 2N4957 transistor. (Courtesy of Motorola Semiconductor Products, Inc.)

The current state of the art for low-noise transistors offers some surprisingly low numbers. The leading edge for room temperature designs at 4 GHz is a NF of about 0.5 dB using gallium arsenide (GaAs) FETs. At 144 MHz, amplifiers with NFs down to 0.3 dB are being employed. The ultimate in low-noise-amplifier (LNA) design utilizes cryogenically cooled circuits (using liquid helium). Noise figures down to about 0.2 dB at microwave frequencies up to about 10 GHz are thereby made possible.

Resistor Combinations

The noise effect of more than one resistor can be determined by the concept of equivalent resistance. This is clearly shown by the following example.

EXAMPLE 1-5

Two resistors, 5 kΩ and 20 kΩ, are at 27°C. Calculate the thermal noise power and voltage for a 10-kHz bandwidth
(a)   For each resistor.
(b)   For their series combination.
(c)   For their parallel combination.

SOLUTION

(a)
$$P_n = kT\ \Delta f$$

$$= 1.38 \times 10^{-23}\ \text{J/K} \times (27 + 273)\ \text{K} \times 10\ \text{kHz} \qquad (1\text{-}2)$$

$$= 4.14 \times 10^{-17}\ \text{W}$$

Note that this is the noise power for all resistor values/combinations.

$$e_n = \sqrt{4kT\ \Delta f\ R}$$

$$= \sqrt{4 \times 4.14 \times 10^{-17} R} \qquad (1\text{-}3)$$

For 5 kΩ,

$$e_n = \sqrt{1.66 \times 10^{-16} \times 5 \text{ k}\Omega} = 0.91 \; \mu\text{V}$$

For 20 kΩ,

$$e_n = \sqrt{1.66 \times 10^{-16} \times 20 \text{ k}\Omega} = 1.82 \; \mu\text{V}$$

(b) In series the equivalent R is 5 kΩ + 20 kΩ = 25 kΩ. Thus,

$$e_n = \sqrt{1.66 \times 10^{-16} \times 25 \text{ k}\Omega} = 2.03 \; \mu\text{V}$$

(c) In parallel the equivalent R is

$$\frac{5 \text{ k}\Omega \times 20 \text{ k}\Omega}{5 \text{ k}\Omega + 20 \text{ k}\Omega} = 4 \text{ k}\Omega$$

Thus,

$$e_n = \sqrt{1.66 \times 10^{-16} \times 4 \text{ k}\Omega} = 0.81 \; \mu\text{V}$$

Reactance Noise Effects

In theory a reactance does not introduce noise to a system. This is true for ideal capacitors and inductors that contain no resistive component. The ideal cannot be attained, but fortunately their resistive elements usually have a negligible effect on system noise considerations compared to semiconductors and other resistances.

The significant effect of reactive circuits on noise is their limitation on frequency response. Our previous discussions on noise have assumed an ideal bandwidth that is rectangular in response. Thus, the 10-kHz bandwidth of Ex. 1-5 implied a total passage within the 10-kHz range and zero effect outside. In practice, RC-, LC-, and RLC-generated passbands are not rectangular but slope off gradually, with the bandwidth defined as a function of half-power frequencies. This is detailed in Sec. 1-6. The equivalent bandwidth (Δf_{eq}) to be used in noise calculations with reactive circuits is given by

$$\Delta f_{eq} = \frac{\pi}{2} \text{BW} \qquad (1\text{-}8)$$

where BW is the 3-dB bandwidth as shown in Sec. 1-6 for RC, LC, or RLC circuits. The fact that the "noise" bandwidth is greater than the "3-dB" bandwidth is not surprising. Significant noise is still being passed through a system beyond the 3-dB cutoff frequency.

Noise Due to Amplifiers in Cascade

We previously specified that the first stage of a system is dominant with regard to noise effect. We are now going to show that effect numerically. *Friiss' formula* is used to provide the overall noise effect of a multistage system.

$$NR = NR_1 + \frac{NR_2 - 1}{P_{G_1}} + \cdots + \frac{NR_n - 1}{P_{G_1} \times P_{G_2} \times \cdots \times P_{G(n-1)}} \qquad (1\text{-}9)$$

where NR = overall noise ratio of *n* stages
P_G = power gain ratio

EXAMPLE 1-6

A three-stage amplifier system has a 3-dB bandwidth of 200 kHz determined by an LC tuned circuit at its input, and operates at 22°C. The first stage has a power gain of 14 dB and a NF of 3 dB. The second and third stages are identical, with power gains of 20 dB and NF = 8 dB. The output load is 300 Ω. The input noise is generated by a 10-kΩ resistor. Calculate

(a)   The noise voltage and power at the input and the output of this system assuming ideal noiseless amplifiers.
(b)   The overall noise figure for the system.
(c)   The actual output noise voltage and power.

SOLUTION

(a)   The effective noise bandwidth is

$$\Delta f_{eq} = \frac{\pi}{2} \, \text{BW}$$

$$= \frac{\pi}{2} \times 200 \ \text{kHz} \qquad (1\text{-}8)$$

$$= 3.14 \times 10^5 \ \text{Hz}$$

Thus, at the input,

$$P_n = kT \, \Delta f$$

$$= 1.38 \times 10^{-23} \ \text{J/K} \times (273 + 22) \ \text{K} \times 3.14 \times 10^5 \ \text{Hz} \qquad (1\text{-}2)$$

$$= 1.28 \times 10^{-15} \ \text{W}$$

and

$$e_n = \sqrt{4kT \, \Delta f \ R}$$

$$= \sqrt{4 \times 1.28 \times 10^{-15} \times 10 \times 10^3} \qquad (1\text{-}3)$$

$$= 7.15 \ \mu\text{V}$$

The total power gain is 14 dB + 20 dB + 20 dB = 54 dB.

$$54 \ \text{dB} = 10 \log P_G$$

Therefore,

$$P_G = 2.51 \times 10^5$$

Assuming perfect noiseless amplifiers,

$$P_{n\,out} = P_{n\,in} \times P_G$$
$$= 1.28 \times 10^{-15}\ \text{W} \times 2.51 \times 10^5$$
$$= 3.22 \times 10^{-10}\ \text{W}$$

Remembering that the output is driven into a 300-Ω load and $P = V^2/R$, we have

$$3.22 \times 10^{-10}\ \text{W} = \frac{(e_{n\,out})^2}{300\ \Omega}$$

$$e_n = 0.311\ \text{mV}$$

Notice that the noise has gone from microvolts to millivolts without considering the noise injected by each amplifier stage.

(b) Recall that to use Friiss' formula, ratios and not decibels must be used. Thus,

$$P_{G_1} = 14\ \text{dB} = 25.1$$
$$P_{G_2} = P_{G_3} = 20\ \text{dB} = 100$$
$$\text{NF}_1 = 3\ \text{dB} \qquad \text{NR}_1 = 2$$
$$\text{NF}_2 = \text{NF}_3 = 8\ \text{dB} \qquad \text{NR}_2 = \text{NR}_3 = 6.31$$
$$\text{NR} = \text{NR}_1 + \frac{\text{NR}_2 - 1}{P_{G_1}} + \cdots + \frac{\text{NR}_n - 1}{P_{G_1} P_{G_2} \cdots P_{G(n-1)}} \tag{1-9}$$
$$= 2 + \frac{6.31 - 1}{25.1} + \frac{6.31 - 1}{25.1 \times 100}$$
$$= 2 + 0.21 + 0.002 = 2.212$$

Thus, the overall noise ratio (2.212) converts into an overall noise figure of $10 \log_{10} 2.212 = 3.45$ dB:

$$\text{NF} = 3.45\ \text{dB}$$

(c)
$$\text{NR} = \frac{S_i/N_i}{S_o/N_o}$$

$$P_G = \frac{S_o}{S_i} = 2.51 \times 10^5$$

Therefore,

$$\text{NR} = \frac{N_o}{N_i \times 2.51 \times 10^5}$$

$$2.212 = \frac{N_o}{1.28 \times 10^{-15} \text{ W} \times 2.51 \times 10^5}$$

$$N_o = 7.11 \times 10^{-10} \text{ W}$$

To get the output noise voltage, since $P = V^2/R$,

$$7.11 \times 10^{-10} \text{ W} = \frac{e_n^2}{300 \ \Omega}$$

$$e_n = 0.462 \text{ mV}$$

Notice that the actual noise voltage (0.462 mV) is about 50% greater than the noise voltage when we did not consider the noise effects of the amplifier stages (0.311 mV).

Equivalent Noise Temperature

Another way of representing noise is by equivalent noise temperature. It is a convenient means of handling noise calculations involved with microwave receivers (1 GHz and above) and their associated antenna system, especially space communication systems. It allows easy calculation of noise power at the receiver using Eq. (1-2) since the equivalent noise temperature (T_{eq}) of microwave antennas and their coupling networks are then simply additive.

The T_{eq} of a receiver is related to its noise ratio, NR, by

$$T_{eq} = T_0(\text{NR} - 1) \tag{1-10}$$

where $T_0 = 290$ K, a reference temperature in kelvin. The use of noise temperature is convenient since microwave antenna and receiver manufacturers usually provide T_{eq} information for their equipment. Additionally, for low noise levels, noise temperature shows greater variation of noise changes than does NF, making the difference easier to comprehend. For example, a NF of 1 dB corresponds to a T_{eq} of 75 K, while 1.6 dB corresponds to 129 K. Verify these comparisons using Eq. (1-10), remembering first to convert NF to NR. Keep in mind that noise temperature is not an actual temperature but is employed because of its convenience.

EXAMPLE 1-7

A satellite receiving system includes a dish antenna ($T_{eq} = 35$ K) connected via a coupling network ($T_{eq} = 40$ K) to a microwave receiver ($T_{eq} = 52$ K referred to its input). What is the noise power to the receiver's input over a 1-MHz frequency range? Determine the receiver's NF.

$$P_n = kT \, \Delta f$$

$$= 1.38 \times 10^{-23} \, \text{J/K} \times (35 + 40 + 52) \, \text{K} \times 1 \, \text{MHz} \qquad (1\text{-}2)$$

$$= 1.75 \times 10^{-15} \, \text{W}$$

$$T_{eq} = T_0(\text{NR} - 1)$$

$$52 \, \text{K} = 290 \, \text{K}(\text{NR} - 1) \qquad\qquad\qquad\qquad\qquad (1\text{-}2)$$

$$\text{NR} = \frac{52}{290} + 1$$

$$= 1.18$$

Therefore, $\text{NF} = 10 \log_{10}(1.18) = 0.716$ dB.

Equivalent Noise Resistance

Manufacturers sometimes represent the noise generated by a device with a fictitious resistance termed the equivalent noise resistance (R_{eq}). It is the resistance that generates the same amount of noise predicted by $\sqrt{4kT \, \Delta f \, R}$ as the device does. The device (or complete amplifier) is then assumed to be noiseless in making subsequent noise calculations. The latest trends in noise analysis have shifted away from the use of equivalent noise resistance in favor of using the noise figure or noise temperatures.

1-4 NOISE MEASUREMENT

Noise measurement has become a very sophisticated process. Specialty noise-measuring instruments that offer many computer-controlled functions are available for thousands of dollars. If you become involved with a large number of measurements, you will become familiar with some of these instruments. In this section we look at some general methods of noise measurement that can be accomplished with relatively standard laboratory instrumentation. A simple and reliable method of noise measurement is the case where the signal is equal to the noise. At some convenient point in the system, a power meter is connected and a reading taken of the noise with no signal input. Then an input signal is raised in power level until the monitored power rises by 3 dB (i.e., doubled). At this point the power level of the signal source is noted. This is equal to the effective input noise level of the system.

Noise Diode Generator

Another noise measurement technique involves using a diode to generate a known amount of noise as predicted by Eq. (1-4); $i_n = \sqrt{2qI_{dc} \, \Delta f}$. You will recall that this is the equation for shot noise. In this technique the output impedance of the diode noise generator circuit is matched into the amplifier under test. In these types of measurements, the amplifier is commonly called the *device under test* or simply *DUT*. The procedure is first to measure the noise power output of the DUT when the dc current to the noise diode is zero. The dc current is then increased until the DUT noise power output is exactly doubled from the original value. The diode dc current is then used in the following equation to determine the noise ratio of the DUT:

$$NR = 20I_{dc}R \qquad (1\text{-}11)$$

where R is the input impedance of the DUT and the temperature is 290 K (approximately room temperature). The reader is referred to "Semiconductor Noise Figure Considerations," Application Note AN-421 from Motorola Semiconductor Products, Inc. for a derivation of this surprisingly simple and useful relationship.

EXAMPLE 1-8

An amplifier has an impedance of 50 Ω. Using a matched-impedance diode noise generator, it is found that the DUT has doubled noise output power when the diode has a dc current of 14 mA. Determine the NR and NF for the DUT.

SOLUTION

$$NR = 20I_{dc}R$$

$$= 20 \times 14 \text{ mA} \times 50 \text{ }\Omega \qquad (1\text{-}11)$$

$$= 14$$

$$NF = 10 \log_{10} NR$$

$$= 10 \log_{10} 14 \qquad (1\text{-}7)$$

$$= 11.46 \text{ dB}$$

Notice that in Ex. 1-8, the NR is numerically equal to the diode's current in mA. This occurs when the DUT has an impedance of 50 Ω—a most convenient situation since many RF amplifier systems are designed with a 50-Ω impedance. Keep in mind that NR is a dimensionless ratio, however, and not measured in mA.

Tangential Noise Measurement Technique

Meters capable of accurately measuring the very low levels involved with noise measurements tend to be expensive and of limited use with regard to other applications. A dual-trace oscilloscope with high sensitivity is an exception to this limitation. Unfortunately, a direct noise reading from the scope results in errors for two reasons:

1. Noise is of a highly random nature and is not sinusoidal. Since rms values are required for noise calculations, the conversion from scope peak-to-peak values by dividing by $2\sqrt{2}$ is not accurate.

2. Since the noise peaks are random, their visibility on the scope is influenced by such things as the scope's intensity setting, the persistence of the CRT's phosphor, and the length of the observation.

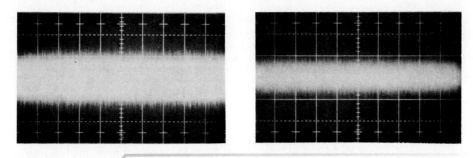

FIGURE 1-7 ■ Scope display of the same noise signal at two different intensity settings. (Courtesy of *Electronic Design.*)

The two displays shown in Fig. 1-7 show exactly the same noise signal at two different intensity settings. The measurement can be erroneous by as much as 6 dB. A specially developed technique, known as the *tangential method,* reduces the possible error to less than 1 dB. The noise signal is connected to both channels of a dual-trace scope with alternate sweep capability. As shown in Fig. 1-8(a), the two displayed signals are set up with both channels identically calibrated. Then their vertical position is adjusted until the dark band between them just disappears [Fig. 1-8(b)]. Now the noise signal input to both channels is removed, and the resulting separation represents twice the rms noise. In this case (with a vertical sensitivity of 20 mV/cm), the rms noise is 0.8 cm × 20 mV/cm ÷ 2,

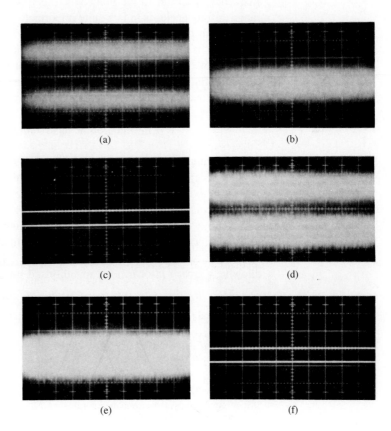

(a)

(b)

FIGURE 1-8 ■ With the tangential method the noise signal is connected to both channels of a dual-channel scope used in the alternate-sweep mode. (a) The offset voltage is adjusted until the traces just merge. (b) The noise signal is then removed. The difference in the noise-free traces is twice the rms noise voltage. (c) This is repeated at a different intensity to show that the method is independent of intensity. (d, e, f) Scope settings are: horizontal = 500 ms/cm, vertical = 20 mV/cm. (Courtesy of *Electronic Design.*)

(c)

(d)

(e)

(f)

or 8 mV rms. Repeating this process with a different scope intensity setting [Fig. 1-8(d), (e), and (f)] yields the same result.

1-5 INFORMATION AND BANDWIDTH

In Sec. 1-1 it was mentioned that there are two basic limitations on the performance of a communications system. By now you should have a good grasp on the noise limitation. Quite simply, if the noise level becomes too high, the information is lost. The other limitation is the bandwidth utilized by the communications system. Stated simply once again, the greater the bandwidth, the greater the information that can be transferred from source to destination. The study of information in communications systems is a science in itself (given the title *information theory*) that uses a highly theoretical method of analysis. It is beyond our intentions here, but if you pursue advanced studies you will hear much more about it. Information theory is the study of information to provide for the most efficient use of a band of frequencies (a *channel*) for electrical communications. Additional information theory is provided in Chapter 9.

It might be asked: Why is efficient channel utilization so important? The band of usable frequencies is limited, and we are living in a world increasingly dependent on electrical communications. Regulatory agencies (the FCC in the United States) allocate the channel that may be used for a given application in a given area. This is done to minimize interference possibilities that will exist with two different signals working at the same frequency. The information explosion of recent years has taxed the total available frequency spectrum to the point where getting the most information from the smallest range of frequencies is in fact quite important.

A formal relationship between bandwidth and information was developed by R. Hartley of Bell Laboratories in 1928 and is called *Hartley's law*. It states that the information that can be transmitted is proportional to the product of the bandwidth utilized times the time of transmission. In simpler terms it means the greater the bandwidth, the more information that can be transmitted. Expressed as an equation, Hartley's law is

$$\text{information} \propto \text{bandwidth} \times \text{time of transmission} \qquad (1\text{-}12)$$

As an example, consider the transmission of a musical performance. The full amount of information available to the human ear is contained in the range of frequencies from just above 0 Hz up to about 15 kHz. The allocated bandwidth of standard AM stations is about 30 kHz. On the other hand, FM stations are allocated a larger bandwidth (200 kHz), which allows the full amount of information (up to 15 kHz) to be reproduced at the receiver. This helps explain the better fidelity available with FM as compared to AM in our two basic commercial radio bands.* This is an example of greater bandwidth allowing a greater information capability and substantiates Hartley's law.

Nonsinusoidal Waveforms

The bandwidth of the AM band imposes limitations on the information capacity of that system, yet it has adequate capacity to please the human ear for most audio applications.

---

*This example has been oversimplified, for reasons that will become obvious as you study AM and FM in subsequent chapters. Its conclusion remains valid, however.

However, a television system would hardly be acceptable if 30 kHz were all that were available for transmission. In fact, the United States allocates a 6-MHz bandwidth per TV channel—a bandwidth 200 times the allowed AM bandwidth! Obviously, TV must require a great deal more information capability than radio to warrant such a large bandwidth. Television signals must not only contain an audio (sound) signal but a video (picture) signal as well. It is the video signal that requires most of the 6-MHz bandwidth. The video signal is mainly a pulse-type waveform and, as will be shown, a pulse waveform at one frequency requires a much larger bandwidth for transmission than a sinusoidal waveform of the same frequency.

Consider the 1000-Hz square wave shown in Fig. 1-9(a). In previous studies you have probably learned that it can be broken down into sinusoidal waveforms at frequencies of 1000, 3000, 5000, 7000, . . . , up to an infinite number of odd multiples of 1000 Hz. In fact, any nonsinusoidal repetitive waveform can be broken down into sinusoidal and/or cosinusoidal components (called *harmonics*) at its basic repetition frequency and its even and/or odd multiples. Thus, to transmit the 1000-Hz square wave exactly would require an infinite bandwidth. This is impractical, but a close approximation of the square wave can be made if a bandwidth allows the fundamental component (1000 Hz) and perhaps the next five harmonics (3000 Hz, 5000 Hz, 7000 Hz, 9000 Hz, 11,000 Hz) to pass. A reference to Fig. 1-9(b) shows the frequency-domain plot of a square wave. This is a common method of representing communication waveforms and is called the *signal spectrum*. It shows the relative amplitude of the different frequency components contained within a complex waveform. Notice that the higher-frequency components of the square wave are decreasing in amplitude. This applies to most waveforms: in general, the higher the harmonic, the lower its magnitude. Even though the square wave theoretically has sinusoidal components out to an infinitely high frequency, they become so small that they can be eliminated by a finite passband without significant signal degradation.

Thus, a 10-kHz bandwidth for transmitting the 1-kHz square wave might be acceptable for many applications. It would result in rounded-off edges, as shown in Fig. 1-10. This nonideal square wave would be acceptable for many applications. If a bandwidth of 30 kHz were used, the resulting square wave would have almost no visible distortion.

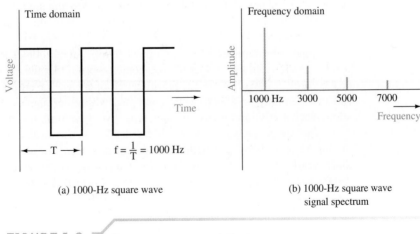

(a) 1000-Hz square wave

(b) 1000-Hz square wave signal spectrum

FIGURE 1-9 ■ Square-wave representations.

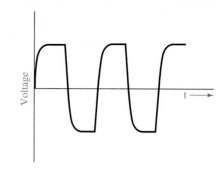

FIGURE 1-10 ■ Nonideal square wave.

Fourier Analysis

A method whereby complex repetitive waveforms can be analyzed is known as *Fourier analysis*. It permits any complex repetitive waveform to be resolved into sine or cosine waves (possibly infinite in number), and a dc component (when necessary). The mathematical tool provided in Fourier analysis helps one to understand the meaning of harmonics and the complex waves they are a part of and also to obtain insight into factors relating to distortion effects.

Figure 1-11(e) shows a complex repetitive waveform: a sawtooth wave. It contains a fundamental sinusoidal component at its basic frequency (as do all complex repetitive waveforms) at (a) as well as a number of higher harmonics at (b), (c), and (d). A detailed study of the effects if it were used to modulate a higher-frequency carrier signal would require us to know the specifics of all its components. Fourier analysis gives us that information and is provided by the equation

$$v = -K\left(\sin\,\omega t + \frac{1}{2}\sin\,2\omega t + \frac{1}{3}\sin\,3\omega t + \frac{1}{4}\sin\,4\omega t + \cdots\right) \qquad (1\text{-}13)$$

The first term ($-K \sin\,\omega t$) is the fundamental component (also known as the first harmonic) and K is a constant and is shown in Fig. 1-11(a). The second harmonic ($-K \times \frac{1}{2} \sin 2\,\omega t$) has a relative amplitude one-half that of the fundamental and is at twice the frequency ($2\omega t$). The third and fourth harmonics are also included and have amplitudes of $\frac{1}{3}$ and $\frac{1}{4}$ and frequencies of 3 and 4 times that of the fundamental, respectively. The higher harmonics are not included, but their amplitudes get smaller and smaller and thus can often be disregarded.

Algebraic addition of these four most significant harmonics results in the composite waveform shown in Fig. 1-11(e). This should be proof that the complex waveform can in fact be broken down into basic sinusoidal components for purposes of further analysis and to determine the effects when combined with other signals. Since Fourier analysis allows such a solution for any complex repetitive waveform, it is indeed a powerful tool.

The composite waveform would be identical to the ideal shown in Fig. 1-11(e) if all the higher-order harmonics (fifth and above) had been included. Reference to Fig. 1-12 more clearly shows that effect. If the original sawtooth were at a frequency of 1 kHz and were passed through a filter [filter A in Fig. 1-12(a)] that fully passed a sine wave at 1 kHz but had no output at 2 kHz, its output would be the fundamental component shown in Fig. 1-12(c). Passing the same sawtooth through filter B of Fig. 1-12(a) would

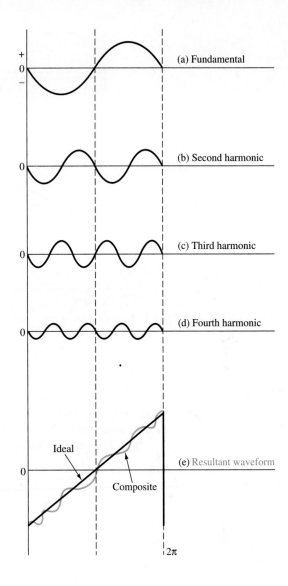

+
0
−

(a) Fundamental

0

(b) Second harmonic

0

(c) Third harmonic

0

(d) Fourth harmonic

Ideal

0

(e) Resultant waveform

Composite

2π

FIGURE 1-11   Formation of a sawtooth waveform.
(By permission, after Matthew Mandl, *Principles of Electronic Communication,* Prentice-Hall, Inc., Englewood Cliffs, N.J., 1973.)

provide the output shown in Fig. 1-12(d), which includes the first four harmonics of the original sawtooth. Passing the sawtooth through filter C allows all of the significant harmonics to pass and provides a nearly perfect replica of the original sawtooth, as shown in Fig. 1-12(e).

A practical conclusion of the previous discussion might be that transmission of the 1-kHz sawtooth with a communication channel of 1-kHz bandwidth would not be adequate. The sawtooth would not be recognizable from a 1-kHz sine wave. On the other hand, using a channel with a 4-kHz bandwidth might be adequate depending on the degree of distortion that could be tolerated. Bear in mind that noise effects have a further degrading effect on the signal that would finally be presented at the receiver's output. Certainly a 100-kHz channel would be adequate, but it would be the wrong choice, for performance (increased noise) and efficiency reasons, if the 4-kHz channel were ade-

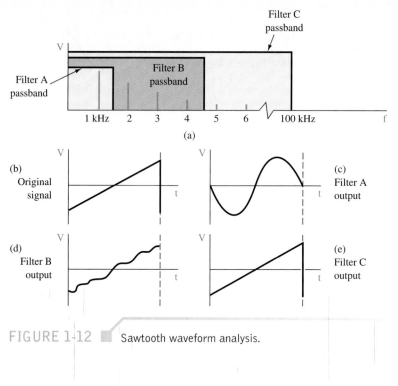

FIGURE 1-12 ■ Sawtooth waveform analysis.

quate. The foregoing considerations (and more) all fall into the category that has been previously defined as the study of *information theory.*

The Fourier expression given in Eq. (1-13) has a predictable nature:

$$v = -K\left(\sin\,\omega t + \frac{1}{2}\sin\,2\omega t + \frac{1}{3}\sin\,3\omega t + \frac{1}{4}\sin\,4\omega t + \cdots\right) \qquad (1\text{-}13)$$

It should be obvious that the next harmonic is $\frac{1}{5}\sin\,5\omega t$. The formal equation to show this is

$$v = \sum_{n=1}^{\infty} -\frac{K}{n}\sin\,2\pi n f_0$$

In this case, the constant K is equal to two times the sawtooth's peak-to-peak amplitude divided by π. Table 1-2 shows this and the Fourier series for some other common repetitive signals. Notice in part (b) that the sawtooth timing is changed, resulting in the alternating sign of each harmonic as compared to part (a). The pulse and full-wave rectifier waveforms at (d) and (f) include a dc component, as you would expect from inspection of the waveform. The reader should consult an advanced-level communications text for details on calculation of the Fourier series for other waveforms (reasonably complex calculus is required for this). It should be understood that Fourier analysis is more than just a useful mathematical way of representing a wave. The component sine or cosine waves are physically real and are actually measurable by using frequency-selective instrumentation. In other words, the sine or cosine waves actually exist within the complex repetitive waveform.

TABLE 1-2 *Fourier Expressions for Selected Periodic Waveforms,* $f = 1/T$, $2\pi f = \omega$

(a) $\quad V = \dfrac{2V}{\pi}\left(\sin \omega t + \dfrac{1}{2}\sin 2\omega t + \dfrac{1}{3}\sin 3\omega t + \dfrac{1}{4}\sin 4\omega t + \cdots\right)$

(b) $\quad V = \dfrac{2V}{\pi}\left(\sin \omega t - \dfrac{1}{2}\sin 2\omega t + \dfrac{1}{3}\sin 3\omega t - \dfrac{1}{4}\sin 4\omega t + \cdots\right)$

(c) $\quad V = \dfrac{4V}{\pi}\left(\sin \omega t + \dfrac{1}{3}\sin 3\omega t + \dfrac{1}{5}\sin 5\omega t + \cdots\right)$

(d) $\quad V = V\dfrac{\tau}{T} + 2V\dfrac{\tau}{T}\left[\dfrac{\sin \pi(\tau/T)}{\pi\tau T}\cos \omega t + \dfrac{\sin 2\pi(\tau/T)}{2\pi(\tau/T)}\cos 2\omega t\right.$

$\left. + \dfrac{\sin 3\pi(\tau/T)}{3\pi(\tau/T)}\cos 3\omega t + \cdots\right]$

(e) $\quad V = \dfrac{8V}{\pi^2}\left[\cos \omega t + \dfrac{1}{(3)^2}\cos 3\omega t + \dfrac{1}{(5)^2}\cos 5\omega t + \cdots\right]$

(f) $\quad V = \dfrac{2V}{\pi}\left[1 + \dfrac{2\cos 2\omega t}{3} - \dfrac{2\cos 4\omega t}{15}\right.$

$\left. + \cdots (-1)^{n/2}\dfrac{2\cos n\omega t}{n^2 - 1}\cdots\right]$ (*n* even)

The remaining sections of this chapter cover some basic characteristics of *LC* circuits and oscillators. This material may be a review to you, but its importance to subsequent communication circuit study merits inclusion at this time.

Practical Inductors and Capacitors

Practical inductors (coils) used at RF frequencies and above have an inductance rating in henrys and a maximum current rating. Similarly, capacitors have a capacitance rating in farads and a maximum voltage rating. When selecting coils and capacitors for use at radio frequencies and above, an additional characteristic must be considered—the *quality* (*Q*) of the component. The *Q* is a ratio of the energy stored to that which is lost in the component.

Inductors store energy in the surrounding magnetic field and lose (dissipate) energy in their winding resistance. A capacitor stores energy in the electric field between its plates and primarily loses energy due to *leakage* between the plates.

For an inductor,

$$Q = \frac{\text{reactance}}{\text{resistance}} = \frac{\omega L}{R} \tag{1-14}$$

where *R* is the series resistance distributed along the coil winding. The required *Q* for a coil varies with circuit application. Values up to about 500 are generally available.

For a capacitor,

$$Q = \frac{\text{susceptance}}{\text{conductance}} = \frac{\omega C}{G} \tag{1-15}$$

where *G* is the value of conductance through the dielectric between the capacitor plates. Good-quality capacitors used in radio circuits have typical *Q* factors of 1000.

At higher radio frequencies (VHF and above—see Table 1-1) the *Q* for inductors and capacitors is generally reduced by such factors as radiation, absorption, lead inductance, and package/mounting capacitance. Occasionally, an inverse term is used rather than *Q*. It is called the component *dissipation* (*D*) and is equal to 1/*Q*. Thus *D* = 1/*Q*, and this is a term used more often in reference to a capacitor.

Resonance

Resonance can be defined as a circuit condition whereby the inductive and capacitive reactance have been balanced ($X_L = X_C$). Consider the series *RLC* circuit shown in Fig. 1-13. In this case, the total impedance, *Z*, is provided by the formula

$$Z = \sqrt{R^2 + (X_L - X_C)^2}$$

An interesting effect occurs at the frequency where X_L is equal to X_C. That frequency is termed the resonant frequency, f_r. At f_r the circuit impedance is equal to the resistor value (which might only be the series winding resistance of the inductor). That result can be shown from the equation above since when $X_L = X_C$, $X_L - X_C$ equals zero, so that

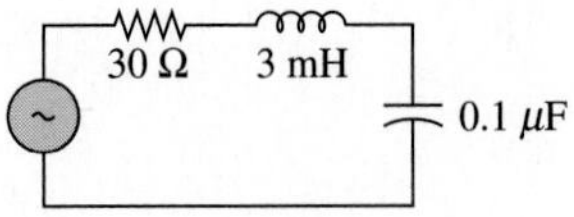

FIGURE 1-13 ■ Series *RLC* circuit.

$Z = \sqrt{R^2 + 0^2} = \sqrt{R^2} = R$. The resonant frequency can be determined by finding the frequency where $X_L = X_C$.

$$X_L = X_C$$

$$2\pi f_r L = \frac{1}{2\pi f_r C}$$

$$f_r^2 = \frac{1}{4\pi^2 LC} \tag{1-16}$$

$$f_r = \frac{1}{2\pi \sqrt{LC}} \text{ (Hz)}$$

EXAMPLE 1-9

Determine the resonant frequency for the circuit shown in Fig. 1-13. Calculate its impedance when f = 12 kHz.

SOLUTION

$$f_r = \frac{1}{2\pi \sqrt{LC}}$$

$$= \frac{1}{2\pi \sqrt{3 \text{ mH} \times 0.1 \text{ }\mu\text{F}}} \tag{1-16}$$

$$= 9.19 \text{ kHz}$$

At 12 kHz,

$$X_L = 2\pi f L$$

$$= 2\pi \times 12 \text{ kHz} \times 3 \text{ mH}$$

$$= 226 \text{ }\Omega$$

$$X_C = \frac{1}{2\pi f_C}$$

$$= \frac{1}{2\pi \times 12 \text{ kHz} \times 0.1 \text{ }\mu\text{F}}$$

$$= 133 \text{ }\Omega$$

$$Z = \sqrt{R^2 + (X_L - X_C)^2}$$

$$= \sqrt{30^2 + (226 - 133)^2}$$

$$= 97.7 \text{ }\Omega$$

This circuit contains more inductive than capacitive reactance at 12 kHz and is therefore said to look inductive.

The impedance of the series *RLC* circuit is minimum at its resonant frequency and equal to the value of *R*. A graph of its impedance, *Z*, versus frequency has the shape of the curve shown in Fig. 1-14(a). At low frequencies the circuit's impedance is very high, since X_C is high. At high frequencies X_L is very high and thus *Z* is high. At resonance, when $f = f_r$, the circuit's $Z = R$ and is at its minimum value. This impedance characteristic can provide a filter effect, as shown in Fig. 1-14(b). At f_r, $X_L = X_C$ and thus

$$e_{out} = e_{in} \times \frac{R_2}{R_1 + R_2}$$

by the voltage-divider effect. At all other frequencies, the impedance of the *LC* combination goes up (from 0 at resonance) and thus e_{out} goes up. The response for the circuit in Fig. 1-14(b) is termed a band-reject or notch filter. A "band" of frequencies is being "rejected" and a "notch" is cut into the output at the resonant frequency, f_r.

Example 1-10 shows that the filter's output increases as the frequency is increased. Calculation of the circuit's output for frequencies below resonance would show a similar increase and is left as an exercise at the end of the chapter. The band-reject or notch filter is sometimes called a trap, since it can "trap" or get rid of a specific range of frequencies near f_r. A trap is commonly used in a television receiver, where rejection of some specific frequencies is necessary for good picture quality.

EXAMPLE 1-10

Determine f_r for the circuit shown in Fig. 1-14(b) when $R_1 = 20\ \Omega$, $R_2 = 1\ \Omega$, $L = 1\ mH$, $C = 0.4\ \mu F$, and $e_{in} = 50\ mV$. Calculate e_{out} at f_r and 12 kHz.

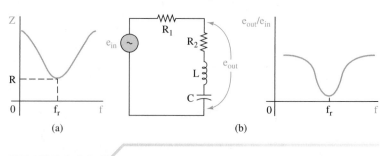

(a)   (b)

FIGURE 1-14 ▇ Series *RLC* circuit effects.

The resonant frequency is

$$f_r = \frac{1}{2\pi\sqrt{LC}} \qquad (1\text{-}16)$$

$$= 7.96 \ \text{kHz}$$

At resonance,

$$e_{out} = e_{in} \times \frac{R_2}{R_2 + R_1}$$

$$= 50 \ \text{mV} \times \frac{1 \ \Omega}{1 \ \Omega + 20 \ \Omega}$$

$$= 2.38 \ \text{mV}$$

At $f = 12$ kHz,

$$X_L = 2\pi f L$$

$$= 2\pi \times 12 \ \text{kHz} \times 1 \ \text{mH}$$

$$= 75.4 \ \Omega$$

and

$$X_C = \frac{1}{2\pi f C}$$

$$= \frac{1}{2\pi \times 12 \ \text{kHz} \times 0.4 \ \mu F}$$

$$= 33.2 \ \Omega$$

Thus,

$$Z_{total} = \sqrt{(R_1 + R_2)^2 + (X_L - X_C)^2}$$

$$= \sqrt{(20 \ \Omega + 1 \ \Omega)^2 + (75.4 \ \Omega - 33.2 \ \Omega)^2}$$

$$= 47.1 \ \Omega$$

and

$$Z_{out} = \sqrt{R_2^2 + (X_L - X_C)^2} = 42.2 \ \Omega$$

$$e_{out} = 50 \ \text{mV} \times \frac{42.2 \ \Omega}{47.1 \ \Omega}$$

$$= 44.8 \ \text{mV}$$

LC Bandpass Filter

If the filter's configuration is changed to that shown in Fig. 1-15(a), it is called a bandpass filter and has a response as shown at Fig. 1-15(b). The term f_{lc} is the low-frequency cutoff where the output voltage has fallen to 0.707 times its maximum value and f_{hc} the high-frequency cutoff. The frequency range between f_{lc} and f_{hc} is called the filter's bandwidth, usually abbreviated BW. The BW is equal to $f_{hc} - f_{lc}$ and it can mathematically be shown that

$$BW = \frac{R}{2\pi L} \tag{1-17}$$

where BW = bandwidth (Hz)
$\quad$ R = total circuit resistance
$\quad$ L = circuit inductance

The filter's quality factor, Q, provides a measure of how selective (narrow) its passband is compared to its center frequency, f_r. Thus,

$$Q = \frac{f_r}{BW} \tag{1-18}$$

As stated earlier, the quality factor, Q, can also be determined as

$$Q = \frac{\omega L}{R} \tag{1-14}$$

where ωL = inductive reactance at resonance
$\quad$ R = total circuit resistance

As Q increases, the filter becomes more selective; that is, a smaller passband (narrower bandwidth) is allowed. A major limiting factor in the highest attainable Q is the resistance factor shown in Eq. (1-14). To obtain a high Q, the circuit resistance must be low. Quite often, the limiting factor becomes the winding resistance of the inductor itself. The turns of wire (and associated resistance) used to make an inductor provide this limiting factor. To obtain the highest Q possible, larger wire (with less resistance) could be used, but then greater cost and physical size to obtain the same amount of inductance is required. Quality factors (Q) approaching 1000 are possible with very high quality inductors.

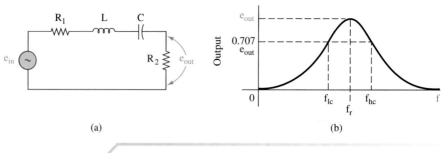

(a) $\qquad\qquad\qquad\qquad\qquad\qquad$ (b)

FIGURE 1-15 ▮ *LC* bandpass filter and response.

EXAMPLE 1-11

A filter circuit of the form shown in Fig. 1-15(a) has a response as shown in Fig. 1-16. Determine

(a)   The bandwidth.
(b)   The Q.
(c)   The value of inductance if C = 0.001 µF.
(d)   The total circuit resistance.

SOLUTION

(a)   From Fig. 1-16, the BW is simply the frequency range between f_{hc} and f_{lc} or 460 kHz − 450 kHz = 10 kHz.

(b)   The filter's peak output occurs at 455 kHz.

$$Q = \frac{f_r}{BW}$$

$$= \frac{455 \ kHz}{10 \ kHz} \tag{1-18}$$

$$= 45.5$$

(c)   Equation (1-16) can be used to solve for L since f_r and C are known.

$$f_r = \frac{1}{2\pi\sqrt{LC}}$$

$$455 \ kHz = \frac{1}{2\pi \ \sqrt{L \times 0.001 \ \mu F}} \tag{1-16}$$

$$L = 0.12 \ mH$$

(d)   Equation (1-17) can be used to solve for total circuit resistance because the BW and L are known.

$$BW = \frac{R}{2\pi L}$$

$$10 \ kHz = \frac{R}{2\pi \times 0.12 \ mH} \tag{1-17}$$

$$R = 10 \times 10^3 \ Hz \times 2\pi \times 0.12 \times 10^{-3} \ mH$$

$$= 7.52 \ \Omega$$

The frequency-response characteristics of LC circuits are affected by the ratio of L and C. Different values of L and C can be used to exhibit resonance at a specific frequency. A high L/C ratio yields a more narrowband response while lower L/C ratios provide a wider frequency response. This effect can be verified by examining the effect of changing L in Eq. (1-17).

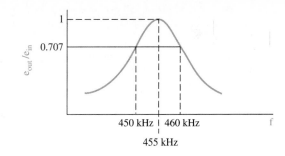

FIGURE 1-16 ■ Response curve for Ex. 1-11.

Parallel *LC* Circuits

A parallel *LC* circuit and its impedance versus frequency characteristic is shown in Fig. 1-17. The only resistance shown for this circuit is the inductor's winding resistance and is effectively in series with the inductor as shown. Notice that the impedance of the parallel *LC* circuit reaches a maximum value at the resonant frequency, f_r, and falls to a low value on either side of resonance. As shown in Fig. 1-17, the maximum impedance is

$$Z_{max} = Q^2 \times R \qquad (1\text{-}19)$$

Equations (1-16) to (1-18) and (1-14) for series *LC* circuits also apply to parallel *LC* circuits when Q is greater than 10 ($Q > 10$), the usual condition.

The parallel *LC* circuit is sometimes called a *tank circuit.* Energy is stored in each reactive element (*L* and *C*), first in one and then released to the other. The transfer of energy between the two elements will occur at a natural rate equal to the resonant frequency and is sinusoidal in form.

EXAMPLE 1-12

A parallel LC tank circuit is made up of an inductor of 3 mH and a winding resistance of 2 Ω. The capacitor is 0.47 μF. Determine

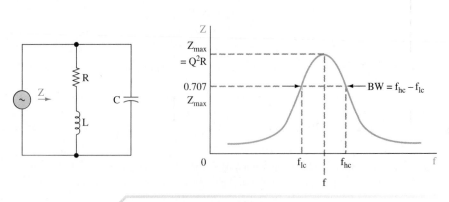

FIGURE 1-17 ■ Parallel *LC* circuit and response.

(a) f_r.

(b) Q.

(c) Z_{max}.

(d) BW.

SOLUTION

(a)
$$f_r = \frac{1}{2\pi\sqrt{LC}}$$

$$= \frac{1}{2\pi\sqrt{3 \text{ mH} \times 0.47\mu\text{F}}} \qquad (1\text{-}16)$$

$$= 4.24 \text{ kHz}$$

(b)
$$Q = \frac{X_L}{R} \qquad (1\text{-}14)$$

where $X_L = 2\pi f L$

$$= 2\pi \times 4.24 \text{ kHz} \times 3 \text{ mH}$$

$$= 79.9 \ \Omega$$

$$Q = \frac{79.9 \ \Omega}{2 \ \Omega}$$

$$= 39.9$$

(c)
$$Z_{max} = Q^2 \times R$$

$$= (39.9)^2 \times 2 \ \Omega \qquad (1\text{-}19)$$

$$= 3.19 \text{ k}\Omega$$

(d)
$$BW = \frac{R}{2\pi L}$$

$$= \frac{2 \ \Omega}{2\pi \times 3 \text{ mH}} \qquad (1\text{-}17)$$

$$= 106 \text{ Hz}$$

1-7 OSCILLATORS

The most basic building block in a communication system is an *oscillator*. An oscillator is a circuit capable of converting energy from a dc form to ac. In other words, an oscillator generates a waveform. The waveform can be of any type but occurs at some repetitive frequency.

A number of different forms of sine-wave oscillators are available for use in electronic circuits. The choice of an oscillator type is based on the following criteria:

1. Output frequency required.
2. Frequency stability required.

3. Is the frequency to be variable, and if so, over what range?
4. Allowable waveform distortion.
5. Power output required.

These performance considerations, combined with economic factors, will dictate the form of oscillator to be used in a given application.

LC Oscillator

The effect of charging the capacitor in Fig. 1-18(a) to some voltage potential and then closing the switch results in the waveform shown in Fig. 1-18(b). The switch closure starts a current flow as the capacitor begins to discharge through the inductor. The inductor, which resists a change in current flow, causes a gradual sinusoidal current buildup that reaches maximum when the capacitor is fully discharged. At this point the potential energy is zero, but since current flow is maximum, the magnetic field energy around the inductor is maximum. The magnetic field no longer maintained by capacitor voltage then starts to collapse, and its counter EMF will keep current flowing in the same direction, thus charging the capacitor to the opposite polarity of its original charge. This repetitive exchange of energy is known as the *flywheel* effect. The circuit losses (mainly the dc winding resistance of the coil) cause the output to become gradually smaller as this process repeats itself after the complete collapse of the magnetic field. The resulting waveform, shown in Fig. 1-18(b), is termed a *damped* sine wave. The energy of the magnetic field has been converted into the energy of the capacitor's electric field, and vice versa. The process repeats itself at the natural or resonant frequency, f_r, as predicted by Eq. (1-16):

$$f_r = \frac{1}{2\pi\sqrt{LC}} \tag{1-16}$$

For an *LC* tank circuit to function as an oscillator, an amplifier is utilized to restore the lost energy to provide a constant-amplitude sine-wave output. The resulting "undamped" waveform is known as a *continuous wave* (CW) in radio work. The most straightforward method of restoring this lost energy is now examined, and the general conditions required for oscillation are introduced.

The *LC* oscillators are basically feedback amplifiers, with the feedback serving to increase or sustain the self-generated output. This is called positive feedback, and it occurs when the fed-back signal is in phase with (reinforces) the input signal. It would

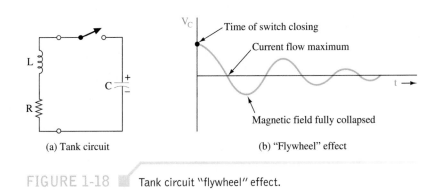

(a) Tank circuit   (b) "Flywheel" effect

FIGURE 1-18 ▮ Tank circuit "flywheel" effect.

seem, then, that the regenerative effects of this positive feedback would cause the output to increase continually with each cycle of fed-back signal. However, in practice, component nonlinearity and power supply limitations limit the theoretically infinite gain.

The criteria for oscillation are formally stated by the *Barkhausen criteria* as follows:

1. The loop gain must equal 1.
2. The loop phase shift must be $n \times 360°$, where $n = 1, 2, 3, \ldots$.

An oscillating amplifier adjusts itself to meet both of these criteria. The initial surge of dc power or noise in the circuit creates a sinusoidal voltage in the tank circuit at its resonant frequency, and it is fed back to the input and amplified repeatedly until the amplifier works into the saturation and cutoff regions. At this time, the flywheel effect of the tank is effective in maintaining a sinusoidal output. This process shows us that too much gain would cause excessive distortion and therefore the gain should be limited to a level that is just greater than or equal to 1.

Hartley Oscillator

Figure 1-19 shows the basic Hartley oscillator in simplified form. The inductors L_1 and L_2 are a single tapped inductor. Positive feedback is obtained by mutual inductance effects between L_1 and L_2 with L_1 in the transistor output circuit and L_2 across the base–emitter circuit. A portion of the amplifier signal in the collector circuit (L_1) is returned to the base circuit by means of inductive coupling from L_1 to L_2. As always in a common-emitter (CE) circuit, the collector and base voltages are 180° out of phase. Another 180° phase reversal between these two voltages occurs because they are taken from opposite ends of an inductor tap that is tied to the common transistor terminal—the emitter. Thus the in-phase feedback requirement is fulfilled and loop gain is of course provided by Q_1. The frequency of oscillation is approximately given by

$$f \simeq \frac{1}{2\pi\sqrt{(L_1 + L_2)C}} \tag{1-20}$$

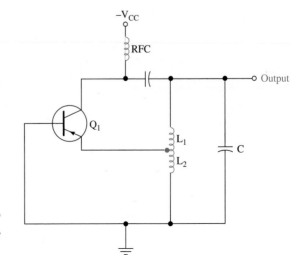

FIGURE 1-19 Simplified Hartley oscillator.

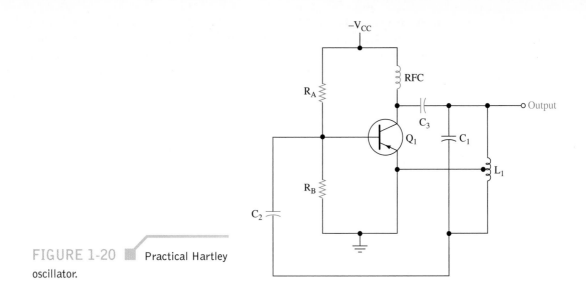

FIGURE 1-20 ▓ Practical Hartley oscillator.

and is influenced slightly by the transistor parameters and amount of coupling between L_1 and L_2.

Figure 1-20 shows a practical Hartley oscillator. A number of additional circuit elements are necessary to make a workable oscillator over the simplified one used for explanatory purposes in Fig. 1-19. Naturally, the resistors R_A and R_B are for biasing purposes. The radio-frequency choke (RFC) is effectively an open circuit to the resonant frequency and thus allows a path for the bias (dc) current but does not allow the power supply to short out the ac signal. The coupling capacitor C_3 prevents dc current from flowing in the tank, and C_2 provides dc isolation between the base and the tank circuit. Both C_2 and C_3 can be considered as short circuits to the oscillator's frequency.

Colpitts Oscillator

Figure 1-21 shows a Colpitts oscillator. It is similar to the Hartley oscillator except that the tank circuit elements have interchanged their roles. The capacitor is now split, so to

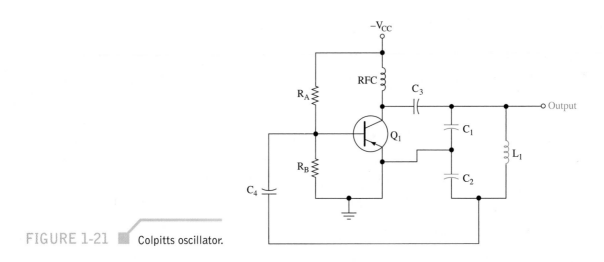

FIGURE 1-21 ▓ Colpitts oscillator.

speak, and the inductor is single-valued with no tap. The details of circuit operation are identical with the Hartley oscillator and therefore will not be explained further. The frequency of oscillation is given approximately by the resonant frequency of L_1 and C_1 in series with the C_2 tank circuit:

$$f \simeq \frac{1}{2\pi\sqrt{[C_1 C_2 / (C_1 + C_2)]L_1}} \qquad (1\text{-}21)$$

The performance differences between these two oscillators' forms are minor, and the choice between them is usually made on the basis of convenience or economics. They may both provide variable oscillator output frequencies by making one of the tank circuit elements variable.

Clapp Oscillator

A variation of the Colpitts oscillator is shown in Fig. 1-22. The Clapp oscillator has a capacitor C_3 in series with the tank circuit inductor. If C_1 and C_2 are made large enough, they will "swamp" out the transistor's inherent junction capacitances, thereby negating transistor variations and junction capacitance changes with temperature. The signal frequency of oscillation is

$$f = \frac{1}{2\pi\sqrt{L_1 C_3}} \qquad (1\text{-}22)$$

and an oscillator with better frequency stability than the Hartley or Colpitts versions results. The Clapp oscillator does not have as much frequency adjustment range, however.

The *LC* oscillators presented in this section are the ones most commonly used. However, many different forms and variations exist and are used for special applications.

Crystal Oscillator

When greater frequency stability than that provided by *LC* oscillators is required, a crystal-controlled oscillator is often utilized. A crystal oscillator is one that uses a piezoelectric crystal as the inductive element of an *LC* circuit. The crystal, usually quartz, also

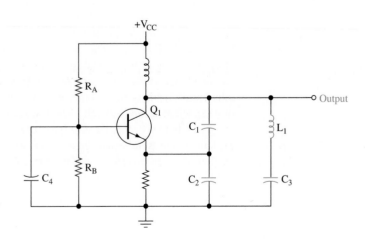

FIGURE 1-22 Clapp oscillator.

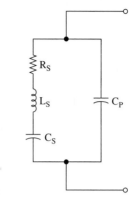

FIGURE 1-23 ■ Electrical equivalent circuit of a crystal.

has a resonant frequency of its own, but optimum performance is obtained when it is coupled with an external capacitance.

The electrical equivalent circuit of a crystal is shown in Fig. 1-23. It represents the crystal by a series resonant circuit (with resistive losses) in parallel with a capacitance C_p. The resonant frequencies of these two resonant circuits (series and parallel) are quite close together (within 1%) and hence the impedance of the crystal varies sharply within a narrow frequency range. This is equivalent to a very high Q circuit, and in fact crystals with a Q-factor of 20,000 are common; a Q of up to 10^6 is possible. This compares to a maximum Q of about 1000 with high-quality LC resonant circuits. For this reason, and because of the good time and temperature stability characteristics of quartz, crystals are capable of maintaining a frequency to ±0.001% over a fairly wide temperature range. The ±0.001% term is equivalent to saying ±10 parts per million (ppm), and this is a preferred way of expressing such very small percentages. Note that 0.001% = 0.00001 = 1/100,000 = 10/1,000,000 = 10 ppm. Over very narrow temperature ranges or by maintaining the crystal in a small temperature-controlled oven, stabilities of ±0.01 ppm are possible.

Crystals are fabricated by "cutting" the crude quartz in a very exacting fashion. The method of "cut" is a science in itself and determines the crystal's natural resonant frequency as well as its temperature characteristics. Crystals are available at frequencies of about 15 kHz and up, with higher frequencies providing the best frequency stability. However, at frequencies above 100 MHz, they become so small that handling becomes a problem.

Crystals may be used in place of the inductors in any of the LC oscillators discussed previously. A circuit especially adapted for crystal oscillators is the Pierce oscillator, shown in Fig. 1-24. The use of an FET is desirable, since its high impedance results in light loading of the crystal, provides for good stability, and does not lower the Q. This circuit is essentially a Colpitts oscillator with the crystal replacing the inductor and the inherent FET junction capacitances functioning as the split capacitor. Because these junction capacitances are generally low, this oscillator is effective only at high frequencies.

In Chapter 8, the use of a basic crystal oscillator in a *frequency synthesizer* will be explained. The synthesizer generates a wide range of frequencies using a single-crystal oscillator as a basic reference. The various output frequencies have the same accuracy and stability as the crystal oscillator.

Crystal oscillators are available in various forms depending upon the frequency stability required. The basic oscillator shown in Fig. 1-24 (often referred to as CXO) may be adequate as a simple clock for a digital system. Increased performance can be attained by

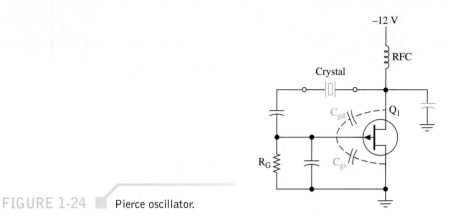

FIGURE 1-24 ■ Pierce oscillator.

adequate as a simple clock for a digital system. Increased performance can be attained by adding temperature compensation circuitry (TCXO). Further improvement is afforded by including microprocessor (digital) control in the crystal oscillator package (DTCXO). The ultimate performance is attained with oven control of the crystal's temperature and sometimes also includes the microprocessor control (OCXO). These obviously require significant power to maintain the oven at some constant elevated temperature. A comparison of the four types of commonly available crystal oscillators is provided in Table 1-3.

Crystal Test

Crystal oscillators may fail to operate due to faulty design or faulty crystals. The circuit shown in Fig. 1-25 works well as a tester for a wide variety of crystals and ceramic res-

TABLE 1-3 *Typical Cost/Performance Comparison for Crystal Oscillators*

	BASIC CRYSTAL OSCILLATOR (CXO)	TEMPERATURE COMPENSATED (TCXO)	DIGITAL TCXO (DTCXO)	OVEN-CONTROLLED CXO (OCXO)
Frequency stability from 0 to 70°C	100 ppm	1 ppm	0.5 ppm	0.05 ppm
Frequency stability for one year at constant temperature	1 ppm	1 ppm	1 ppm	1 ppm
Price	$2–$10	$10–$50	$25–$100	$100 and up

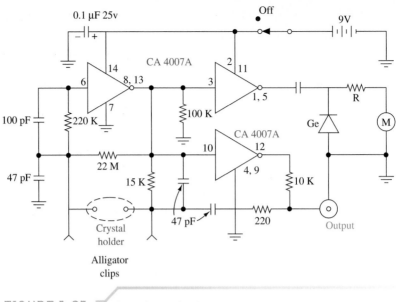

FIGURE 1-25 ■ Crystal test circuit.

onators over the 40-kHz to 20-MHz range. See Sec. 4-3 for a discussion of ceramic resonators.

The oscillator in Fig. 1-25 is a Pierce type that operates at the crystal's parallel resonant frequency and presents about 30 pF capacitance to the crystal. The CA4007A contains three pairs of complementary MOSFETs with the first (input at pin 6) functioning as the Pierce oscillator. The second (input at pin 3) drives a 200–500 μA meter movement. The resistor R is selected to provide about 90% deflection with an active (good) crystal. The "tuning" meter from a discarded stereo is usually ideal for this application. The other complementary pair (input at pin 10) provides a low-impedance output that can drive a frequency counter or provide a connection for an oscilloscope.

The crystal being tested can be inserted in the crystal holder or connected with alligator clips. The input MOSFETs are well protected from electrostatic and leakage damage.

TROUBLESHOOTING

Due to the increasing complexity of electronic communications equipment, the technician must have a good understanding of communication circuits and concepts. To be an effective troubleshooter, the technician must also be able to quickly isolate faulty components and repair the defective circuit. Recognizing the way a circuit may malfunction is a key factor to speedy repair procedures.

After completing this section you should be able to

- Explain general troubleshooting techniques
- Recognize major types of circuit failures
- List the four troubleshooting techniques
- Test for a defective crystal
- Test for defective capacitors and inductors

General Troubleshooting Techniques

Troubleshooting requires the technician to ask questions such as: What could cause this to happen? Why is this voltage so low/high? Or, if this resistance were open/shorted, what effect would it have on the operation of the circuit I'm working on? Each question calls for measurements to be made and tests to be performed. The defective component(s) is isolated when measurements give results far different than they would be in a properly operating unit, or when a test fails. The ability to ask the right questions makes a good troubleshooter. Obviously, the more you know about the circuit or system being worked on, the quicker the problem will be corrected.

Always start troubleshooting by doing the easy things first:

- be sure unit is plugged in and turned on
- check fuses
- check if all connections are made
- ask yourself, "Am I forgetting something?"

Basic troubleshooting test equipment:

- digital multimeter (DMM) capable of reading at the frequencies you intend to work at
- broadband oscilloscope, preferably dual trace
- signal generators, both audio and RF. The RF generator should have internal modulation capabilities.
- a collection of probes and clip leads

Advanced test equipment would include a spectrum analyzer to observe frequency spectra and a logic analyzer for digital work.

Time spent learning your test equipment, its capabilities and limitations, and how to use it will pay off with faster troubleshooting.

Always be aware of any possible effects the test equipment you connect to a circuit may have on the operation of that circuit. Don't let the measuring equipment change that which is being measured. For example, a scope's test lead may have a capacitance of several hundred picofarads. Should that lead be connected across the output of an oscillator, the oscillator's frequency could be changed to the point that any measurements are worthless.

If the equipment you're troubleshooting employs dangerous voltages, do not work alone. Turn off all power switches before entering equipment. See additional comments on safe procedures in the Troubleshooting section of Chapter 2.

Keep all manuals that came with the equipment. Such manuals usually include troubleshooting procedures. Check them before trying any other approaches.

Maintain clear, up-to-date records of all changes made to equipment.

Replace a suspicious unit with a known good one—this is one of the best, most commonly used troubleshooting techniques.

Test points are often built into electronic equipment. They provide convenient connections to the circuitry for adjustment and/or testing. There are various types, from jacks or sockets to short, stubby wires sticking up from PC boards. Equipment manuals will diagram the location of each test point and describe and sometimes illustrate the condition and/or signal that should be found there. The better manuals indicate the proper test equipment to use.

Plot a "game plan" or strategy with which you will troubleshoot a problem (just as you might with a car problem).

Use all your senses when troubleshooting:

Look—for discolored or charred components that might indicate overheating.

Smell—some components, especially transformers, that emit characteristic odors when overheated.

Feel—for hot components. Wiggle components to find broken connections.

Listen—for "frying" noises that indicate a component is about to fail.

Reasons Electronic Circuits Fail

Electronic circuits fail in many ways. Let's look at some major types of failures that you will encounter as an electronics technician.

1. COMPLETE FAILURES Complete failures cause the piece of equipment to go totally dead. Equipment with some circuits still operating has not completely failed. Normally this type of failure is the result of a major circuit path becoming open. Blown (open) fuses, open power resistors, defective power supply rails, and bad regulator transistors in the power supply can cause complete failures. Complete failures are often the easiest problems to repair.

2. INTERMITTENT FAULTS Intermittent faults are characterized by sporadic circuit operation. The circuit works for awhile and then quits working. It works one moment and doesn't work the next. Keeping the circuit in a failed condition can be quite difficult. Loose wires and components, poor soldering, and effects of temperature on sensitive components can all contribute to intermittent operation in a piece of communication equipment. Intermittent faults are usually the most difficult to repair since troubleshooting can only be done when the equipment is malfunctioning.

3. POOR SYSTEM PERFORMANCE Equipment that is functioning below specified operational standards is said to have poor system performance characteristics. For example, a transmitter is showing poor performance if the specifications call for 4 W of output power but it is only putting out 2 W. Degradation of equipment performance takes place over a period of time due to deteriorating components (components change in value), poor alignments, and weakening power components. Regular performance checks are necessary for critical communications systems. Commercial radio transmitters require performance checks to be done on a regular basis.

4. INDUCED FAILURES Induced failures often come from equipment abuse. Unauthorized modifications may have been performed on the equipment. An inexperienced technician without supervision may have attempted repairs and damaged the equipment. Induced failures can be eliminated by exercising proper equipment care. Repairs should be done or supervised by experienced technicians.

Troubleshooting Plan

Experienced technicians have developed a method for troubleshooting. They follow certain logical steps when looking for a defect in a piece of equipment. The following four troubleshooting techniques are popular and widely used to find defects in communication equipment.

1. SYMPTOMS AS CLUES TO FAULTY STAGES   This technique relates a particular fault to a circuit function in the electronic equipment. For example, if a white horizontal line were displayed on the TV screen of a set brought in for repair, the service technician would associate this symptom to the vertical output section. Troubleshooting would begin in that section of the TV set. As you gain experience in troubleshooting you will start associating symptoms to specific circuit functions.

2. SIGNAL TRACING AND SIGNAL INJECTION   Signal injection is supplying a test signal at the input of a circuit and looking for the test signal at the circuit's output or listening for an audible tone at the speaker (Fig. 1-26). This test signal is usually composed of an RF signal modulated with an audible frequency. If the signal is good at the circuit's output, then move to the next stage down the line and repeat the test. Signal tracing, as illustrated in Fig. 1-27, is actually checking for the normal output signal from a stage. An oscilloscope is used to check for these signals. However, other test equipment is available that can be used to detect the presence of output signals. Signal tracing is monitoring the output of a stage for the presence of the expected signal. If the signal is there, then the next stage in line is checked. The malfunctioning stage precedes the point where the output signal is missing.

3. VOLTAGE AND RESISTANCE MEASUREMENTS   Voltage and resistance measurements are made with respect to chassis ground. Using the DMM (digital multimeter), measurements at specific points in the circuit are compared to those found in the equipment's service manual. Service manuals furnish equipment voltage and resistance charts or print the values right on the schematic diagram. Voltage and resistance checks are done to isolate defective components once the trouble has been pinpointed to a specific stage of the equipment. Remember that resistance measurements are done on circuits with the power turned off.

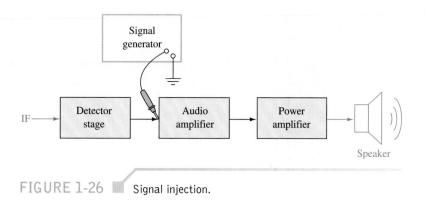

FIGURE 1-26 ▦ Signal injection.

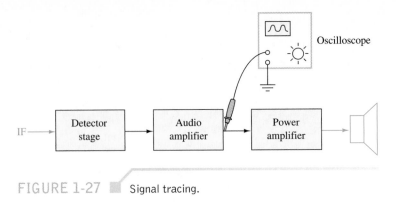

FIGURE 1-27 ■ Signal tracing.

4. SUBSTITUTION   Another method often used to troubleshoot electronic circuits is to swap a known good component for the suspected bad component. A warning is in order here: the good component could get damaged in the substitution process. Don't get into the habit of indiscriminately substituting parts. This method works best when you have narrowed the failure down to a specific component.

Testing a Crystal

An oscillator with a bad crystal may not oscillate at all, may be erratic, or may not oscillate at the correct frequency. One common crystal failure mode is a broken or corroded internal connection. Or if the crystal has been dropped, it may be cracked.

Figure 1-28 shows how to make a simple test to quickly determine the condition of the crystal. Normally, a crystal oscillator will oscillate at a slightly higher frequency than the crystal's series resonant point. If you can find the series resonant point of the crystal, you know the crystal is good.

Recall that at the series resonant point, the crystal should have a very low resistance in the order of 100 ohms. At other frequencies, the crystal impedance should be quite high.

The generator should be very carefully tuned across the specified frequency of the crystal. If the crystal is operating properly, the voltmeter will show a dramatic dip at the series resonant point. Remember that the crystal is a very high-Q device, and tuning the signal generator will have to be done very carefully.

Because the impedance of the crystal is very high at the parallel or anti-resonant point, perhaps 50,000 ohms, there should be a peak on the voltmeter at a frequency just slightly above the series resonant point. You should look for the series resonant point first because it is easier to find.

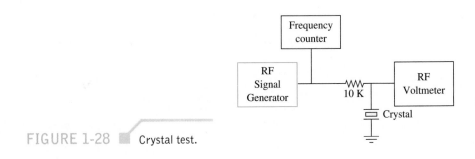

FIGURE 1-28 ■ Crystal test.

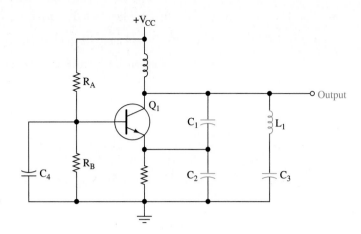

FIGURE 1-29 ▦ Clapp oscillator.

The voltage across a broken crystal will not change much as the generator frequency is varied. Internal connection problems could cause erratic operation. Corrosion problems will cause the resonant frequency to shift from the specified value.

Testing Oscillator Capacitors

The capacitors associated with the crystal or inductor together with the inductor determine the exact frequency of oscillation. This type of capacitor will seldom show a short, but it can become sensitive to temperature and shock or change value with age.

In the Clapp circuit shown in Fig. 1-29, C3 is primarily responsible for setting the frequency. While observing the frequency with a counter, cool the capacitor with an aerosol spray sold for cooling electronic equipment. Defective capacitors will generally change value suddenly and shift the frequency a good bit when cooled. If C3 is open the circuit probably will not oscillate at all.

In the Clapp circuit, C1 and C2 are primarily responsible for providing the proper amount of feedback to allow oscillation. If either of these capacitors fails the oscillator will not work. An oscilloscope connected to the collector of Q1 should show a high quality sine wave. C1 and C2 do have some effect on the frequency and should not be excluded from suspicion if the frequency is not correct.

Testing Oscillator Inductors

A shorted or open inductor will completely kill an oscillator. Inductors can be easily checked for an open circuit with an ohmmeter, though the ohmmeter will not detect a shorted turn. A short in the inductor is best detected with a Q-meter or impedance bridge.

SUMMARY

In Chapter 1 the concept of a communication system was introduced. The effects of electrical noise were explained, and *LC* circuits and oscillators were discussed. The major topics that the student should now understand include:

- the function and basic building blocks of a communication system

- an understanding of the need for modulation/demodulation in a communication system

- the difference between the carrier wave and intelligence wave and their importance

- the effects and analysis of electrical noise in a communication system

- the performance of signal-to-noise ratio and noise figure calculations

- the performance of electrical noise measurements on a communication system

- the makeup of nonsinusoidal waveforms

- the mathematical analysis of waveforms using Fourier analysis

- the analysis of *LC* filters

- the understanding of common oscillator types, including Hartley, Colpitts, Clapp, and crystal varieties

QUESTIONS AND PROBLEMS

SECTION 1

1. Define *modulation.*
*2. What is *carrier frequency*?
3. Describe the two reasons that modulation is used for communications transmissions.
4. List the three parameters of a high-frequency carrier that may be varied by a low-frequency intelligence signal.
5. What are the frequency ranges included in the following frequency subdivisions: MF (medium frequency), HF (high frequency), VHF (very high frequency), UHF (ultra high frequency), and SHF (super high frequency)?
6. List the two basic limitations on the performance of a communications system.

SECTION 2

7. Define *electrical noise,* and explain why it is so troublesome to a communications receiver.
8. Explain the difference between external and internal noise.
9. List and briefly explain the various types of external noise.
10. Provide two other names for Johnson noise and calculate the noise voltage output of a 1-MΩ resistor at 27°C over a 1-MHz frequency range. (128.7 μV)
11. The noise produced by a resistor is to be amplified by a noiseless amplifier having a voltage gain of 75 and a bandwidth of 100 kHz. A sensitive meter at the output reads 240 μV rms. Assuming operation at 37°C, calculate the resistor's resistance. If the bandwidth were cut to 25 kHz, determine the expected output meter reading. (5.985 kΩ, 120 μV)
12. Explain the term *low-noise resistor.*
13. Calculate the noise power and voltage outputs for the following:
    (a) A 1-kΩ resistor at 27°C over a 100-kHz bandwidth. (4.14×10^{-16} W, 1.287 μV)
    (b) A 100-kΩ resistor at 27°C over a 1-kHz bandwidth. (4.14×10^{-18} W, 1.287 μV)

*An asterisk preceding a number indicates a question that has been provided by the FCC as a study aide for licensing examinations.

(b) A 100-kΩ resistor at 27°C over a 1-kHz bandwidth. (4.14×10^{-18} W, 1.287 μV)

(c) A 100-kΩ resistor at 27°C over a 100-kHz bandwidth. (4.14×10^{-16} W, 12.87 μV)

14. Determine the noise current and equivalent noise voltage for a diode with a forward bias of 1 mA over a 1-MHz bandwidth. (17.9 nA, 0.465 μV)

15. The diode in Problem 14 is in a circuit with a series resistance of 1 kΩ. Determine the output noise voltage. Assume that $T = 27$°C. (0.477 μV)

SECTION 3

16. Calculate the S/N ratio for a receiver output of 4 V signal and 0.48 V noise both as a ratio and in decibel form. (69.44, 18.42 dB)

17. The receiver in Problem 16 has a S/N ratio of 110 at its input. Calculate the receiver's noise figure (NF) and noise ratio (NR). (1.998 dB, 1.584)

18. An amplifier with NF = 6 dB has S_i/N_i of 25 dB. Calculate the S_o/N_o in dB and as a ratio. (19 dB, 79.4)

19. A single-stage amplifier has a 200-kHz bandwidth and a voltage gain of 100 at room temperature. Assume that the external noise is negligible and that a 1-mV signal is applied to the amplifier's input. Calculate the output noise voltage if the amplifier has a 5-dB NF and the input noise is generated by a 2-kΩ resistor. (458 μV)

20. Calculate the noise voltage of a 1-kΩ resistor at 17°C over a 1-MHz frequency range. Repeat for the series and parallel combination of three 1-kΩ resistors. (4 μV, 6.93 μV, 2.31 μV)

21. Calculate the noise voltage for a 1-kΩ resistor at 17°C "tuned" by an LC circuit with a BW of 1 MHz. (5.01 μV)

22. A three-stage amplifier has an input stage with noise ratio (NR) = 5 and power gain (P_G) = 50. Stages 2 and 3 have NR = 10 and P_G = 1000. Calculate the NF for the overall system. (8.33 dB)

23. A two-stage amplifier has a 3-dB bandwidth of 150 kHz determined by an LC circuit at its input and operates at 27°C. The first stage has P_G = 8 dB and NF = 2.4 dB. The second stage has P_G = 40 dB and NF = 6.5 dB. The output is driving a load of 300 Ω. In testing this system, the noise of a 100-kΩ resistor is applied to its input. Calculate the input and output noise voltage and power and the system noise figure. (19.8 μV, 0.206 mV, 9.75×10^{-16} W, 1.4×10^{-10} W, 3.6 dB)

24. A microwave antenna (T_{eq} = 25 K) is coupled through a network (T_{eq} = 30 K) to a microwave receiver with T_{eq} = 60 K referred to its input. Calculate the noise power at its input for a 2-MHz bandwidth. Determine the receiver's NF. (3.17×10^{-15} W, 0.817 dB)

SECTION 4

25. Explain the meaning of *equivalent noise resistance* and *equivalent noise temperature*.

26. Calculate the noise power at the input of a microwave receiver with an equivalent noise temperature of 45 K. It is fed from an antenna with a 35 K equivalent noise temperature and operates over a 5-MHz bandwidth. (5.52×10^{-15} W)

27. Calculate the minimum signal power for the receiver described in Problem 26 needed for good reception if the signal-to-noise ratio must be not less than 100:1. $(5.52 \times 10^{-13} \text{ W})$

28. Calculate the NF and T_{eq} for an amplifier that has $Z_{in} = 300 \ \Omega$. It is found that when driven from a matched-impedance diode noise generator, its output noise is doubled (as compared to no input noise) when the diode is forward biased with 0.3 mA. (2.55 dB, 232 K)

29. Describe the procedure used for noise measurement using the noise diode generator.

30. Describe the procedure for noise measurement using the tangential technique.

SECTION 5

31. Define *information theory*.

32. What is Hartley's law? Explain its significance.

*33. What is a *harmonic?*

*34. What is the seventh harmonic of 360 kHz? (2520 kHz)

35. Why does transmission of a 2-kHz square wave require greater bandwidth than a 2-kHz sine wave?

36. Draw time- *and* frequency-domain sketches for a 2-kHz square wave. The time-domain sketch is a standard oscilloscope display while the frequency domain is provided by a spectrum analyzer.

37. Explain the function of Fourier analysis.

38. A 2-kHz square wave is passed through a filter with a 0- to 10-kHz frequency response. Sketch the resulting signal, and explain why the distortion occurs.

39. A triangle wave of the type shown in Table 1-2(e) has a peak-to-peak amplitude of 2 V and $f = 1$ kHz. Write the expression $v(t)$, including the first five harmonics. Graphically add the harmonics to show the effects of passing the wave through a low-pass filter with cutoff frequency equal to 6 kHz.

SECTION 6

40. Explain the makeup of a practical inductor and capacitor. Include the quality and dissipation in your discussion.

41. Calculate an inductor's Q at 100 MHz. It has an inductance of 6 mH and a series resistance of 1.2 kΩ. Determine its dissipation. $(3.14 \times 10^3, 0.318 \times 10^{-3})$

42. Calculate a capacitor's Q at 100 MHz given 0.001 μF and a leakage resistance of 0.7 MΩ. Calculate D for the same capacitor. $(4.39 \times 10^5, 2.27 \times 10^{-6})$

43. The inductor and capacitor for Problems 41 and 42 are put in series. Calculate the impedance at 100 MHz. Calculate the frequency of resonance (f_r) and the impedance at that frequency. (37.5 kΩ, 65 kHz, 1200 Ω)

44. Calculate the output voltage for the circuit shown in Fig. 1-14 at 6 kHz and 4 kHz. Graph these results together with those of Ex. 1-10 versus frequency. Use the circuit values given in Ex. 1-10.

45. Sketch the e_{out}/e_{in} versus frequency characteristic for a *LC* bandpass filter. Show f_{lc} and f_{hc} on the sketch and explain how they are defined. On this sketch, show the bandwidth (BW) of the filter and explain how it is defined.

46. Define the quality factor (Q) of an LC bandpass filter. Explain how it relates to the "selectivity" of the filter. Describe the major limiting value on the Q of a filter.

47. An FM radio receiver uses an LC bandpass filter with $f_r = 10.7$ MHz and requires a BW of 200 kHz. Calculate the Q for this filter. (53.5)

48. The circuit described in Problem 47 is shown in Fig. 1-17(a). If $C = 0.1$ nF (0.1×10^{-9} F), calculate the required inductor value and the value of R. (2.21 μH, 2.78 Ω)

49. A parallel LC tank circuit has a Q of 60 and coil winding resistance of 5 Ω. Determine the circuit's impedance at resonance. (18 kΩ)

50. A parallel LC tank circuit has $L = 27$ mH, $C = 0.68$ μF, and a coil winding resistance of 4 Ω. Calculate f_r, Q, Z_{max}, the BW, f_{lc}, and f_{hc}. (1175 Hz, 49.8, 9.93 kΩ, 23.6 Hz, 1163 Hz, 1187 Hz)

SECTION 7

51. Explain the Barkhausen criteria for oscillation. Describe the relationship to positive feedback in your explanation.

52. Draw schematics for Hartley and Colpitts oscillators. Briefly explain their operation and differences.

53. Describe the reason that a Clapp oscillator has better frequency stability than the Hartley or Colpitts oscillators.

54. List the major advantages of crystal oscillators over the LC varieties. Draw a schematic for a Pierce oscillator.

55. The crystal oscillator time base for a digital wristwatch yields an accuracy of ± 15 s/month. Express this accuracy in parts per million (ppm). (± 5.787 ppm)

AMPLITUDE MODULATION: TRANSMISSION

Spectrum Analyzers

EMI Test Receivers

Network Analyzers

Communication Service Monitors

RF Power Meters

RF Signal Generators

Portable Field Analyzers

OBJECTIVES

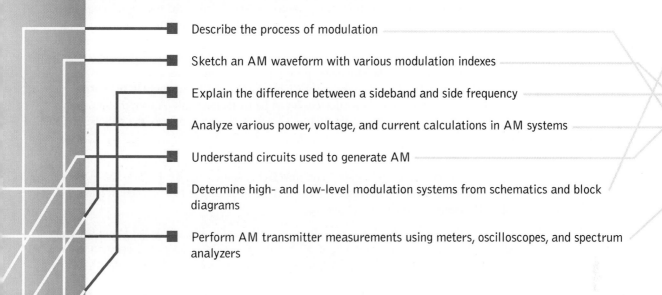

- Describe the process of modulation

- Sketch an AM waveform with various modulation indexes

- Explain the difference between a sideband and side frequency

- Analyze various power, voltage, and current calculations in AM systems

- Understand circuits used to generate AM

- Determine high- and low-level modulation systems from schematics and block diagrams

- Perform AM transmitter measurements using meters, oscilloscopes, and spectrum analyzers

2-1 INTRODUCTION

The reasons that modulation is used in electronic communications have previously been explained as:

1. Direct transmission of intelligible signals would result in catastrophic interference problems, since the resulting radio waves would be at approximately the same frequency.

2. Most intelligible signals occur at relatively low frequencies. Efficient transmission and reception of radio waves at low frequencies is not practical due to the large antennas required.

The process of impressing a low-frequency intelligence signal onto a higher-frequency "carrier" signal may be defined as *modulation.* The higher-frequency "carrier" signal will hereafter be referred to as simply the carrier. It is also termed the radio-frequency (RF) signal, since it is at a high-enough frequency to be transmitted through free space as a radio wave. The low-frequency intelligence signal will subsequently be termed the "intelligence." It may also be identified by terms such as modulating signal, information signal, audio signal, or modulating wave.

Three different characteristics of a carrier can be modified so as to allow it to "carry" intelligence. Either the amplitude, frequency, or phase of a carrier are altered by the intelligence signal. Varying the carrier's amplitude to accomplish this goal is the subject of this chapter.

Combining two widely different sine-wave frequencies such as a carrier and intelligence in a linear fashion results in their simple algebraic addition, as shown in Fig. 2-1. A circuit that would perform this function is shown in Fig. 2-1(a)—the two signals combined in a linear device such as a resistor. Unfortunately, the resultant [Fig. 2-1(d)] is *not* suitable for transmission as an AM waveform. If it were transmitted, the receiving antenna would be detecting just the carrier signal [Fig. 2-1(c)], because the low-frequency intelligence component cannot be efficiently propagated as a radio wave.

The method utilized to produce a usable AM signal is to combine the carrier and intelligence through a *nonlinear* device. It can be mathematically proven that the combination of any two sine waves through a nonlinear device produces the following frequency components:

1. A dc level
2. Components at each of the two original frequencies
3. Components at the sum and difference frequencies of the two original frequencies
4. Harmonics of the two original frequencies

Figure 2-2 shows this process pictorially with the two sine waves, labeled f_c and f_i, to represent the carrier and intelligence. If all but the $f_c - f_i$, f_c, and $f_c + f_i$ components are removed (perhaps with a bandpass filter), the three components left form an AM waveform. They are referred to as:

1. The *lower side frequency* $(f_c - f_i)$
2. The *carrier frequency* (f_c)
3. The *upper side frequency* $(f_c + f_i)$

Mathematical analysis of this process is provided in Sec. 2-4.

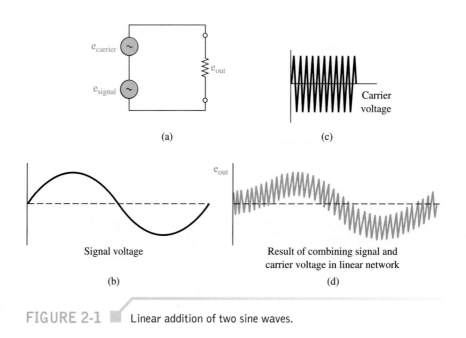

$e_{carrier}$ e_{out}

e_{signal}

(a)

Carrier voltage

(c)

Signal voltage

(b)

e_{out}

Result of combining signal and carrier voltage in linear network

(d)

FIGURE 2-1   Linear addition of two sine waves.

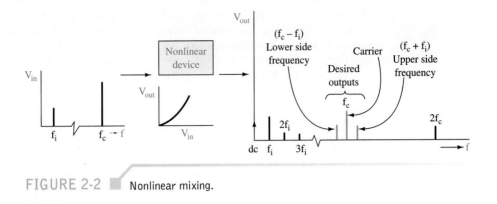

FIGURE 2-2 ■ Nonlinear mixing.

AM Waveforms

Figure 2-3 shows the actual AM waveform under varying conditions of the intelligence signal. Note in Fig. 2-3(a) that the resultant AM waveform is basically a signal at the carrier frequency whose amplitude is changing at the same rate as the intelligence frequency. As the intelligence amplitude reaches a maximum positive value, the AM waveform has a maximum amplitude. The AM waveform reaches a minimum value when the intelligence amplitude is at a maximum negative value. In Fig. 2-3(b), the intelligence frequency remains the same, but its amplitude has been increased. The resulting AM waveform reacts by reaching a larger maximum value and smaller minimum value. In Fig. 2-3(c),

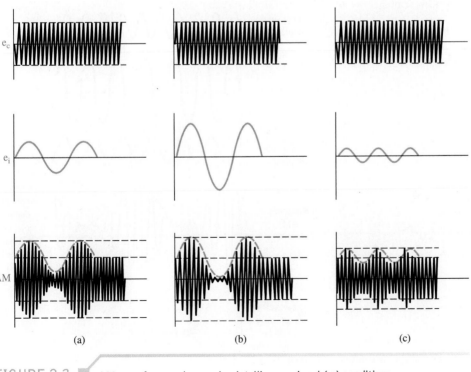

FIGURE 2-3 ■ AM waveform under varying intelligence signal (e_i) conditions.

the intelligence amplitude is reduced and its frequency has gone up. The resulting AM waveform, therefore, has reduced maximums and minimums, and the rate at which it swings between these extremes has increased to the same frequency as the intelligence signal.

It may now be correctly concluded that both the top and bottom envelopes of an AM waveform are replicas of the frequency and amplitude of the intelligence (notice the 180° phase shift). However, the AM waveform does *not* include any component at the intelligence frequency. If a 1-MHz carrier were modulated by a 5-kHz intelligence signal, the AM waveform would include the following components:

$$1 \text{ MHz} + 5 \text{ kHz} = 1,005,000 \text{ Hz (upper side frequency)}$$

$$1 \text{ MHz} = 1,000,000 \text{ Hz (carrier frequency)}$$

$$1 \text{ MHz} - 5 \text{ kHz} = 995,000 \text{ Hz (lower side frequency)}$$

This process is shown in Fig. 2-4. Thus, even though the AM waveform has envelopes that are replicas of the intelligence signal, it *does not* contain a frequency component at the intelligence frequency.

The intelligence envelope is shown in the resultant waveform and results from connecting a line from each RF peak value to the next one for both the top and bottom halves of the AM waveform. The drawn-in envelope is not really a component of the waveform and would not be seen on an oscilloscope display. In addition, the top and bottom envelopes are *not* the upper and lower side frequencies, respectively. The envelopes result

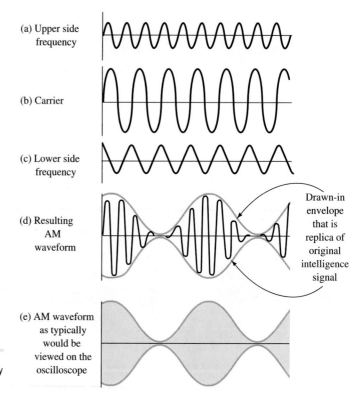

(a) Upper side frequency

(b) Carrier

(c) Lower side frequency

(d) Resulting AM waveform

Drawn-in envelope that is replica of original intelligence signal

(e) AM waveform as typically would be viewed on the oscilloscope

FIGURE 2-4 ■ Carrier and side-frequency components result in AM waveform.

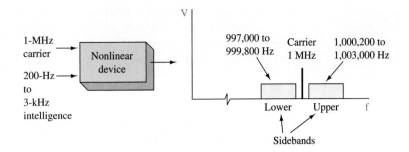

FIGURE 2-5 ▊ Modulation by a band of intelligence frequencies.

from nonlinear combination of a carrier with two lower-amplitude signals spaced in frequency equal amounts above and below the carrier frequency. The increase and decrease in the AM waveform's amplitude is caused by the frequency difference in the side frequencies, which allows them to alternately add to and subtract from the carrier amplitude, depending on their instantaneous phase relationships.

The AM waveform in Fig. 2-4(d) does not show the relative frequencies to scale. The ratio of f_c to the envelope frequency (which is also f_i) is 1 MHz to 5 kHz, or 200:1. Thus, the fluctuating RF should show 200 cycles for every cycle of envelope variation. To do that in a sketch is not possible, and an oscilloscope display of this example, and most practical AM waveforms, results in a well-defined envelope but with so many RF variations that they appear as a blur, as shown in Fig. 2-4(e).

Modulation of a carrier with a pure sine-wave intelligence signal has thus far been shown. However, in most systems the intelligence is a rather complex waveform that contains many frequency components. For example, the human voice contains components from roughly 200 Hz to 3 kHz and has a very erratic shape. If it were used to modulate the carrier, a whole *band* of side frequencies would be generated. The band of frequencies thus generated above the carrier is termed the *upper sideband,* while those below the carrier are called the *lower sideband.* This situation is illustrated in Fig. 2-5 for a 1-MHz carrier modulated by a whole band of frequencies, which range from 200 Hz up to 3kHz. The upper sideband is from 1,000,200 to 1,003,000 Hz, and the lower sideband ranges from 997,000 to 999,800 Hz.

EXAMPLE 2-1

A 1.4-MHz carrier is modulated by a music signal that has frequency components from 20 Hz to 15 kHz. Determine the range of frequencies generated for the upper and lower sidebands.

SOLUTION

The upper sideband is equal to the sum of carrier and intelligence frequencies. Therefore, the upper sideband (usb) will include the frequencies from

$$1,400,000 \text{ Hz} + 20 \text{ Hz} = 1,400,020 \text{ Hz}$$

to

$$1,400,000 \text{ Hz} + 15,000 \text{ Hz} = 1,415,000 \text{ Hz}$$

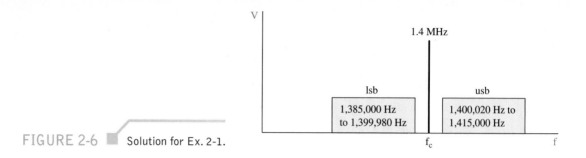

FIGURE 2-6 ■ Solution for Ex. 2-1.

The lower sideband (lsb) will include the frequencies from

$$1,400,000 \text{ Hz} - 15,000 \text{ Hz} = 1,385,000 \text{ Hz}$$

to

$$1,400,000 \text{ Hz} - 20 \text{ Hz} = 1,399,980 \text{ Hz}$$

This result is shown in Fig. 2-6 with a frequency spectrum of the AM modulator's output.

2-3 PERCENTAGE MODULATION

In Sec. 2-2 it was determined that an increase in intelligence amplitude resulted in an AM signal with larger maximums and smaller minimums. It is helpful to have a mathematical relationship between the relative amplitude of the carrier and intelligence signals. The *percentage modulation* provides this, and it is a measure of the extent to which a carrier voltage is varied by the intelligence. The percentage modulation is also referred to as *modulation index* or *modulation factor,* and they are symbolized by *m*.

Figure 2-7 illustrates the two most common methods for determining the percentage modulation. Notice that when the intelligence signal is zero, the carrier is unmodulated and has a peak amplitude labeled as E_c. When the intelligence reaches its first peak value (point *w*), the AM signal reaches a peak value labeled E_i (the increase from E_c). Percentage modulation is then given as

$$\%m = \frac{E_i}{E_c} \times 100\% \qquad (2\text{-}1)$$

or expressed simply by a ratio:

$$m = \frac{E_i}{E_c} \qquad (2\text{-}2)$$

The same result can be obtained by utilizing the maximum peak-to-peak value of the AM waveform (point *w*), which is shown as *B,* and the minimum peak-to-peak value (point *x*), which is *A* in the following equation:

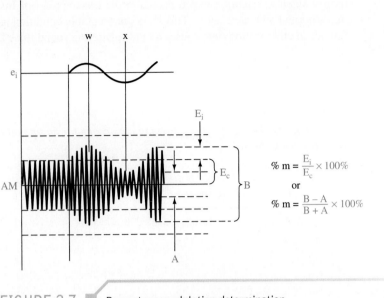

$$\% \, m = \frac{E_i}{E_c} \times 100\%$$

or

$$\% \, m = \frac{B - A}{B + A} \times 100\%$$

FIGURE 2-7   Percentage modulation determination.

$$\%m = \frac{B - A}{B + A} \times 100\% \qquad\qquad (2\text{-}3)$$

This method is usually more convenient in graphical (oscilloscope) solutions.

Overmodulation

If the AM waveform's minimum value *A* falls to zero as a result of an increase in the intelligence amplitude, the percentage modulation becomes

$$\%m = \frac{B - A}{B + A} \times 100\% = \frac{B - O}{B + O} \times 100\% = 100\%$$

This is the maximum possible degree of modulation. In this situation the carrier is being varied between zero and double its unmodulated value. Any further increase in the intelligence amplitude will cause a condition known as *overmodulation* to occur. If this does occur, the modulated carrier will go to more than double its unmodulated value but will fall to zero for an interval of time, as shown in Fig. 2-8. This "gap" produces distortion

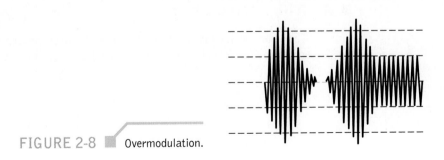

FIGURE 2-8   Overmodulation.

termed *sideband splatter,* which results in the transmission of frequencies outside a station's normal allocated range. This is an unacceptable condition as it causes severe interference to other stations and causes a loud splattering sound to be heard at the receiver.

EXAMPLE 2-2

Determine the %m for the following conditions for an unmodulated carrier of 80 V peak-to-peak (p-p).

	MAXIMUM P-P CARRIER (V)	MINIMUM P-P CARRIER (V)
(a)	100	60
(b)	125	35
(c)	160	0
(d)	180	0
(e)	135	35

SOLUTION

(a)
$$\%m = \frac{B - A}{B + A} \times 100\%$$

(2-3)

$$= \frac{100 - 60}{100 + 60} \times 100\% = 25\%$$

(b)
$$\%m = \frac{125 - 35}{125 + 35} \times 100\% = 56.25\%$$

(c)
$$\%m = \frac{160 - 0}{160 + 0} \times 100\% = 100\%$$

(d) This is a case of overmodulation since the modulated carrier reaches a value more than twice its unmodulated value.

(e) The increase is greater than the decrease in the carrier's amplitude. This is a distorted AM wave.

2-4 AM ANALYSIS

The equation for the amplitude of an AM waveform can be written as the carrier peak amplitude, E_c, plus the intelligence signal, e_i. Thus, the amplitude E is

$$E = E_c + e_i$$

but $e_i = E_i \sin \omega_i t$, so that

$$E = E_c + E_i \sin \omega_i t$$

From Eq. (2-2), $E_i = mE_c$, so that

$$E = E_c + mE_c \sin \omega_i t$$

$$= E_c(1 + m \sin \omega_i t)$$

The instantaneous value of the AM wave is the amplitude term E just developed times $\sin \omega_c t$. Thus,

$$e = E \sin \omega_c t$$

$$= E_c(1 + m \sin \omega_i t) \sin \omega_c t$$

Notice that the AM wave (e) is the result of the product of two sine waves. This can be expanded with the help of the trigonometric relation $\sin x \sin y = \frac{1}{2}[\cos (x - y) - \cos (x + y)]$. Therefore,

$$e = \overbrace{E_c \sin \omega_c t}^{①} + \overbrace{\frac{mE_c}{2} \cos (\omega_c - \omega_i)t}^{②} - \overbrace{\frac{mE_c}{2} \cos (\omega_c + \omega_i)t}^{③}$$

The preceding equation proves that the AM wave contains the three terms previously listed: the carrier ①, the upper sideband at $f_c + f_i$ ③, and the lower sideband at $f_c - f_i$ ②. It also proves that the instantaneous amplitude of the side frequencies is $mE_c/2$. It shows conclusively that the bandwidth required for AM transmission is twice the highest intelligence frequency.

In the case where a carrier is modulated by a pure sine wave, it can be shown that at 100% modulation, the upper and lower side frequencies are one-half the amplitude of the carrier. In general, as just developed,

$$E_{\text{SF}} = \frac{mE_c}{2} \qquad (2\text{-}4)$$

where E_{SF} = side-frequency amplitude
$\quad m$ = modulation index
$\quad E_c$ = carrier amplitude

In an AM transmission, the carrier amplitude and frequency always remain constant while the sidebands are usually changing in amplitude and frequency. The carrier contains no information since it never changes. However, it does contain the most power since its amplitude is always at least double (when $m = 100\%$) the sideband's amplitude. It is the sidebands that contain the information.

EXAMPLE 2-3

Determine the maximum sideband power if the carrier output is 1 kW and calculate the total maximum transmitted power.

Since

$$E_{SF} = \frac{mE_c}{2} \qquad (2\text{-}4)$$

it is obvious that the maximum sideband power occurs when $m = 1$ or 100%. At that percentage modulation, each side frequency is $\frac{1}{2}$ the carrier amplitude. Since power is proportional to the square of voltage, each sideband has $\frac{1}{4}$ of the carrier power or $\frac{1}{4} \times 1$ kW, or 250 W. Therefore, the total sideband power is 250 W $\times$ 2 = 500 W and the total transmitted power is 1 kW + 500 W, or 1.5 kW.

Importance of High-Percentage Modulation

It is important to use as high a percentage modulation as possible while ensuring that overmodulation does not occur. The sidebands contain the information and have maximum power at 100% modulation. For example, if 50% modulation were used in Ex. 2-3, the sideband amplitudes are $\frac{1}{4}$ the carrier amplitude, and since power is proportional to E^2, we have $(\frac{1}{4})^2$, or $\frac{1}{16}$ the carrier power. Thus, total sideband power is now $\frac{1}{16} \times 1$ kW $\times 2$, or 125 W. The actual transmitted intelligence is thus only $\frac{1}{4}$ of the 500 W sideband power transmitted at full 100% modulation. These results are summarized in Table 2-1. Even though the total transmitted power has only fallen from 1.5 kW to 1.125 kW, the effective transmission has only $\frac{1}{4}$ the strength at 50% modulation as compared to 100%. Because of these considerations, most AM transmitters attempt to maintain between 90 and 95% modulation as a compromise between efficiency and the chance of drifting into overmodulation.

A valuable relationship for many AM calculations is

$$P_t = P_c\left(1 + \frac{m^2}{2}\right) \qquad (2\text{-}5)$$

where P_t = total transmitted power (sidebands and carrier)
$\quad\ P_c$ = carrier power
$\quad\ m$ = modulation index

TABLE 2-1 *Effective Transmission at 50% versus 100% Modulation*

MODULATION INDEX, m	CARRIER POWER (kW)	POWER IN ONE SIDEBAND (W)	TOTAL SIDEBAND POWER (W)	TOTAL TRANSMITTED POWER, P_t (kW)
1.0	1	250	500	1.5
0.5	1	62.5	125	1.125

Equation (2-5) can be manipulated to utilize current instead of power. This is a useful relationship since current is often the most easily measured parameter of a transmitter's output to the antenna.

$$I_t = I_c \sqrt{1 + \frac{m^2}{2}}$$

(2-6)

where I_t = total transmitted current
I_c = carrier current
m = modulation index

Equation (2-6) can also be used with E substituted for I ($E_t = E_c \sqrt{1 + m^2/2}$).

EXAMPLE 2-4

A 500-W carrier is to be modulated to a 90% level. Determine the total transmitted power.

$$P_t = P_c \left(1 + \frac{m^2}{2}\right)$$

$$= 500 \text{ W} \left(1 + \frac{0.9^2}{2}\right)$$

(2-5)

$$= 702.5 \text{ W}$$

EXAMPLE 2-5

An AM broadcast station operates at its maximum allowed total output of 50 kW and at 95% modulation. How much of its transmitted power is intelligence (sidebands)?

SOLUTION

$$P_t = P_c \left(1 + \frac{m^2}{2}\right)$$

$$50 \text{ kW} = P_c \left(1 + \frac{0.95^2}{2}\right)$$

(2-5)

$$P_c = \frac{50 \text{ kW}}{1 + (0.95^2/2)}$$

$$= 34.5 \text{ kW}$$

Therefore, the total intelligence signal is

$$P_i = P_t - P_c = 50 \text{ kW} - 34.5 \text{ kW}$$

$$= 15.5 \text{ kW}$$

EXAMPLE 2-6

The antenna current of an AM transmitter is 12 A when unmodulated but increases to 13 A when modulated. Calculate %m.

$$I_t = I_c \sqrt{1 + \frac{m^2}{2}}$$

$$13 \text{ A} = 12 \text{ A} \sqrt{1 + \frac{m^2}{2}}$$

$$1 + \frac{m^2}{2} = \left(\frac{13}{12}\right)^2 \qquad (2\text{-}6)$$

$$m^2 = 2\left[\left(\frac{13}{12}\right)^2 - 1\right]$$

$$= 0.34$$

$$m = 0.59$$

$$\%m = 0.59 \times 100\% = 59\%$$

EXAMPLE 2-7

An intelligence signal is amplified by a 70% efficient amplifier before being combined with a 10-kW carrier to generate the AM signal. If it is desired to operate at 100% modulation, what is the dc input power to the final intelligence amplifier?

You may recall that the efficiency of an amplifier is the ratio of ac output power to dc input power. To fully modulate a 10-kW carrier requires 5 kW of intelligence. Therefore, to provide 5 kW of sideband (intelligence) power through a 70% efficient amplifier requires a dc input of

$$\frac{5 \text{ kW}}{0.70} = 7.14 \text{ kW}$$

If a carrier is modulated by more than a single sine wave, the effective modulation index is given by

$$m_{\text{eff}} = \sqrt{m_1^2 + m_2^2 + m_3^2 + \cdots} \qquad (2\text{-}7)$$

The total effective modulation index must not exceed 1 or distortion (as with a single sine wave) will result. The term m_{eff} can be used in all previously developed equations using m.

EXAMPLE 2-8

A transmitter with a 10-kW carrier transmits 11.2 kW when modulated with a single sine wave. Calculate the modulation index. If the carrier is simultaneously modulated with another sine wave at 50% modulation, calculate the total transmitted power.

SOLUTION

$$P_t = P_c\left(1 + \frac{m^2}{2}\right)$$

$$11.2 \text{ kW} = 10 \text{ kW} \left(1 + \frac{m^2}{2}\right) \tag{2-5}$$

$$m = 0.49$$

$$m_{\text{eff}} = \sqrt{m_1^2 + m_2^2} \tag{2-7}$$

$$= \sqrt{0.49^2 + 0.5^2}$$

$$= 0.7$$

$$P_t = P_c\left(1 + \frac{m^2}{2}\right)$$

$$= 10 \text{ kW} \left(1 + \frac{0.7^2}{2}\right)$$

$$= 12.45 \text{ kW}$$

2-5 CIRCUITS FOR AM GENERATION

Amplitude modulation is generated by combining carrier and intelligence frequencies through a nonlinear device. Diodes have nonlinear areas, but they are not often used because, being passive devices, they offer no gain. Transistors offer nonlinear operation (if properly biased) and provide amplification, thus making them ideal for this application. Figure 2-9(a) shows an input/output relationship for a typical bipolar junction transistor (BJT). Notice that at both low and high values of current, nonlinear areas exist. Between these two extremes is the linear area that should be used for normal amplification. One of the nonlinear areas must be used to generate AM.

Figure 2-9(b) shows a very simple transistor modulator. It operates with no base bias and thus depends on the positive peaks of e_c and e_i to bias it into the first nonlinear area shown in Fig. 2-9(a). Proper adjustment of the levels of e_c and e_i is necessary for good operation. Their levels must be low to stay in the first nonlinear area, and the intelligence power must be one-half the carrier power (or less) for 100% modulation (or less). In the collector a parallel resonant circuit, tuned to the carrier frequency, is used to tune into the three desired frequencies—the upper and lower side bands and the carrier. The resonant circuit presents a high impedance to the carrier (and any other close frequencies such as the sidebands) and thus allows a high output to those components, but its very low impedance to all other frequencies effectively shorts them out. Recall that the mixing of two fre-

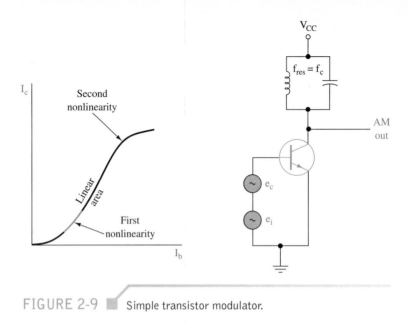

FIGURE 2-9 ■ Simple transistor modulator.

quencies through a nonlinear device generates more than just the desired AM components, as illustrated in Fig. 2-2. The tuned circuit then "sorts" out the three desired AM components and serves to provide good sinusoidal components by the flywheel effect.

In practice, amplitude modulation can be obtained in a number of ways. For descriptive purposes, the point of intelligence injection is utilized. For example, in Fig. 2-9(b) the intelligence is injected into the base, hence it is termed *base modulation*. *Collector* and *emitter modulation* are also used. In previous years, when vacuum tubes were widely used, the most common form was *plate modulation,* but *grid, cathode,* and (for pentodes) *suppressor-grid* and *screen-grid* modulation schemes were also utilized.

High- and Low-Level Modulation

Another common designator for modulators involves whether or not the intelligence is injected at the last possible place or not. For example, the plate-modulated circuit shown in Fig. 2-10 has the intelligence added at the last possible point before the transmitting antenna and is termed a *high-level* modulation scheme. If the intelligence was injected at any previous point, such as a base, emitter, grid, or cathode, or even a previous stage, it would be termed *low-level* modulation. The designer's choice between high- and low-level systems is made largely on the basis of the required power output. For high-power applications such as standard radio broadcasting, where outputs are measured in terms of kilowatts instead of watts, high-level modulation is the most economical approach. Recall that class C bias (device conduction for less than 180°) allows for the highest possible efficiency. It realistically provides 70 to 80% efficiency as compared to about 50 to 60% for the next best configuration, a class B (linear) amplifier. However, class C amplification cannot be used for reproduction of the complete AM signal, and hence large amounts of intelligence power must be injected at the final output to provide a high percentage modulation.

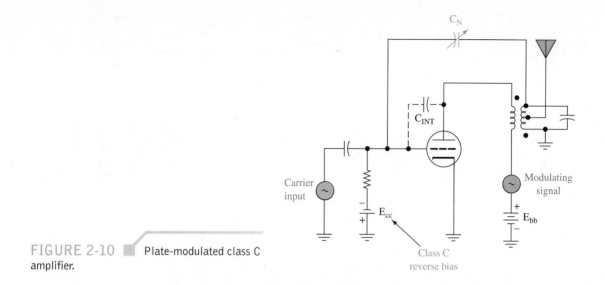

FIGURE 2-10 Plate-modulated class C
amplifier.

The modulation process is accomplished in a nonlinear device, but all circuitry that
follows must be linear. This is required to provide reproduction of the AM signal without
distortion. The class C amplifier is not linear but can reproduce (and amplify) the single
frequency carrier. However, it would distort the carrier and sidebands combination of the
AM signal. This is due to their changing amplitude that would be distorted by the flywheel effect in the class C tank circuit. Block diagrams for typical high- and low-level

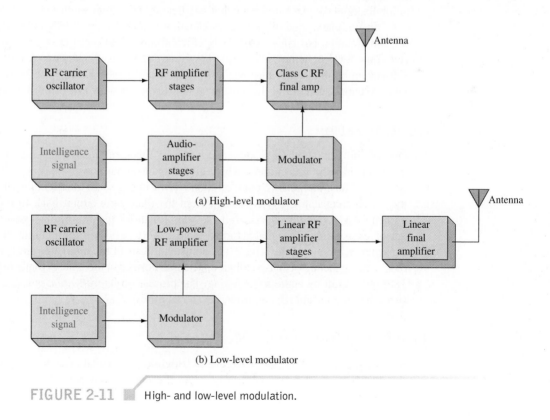

FIGURE 2-11 High- and low-level modulation.

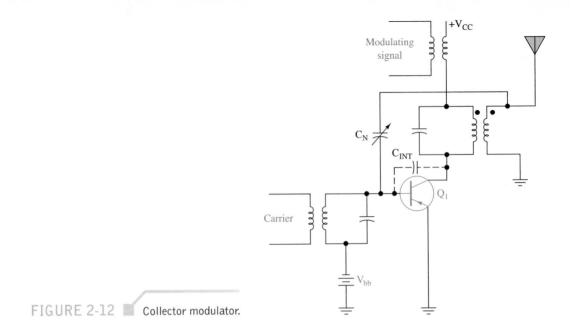

FIGURE 2-12 ▮ Collector modulator.

modulator systems are shown in Fig. 2-11(a) and (b), respectively. Note that in the high-level modulation system [Fig. 2-11(a)] the majority of power amplification takes place in the highly efficient class C amplifier. The low-level modulation scheme has its power amplification take place in the much less efficient linear final amplifier.

In summary, then, high-level modulation requires larger intelligence power to produce modulation but allows extremely efficient amplification of the higher-powered carrier. Low-level schemes allow low-powered intelligence signals to be used, but all subsequent output stages must use less efficient linear (not class C) configurations. Low-level systems usually offer the most economical approach for low-power transmitters.

Neutralization

One of the last remaining applications where tubes offer advantages over solid-state devices is in radio transmitters, where kilowatts of output power are required at high frequencies. Thus, the general configuration shown in Fig. 2-10 is still being utilized. Note the variable capacitor, C_N, connected from the plate tank circuit back to the grid. It is termed the *neutralizing* capacitor. It provides a path for the return of a signal that is 180° out of phase with the signal returned from plate to grid via the internal interelectrode capacitance, (C_{INT}) of the tube. C_N is adjusted to cancel the internally fed-back signal to reduce the tendency of self-oscillation. The transformer in the plate is made to introduce a 180° phase shift by appropriate wiring. The process of *neutralization* is also used in transistor modulators and RF amplifiers, as shown in Fig. 2-12.

Transistor High-Level Modulator

Figure 2-12 shows a transistorized class C, high-level modulation scheme. Class C operation provides an abrupt nonlinearity when the device switches "on" and "off," which allows for the generation of sum and difference frequencies. This is in contrast to the use of the gradual nonlinearities offered by a transistor at high and low levels of class A bias, as previously shown in Fig. 2-9(a). Generally, the operating point is established so as to allow half the maximum ac output voltage to be supplied at the collector when the intelligence signal is zero. The V_{bb} supply provides a reverse bias for Q_1 so that it conducts on only the positive peak of the input carrier signal. This, by definition, is class C bias, since Q_1 conducts for less than $180°$ per cycle. The tank circuit in Q_1's collector is tuned to resonate at f_c, and thus the full carrier sine wave is reconstructed there by the flywheel effect at the extremely high efficiency afforded by class C operation.

The intelligence (modulating) signal for the collector modulator of Fig. 2-12 is added directly in series with the collector supply voltage. The net effect of the intelligence signal is to vary the energy available to the tank circuit each time Q_1 conducts on the positive peaks of carrier input. This causes the output to reach a maximum value when the intelligence is at its peak positive value and a minimum value when the intelligence is at its peak negative value. Since the circuit is biased so as to provide one-half of the maximum possible carrier output when the intelligence is zero, theoretically an intelligence signal level exists where the carrier will swing between twice its static value and zero. This is a fully modulated (100% modulation) AM waveform. In practice, however, the collector modulator cannot achieve 100% modulation, because the transistor's knee in its characteristic curve changes at the intelligence frequency rate. This limits the region over which the collector voltage can vary, and slight collector modulation of the preceding stage is necessary to allow the high modulation indexes that are usually desirable. This is sometimes not a necessary measure in the tube-type high-level modulators.

Figure 2-13(a) shows an intelligence signal for a collector modulator, and Fig. 2-13(b) shows its effect on the collector supply voltage. In Fig. 2-13(c), the resulting collec-

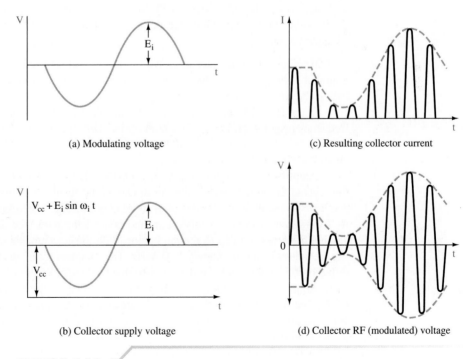

(a) Modulating voltage

(c) Resulting collector current

(b) Collector supply voltage

(d) Collector RF (modulated) voltage

FIGURE 2-13 Collector modulator waveforms.

tor current variations that are in step with the available supply voltages are shown. Fig. 2-13(d) shows the collector voltage produced by the flywheel effect of the tank circuit as a result of the varying current peaks that are flowing through the tank.

Linear-Integrated-Circuit Modulators

The process of generating high-quality AM signals economically is greatly simplified by the availability of low-cost specialty linear integrated circuits (LIC). This is especially true for low-power systems, where low-level modulation schemes are attractive. As an example, the RCA CA3080 operational transconductance amplifier (OTA) can be used to provide AM with an absolute minimum of design considerations. The OTA is similar to conventional operational amplifiers inasmuch as they employ the usual differential input terminals, but its output is best described in terms of the output current, rather than voltage, that it can supply. In addition, it contains an extra control terminal that enhances flexibility for use in a variety of applications, including AM generation.

Figure 2-14(a) shows the CA3080 connected as an amplitude modulator. The gain of the OTA to the input carrier signal is controlled by variation of the amplifier-bias current at pin 5 (I_{ABC}), because the OTA transconductance (and hence gain) is directly proportional to this current. The level of the unmodulated carrier output is determined by the quiescent I_{ABC} current, which is set by the value of R_M. The 100-kΩ potentiometer is adjusted to set the output voltage symmetrically about zero, thus nulling the effects of amplifier input offset voltage. Figure 2-14(b) shows the following:

Top trace: the original intelligence signal superimposed on the upper AM envelope, which gives an indication of the high quality of this AM generator

Center trace: the AM output

Lower trace: the AM output with the scope's vertical sensitivity greatly expanded to show the ability to provide high degrees of modulation (99% in this case) with a high degree of quality

Another LIC modulator is shown in Fig. 2-15(a). This circuit uses an HA-2735 programmable dual op amp—half of it to generate the carrier frequency (A_1) and the other half as the AM generator (A_2).

In the HA-2735 op amp, the set current (pin 1 for A_1 and pin 13 for A_2) controls the frequency response and gain of each amplifier. This "programmable" function is unnecessary for the oscillator circuit, and thus $I_{set\,1}$ is fixed by the 147-kΩ resistor. Carriers up to about 2 MHz can be generated with A_1.

Amplifier A_1 operates as a Wien-bridge oscillator. Amplitude control of the oscillator is achieved with 1N914 clamping diodes in the feedback network. If the output voltage tends to increase, the diodes offer more conductance, which lowers the gain. Resistor R_1 is adjusted to minimize distortion and control the gain (see Sec. 1-7), and thus the amount of carrier applied to the modulator. With the components in Fig. 2-15(a), the carrier frequency is approximately 1.33 MHz. This frequency can be changed by selection of different RC combinations in the Wien-bridge feedback circuit (1 kΩ and 120 pF).

Amplifier A_2's open-loop response is controlled by the modulating voltage applied to R_2. The percentage of modulation is directly proportional to the modulating voltage. When sinusoidal modulation is applied to R_2, the circuit gain varies from a maximum, A_H, to a minimum, A_L, as A_2's frequency response to the carrier frequency, f_c, is modulated by the set current [Fig. 2-15(b)]. This results in a very distortion-free AM output at pin

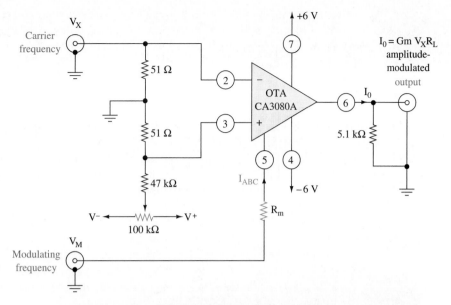

(a) Amplitude modulator circuit using the OTA

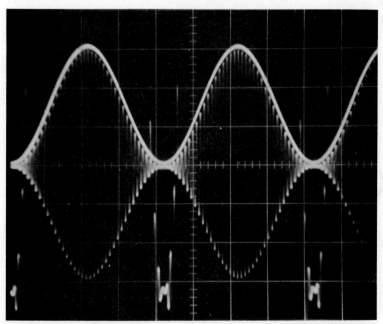

(b) Top trace: modulation frequency input
    ≈ 20 V p-p and 50 μsec/div
    Center trace: amplitude modulation/output
    500 mV/div and 50 μsec/div

Bottom trace: expanded output to show
depth of modulation 20 mV/div
and 50 μsec/div

FIGURE 2-14 ▇ LIC amplitude modulator and resulting waveforms. (Courtesy of RCA Solid State Division.)

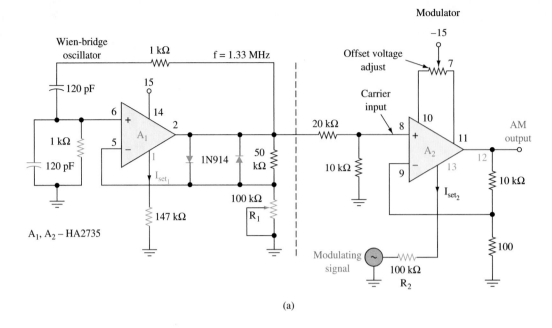

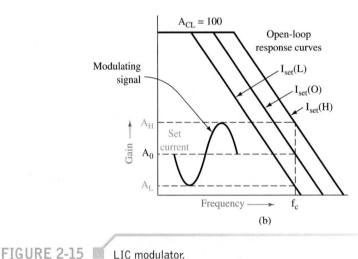

FIGURE 2-15 ■ LIC modulator.

12 of A_2. This circuit makes an ideal test generator for troubleshooting AM systems with carrier frequencies to 2 MHz. If a crystal oscillator were used instead of the Wien-bridge circuit, a high-quality AM transmitter could be fabricated with this AM generator.

2-6 AM TRANSMITTER SYSTEMS

Section 2-5 dealt with specific circuits to generate AM. Those circuits are only one element of a transmitting system. It is important to obtain a good understanding of a complete transmitting unit, and that is the goal of this section.

Figure 2-16 provides block diagrams of simple high- and low-level AM transmitters. The oscillator that generates the carrier signal will invariably be crystal-controlled to

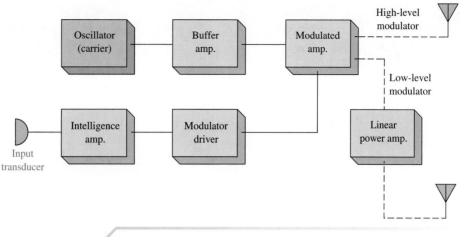

FIGURE 2-16 ■ Simple AM transmitter block diagram.

maintain the high accuracy required by the FCC. It is followed by the "buffer" amplifier, which provides a high-impedance load for the oscillator to minimize drift. It also provides enough gain to sufficiently drive the modulated amplifier. Thus, the buffer amplifier could be a single stage, or however many stages are necessary to drive the following stage, the modulated amplifier.

The intelligence amplifier receives its signal from the input transducer (often a microphone) and contains whatever stages of intelligence amplification are required except for the last one. The last stage of intelligence amplification is called the *modulator,* and its output is mixed in the following stage with the carrier to generate the AM signal. The stage that generates this signal is termed the *modulated amplifier.* This is also the output stage for high-level systems, but low-level systems have whatever number (one or more) of linear power amplifier stages required. Recall that these stages are now amplifying the AM signal and must, therefore, be linear (class A or B), as opposed to the more efficient but nonlinear class C amplifier that can be used as an output stage in high-level schemes.

Citizen's Band Transmitter

Figure 2-17 provides a typical AM transmitter configuration for use on the 27-MHz class D citizen's band. It is taken from the Motorola Semiconductor Products, Inc., Applications Note AN596. It is designed for 13.6-V dc operation, which is the typical voltage level in standard 12-V automotive electrical systems. It employs low-cost plastic transistors and features a novel high-level collector modulation method using two diodes and a double-pi output filter network for matching to the antenna impedance and harmonic suppression.

The first stage uses a MPS 8001 transistor in a common-emitter crystal oscillator configuration. Notice the 27-MHz crystal, which provides excellent frequency stability with respect to temperature and supply voltage variations (well within the 0.005% allowance by FCC regulations for this band). This RF oscillator delivers about 100 mW of 27-MHz carrier power through the L_1 coupling transformer into the buffer (sometimes termed the *driver*) amplifier, which uses a MPS8000 transistor in a common-emitter configuration. Information that allows fabrication of the coils used in this transmitter is provided in Fig. 2-18. The use of coils such as these is a necessity in transmitters and allows for required impedance transformations, interstage coupling, and "tuning" into desired frequencies when combined with the appropriate capacitance to form electrical resonance.

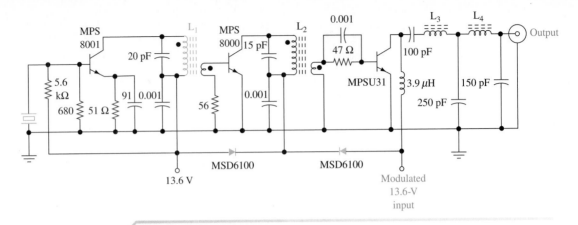

FIGURE 2-17 ▦ Class D citizen's band transmitter. (Courtesy of Motorola Semiconductor Products, Inc.)

The buffer drives about 350 mW into the modulated amplifier. It uses a MPSU31 RF power transistor and can subsequently drive 3.5 W of AM signal into the antenna. This system uses high-level collector modulation on the MPSU31 final transistor, but to obtain a high modulation percentage it is necessary to collector-modulate the previous MPS8000 transistor. This is accomplished with the aid of the MSD6100 dual diode shown in Fig. 2-17. The point labeled "modulated 13.6 V" is the injection point for the intelligence signal riding on the dc supply level of 13.6 V dc. A complete system block diagram for this transmitter is shown in Fig. 2-19. To obtain 100% modulation requires about 2.5 W of intelligence power. Thus, the audio amplification blocks between the microphone and the coupling transformer could easily be accomplished by a single low-

FIGURE 2-18 ▦ Coil description for transmitter shown in Fig. 2-17: Conventional transformer coupling is employed between the oscillator and driver stages (L_1) and the driver and final stages (L_2). To obtain good harmonic suppression, a double-pi matching network consisting of (L_3) and (L_4) is utilized to couple the output to the antenna. All coils are wound on standard $\frac{1}{4}$-in. coil forms with No. 22 AWG wire. Carbonyl J $\frac{1}{4}$ - $\times \frac{3}{8}$ -in.-long cores are used in all coils. Secondaries are overwound on the bottom of the primary winding. The "cold" end of both windings is the start (bottom) and both windings are wound in the same direction. L_1—Primary: 12 turns (close wound); secondary: 2 turns overwound on bottom of primary winding. L_2—Primary: 18 turns (close wound); secondary: 2 turns overwound on bottom of primary winding. L_3—7 turns (close wound). L_4—5 turns (close wound).

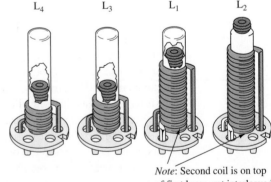

Note: Second coil is on top of first layer, not interleaved

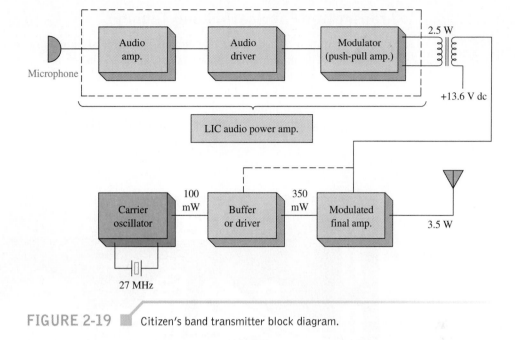

FIGURE 2-19 ■ Citizen's band transmitter block diagram.

cost LIC audio power amplifier (shown in dashed lines) capable of 2.5 W of output. The audio output is combined with the dc by the coupling transformer, as is required to modulate the 27-MHz carrier.

Antenna Coupler

Once the 3.5 W of AM signal is obtained at the final stage (MPSU31), it is necessary to "couple" this signal into the antenna. The coupling network for this system is comprised of L_3, L_4, and the 250- and 150-pF capacitors in Fig. 2-17. This filter configuration is termed a *double-pi network*. To obtain maximum power transfer to the antenna, it is necessary that the transmitter's output impedance be properly matched to the antenna's input impedance. This means equality in the case of a resistive antenna or the complex conjugate in the case of a reactive antenna input. If the transmitter was required to operate at a number of different carrier frequencies, the coupling circuit is usually made variable to obtain maximum transmitted power at each frequency. Coupling circuits are also required to perform some filtering action (to eliminate unwanted frequency components), in addition to their efficient energy transfer function. Conversely, a filter invariably performs a coupling function, and hence the two terms (filter and coupler) are really interchangeable, with what they are called generally governed by the function considered of major importance.

The double-pi network used in the citizen's band transmitter is very effective in suppressing (i.e., filtering out) the second and third harmonics, which would otherwise interfere with communications at 2×27 MHz and 3×27 MHz. It typically offers 37-dB second harmonic suppression and 55-dB third harmonic suppression. The capacitors and inductors in the double-pi network are resonant so as to allow frequencies in the 27-MHz region (the carrier and sidebands) to pass, but all other frequencies are severely attenuated. The ratios of the values of the two capacitors determines what part of the total impedance across L_4 is coupled to the antenna, and the value of the 150-pF capacitor has a direct effect on the output impedance.

Transmitter Fabrication and Tuning

The fabrication of high-frequency circuits is much less straightforward than for low frequencies. The minimal inductance of a simple conductor or capacitance between two adjacent conductors can play havoc at high frequencies. Common sense and experimentation generally yield a suitable configuration. The information contained in Fig. 2-20 provides a suggested printed-circuit-board layout and component mounting photograph for the high-frequency sections (shown schematically in Fig. 2-17).

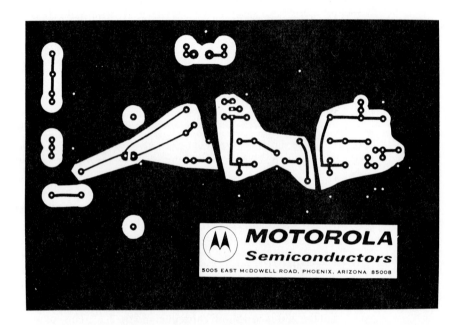

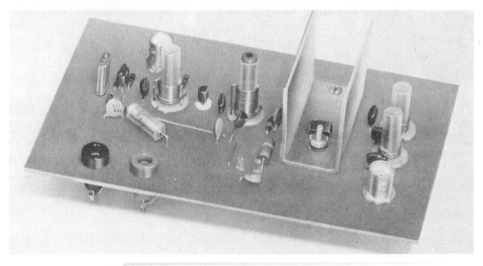

FIGURE 2-20 ■ Citizen's band transmitter PC board layout and complete assembly pictorial. (Courtesy of Motorola Semiconductor Products, Inc.)

After assembly it is necessary to go through a "tune-up" procedure to get the transmitter on the air. Initially, L_1's variable core must be adjusted to get the oscillator to oscillate. This is necessary to get its inductance precisely adjusted so that, in association with its shunt capacitance, it will resonate at the precise 27-MHz resonant frequency of the crystal. The tune-up procedure starts by adjusting the cores of all four coils one-half turn out of the windings. Then tune L_1 clockwise until the oscillator starts and continue for one additional turn. Ensure that the oscillator starts every time by turning the dc on and off (a process termed *keying*) a number of times. If it does not reliably start every time, turn L_1 clockwise one-quarter turn at a time until it does. Then tune the other coils in order, with the antenna connected, for maximum power output. Apply nearly 100% sine-wave intelligence modulation and retune L_2, L_3, and L_4 once again for maximum power output while observing the output on an oscilloscope to ensure that overmodulation and/or distortion do not occur.

2-7 STEREO BROADCASTING

It is known that the reproduction of music with two separate channels can enrich and add to its realism. Broadcast AM has started to move into stereo broadcasts since a number of schemes were advanced in the late 1970s. Unfortunately, the FCC decided to let the marketplace decide on the best system. This led to confusion and no clear favorite. At this juncture we find that the Motorola system has become the de facto standard. It is no wonder that AM stereo has not become a favorite mode of broadcast as has FM radio, where a single approved system led to essentially total market coverage.

Two of the four systems proposed are quite similar in approach. They are Motorola and Harris systems, and the block diagram in Fig. 2-21 helps to illustrate their operation.

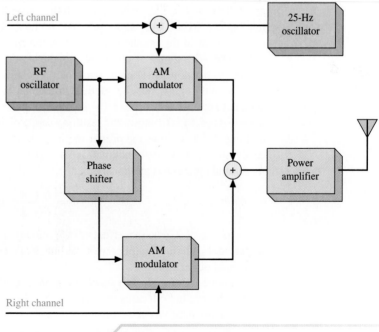

FIGURE 2-21   AM stereo block diagram.

The carrier is phase-shifted so that essentially two carrier signals are developed. The two audio signals (left and right channels) are used to modulate the two carriers individually. Note that a reference 25-Hz signal also modulates one of the carrier signals. When the receiver detects the 25-Hz tone, it lights up an indicator to indicate stereo reception. The two AM signals are summed out of the modulator for final amplification and transmission. Regular receivers simply detect the left-plus-right signals for normal monaural reception. A specially equipped stereo receiver is able to differentiate between the two out-of-phase carriers and thereby develop the two separate audio signals.

The Kahn–Hazeltine system was the first and remains the simplest. In it the two audio signals are processed into left plus right (L + R) and left minus right (L − R) components. The L + R (monaural) signal modulates the carrier as in a normal (nonstereo) transmitter. The L − R component phase modulates the carrier and generates two principal frequency modulation sidebands that are of opposite phase. The regular AM sidebands are in phase. The left channel then winds up on one sideband and the right on the other since (L + R) + (L − R) = 2L and (L + R) − (L − R) = 2R. To receive stereo the listener tunes two radios with one a few kHz high and the other a few kHz low. The system by Magnavox employs a similar phase modulation scheme but allows reception with a single specially designed receiver.

2-8 TRANSMITTER MEASUREMENTS

Trapezoid Patterns

A number of techniques are available to make operational checks on a transmitter's performance. A standard oscilloscope display of the transmitted AM signal will indicate any gross deficiencies. This technique is all the better if a dual-trace scope is used to allow the intelligence signal to be superimposed on the AM signal, as illustrated in Fig. 2-14. An improvement in this method is known as the *trapezoidal pattern*. It is illustrated in Fig. 2-22. The AM signal is connected to the vertical input and the intelligence signal is applied to the horizontal input with the scope's internal sweep disconnected. The intelligence signal usually must be applied through an adjustable *RC* phase-shifting network, as shown, to ensure that it is exactly in phase with the modulation envelope of the AM waveform. Figure 2-22(b) shows the resulting scope display with improper phase relationships, and Fig. 2-22(c) shows the proper in-phase trapezoidal pattern for a typical AM signal. It easily allows percentage modulation calculations by applying the B and A dimensions to the formula presented previously,

$$\%m = \frac{B - A}{B + A} \times 100\% \tag{2-3}$$

In Fig. 2-22(d), the effect of 0% modulation (just the carrier) is indicated. The trapezoidal pattern is simply a vertical line, since there is no intelligence signal to provide horizontal deflection.

Figure 2-22(e) and (f) show two more trapezoidal displays indicative of some common problems. In both cases the trapezoid's sides are not straight (linear). The concave curvature at (e) indicates poor linearity in the modulation stage, which is often caused by improper neutralization or by stray coupling in a previous stage. The convex curvature at (f) is usually caused by improper bias or low carrier signal power (often termed low *excitation*).

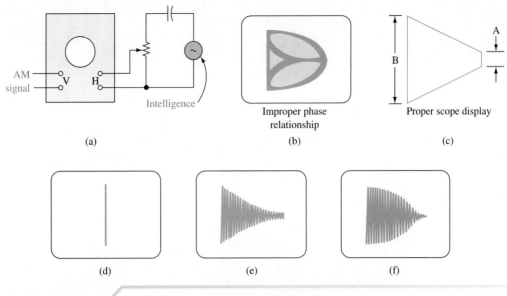

(a)

Improper phase
relationship

(b)

Proper scope display

(c)

(d)

(e)

(f)

FIGURE 2-22 ▮ Trapezoidal pattern connection scheme and displays.

Meter Measurement

It is possible to make some meaningful transmitter checks with a dc ammeter in the collector (or plate) of the modulated stage. If the operation is correct, this current should not change as the intelligence signal is varied between zero and the point where full modulation is attained. This is true since the increase in current during the crest of the modulated wave should be exactly offset by the drop during the trough. A distorted AM signal will usually cause a change in dc current flow. In the case of overmodulation, the current will increase further during the crest but cannot decrease below zero at the trough, and a net increase in dc current will occur. It is also common for this current to decrease as modulation is applied. This malfunction is termed *downward modulation* and is usually the result of insufficient excitation. The current increase during the modulation envelope crest is minimized, but the decrease during the trough is nearly normal.

Spectrum Analyzers

The use of spectrum analyzers has become widespread in all fields of electronics, but especially in the communications industry. A *spectrum analyzer* visually displays (on a CRT) the amplitude of the components of a wave as a function of frequency. This can be contrasted with an oscilloscope display, which shows the amplitude of the total wave (all components) versus time. Thus, an oscilloscope shows us the "time" domain while the spectrum analyzer shows the "frequency" domain. In Fig. 2-23(a) the frequency domain for a 1-MHz carrier modulated by a 5-kHz intelligence signal is shown. Proper operation is indicated since only the carrier and upper and lower side frequencies are present. During malfunctions, and to a lesser extent even under normal conditions, transmitters will often generate *spurious frequencies* as shown in (b), where components other than just the three desired are present. These spurious undesired components are usually called

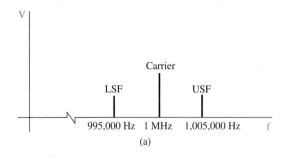

(a)

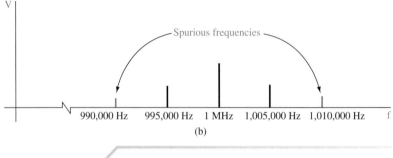

(b)

FIGURE 2-23 Spectrum analysis of AM waveforms.

spurs, and their amplitude is tightly controlled by FCC regulation to minimize interference on adjacent channels. The coupling stage between the transmitter and its antenna is designed to attenuate the spurs, but the transmitter's output stage must also be carefully designed to keep the spurs to a minimum level. The spectrum analyzer is obviously a very handy tool for use in evaluating a transmitter's performance.

The spectrum analyzer is, in effect, an automatic frequency-selective voltmeter that provides both frequency and voltage on its CRT display. It can be thought of as a radio receiver with broad frequency-range coverage and sharp sweep tuning, narrow-bandwidth circuits. The more sophisticated units are calibrated to read signals in dB or dBm. This provides better resolution between low-level sideband signals and the carrier and of course allows direct reading of power levels without resorting to calculation from voltage levels. Most recent spectrum analyzers utilize microprocessor principles (software programming) for ease of operation.

Figure 2-24 shows a typical spectrum analyzer. Available software packages link the waveform analyzer and a computer that enables simplification and automation of complex operations and measurements. For example, the software package shown on page 84 can be used to spot spurious signals. Many different applications/utility routines are selected through menu-driven operation. This allows immediate access to operations such as swept frequency measurements, identifying true signals from false responses, and performing signal analysis, including measurements of filter response, harmonic distortion, and signal-to-noise ratio. A typical display on the computer's CRT is shown in Fig. 2-24. It shows an AM carrier that has four spurious outputs. Notice the noise between the spurs and the carrier. This is commonly referred to as the *noise floor* of the system under test.

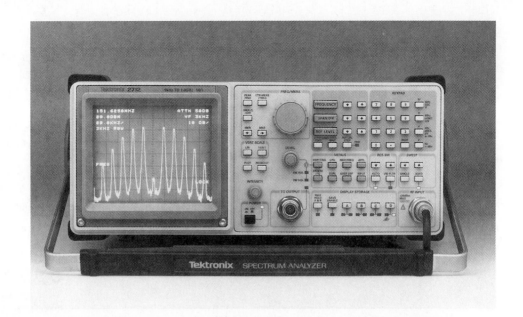

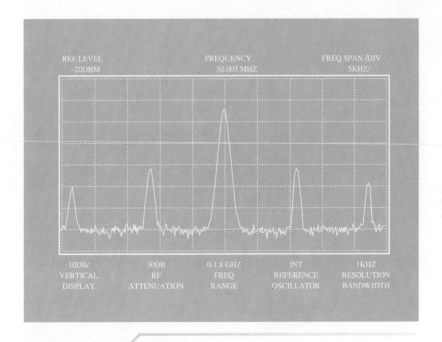

FIGURE 2-24 Spectrum analyzer and typical display. (Courtesy of Tektronix, Inc.)

Mini-Spur™ 1.0
turns your PC into a Spectrum Analyzer *

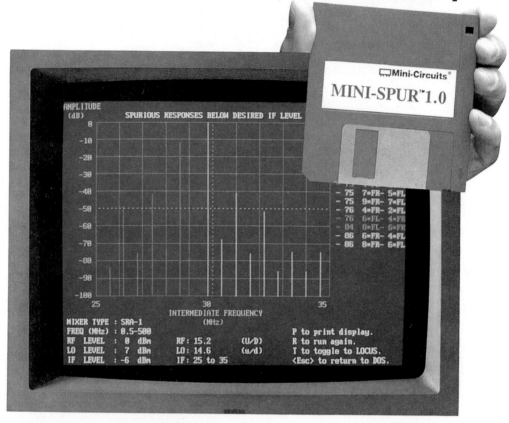

Spot Spurious Signals Easily...
choose the best mixer for your design problems

Introducing Mini-Spur™, the software simulation program for analysis of system spurious responses. Using actual data on Mini-Circuits mixers, spurious signal levels are calculated and then displayed.

Operation is simple. The user defines the input frequency and power level, the program then graphically displays the various outputs including all the spurs (up to 10xLO ± 10xRF) falling within the user-defined IF filter bandwidth. As the user tunes the frequency, the output spectrum scrolls across the screen just like that of a sophisticated spectrum analyzer.

Required hardware; IBM AT or compatible with 640k memory, and EGA or VGA display. Optional, dot matrix, laser printer or plotter.

So maximize design efficiency... use Mini-Spur™ only from Mini-Circuits.

* PC simulates measured results. Does not become a measuring instrument.

Mini-Spur software package. (Courtesy of Mini-Circuits.)

Harmonic Distortion Measurements

Harmonic distortion measurements can easily be made by applying a spectrally pure signal source to the device under test (DUT). The quality of the measurement is dependent on the harmonic distortion of both the signal source and spectrum analyzer. The source provides a signal to the DUT and the spectrum analyzer is used to monitor the output. Figure 2-25 shows the results of a typical harmonic distortion measurement. The distortion can be specified by expressing the fundamental with respect to the largest harmonic in dB. This is termed the *relative harmonic distortion.*

If the fundamental in Fig. 2-25 is 1V and the harmonic at 3 kHz (the largest distortion component) is 0.05V, the relative harmonic distortion is

$$20 \log \frac{1V}{0.05V} = 26 \text{ dB}$$

A somewhat more descriptive distortion specification is *total harmonic distortion* (THD). THD takes into account the power in all the significant harmonics:

$$\text{THD} = \sqrt{(V_2^2 + V_3^2 + \cdots)} / V_1 \qquad (2\text{-}4)$$

where V_1 is the rms voltage of the fundamental and V_2, V_3, . . . are the rms voltages of the harmonics. An infinite number of harmonics are theoretically created, but in practice the amplitude falls off for the higher harmonics. Virtually no error is introduced if the calculation does not include harmonics less than one-tenth of the largest harmonic.

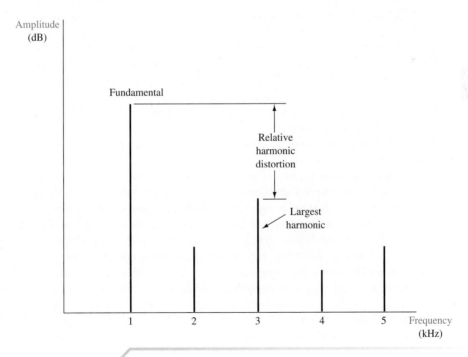

FIGURE 2-25 Relative harmonic distortion.

EXAMPLE 2-9

Determine the THD if the spectrum analyzer display in Fig. 2-25 has $V_1 = 1$ v, $V_2 = 0.03$ v, $V_3 = 0.05$ v, $V_4 = 0.02$ v, and $V_5 = 0.04$ v.

SOLUTION

$$\begin{aligned}
\text{THD} &= \sqrt{(V_2^2 + V_3^2 + V_4^2 + V_5^2)}\,/\,V_1 \\
&= \sqrt{(0.03^2 + 0.05^2 + 0.02^2 + 0.04^2)}\,/\,1 \\
&= 0.07348 \\
&= 7.35\%
\end{aligned}$$

$(2\text{-}4)$

THD calculations are somewhat tedious when a large number of significant harmonics exist. Some spectrum analyzers include an automatic THD function that does all the work and prints out the THD percentage.

Special RF Signal Measurement Precautions

The frequency-domain measurements of the spectrum analyzer provide a more thorough "reading" of RF frequency signals than does the time-domain oscilloscope. The high cost and additional setup time of the spectrum analyzer dictates the continued use of more standard measurement techniques—mainly voltmeter and oscilloscope usage. Whatever the means of measurement, certain effects must be understood when testing RF signals as compared to the audio frequencies with which you are probably more familiar. These effects are the loading of high-Q parallel-resonant circuits by a relatively low impedance instrument and the frequency response shift that can be caused by test lead and instrument input capacitance.

The consequence of connecting a 50-Ω signal generator into an RF tuned circuit that has a Z_p in the kilohm region would be a drastically reduced Q and increased bandwidth. The same loading effect would result if a low-impedance detector were used to make RF impedance measurements. This loading is minimized by using resistors, capacitors, or transformers in conjunction with the measurement instrument.

The consequence of test lead or instrument capacitance is to shift the circuit's frequency response. If you were looking at a 10-MHz AM signal that had its resonant circuit shifted to 9.8 MHz by the measurement capacitance, an obvious problem has resulted. Besides some simple attenuation, the equal amplitude relationship between the USB and LSB would be destroyed and waveform distortion results. This effect is minimized by using low-valued series-connected coupling capacitors, or canceled with small series-connected inductors. If more precise readings with inconsequential loading are necessary, specially designed resonant matching networks are required. They can be used as an add-on with the measuring instrument or built into the RF system at convenient test locations.

When testing a transmitter the use of a "dummy antenna" is often a necessity. A *dummy antenna* is a resistive load used in place of the regular antenna. It is used to prevent undesired transmissions that may otherwise occur. The dummy antenna also prevents damage to the output circuits that may occur under unloaded conditions.

TROUBLESHOOTING

When traveling uncharted roads, having a map can make the difference between getting lost and finding your way. A plan for troubleshooting, like a map, can help guide you to an equipment malfunction. Made up of a logical sequence of troubleshooting steps, this troubleshooting map can help the technician hunt down the defect in a piece of electronic communication equipment. By developing and using this strategy, the technician can become very proficient at locating electronic problems and repairing them.

After completing this section you should be able to

- Describe the purpose of the inspection
- State the sequence of troubleshooting steps
- Troubleshoot an RF amplifier and oscillator
- Check for transmitter operation on proper frequency
- Correct for low transmitter output power

Inspection

The first phase of any repair action is to do a visual inspection of the defective equipment. During this inspection look for broken wires, loose connections, discolored or burned resistors, and exploded capacitors. Burned resistors are easily seen and will give off a distinguishing odor. Equipment that has been dropped may have a broken PCB (printed circuit board). Connectors may have been knocked loose. Look for bad soldering and cold solder joints. Cold solder connections look dull and dingy as opposed to nice shiny good ones. Intermittent faults are usually the result of cold solder joints. Components hot to the touch after the equipment has been on for a few minutes can indicate shorted components. Listen for unusual sounds when the equipment is turned on. Unusual sounds could lead you to the malfunctioning component. Many defects are found during this inspection phase and the equipment frequently gets fixed without further troubleshooting.

Strategy for Repair

1. VERIFY THAT A PROBLEM DOES EXIST   Always verify that the reported problem exists before troubleshooting the equipment. By confirming the problem you will save time that could get wasted looking for a nonexistent defect. If the equipment is not completely dead, try to localize the symptom to a particular stage. Check the service literature for troubleshooting hints. Some manufacturers provide diagnostic charts to help pinpoint malfunctions. The more clues you gather the more apt you are to successfully associate a fault to a particular circuit function. Another reason for confirming that the problem exists is to rule out operator error. The equipment owner or operator may not know all the equipment's functions and may consider the unit bad if it can't do something it wasn't designed for. Check the operating manual when in doubt about the equipment functions.

2. ISOLATE THE DEFECTIVE STAGE   When a problem has been verified in a piece of electronic equipment, begin troubleshooting to isolate the defect to a particular stage. Normally the defective stage can be found by signal tracing or signal injection. The oscilloscope can be used with either the signal tracing or signal injection method. An example of this is in Fig. 2-22, where an AM transmitter's modulation is being monitored by an oscilloscope. As shown, the test pattern appears on the screen and represents over- or undermodulation of the AM transmitted signal. If the transmitted signal were incorrect, the modulator would be adjusted for the proper signal display.

A signal generator or a function generator are used to supply a test signal at the input of the specific stage under examination. Figure 2-26 illustrates a dual trace oscilloscope connected to an amplifier stage to show the input test signal and the output signal. The input signal is easily compared to the output signal with this test setup. If the output signal is missing or distorted, then the defective stage has been located.

3. ISOLATE THE DEFECTIVE COMPONENT   Once the defective stage has been located, the next step is to find the bad component or components. Voltage and resistance measurements should be used to locate the defective component. Remember, voltage measurements are made with respect to ground, and resistance measurements are done with the power off. The voltage and resistance measurements are compared to specified values stated in the service literature. Incorrect readings usually pinpoint the defective component. It is very possible that there may be more than one bad component in a malfunctioning circuit.

4. REPLACE THE DEFECTIVE COMPONENT AND HOT CHECK   Before replacing the bad component in a circuit, make sure that another component is not causing it to go bad. For example, a shorted diode or transistor can cause resistors to burn up. Diodes and transistors should be checked for shorted conditions before associated components are replaced. Replace defective components with exact replacement parts. Once the defective component is replaced, ensure that the circuit is operating normally. You may need to do a

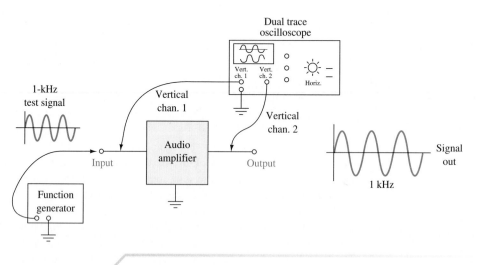

FIGURE 2-26   Comparing input and output signals.

few more voltage checks to confirm things are back to normal. Burn-in the equipment by turning it on and letting it operate a number of hours on the test bench. This burn-in is a vital part of equipment repair. If failure is going to reoccur, it will usually show up during this burn-in period.

RF Amplifier Troubleshooting

BIAS SUPPLY   Many RF amplifiers utilize power from the previous stage to provide dc bias.

Figure 2-27 shows how bias for the transistor Q1 is developed. RF from the previous stage is rectified by the base–emitter junction of Q1. The current flows through R1 and the transformer to ground. The reactance of C1 is low at RF, so the RF bypasses the resistor. C1 also serves to filter the RF pulses and develop a dc voltage across R1. At the base of Q1, this dc voltage is negative with respect to ground. Therefore, Q1 will be a class C amplifier conducting only on positive RF peaks. Figure 2-28 shows the instantaneous voltage at the base of Q1 that you can observe with an oscilloscope.

SHORTED C1   If C1 were to short, excessive drive would reach Q1. No negative bias for Q1 could be developed. This would cause Q1 to draw excessive current and destroy itself. If Q1 is bad, always check all components ahead of Q1 before replacing it.

OPEN C1   If C1 were open, the drive reaching Q1 would be greatly reduced. Bias voltage would be low and Q1 would not develop full power output.

OPEN R1   Resistors in these circuits may overheat and fail open. C1 will charge to the negative peak of the RF drive voltage due to the rectifier action of the base–emitter junction. This will cut Q1 off and there will be no power output.

OUTPUT NETWORK   Now consider possible faults in components on the output side of Q1. Common faults are shorted blocking capacitors, overheated tuning capacitors, and open chokes.

SHORTED BLOCKING CAPACITOR   Consider the circuit in Fig. 2-29. Assume that capacitor Cb has shorted. If this amplifier is connected to an antenna that is not dc grounded, there will be no effect at all. Cb is not part of any tuning circuit; its job is to block the dc power supply from the succeeding stage or antenna.

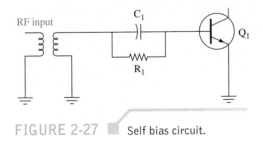

FIGURE 2-27   Self bias circuit.

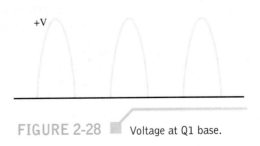

FIGURE 2-28   Voltage at Q1 base.

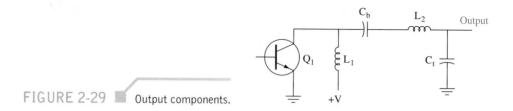

FIGURE 2-29 ■ Output components.

Many antennas show a short circuit to dc. In this case, excessive current would flow through L1 and L2, possibly damaging them and the power supply. If a shorted blocking capacitor is found, it is wise to check for damage to the wiring or printed circuit board and the power supply.

FAULTY TUNING CAPACITOR   The ac load impedance presented to the transistor Q1 is dependent on Ct and L2 forming a tuned circuit that transforms the antenna impedance to the correct value. Assume that Ct is shorted. In this case, the load impedance would probably be too low and the Q1 would draw excessive current. If Ct were open, the opposite would happen. In either case, power output will be very low.

ADJUSTMENTS   Assume that Ct is simply not properly adjusted (it is usually variable). Power output will be too high or too low depending on the direction of error. High-power output will be accompanied by overheating of Q1 due to excessive collector current.

L2 is also usually adjustable. You must alternately adjust both Ct and L2 to obtain the proper impedance match. Look for minimum collector current and maximum power output. Use a spectrum analyzer to be certain the amplifier is not tuned to a harmonic. Some amplifiers will happily tune to the second or third harmonic. Others will break into self-oscillation on many frequencies at once. The spectrum analyzer will reveal many of the bad habits an amplifier might have.

Checking a Transmitter

A word of caution before starting. High-power transmitters (perhaps anything greater than 10 kW) frequently employ vacuum tubes. These are high-voltage devices using voltages of 5 KV or more. *THESE VOLTAGES CAN KILL YOU.* Therefore, troubleshoot with extreme caution. Get experienced help until the following rules become second nature:

Before entering the transmitter cabinet turn off all power switches. Better yet, remove power plug from socket. There may be no plug; the transmitter power input lines may be hard-wired to a fuse panel. If so, there will probably be a main switch; turn it off.

Even though you have turned off the power, make absolutely certain the high voltage is off before touching any circuits within the transmitter cabinet. The best and only sure test of this is to attach a bare, uninsulated piece of 12-gage or larger copper wire (no insulation; you must be able to see that the entire length of the wire is intact) to a non-conducting wooden or plastic rod perhaps 2-feet long. Ground one end of this wire, that is, connect it to the metal chassis of the transmitter. Holding the end of the rod furthest from the wire, move the rod so that the ungrounded end of the wire touches those points which would have high voltage on them if the high voltage was still on. You'll see arcing if the voltage is still on.

Do not trust automatic switches (interlocks) that are supposed to turn off high-voltage circuits automatically.

Remember, charged capacitors such as those used for power supply filters can store a lethal charge. Short these units with your bare wire on a rod.

TRANSMITTER NOT OPERATING ON PROPER FREQUENCY   The simplest way to determine a transmitter's operating frequency is to listen to its output signal on a receiver with a calibrated readout that accurately indicates the frequency to which the receiver is tuned. Such receivers may have a built-in crystal oscillator called a crystal calibrator. Calibrator oscillators are specially designed to have rich harmonic output. If, for example, a receiver had a calibrator operating at 100 kHz, signals would be heard on the receiver at harmonics or multiples of 100 kHz over a broad range. Tune the receiver to the 100 kHz multiple nearest the frequency of the transmitter under test. Compare the frequency shown on the receiver's readout to the known multiple of 100 kHz and determine the error between the two. Then, tune to the transmitter frequency and adjust the receiver's readout up or down according to the error. For greater accuracy, the calibrator can be set to the frequency of radio station WWV, operated by The National Institute of Standards and Technology, Fort Collins, Colorado. This station broadcasts on accurate frequencies of 2.5, 5, 10, 15, and 20 MHz on the shortwave bands. Canadian station CHU, broadcasting from Ottawa, can also be used for calibration. It is found at 3330, 7335, and 14,670 kHz on the shortwave bands.

A transmitter's operating frequency can also be determined with a spectrum analyzer or a frequency meter, as shown in Fig. 2-30. It may be necessary to connect a short antenna, perhaps a few feet long, to the input terminal of the measuring equipment if the transmitter has low output power.

If the transmitter is not operating on the correct frequency, adjust the carrier oscillator to the proper frequency. Check the transmitter's maintenance manual for instructions. See additional comments for oscillators in the Chapter 1 troubleshooting section.

MEASURING TRANSMITTER OUTPUT POWER   Figure 2-31 shows the circuit to be used when measuring and troubleshooting the output power of a transmitter. The dummy load acts like an antenna in that it absorbs the energy output from the transmitter without

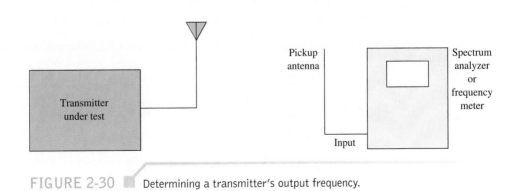

FIGURE 2-30 ▪ Determining a transmitter's output frequency.

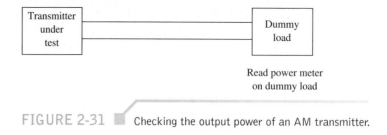

FIGURE 2-31 ■ Checking the output power of an AM transmitter.

allowing that energy to radiate and interfere with other stations. Its input impedance must match (be equal to) the transmitter's output impedance; this is usually 50 ohms.

Suppose we are checking a low-power commercial transmitter that is rated at 250-W output (the dummy load and wattmeter must be rated for this level). If the output power is greater than the station license allows, the drive control must be adjusted to bring the unit within specs. What if the output is below specs? Let us consider possible causes.

Remember the suggestion in Chapter 1: do the easy things first. Perhaps the easiest thing to do in the case of low-power output is to check the drive control: is it set correctly? Assuming it is, check the power supply voltage: is it correct? Observe that voltage on a scope: is it good, pure dc or has a rectifier shorted or opened?

Once the easy things have been done, check the tuning of each amplifier stage between the carrier oscillator and the last or final amplifier driving the antenna. If the output power is still too low after peaking the tuning controls, use an oscilloscope to check the output voltage of each stage to see if they are up to specs. Are the signals good sinusoids? If a stage has a clean, undistorted input signal and a distorted output, there may be a defective component in the bias network. Or perhaps the tube/transistor needs checking and/or replacement.

SUMMARY

In Chapter 2 we studied the concept of amplitude modulation as it is specifically utilized in a transmitter. The major topics the student should now understand include:

- the fundamental concept of amplitude modulation

- the meaning of modulation index and its use in AM calculations

- the cause of overmodulation and why it must be avoided

- the mathematical analysis of AM and the effect of modulation index on sideband amplitude

- the elements of a simple transistor AM generator and the analysis of its operation

- the understanding of high- and low-level modulation

- the analysis of a high-level transistor modulator

- the analysis and operation of various linear integrated circuit modulators

- the caution necessary when working with high-powered transmitters

1. Explain why the *linear* combination of a low-frequency intelligence signal and a high-frequency carrier signal is *not* effective as a radio transmission.
2. A 1500-kHz carrier and 2-kHz intelligence signal are combined in a *nonlinear* device. List *all* the frequency components produced.
*3. If a 1500-kHz radio wave is modulated by a 2-kHz sine-wave tone, what frequencies are contained in the modulated wave (the actual AM signal)?
*4. If a carrier is amplitude-modulated, what causes the sideband frequencies?
*5. What determines the bandwidth of emission for an AM transmission?
6. Describe the significance of the upper and lower envelope of an AM waveform.
7. Explain the difference between a sideband and a side frequency.

SECTION 3

*8. Draw a diagram of a carrier wave envelope when modulated 50% by a sinusoidal wave. Indicate on the diagram the dimensions from which the percentage of modulation is determined.
*9. What are some of the possible results of overmodulation?
10. An unmodulated carrier is 300 V p-p. Calculate %m when its maximum p-p value reaches 400, 500, and 600 V. (33.3%, 66.7%, 100%)

SECTION 4

11. Given that the amplitude of an AM waveform can be expressed as the sum of the carrier peak amplitude and intelligence signal, derive the expression for an AM signal that shows the existence of carrier and side frequencies.
12. A 100-V carrier is modulated by a 1-kHz sine wave. Determine the side-frequency amplitudes when $m = 0.75$. (37.5 V)
13. A 1-MHz, 40-V-peak carrier is modulated by a 5-kHz intelligence signal such that $m = 0.7$. This AM signal is fed to a 50-Ω antenna. Calculate the power of each spectral component fed to the antenna. ($P_c = 16$ W, $P_{usb} = P_{lsb} = 1.96$ W)
14. Calculate the carrier and sideband power if the total transmitted power is 500 W in Problem 12. (390 W, 110 W)
15. The ac rms antenna current of an AM transmitter is 6.2 A when unmodulated and rises to 6.7 A when modulated. Calculate %m. (57.9%)
*16. Why is a high percentage of modulation desirable?
*17. During 100% modulation, what percentage of the average output power is in the sidebands? (33.3%)
18. An AM transmitter at 27 MHz develops 10 W of carrier power into a 50-Ω load. It is modulated by a 2-kHz sine wave between 20 and 90% modulation. Determine:
(a) Component frequencies in the AM signal.
(b) Maximum and minimum waveform voltage of the AM signal at 20% and 90% modulation. (25.3 to 37.9 V peak, 3.14 to 60.1 V peak)
(c) Sideband signal voltage and power at 20% and 90% modulation. (2.24 V, 0.1 W, 10.06 V, 2.025 W)

(d) Load current at 20% and 90% modulation. (0.451 A, 0.530 A)

19. An AM transmitter has a 1-kW carrier and is modulated by three different sine waves having equal amplitudes. If $m_{eff} = 0.8$, calculate the individual values of m and the total transmitted power. (0.462, 1.32 kW)

20. A 50-V rms carrier is modulated by a square wave as shown in Table 1-2(c). If only the first four harmonics are considered and $V = 20$ V, calculate m_{eff}. (0.77)

SECTION 5

21. Describe two possible ways that a transistor can be used to generate an AM signal.
*22. What is *low-level modulation?*
*23. What is *high-level modulation?*
24. Explain the relative merits of high- and low-level modulation schemes.
*25. Why must some radio-frequency amplifiers be neutralized?
26. Draw a schematic of a class C transistor modulator and explain its operation.
*27. What is the principal advantage of a class C amplifier?
*28. What is the function of a quartz crystal in a radio transmitter?

SECTION 6

*29. Draw a block diagram of an AM transmitter.
*30. What is the purpose of a buffer amplifier stage in a transmitter?
31. Describe the means by which the transmitter shown in Fig. 2-16 is modulated.
*32. Draw a simple schematic diagram showing a method of coupling the radio-frequency output of the final power amplifier stage of a transmitter to an antenna.
33. Describe the functions of an antenna coupler.
*34. A ship radio-telephone transmitter operates on 2738 kHz. At a certain point distant from the transmitter the 2738-kHz signal has a measured field of 147 mV/m. The second harmonic field at the same point is measured as 405 μV/m. To the nearest whole unit in decibels, how much has the harmonic emission been attenuated below the 2738-kHz fundamental? (51.2 dB)
35. What is a *tune-up procedure?*

SECTION 7

36. Discuss the basic principles of the four systems for stereo AM transmission.
37. List the required circuit changes to change the transmission frequency of the IC transmitter shown in Fig. 2-20.

SECTION 8

*38. Draw a sample sketch of the trapezoidal pattern on a cathode-ray oscilloscope screen indicating low percent modulation without distortion.
39. Explain the advantages of the trapezoidal display over a standard oscilloscope display of AM signals.

40. Compare the display of an oscilloscope to that of a spectrum analyzer.

41. A spectrum analyzer display shows that a signal is made up of three components only: 960 kHz at 1 V, 962 kHz at $\frac{1}{2}$ V, 958 kHz at $\frac{1}{2}$ V. What is the signal and how was it generated?

42. Define *spur.*

43. Provide a sketch of the display of a spectrum analyzer for the AM signal described in Problem 18 at both 20% and 90% modulation. Label the amplitudes in dBm.

44. The spectrum analyzer display shown in Fig. 2-24 is calibrated at 10 dB/vertical division and 5 kHz/horizontal division. The 50.0034-MHz carrier is shown riding on a −20-dBm noise floor. Calculate the carrier power, the frequency, and the power of the spurs. (2.51 W, 50.0149 MHz, 49.9919 MHz, 50.0264 MHz, 49.9804 MHz, 6.3 mW, 1 mW)

*45. What is the purpose of a *dummy antenna?*

46. An amplifier has a spectrally pure sine wave input of 50 mv. It has a voltage gain of 60. A spectrum analyzer shows harmonics of 0.035 v, 0.027 v, 0.019 v, 0.011 v, and 0.005 v. Calculate the total harmonic distortion (THD) and the relative harmonic distortion. (2.864%, 38.66 dB)

47. An additional harmonic ($V_6 = 0.01$ v) was neglected in the THD calculation shown in Ex. 2-9. Determine the percentage error introduced by this omission. (0.91%)

AMPLITUDE MODULATION: RECEPTION

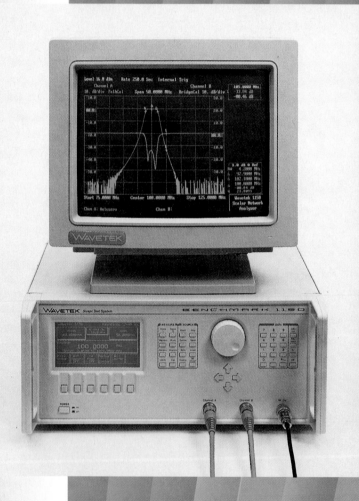

OBJECTIVES

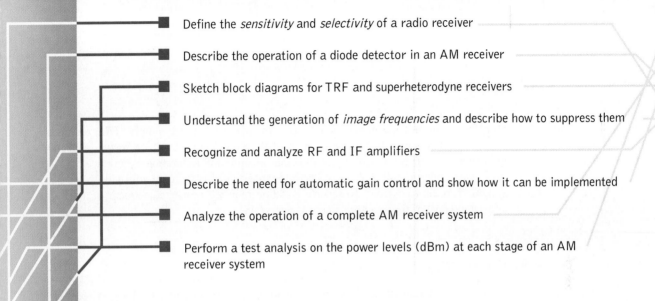

- Define the *sensitivity* and *selectivity* of a radio receiver

- Describe the operation of a diode detector in an AM receiver

- Sketch block diagrams for TRF and superheterodyne receivers

- Understand the generation of *image frequencies* and describe how to suppress them

- Recognize and analyze RF and IF amplifiers

- Describe the need for automatic gain control and show how it can be implemented

- Analyze the operation of a complete AM receiver system

- Perform a test analysis on the power levels (dBm) at each stage of an AM receiver system

3-1 RECEIVER CHARACTERISTICS

If you were to logically envision a block diagram for a radio receiver, you would probably go through the following thought process:

1. The signal from the antenna is usually very small—therefore, amplification is necessary. This amplifier should have low-noise characteristics and should be "tuned" to accept only the desired carrier and sideband frequencies to avoid interference from other stations and to minimize the received noise. Recall that noise is proportional to bandwidth.

2. After sufficient amplification, a circuit to detect the intelligence from the radio frequency is required.

3. Following the detection of the intelligence, further amplification is necessary to give it sufficient power to drive a loudspeaker.

This logical train of thought leads to the block diagram shown in Fig. 3-1. It consists of an RF amplifier, detector, and audio amplifier. The first radio receivers for broadcast AM

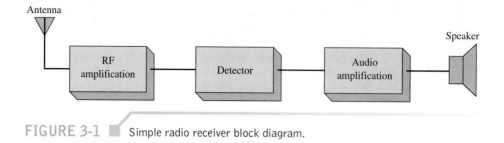

FIGURE 3-1 Simple radio receiver block diagram.

took this form and are called *tuned radio frequency* or, more simply, *TRF receivers.* These receivers generally had three stages of RF amplification, with each stage preceded by a separate variable-tuned circuit. You can imagine the frustration experienced by the user when tuning to a new station. The three tuned circuits were all adjusted by separate variable capacitor controls. To receive a station required proper adjustment of all three, and a good deal of time and practice was necessary.

Sensitivity and Selectivity

Two major characteristics of any receiver are its sensitivity and selectivity. A receiver's *sensitivity* may be defined as its ability to drive the output transducer (e.g., speaker) to an acceptable level. A more technical definition is the minimum input signal (usually expressed as a voltage) required to produce a specified output signal or sometimes to just provide a discernible output. The range of sensitivities for communication receivers varies from the millivolt region for low-cost AM receivers down to the nanovolt region for ultra-sophisticated units for more exacting applications. In essence, a receiver's sensitivity is determined by the amount of gain provided and, more important, its noise characteristics. In general, the input signal must be somewhat greater than the noise at the receiver's input. This input noise is termed the *noise floor* of the receiver. It is not difficult to insert more gain in a radio, but getting noise figures below a certain level presents a more difficult challenge.

Selectivity may be defined as the extent to which a receiver is capable of differentiating between the desired signal and other frequencies (unwanted radio signals and noise). A receiver can also be overly selective. For instance, on commercial broadcast AM, we have seen that the transmitted signal can include intelligence signals up to about a maximum of 15 kHz, which subsequently generates upper and lower sidebands extending 15 kHz above and below the carrier frequency. The total signal has a 30-kHz bandwidth. Optimum receiver selectivity is thus 30 kHz, but if a 5-kHz bandwidth were "selected," the upper and lower sidebands would only extend 2.5 kHz above and below the carrier. The radio's output would suffer from a lack of the full possible "fidelity," since the output would include intelligence up to a maximum of 2.5 kHz. On the other hand, an excessive selectivity of 50 kHz results in the reception of unwanted adjacent radio signals and the additional external noise that is directly proportional to the bandwidth selected. Unfortunately, TRF receivers did suffer from selectivity problems. This led to their replacement by the superheterodyne receiver.

As has been stated, broadcast AM can extend to about 30 kHz bandwidth. As a practical matter, however, many stations and receivers use a more limited bandwidth. The lost fidelity is often not detrimental, due to the talk-show format of many AM stations. For instance, a 10-kHz bandwidth receiver provides audio output up to 5 kHz, which more than handles the human voice range. Musical reproduction with a 5-kHz maximum frequency is not "high fidelity" but certainly adequate for casual listening.

TRF Selectivity

Consider a standard AM broadcast band receiver that spans the frequency range from 550 to 1550 kHz. If the approximate center of 1000 kHz is considered, we find that for a desired 10-kHz BW, a Q of 100 is required.

$$Q = \frac{f_r}{BW}$$

$$= \frac{1000 \text{ kHz}}{10 \text{ kHz}} \tag{1-18}$$

$$= 100$$

Now, since the Q of a tuned circuit remains fairly constant as its capacitance is varied, a change to 1550 kHz will change the BW to 15.5 kHz.

$$Q = \frac{f_r}{BW} \tag{1-18}$$

Therefore,

$$BW = \frac{f_r}{Q}$$

$$= \frac{1550 \text{ kHz}}{100}$$

$$= 15.5 \text{ kHz}$$

The receiver's BW is now too large, and it will suffer from increased noise. On the other hand, the opposite problem is encountered at the lower end of the frequency range. At 550 kHz, the BW is 5.5 kHz.

$$BW = \frac{f_r}{Q}$$

$$= \frac{550 \text{ kHz}}{100}$$

$$= 5.5 \text{ kHz}$$

The fidelity of reception is now impaired. The maximum intelligence frequency possible is 5.5 kHz/2, or 2.75 kHz, instead of the full 5 kHz transmitted. It is this selectivity problem that led to the general use of the superheterodyne receiver (see Sec. 3-3) in place of TRF designs.

EXAMPLE 3-1

A TRF receiver is to be designed with a single tuned circuit using a 10-μH inductor.
(a) Calculate the capacitance range of the variable capacitor required to tune from 550 to 1550 kHz.
(b) The ideal 10-kHz BW is to occur at 1100 kHz. Determine the required Q.
(c) Calculate the BW of this receiver at 550 kHz and 1550 kHz.

SOLUTION

(a) At 550 kHz, calculate C.

$$f_r = \frac{1}{2\pi\sqrt{LC}} \qquad (1\text{-}16)$$

$$550 \text{ kHz} = \frac{1}{2\pi\sqrt{10\mu H \times C}}$$

$$C = 8.37 \text{ nF}$$

At 1550 kHz,

$$1550 \text{ kHz} = \frac{1}{2\pi\sqrt{10\mu H \times C}}$$

$$C = 1.06 \text{ nF}$$

Therefore, the required range of capacitance is from

$$1.06 \text{ to } 8.37 \text{ nF}$$

(b)
$$Q = \frac{f_r}{BW} \qquad (1\text{-}18)$$

$$= \frac{1100 \text{ kHz}}{10 \text{ kHz}}$$

$$= 110$$

(c) At 1550 kHz,

$$BW = \frac{f_r}{Q}$$

$$= \frac{1550 \text{ kHz}}{110}$$

$$= 14.1 \text{ kHz}$$

At 550 kHz,

$$BW = \frac{550 \text{ kHz}}{110}$$

$$= 5 \text{ kHz}$$

3-2 AM DETECTION

The process of detecting the intelligence out of the carrier and sidebands (the AM signal) has thus far been mentioned but not explained. In fact, the detection process can easily be accomplished. Recall our discussions about generating AM. We said if two different frequencies were passed through a nonlinear device, sum and difference components would be generated. The carrier and sidebands of the AM signal are separated in frequency by an amount equal to the intelligence frequency. If the AM signal is passed through a nonlinear device, difference frequencies between the carrier and sidebands will be generated and these frequencies are, in fact, the intelligence. It follows that passing the AM signal through a nonlinear device will provide detection, just as passing the carrier and intelligence through a nonlinear device enables AM generation. The mathematical proof for this was given in Sec. 2-4.

Detection of amplitude-modulated signals requires a nonlinear electrical network. An ideal nonlinear curve for this is one that affects the positive half-cycle of the modulated wave differently than the negative half-cycles. This distorts an applied voltage wave of zero average value so that the average resultant current varies with the intelligence signal amplitude. The curve shown in Fig. 3-2(a) is called an *ideal curve* because it is linear on each side of the operating point *P* and does not introduce harmonic frequencies.

When the input to an ideal nonlinear device is a carrier and its sidebands, the output contains the following frequencies:

1. The carrier frequency ⎫
2. The upper sideband   ⎬   original components
3. The lower sideband   ⎭

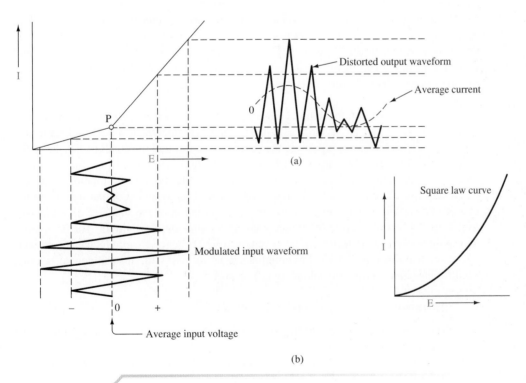

(a)

(b)

FIGURE 3-2   Nonlinear device used as detector.

4. A dc component
5. A frequency equal to the carrier minus the lower sideband and the upper sideband minus the carrier, which is the original signal frequency

The detector reproduces the signal frequency by producing a distortion of a desirable kind in its output. When the output of the detector is impressed upon a low-pass filter, the radio frequencies are suppressed and only the low-frequency intelligence signal and dc components are left. This is shown as the dashed average current curve in Fig. 3-2(a).

In some practical detector circuits, the nearest approach to the ideal curve is the square-law curve shown in Fig. 3-2(b). The output of a device using this curve contains, in addition to all the frequencies that were listed, the harmonics of each of these frequencies. The harmonics of radio frequencies can be filtered out, but the harmonics of the sum and difference frequencies, even though they produce an undesirable distortion, may have to be tolerated, since they can be in the audio-frequency range.

Diode Detector

One of the simplest and most effective types of detectors, and one with nearly an ideal nonlinear resistance characteristic, is the diode detector circuit shown in Fig. 3-3(a). Notice the *I–V* curve in Fig. 3-3(b). This is the type of curve on which the diode detector at (a) operates. The curved part of its response is the region of low current and indicates that for small signals the output of the detector will follow the square law. For input signals with large amplitudes, however, the output is essentially linear. Therefore, harmonic outputs are limited. The abrupt nonlinearity occurs for the negative half-cycle as shown in Fig. 3-3(b).

The modulated carrier is introduced into the tuned circuit made up of LC_1 in Fig. 3-3(a). The waveshape of the input to the diode is shown in Fig. 3-3(c). Since the diode conducts only during half-cycles, this circuit removes all the negative half-cycles and gives the result shown in Fig. 3-3(d). The average output is shown at (e). Although the average input voltage is zero, the average output voltage across R always varies above zero.

The low-pass filter, made up of capacitor C_2 and resistor R, removes the RF (carrier frequency), which, so far as the rest of the receiver is concerned, serves no useful purpose. Capacitor C_2 charges rapidly to the peak voltage through the small resistance of the conducting diode, but discharges slowly through the high resistance of R. The sizes of R and C_2 normally form a rather short time constant at the intelligence (audio) frequency and a very long time constant at the radio frequencies. The resultant output with C_2 in the circuit is a varying voltage that follows the peak variation of the modulated carrier [see Fig. 3-3(f)]. It is for this reason often termed an *envelope detector* circuit. The dc component produced by the detector circuit is removed by capacitor C_3, producing the ac voltage waveshape in Fig. 3-3(g). In communications receivers the dc component is often used for providing automatic volume (gain) control.

Advantages of diode detectors are as follows:

1. They can handle relatively high power signals. There is no practical limit to the amplitude of the input signal.
2. Distortion levels are acceptable for most AM applications. Distortion decreases as the amplitude increases.

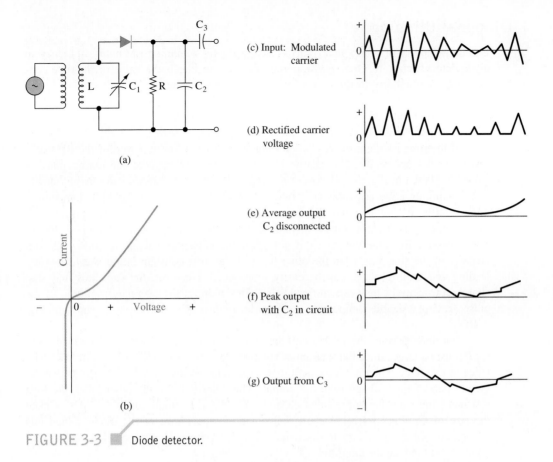

(c) Input: Modulated carrier

(d) Rectified carrier voltage

(e) Average output C_2 disconnected

(f) Peak output with C_2 in circuit

(g) Output from C_3

FIGURE 3-3 Diode detector.

3. They are highly efficient. When properly designed, 90% efficiency is obtainable.
4. They develop a readily usable dc voltage for the automatic gain control circuits.

Disadvantages of the diode detectors are:

1. Power is absorbed from the tuned circuit by the diode circuit. This reduces the Q and selectivity of the tuned input circuit.
2. No amplification occurs in a diode detector circuit.

Detector Diode Types

In non-critical applications, the standard pn junction diode can be used. It is usually adequate for the LF, HF, and low VHF bands. Low-cost silicon switching diodes such as the 1N914 and 1N4148 are frequently used. For higher frequencies, point contact diodes are often used. These diodes have the pn junction on the surface of the substrate and make contact to the p-type material via a small wire. The junction is very small, yielding very low capacitance that makes them useful for microwave operation up to 40 GHz. The commonly used point contact varieties are the 1N21, 1N23, and 1N34.

An important specification for detector diodes is voltage sensitivity. This is a measure of detector output per unit of RF input power. It is specified as V/mW or mV/μW at some specified dc bias current.

Diagonal Clipping

Careful selection of component parts is necessary for obtaining optimum efficiency in diode detector circuits. One very important fact to consider is the value of the time constant RC_2, particularly in the case of pulse modulation. When a carrier modulated by a square pulse [Fig. 3-4(b)] is applied to an ideal diode detector, the waveshape shown in Fig. 3-4(c) is produced. Notice that for clarity, the amplitude of the wave at (c) is exaggerated in comparison to the high-frequency carrier shown at (b).

If the time constant of RC_2 is too long as compared to the period of the RF wave, several cycles are required to charge C_2, and the leading edge of the output pulse is sloped as shown in Fig. 3-4(d). After the pulse passes by, the capacitor discharges slowly and the trailing edge is exponential rather than square as desired. This phenomenon is often referred to as "diagonal clipping." The diagonal clipping effect from a sine-wave intelligence signal is shown at (e). Notice that the detected sine wave at (e) is distorted. The excessive RC time constant did not allow the capacitor voltage to follow the full changes of the sine wave. On the other hand, if the time constant is too short, both the leading and trailing edges can be easily reproduced. However, the capacitor may discharge considerably between carrier cycles. This reduces the average amplitude of the pulse, leaving a sizable component of the carrier frequency in the output, as shown in Fig. 3-4(f).

For these reasons the selection of the time constant is a compromise. The load resistor R must be large, since the total input voltage is divided across R and the internal resistance of the diode when it is conducting. A large value of load resistance ensures that the greater part of this voltage is in the output, where desired. On the other hand, the load resistance must not be so high that capacitor C_2 becomes small enough to approximate the size of C_1 [Fig. 3-4(a)], the internal junction capacitance of the diode. When this occurs, capacitor C_2 will try to discharge through C_1 during the nonconducting periods, which would reduce the amplitude of the detector output.

Synchronous Detection

Diode detectors are used in the vast majority of AM detection schemes. Since "high fidelity" is usually not an important aspect in AM, the distortion levels of several percentage points or more from a diode detector can easily be tolerated. In applications demanding greater performance, the use of a synchronous detector offers the following advantages:

1. Low distortion—well under 1%
2. Greater ability to follow fast-modulation waveforms, as in pulse-modulation or high-fidelity applications
3. The ability to provide gain instead of attenuation, as in diode detectors

Synchronous detectors are also called *product* or *heterodyne* detectors. The principle of operation involves mixing in a nonlinear fashion just as in AM generation. Imagine receiving a transmission at 900 kHz. If it contained a 1-kHz tone, the reception consists of three components:

1. The carrier at 900 kHz
2. The USB at 901 kHz
3. The LSB at 899 kHz

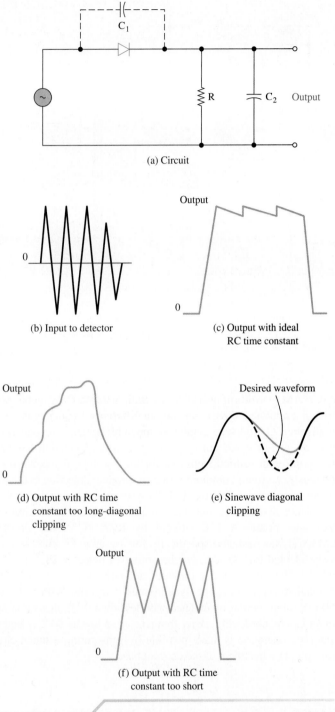

(a) Circuit

(b) Input to detector

(c) Output with ideal
RC time constant

(d) Output with RC time
constant too long-diagonal
clipping

(e) Sinewave diagonal
clipping

(f) Output with RC time
constant too short

FIGURE 3-4 ▮ Diode detector component considerations. Dashed lines indicate average voltage during pulse.

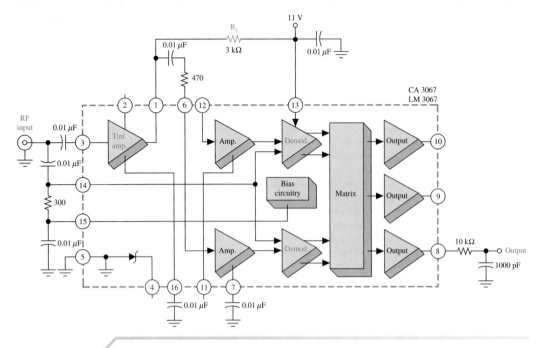

FIGURE 3-5 ■ Synchronous AM detection.

If this AM waveform were mixed with an internally generated 900-kHz sine wave through a nonlinear device, a resulting difference frequency is 1 kHz—the desired output intelligence. Of course, a number of much higher sum frequencies are also generated, but they are easily filtered out by a low-pass filter. Detection has been achieved in a completely different fashion from that for the envelope detector (diode detection) discussed previously. A circuit commonly used for product detection is the balanced modulator. It is widely used in single-sideband (SSB) systems and is detailed in Chapter 4.

Since synchronous detectors require rather complex circuitry, the use of a LIC is most appropriate. A LIC offered by RCA (CA3067) or National Semiconductor (LM3067) was designed specifically for use as a TV chroma (color) demodulator. It is easily adapted for use as a synchronous AM detector. Figure 3-5 shows the circuit connection. The tint amplifier provides an initial stage of gain with the double-balanced demodulators providing the synchronous detection. With a 35-mV AM signal input, a 450-mV audio output is obtained with less than 0.7% distortion at 80% modulation. The circuit can be used with carrier frequencies as low as 10 kHz and up to 10 MHz. Higher-frequency operation is made possible by substituting a tuned circuit for R_1, which provides proper adjustment of the carrier phase.

3-3 SUPERHETERODYNE RECEIVERS

The basic variable-selectivity problem in TRF systems led to the development and general usage of the superheterodyne receivers in the early 1930s. This basic receiver configuration is still dominant after all these years, an indication of its utility. A block diagram for a superheterodyne receiver is provided in Fig. 3-6. The first stage is a standard RF amplifier that may or may not be required, depending on factors to be discussed later. The next stage is the mixer, which accepts two inputs, the output of the RF amplifier (or antenna input when an RF amplifier is omitted) and a steady sine wave from the local oscillator (LO). The mixer is yet another nonlinear device utilized in AM. Its function is to mix the AM signal with a sine wave to generate a new set of sum and difference frequencies. Its output, as will be shown, is an AM signal with a constant carrier frequency regardless of the transmitter's frequency. The next stage is the intermediate-frequency (IF) amplifier, which provides the bulk of radio-frequency signal amplification at a *fixed* frequency. This allows for a constant BW over the entire band of the receiver and is the key to the superior selectivity of the superheterodyne receiver. Additionally, since the IF frequency is usually lower than the RF, voltage gain of the signal is more easily attained at the IF frequency. Following the IF amplifiers is the detector, which extracts the intelligence from the radio signal. It is subsequently amplified by the audio amplifiers into the speaker. A dc level proportional to the received signal's strength is extracted from the detector stage and fed back to the IF amplifiers and sometimes to the mixer and/or the RF amplifier. This is the automatic gain control (AGC) level, which allows relatively constant receiver output for widely variable received signals. Detail on AGC is provided in Sec. 3-6.

Frequency Conversion

It has been stated that the mixer performs a frequency conversion process. Consider the situation shown in Fig. 3-7. The AM signal into the mixer is a 1000-kHz carrier that has been modulated by a 1-kHz sine wave, thus producing side frequencies at 999 kHz and

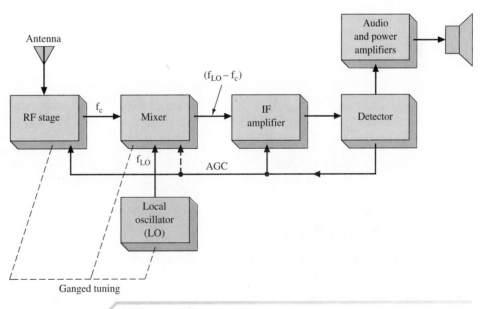

FIGURE 3-6 ■ Superheterodyne receiver block diagram.

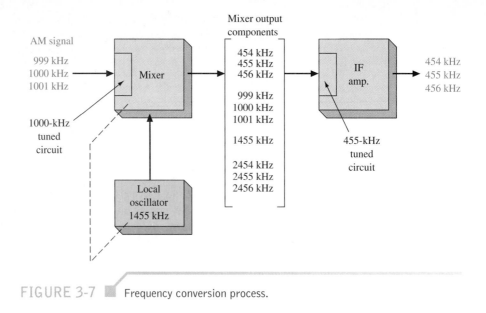

FIGURE 3-7 ■ Frequency conversion process.

1001 kHz. The LO input is a 1455-kHz sine wave. The mixer, being a nonlinear device, will generate the following components:

1. Frequencies at all of the original inputs: 999 kHz, 1000 kHz, 1001 kHz, and 1455 kHz.
2. Sum and difference components of all the original inputs: 1455 kHz ±(999 kHz, 1000 kHz, and 1001 kHz). This means outputs at 2454 kHz, 2455 kHz, 2456 kHz, 454 kHz, 455 kHz, and 456 kHz.
3. Harmonics of all the frequency components listed in 1 and 2 and a dc component.

The IF amplifier has a tuned circuit that only accepts components near 455 kHz, in this case 454 kHz, 455 kHz, and 456 kHz. Since the mixer maintains the same amplitude proportion that existed with the original AM signal input at 999 kHz, 1000 kHz, and 1001 kHz, the signal now passing through the IF amplifiers is a replica of the original AM signal. The only difference is that now its carrier frequency is 455 kHz. Its envelope is identical to that of the original AM signal. A frequency conversion has occurred that has translated the carrier from 1000 kHz to 455 kHz—a frequency intermediate to the original carrier and intelligence frequencies—which led to the terminology "intermediate-frequency amplifier," or IF amplifier. Since the mixer and detector both have nonlinear characteristics, the mixer is often referred to as the *first detector.*

Tuned-Circuit Adjustment

Now consider the effect of changing the tuned circuit at the front end of the mixer to accept a station at 1600 kHz. This means a reduction in either its inductance or capacitance (usually the latter) to change its center frequency from 1000 kHz to 1600 kHz. If the capacitance in the local oscillator's tuned circuit were simultaneously reduced so that its frequency of oscillation went up by 600 kHz, the situation shown in Fig. 3-8 would

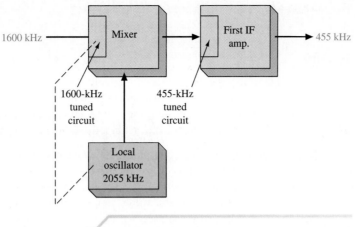

1600 kHz ——→ Mixer ——→ First IF amp. ——→ 455 kHz

1600-kHz tuned circuit

455-kHz tuned circuit

Local oscillator 2055 kHz

FIGURE 3-8 ■ Frequency conversion.

now exist. The mixer's output still contains a component at 455 kHz (among others), as in the previous case when we were tuned to a 1000-kHz station. Of course, the other frequency components at the output of the mixer are not accepted by the selective circuits in the IF amplifiers.

Thus, the key to superheterodyne operation is to make the LO frequency "track" with the circuit or circuits that are tuning the incoming radio signal such that their difference is a constant frequency (the IF). For a 455-kHz IF frequency, the most common case for broadcast AM receivers, this means the LO should always be at a frequency 455 kHz above the incoming carrier frequency. The receiver's "front-end" tuned circuits are usually made to track together by mechanically linking (ganging) the capacitors in these circuits on a common variable rotor assembly, as shown in Fig. 3-9. Note that this ganged capacitor has three separate capacitor elements.

Trimmers

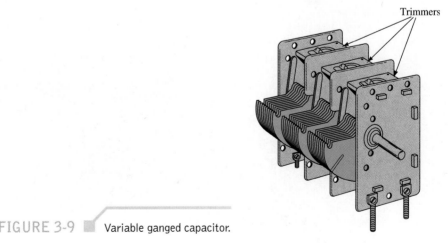

FIGURE 3-9 ■ Variable ganged capacitor.

Tracking

It is not possible to make a receiver track perfectly over an entire wide range of frequencies. The perfect situation occurs when the RF amplifier and mixer tuned circuits are exactly together and the LO is above these two by an amount exactly equal to the IF frequency. To obtain a practical degree of tracking, the following steps are employed:

1. A small variable capacitance in parallel with each section of the ganged capacitor, called the *trimmer,* is adjusted for proper operation at the highest frequency. The trimmer capacitors are shown in Fig. 3-9. The highest frequency requires the main capacitor to be at its minimum value (i.e., the plates all the way open). The trimmers are then adjusted to balance out the remaining stray capacitances to provide perfect tracking at the highest frequency.

2. At the lowest frequency, when the ganged capacitors are fully meshed (maximum value), a small variable capacitor known as the *padder* capacitor is put in series with the tank inductor. The "padders" are adjusted to provide tracking at the low frequency in the band.

3. The final adjustment is made at midfrequency by slight adjustment of the inductance in each tank.

The curve in Fig. 3-10(a) shows that performing the steps above, and then rechecking them once again to allow for interaction effects, provides nearly perfect tracking at three points. The minor imperfections between these points are generally of an acceptable nature. Figure 3-10(b) shows the circuit for each tank circuit and summarizes the adjustment procedure.

Electronic Tuning

The bulk and cost of ganged capacitors have led to their gradual replacement by a technique loosely called "electronic tuning." The majority of new designs use electronic frequency synthesis (see Chapter 8). Another electronic method relies on the capacitance

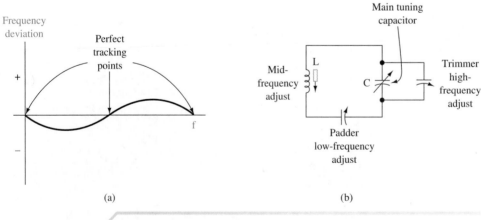

FIGURE 3-10 Tracking considerations.

RF inductors. (Courtesy of Coilcraft, Inc., Cary, Illinois.)

offered by a reverse-biased diode. Since this capacitance varies with the amount of reverse bias, a potentiometer can be used to provide the variable capacitance required for tuning. Diodes that have been specially fabricated to enhance this variable capacitance versus reverse bias characteristic are referred to as *varactor diodes, varicap diodes,* or *VVC diodes.* Figure 3-11 shows the two generally used symbols for these diodes and a typical capacitance versus reverse bias characteristic.

The amount of capacitance exhibited by a reverse-biased silicon diode, C_d, can be approximated as

$$C_d = \frac{C_0}{\left(1 + 2|V_R|\right)^{\frac{1}{2}}} \tag{3-1}$$

where C_0 = diode capacitance at zero bias
$\quad\quad V_R$ = diode reverse bias voltage

The use of Eq. (3-1) allows the designer to determine accurately the amount of reverse bias needed to provide the necessary tuning range. The varactor diode can also be used to generate FM, as explained in Chapter 5.

Figure 3-12 shows the front end of a broadcast-band receiver. It does not incorporate an RF amplifier, and Q_1 performs the dual function of mixer and local oscillator. The varactor diode D_1 provides the variable capacitance necessary to tune the radio signal

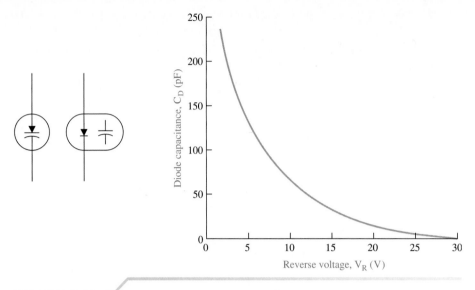

FIGURE 3-11 ■ Varactor diode symbols and *C/V* characteristic.

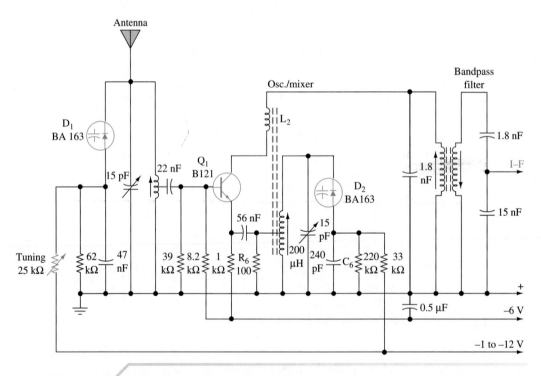

FIGURE 3-12 ■ Broadcast-band AM receiver front end with electronic tuning.

from the antenna while D_2 allows for the variable LO frequency. The –1- to –12-V supply comes from the tuning potentiometer and provides the necessary variable reverse voltage for both varactor diodes. The matched diode characteristics required for good tracking often lead to the use of varactor diodes fabricated on a common silicon chip and provided in a single package.

3-5 SUPERHETERODYNE ANALYSIS

Image Frequency

The superheterodyne receiver has been shown to have that one great advantage over the TRF—constant selectivity over a wide range of received frequencies. This was shown to be true since the bulk of the amplification in a superheterodyne receiver occurs in the IF amplifiers at a fixed frequency, and this allows for relatively simple and yet highly effective frequency selective circuits. A disadvantage does exist, however, other than the obvious increase in complexity. The frequency conversion process performed by the mixer–oscillator combination sometimes will allow a station other than the desired one to be fed into the IF. Consider a receiver tuned to receive a 20-MHz station that uses a 1-MHz IF. The LO would, in this case, be at 21 MHz to generate a 1-MHz frequency component at the mixer output. This situation is illustrated in Fig. 3-13. If an undesired station at 22 MHz were also on the air, it is possible for it to also get into the mixer. Even though the tuned circuit at the mixer's front end is "selecting" a center frequency of 20 MHz, a look at its response curve in Fig. 3-13 shows that it will not fully attenuate the undesired station at 22 MHz. As soon as the 22-MHz signal is fed into the mixer, we have a problem. It mixes with the 21-MHz LO signal and one of the components produced is 22 MHz – 21 MHz = 1 MHz—the IF frequency! Thus, we now have a desired 20-MHz station and an undesired 22 MHz station, which both look correct to the IF. Depending on the strength of the undesired station, it can either interfere with or even completely override the desired station.

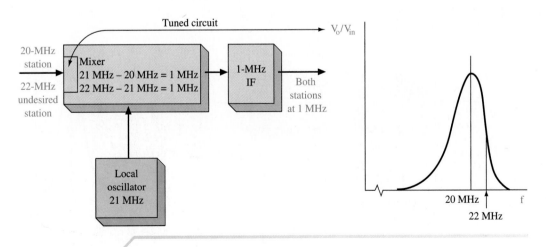

FIGURE 3-13   Image frequency illustration.

EXAMPLE 3-2

Determine the image frequency for a standard broadcast band receiver using a 455-kHz IF and tuned to a station at 620 kHz.

SOLUTION

The first step is to determine the frequency of the LO. The LO frequency minus the desired station's frequency of 620 kHz should equal the IF of 455 kHz. Hence,

$$LO - 620 \text{ kHz} = 455 \text{ kHz}$$

$$LO = 620 \text{ kHz} + 455 \text{ kHz}$$

$$= 1075 \text{ kHz}$$

Now determine what other frequency, when mixed with 1075 kHz, yields an output component at 455 kHz.

$$X - 1075 \text{ kHz} = 455 \text{ kHz}$$

$$X = 1075 \text{ kHz} + 455 \text{ kHz}$$

$$= 1530 \text{ kHz}$$

Thus, 1530 kHz is the image frequency in this situation.

In the preceding discussion, the undesired 22-MHz station is called the *image frequency*. Designing superheterodyne receivers with a high degree of image frequency rejection is obviously an important consideration.

Image frequency rejection on the standard broadcast band is not a major problem. A glance at Fig. 3-14 serves to illustrate this point. This tuned circuit at the mixer's input comes fairly close to fully attenuating the image frequency, in this case, since 1530 kHz

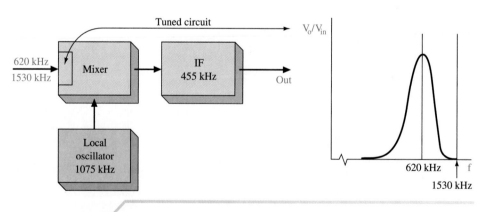

FIGURE 3-14 ▪ Image frequency not a problem.

is so far away from the tuned circuit's center frequency of 620 kHz. Unfortunately, this situation is not so easy to attain at the higher frequencies used by many communication receivers. In these cases, a technique known as *double conversion* is employed to solve the image frequency problems. This process is described in Chapter 8.

The use of an RF amplifier with its own input tuned circuit also helps to minimize this problem. Now the image frequency must pass through two tuned circuits (tuned to the desired frequency) before it is mixed. These tuned circuits at the input of the RF and mixer stages obviously serve to attenuate the image frequency to a greater extent than can the single tuned circuit in receivers without an RF stage.

RF Amplifiers

The use of RF amplifiers in superheterodyne receivers varies from none in undemanding applications up to three or even four stages in sophisticated communication receivers. Even inexpensive AM broadcast receivers not having an RF amplifier do contain an RF section—the tuned circuit at the mixer's input. The major benefits of using RF amplification are the following:

1. Improved image frequency rejection
2. More gain and thus better sensitivity
3. Improved noise characteristics

The first two are self-explanatory at this point, but the last advantage requires further elaboration. Mixer stages require devices to be operated in a nonlinear area in order to generate the required difference frequency at their output. This process is inherently more noisy (i.e., higher NF) than normal class A linear bias. The use of RF amplification stages to bring the signal up to appreciable levels minimizes the effect of mixer noise.

The RF amplifier usually employs a FET as its active component. While BJTs certainly can be utilized, the following advantages of FETs have led to their general usage in RF amplifiers.

1. Their high input impedance does not "load" down the Q of the tuned circuit preceding the FET. It thus serves to keep the selectivity at the highest possible level.
2. The availability of dual-gate FETs provides an isolated injection point for the AGC signal.
3. Their input/output square-law relationship allows for lower distortion levels.

The distortion referred to in the last item is called *cross-modulation* and is explained in Sec. 6-2.

A typical MOSFET RF amplifier stage is shown in Fig. 3-15. It is a dual-gate unit, with the AGC level applied to gate 2 to provide for automatically variable gain. The received antenna signal is fed via a tuned coupling circuit to gate 1. The gate 1 and output drain connections are tapped down on their respective coupling networks, which keeps the device from self-oscillation without the need for a neutralizing capacitor. Notice the built-in transient protection shown within the symbol for the 40673 MOSFET. Zener diodes between the gates and source/substrate connections provide protection from up to 10-V p-p transient voltages. This is a valuable safeguard because of the extreme fragility of the MOSFET gate/channel junction. Additional RF amplifier information is included in Sec. 6-2.

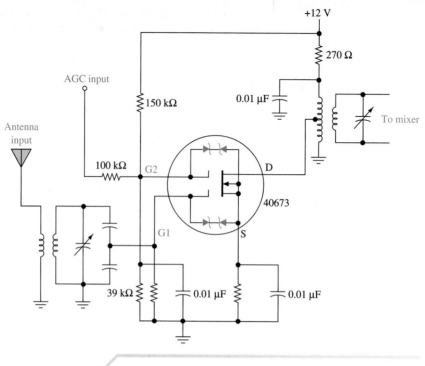

FIGURE 3-15 ■ Dual-gate MOSFET RF amplifier.

Mixer/LO

The frequency conversion accomplished by the mixer/LO combination can be accomplished in a number of ways. The circuits shown in Fig. 3-16 illustrate some of the possibilities. They all make use of a device's nonlinearity to generate sum and difference frequencies between the RF and local oscillator signals. This process generates output components at the IF frequency. The circuit in Fig. 3-16(a) utilizes a BJT and requires a

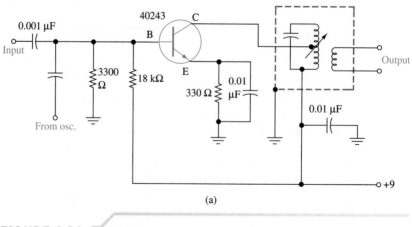

(a)

FIGURE 3-16 ■ Typical mixer circuits.

separate oscillator (often a Hartley oscillator) input as does the circuit at (c), which uses a MOSFET as the nonlinear element required to generate sum and difference frequencies.

The mixer at Fig. 3-16(b) incorporates a monolithic IC. Utilizing the Analog Device AD831, it provides a low distortion, wide dynamic range mixer for applications of

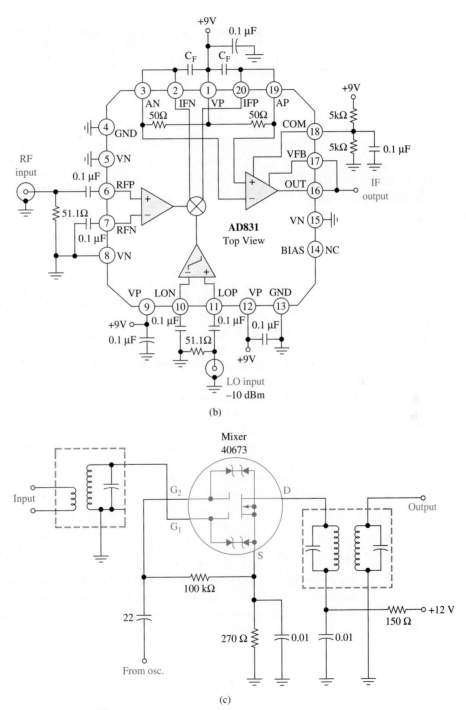

(b)

(c)

FIGURE 3-16 ■ *(Continued)*

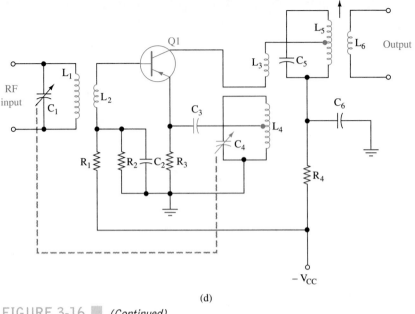

(d)

FIGURE 3-16 ■ (Continued)

RF and LO inputs up to 500 MHz. An integral low-noise amplifier provides an output (pin 16) that can directly drive low-impedance loads such as filters, 50-Ω amplifier inputs, and A/D converters. The circuit shown at (d) is a *self-excited mixer* in that a single device does the mixing and generates the LO frequency. Self-excited mixers are sometimes referred to as *autodyne mixers.* The oscillator-tuned circuit of C_4 and L_4 provides a positive feedback signal to maintain oscillation via coil L_3, which is magnetically coupled to L_4. The oscillator signal is injected into Q_1's emitter via C_3 and the RF signal into its base via L_1–L_2 transformer action. The "mixed" output at Q_1's collector is fed to the C_5–L_5 tank circuit, which "tunes" in the desired frequency for the IF amplifiers. Recall that mixing signals through a nonlinear device generates many frequency components; the tuned circuit is used to select the desired ones for the IF amplifiers.

A diode could be used as the mixer element but seldom is because the use of transistors provides gain as well as frequency conversion. Mixers are also referred to as *converters* and the *first detector.*

IF Amplifiers

The IF amplifiers provide the bulk of a receiver's gain (and thus are a major influence on its sensitivity) and selectivity characteristics. An IF amplifier is not a whole lot different than an RF stage except it operates at a fixed frequency. This allows the use of fixed double-tuned inductively coupled circuits to allow for the sharply defined bandpass response characteristic of superheterodyne receivers.

The number of IF stages in any given receiver varies, but from two to four is typical. Some typical IF amplifiers are shown in Fig. 3-17. The circuit at (a) uses the 40673 dual-gate MOSFET while the other two use LICs specially made for IF amplifier applica-

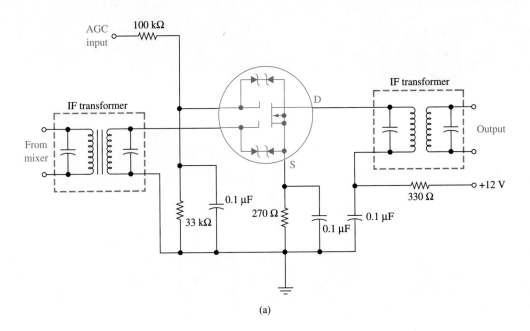

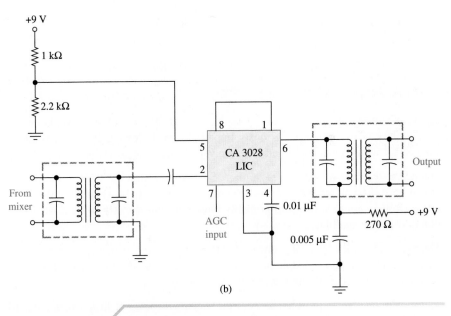

FIGURE 3-17 Typical IF amplifiers.

tions. Notice the double-tuned LC circuits at the input and output of all three circuits. They are shown within dashed lines to indicate they are one complete assembly. They can be economically purchased for all common IF frequencies and have a variable slug in the transformer core for fine tuning their center frequency. All three of the circuits have provision for the AGC level. Not all receivers utilize AGC to control the gain of mixer and/or RF stages, but they invariably do control the gain of IF stages.

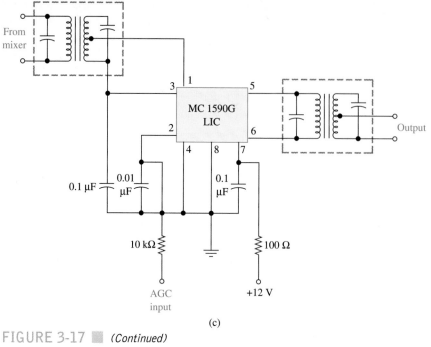

From mixer

MC 1590G
LIC

Output

0.1 μF
0.01 μF
0.1 μF

10 kΩ

100 Ω

AGC
input

+12 V

(c)

FIGURE 3-17 ■ *(Continued)*

3-6 AUTOMATIC GAIN CONTROL

The purpose of automatic gain control (AGC) has already been explained. Without this function, a receiver's usefulness is seriously impaired. The following list gives some of the problems that would be encountered in a receiver without this provision:

1. Tuning the receiver would be a nightmare. So as to not miss the weak stations, you would have the volume control (in the non-AGC set) turned way up. As you tuned into a strong station, you would probably blow out your speaker while a weak station may not be audible.

2. The received signal from any given station is constantly changing as a result of changing weather and ionospheric conditions. The AGC allows you to listen to a station without constantly monitoring the volume control.

3. Many radio receivers are utilized under mobile conditions. For instance, a standard broadcast AM car radio would be virtually unusable without a good AGC to compensate for the signal variation in different locations.

Obtaining the AGC Level

Most AGC systems obtain the AGC level just following the detector. Recall that following the detector diode, an *RC* filter removes the high frequency but hopefully leaves the low-frequency envelope intact. By simply increasing that *RC* time constant, a slowly varying dc level is obtained. The dc level changes with variations in the strength of the overall received signal.

Figure 3-18(a) shows the output from a diode detector with no filtering. In this case, the output is simply the AM waveform with the positive portion rectified out for two different levels of received signal into the diode. At (b), the addition of a filter has provided the two different envelope levels while filtering out the high-frequency content. These signals correspond to an undesired change in volume of two different received stations.

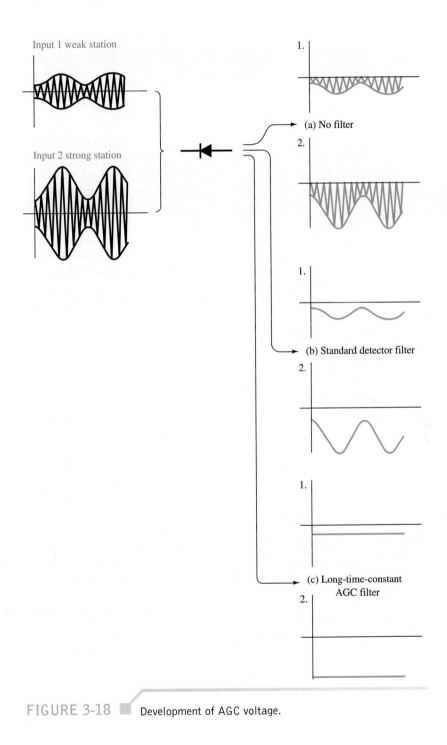

FIGURE 3-18 ■ Development of AGC voltage.

At (c), a much longer time constant filter has actually filtered the output into a dc level. Notice that the dc level changes, however, with the two different levels of input signal. This is a typical AGC level that is subsequently fed back to control the gain of IF stages and/or the mixer and RF stages.

In this case, the larger negative dc level at C_2 would cause the receiver's gain to be decreased such that the ultimate speaker output is roughly the same for either the weak or strong station. It is important that the AGC time constant be long enough so that desired radio signal level changes that constantly occur do *not* cause a change in receiver gain. The AGC should only respond to average signal strength changes, and as such usually has a time constant of about a full second.

Controlling the Gain of a Transistor

Figure 3-19 illustrates a method whereby the variable dc AGC level can be used to control the gain of a common emitter (CE) transistor amplifier stage. In the case of a strong received station, the AGC voltage developed across the AGC filter capacitor (C_{AGC}) is a large negative value that subsequently lowers the forward bias on Q_1. It causes more dc current to be drawn through R_2, and hence less is available for the base of Q_1, since R_1, which supplies current for both, can only supply a relatively constant amount. The voltage gain of a CE stage with an emitter bypass capacitor (C_E) is nearly directly proportional to dc bias current, and therefore the strong station reduces the gain of Q_1. The reception of very weak stations would reduce the gain of Q_1 very slightly, if at all. The introduction of AGC back in the 1920s marked the first major use of an electronic feedback control system. The AGC feedback path is called the AGC bus, because in a full receiver it is usually "bussed" back into a number of stages to obtain a large amount of gain control. Some receivers require more elaborate AGC schemes, and they will be examined in Chapter 8.

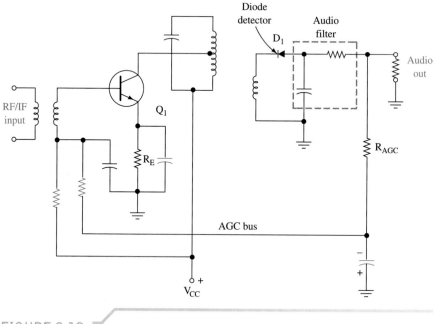

FIGURE 3-19 ■ AGC circuit illustration.

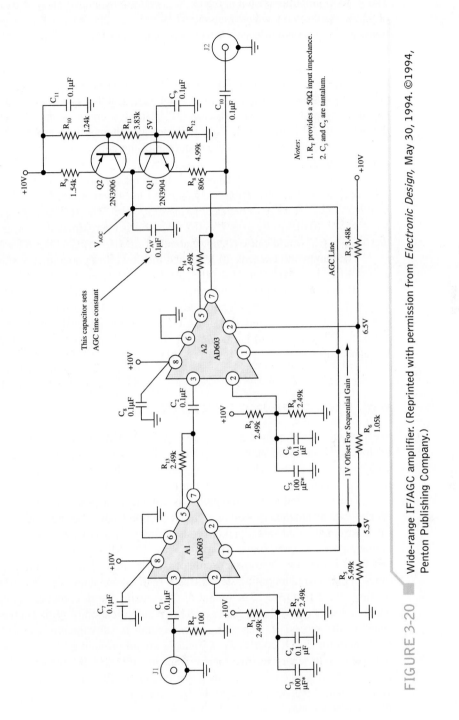

FIGURE 3-20 Wide-range IF/AGC amplifier. (Reprinted with permission from *Electronic Design*, May 30, 1994. ©1994, Penton Publishing Company.)

IF/AGC Amplifier

The IF/AGC amplifier shown in Fig. 3-20 operates over an extremely wide input (J1) range of 82 dB. It uses two low-cost transistors (2N3904 and 2N3906) as peak detectors. Q2 functions as a temperature-dependent current source and Q1 as a half-wave detector. Q2 is biased for a collector current of 300 μA at 27°C with a 1 μA/°C temperature coefficient.

The current into capacitor C_{AV} is the difference in the Q1, Q2 collector currents, which is proportional to the output signal at J2. The AGC voltage (V_{AGC}) is the time integral of this difference current. To insure that V_{AGC} is not sensitive to the short-term output signal changes, the rectified current in Q1 must, on average, balance the current in Q2. If the output of A2 is too small, V_{AGC} will increase, thereby increasing the gain of A1 and A2. This will cause Q1 to further conduct until the current through Q1 balances the current through Q2.

The gain of ICs A1 and A2 is set at 41 dB maximum for a total possible 82-dB gain. They operate sequentially in that the gain of A1 goes from minimum to maximum first and then A2's does the same as dictated by the AGC level. The full range of gain occurs from $V_{AGC} \cong 5$ V (0 dB) to $V_{AGC} \cong 7$ V (82 dB). This is approximately a linear relationship so that $V_{AGC} = 6$ V would cause a gain of about 41 dB [(6 − 5)/(7 − 5) × 82 = 41].

The bandwidth exceeds 40 MHz and thereby allows operation at standard IFs such as 455 kHz, 10.7 MHz, or 21.4 MHz. At 10.7 MHz the AGC threshold is 100 μV rms (−67 dBm) and the output is 1.4 V rms (3.9 V p-p). This corresponds to a gain of 83 dB (20 log 1.4 V/100 μV). The output holds steady at 1.4 V rms for inputs from −67 dBm to as high as +15 dBm, giving an 83-dB AGC range. Input signals above 15 dBm overdrive the device. The undesired harmonic outputs are typically at least 34 dB down from the fundamental.

3-7 AM RECEIVER SYSTEMS

We have thus far examined the various sections of AM receivers. It is now time to "put it all together" and look at the complete system. Figure 3-21 shows the schematic of a widely used circuit for a low-cost AM receiver. In the schematic shown in Fig. 3-21, the push-pull audio power amp, which requires two more transistors, has been omitted.

The L_1–L_2 inductor combination is wound on a powdered-iron (ferrite) core and functions as an antenna as well as an input coupling stage. Ferrite-core loopstick antennas offer extremely good signal pickup, considering their small size, and are adequate for the strong signal strengths found in urban areas. The RF signal is then fed into Q_1, which functions as the mixer and local oscillator (self-excited). The ganged tuning capacitor, C_1, tunes to the desired incoming station (the B section) and adjusts the LO (the D section) to its appropriate frequency. The output of Q_1 contains the IF components, which are tuned and coupled to Q_2 by the T_1 package. The IF amplification of Q_2 is coupled via the T_2 IF "can" to the second IF stage, Q_3, whose output is subsequently coupled via T_3 to the diode detector E_2. Of course, T_1, T_2, and T_3 are all providing the very good superheterodyne selectivity characteristics at the standard 455-kHz IF frequency. The E_2 detector diode's output is filtered by C_{11} such that just the intelligence envelope is fed via the R_{12} volume control potentiometer into the Q_4 audio amplifier. The AGC filter, C_4, then allows for a fed-back control level into the base of Q_2.

This receiver also illustrates the use of an *auxiliary AGC diode* (E_1). Under normal signal conditions, E_1 is reverse biased and has no effect on the operation. At some predetermined high signal level, the regular AGC action causes the dc level at E_1's cathode to decrease to the point where E_1 starts to conduct (forward bias), and it loads down the T_1

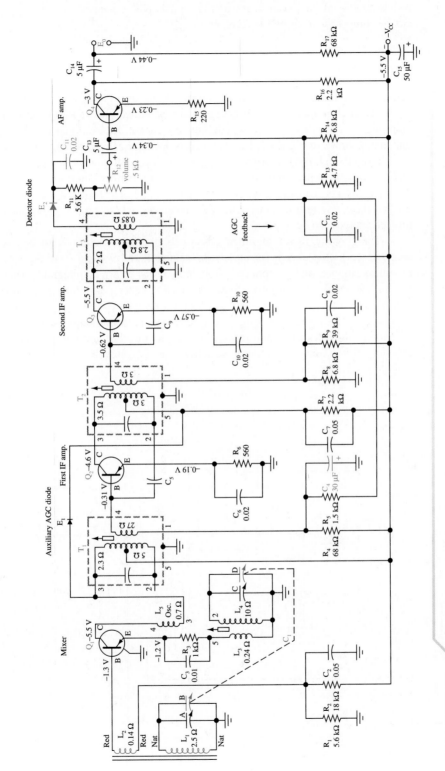

FIGURE 3-21 ■ AM broadcast superheterodyne receiver.

tank circuit, thus reducing the signal coupled into Q_2. The auxiliary AGC diode thus furnishes additional gain control for strong signals and enhances the range of signals that can be compensated for by the receiver.

LIC AM Receiver

The complete function of a superheterodyne AM receiver can be accomplished with LICs. The only hitch is that the tuned circuits must be added on externally. A number of AM chips are available from the various IC manufacturers. The National LM1820 is a typical unit and is shown in Fig. 3-22. Notice that this design uses variable inductors for tuning.

Even though the use of the LIC greatly reduces component count, the physical size and cost are not appreciably affected, since they are mainly determined by the frequency selective circuits. Thus, LIC AM radios are not widely used for low-cost applications but do find their way into higher-quality AM receivers, where certain performance and feature advantages can be realized.

The limiting factor of tuned circuits is the only roadblock to having complete receivers on a chip except for the station selection and volume controls. Alternatives to LC-tuned circuits, such as ceramic filters, may in the future be integrable. (See Chapter 8 for further detail on alternative filter circuits.) Another possibility is the use of phase-locked-loop (PLL) technology in providing a nonsuperheterodyne type of receiver. (See Chapter 6 for PLL theory.) Using this approach, it is theoretically possible to fabricate a functional AM broadcast-band receiver using just the chip and two external potentiometers (for volume control and station selection) and the antenna.

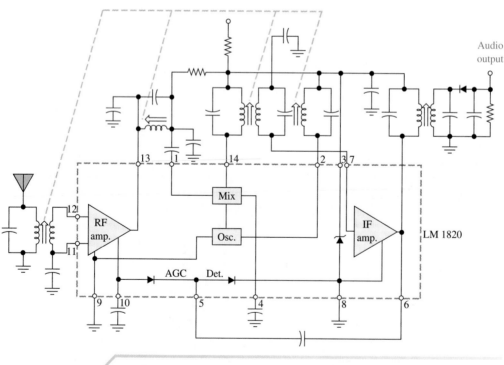

FIGURE 3-22 LM1820 AM radio system. (Courtesy of National Semiconductor Corp.)

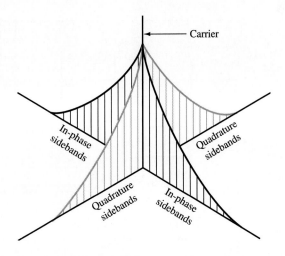

FIGURE 3-23 Phase relationships in AM stereo.

AM Stereo Receiver

As mentioned in Chapter 2, the Motorola C-Quam AM stereo system has become the de facto standard. In it, two sets of sidebands are generated 90° out of phase to carry the additional information. Figure 3-23 provides a pictorial representation of this condition. This is an example of combining two separate signals (left and right channels) into one frequency band, a practice that is called *frequency multiplexing.* Additional information on this concept is provided in subsequent chapters.

A block diagram of a C-Quam AM stereo receiver is shown in Fig. 3-24. A MC13024 IC is the basis of this system and the required "external" components are also shown. This circuit provides the complete receiver function requiring only a stereo audio power amplifier for the left and right channel outputs at pins 23 and 20. As you study this block diagram you will not understand some of the "blocks." For instance, instead of a local oscillator input to the mixer, a VCLO (voltage-controlled local oscillator) is provided. It is controlled by an AFC (automatic frequency control) signal at pin 7 that is the result of a PLL (phase-locked loop). All of these devices will be explained in subsequent chapters, so do not be alarmed at this point.

Receiver Analysis

It is convenient to consider power gain or attenuation of various receiver stages in terms of decibels related to a reference power level. The most often used references are with respect to 1 milliwatt (dBm) and 1 watt (dBW). In equation form,

$$dBm = 10\log_{10}\frac{P}{1\ mW} \tag{3-2}$$

$$dBW = 10\log_{10}\frac{P}{1\ W} \tag{3-3}$$

A dBm or dBW is an actual amount of power, whereas a dB represents a ratio of power. When dealing with a system that has a number of stages, the effect of dB and dBm can easily be dealt with. The following example shows this process.

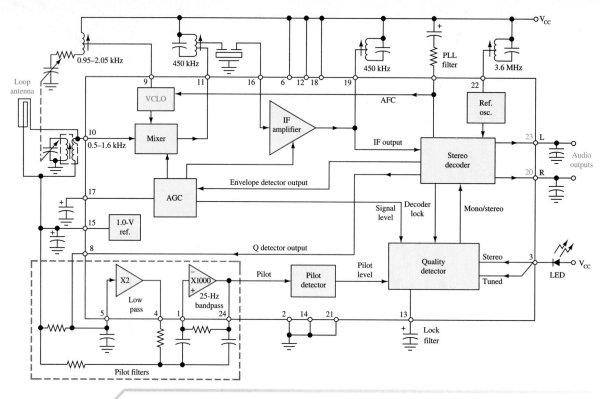

FIGURE 3-24 ▤ C-Quam receiver system. (Courtesy of Motorola Inc.)

EXAMPLE 3-3

Consider the radio receiver shown in Fig. 3-25. The antenna receives an 8-µV signal into its 50-Ω input impedance. Calculate the input power in watts, dBm, and dBW. Calculate the power driven into the speaker.

SOLUTION

$$P = \frac{V^2}{R} = \frac{(8\ \mu V)^2}{50\ \Omega} = 1.28 \times 10^{-12}\,\text{W}$$

$$\text{dBm} = 10 \log_{10} \frac{P}{1\ \text{mW}} \tag{3-2}$$

$$= 10 \log_{10} \frac{1.28 \times 10^{-12}}{1 \times 10^{-3}} = -89\ \text{dBm}$$

$$\text{dBW} = 10 \log_{10} \frac{P}{1\ \text{W}} \tag{3-3}$$

$$= 10 \log_{10} \frac{1.28 \times 10^{-12}}{1} = -119\ \text{dBW}$$

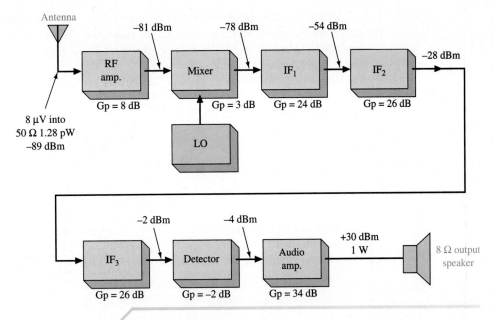

FIGURE 3-25 ▮ Receiver block diagram.

Notice that dBm and dBW are separated by 30 dB—this is always the case since 30 dB represents a 1000:1 power ratio. To determine the power driven into the speaker, simply add the gains and subtract the losses (in dB) all the way through the system. The −89 dBm at the input is added to the 8-dB gain of the RF stage to give −81 dBm. Notice that the 8-dB gain is simply added to the dBm input to give −81 dBm. This, and all subsequent stages, is shown in Fig. 3-24. Thus,

$$P_{\text{out}_{\text{dBm}}} = -89 \text{ dBm} + 8 \text{ dB} + 3 \text{ dB} + 24 \text{ dB} + 26 \text{ dB}$$

$$+ 26 \text{ dB} - 2 \text{ dB} + 34 \text{ dB}$$

$$= 30 \text{ dBm into speaker}$$

$$30 \text{ dBm} = 10 \log_{10} \frac{P_{\text{out}}}{1 \text{ mW}}$$

$$3 = \log_{10} \frac{P_{\text{out}}}{1 \text{ mW}}$$

Therefore,

$$1000 = \frac{P_{\text{out}}}{1 \text{ mW}}$$

$$P_{\text{out}} = 1 \text{ W}$$

Example 3-3 assumed that the receiver's AGC was operating at some fixed level based on the input signal's strength. As previously explained, the AGC system will attempt to maintain that same output level over some range of input signal. *Dynamic range* is the decibel difference between the largest tolerable receiver input signal (without causing audible distortion in the output) and its sensitivity (usually the minimum discernible signal). Dynamic ranges of up to about 100 dB represent current state-of-the-art receiver performance.

TROUBLESHOOTING

In this section we are going to analyze and troubleshoot the AM converter circuit. The converter, also known as an autodyne circuit, is a combination of the local oscillator and the mixer in a single stage. We're also going to discuss the power supply and audio amplifier problems in this section.

Upon completing this section you should be able to

- Troubleshoot an AM converter circuit
- Identify an open input circuit
- Identify a dead or intermittent local oscillator circuit
- Identify causes for a dead and intermittent local oscillator
- Troubleshoot the receiver's power supply
- Troubleshoot the receiver's audio amplifier

The Converter Circuit

In Sec. 3-3 it was shown that the local oscillator and the mixer play a very important part in AM reception. Figure 3-26 shows the converter stage (autodyne circuit) of an AM radio. The received RF input signal is fed into the base of Q1 from coils L1 and L2. The input AM radio signal is selected by tuning C1; notice also that C1 and C4 are ganged. When C1 is adjusted, C4 will be adjusted by the same amount. The local oscillator portion of the converter stage is made up of L3, L4, and C4. As C4 is adjusted the oscillator frequency changes to maintain a difference frequency of 455 kHz above the received AM signal. The feedback capacitor C3 sends a portion of the oscillator signal from a tap on L4 back to the emitter of Q1. The received RF signal and the oscillator signal are mixed in Q1 to produce the IF, which is sent to L5. All frequencies except the 455-kHz IF signal are filtered out by the tuned circuit's L5 and C5. Resistors R1 and R2 form a voltage divider network to bias the transistor's base–emitter circuit. Resistor R3 acts as a dc stabilizer for the emitter circuit. Capacitor C2 is a decoupling capacitor to keep the IF frequency from being fed back to the base of Q1. Any IF signal present at the base would be shorted to ground. The transistor's collector dc voltage is supplied by R4.

No AM RF Signal

If the received AM RF signal does not reach the base of Q1, no audio will be heard from the speaker. Noise may be heard when the tuning dial is moved across the band, but no stations will come in. An exception might be where a strong AM radio station in close proximity bleeds through into the converter transistor. A good indication of a working converter stage is to monitor the emitter voltage using a DMM as depicted in Fig. 3-26. As the radio is tuned across the AM band, the voltage reading on DMM will change. An open winding in coil L1 will cause the received AM signal to be lost. If a test signal were

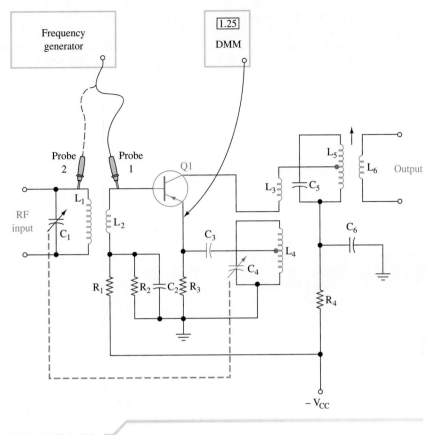

FIGURE 3-26 Troubleshooting a self-excited mixer.

injected at the base of Q1 (refer to Fig. 3-26 signal generator probe 1), it would be heard from the speaker. If the test signal were applied to L1 (Fig. 3-26 probe 2), no signal would be heard. If the coil L2 were open, AM reception would be lost. In addition, an open L2 will isolate the base from the resistor voltage divider network. As a result, the base–emitter bias would be removed and the transistor would cut off. Coils L1 and L2 in most AM receivers are part of the antenna system. The antenna consists of a ferrite metal stick with very fine wires making up the two coils. These fine wires very often break at the antenna or come loose from the printed circuit board, causing L1 or L2 to become open. Also, the wires from radio transformers usually break at the base of the transformer, where they are connected to the PCB. A dead converter stage can also result from a defective transistor.

Dead Local Oscillator Portion of Converter

Measuring the voltage at the emitter with a DMM and tuning the radio across the AM band is a good indication of oscillator operation. If this voltage changes as the radio is tuned, the oscillator can be assumed to be functioning. An oscilloscope at the emitter of transistor Q1 will show the oscillator waveform if it is present. If the local oscillator is dead (not operating), the signal will be missing and no voltage change will be detected by the DMM at the emitter of Q1. An open L4 will shut down the oscillator operation. The same is true for an open C4.

Poor AM Reception

A leaky capacitor C3 can cause erratic operation of the local oscillator circuit. Received radio stations will fade in and fade out as a result of this erratic operation of the oscillator. A station may fade out altogether and the converter quit working from a severely leaking capacitor. This is due to the loading effect on the emitter circuit of Q1. A local oscillator with poor tracking will affect radio reception at the high end or the low end of the AM band. A faulty C4 or C1 are likely suspects if poor tracking occurs.

Symptoms and Likely Causes

The following table lists symptoms and the likely circuit components that can cause them.

TABLE 3-1   *Converter Troubleshooting Chart*

Symptom	Troubleshooting Checks	Likely Trouble
No reception	Power OK; converter working	No input signal at base of Q1; L1 or L2 open; transistor bad
Stations fade in and out	Q1's emitter voltage fluctuates	Converter operation erratic; C3 leaky or open
No stations heard from mid to low AM band	DMM voltage changes when radio is tuned	LO not tracking across AM band; C1 or C4 faulty
No stations heard from mid to high AM band	DMM voltage changes when radio is tuned	LO not tracking across AM band; C1 or C4 faulty

Suspected faulty capacitors should be tested. The best method for testing capacitors is to use a capacitor checker. Some DMMs on the market today have a capacitor check function. The capacitor values are small in the converter circuit and should be tested out of the circuit. Open coils can be found using the ohmmeter setting of the DMM. A good coil will measure a low resistance and an open coil will measure a very high resistance. Coils can usually be measured without removing them from the circuit. If the converter transistor is suspect, test it with a transistor tester. Modern DMMs are equipped with this function. An open or shorted transistor can be tested with the DMM diode check setting or the ohmmeter setting.

Troubleshooting the Power Supply

If the receiver is completely dead, that is, no sound comes from the speaker, you should immediately suspect the power supply. This is one part of a receiver where the average technician can often easily find and repair a problem.

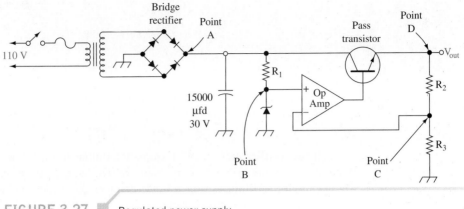

FIGURE 3-27 ▦ Regulated power supply.

Receivers are powered by batteries or a transformer-rectifier supply connected to the 110-V lines. Batteries usually power portable radios. A 9-V battery is most common. To check its output voltage, turn the radio on (to load the battery) and measure the battery's terminal voltage. If it is significantly below 9 V, perhaps 8 V or less, replace the battery and recheck the unit. Also check for corroded terminals.

Some radios employ a group of cells to obtain the necessary voltage. These must be connected in a series-aiding configuration of the positive terminal of one cell to the negative terminal of the next, and so on. The battery compartment has a diagram with battery symbols and plus and minus signs molded into the plastic to help us install the cells properly. Should one cell be placed in the compartment backwards, it would cancel the voltage of two cells, thereby dropping the total voltage to the point where the radio would not work. Check for proper installation of all cells. Then perform the "loaded" test described above for 9-V batteries.

Stereos and communications receivers will most likely use a regulated power supply similar to that shown in Fig. 3-27. Start troubleshooting by checking the output voltage with a DMM connected between point D and ground. If the voltage is correct (per manual specs), your problem lies elsewhere. If not, test the fuse for continuity and be sure the power plug is connected to a "hot" outlet and the switch is on. Next, check the rectifier output waveform at point A with an oscilloscope per the diagram in Fig. 3-28. The waveform should be similar to the one illustrated.

Output waveform at point D
should be almost perfect dc

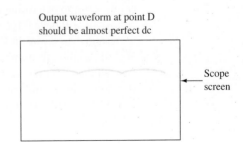

Scope screen

FIGURE 3-28 ▦ Bridge rectifier and filter operating properly.

FIGURE 3-29 ▊ Ripple increase caused by open diode.

If the rectifier output waveform is not similar to that shown in Fig. 3-28, one or more diodes in the bridge have probably failed. Diodes fail in one of two ways, either by opening or shorting. An open diode changes the bridge rectifier from full-wave to half-wave. As a result, ripple increases dramatically (see Fig. 3-29). A shorted diode causes heavy currents that should blow the fuse or, at the very least, cause overheated components.

Bridge rectifiers are usually encapsulated (you cannot get at the individual diodes). The unit must therefore be replaced should problems be found.

If the filter capacitor (from point A to ground) opens, the bridge output will be unfiltered, making it more difficult for the voltage regulator to eliminate ripple. It's difficult to say exactly what the waveform would look like; check the maintenance manual for details.

A shorted filter capacitor shorts the rectifier output, causing, at best, a blown fuse, and at worst, a burned-out rectifier and/or power transformer. In either case, open or shorted, replace the capacitor.

Assuming the rectifier and filter capacitor pass the tests discussed above, measure the zener reference voltage at point B and compare with specs per manual. Measure the voltage at point C, the feedback voltage to the inverting input of the op-amp. It should be within a tenth of a volt or so of the zener voltage. The point C voltage can be calculated using the voltage division formula:

$$V \text{ at point C} = V_{\text{out}}(R3)/(R2 + R3)$$

Measure the emitter–collector voltage of the pass transistor. It should be approximately 5 to 7 V depending on power supply load. If this voltage is a few tenths of a volt or less, the transistor is shorted and must be replaced.

Note: The above comments on power supply troubleshooting apply for any piece of equipment using a regulated power supply, not just superheterodyne receivers.

Troubleshooting the Audio Amplifier

A quick test to determine whether an audio amplifier is working is to first find the volume control. It will have three terminals on it. Touch a screwdriver or piece of wire to the center terminal as shown in Fig. 3-30. If the amplifier is working, you should hear a loud 60-Hz hum coming from the loudspeaker.

If the amplifier fails this quick test, do a dc check of voltages throughout the circuit. If nothing shows up, connect an audio generator via a 0.1-μfd capacitor to the center terminal of the volume control. Set the generator to approximately 1 kHz at perhaps 500-mV amplitude. Using an oscilloscope, observe the signal at each collector and base between the volume control and loudspeaker. Should the signal be present at one point and not the next, find the defective component causing this and replace it.

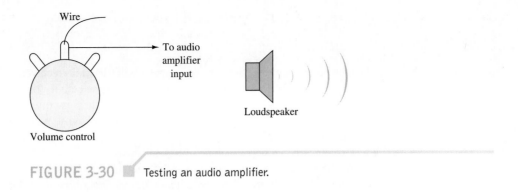

To audio
amplifier
input

Wire

Loudspeaker

Volume control

FIGURE 3-30   Testing an audio amplifier.

Troubleshooting the RF Portions of a Superhet Receiver

In general, troubleshooting a receiver's RF sections is done using the time-tested method of signal injection and tracing. The approach is the same discussed above for audio amplifiers except that now a high-frequency RF signal, usually modulated, is being used. This signal is connected to or injected into the receiver's antenna input terminals. We call this signal injection. The signal tracer, which can be either a scope or a RF probe on a voltmeter, is then connected to the inputs and outputs of each amplifier stage, one after the other, until the signal is lost. In this way, the defect is isolated and located with further tests.

SUMMARY

In Chapter 3 the basics of AM receivers were introduced. The development of receivers from the simplest to superheterodyne systems was discussed. The major topics the student should now understand include:

- the basics of a simple radio receiver
- the fundamental concepts of sensitivity and selectivity
- the functional blocks of a tuned radio frequency receiver (TRF)
- the input/output characteristics of a nonlinear device used as an AM detector
- the characteristics, operation, types, and design considerations of diode detectors
- the advantages of synchronous detection over the basic diode detector
- a complete analysis of superheterodyne receiver operation
- the tuning and tracking of a superheterodyne receiver
- an analysis of image frequency and methods for its attenuation
- the operation and typical circuits of the functional blocks in a superheterodyne receiver
- the need for automatic gain control (AGC) in a receiver and the description of a typical circuit and its operation
- the description of various superheterodyne receiver systems with power gain analysis

QUESTIONS AND PROBLEMS

SECTION 1

*1. Draw a diagram of a tuned radio-frequency (TRF) radio receiver.
*2. Explain the following: sensitivity of a receiver; selectivity of a receiver. Why are these important characteristics? In what units are they usually expressed?
 3. List the two factors that determine a receiver's sensitivity. Which is more important, and why?
 4. Explain why a receiver can be overly selective.
 5. A TRF receiver is to be tuned over the range 550 to 1550 kHz with a 25-μH inductor. Calculate the required capacitance range. Determine the tuned circuit's necessary Q if a 10-kHz bandwidth is desired at 1000 kHz. Calculate the receiver's selectivity at 550 and 1550 kHz. (0.422 to 3.35 nF, 100, 5.5 kHz, 15.5 kHz)

SECTION 2

 6. Why does passing an AM signal through a nonlinear device allow recovery of the low-frequency intelligence signal when the AM signal contains only high frequencies?
*7. Explain the operation of a diode detector.
 8. Describe the advantages and disadvantages of a diode detector.
 9. Explain how diagonal clipping occurs in a diode detector.
 10. Provide the advantages of a synchronous detector compared to a diode detector. Explain its principle of operation.

SECTION 3

*11. Draw a block diagram of a superheterodyne AM receiver. Assume an incident signal, and explain briefly what occurs in each stage.
*12. What type of radio receivers contains intermediate-frequency transformers?
 13. Explain how the superheterodyne receiver allows for constant selectivity over an entire band of received frequencies.
 14. The AM signal into a mixer is a 1.1-MHz carrier that was modulated by a 2-kHz sine wave. The local oscillator is at 1.555 MHz. List all mixer output components and indicate those "accepted" by the IF amplifier stage.
*15. Explain the purpose and operation of the first detector in a superhet receiver.
 16. Explain how the variable tuned circuits in a superheterodyne receiver are adjusted with a single control.

SECTION 4

 17. Provide an adjustment procedure whereby adequate tracking characteristics are obtained in a superheterodyne receiver.
 18. Draw a schematic that illustrates *electronic* tuning using a varactor diode.
 19. A silicon varactor diode exhibits a capacitance of 200 pF at zero bias. If it is in parallel with a 60-pF capacitor and 200-μH inductor, calculate the range of resonant frequency as the diode varies through a reverse bias of 3–15 V. (966 kHz, 1.15 MHz)
 20. A varactor diode has C_0 equal to 320 pF. Plot a curve of capacitance versus V_R from 0 to 20 V. The diode is used with a 200-μH coil. Plot the resonant frequency versus V_R from 0 to 20 V and suggest how the response could be linearized.

*21. If a superheterodyne receiver is tuned to a desired signal at 1000 kHz and its conversion (local) oscillator is operating at 1300 kHz, what would be the frequency of an incoming signal that would possibly cause *image* reception? (1600 kHz)

22. A receiver tunes from 20 to 30 MHz using a 10.7-MHz IF. Calculate the required range of oscillator frequencies and the range of image frequencies.

23. A superheterodyne receiver tunes the band of frequencies from 4 to 10 MHz with an IF of 1.8 MHz. The double-ganged capacitor used has a 325 pF maximum capacitance per section. Calculate the required RF and local oscillator coil inductance and the required tuning capacitor values when the receiver is tuned to receive 4 MHz. (4.87 μH, 2.32 μH, 52 pF, 78.5 pF)

24. Show why image frequency rejection is not a major problem for the standard AM broadcast band.

*25. What are the advantages to be obtained from adding a tuned radio-frequency amplifier stage ahead of the first detector (converter) stage of a superheterodyne receiver?

*26. If a transistor in the only radio-frequency stage of your receiver shorted out, how could temporary repairs or modifications be made?

27. What advantages do dual-gate MOSFETs have over BJTs for use as RF amplifiers?

*28. What is the *mixer* in a superheterodyne receiver?

29. Describe the advantage of an autodyne mixer over a standard mixer.

30. Why is the bulk of a receiver's gain and selectivity obtained in the IF amplifier stages?

SECTION 6

31. Describe the difficulties in listening to a receiver without AGC.

*32. How is *automatic volume control* accomplished in a radio receiver?

33. Explain how the ac gain of a transistor can be controlled by a dc AGC level.

34. The IF/AGC system in Fig. 3-20 has an AGC level of 5.5 V (V_{AGC} = 5.5 V). Determine the rms output voltage and the gain of the A1, A2 amplifier combination. Calculate the rms input voltage. (1.4 V rms, 20.5 dB, 0.132 V rms)

SECTION 7

35. Describe the function of auxiliary AGC.

36. What is the major limiting function with respect to manufacturing a complete superheterodyne receiver on a LIC chip?

37. A superhet receiver tuned to 1 MHz has the following specifications:
    RF amplifier: P_G = 6.5 dB, R_{in} = 50 Ω *Detector:* 4-dB attenuation
    Mixer: P_G = 3 dB *Audio amplifier:* P_G = 13 dB
    3 IFs: P_G = 24 dB each at 455 kHz
    The antenna delivers a 21-μV signal to the RF amplifier. Calculate the receiver's image frequency and input/output power in watts and dBm. Draw a block diagram of the receiver and label dBm power throughout. (1.91 MHz, 8.82 pW, −80.5 dBm, 10 mW, 10 dBm)

38. A receiver has a dynamic range of 81 dB. It has 0.55 nW sensitivity. Determine the maximum allowable input signal. (0.0692 W)

SINGLE-SIDEBAND COMMUNICATIONS

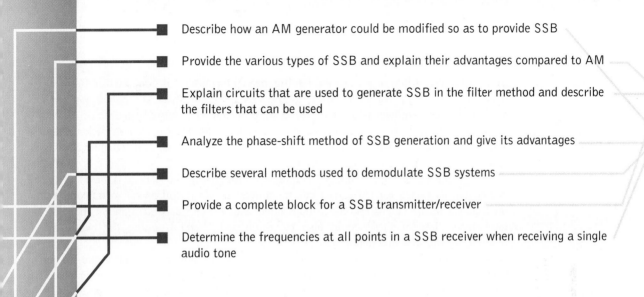

OBJECTIVES

- Describe how an AM generator could be modified so as to provide SSB

- Provide the various types of SSB and explain their advantages compared to AM

- Explain circuits that are used to generate SSB in the filter method and describe the filters that can be used

- Analyze the phase-shift method of SSB generation and give its advantages

- Describe several methods used to demodulate SSB systems

- Provide a complete block for a SSB transmitter/receiver

- Determine the frequencies at all points in a SSB receiver when receiving a single audio tone

4-1 SINGLE-SIDEBAND CHARACTERISTICS

The basic concept of single-sideband (SSB) communications was understood as early as 1914. It was first realized through mathematical analysis of an amplitude-modulated RF carrier. Recall that when a carrier is amplitude modulated by a single sine wave it generates three different frequencies: (1) the original carrier with amplitude unchanged; (2) a frequency equal to the difference between the carrier and the modulating frequencies with an amplitude up to one-half (at 100% modulation) the modulating signal; and (3) a frequency equal to the sum of the carrier and the modulating frequencies, with an amplitude also equal to a maximum of one-half that of the modulating signal. The two new frequencies of course are the side frequencies.

Upon recognition of the fact that sidebands existed, further investigation showed that after the carrier and one of the sidebands were eliminated, the other sideband could be used to transmit the intelligence. Since its amplitude and frequency never change, there is no information contained in the carrier. Further experiments proved that both sidebands could be transmitted, each containing different intelligence, with a suppressed or completely eliminated carrier.

By 1923, the first patent for this system had been granted, and a successful SSB communications system was established between the United States and England. Today,

SSB communications play a vital role in radio communications because of their many advantages over standard AM systems. The FCC, recognizing these advantages, further increased their use by requiring most transmissions in the overcrowded 2- to 30-MHz range to be SSB starting in 1977.

Power Distribution

You should recall that in AM all the intelligence (information) is contained in the sidebands, but two-thirds (or more) of the total power is in the carrier. It would appear that a great amount of power is wasted during transmission. The basic principle of single-sideband transmission is to eliminate or greatly suppress the high-energy RF carrier. This can be accomplished without affecting the fidelity of the emitted intelligence, since the carrier contains no information.

If a means of suppressing or completely eliminating the carrier is devised, the power that was used for the carrier can be converted into useful power to transmit the intelligence in the sidebands. Since both upper and lower sidebands contain the same intelligence, one of these could also be eliminated, thereby cutting the bandwidth required for transmission in half.

The total power output of a conventional AM transmitter is equal to the carrier power plus the sideband power. Conventional AM transmitters are rated in carrier power output. Consider a low-power AM system operating at 100% modulation. The carrier is 4 W and therefore each sideband is 1 W. The total transmitted power at 100% modulation is 6 W (4 W + 1 W + 1 W), but the AM transmitter is rated as a 4 W (just the carrier power) transmitter. If this system were converted to SSB, just a sideband at 1 W would be transmitted. This, of course, assumes a sine-wave intelligence signal. SSB systems are most often used for voice communications, which certainly do not generate a sinusoidal waveform.

SSB transmitters (and linear power amplifiers in general) are usually rated in terms of *peak envelope power* (PEP). To calculate PEP, the maximum (peak) envelope voltage is multiplied by 0.707, square the result, and divide by the load resistance. For instance, a SSB signal with a maximum level (over time) of 150 V p-p driven into a 50-Ω antenna results in a PEP rating of $(150/2 \times 0.707)^2 \div 50\ \Omega = 56.2$ W. This is the same power rating that would be given to the 150-V p-p sine wave, but there is a difference. The 150-V p-p level in the SSB voice transmission may occur only occasionally, while for the sine wave it occurs every cycle. These calculations are valid no matter what type of waveform the transmitter is providing. This could range from a series of short spikes with low average power (perhaps 5 W out of the PEP of 56.2 W) to a sine wave that would yield 56.2 W of average power. With a normal voice signal an SSB transmitter develops an average power of only one-fourth to one-third its PEP rating. Most transmitters cannot deliver an average power output equal to their peak envelope power capability. This is because their power supplies and/or components in the output stage are designed for a lower average power (voice operation) and cannot continuously operate at higher power levels.

Types of Sideband Transmission

A number of single-sideband systems have been developed. The major types include:

1. The carrier and one of the sidebands are completely eliminated at the transmitter; only one sideband is transmitted. This is standard single sideband, or simply SSB;

it is quite popular with amateur radio operators. The chief advantage of this system is maximum transmitted signal range with minimum transmitter power.

2. Another system eliminates one sideband and suppresses the carrier to a desired level. The suppressed carrier can then be used at the receiver for a reference, AGC, automatic frequency control (AFC), and, in some cases, demodulation of the intelligence-bearing sideband. This is called a single-sideband suppressed carrier (SSSC). The suppressed carrier is sometimes called a *pilot carrier.*

3. The type of system often used in military communications is referred to as *twin-sideband suppressed carrier,* or *independent sideband* (ISB) *transmission.* This system involves the transmission of two independent sidebands, each containing different intelligence, with the carrier suppressed to a desired level.

4. Vestigial sideband is used for television video transmissions. In it, a vestige (trace) of the unwanted sideband and the carrier is included with one full sideband. It is explained with television analysis in Chapter 7.

5. A more recently developed system is called amplitude-compandored single sideband (ACSSB). It is actually a type of SSSC in that a pilot carrier is usually included. In ACSSB the amplitude of the speech signal is compressed at the transmitter and expanded at the receiver. Performance gains of ACSSB systems over SSB are explained in Sec. 4-4.

Advantages of SSB

The most important advantage of SSB systems is a more effective utilization of the available frequency spectrum. The bandwidth required for the transmission of one conventional AM signal contains two equivalent SSB transmissions. This type of communications is, therefore, especially adaptable to the already overcrowded high-frequency spectrum.

A second advantage of this system is that it is less subject to the effects of selective fading. This is because there is *no* definite phase relationship between the upper and lower sidebands and the carrier as there is in conventional AM. In the propagation of conventional AM transmissions, if the upper-sideband frequency strikes the ionosphere and is refracted back to earth at a different phase angle from that of the carrier and lower-sideband frequencies, distortion is introduced at the receiver. Under extremely bad conditions, complete signal cancellation may result. The two sidebands should be identical in phase with respect to the carrier so that when passed through a nonlinear device (i.e., a diode detector), the difference between the sidebands and carrier is identical. That difference is the intelligence and will be distorted in AM systems if the two sidebands have a phase difference.

Another major advantage realized by all types of SSB systems is that a higher percentage of power is in the radiated intelligence. The resultant lower power requirements and weight reduction are especially important in mobile communication systems.

The SSB system has a noise advantage over AM due to the bandwidth reduction (one-half). Taking into account the selective fading improvement, noise reduction, and power savings, SSB offers about a 10- to 12-dB advantage over AM. This means that to have the same overall effectiveness, an AM system must transmit 10 to 12 dB more power than SSB. Some controversy exists on this issue because of the many variables that affect the savings. Suffice it to say that a 10-W SSB transmission is at least equivalent to the 100-W AM transmission (10-dB difference).

4-2 SIDEBAND GENERATION: THE BALANCED MODULATOR

A single-sideband transmission requires: (1) carrier elimination or suppression; and (2) elimination of one sideband. Once this has been accomplished, the selected sideband can be applied to the final power amplifier for transmission.

Balanced modulation is a system of adding intelligence to a carrier whereby only the sidebands are produced; the carrier is eliminated. The balanced modulator resembles the conventional push-pull amplifier in circuitry but not in operation. Figure 4-1 shows a conventional push-pull amplifier. For proper operation as an amplifier, the signals applied to the gates of Q_1 and Q_2 must be 180° out of phase. When the positive cycle is applied to the gate of Q_1, drain current flowing through T_2 will produce the positive portion of the output voltage. When the gate of Q_1 goes negative, the gate of Q_2 goes positive and drain current from Q_2 flows in the opposite direction through the primary of T_2, producing the negative-going alternation of the output waveshape.

By slightly modifying the push-pull amplifier, as shown in Fig. 4-2, no output signal will be obtained when an input signal is applied. Notice that the input signal applied to the gates of Q_1 and Q_2 in Fig. 4-2 has the same polarity. Both gates are positive at the same time. A positive-going gate in each FET causes the drain current to increase simultaneously in each FET. Since the drain current of Q_1 is flowing down through the primary of T_2 at the time the drain current of Q_2 is flowing up, the two magnetic fields produced in T_2 will effectively cancel. Thus, no output signal will be developed. At first glance, this particular circuit is apparently not performing any useful function. On the other hand, assume the input signal to be the carrier frequency. When the carrier frequency is applied in phase, no output signal will be developed in the output; thus, the carrier frequency has been suppressed, or eliminated, in the output.

Again, modify the push-pull amplifier circuit as shown in Fig. 4-3 (essentially a combination of Figs. 4-1 and 4-2). In this simplified push-pull balanced modulator, the

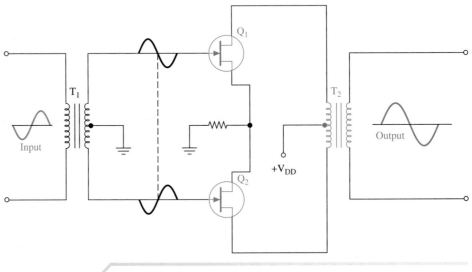

FIGURE 4-1 Conventional push-pull amplifier.

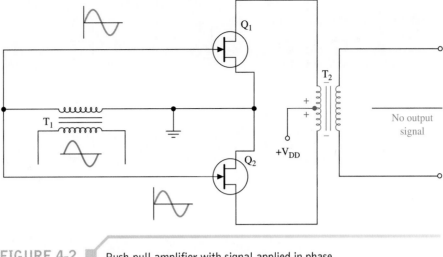

FIGURE 4-2 ■ Push-pull amplifier with signal applied in phase.

carrier input is applied to the gates of both Q_1 and Q_2 in phase; when the RF voltage is going positive, both gates are driven positive. The modulating voltage is applied to the two gates in the conventional push-pull manner; when the modulating signal drives the gate of Q_1 in a positive direction, the gate of Q_2 is driven in a negative direction.

Now consider the effect of the RF voltage on the stage. Since the carrier frequency causes both gates to be driven positive at the same time, the drain current of both Q_1 and Q_2 increases at the same time. Consequently, the magnetic fields caused by the two equal drain currents are effectively canceled in the output transformer T_2, which is center-tapped to $+V_{DD}$. On the negative swing of the RF voltage, the drain current of both FETs decreases at the same time. Again, the magnetic fields cancel. The carrier frequency, therefore, does not appear in the output.

The effect of the modulating voltage is different. When the positive swing of the modulating voltage is applied, the gate of Q_2 is driven negative. On the negative alternation of the modulating voltage, the gate of Q_1 is driven negative, but the gate of Q_2 is driven positive. Because of this push-pull arrangement, drain current flows on both the positive and negative swings, first through one FET and then through the other. There is no cancellation in the output.

When the modulating and RF carrier signals are applied at the same time, first one FET conducts heavily and then the other, depending on which gate is driven positive by the modulating voltage. The modulating voltage acts as a varying bias. Assuming nonlinear operation, when the gate of Q_1 is positive, the following components flow through Q_1:

1. The carrier frequency
2. The modulating frequency
3. The sum and difference frequencies of 1 and 2

The carrier frequency currents of Q_1 are effectively canceled by the carrier frequency currents from Q_2 in the common drain load, as explained before. The sum and difference frequencies cause induced voltages in the secondary of transformer T_2, and the modulating frequency is so low that it is highly attenuated at the output of T_2.

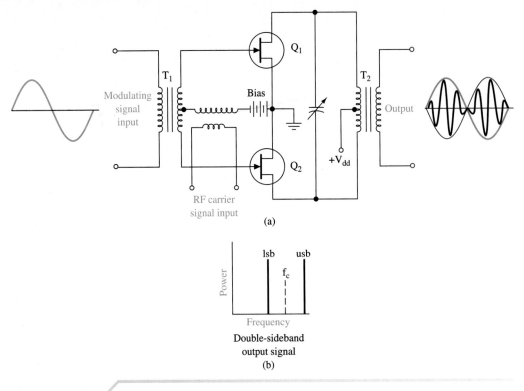

FIGURE 4-3 ■ Simplified push-pull modulator.

The output signal of the balanced modulator thus consists of the upper and lower sidebands; the carrier and modulating frequencies have been eliminated. This is shown in Fig. 4-3(b). As in AM generation, the generation of sum and difference frequencies relies on nonlinear operation. A tuned circuit is required to select the desired output frequencies from the other components generated by the heterodyning process.

Notice the dark blue envelope of the balanced modulator output in Fig. 4-3. It is a replica of the input intelligence signal. This is in contrast with the AM modulator, which replicates the intelligence in its positive or negative envelope.

Balanced Ring Modulator

Another commonly used means of eliminating the carrier is called the *balanced ring modulator*. Figure 4-4 shows a ring modulator schematically. Consider the carrier with the instantaneous conventional current flow as indicated by the arrows. The current flow through both halves of L_5 is equal but opposite, and thus the carrier is canceled in the output. This is also true on the carrier's other half-cycle, only now diodes B and C conduct instead of A and D.

Considering just the modulating signal, current flow occurs from winding L_2 through diodes C and D or A and B but not through L_5. Thus, there is no output of the modulating signal either. Now with both signals applied, but with the carrier amplitude

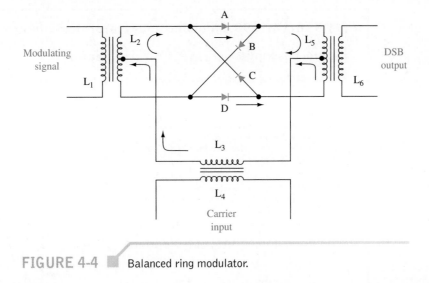

FIGURE 4-4 ■ Balanced ring modulator.

much greater than the modulating signal, the conduction is determined by the polarity of the carrier. The modulating signal either aids or opposes this conduction. When the modulating signal is applied, current will flow from L_2 and diode D will conduct more than A, and the current balance in winding L_5 is upset. This causes outputs of the desired sidebands but continued suppression of the carrier. This modulator is capable of 60 dB carrier suppression when carefully matched diodes are utilized. It relies on the nonlinearity of the diodes to generate the sum and difference sideband signals.

LIC Balanced Modulator

A balanced modulator of either type previously explained requires extremely well matched components to provide good suppression of the carrier (40 or 50 dB suppression is usually adequate). This suggests the use of LICs because of the superior component matching characteristics obtainable when devices are fabricated on the same silicon chip. A number of devices specially formulated for balanced modulator applications are available. A data sheet for the device used most often, the 1496/1596 unit, is provided in Fig. 4-5. As shown in the first page of the data sheets, this approach does not require the use of transformers or tuned circuits. The balanced modulator function is achieved with matched transistors in the differential amplifiers, with the modulating signal controlling the emitter current of the "diff-amps." The carrier signal is applied so as to switch the diff-amps' bases, resulting in a mixing process with the mixing product signals out of phase at the collectors.

This is an extremely versatile device since it can be used not only as a balanced modulator but also as an amplitude modulator, synchronous detector, FM detector, or frequency doubler. Its use as a SSB detector is shown at the end of the data sheets. The first graph on the second page provides an indication of the amount of carrier suppression versus carrier input level at two different carrier frequencies. Notice that 50- to 60-dB attenuation can be attained with relative ease.

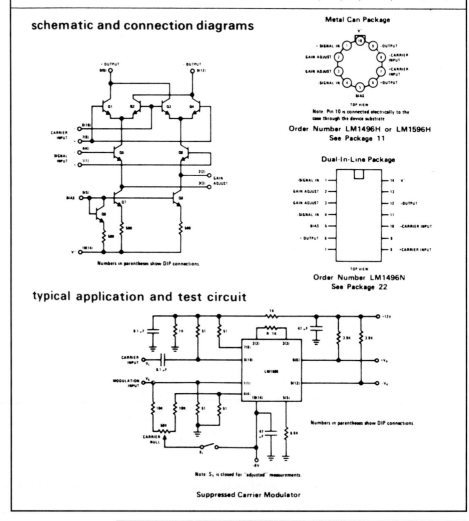

LM1596/LM1496 balanced modulator-demodulator

general description

The LM1596/LM1496 are double balanced modulator-demodulators which produce an output voltage proportional to the product of an input (signal) voltage and a switching (carrier) signal. Typical applications include suppressed carrier modulation, amplitude modulation, synchronous detection, FM or PM detection, broadband frequency doubling and chopping.

The LM1596 is specified for operation over the −55°C to +125°C military temperature range. The LM1496 is specified for operation over the 0°C to +70°C temperature range.

features

■ Excellent carrier suppression
  65 dB typical at 0.5 MHz
  50 dB typical at 10 MHz

■ Adjustable gain and signal handling

■ Fully balanced inputs and outputs

■ Low offset and drift

■ Wide frequency response up to 100 MHz

schematic and connection diagrams

Metal Can Package

Note: Pin 10 is connected electrically to the case through the device substrate

Order Number LM1496H or LM1596H
See Package 11

Dual-In-Line Package

Order Number LM1496N
See Package 22

typical application and test circuit

Note: S₁ is closed for "adjusted" measurements.

Suppressed Carrier Modulator

FIGURE 4-5 ■ Balanced modulator LIC. (Courtesy of National Semiconductor Corp.)

typical performance characteristics

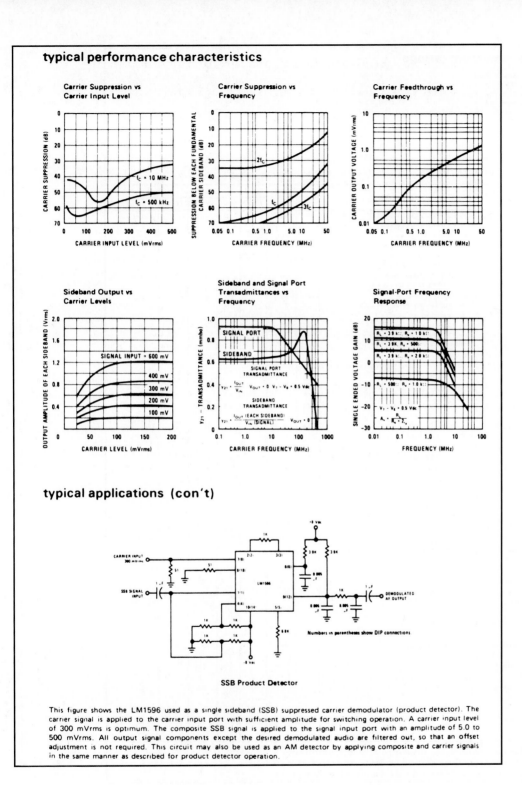

Carrier Suppression vs Carrier Input Level

Carrier Suppression vs Frequency

Carrier Feedthrough vs Frequency

Sideband Output vs Carrier Levels

Sideband and Signal Port Transadmittances vs Frequency

Signal-Port Frequency Response

typical applications (con't)

SSB Product Detector

This figure shows the LM1596 used as a single sideband (SSB) suppressed carrier demodulator (product detector). The carrier signal is applied to the carrier input port with sufficient amplitude for switching operation. A carrier input level of 300 mVrms is optimum. The composite SSB signal is applied to the signal input port with an amplitude of 5.0 to 500 mVrms. All output signal components except the desired demodulated audio are filtered out, so that an offset adjustment is not required. This circuit may also be used as an AM detector by applying composite and carrier signals in the same manner as described for product detector operation.

FIGURE 4-5 ■ *(Continued)*

absolute maximum ratings

Internal Power Dissipation (Note 1)	500 mW
Applied Voltage (Note 2)	30V
Differential Input Signal ($V_7 - V_8$)	±5.0V
Differential Input Signal ($V_4 - V_1$)	$±(5+I_5 R_e)$ V
Input Signal ($V_2 - V_1, V_3 - V_4$)	5.0V
Bias Current (I_5)	12 mA
Operating Temperature Range LM1596	$-55°C$ to $+125°C$
LM1496	$0°C$ to $+70°C$
Storage Temperature Range	$-65°C$ to $+150°C$
Lead Temperature (Soldering, 10 sec)	$300°C$

electrical characteristics ($T_A = 25°C$, unless otherwise specified, see test circuit)

PARAMETER	CONDITIONS	LM1596			LM1496			UNITS
		MIN	TYP	MAX	MIN	TYP	MAX	
Carrier Feedthrough	$V_C = 60$ mVrms sine wave $f_C = 1.0$ kHz, offset adjusted		40			40		μVrms
	$V_C = 60$ mVrms sine wave $f_C = 10$ MHz, offset adjusted		140			140		μVrms
	$V_C = 300$ mV$_{pp}$ square wave $f_C = 1.0$ kHz, offset adjusted		0.04	0.2		0.04	0.2	mVrms
	$V_C = 300$ mV$_{pp}$ square wave $f_C = 1.0$ kHz, offset not adjusted		20	100		20	150	mVrms
Carrier Suppression	$f_S = 10$ kHz, 300 mVrms $f_C = 500$ kHz, 60 mVrms sine wave offset adjusted	50	65		50	65		dB
	$f_S = 10$ kHz, 300 mVrms $f_C = 10$ MHz, 60 mVrms sine wave offset adjusted		50			50		dB
Transadmittance Bandwidth	$R_L = 50Ω$ Carrier Input Port, $V_C = 60$ mVrms sine wave $f_S = 1.0$ kHz, 300 mVrms sine wave		300			300		MHz
	Signal Input Port, $V_S = 300$ mVrms sine wave $V_7 - V_8 = 0.5$Vdc		80			80		MHz
Voltage Gain, Signal Channel	$V_S = 100$ mVrms, f = 1.0 kHz $V_7 - V_8 = 0.5$Vdc	2.5	3.5		2.5	3.5		V/V
Input Resistance, Signal Port	f = 5.0 MHz $V_7 - V_8 = 0.5$ Vdc		200			200		kΩ
Input Capacitance, Signal Port	f = 5.0 MHz $V_7 - V_8 = 0.5$ Vdc		2.0			2.0		pF
Single Ended Output Resistance	f = 10 MHz		40			40		kΩ
Single Ended Output Capacitance	f = 10 MHz		5.0			5.0		pF
Input Bias Current	$(I_1 + I_4)/2$		12	25		12	30	μA
Input Bias Current	$(I_7 + I_8)/2$		12	25		12	30	μA
Input Offset Current	$(I_1 - I_4)$		0.7	5.0		0.7	5.0	μA
Input Offset Current	$(I_7 - I_8)$		0.7	5.0		5.0	5.0	μA
Average Temperature Coefficient of Input Offset Current	$(-55°C < T_A < +125°C)$ $(0°C < T_A < +70°C$		2.0			2.0		nA/°C nA/°C
Output Offset Current	$(I_6 - I_9)$		14	50		14	60	μA
Average Temperature Coefficient of Output Offset Current	$(-55°C < T_A < +125°C)$ $(0°C < T_A < +70°C)$		90			90		nA/°C nA/°C
Signal Port Common Mode Input Voltage Range	$f_S = 1.0$ kHz		5.0			5.0		V$_{p-p}$
Signal Port Common Mode Rejection Ratio	$V_7 - V_8 = 0.5$ Vdc		−85			−85		dB
Common Mode Quiescent Output Voltage			8.0			8.0		Vdc
Differential Output Swing Capability			8.0			8.0		V$_{p-p}$
Positive Supply Current	$(I_6 + I_9)$		2.0	3.0		2.0	3.0	mA
Negative Supply Current	(I_{10})		3.0	4.0		3.0	4.0	mA
Power Dissipation			33			33		mW

Note 1: LM1596 rating applies to case temperatures to +125°C; derate linearly at 6.5 mW/°C for ambient temperature above 75°C. LM1496 rating applies to case temperatures to +70°C.

Note 2: Voltage applied between pins 6-7, 8-1, 9-7, 9-8, 7-4, 7-1, 8-4, 6-8, 2-5, 3-5.

FIGURE 4-5 ■ *(Continued)*

4-3 SSB FILTERS

Once the carrier has been eliminated, it is necessary to cancel one of the sidebands without affecting the other one. This requires a sharply defined filter, as Fig. 4-6 helps illustrate. Voice transmission requires audio frequencies from about 100 Hz to 3 kHz. Therefore, the upper and lower sidebands generated by the balanced modulator are separated by 200 Hz, as shown in Fig. 4-6.

The required Q depends on the center or carrier frequency, f_c; the separation between the two sidebands, Δf; and the desired attenuation level of the unwanted sideband. It can be calculated from

$$Q = \frac{f_c \left(\log^{-1} \text{dB} / 20 \right)^{1/2}}{4 \Delta f} \tag{4-1}$$

where dB is the suppression of the unwanted sideband.

EXAMPLE 4-1

Calculate the required Q for the situation depicted in Fig. 4-6 for
(a)   A 1-MHz carrier and 80-dB sideband suppression.
(b)   A 100-kHz carrier and 80-dB sideband suppression.

SOLUTION

(a)
$$Q = \frac{f_c \left(\log^{-1} \text{dB} / 20 \right)^{1/2}}{4 \Delta f} \tag{4-1}$$

$$= \frac{1 \text{ MHz} \left(\log^{-1} 80 / 20 \right)^{1/2}}{4 \times 200 \text{ Hz}} = \frac{1 \times 10^6 \left(10^4 \right)^{1/2}}{800}$$

$$= \frac{1 \times 10^8}{8 \times 10^2} = 125,000$$

(b)
$$Q = \frac{100 \text{ kHz} \left(\log^{-1} 80 / 20 \right)^{1/2}}{4 \times 200 \text{ Hz}}$$

$$= \frac{10^7}{8 \times 10^2} = 12,500$$

A practical consequence of the preceding example is that the SSB signal would be generated around the lower 100-kHz carrier in conjunction with a crystal filter. Then, after removing one sideband, an additional frequency translation is usually employed to get the sideband up to the desired frequency range. This is accomplished with a mixer circuit.

Both SSB transmitters and receivers require extremely sensitive bandpass filters in the region of 100 to 500 kHz. In receivers a high order of adjacent channel rejection is required if channels are to be closely spaced in order to conserve spectrum space. The fil-

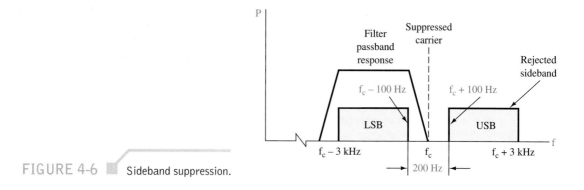

FIGURE 4-6   Sideband suppression.

ter used, therefore, must have very steep skirt characteristics (fast roll-off) and a flat
bandpass characteristic in order to pass all frequencies in the band equally well. These filter requirements are met by crystal filters, ceramic filters, and mechanical filters. A fourth
type of high-Q filter of more recent popularity is the *surface acoustic wave* (SAW)
device. It is most often used in TV and radar applications and is treated in Chapter 7. It is
most applicable to higher frequencies than are used in most SSB systems.

Crystal Filters

The crystal filter is commonly used in single-sideband systems to attenuate the unwanted
sideband. Because of its very high Q, the crystal filter passes a much narrower band of
frequencies than the best LC filter. Crystals with a Q up to about 50,000 are available.
    The equivalent circuit of the crystal and crystal holder is illustrated in Fig. 4-7(a).
Recall that the basics of crystal operation were introduced in Chapter 1. The components
L_s, C_s, and R_s represent the series resonant circuit of the crystal itself. C_p represents the
parallel capacitance of the crystal holder. The crystal offers a very low-impedance path to
the frequency to which it is resonant and a high-impedance path to other frequencies.
However, the crystal holder capacitance, C_p, shunts the crystal and offers a path to other
frequencies. For the crystal to operate as a bandpass filter, some means must be provided
to counteract the shunting effect of the crystal holder. This is accomplished by placing an
external variable capacitor in the circuit [C_1 in Fig. 4-7(b)].

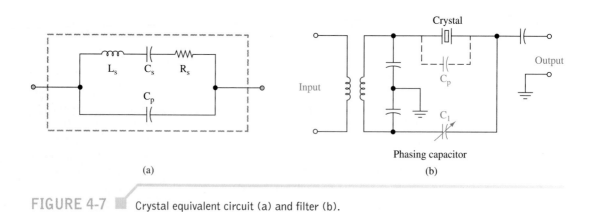

FIGURE 4-7   Crystal equivalent circuit (a) and filter (b).

In Fig. 4-7(b), a simple bandpass crystal filter is shown. The variable capacitor C_1, called the *phasing capacitor*, counteracts holder capacitance C_p. C_1 can be adjusted so that its capacitance equals the capacitance of C_p. Then both C_p and C_1 pass undesired frequencies equally well. Because of the circuit arrangement, the voltages across C_p and C_1 due to undesired frequencies are equal and 180° out of phase. Therefore, undesirable frequencies cancel and do not appear in the output. This cancellation effect is called the *rejection notch*.

For circuit operation, assume that a lower sideband with a maximum frequency of 99.9 kHz and an upper sideband with a minimum frequency of 100.1 kHz are applied to the input of the crystal filter in Fig. 4-7(b). Assume that the upper sideband is the unwanted sideband. By selecting a crystal that will provide a low-impedance path (series resonance) at about 99.9 kHz, the lower-sideband frequency will appear in the output. The upper sideband, as well as all other frequencies, will have been attenuated by the crystal filter. Improved performance is possible when two or more crystals are combined in a single filter circuit.

Ceramic Filters

Ceramic filters utilize the piezoelectric effect just as do crystals. However, they are normally constructed from lead zirconate–titanate. While ceramic filters do not offer Qs as high as a crystal, they do outperform LC filters in that regard. A Q of up to 2000 is practical with ceramic filters. They are lower in cost, more rugged, and smaller in size than crystal filters. They are used not only as sideband filters but also as replacements for the tuned IF transformers for superheterodyne receivers.

The circuit symbol for a ceramic filter is shown in Fig. 4-8(a) and a typical attenuation response curve is shown at (b). Note that the bandwidths at 60 dB and at 6 dB are shown. The ratio of these two bandwidths (8 kHz/6.8 kHz = 1.18) is defined as the *shape factor*. The shape factor (60-dB BW divided by 6-dB BW) provides an indication of the filter's selectivity. The ideal value of 1 would indicate a vertical slope at both frequency extremes. The ideal filter would have a horizontal slope within the passband with zero attenuation. The practical case is shown in Fig. 4-8(b), where a variation is illustrated. This variation is termed the *peak-to-valley ratio* or *ripple amplitude*. The shape factor and ripple amplitude characteristics also apply to the mechanical filters discussed next.

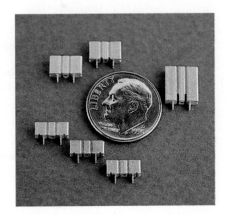

Ceramic filters. (Courtesy of Integrated Microwave, San Diego, CA.)

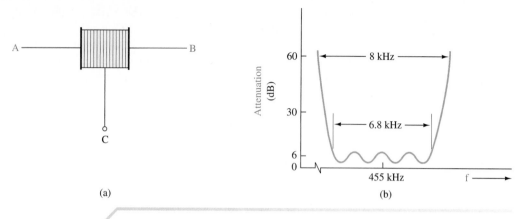

(a)

(b)

FIGURE 4-8 ▪ Ceramic filter and response curve.

Mechanical Filters

Mechanical filters have been used in single-sideband equipment since the 1950s. Some of the advantages of mechanical filters are their excellent rejection characteristics, extreme ruggedness, size small enough to be compatible with the miniaturization of equipment, and a Q in the order of 10,000, which is about 50 times that obtainable with LC filters.

The mechanical filter is a device that is mechanically resonant; it receives electrical energy, converts it to mechanical vibration, then converts this mechanical energy back into electrical energy at the output. Figure 4-9 shows a cutaway view of a typical unit. There are four elements constituting a mechanical filter: (1) an input transducer that converts the electrical energy at the input into mechanical vibrations, (2) metal disks that are manufactured to be mechanically resonant at the desired frequency, (3) rods that couple the metal disks, and (4) an output transducer that converts the mechanical vibrations back into electrical energy.

Not all the disks are shown in the illustration. The shields around the transducer coils have been cut away to show the coil and magnetostrictive driving rods. As you can see by its symmetrical construction, either end of the filter may be used as the input.

Figure 4-10 is the electrical equivalent of the mechanical filter. The disks of the mechanical filter are represented by the series resonant circuits L_1C_1 while C_2 represents

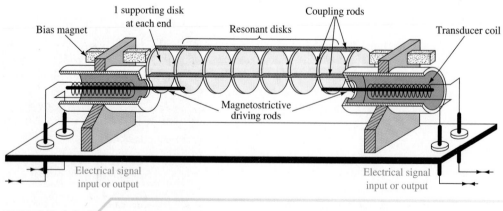

FIGURE 4-9 ▪ Mechanical filter.

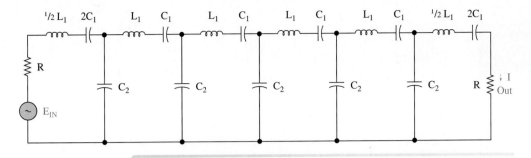

FIGURE 4-10 Electrical analogy of a mechanical filter.

the coupling rods. The resistance R in both the input and output represents the matching mechanical loads.

Let us assume that the mechanical filter of Fig. 4-9 has disks tuned to pass the frequencies of the desired sideband. The input to the filter contains both sidebands, and the transducer driving rod applies both sidebands to the first disk. The vibration of the disk will be greater at a frequency to which it is tuned (resonant frequency), which is the desired sideband, than at the undesired sideband frequency. The mechanical vibration of the first disk is transferred to the second disk, but a smaller percentage of the unwanted sideband frequency is transferred. Each time the vibrations are transferred from one disk to the next, there is a smaller amount of the unwanted sideband. At the end of the filter there is practically none of the undesired sideband left. The desired sideband frequencies are taken off the transducer coil at the output end of the filter.

Varying the size of C_2 in the electrical equivalent circuit in Fig. 4-10 varies the bandwidth of the filter. Similarly, by varying the mechanical coupling between the disks (Fig. 4-9), that is, by making the coupling rods either larger or smaller, the bandwidth of the mechanical filter is varied. Because the bandwidth varies approximately as the total cross-sectional area of the coupling rods, the bandwidth of the mechanical filter can be increased by using either larger coupling rods or more coupling rods. Mechanical filters with bandwidths as narrow as 500 Hz and as wide as 35 kHz are practical in the range 100 to 500 kHz.

4-4 SSB TRANSMITTERS

Filter Method

Figure 4-11 is a block diagram of a single-sideband transmitter using a balanced modulator to generate DSB and the filter method of eliminating one of the sidebands. For illustrative purposes, a single-tone 2000-Hz intelligence signal is used, but it is normally a complex intelligence signal, such as produced by the human voice.

The 2-kHz signal is amplified and mixed with a 100-kHz carrier (*conversion frequency*) in the balanced modulator. Remember, neither the carrier nor audio frequencies appear in the output of the balanced modulator; the sum and difference frequencies (98 kHz and 102 kHz) are its output. As illustrated in Fig. 4-11, the two sidebands from the balanced modulator are applied to the filter. Only the desired upper sideband is passed. The dashed lines show that the carrier and lower sideband have been removed.

Because the remaining sideband (containing the intelligence) is too low in frequency to transmit efficiently, it must be mixed again with a new conversion frequency to raise

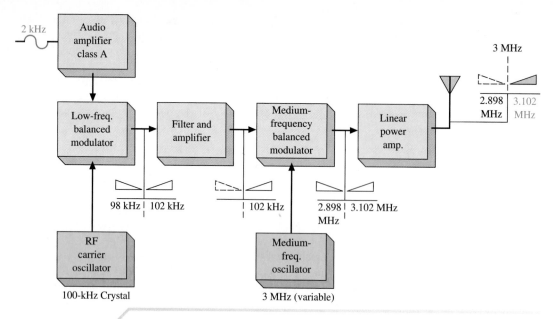

FIGURE 4-11 ■ SSB transmitter block diagram.

it to the desired transmitter frequency. As mentioned previously, sideband elimination at an initial low frequency allows for the use of reasonably low-Q filter circuits. It also allows for the use of mechanical filters that have an upper-frequency limit of about 500 kHz. In addition, crystals and ceramic units generally have better frequency stability at lower frequencies. The 3-MHz oscillator applies a signal to another balanced modulator. Again, the balanced modulator, after mixing the two inputs to get two new sidebands, removes the new 3-MHz carrier and applies the two new sidebands (3102 kHz and 2898 kHz) to a tunable linear power amplifier.

EXAMPLE 4-2

For the transmitter system shown in Fig. 4-11, determine the filter Q required in the linear power amplifier.

SOLUTION

The second balanced modulator created another DSB signal from the SSB signal of the preceding high-Q filter. However, the frequency translation of the second balanced modulator means that a low-quality filter can be used to once again create SSB. The new DSB signal is at about 2.9 MHz and 3.1 MHz. The required filter Q is

$$\frac{3 \text{ MHz}}{3.1 \text{ MHz} - 2.9 \text{ MHz}} = \frac{3 \text{ MHz}}{0.2 \text{ MHz}} = 15$$

The input and output circuits of the linear power amplifier are tuned to reject one sideband and pass the other to the antenna for transmission. A standard *LC* filter is now adequate to remove one of the two new sidebands. The new sidebands are about 200 kHz apart ($\approx$3100 kHz − 2900 kHz), so the required Q is quite low. (See Ex. 4-2 for further illustration.) The high-frequency oscillator is variable so that the transmitter output frequency can be varied over a range of transmitting frequencies. Since both the carrier and one sideband have been eliminated, all the transmitted energy is in the single sideband.

Filter SSB Generator

The circuit shown in Fig. 4-12 provides a complete, practical SSB generator. Its output is at 9 MHz and can be heterodyned to any desired frequency. The audio signal is amplified by a 741 op amp (U1). Its output is applied to the gates of Q_1. The balanced modulator is formed by the Q_1–Q_2 combination. The balance for maximum carrier suppression is made by adjusting R_2. The required 180° phase difference for the drains of Q_1 and Q_2 is provided by transformer T_1. It also couples the balanced modulator output into the IF preamplifier, Q_3. The SSB output is filtered by a prepackaged crystal-lattice filter at the collector of Q_3. A *crystal-lattice* filter contains at least two, but usually four crystals. It offers a wider possible passband than a single-crystal filter.

The carrier is generated with either crystal Y_1 (USB) or Y_2 (LSB). The oscillator's output at the drain of Q_5 is amplified by Q_6, which allows 4-V p-p injection into the sources of balanced modulation transistors Q_1 and Q_2. Fine adjustment of carrier frequency is accomplished with trimmer capacitors C_1 or C_2. They are adjusted to just "fit" the desired sideband into the passband of FL$_1$ and thus attenuate the undesired sideband and any vestige of the carrier that is already suppressed 45 to 50 dB by the balanced modulator.

Continuous wave (CW) operation is also possible with this system. CW is telegraphy by on–off keying of a carrier. It is the oldest radio modulation system and simply means either to transmit a carrier or not, representing a mark or space in telegraphy. For CW operation, S_2 is switched to the CW position, which activates Q_4 as a variable dc attenuator. R_3 is varied to change the bias of Q_4, which then shifts Q_2's source voltage to permit carrier insertion.

Phase Method

The phase method of SSB generation offers the following advantages over the filter method:

1. The bulk and expense of high-Q filters are eliminated.
2. There is greater ease in switching from one sideband to the other.
3. SSB can be generated directly at the desired transmitting frequency, which means that intermediate balanced modulators are not necessary.
4. Lower intelligence frequencies can be economically used since a high-Q filter is not necessary.

Despite these advantages, the filter method is rather firmly entrenched for many systems because of adequate performance and the complexity of the phase method. The increased availability of special LICs in recent years has increased SSB designs using the phase method.

The phase method of SSB generation relies on the fact that the upper and lower sidebands of an AM signal differ in the sign of their phase angles. This means that phase discrimination may be used to cancel one sideband of the DSB signal.

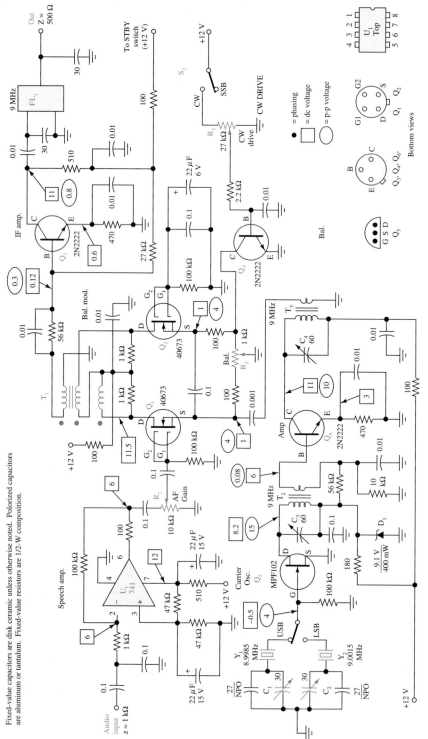

FIGURE 4-12 ■ SSB generator-filter method. (From the *ARRL Handbook*, courtesy of the American Radio Relay League.)

Consider a modulating signal $f(t)$ to be a pure cosine wave. A resulting balanced modulator output (DSB) can then be written as

$$f_{\text{DSB}1}(t) = (\cos \omega_i t)(\cos \omega_c t) \qquad (4\text{-}2)$$

where $\cos \omega_i t$ is the intelligence signal and $\cos \omega_c t$ the carrier. The term $\cos A \cos B$ is equal to $\frac{1}{2}[\cos (A + B) + \cos(A - B)]$ by trigonometric identity, and therefore Eq. (4-2) can be rewritten as

$$f_{\text{DSB}1}(t) = \frac{1}{2}\left[\cos(\omega_c + \omega_i)t + \cos(\omega_c - \omega_i)t\right] \qquad (4\text{-}3)$$

If another signal,

$$f_{\text{DSB}2}(t) = \frac{1}{2}\left[\cos(\omega_c - \omega_i)t - \cos(\omega_c + \omega_i)t\right] \qquad (4\text{-}4)$$

were added to Eq. (4-3), the upper sideband would cancel, leaving just the lower sideband,

$$f_{\text{DSB}1}(t) + f_{\text{DSB}2}(t) = \cos(\omega_c - \omega_i)t$$

Since the signal in Eq. (4-4) is equal to

$$\sin \omega_i t \, \sin \omega_c t$$

by trigonometric identity, it can be generated by shifting the phase of the carrier and intelligence signal by exactly 90° and then feeding them into a balanced modulator. Recall that sine and cosine waves are identical except for a 90° phase difference.

A block diagram for the system just described is shown in Fig. 4-13. The upper balanced modulator receives the carrier and intelligence signals directly, while the lower bal-

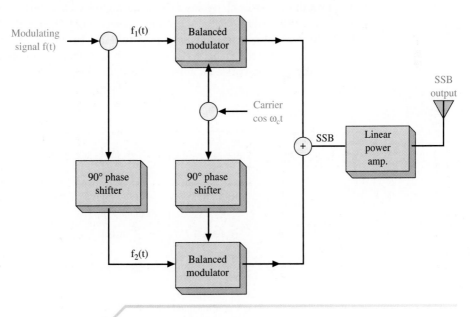

FIGURE 4-13   Phase-shift SSB generator.

anced modulator receives both of them shifted in phase by 90°. Thus, combining the outputs of both balanced modulators in the adder results in a SSB output that is subsequently amplified and then driven into the transmitting antenna.

A major disadvantage of this system is the 90° phase-shifting network required for the intelligence signal. The *carrier* 90° phase shifter is easily accomplished because of its single-frequency nature, but the audio signal covers a wide range of frequencies. To obtain exactly 90° of phase shift for a complete range of frequencies is difficult. The system is critical inasmuch as an 88° phase shift (2° error) for a given audio frequency results in about 30 dB of unwanted sideband suppression instead of the desired complete suppression obtained at 90° phase shift. The difficulty in obtaining adequate performance of the intelligence phase-shifting network is becoming less of a problem with the newer LICs designed to address this situation.

ACSSB Systems

Amplitude compandoring (*com*pression-ex*pandor*) single-sideband (ACSSB) systems are now allowing narrowband voice communications with the performance of FM systems for the land-mobile communications industry. This equivalent performance is provided with less than one-third the bandwidth of the comparable FM systems. The basis of ACSSB is to compress the audio before modulation and to expand it following demodulation at the receiver. A commonly used method to achieve this is use of the NE571N compandor LIC. It is shown in Fig. 4-14 connected as an expandor. This IC has a unity gain for a 0-dBm input. When used as a compressor, all negative dBm power levels are increased and positive dBm powers are decreased. For example, −40 dBm becomes −20 dBm, +15 dBm becomes +7.5 dBm, and so on. The expandor reverses the process to restore the signal's original dynamic range. Thus, a −20 dBm to +7.5 dBm signal becomes −40 dBm to +15 dBm at the expandor's output. The only signals not changed by the 571 IC are those at 0 dBm.

This system significantly cuts down the dynamic range that must be dealt with. It allows the lower-level signals to be transmitted with greater power while remaining within the PEP ratings of the transmitter power amplifier for the highest-level signals. Thus the *S/N* ratio is significantly improved at the lower end, while the somewhat increased noise for the louder passages (due to their reduced amplitude) is not a problem. At the receiver, the expandor restores the demodulated output to its original dynamic range.

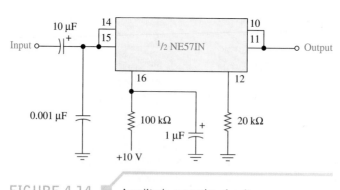

FIGURE 4-14 ▦ Amplitude expandor circuit.

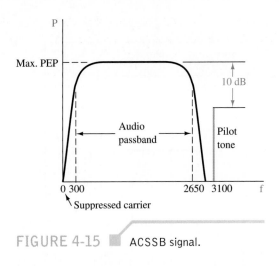

FIGURE 4-15 ■ ACSSB signal.

These ACSSB systems also include a pilot carrier signal as illustrated in Fig. 4-15. It is shown added to the audio signal sufficiently separated so that the receiver can ultimately distinguish between the two. The audio passband for voice transmission is fully attenuated by 3 kHz, and a pilot tone at 3.1 kHz above the eliminated carrier is the norm. It is suppressed by 10 dB from the maximum PEP as shown in Fig. 4-15. Thus the transmitter will have output power of about $\frac{1}{10}$ (−10 dB) of the maximum when there is no voice modulation. At the receiver, the pilot tone is usually compared to a reference oscillator in a phase-locked-loop (PLL) circuit. The PLL difference voltage is used to shift the receiver oscillator until error is eliminated. The pilot tone can also be used for AGC and squelch circuits at the receiver. A complete introduction to the PLL is provided in Chapter 6.

Transmitter Linear Power Amplifier

Once a SSB signal has been generated, a linear power amplifier is necessary to obtain significant power levels for transmission. The circuit in Fig. 4-16 provides 140 W of PEP nominal output power from 2 to 30 MHz when supplied with about 3 W of signal input. Fairly linear outputs up to 200 W PEP are possible with increased input drive. Its simplicity and use of low-cost components makes it an attractive design for mobile transmitters. It operates on the standard 13.6 V dc available from automotive electrical systems.

The amplifier is a class AB push-pull design. The quiescent current for each MRF454 power transistor (Q_3 and Q_4) is about 500 mA. This amount of bias is needed to prevent crossover distortion under high-output conditions. Diode D_2 is mounted on the same heat sink with the power transistors and "temperature tracks" them to provide bias adjustment with temperature changes. The relay K_1 and associated control circuitry, including Q_1 and Q_2, serve to "engage" the power amplifier only when RF input is present. The student is asked to analyze this function in an end-of-chapter question. Further details and circuit construction information can be obtained by requesting engineering bulletin EB63 from Motorola Semiconductor Products, Inc., P.O. Box 20912, Phoenix, AZ 85036.

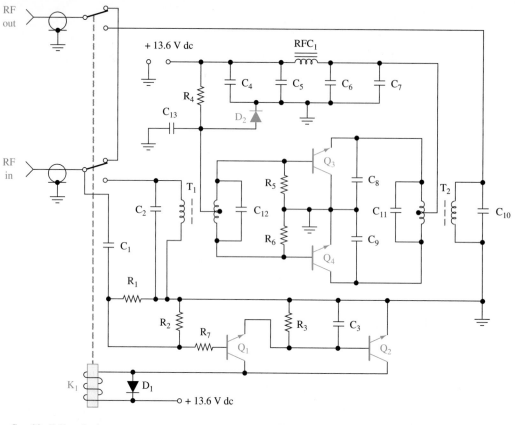

C₁—33 pF dipped mica
C₂—18 pF dipped mica
C₃—10 μF, 35 V dc for AM operation
    100 μF, 35 V dc for SSB operation
C₄—0.1 μF Erie
C₅—10 μF, 35 V dc electrolytic
C₆—1 μF tantalum
C₇—0.001 μF Erie disk
C₈, C₉—330 pF dipped mica
C₁₀—24 pF dipped mica
C₁₁—910 pF dipped mica
C₁₂—1100 pF dipped mica
C₁₃—500 μF, 3 V dc electrolytic

R₁—100 kΩ, 0.25 W
R₂—10 kΩ, 0.25 W
R₃—10 kΩ, 0.25 W
R₄—33 Ω, 5 W wirewound
R₅, R₆—10 Ω, 0.5 W
R₇—100 Ω, 0.25 W
RFC₁—9 ferroxcube beads on No. 18 AWG wire
D₁—1N4001
D₂—1N4997
Q₁, Q₂—2N4401
Q₃, Q₄—MRF454
T₁, T₂—16:1 transformers
K₁—Potter & Brumfield KT11A 12 V dc relay or equivalent

FIGURE 4-16 ▨ Linear power amplifier. (Courtesy of *Microwaves and RF,* September 1985.)

4-5 SSB DEMODULATION

One of the major advantages of SSB has been shown to be the elimination of the trans-mitted carrier. We have shown that this allows an increase in effective radiated power (erp), since the sidebands contain the information, and the never-changing carrier is redundant. Unfortunately, even though the carrier is redundant (contains no information), it *is* needed at the receiver! Recall that the intelligence in an AM system is equal in frequency to the difference of the sideband and carrier frequencies.

Waveforms

Figure 4-17(a) shows three different sine-wave intelligence signals; at (b) the resulting AM waveforms are shown, and (c) shows the DSB (no carrier) waveform. Notice that the DSB envelope (drawn in for illustrative purposes) looks like a full-wave rectification of the corresponding AM waveforms envelope. It is double the frequency of the AM envelope. At (d) the SSB waveforms are simply pure sine waves. This is precisely what is transmitted in the case of a sine-wave modulating signal. These waveforms are either at the carrier plus the intelligence frequency (USB) or carrier minus intelligence frequency (LSB). A SSB receiver would have to somehow "reinsert the carrier" to enable detection of the original audio or intelligence signal. A simple way to form a SSB detector is to use a mixer stage identical to a standard AM receiver mixer. The mixer is a nonlinear device, and the local oscillator input should be equivalent to the desired carrier frequency.

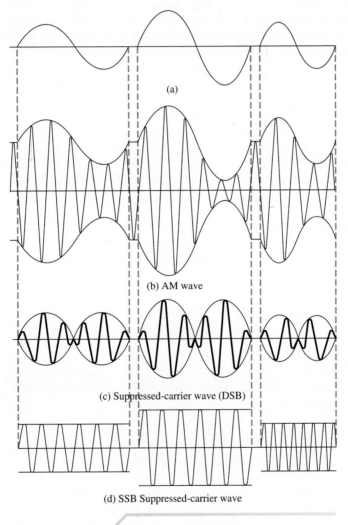

(a)

(b) AM wave

(c) Suppressed-carrier wave (DSB)

(d) SSB Suppressed-carrier wave

FIGURE 4-17 ■ AM, DSBB, and SSB waves from sinusoidal modulating signals.

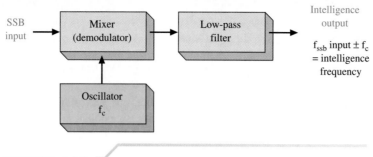

FIGURE 4-18 ▮ Mixer used as SSB demodulator.

Mixer SSB Demodulator

Figure 4-18 shows this situation pictorially. Consider a 500-kHz carrier frequency that has been modulated by a 1-kHz sine wave. If the upper sideband were transmitted, the receiver's demodulator would see a 501-kHz sine wave at its input. Therefore, a 500-kHz oscillator input will result in a mixer output frequency component of 1 kHz, which is the desired result. If the 500-kHz oscillator is not exactly 500 kHz, the recovered intelligence will not be exactly 1 kHz. If the receiver is to be used on several specific channels, a crystal for each channel will provide the necessary stability. If the receiver is to be used over a complete band of frequencies, the variable frequency oscillator (VFO), often called the beat frequency oscillator (BFO), must have some sort of automatic frequency control (AFC) to provide adequate quality reception. This can be accomplished by including a *pilot carrier* signal with the transmitted SSB signal. The pilot carrier can then be used to calibrate the receiver's oscillator at periodic intervals. Another approach is to utilize rather elaborate AFC circuits completely at the receiver, and the third possibility is the use of frequency synthesizers. They are covered in Chapter 8.

BFO Drift Effect

In any event, even minor drifts in BFO frequency can cause serious problems in SSB reception. If the oscillator drifts 15 Hz, a 1-kHz intelligence signal would be detected

EXAMPLE 4-3

At one instant of time, a SSB music transmission consists of a 256-Hz sine wave and its second and fourth harmonics, 512 Hz and 1024 Hz. If the receiver's demodulator oscillator has drifted 5 Hz, determine the resulting speaker output frequencies.

SOLUTION

The 5-Hz oscillator drift means that the detected audio will be 5 Hz in error, either up or down, depending on whether it is a USB or LSB transmission and on the direction of the oscillator's drift. Thus, the output would be either 251, 507, and 1019 Hz or 261, 517, and 1029 Hz. The speaker's output is no longer harmonic (exact frequency multiples), and even though it is just slightly off, the human ear would be offended by the new "music."

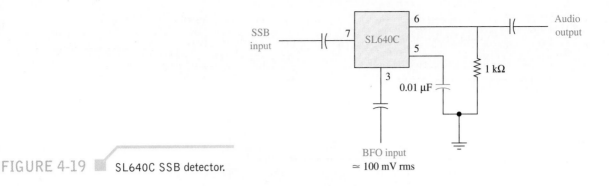

FIGURE 4-19 SL640C SSB detector.

either as 1015 Hz or 985 Hz. Speech transmission requires less than a 15-Hz shift or the talker starts sounding like Donald Duck and becomes completely unintelligible with a 40- to 50-Hz drift. Obtaining good-quality SSB reception of music and digital signals requires stabilities of several hertz.

LIC SSB Demodulator

Other forms of SSB demodulators include modified versions of the balanced modulators used to create DSB and a simple diode detector. In any event, the frequency stability requirements of the reinjected carrier signal serve to make SSB receivers much more complex and costly than standard AM receivers.

Figure 4-19 shows the Plessey Semiconductors SL640C LIC used as a SSB detector. This device is also used for the generation of DSB. This dual function is analogous to the 1496/1596 LIC introduced in Sec. 4-2. The capacitor connected to output pin 5 forms the low-pass filter to allow just the audio (low)-frequency component to appear in the output. The simplicity of this demodulator makes its desirability clear.

4-6 SSB RECEIVERS

To see the relationship of the parts in a single-sideband receiver, observe the block diagram in Fig. 4-20. Basically, the receiver is similar to an ordinary AM superheterodyne receiver; that is, it has RF and IF amplifiers, a mixer, a detector, and audio amplifiers.

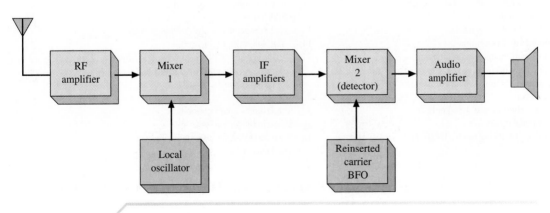

FIGURE 4-20 SSB receiver block diagram.

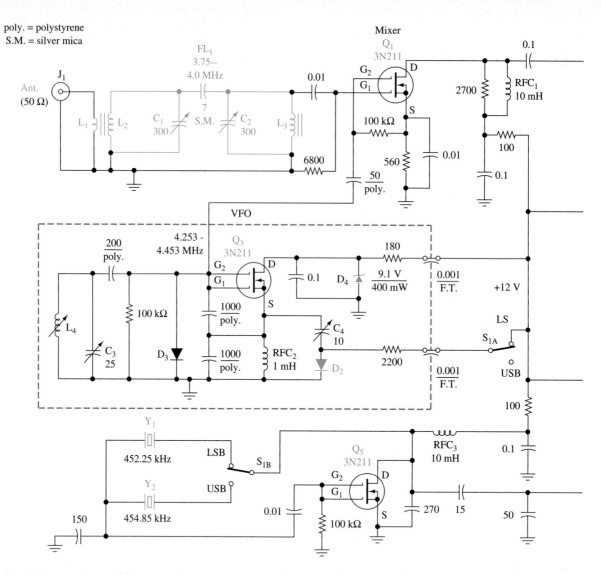

poly. = polystyrene
S.M. = silver mica

Fixed-value capacitors are disk ceramic unless noted otherwise. Polarized capacitors are electrolytic.
Fixed-value resistors are $^1/_4$- or $^1/_2$-Watt composition.

C_1, C_2—Mica compression trimmer, 300 pF max. Arco 427 or equiv.

C_3—Miniature 25-pF air variable. Hammarlund HF-25 or similar.

C_4—Circuit-board mount subminiature air variable or glass piston trimmer, 10 pF max. NPO miniature ceramic trimmer suitable as second choice.

D_1—LED, any color or size. Used only as 1.5-V reference diode.

D_2, D_3—Silicon switching diode, 1N914 or equiv.

D_4—Polarity-guarding diode. Silicon rectifier, 50 PIV, 1A.

D_5—Zener diode, 9.1 V, 400 mW or 1 Watt.

FL_1—Bandpass filter (see text).

FL_2—Collins Radio CB-type mechanical filter, Rockwell International No. 5269939010, 453.33-kHz center freq.

J_1—SO–239

J_2—Single-hole-mount phono jack.

J_3—Two-circuit phone jack.

L_1—Two turns No. 24 insulated wire over ground end of L2.

FIGURE 4-21 ■ SSB receiver. (From the *ARRL Handbook,* courtesy of the American Radio Relay League.)

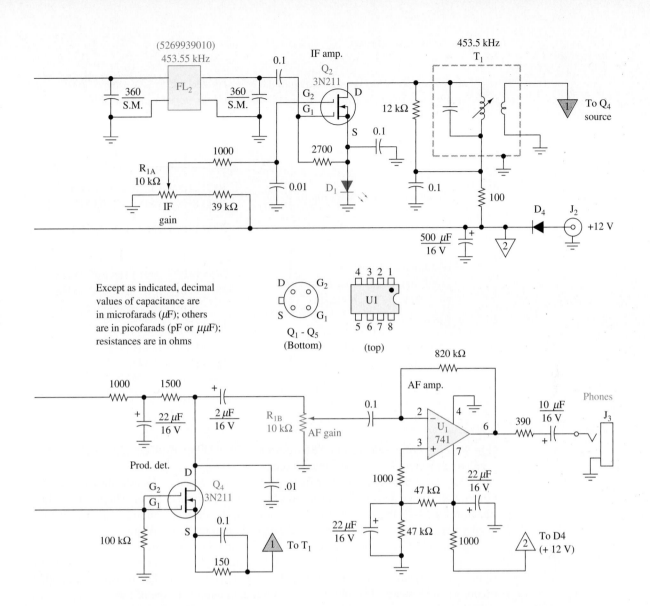

Except as indicated, decimal values of capacitance are in microfarads (μF); others are in picofarads (pF or μμF); resistances are in ohms

L₂, L₃—40 turns No. 24 enameled wire on T68-6 toroid core.

L₄—Slug-tuned inductor, 3.6- to 8.5-μH range, J. W. Miller 42A686CBI or equivalent suitable. Substitutes should have Q of 100 or greater at 4 MHz and be mechanically rigid.

Q₁–Q₅, incl. —Texas Instruments 3N211 FET,

R₁—Dual control, 10-kΩ per section, linear taper. Allen-Bradley type JD1N200P or similar. Separate controls can be used by providing extra hole in front panel.

RFC₁, RFC₃ — 10-mH miniature RF choke, J. W. Miller 70F 102AI or equiv.

RFC₂—1-mH miniature RF choke, J. W. Miller 70F 103AI or equiv.

S₁—Two-pole, two-position phenolic or ceramic wafer switch.

T₁—455-kHz miniature IF transformer (see text). J. W. Miller No. 2067.

U₁—8-pin dual-in-line 741 op amp.

Y₁, Y₂—International Crystal Co. type GP, 30-pF load capacitance, HC-6/U style of holder. LSB 452.25 kHz, and USB 454.85 kHz.

FIGURE 4-21 ▮ (Continued)

EXAMPLE 4-4

The SSB receiver in Fig. 4-20 has outputs at 1 kHz and 3 kHz. The carrier used and suppressed at the transmitter was 2 MHz, and the upper sideband was utilized. Determine the exact frequencies at all stages for a 455-kHz IF frequency.

SOLUTION

RF amp and first mixer input }	2000 kHz + 1 kHz = 2001 kHz 2000 kHz + 3 kHz = 2003 kHz
Local oscillator	2000 kHz + 455 kHz = 2455 kHz
First mixer output: IF amp and second mixer input (the other components attenuated by tuned circuits }	2455 kHz − 2001 kHz = 454 kHz 2455 kHz − 2003 kHz = 452 kHz
BFO	455 kHz
Second mixer output and audio amp }	455 kHz − 454 kHz = 1 kHz 455 kHz − 452 kHz = 3 kHz

However, to permit satisfactory SSB reception, an additional mixer (demodulator) and oscillator must replace the conventional diode detector.

As shown before, the carrier frequency was suppressed at the transmitter; thus, for proper intelligence detection a carrier must be inserted by the receiver. The receiver illustrated in Fig. 4-20 inserts a carrier frequency into the detector, although the carrier frequency may be inserted at any point in the receiver before demodulation.

When the SSB signal is received at the antenna, it is amplified by the RF amplifier and applied to the first mixer. By mixing the output of the local oscillator with the input signal (heterodyning), a difference frequency, or IF, is obtained. The IF is then amplified by one or more stages. Of course, this is dependent upon the type of receiver. Up to this point it is identical to an AM superheterodyne receiver. The IF output is applied to the second mixer (detector). The detector output is applied to the audio amplifier and then on to the output speaker.

Tuning the sideband receiver is somewhat more difficult than in a regular AM receiver. The carrier injection oscillator must be precisely adjusted to simulate the carrier frequency at all times. As previously explained, any tendency to drift within the oscillator will cause the output intelligence to be distorted.

Basic SSB Receiver

A basic SSB receiver is shown schematically in Fig. 4-21 (pages 164-165). This superhet design functions well without an RF amplifier. The input signal comes in to a fixed Butterworth front-end filter (FL$_1$) that passes 3.75 to 4.0 MHz without tuning. A *Butterworth filter* exhibits a very flat response in the passband and approaches a 6-dB slope per octave. Individual channels in this amateur radio band are tuned by varying the local oscillator (Q_3) frequency. Notice that this stage is labeled with VFO, or *variable frequency oscillator*, and has a range from 4.253 to 4.453 MHz. This signal and the received signal are applied to the two gates of the mixer (Q_1). This 3N211 MOSFET provides high gain ($g_m \simeq 30,000\ \mu S$) and its output is applied to a mechanical filter, FL$_2$. The specified filter has a 2.2-kHz bandwidth at the 3-dB points and has a 5.5-kHz bandwidth at −60 dB.

The mechanical filter output is applied to the IF amplifier, Q_2, which is another 3N211 MOSFET. Its gain, and that of the audio amplifier U_1, are manually controlled by ganged potentiometers R_{1A} and R_{1B}. The bias at gate 2 of Q_2 is varied by R_{1A}. To obtain a wide range of control it is necessary to have gate 2 a volt or two less than gate 1. This is done by "bootstrapping" this stage with an LED, D_1, that conducts at about 1.5 V. Thus, when R_{1A} has its arm at ground, gate 2 is effectively at −1.5 V and minimum gain for Q_2 occurs.

Another 3N211 device is used as the LO (or VFO if you prefer). Gates 1 and 2 of Q_3 are tied together. The oscillator signal is applied from the gate of Q_3 to gate 2 of the mixer, Q_1. A pure 3-V p-p sine wave is thereby available. D_2 is used as a switching diode to offset the VFO frequency when changing from USB to LSB.

The product detector stage (Q_4) is fed from the IF amp into the source of Q_4. The beat frequency oscillator (Q_5) is a switchable crystal oscillator. S1B selects either crystal Y_1 or Y_2 for LSB or USB, respectively. The product detector output (at Q_4's drain) is applied to a 741-op amp audio amplifier that offers up to 40-dB gain. Its output is sufficient to drive headphones, or an IC power amp could be added if a speaker is needed.

TROUBLESHOOTING

There are two ways to generate SSB signals, but modern manufacturing methods have reduced the cost of filters to the point that nearly all generate the SSB signal with balanced modulators and filters. Most radios even use separate filters to select the upper or lower sideband as desired instead of switching oscillators.

A spectrum analyzer is an extremely desirable tool, but if one is not available, a good general coverage shortwave receiver is the next best tool. In troubleshooting SSB generators, you will be mainly looking for the presence or absence of various oscillations. It is also desirable to have a frequency counter to measure the exact frequency of the oscillators.

What do we do when faced with a radio receiver that has no reception? Where do we start to look for the trouble? When faced with this kind of problem, how does the technician proceed in formulating a plan of action? This section will show you a popular method used for finding the problem in a receiver with no reception. After completing this section you should be able to

- Troubleshoot SSB generators and demodulators
- Test for carrier leakthrough with an oscilloscope or spectrum analyzer
- Identify a defective stage in a SSB receiver
- Describe the signal injection method of troubleshooting

Balanced Modulators (Figure 4-22)

Things to look for and do:

1. With no audio input, there should be no RF output. An oscilloscope will be helpful here.

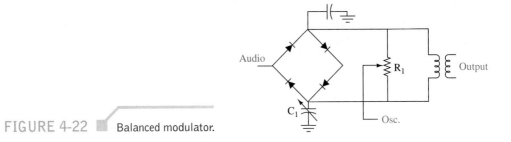

FIGURE 4-22 ▮ Balanced modulator.

2. The voltage from the oscillator must be 6 to 8 times the peak audio voltage. There should be several volts of RF and a few tenths of a volt of audio.

3. The diodes should be well matched. An ohmmeter can be used to select matched pairs or quads.

4. You should be able to null the carrier at the output by adjusting R1 and C1. It may be necessary to adjust each control several times alternately to secure optimum carrier suppression. Further detail on testing for "carrier leakthrough" is provided in the next few paragraphs.

Testing for Carrier Leakthrough

The purpose of the balanced modulator is to suppress or cancel out the carrier. An exactly balanced modulator would totally suppress or remove the carrier. This is an impossibility because there are always imbalances—one diode conducts a little more current than another, perhaps. To achieve a circuit's maximum suppression, balanced modulators usually include one or more balance controls, as was shown in Fig. 4-22.

The first thing to do when troubleshooting a balanced modulator is to check the condition of its balance. This can be done by looking for "carrier leakthrough" with an oscilloscope. The circuit for this test is seen in Fig. 4-23. *Carrier leakthrough* simply means the amount of carrier not suppressed by the balanced modulator.

The audio signal generator is set to some frequency within the normal audio range of the xmtr, perhaps 1 or 1.5 kHz. Check the manual for the correct RF and audio signal levels into the modulator. (Test points may be included for measuring these.) Typically, the RF input (oscillator output) will be about 4 to 6 times the audio level for proper diode switching. Any DMM can be used to measure the audio, but the meter will require a RF probe for the oscillator output.

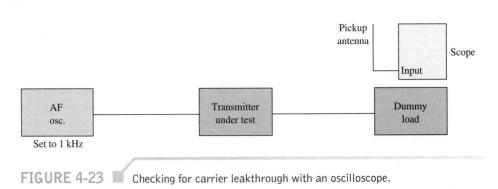

FIGURE 4-23 ▮ Checking for carrier leakthrough with an oscilloscope.

(a)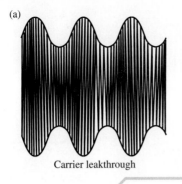

(b)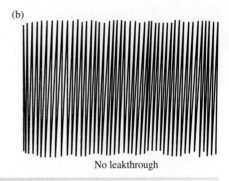

Carrier leakthrough

No leakthrough

FIGURE 4-24 ■ Single-sideband signal with and without carrier leakthrough.

Part A of Fig. 4-24 shows the transmitter signal when there is carrier leakthrough, that is, the carrier is not fully suppressed. Note the similarity to a partially modulated AM signal. Part B shows the signal as it should be, a single tone signal; the carrier is fully suppressed.

One of two conditions could cause carrier leakthrough: either the circuit has become unbalanced or there are defective components. To check for unbalance, adjust the balance control(s) for minimum carrier amplitude. Should there be more than one control, it may be necessary to go back and forth more than once between controls because the setting of one often affects the setting of another.

If there is no balance problem and the input signal levels are correct, there is a defective component, most likely one of the diodes in the bridge assembly. Such bridges are usually a sealed unit; you cannot get at individual diodes. If this is the case, replace the suspect unit with a known good one and recheck for proper operation. Be sure to check balance again with the new unit in place.

Figure 4-25 shows the circuit for checking carrier leakthrough and suppression with a spectrum analyzer.

Having determined that the RF and audio frequencies and levels are correct, observe the screen of the spectrum analyzer. If the modulator is operating correctly, you will see the display in part A of Fig. 4-26. Note the location of the suppressed carrier. Part

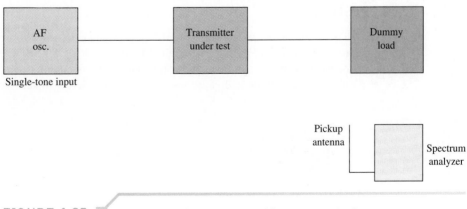

FIGURE 4-25 ■ Checking carrier suppression with a spectrum analyzer.

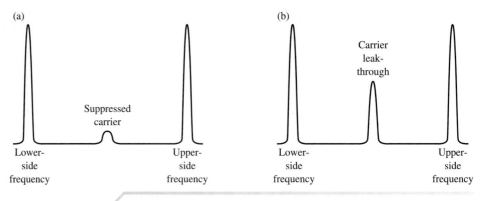

(a)

Suppressed carrier

Lower-side frequency

Upper-side frequency

(b)

Carrier leak-through

Lower-side frequency

Upper-side frequency

FIGURE 4-26 ▪ Carrier suppression as seen on a spectrum analyzer.

B shows the same signal with some carrier leakthrough. As before, adjust the balance controls for minimum carrier amplitude (maximum suppression).

An advantage of the spectrum analyzer is that carrier suppression can be measured in dB directly on the analyzer's log scale. Check the transmitter's manual for the suppression figure; it will be in the neighborhood of −60 to −70 dB. The instruction manual for the spectrum analyzer will tell you how to set the unit's controls for logarithmic measurements.

Testing Filters

Filters designed for SSB service can be ceramic, crystal, or mechanical, but test methods are the same. The filter can have 6 to 10 dB of loss in the pass band. Figure 4-27 shows how you would set up the equipment to test a filter.

The technician should slowly sweep the generator frequency across the passband of the filter. If the sweep speed is too fast, the filter's time delay will cause misleading results. Response of the filter should fall off rapidly at the band edges. Two or three dB of ripple in the passband is normal. If the ripple is as much as 6 to 10 dB, the signal passing through the filter will be badly distorted.

Testing Linear Amplifiers

The two-tone test is generally used to check amplifier linearity. For example, a 400-Hz tone and a 2500-Hz tone are applied to the input of a SSB transmitter. The output is observed with a spectrum analyzer tuned to the transmitter's carrier frequency. Nonlinearity in the amplifier will cause the amplifier to generate mixer-like products. The proper

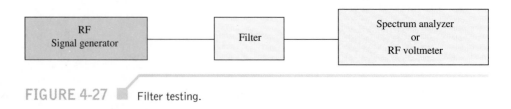

RF Signal generator

Filter

Spectrum analyzer or RF voltmeter

FIGURE 4-27 ▪ Filter testing.

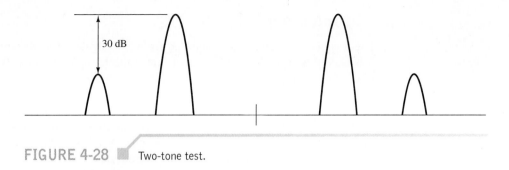

30 dB

response is shown in Fig. 4-28. The carrier will not be present if the balanced modulator is properly adjusted. In a good linear amplifier, the distortion products will be at least 30 dB below the two desired tones. If the amplifier is not linear, several spurious outputs will appear and may be only a few dB below the desired signals. Amplifier nonlinearity is usually caused by improper bias points in the amplifier.

A qualitative check on linearity can be made with the two-tone test by observing the RF output on an oscilloscope. The output should appear as in Fig. 4-17. The output will approach a perfect sine wave if the amplifier is linear. Nonlinearity shows up as flat-topping of the sine wave.

Testing the SSB Receiver System

It's always best to start by having proper servicing material available for the set being repaired. This material includes the service manual, block diagrams, and schematic diagrams. The bare minimum would be the schematic diagrams. These items enhance the troubleshooting job tremendously. With today's complex electronic circuits, it is very difficult to attempt a service job without the service literature.

From the antenna of the receiver illustrated in Fig. 4-29 radio frequency signals are amplified by the RF amplifier. The radio signal can be traced from the antenna to where it is finally heard at the speaker. The RF amplifier boosts the signal before it goes to mixer one. As stated in our Chapter 3 Troubleshooting section, the local oscillator can kill the received signal if it is not working. So the local oscillator is included in the signal chain as a possible cause of no reception. Next, the converted RF signal is amplified in the IF amplifiers. From the IF strip, the signal is applied to mixer two, the detector. Mixer two is responsible for recovering the original intelligence signal. To recover this original intelligence, the BFO reinserts the missing carrier. If the carrier were not reinserted, the radio would still produce sound at the speaker. The sound is garbled without this reinserted carrier, but the reception is not otherwise hindered. However, a problem in mixer two would interfere with the signal reception. The last stage before the speaker is the audio amplifier. In summary, the stages that can interfere with reception are the antenna, the RF amplifier, mixer one, the local oscillator, the IF amplifiers, mixer two, the audio amplifier section, the speaker, and the power supply.

The speaker and power supply should be checked first in the sequence of troubleshooting events. Check the power supply voltages using the DMM and compare these measurements to the specified values given in the service literature. If the voltage measurements are correct, the power supply is good. The speaker can be checked by inserting a tone across its terminals. If the tone is heard then the speaker is working.

Signal Injection

Use a signal generator that is adjusted for a 1-V signal set to the center IF frequency. Modulate this IF with a 1-kHz signal. Inject the IF signal at point B as shown on the block diagram in Fig. 4-29. The 1-kHz test tone should be heard at the speaker output if mixer two and the audio amplifier are operating correctly. This would mean that the trouble lies toward the antenna. If the tone is not heard, inject a pure 1-kHz signal at point A, just ahead of the audio amplifier. A tone from the speaker would now indicate a problem in mixer two.

Supposing that the tone was heard when the signal was injected at point B, we would then move the probe to point C. A tone heard from the speaker indicates the IF amplifier is functioning correctly. Before moving to point D and applying the test signal, readjust the signal generator for a received RF frequency with the modulated 1-kHz signal. Set the generator's output voltage to 20 mV (check the service literature for exact signal levels). If the test tone is heard from the speaker at this point, then the RF amplifier or the antenna can be considered faulty. Decrease the output amplitude of the signal generator to around 2 mV and inject the test signal at point E. If the RF amplifier is bad, no tone will reach the speaker.

The following table will guide you through the signal injecting procedure.

TABLE 4-1

Signal Injected	Test Point	Tone Present	Analysis
1. 1-kHz modulated IF	B	Y (yes)	Mixer 2 & audio amplifier good; go to step three
		N (no)	Go to step two
2. 1-kHz signal	A	Y	Mixer 2 stage bad, audio amplifier good
		N	Audio amplifier bad
3. Modulated IF	C	Y	IF amplifier ok; go to step four
		N	IF amplifier bad
4. Modulated RF	D	Y	Mixer 1 and localoscillator good; go to step five
		N	Problem resides in mixer 1 or the LO
5. Modulated RF	E	Y	Trouble lies beyond the RF amplifier; check antenna, cabling, connectors,or coupling
		N	RF amplifier bad

Using the table above, you can isolate a defective stage rather quickly. A variation on the above technique is to use an oscilloscope to look at the test signal at the output of each stage instead of relying on hearing the signal.

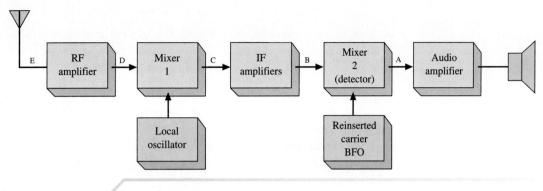

FIGURE 4-29   SSB receiver block diagram.

SUMMARY

In Chapter 4 we introduced single-sideband (SSB) systems and explained their various advantages over standard AM systems. The major topics the student should now understand include:

- the advantages of SSB systems, including the utilization of available frequency bandwidth, noise reduction, power requirements, and selective fading effects

- the various SSB systems and their general characteristics

- an analysis of the function of a balanced modulator

- the development and operation of a simplified push-pull balanced modulator

- the operation of a balanced ring modulator

- the application of linear integrated circuit balanced modulators

- the need for high-Q bandpass filters and the description of mechanical, ceramic, and crystal varieties

- the analysis of SSB transmission systems, including the filter and phase methods

- an understanding of the need for amplitude compandoring and a method of implementation

- the description and operation of a class AB push-pull linear power amplifier

- the analysis of SSB demodulation techniques

- the analysis of a complete SSB receiver

QUESTIONS AND PROBLEMS

SECTION 1

1. Explain why two components of an AM signal (carrier and one sideband) may be eliminated and still result in a usable transmission.
2. An AM transmission of 1000 W is fully modulated. Calculate the power transmitted if it is transmitted as a SSB signal. (167 W)

3. A SSB transmission drives 121 V peak into a 50-Ω antenna. Calculate the PEP. (146 W)
4. Explain the difference between rms and PEP designations.
5. Provide detail on the differences between ACSSB, SSB, SSSC, and ISB transmissions.
*6. Explain the principles involved in a single-sideband, suppressed-carrier (SSSC) emission. How does its bandwidth of emission and required power compare with that of full carrier and sidebands?
7. List and explain the advantages of SSB over conventional AM transmissions. Are there any disadvantages?

Section 2

8. What typically are the inputs and outputs for a balanced modulator?
9. Draw a schematic of a balanced modulator using two JFETs, and briefly explain its operation.
10. What disadvantage might a balanced ring modulator have compared to the circuit used in Problem 9?
11. Referring to the specifications for the 1496/1596 LIC balanced modulator in Fig. 4-5, determine its typical dc power dissipation. Determine the typical carrier output with adjusted offset for a 60-mV rms, 10-MHz sine-wave carrier. Calculate the carrier suppression in dB. (33 mW, 140 μV, 52.6 dB)

Section 3

12. Calculate a filter's required Q to convert DSB to SSB, given that the two sidebands are separated by 200 Hz. The supressed carrier (40 dB) is 2.9 MHz. Explain how this required Q could be greatly reduced. (36,250)
*13. Draw the approximate equivalent circuit of a quartz crystal.
14. What are the undesired effects of the crystal holder capacitance in a crystal filter, and how are they overcome?
*15. What crystalline substance is widely used in crystal oscillators (and filters)?
16. Using your library or some other source, provide a schematic for a four-element crystal lattice filter and explain its operation.
*17. What are the principal advantages of crystal control over tuned circuit oscillators (or filters)?
18. Explain the operation of a ceramic filter. What is the significance of a filter's shape factor?
19. A bandpass filter has a 3-dB ripple amplitude. Explain this specification.
20. Explain the operation and use of mechanical filters.
21. Why are SAW filters not often used in SSB equipment?
22. A SSB signal is generated around a 200-kHz carrier. Before filtering, the upper and lower sidebands are separated by 200 Hz. Calculate the filter Q required to obtain 40-dB suppression. (2500)

Section 4

23. Determine the carrier frequency for the transmitter shown in Fig. 4-11. (It is *not* 3 MHz.)

24. Draw a detailed block diagram of the SSB generator shown in Fig. 4-12. Label frequencies involved at each stage if the intelligence is a 2-kHz tone and the USB is utilized.

25. The sideband filter (FL$_1$) in Fig. 4-12 has a 5-dB *insertion loss* (i.e., a 5-dB signal loss from input to output). Calculate the filter's output voltage assuming equal impedances at its input and output. (0.45 V p-p)

26. Calculate the total impedance in the collector of Q_3 in Fig. 4-12. (54.3 Ω)

27. List the advantages of the phase versus filter method of SSB generation. Why isn't the phase method more popular than the filter method?

28. Mathematically show how a DSB signal, cos $\omega_i t$ cos $\omega_c t$, can be manipulated to provide SSB.

29. Explain the operation of the phase-shift SSB generator illustrated in Fig. 4-13. Why is the carrier phase shift of 90° not a problem, whereas that for the audio signal is?

30. Explain the operation and need for the control circuitry (K, Q_1, Q_2) in the linear power amplifier shown in Fig. 4-16.

31. The PEP transmitted by an ACSSB system is 140 W. It uses a NE571N compandor LIC. Calculate the power transmitted under the no-modulation condition. The audio signal ranges from −28 dBm to +34 dBm before compression. Determine the compressor's output range. (14 W, −14 dBm to +17 dBm)

32. Explain how an ACSSB system is able to provide improved noise performance compared to a regular SSB system.

SECTION 5

33. List the components of an AM signal at 1 MHz when modulated by a 1-kHz sine wave. What is the component(s) if it is converted to a USB transmission? If the carrier is redundant, why must it be "reinserted" at the receiver?

34. Explain why the BFO in a SSB demodulator has such stringent accuracy requirements.

35. If, in an emergency, you had to use an AM receiver to receive a SSB broadcast, what modifications to the receiver would be appropriate?

SECTION 6

36. Draw a block diagram for the receiver shown schematically in Fig. 4-21. Suggest a change to the schematic that you feel would improve its performance and explain why.

CHAPTER

5

FREQUENCY MODULATION: TRANSMISSION

OBJECTIVES

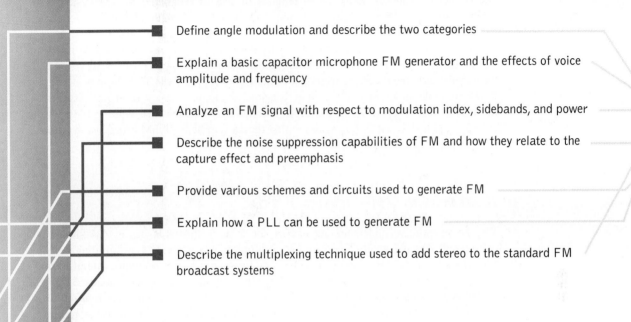

■ Define angle modulation and describe the two categories

■ Explain a basic capacitor microphone FM generator and the effects of voice amplitude and frequency

■ Analyze an FM signal with respect to modulation index, sidebands, and power

■ Describe the noise suppression capabilities of FM and how they relate to the capture effect and preemphasis

■ Provide various schemes and circuits used to generate FM

■ Explain how a PLL can be used to generate FM

■ Describe the multiplexing technique used to add stereo to the standard FM broadcast systems

5-1 ANGLE MODULATION

As has been stated previously, there are three parameters of a sine-wave carrier that can be varied to allow it to "carry" a low-frequency intelligence signal. They are its amplitude, frequency, and phase. The latter two, frequency and phase, are actually interrelated, as one cannot be changed without changing the other. They both fall under the general category of *angle modulation. Angle modulation* is defined as modulation where the angle of a sine-wave carrier is varied from its reference value. Angle modulation has two subcategories, phase modulation and frequency modulation, with the following definitions:

Phase modulation (PM): angle modulation where the phase angle of a carrier is caused to depart from its reference value by an amount proportional to the modulating signal amplitude.

Frequency modulation (FM): angle modulation where the instantaneous frequency of a carrier is caused to vary by an amount proportional to the modulating signal amplitude.

The key difference between these two similar forms of modulation is that in PM the amount of phase change is proportional to intelligence amplitude, while in FM it is

Photo: Final power stage of 20-kW amplifier uses a tube as active element.
(Courtesy of Ehrhorn Technological Operations, Inc.)

177

the frequency change that is proportional to intelligence amplitude. As it turns out, PM is *not* directly used as the transmitted signal in communication systems but does have importance since it is often used to help generate FM, *and* a knowledge of PM helps us to understand the superior noise characteristics of FM as compared to AM systems. In recent years, it has become fairly common practice to denote angle modulation simply as FM instead of specifically referring to FM and PM.

The concept of FM was first practically postulated as an alternative to AM in 1931. At that point, commercial AM broadcasting had been in existence for over 10 years, and the superheterodyne receivers were just beginning to supplant the TRF designs. The goal of research into an alternative to AM at that time was to develop a system less susceptible to external noise pickup. Major E. H. Armstrong developed the first working FM system in 1936, and in July 1939, he began the first regularly scheduled FM broadcast in Alpine, New Jersey.

5-2 A SIMPLE FM GENERATOR

To gain an intuitive understanding of FM, the system illustrated in Fig. 5-1 should be considered. This is actually a very simple, yet highly instructive, FM transmitting system. It consists of an *LC* tank circuit, which, in conjunction with an oscillator circuit, generates a sine-wave output. The capacitance section of the *LC* tank is not a standard capacitor but is a capacitor microphone. This popular type of microphone is often referred to as a condenser mike and is, in fact, a variable capacitor. When no sound waves reach its plates, it presents a constant value of capacitance at its two output terminals. However, when sound waves reach the mike, they alternately cause its plates to move in and out. This causes its capacitance to go up and down around its center value. The *rate* of this capacitance change is equal to the frequency of the sound waves striking the mike, *and* the *amount* of capacitance change is proportional to the amplitude of the sound waves.

Since this capacitance value has a direct effect on the oscillator's frequency, the following two *important* conclusions can be made concerning the system's output frequency:

1. The frequency of impinging sound waves determines the *rate* of frequency change.
2. The amplitude of impinging sound waves determines the *amount* of frequency change.

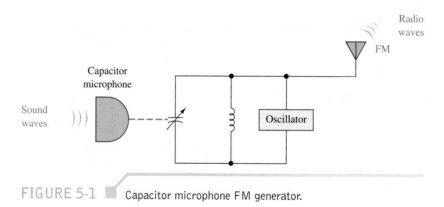

FIGURE 5-1 ▮ Capacitor microphone FM generator.

Consider the case of the sinusoidal sound wave (the intelligence signal) shown in Fig. 5-2(a). Up until time T_1 the oscillator's waveform at (b) is a constant frequency with constant amplitude. This corresponds to the carrier frequency (f_c) or *rest* frequency in FM systems. At T_1 the sound wave at (a) starts increasing sinusoidally and reaches a maximum positive value at T_2. During this period, the oscillator frequency is gradually increasing and reaches its highest frequency when the sound wave has maximum amplitude at time T_2. From time T_2 to T_4 the sound wave goes from maximum positive to maximum negative and the resulting oscillator frequency goes from a maximum frequency *above* the rest value to a maximum value *below* the rest frequency. At time T_3 the sound wave is passing through zero, and therefore the oscillator output is instantaneously equal to the carrier frequency.

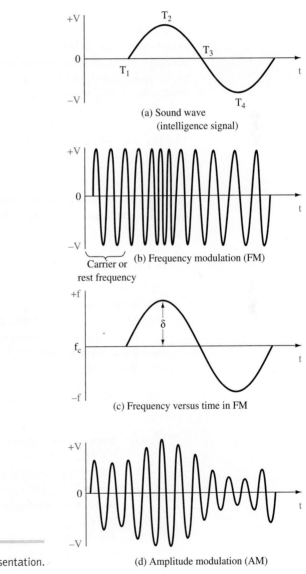

(a) Sound wave
(intelligence signal)

(b) Frequency modulation (FM)

(c) Frequency versus time in FM

(d) Amplitude modulation (AM)

FIGURE 5-2　FM representation.

The Two Major Concepts

The amount of oscillator frequency increase and decrease around f_c is called the *frequency deviation,* δ. This deviation is shown in Fig. 5-2(c) as a function of time. It is ideally shown as a sine-wave replica of the original intelligence signal. It shows that the oscillator output is indeed an FM waveform. Recall that FM is defined as a sine-wave carrier that changes in frequency by an *amount* proportional to the instantaneous value of the intelligence wave and at a *rate* equal to the intelligence frequency.

Figure 5-2(d) shows the AM wave resulting from the intelligence signal shown at (a). This should help you to see the difference between an AM and FM signal. In the case of AM, the carrier's amplitude is varied (by its sidebands) in step with the intelligence, while in FM, the carrier's frequency is varied in step with the intelligence.

The capacitor microphone FM generation system is seldom used in practical applications; its importance is derived from its relative ease of providing an understanding of FM basics. If the sound-wave intelligence striking the microphone were doubled in frequency from 1 kHz to 2 kHz with constant amplitude, the rate at which the FM output swings above and below the center (f_c) frequency would change from 1 kHz to 2 kHz. However, since the intelligence amplitude was not changed, the *amount* of frequency deviation (δ) above and below f_c will remain the same. On the other hand, if the 1-kHz intelligence frequency were kept the same but its amplitude were doubled, the *rate* of deviation above and below f_c would remain at 1 kHz, but the *amount* of frequency deviation would double.

As you continue through your study of FM, whenever you start getting bogged down on basic theory, it will often be helpful to review the capacitor mike FM generator. *Remember:*

1. The intelligence amplitude determines the *amount* of carrier frequency deviation.
2. The intelligence frequency (f_i) determines the *rate* of carrier frequency deviation.

EXAMPLE 5-1

An FM signal has a center frequency of 100 MHz but is swinging between 100.001 MHz and 99.999 MHz at a rate of 100 times per second. Determine
(a) The intelligence frequency f_i.
(b) The intelligence amplitude.
(c) What happened to the intelligence amplitude if the frequency deviation changed to between 100.002 and 99.998 MHz.

SOLUTION

(a) Since the FM signal is changing frequency at a 100-Hz rate, $f_i = 100$ Hz.

(b) There is no way of determining the actual amplitude of the intelligence signal. Every FM system has a different proportionality constant between the intelligence amplitude and the amount of deviation it causes.

(c) The frequency deviation has now been doubled, which means that the intelligence amplitude is now double whatever it originally was.

The complete mathematical analysis of angle modulation requires the use of high-level mathematics. For our purposes, it will suffice to simply give the solutions and discuss them. For phase modulation (PM), the equation for the instantaneous voltage is

$$e = A \sin\left(\omega_c t + m_p \sin \omega_i t\right) \qquad (5\text{-}1)$$

where $e =$ instantaneous voltage
$A =$ peak value of original carrier wave
$\omega_c =$ carrier angular velocity $(2\pi f_c)$
$m_p =$ maximum phase shift caused by the intelligence signal (radians)
$\omega_i =$ modulating (intelligence) signal angular velocity $(2\pi f_i)$

The maximum phase shift caused by the intelligence signal, m_p, is defined as the *modulation index* for PM.

The following equation provides the equivalent formula for FM:

$$e = A \sin\left(\omega_c t + m_f \sin \omega_i t\right) \qquad (5\text{-}2)$$

All the terms in Eq. (5-2) are defined as they were for Eq. (5-1), with the exception of the new term, m_f. In fact, the two equations are identical except for that term. It is defined as the modulation index for FM, m_f. It is equal to

$$m_f = \text{FM modulation index} = \frac{\delta}{f_i} \qquad (5\text{-}3)$$

where $\delta =$ maximum frequency shift caused by the intelligence signal (deviation)
$f_i =$ frequency of the intelligence (modulating) signal

Comparison of Eqs. (5-1) and (5-2) points out the only difference between PM and FM. The equation for PM shows that the phase of the carrier varies with the modulating signal amplitude (since m_p is determined by this), and in FM the carrier phase is determined by the ratio of intelligence signal amplitude (which determines δ) to the intelligence frequency (f_i). Thus, FM is *not* sensitive to the modulating signal frequency but PM *is*. The difference between them is subtle—in fact, if the intelligence signal is integrated and then allowed to phase-modulate the carrier, an FM signal is created. This is the method used in the Armstrong indirect FM system, as will be explained in Sec. 5-6. In FM the amount of deviation produced is not dependent on the intelligence frequency, as it is for PM. The amount of deviation is proportional to the intelligence signal amplitude for both PM and FM. These conditions are shown in Fig. 5-3.

FM Mathematical Solution

The FM formula [Eq. (5-2)] is really more complex than it looks because it contains the sine of a sine. To solve for the frequency components of an FM wave requires the use of a high-level mathematical tool, *Bessel functions*. They show that frequency-modulating a carrier with a pure sine wave actually generates an infinite number of sidebands (components) spaced at multiples of the intelligence frequency, f_i, above and below the carrier!

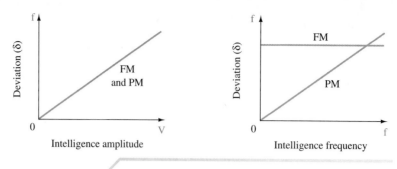

FIGURE 5-3 ■ Deviation effects on FM/PM by intelligence parameters.

Fortunately, the amplitude of these sidebands approaches a negligible level the farther away they are from the carrier, which allows FM transmission within finite bandwidths. The Bessel function solution to the FM equation is

$$f_c(t) = J_0(m_f)\cos \omega_c t - J_1(m_f)[\cos(\omega_c - \omega_i)t - \cos(\omega_c + \omega_i)t]$$
$$+ J_2(m_f)[\cos(\omega_c - 2\omega_i)t + \cos(\omega_c + 2\omega_i)t]$$
$$- J_3(m_f)[\cos(\omega_c - 3\omega_i)t + \cos(\omega_c + 3\omega_i)t]$$
$$+ \cdots$$

(5-4)

where

$$f_c(t) = \text{FM frequency components}$$
$$J_0(m_f)\cos \omega_c t = \text{carrier component}$$
$$J_1(m_f)\cos(\omega_c - \omega_i)t + \cos(\omega_c + \omega_i)t = \text{component at } \pm f_i \text{ around the carrier}$$
$$J_2(m_f)\cos(\omega_c - 2\omega_i)t + \cos(\omega_c + 2\omega_i)t = \text{component at } \pm 2f_i \text{ around}$$
$$\text{the carrier, etc.}$$

To solve for the amplitude of any side-frequency component, J_n, the following equation should be applied:

$$J_N(m_f) = \left(\frac{m_f}{2}\right)^n \left[\frac{1}{n!} - \frac{(m_f/2)^2}{1!(n+1)!} + \frac{(m_f/2)^4}{2!(n+2)!} - \frac{(m_f/2)^6}{3!(n+3)!} + \cdots \right]$$

(5-5)

Thus, solving for these amplitudes is a very tedious process and strictly dependent on the modulation index, m_f. Table 5-1 gives the solution for a number of modulation indexes. Notice that for no modulation ($m_f = 0$), the carrier (J_0) is the only frequency present and exists at its full value of 1. However, as the carrier becomes modulated, energy is shifted from the carrier and into the sidebands. For $m_f = 0.25$, the carrier amplitude has dropped to 0.98, and the first side frequencies at $\pm f_i$ around the carrier (J_1) have an amplitude of 0.12. As indicated previously, FM generates an infinite number of sidebands, but in this case, J_2, J_3, J_4, . . . all have negligible value. Thus, an FM transmission with $m_f = 0.25$ requires the same bandwidth ($2f_i$) as an AM broadcast. Computer programs are also used for Bessel function solutions.

TABLE 5-1 FM Side Frequencies from Bessel Functions

x (CARRIER) (m_f)	J_0	J_1	J_2	J_3	J_4	J_5	J_6	J_7	J_8	J_9	J_{10}	J_{11}	J_{12}	J_{13}	J_{14}	J_{15}	J_{16}
0.00	1.00	—	—	—	—	—	—	—	—	—	—	—	—	—	—	—	—
0.25	0.98	0.12	—	—	—	—	—	—	—	—	—	—	—	—	—	—	—
0.5	0.94	0.24	0.03	—	—	—	—	—	—	—	—	—	—	—	—	—	—
1.0	0.77	0.44	0.11	0.02	—	—	—	—	—	—	—	—	—	—	—	—	—
1.5	0.51	0.56	0.23	0.06	0.01	—	—	—	—	—	—	—	—	—	—	—	—
2.0	0.22	0.58	0.35	0.13	0.03	—	—	—	—	—	—	—	—	—	—	—	—
2.5	-0.05	0.50	0.45	0.22	0.07	0.02	—	—	—	—	—	—	—	—	—	—	—
3.0	-0.26	0.34	0.49	0.31	0.13	0.04	0.01	—	—	—	—	—	—	—	—	—	—
4.0	-0.40	-0.07	0.36	0.43	0.28	0.13	0.05	0.02	—	—	—	—	—	—	—	—	—
5.0	-0.18	-0.33	0.05	0.36	0.39	0.26	0.13	0.05	0.02	—	—	—	—	—	—	—	—
6.0	0.15	-0.28	-0.24	0.11	0.36	0.36	0.25	0.13	0.06	0.02	—	—	—	—	—	—	—
7.0	0.30	0.00	-0.30	-0.17	0.16	0.35	0.34	0.23	0.13	0.06	0.02	—	—	—	—	—	—
8.0	0.17	0.23	-0.11	-0.29	-0.10	0.19	0.34	0.32	0.22	0.13	0.06	0.03	—	—	—	—	—
9.0	-0.09	0.24	0.14	-0.18	-0.27	-0.06	0.20	0.33	0.30	0.21	0.12	0.06	0.03	0.01	—	—	—
10.0	-0.25	0.04	0.25	0.06	-0.22	-0.23	-0.01	0.22	0.31	0.29	0.20	0.12	0.06	0.03	0.01	—	—
12.0	0.05	-0.22	-0.08	0.20	0.18	-0.07	-0.24	-0.17	0.05	0.23	0.30	0.27	0.20	0.12	0.07	0.03	0.01
15.0	-0.01	0.21	0.04	-0.19	-0.12	0.13	0.21	0.03	-0.17	-0.22	-0.09	0.10	0.24	0.28	0.25	0.18	0.12

n OR ORDER

Source: E. Cambi, *Bessel Functions*, Dover Publications, Inc., New York, 1948. Courtesy of the publisher.

EXAMPLE 5-2

Determine a bandwidth required to transmit an FM signal with $f_i = 10$ kHz and a maximum deviation $\delta = 20$ kHz.

SOLUTION

$$m_f = \frac{\delta}{f_i} = \frac{20 \text{ kHz}}{10 \text{ kHz}} = 2 \qquad (5\text{-}3)$$

From Table 5-1 with $m_f = 2$, the following significant components are obtained:

$$J_0, J_1, J_2, J_3, J_4$$

This means that besides the carrier, J_1 will exist ± 10 kHz around the carrier, J_2 at ± 20 kHz, J_3 at ± 30 kHz, and J_4 at ± 40 kHz. Therefore, the total required bandwidth is 2×40 kHz = 80 kHz.

EXAMPLE 5-3

Repeat Ex. 5-2 with f_i changed to 5 kHz.

SOLUTION

$$m_f = \frac{\delta}{f_i} \qquad (5\text{-}3)$$

$$= \frac{20 \text{ kHz}}{5 \text{ kHz}}$$

$$= 4$$

Referring to Table 5-1 with $m_f = 4$ shows that the highest significant side-frequency component is J_7. Since J_7 will be at $\pm 7 \times 5$ kHz around the carrier, the required BW is 2×35 kHz = 70 kHz.

Examples 5-2 and 5-3 point out a confusing aspect about FM analysis. Deviation and bandwidth are related *but different.* They are related in that deviation determines modulation index, which in turn determines significant sideband pairs. The bandwidth, however, is computed by sideband pairs and *not* deviation frequency. Notice in Ex. 5-2 that the maximum deviation was ± 20 kHz, yet the bandwidth was 80 kHz. The deviation is *not* the bandwidth but *does* have an effect on the bandwidth.

An approximation known as *Carson's rule* is often used to predict the bandwidth necessary for an FM signal.

$$\text{BW} \approx 2\left(\delta_{\max} + f_{i_{\max}}\right) \qquad (5\text{-}6)$$

EXAMPLE 5-4

An FM signal, $2000 \sin (2\pi \times 10^8 t + 2 \sin \pi \times 10^4 t)$, is applied to a 50-$\Omega$ antenna. Determine

(a) The carrier frequency.
(b) The transmitted power.
(c) m_f.
(d) f_i.
(e) BW (by two methods).
(f) Power in the largest and smallest sidebands predicted by Table 5-1.

SOLUTION

(a) By inspection of the FM equation, $f_i = (2\pi \times 10^8)/2\pi = 10^8 = 100$ MHz

(b) The peak voltage is 2000 V. Thus,

$$P = \frac{\left(2000/\sqrt{2}\right)^2}{50 \ \Omega} = 40 \ \text{kW}$$

(c) By inspection of the FM equation, we have

$$m_f = 2$$

(d) The intelligence frequency, f_i, is derived from the $\sin \pi \ 10^4 t$ term [Eq. (5-2)]. Thus,

$$f_i = \frac{\pi \times 10^4}{2\pi} = 5 \ \text{kHz}$$

(e)
$$m_f = \frac{\delta}{f_i} \qquad (5\text{-}3)$$

$$2 = \frac{\delta}{5 \ \text{kHz}}$$

$$\delta = 10 \ \text{kHz}$$

From Table 5-1 with $m_f = 2$, significant sidebands exist to J_4 (4×5 kHz = 20 kHz). Thus, BW = 2×20 kHz = 40 kHz. Using Carson's rule yields

$$\text{BW} \simeq 2\left(\delta_{max} + f_{i_{max}}\right) \qquad (5\text{-}6)$$

$$= 2\left(10 \ \text{kHz} + 5 \ \text{kHz}\right) = 30 \ \text{kHz}$$

(f) From Table 5-1, J_1 is the largest sideband at 0.58 times the unmodulated carrier amplitude.

$$P = \frac{\left(0.58 \times 2000/\sqrt{2}\right)^2}{50 \ \Omega} = 13.5 \ \text{kW}$$

or 2×13.5 kW = 27 kW for the two sidebands at ± 5 kHz from the carrier. The smallest sideband, J_4, is 0.03 times the carrier or $(0.03 \times 2000/\sqrt{2})^2/50 \ \Omega = 36$ W.

This approximation includes about 98% of the total power; that is, about 2% of the power is in the sidebands outside its predicted BW. Reasonably good fidelity results when limiting the BW to that predicted by Eq. (5-6). Referring back to Ex. 5-2, Carson's rule predicts a BW of 2(20 kHz + 10 kHz) = 60 kHz versus the 80 kHz shown. In Ex. 5-3, BW = 2(20 kHz + 5 kHz) = 50 kHz versus 70 kHz. It should be remembered that even the 70-kHz prediction does not include all of the sidebands created.

Zero Carrier Amplitude

Figure 5-4 shows the FM frequency spectrum for various levels of modulation while keeping the modulation frequency constant. The relative amplitude of all components is obtained from Table 5-1. Notice from the table that between $m_f = 2$ and $m_f = 2.5$, the carrier goes from a plus to a minus value. The minus sign simply indicates a phase reversal, but when $m_f = 2.4$, the carrier component has zero amplitude and all the energy is contained in the side frequencies. This also occurs when $m_f = 5.5$, 8.65, and between 10 and 12, and 12 and 15.

The zero-carrier condition suggests a convenient means of determining the deviation produced in an FM modulator. A carrier is modulated by a single sine wave at a known frequency. The modulating signal's amplitude is varied while observing the generated FM on a spectrum analyzer. At the point where the carrier amplitude goes to zero, the modulation index, m_f, is determined based on the number of sidebands displayed. If four or five sidebands appear on both sides of the nulled carrier, you can assume that $m_f = 2.4$. The deviation, δ, is then equal to $2.4 \times f_i$. The modulating signal could be increased in amplitude, and the next carrier null should be at $m_f = 5.5$. A check on modulator linearity is thereby possible since the frequency deviation should be directly proportional to the modulating signal's amplitude.

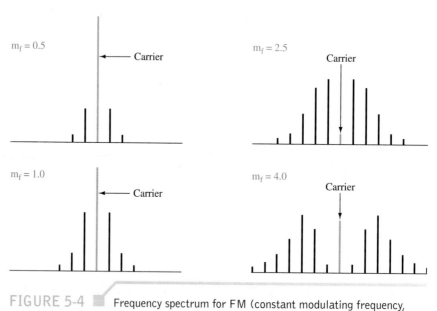

FIGURE 5-4   Frequency spectrum for FM (constant modulating frequency, variable deviation).

Broadcast FM

Standard broadcast FM uses a 200-kHz bandwidth for each station. This is a very large allocation when one considers that one FM station has a bandwidth that could contain many standard AM stations. Broadcast FM, however, allows for a true high-fidelity modulating signal up to 15 kHz and offers superior noise performance (see Sec. 5-4).

Figure 5-5 shows the FCC allocation for commercial FM stations. The maximum allowed deviation around the carrier is ±75 kHz, and 25-kHz *guard* bands at the upper and lower ends are also provided. The carrier is required to maintain a ±2-kHz stability. Recall that an infinite number of side frequencies are generated during frequency modulation, but their amplitude gradually decreases as you move away from the carrier. In other words, the significant side frequencies exist up to ±75 kHz around the carrier, and the guard bands ensure that adjacent channel interference will not be a problem.

Since full deviation (δ) is 75 kHz, that is 100% modulation. By definition, 100% modulation in FM is when the deviation is the full permissible amount. Recall that the modulation index, m_f, is

$$m_f = \frac{\delta}{f_i} \tag{5-3}$$

so that the actual modulation index at 100% modulation varies inversely with the intelligence frequency, f_i. This is in contrast with AM, where full or 100% modulation means a modulation index of 1 regardless of intelligence frequency.

Narrowband FM

Frequency modulation is also widely used in communication (i.e., not to entertain) systems such as police, aircraft, taxicabs, weather service, and private industry networks. These applications are often voice transmissions, which means that intelligence frequency maximums of 3 kHz are the norm. These are *narrowband* FM systems in that FCC bandwidth allocations of 10 to 30 kHz are provided.

The modulation index when δ and f_i are the maximum possible value (75 kHz and 15 kHz, respectively, for broadcast FM) is often called the *deviation ratio*. It has a value of 75 kHz/15 kHz, or 5, for broadcast FM.

The result of Ex. 5-6 is predictable. In FM, the transmitted waveform never varies in amplitude, just frequency. Therefore, the total transmitted power must remain constant regardless of the level of modulation. It is thus seen that whatever energy is contained in the side frequencies has been obtained from the carrier. No additional energy is added during the modulation process. The carrier in FM is not redundant as in AM, since its (the carrier) amplitude is dependent on the intelligence signal.

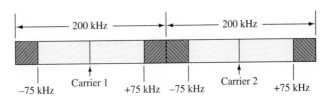

FIGURE 5-5 ■ Commercial FM bandwidth allocations for two adjacent stations.

EXAMPLE 5-5

(a) Determine the permissible range in maximum modulation index for commercial FM that has 30-Hz to 15-kHz modulating frequencies.

(b) Repeat for a narrowband system that allows a maximum deviation of 10 kHz and 100-Hz to 3-kHz modulating frequencies.

SOLUTION

(a) The maximum deviation in broadcast FM is 75 kHz.

$$m_f = \frac{\delta}{f_i} \qquad (5\text{-}3)$$

$$= \frac{75 \text{ kHz}}{30 \text{ Hz}} = 2500$$

to

$$\frac{75 \text{ kHz}}{15 \text{ kHz}} = 5$$

(b)

$$m_f = \frac{\delta}{f_i} = \frac{10 \text{ kHz}}{100 \text{ Hz}} = 100$$

to

$$\frac{10 \text{ kHz}}{3 \text{ kHz}} = 3\tfrac{1}{3}$$

EXAMPLE 5-6

Determine the relative total power of the carrier and side frequencies when $m_f = 0.25$ for a 10-kW FM transmitter.

SOLUTION

For $m_f = 0.25$, the carrier is equal to 0.98 times its unmodulated amplitude and the only significant sideband is J_1, with a relative amplitude of 0.12 (from Table 5-1). Therefore, since power is proportional to the voltage squared, the carrier power is

$$(0.98)^2 \times 10 \text{ kW} = 9.604 \text{ kW}$$

and the power of each sideband is

$$(0.12)^2 \times 10 \text{ kW} = 144 \text{ W}$$

The total power is

$$9604 \text{ W} + 144 \text{ W} + 144 \text{ W} = 9.892 \text{ kW}$$

$$\cong 10 \text{ kW}$$

5-4 NOISE SUPPRESSION

The most important advantage of FM over AM is the superior noise characteristics. You are probably aware that static noise is rarely heard on FM, although it is quite common in AM reception. You may be able to guess a reason for this improvement. The addition of noise to a received signal causes a change in its amplitude. Since the amplitude changes in AM contain the intelligence, any attempt to get rid of the noise adversely affects the received signal. However, in FM, the intelligence is *not* carried by amplitude changes but instead by frequency changes. The spikes of external noise picked up during transmission are "clipped" off by a *limiter* circuit and/or through the use of detector circuits that are insensitive to amplitude changes. Chapter 6 provides more detailed information on these FM receiver circuits.

Figure 5-6(a) shows the noise removal action of an FM limiter circuit, while at Fig. 5-6(b) the noise spike feeds right through to the speaker in an AM system. The advantage for FM is clearly evident; in fact, you may think that the limiter removes all the effects of this noise spike. Unfortunately, while it is possible to clip the noise spike off, it still causes an undesired phase shift and thus frequency shift of the FM signal, and this frequency shift *cannot* be removed.

The noise signal frequency will be close to the frequency of the desired FM signal due to the selective effect of the tuned circuits in a receiver. In other words, if you are tuned to an FM station at 96 MHz, the receiver's selectivity provides gain only for frequencies near 96 MHz. The noise that will affect this reception must, therefore, also be around 96 MHz, since all other frequencies will be greatly attenuated. The effect of adding the desired *and* noise signals will give a resultant signal with a different phase angle than the desired FM signal alone. Therefore, the noise signal, even though it is clipped off in amplitude, will cause phase modulation (PM), which indirectly causes undesired FM. The amount of frequency deviation (FM) caused by PM is

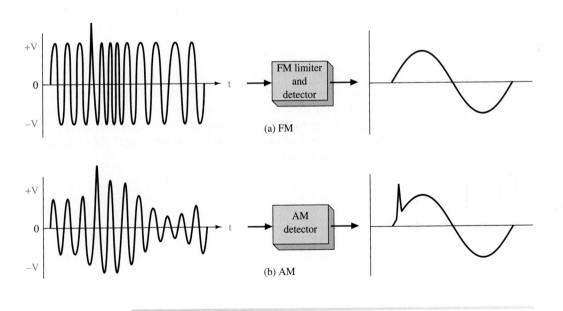

(a) FM

(b) AM

FIGURE 5-6 FM, AM noise comparison.

$$\delta = \phi \times f_i \tag{5-7}$$

where δ = frequency deviation
ϕ = phase shift (radians)
f_i = frequency of intelligence signal

FM Noise Analysis

The phase shift caused by the noise signal results in a frequency deviation that is predicted by Eq. (5-7). Consider the situation illustrated in Fig. 5-7. Here the noise signal is one-half the desired signal amplitude, which provides a voltage *S/N* ratio of 2:1. This is an intolerable situation in AM but, as the following analysis will show, not so bad in FM.

Since the noise (*N*) and desired signal (*S*) are at different frequencies (but in the same range, as dictated by a receiver's tuned circuits), the noise is shown as a rotating vector using the *S* signal as a reference. The phase shift of the resultant (*R*) is maximum when *R* and *N* are at right angles to one another. At this worst-case condition

$$\phi = \sin^{-1}\frac{N}{S} = \sin^{-1}\frac{1}{2}$$

$$= 30°$$

or $30°/(57.3°$ per radian$) = 0.52$ rad, or about $\frac{1}{2}$ rad.

If the intelligence frequency, f_i, were known, then the deviation (δ) caused by this severe noise condition could now be calculated using Eq. (5-7). Since $\delta = \phi \times f_i$ the worst-case deviation occurs for the maximum intelligence frequency. Assuming an f_i maximum of 15 kHz, the absolute worst-case δ due to this severe noise signal is

$$\delta = \phi \times f_i = 0.5 \times 15 \text{ kHz} = 7.5 \text{ kHz}$$

In standard broadcast FM, the maximum modulating frequency is 15 kHz and the maximum allowed deviation is 75 kHz above and below the carrier. Thus, a 75-kHz deviation corresponds to maximum modulating signal amplitude and full volume at the receiver's output. The 7.5-kHz worst-case deviation output due to the *S/N* = 2 condition is

$$\frac{7.5 \text{ kHz}}{75 \text{ kHz}} = \frac{1}{10}$$

and, therefore, the 2:1 signal-to-noise ratio results in an output signal-to-noise ratio of 10:1. This result assumes that the receiver's internal noise is negligible. Thus, FM is seen to exhibit a very strong capability to nullify the effects of noise! In AM, a 2:1 signal-to-noise ratio at the input essentially results in the same ratio at the output. Thus, FM is seen to have an inherent noise reduction capability not possible with AM.

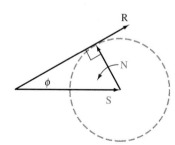

FIGURE 5-7   Phase shift (ϕ) as a result of noise.

EXAMPLE 5-7

Determine the worst-case output S/N for a broadcast FM program that has a maximum intelligence frequency of 5 kHz. The input S/N is 2.

SOLUTION

The input $S/N = 2$ means that the worst-case deviation is about $\frac{1}{2}$ rad (see the preceding paragraphs). Therefore,

$$\delta = \phi \times f_i \tag{5-7}$$
$$= 0.5 \times 5 \text{ kHz} = 2.5 \text{ kHz}$$

Since full volume in broadcast FM corresponds to a 75-kHz deviation, this 2.5-kHz worst-case noise deviation means that the output *S/N* is

$$\frac{75 \text{ kHz}}{2.5 \text{ kHz}} = 30$$

Example 5-7 shows that the inherent noise reduction capability of FM is improved when the maximum intelligence (modulating) frequency is reduced. A little thought shows that this capability can also be improved by increasing the maximum allowed frequency deviation from the standard 75-kHz value. An increase in allowed deviation means that increased bandwidths for each station would be necessary, however. In fact,

EXAMPLE 5-8

Determine the worst-case output S/N for a narrowband FM receiver with $\delta_{max} = 10$ kHz and a maximum intelligence frequency of 3 kHz. The S/N input is 3:1.

SOLUTION

The worst-case phase shift (ϕ) due to the noise occurs when $\phi = \sin^{-1} (N/S)$.

$$\phi = \sin^{-1} \tfrac{1}{3} = 19.5°, \text{ or } 0.34 \text{ rad}$$

and

$$\delta = \phi \times f_i \tag{5-7}$$
$$= 0.34 \times 3 \text{ kHz} \cong 1 \text{ kHz}$$

The *S/N* output will be

$$\frac{10 \text{ kHz}}{1 \text{ kHz}} = 10$$

and thus the input signal-to-noise ratio of 3 is transformed to 10 or higher at the output.

many FM systems utilized as communication links operate with decreased bandwidths—narrowband FM systems. It is typical for them to operate with a 10-kHz maximum deviation. The inherent noise reduction of these systems is reduced by the lower allowed δ but is somewhat offset by the lower maximum modulating frequency of 3 kHz usually used for voice transmissions.

Capture Effect

This inherent ability of FM to minimize the effect of undesired signals (noise in the preceding paragraphs) also applies to the reception of an undesired station operating at the same or nearly the same frequency as the desired station. This is known as the *capture effect.* You may have noticed when riding in a car that an FM station is suddenly replaced by a different one. You may even find that the receiver alternates abruptly back and forth between the two. This occurs because the two stations are presenting a variable signal as you drive. The capture effect causes the receiver to lock on the stronger signal by suppressing the weaker but can fluctuate back and forth when the two are nearly equal. However, when they are not nearly equal, the inherent FM noise suppression action is very effective in preventing the interference of an unwanted (but weaker) station. The weaker station is suppressed just as noise was in the preceding noise discussion. FM receivers typically have a *capture ratio* of 1 dB—this means suppression of a 1-dB (or more) weaker station is accomplished. In AM, it is not uncommon to hear two separate broadcasts at the same time, but this is certainly a rare occurrence with FM.

The capture effect can also be illustrated by referring to Fig. 5-8. Notice that the *S/N* before and after demodulation for SSB and AM is linear. Assuming noiseless demodulation schemes, SSB (and DSB) has the same *S/N* at the detector's input and output. The degradation shown for AM is due to so much of the signal's power being wasted in the redundant carrier. FM systems with m_f greater than 1 show an actual improvement in *S/N,* as illustrated in Exs. 5-7 and 5-8. For example, consider $m_f = 5$ in Fig. 5-8. When *S/N* before demodulation is 20, the *S/N* after demodulation is about 38—a significant improvement.

Insight into the capture effect is provided by consideration of the inflection point (often termed *threshold*) shown in Fig. 5-8. Notice that a rapid degradation in *S/N* after demodulation results when the noise approaches the same level as the desired signal. This threshold situation is noticeable when driving in a large city. The "fluttering" noise often heard is caused when the FM signal is reflecting off various structures. The signal

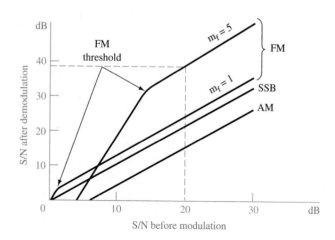

FIGURE 5-8 ■ *S/N* for basic modulation schemes.

strength fluctuates widely due to the additive or subtractive effects on the total received signal. The effect can cause the output to totally blank out and resume at a rapid rate, as the *S/N* before demodulation moves back and forth through the threshold level.

Preemphasis

The noise suppression ability of FM has been shown to decrease with higher intelligence frequencies. This is unfortunate since the higher intelligence frequencies tend to be of lower amplitude than the low frequencies. Thus, a high-pitched violin note that the human ear may perceive as the same "sound" level as the crash of a bass drum may have only half the electrical amplitude as the low-frequency drum signal. In FM, half the amplitude means half the deviation and, subsequently, half the noise reduction capability. To counteract this effect, virtually all FM transmissions provide an artificial boost to the electrical amplitude of the higher frequencies. This process is termed *preemphasis*.

By definition, *preemphasis* involves increasing the relative strength of the high-frequency components of the audio signal before it is fed to the modulator. Thus, the relationship between the high-frequency intelligence components and the noise is altered. While the noise remains the same, the desired signal strength is increased.

A potential disadvantage, however, is that the natural balance between high- and low-frequency tones at the receiver would be altered. A *deemphasis* circuit in the receiver, however, corrects this defect by reducing the high-frequency audio the same amount as the preemphasis circuit increased it, thus regaining the original tonal balance. In addition, the deemphasis network operates on both the high-frequency signal and the high-frequency noise; therefore, there is no change in the improved signal-to-noise ratio. The main reason for the preemphasis network, then, is to prevent the high-frequency components of the transmitted intelligence from being degraded by noise that would otherwise have more effect on the higher than on the lower intelligence frequencies.

The deemphasis network is normally inserted between the detector and the audio amplifier in the receiver. This ensures that the audio frequencies are returned to their original relative level before amplification. The preemphasis characteristic curve is flat up to 500 Hz, as shown in Fig. 5-9. From 500 to 15,000 Hz there is a sharp increase in gain up

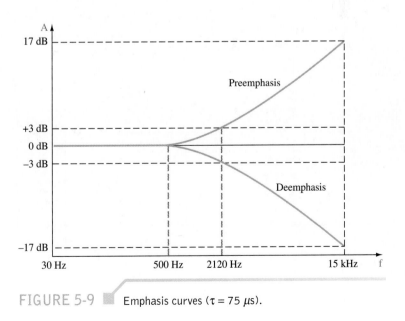

FIGURE 5-9 ■ Emphasis curves (τ = 75 μs).

to approximately 17 dB. The gain at these frequencies is necessary to maintain the signal-to-noise ratio at high audio frequencies. The frequency characteristic of the deemphasis network is directly opposite to that of the preemphasis network. The high-frequency response decreases in proportion to its increase in the preemphasis network. The characteristic curve of the deemphasis circuit should be a mirror image of the preemphasis characteristic curve. Figure 5-9 shows the pre- and deemphasis curves as used by standard FM broadcast in the United States. As shown, the 3-dB points occur at 2120 Hz, as predicted by the RC time constant (τ) of 75 μs used to generate them.

$$f = \frac{1}{2\pi RC} = \frac{1}{2\pi \times 75\mu s} = 2120 \text{ Hz}$$

Figure 5-10(a) shows a typical preemphasis circuit. The impedance to the audio voltage is mainly that of the parallel circuit of C and R_1, as the effect of R_2 is small in comparison to that of either C or R_1. Since capacitive reactance is inversely proportional to frequency, audio frequency increases cause the reactance of C to decrease. This decrease of X_c provides an easier path for high frequencies as compared to R. Thus, with an increase of audio frequency, there is an increase in signal voltage. The result is a larger voltage drop across R_2 (the amplifier's input) at the higher frequencies and thus greater output.

Figure 5-10(b) depicts a typical deemphasis network. Note the physical position of R and C in relation to the base of the transistor. As the frequency of the audio signal increases, the reactance of capacitor C decreases. The voltage division between R and C now provides a smaller drop across C. The audio voltage applied to the base decreases; therefore, a reverse of the preemphasis circuit is accomplished. For the signal to be exactly the same as before preemphasis and deemphasis, the time constants of the two circuits must be equal to each other.

Dolby System

The FCC has ruled that FM broadcast stations can, if desired, use a 25-μs time constant. They then must use the *Dolby* system, which works like preemphasis but in a dynamic fashion. The amount of preemphasis (and subsequent deemphasis in a "Dolbyized"

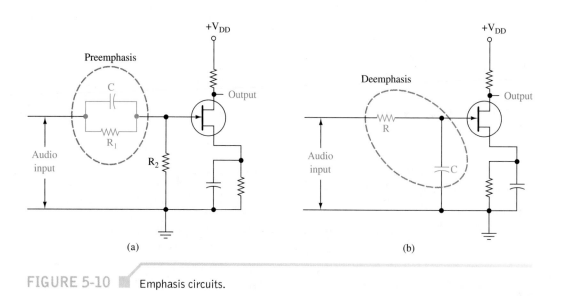

(a)   (b)

FIGURE 5-10 Emphasis circuits.

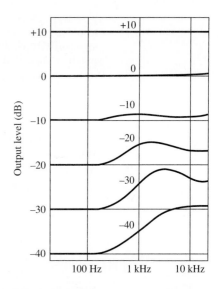

FIGURE 5-11 ▮ Dolby dynamic preemphasis.

receiver) varies depending on the loudness level at any instant. Maximum noise reduction of this system is realized by listeners in "fringe" areas where noise effects are most detrimental. Those people with regular 75-μs, non-Dolbyized receivers do not seem to be adversely affected but will usually turn down their treble control somewhat.

The Dolby system used by FM transmitters is not concerned primarily with noise reduction, however. Standard 75-μs preemphasis provides the same amount of "boost" to both weak and strong high-frequency signals. This created no real problem years ago because few, if any, high frequencies were strong anyway. The better-quality recordings of today do offer some fairly strong high frequencies. Their strength, combined with 75-μs preemphasis, now causes excessive bandwidth of the transmitted FM signal, and thus a station is forced to lower the strength of all signals. The transmitting station, to counteract this effect, will artificially reduce the strength of high-frequency signals, resulting in a less-than-true signal reproduction (and less dynamic range) at the receiver.

The Dolby solution to this problem is to provide varying degrees of high-frequency boost, as depicted in Fig. 5-11. The Dolby receiver must reverse these characteristics and thus requires some relatively complex circuitry. Notice how the weak high frequencies are given a significantly greater boost than the stronger ones. The overall result of this system is a stronger received FM signal with greater "dynamic" range as compared to a broadcast signal, which unnaturally attenuates the high frequencies. The Dolby system is also highly effective in minimizing high-frequency tape hiss noise in tape players.

5-5 DIRECT FM GENERATION

The capacitance microphone system explained in Sec. 5-2 can be used to generate FM directly. Recall that the capacitance of the microphone varies in step with the sound wave striking it. Reference to Fig. 5-1 shows that if the amplitude of sound striking it is increased, the amount of deviation of the oscillator's frequency is increased. If the frequency of sound waves is increased, the rate of the oscillator's deviation is increased. This system is useful in explaining the principles of an FM signal but is not normally used to generate FM in practical systems. It is not able to produce enough deviation as required in actual applications.

Varactor Diode

A varactor diode may be used to generate FM directly. All reverse-biased diodes exhibit a junction capacitance that varies inversely with the amount of reverse bias, as was shown in Fig. 3-11. A diode that is physically constructed so as to enhance this characteristic is termed a *varactor diode.* Figure 5-12 shows a schematic of a varactor diode modulator. With no intelligence signal (E_i) applied, the parallel combination of C_1, L_1, and D_1's capacitance forms the resonant carrier frequency. The coupling capacitor, C_c, isolates the dc levels and intelligence signal while looking like a short to the high-frequency carrier. When the intelligence signal, E_i, is applied to the varactor diode, its reverse bias is varied, which causes the diode's junction capacitance to vary in step with E_i. The oscillator frequency is subsequently varied as required for FM, and the FM signal is available at Q_1's collector. For simplicity, the dc bias and oscillator feedback circuitry is not shown in Fig. 5-12.

Reactance Modulator

While the varactor diode modulator can be called a *reactance modulator,* the term is usually applied to those in which an active device is made to look like a variable reactance. The reactance modulator is a very popular means of FM generation. In order to determine how an active device can be made to look like a reactance, consider the JFET in Fig. 5-13. The impedance, z, looking back into the JFET's drain will be shown to be reactive in this discussion. Assuming the JFET's gate current to be nearly zero, we obtain

$$e_g = i_1 R \tag{5-8}$$

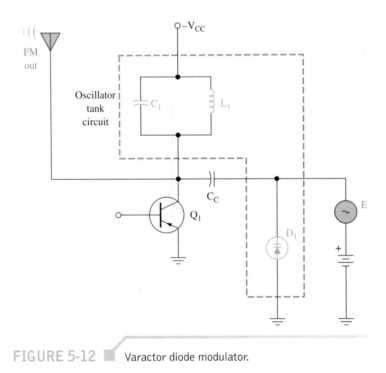

FIGURE 5-12 ■ Varactor diode modulator.

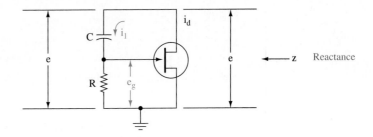

FIGURE 5-13 Reactance circuit.

but i_1 is

$$i_1 = \frac{e}{R - jX_c} \tag{5-9}$$

and substituting Eq. (5-9) into Eq. (5-8), we have

$$e_g = \frac{R \times e}{R - jX_c} \tag{5-10}$$

The JFET drain current, i_d, is

$$i_d = g_m e_g \tag{5-11}$$

(where g_m is the JFET's transconductance), which, using Eq. (5-10), can be rewritten as

$$i_d = \frac{g_m \times R \times e}{R - jX_c} \tag{5-12}$$

Therefore, the impedance z seen from the drain to ground will be

$$z = \frac{e}{i_d} = e \div \frac{g_m \times R \times e}{R - jX_c} \tag{5-13}$$

which can be rewritten as

$$z = \frac{R - jX_c}{g_m R} = \frac{1}{g_m} - \frac{jX_c}{g_m R} \tag{5-14}$$

If the values of R and C are chosen so that $R \ll X_c$, Eq. (5-14) reduces to

$$z = -j\frac{X_c}{g_m R} = \frac{-j}{2\pi f g_m RC} \tag{5-15}$$

which can be rewritten as

$$z = \frac{-j}{2\pi f C_{eq}} \tag{5-16}$$

where

$$C_{eq} = g_m RC \qquad (5\text{-}17)$$

Thus, the impedance z has been made to look like a capacitance. If a modulating signal is applied to the JFET gate in Fig. 5-13, the amount of capacitance will vary in step because the JFET transconductance, g_m, is varied by an applied gate voltage. All that is necessary to generate FM, then, is to connect a JFET's drain (or BJT's collector) to ground terminals across an oscillator's tank circuit to provide an FM generator, as shown in Fig. 5-14. The active device can be either a FET or BJT. Interchanging the position of R and C causes the variable impedance to look inductive rather than capacitive but still allows the generation of FM.

LIC VCO FM Generation

A *voltage-controlled oscillator* (VCO) produces an output frequency that is directly proportional to a control voltage level. The circuitry necessary to produce such an oscillator with a high degree of linearity between control voltage and frequency was formerly prohibitive on a discrete component basis, but now that low-cost monolithic LIC VCOs are available, they make FM generation extremely simple. Figure 5-15 provides the specifications for the Signetics 566 VCO. The circuit shown at the end of the specifications (Signetics Figure 1) provides a high-quality FM generator with the modulating voltage applied to C_2. The FM output can be taken at pin 4 (triangle wave) or pin 3 (square wave). Feeding either of these two outputs into an *LC* tank circuit resonant at the VCO center frequency (i.e., carrier) will subsequently provide a standard sinusoidal FM signal by the flywheel effect.

Crosby Modulator

Now that three practical methods of FM generation have been shown—varactor diode, reactance modulator, and the VCO—it is time to consider the weakness of these methods. Notice that in no case was a crystal oscillator used as the basic reference or carrier fre-

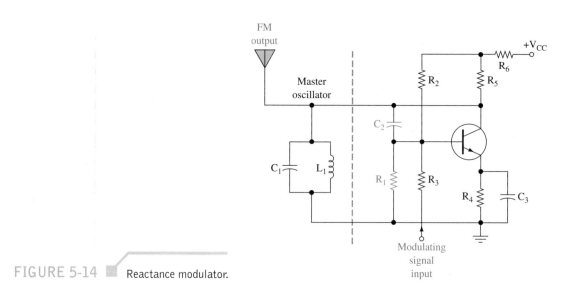

FIGURE 5-14 ■ Reactance modulator.

LINEAR INTEGRATED CIRCUITS

DESCRIPTION

The SE/NE 566 Function Generator is a voltage controlled oscillator of exceptional stability and linearity with buffered square wave and triangle wave outputs. The frequency of oscillation is determined by an external resistor and capacitor and the voltage applied to the control terminal. The oscillator can be programmed over a ten to one frequency range by proper selection of an external resistance and modulated over a ten to one range by the control voltage, with exceptional linearity.

FEATURES

- **WIDE RANGE OF OPERATING VOLTAGE** (10 to 24 volts)
- **VERY HIGH LINEARITY OF MODULATION**
- **EXTREME STABILITY OF FREQUENCY** (100 ppm/°C typical)
- **HIGHLY LINEAR TRIANGLE WAVE OUTPUT**
- **HIGH ACCURACY SQUARE WAVE OUTPUT**
- **FREQUENCY PROGRAMMING BY MEANS OF A RESISTOR, CAPACITOR, VOLTAGE OR CURRENT**
- **FREQUENCY ADJUSTABLE OVER 10 TO 1 RANGE WITH SAME CAPACITOR**

APPLICATIONS

TONE GENERATORS
FREQUENCY SHIFT KEYING
FM MODULATORS
CLOCK GENERATORS
SIGNAL GENERATORS
FUNCTION GENERATORS

PIN CONFIGURATION (Top View)

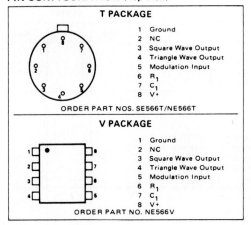

T PACKAGE

1 Ground
2 NC
3 Square Wave Output
4 Triangle Wave Output
5 Modulation Input
6 R_1
7 C_1
8 V+

ORDER PART NOS. SE566T/NE566T

V PACKAGE

1 Ground
2 NC
3 Square Wave Output
4 Triangle Wave Output
5 Modulation Input
6 R_1
7 C_1
8 V+

ORDER PART NO. NE566V

BLOCK DIAGRAM

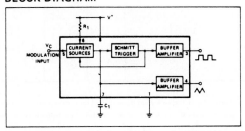

EQUIVALENT CIRCUIT

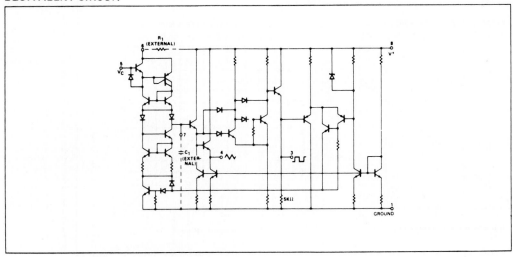

FIGURE 5-15 ■ 566 VCO specifications. (Courtesy of Signetics.)

ABSOLUTE MAXIMUM RATINGS (Limiting values above which serviceability may be impaired)

Maximum Operating Voltage	26V
Storage Temperature	65°C to 150° C
Power Dissipation	300mW

ELECTRICAL CHARACTERISTICS (25°C, 12 Volts, unless otherwise stated)

CHARACTERISTICS	SE566			NE566			UNITS
	MIN.	TYP.	MAX.	MIN.	TYP.	MAX.	
GENERAL							
Operating Temperature Range	55		125	0		70	°C
Operating Supply Voltage			24			24	Volts
Operating Supply Current		7	12.5		7	12.5	mA
VCO (Note 1)							
Maximum Operating Frequency		1			1		MHz
Frequency Drift with Temperature		100			200		ppm/°C
Frequency Drift with Supply Voltage		1			2		%/volt
Control Terminal Input Impedance (Note 2)		1			1		MΩ
FM Distortion (±10% Deviation)		0.2	0.75		0.2	1.5	%
Maximum Sweep Rate		1			1		MHz
Sweep Range		10:1			10:1		
OUTPUT							
Triangle Wave Output ·							
Impedance		50			50		Ω
Voltage	2	2.4		2	2.4		Volts pp
Linearity		0.2			0.5		%
Square Wave Output ·							
Impedance		50			50		Ω
Voltage	5	5.4		5	5.4		Volts pp
Duty Cycle	45	50	55	40	50	60	%
Rise Time		20			20		nsec
Fall Time		50			50		nsec

NOTES:

1. The external resistance for frequency adjustment (R_1) must have a value between 2KΩ and 20KΩ.

2. The bias voltage (Vc) applied to the control terminal (pin 5) should be in the range $3/4\,V^+ \leqslant V_C \leqslant V^+$

TYPICAL PERFORMANCE CHARACTERISTICS

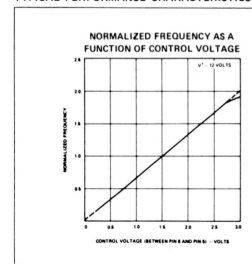

NORMALIZED FREQUENCY AS A
FUNCTION OF CONTROL VOLTAGE

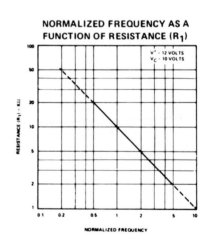

NORMALIZED FREQUENCY AS A
FUNCTION OF RESISTANCE (R_1)

FIGURE 5-15 ▓ (Continued)

TYPICAL PERFORMANCE CHARACTERISTICS (Cont'd)

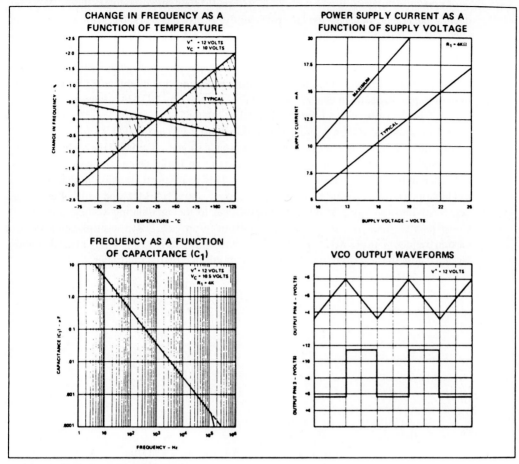

OPERATING INSTRUCTIONS

The SE/NE 566 Function Generator is a general purpose voltage controlled oscillator designed for highly linear frequency modulation. The circuit provides simultaneous square wave and triangle wave outputs at frequencies up to 1 MHz. A typical connection diagram is shown in Figure 1. The control terminal (pin 5) must be biased externally with a voltage (V_C) in the range

$$3/4 \; V^+ \leqslant V_C \leqslant V^+$$

where V_{CC} is the total supply voltage. In Figure 1, the control voltage is set by the voltage divider formed with R_2 and R_3. The modulating signal is then ac coupled with the capacitor C_2. The modulating signal can be direct coupled as well, if the appropriate dc bias voltage is applied to the control terminal. The frequency is given approximately by

$$f_o \simeq \frac{2(V^+ - V_C)}{R_1 C_1 V^+}$$

and R_1 should be in the range $2K < R_1 < 20K\Omega$.

A small capacitor (typically $0.001\mu f$) should be connected between pins 5 and 6 to eliminate possible oscillation in the control current source.

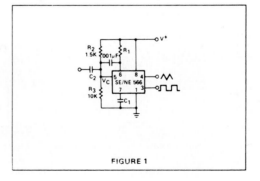

FIGURE 1

FIGURE 5-15 ■ (Continued)

quency. The stability of the carrier frequency is very tightly controlled by the FCC, and that stability is not attained by any of the methods described thus far. Because of the high Q of crystal oscillators, it is not possible to directly frequency-modulate them—their frequency cannot be made to deviate sufficiently to provide workable wideband FM systems. It is possible to directly modulate a crystal oscillator in some narrowband applications. If a crystal is modulated to a deviation of ±50 Hz around a 5-MHz center frequency and both are multiplied by 100, a narrowband system with a 500-MHz carrier ±5-kHz deviation results. One method of circumventing this dilemma for wideband systems is to provide some means of *automatic frequency control* (AFC) to correct any carrier drift by comparing it to a reference crystal oscillator.

FM systems utilizing direct generation with AFC are called *Crosby* systems. A Crosby direct FM transmitter for a standard broadcast station at 90 MHz is shown in Fig. 5-16. Notice that the reactance modulator starts at an initial center frequency of 5 MHz and has a maximum deviation of ±4.167 kHz. This is a typical situation in that reactance modulators cannot provide deviations exceeding about ±5 kHz and still offer a high degree of linearity (i.e., Δf directly proportional to the modulating voltage amplitude). Consequently, *frequency multipliers* are utilized to provide a ×18 multiplication up to a carrier frequency of 90 MHz (18 × 5 MHz) with a ±75-kHz (18 × 4.167 kHz) deviation. Notice that both the carrier and deviation are multiplied by the multiplier.

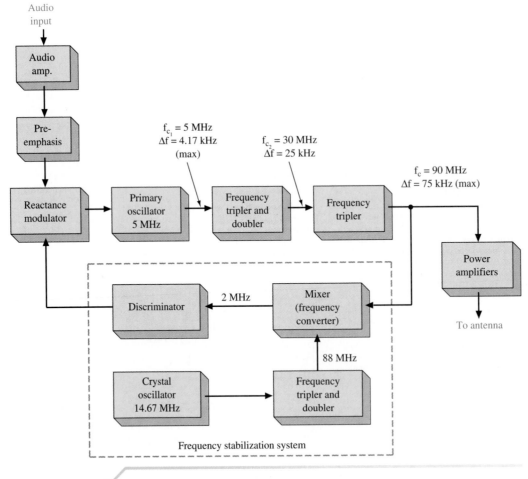

FIGURE 5-16 ■ Crosby direct FM transmitter.

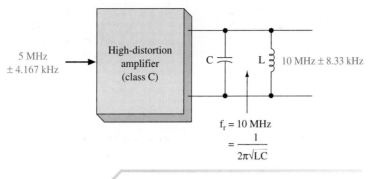

$$f_r = 10 \text{ MHz}$$
$$= \frac{1}{2\pi\sqrt{LC}}$$

FIGURE 5-17 ■ Frequency multiplication (doubler).

Frequency multiplication is normally obtained in steps of ×2 or ×3 (doublers or triplers). The principle involved is to feed a frequency rich in harmonic distortion (i.e., from a class C amplifier) into an *LC* tank circuit tuned to two or three times the input frequency. The harmonic is then the only significant output, as illustrated in Fig. 5-17.

After the ×18 multiplication (3 × 2 × 3) shown in Fig. 5-16, the FM *exciter* function is complete. The term *exciter* is often used to denote the circuitry that generates the modulated signal. The excited output goes to the power amplifiers for transmission *and* to the frequency stabilization system. The purpose of this system is to provide a control voltage to the reactance modulator whenever it drifts from its desired 5-MHz value. The control (AFC) voltage then varies the reactance of the primary 5-MHz oscillator slightly to bring it back on frequency.

The mixer in Fig. 5-16 has the 90-MHz carrier and 88-MHz crystal oscillator signal as inputs. The mixer output only accepts the difference component of 2 MHz, which is fed to the discriminator. A *discriminator* is the opposite of a VCO, in that it provides a dc level output based on the frequency input. The discriminator output in Fig. 5-16 will be zero if it has an input of exactly 2 MHz, which occurs when the transmitter is at precisely 90 MHz. Any carrier drift up or down causes the discriminator output to go positive or negative, resulting in the appropriate primary oscillator readjustment. Further detail on discriminator circuits is provided in Chapter 6.

5-6 INDIRECT FM GENERATION

If the phase of a crystal oscillator's output is varied, phase modulation (PM) will result. As discussed previously, changing the phase of a signal indirectly causes its frequency to be changed. We thus find that direct modulation of a crystal is possible via PM, which indirectly creates FM. This indirect method of FM generation is usually referred to as the *Armstrong* type, after its originator, E. H. Armstrong. It permits modulation of a stable crystal oscillator without the need for the cumbersome AFC circuitry and also provides carrier accuracies identical to the crystal accuracy, as opposed to the slight degradation of the Crosby system's accuracy.

A simple Armstrong modulator is depicted in Fig. 5-18. The JFET is biased in the ohmic region by keeping V_{DS} low. In that way it presents a resistance from drain to source that is made variable by the gate voltage (the modulating signal). Notice that the modulating signal is first given the standard preemphasis and then applied to a frequency-correcting network. This network is a low-pass *RC* circuit (an integrator) that makes the audio output amplitude inversely proportional to its frequency. This is necessary because in

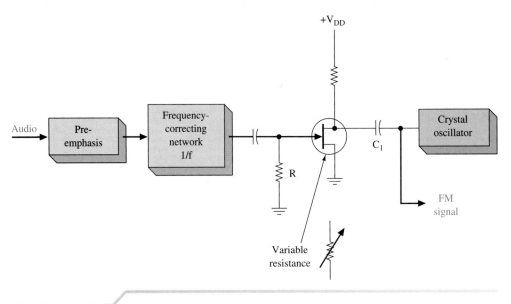

FIGURE 5-18   Indirect FM via PM (Armstrong modulator).

phase modulation, the frequency deviation created is not only proportional to modulating signal amplitude (as desired for FM) but also to the modulating signal frequency (undesired for FM). Thus, in PM if a 1-V, 1-kHz modulating signal caused a 100-Hz deviation, a 1-V, 2-kHz signal would cause a 200-Hz deviation instead of the same deviation of 100 Hz if that signal were applied to the $1/f$ network.

In summary, the Armstrong modulator of Fig. 5-18 indirectly generates FM by changing the phase of a crystal oscillator's output. That phase change is accomplished by varying the phase angle of an RC network (C_1 and the JFET's resistance), in step with the frequency-corrected modulating signal.

Obtaining Wideband Deviation

The indirectly created FM is not capable of much frequency deviation. A typical deviation is 50 Hz out of 1 MHz (50 ppm). Thus, even with a ×90 frequency multiplication, a 90-MHz station would have a deviation of 90 × 50 Hz = 4.5 kHz. This may be adequate for narrowband communication FM but falls far short of the 75-kHz deviation required for broadcast FM. A complete Armstrong FM system providing a 75-kHz deviation is shown in Fig. 5-19. It uses a balanced modulator and 90° phase shifter to phase-modulate a crystal oscillator. Sufficient deviation is obtained by a combination of multipliers and mixing. The ×81 multipliers (3 × 3 × 3 × 3) raise the initial 400-kHz ± 14.47-Hz signal to 32.4 MHz ± 1172 Hz. The carrier *and* deviation are multiplied by 81. Applying this signal to the mixer, which also has a crystal oscillator signal input of 33.81 MHz, provides an output component (among others) of 33.81 MHz − (32.4 MHz ± 1172 Hz), or 1.41 MHz ± 1172 Hz. Notice that the mixer output changes the center frequency *without* changing the deviation. Following the mixer, the ×64 multipliers accept only the mixer difference output component of 1.41 MHz ± 1172 Hz and raise that to (64 × 1.41 MHz) ± (64 × 1172 Hz), or the desired 90 MHz ± 75 kHz.

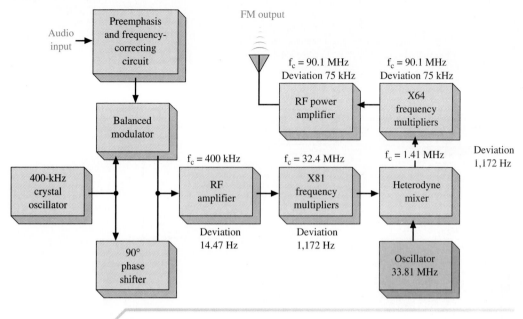

FIGURE 5-19 ▓ Wideband Armstrong FM.

5-7 PHASE-LOCKED-LOOP FM TRANSMITTER

Block Diagram

The block diagram shown in Fig. 5-20 provides a very practical way to fabricate an FM transmitter. The amplified audio signal is used to frequency-modulate a crystal oscillator. The crystal frequency is "pulled" slightly by the variable capacitance exhibited by the varactor diode. The approximate ±200-Hz deviation possible in this fashion is adequate for narrowband systems. The FM output from the crystal oscillator is then divided by 2 and applied as one of the inputs to the phase detector of a phase-locked-loop (PLL) system. As indicated in Fig. 5-20, the other input to the phase detector is the same, and its output is therefore (in this case) the original audio signal. The input control signal to the VCO is therefore the same audio signal, and its output will be its free-running value of 125 MHz ± 5 kHz, which is set up to be exactly 50 times the 2.5-MHz value of the divided-by-2 crystal frequency of 5 MHz.

The FM output signal from the VCO is given power amplification and then driven into the transmitting antenna. This output is also sampled by a ÷50 network, which provides the other FM signal input to the phase detector. The PLL system effectively provides the required ×50 multiplication but, more important, provides the transmitter's required frequency stability. Any drift in the VCO center frequency causes an input to the phase detector (input 2 in Fig. 5-20) that is slightly different from the exact 2.5-MHz crystal reference value. The phase detector output therefore develops an error signal that corrects the VCO center frequency output back to exactly 125 MHz. This dynamic action of the phase detector/VCO and feedback path is the basis of a PLL. More detail on PLL action is provided in Chapter 6.

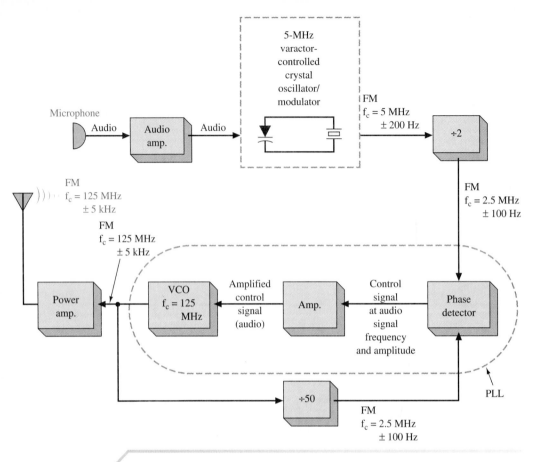

FIGURE 5-20 ▤ PLL FM transmitter block diagram.

Circuit Description

The circuit schematic shown in Fig. 5-21 is a practical working system for the block diagram of Fig. 5-20. The crystal frequency is labeled 5 MHz to help you correlate to the block diagram in Fig. 5-20. The actual crystal frequency is to be 5.76–5.92 MHz for the circuit values shown in Fig. 5-21. Similarly, the VCO frequency is really 146 MHz, rather than the 125 MHz shown. This circuitry approach eliminates the cumbersome oscillator–multiplier chain approach and allows for a very compact transmitter. The microphone input is amplified by U_5, which is an RCA CA3130 IC. Its output is used to "pull" the varactor-controlled crystal oscillator package comprised of Y_1, CR_3, and Q_3. Its output is amplified by Q_4 and then divided by 2 by the U_{3A} IC. Refer back to the block diagram in Fig. 5-20 as an aid in this circuit description. The output of the U_{3A} is applied as one of the inputs to the U_4 phase detector amplifier IC. The U_4 output at pin 8 is applied to the VCO made up of varactor diode CR_2 and Q_1. Its output (about 200 mW) is applied to the power amplifier stage (Q_2), which provides about 2 W into the antenna. Its output is sampled by the U_2IC, which is an emitter-coupled logic (ECL) device that provides a ÷10 function at frequencies up to 250 MHz. Its output is then ÷5 by the TTL U_{3B} IC and applied as the other input to the U_4 phase detector/amplifier IC. The values in this schematic are selected for operation on the 146-MHz amateur band, but operation can be accomplished at up to 250 MHz, with the U_2 ECL divider IC being the limiting upper frequency factor.

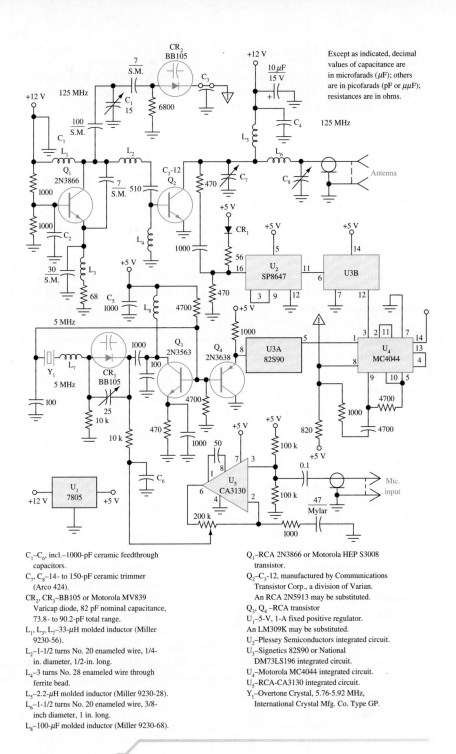

Except as indicated, decimal values of capacitance are in microfarads (μF); others are in picofarads (pF or $\mu\mu$F); resistances are in ohms.

C_1–C_6, incl.–1000-pF ceramic feedthrough capacitors.

C_7, C_8–14- to 150-pF ceramic trimmer (Arco 424).

CR_2, CR_3–BB105 or Motorola MV839 Varicap diode, 82 pF nominal capacitance, 73.8- to 90.2-pF total range.

L_1, L_3, L_7–33-μH molded inductor (Miller 9230-56).

L_2–1-1/2 turns No. 20 enameled wire, 1/4-in. diameter, 1/2-in. long.

L_4–3 turns No. 28 enameled wire through ferrite bead.

L_5–2.2-μH molded inductor (Miller 9230-28).

L_6–1-1/2 turns No. 20 enameled wire, 3/8-inch diameter, 1 in. long.

L_8–100-μF molded inductor (Miller 9230-68).

Q_1–RCA 2N3866 or Motorola HEP S3008 transistor.

Q_2–C_3-12, manufactured by Communications Transistor Corp., a division of Varian. An RCA 2N5913 may be substituted.

Q_3, Q_4 –RCA transistor

U_1–5-V, 1-A fixed positive regulator. An LM309K may be substituted.

U_2–Plessey Semiconductors integrated circuit.

U_3–Signetics 82S90 or National DM73LS196 integrated circuit.

U_4–Motorola MC4044 integrated circuit.

U_5–RCA-CA3130 integrated circuit.

Y_1–Overtone Crystal, 5.76-5.92 MHz, International Crystal Mfg. Co. Type GP.

FIGURE 5-21 ▇ PLL FM transmitter schematic.

Alignment and Operation

The reference crystal frequency is determined by dividing the desired operating frequency by 25. Varactor voltage is monitored with a DMM or oscilloscope while C_1 is varied through its range. If the loop is locked, the varactor voltage will vary with adjustment of C_1 and should be adjusted to 2.5 V. The transmitter should be terminated in a nonreactive 50-Ω load and the RF amplifier adjusted for maximum power output. Some means of determining deviation will be necessary, and the transmitter will then be ready for use.

Operation on Other Bands

A transmitter may be constructed for use on the 200-MHz band by redesigning the oscillator and RF amplifier tuned circuits to resonate in that band. Q_1 and Q_2 will operate efficiently at frequencies up to 400 MHz. Crystal frequency is determined in the manner indicated previously. If separate oscillator–amplifier modules are constructed for 144 and 200 MHz, or perhaps even 50 MHz, and switched electronically with ECL gates, it is possible to operate on several bands with the same phase-locked-loop components, at a considerable cost savings. It is also possible to select a low-power oscillator and an unmodulated crystal oscillator to generate the LO signal for a receiver. ECL dividers are available that allow application of this circuit at higher frequencies, but a frequency division of more than 50 is required in order that the maximum operating frequency of U_4 not be exceeded.

5-8 STEREO FM

The advent of stereo records and tapes and the associated high-fidelity playback equipment in the 1950s led to the development of stereo FM transmissions as authorized by the FCC in 1961. Stereo systems involve generating two separate signals, as from the left and right sides of a concert hall performance. When played back on left and right speakers, the listener gains greater spatial dimension or directivity.

A stereo radio broadcast requires that two separate 30-Hz to 15-kHz signals be used to modulate the carrier in such a way that the receiver can extract the "left" and "right" channel information and separately amplify them into their respective speakers. In essence, then, the amount of information to be transmitted is doubled in a stereo broadcast. Hartley's law (Chapter 1) tells us that either the bandwidth or time of transmission must therefore be doubled, but this is not practical. The problem was solved by making more efficient use of the available bandwidth (200 kHz) by *frequency multiplexing* the two required modulating signals. *Multiplex operation* is the simultaneous transmission of two or more signals on one carrier.

Modulating Signal

The system approved by the FCC is *compatible* in that a stereo broadcast received by a normal FM receiver will provide an output equal to the sum of the left plus right channels (L + R), while a stereo receiver can provide separate left and right channel signals. The stereo transmitter has a modulating signal, as shown in Fig. 5-22. Notice that the sum of L + R modulating signal extends from 30 Hz to 15 kHz just as does the full audio signal used to modulate the carrier in standard FM broadcasts. However, a signal corresponding

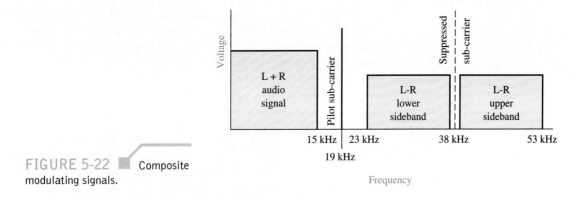

FIGURE 5-22 ■ Composite modulating signals.

to the left channel minus right channel (L − R) extends from 23 to 53 kHz. In addition, a 19-kHz pilot subcarrier is included in the composite stereo modulating signal.

The reasons for the peculiar arrangement of the stereo modulating signal will become more apparent when the receiver portion of stereo FM is discussed in Chapter 6. For now, suffice it to say that two different signals (L + R and L − R) are used to modulate the carrier. The signal is an example of *frequency-division multiplexing,* in that two different signals are multiplexed together by having them exist in two different frequency ranges.

FM Stereo Generation

The block diagram in Fig. 5-23 shows the method whereby the composite modulating signal is generated and applied to the FM modulator for subsequent transmission. The left and right channels are picked up by their respective microphones and individually preem-

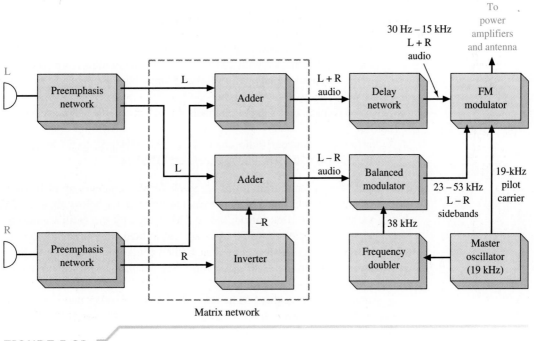

FIGURE 5-23 ■ Stereo FM transmitter.

phasized. They are then applied to a *matrix* network that inverts the right channel, giving a −R signal, and then combines (adds) L and R to provide an (L + R) signal and also combines L and −R to provide the (L − R) signal. The two outputs *are still* 30-Hz to 15-kHz audio signals at this point. The (L − R) signal and a 38-kHz carrier signal are then applied to a balanced modulator that suppresses the carrier but provides a double-side-band (DSB) signal at its output. The upper and lower sidebands extend from 30 Hz to 15 kHz above and below the suppressed 38-kHz carrier and therefore range from 23 kHz (38 kHz − 15 kHz) up to 53 kHz (38 kHz + 15 kHz). Thus, the (L − R) signal has been translated from audio up to a higher frequency so as to keep it separate from the 30-Hz to 15-kHz (L + R) signal. The (L + R) signal is given a slight delay so that both signals are applied to the FM modulator in time phase due to the slight delay encountered by the (L − R) signal in the balanced modulator. The 19-kHz master oscillator in Fig. 5-23 is applied directly to the FM modulator and also doubled in frequency to 38 kHz for the balanced modulator carrier input.

Stereo FM is more prone to noise than are monophonic broadcasts. The (L − R) signal is weaker than the (L + R) signal, as shown in Fig. 5-22. The (L − R) signal is also at a higher modulating frequency (23 to 53 kHz), and both of these effects cause poorer noise performance. The net result to the receiver is a *S/N* of about 20 dB less than the monophonic signal. Because of this, some receivers have a mono/stereo switch so that a noisy (weak) stereo signal can be changed to monophonic for improved reception. A stereo signal received by a monophonic receiver is only about 1 dB worse (*S/N*) than an equivalent monophonic broadcast, due to the presence of the 19-kHz pilot carrier. The reader may wish to refer to Sec. 6-6 to continue this stereo FM discussion into the receiver section at this time.

5-9 FM TRANSMISSIONS

There are five major categories in which FM is used:

1. Noncommercial broadcast at 88 to 90 MHz
2. Commercial broadcast with 200-kHz channel bandwidths from 90 to 108 MHz
3. Television audio signals with 50-kHz channel bandwidths at 54 to 88 MHz, 174 to 216 MHz, and 470 to 806 MHz
4. Narrowband public service channels from 108 to 174 MHz and in excess of 806 MHz
5. Narrowband amateur radio channels at 29.6 MHz, 52 to 53 MHz, 144 to 147.99 MHz, 440 to 450 MHz, and in excess of 902 MHz

The output powers range from milliwatt levels for the amateurs up to 100 kW for broadcast FM. You will note that FM is not used at frequencies below about 30 MHz. This is due to the phase distortion introduced to FM signals by the earth's ionosphere to these frequencies. Frequencies above 30 MHz are transmitted "line-of-sight" and are not significantly affected by the ionosphere. The limited range (normally, 70 to 80 mi) for FM transmission is due to the earth's curvature. See Chapter 13 for a more complete discussion of these effects.

Another advantage that FM has over SSB and AM, other than superior noise performance, is the fact that *low-level modulation* (see Sec. 2-5) can be used with subsequent highly efficient class C power amplifiers. Since the FM waveform does not vary in amplitude, the intelligence is not lost by class C power amplification as it is for AM and SSB.

Recall that a class C amplifier tends to provide a constant output amplitude due to the *LC* tank circuit flywheel effect. Thus, there is no need for high-power audio amplifiers in an FM transmitter and, more important, all the power amplification takes place at about 90% efficiency (class C), as compared to a maximum of about 70% for linear power amplifiers.

TROUBLESHOOTING

The most likely types of FM radio a technician will be called to service are an automotive mobile, a fixed base station, or a hand-held portable. Either the mobile or the fixed base station can have power outputs as high as 150 W. Some of these transmitters can be damaged if not operated into the proper load impedance, usually 50 Ω. Some contain automatic circuitry to shut the transmitter off if it is not connected to a proper load. Dummy loads are made for this purpose, so make sure you have one rated for sufficient power.

In an FM transmitter, an oscillator is typically controlled so that its frequency changes when an intelligence signal changes. This control is provided by a modulating circuit. Recall in Sec. 5-5 there are several ways that this modulator can work. In this section we will learn some troubleshooting techniques for a reactance modulator circuit.

Frequency-modulated transmitters can be roughly divided into two categories. The first is wideband FM. More popular than AM broadcasting, it is the FM we listen to on our car radios and home stereos (it also produces the audio portion of a TV signal). The second type of FM transmitters is called narrowband FM. It is used in police and fire department radios as well as taxicabs and VHF boat radios and the popular hand-held transceivers called handy talkies. We'll look at testing a wideband FM generator in this section.

After completing this section you should be able to

- Troubleshoot FM transmitter systems
- Describe the operation of the reactance modulator
- Locate the master oscillator section
- Locate the reactance circuit section

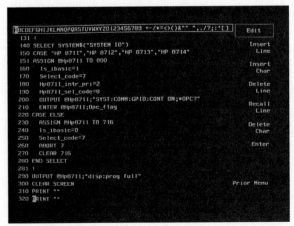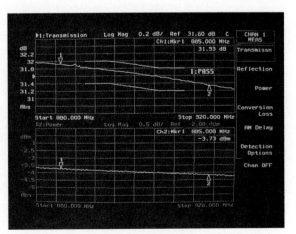

Spectrum/network analyzer used in testing amplifier performance. Left: Keystroke recording allows users without programming experience to automate test processes. Right: An amplifier measurement in a split display of transmission and power. User-defined limit lines and markers show pass/fail status. (Courtesy of Hewlett-Packard.)

Spectrum/network analyzer.
(Courtesy of Hewlett-Packard.)

- Recognize the difference between no modulator output, low output, or oscillator output without FM
- Troubleshoot a stereo/SCA FM generator
- Measure an FM transmitter's carrier frequency and deviation

FM Transmitter Systems

1. NO RF OUTPUT   In this case, one should first verify that the oscillator is running. Most of these transmitters multiply the oscillator by 12 or 18 to obtain the output frequency. The service manual will tell you what the multiplier is. It is best to have a spectrum analyzer, but a shortwave receiver will do.

Hints on oscillator problems are in Chapter 1, so let's assume that the oscillator is running and move on to the first multiplier stage. Figure 5-24 shows a simplified schematic of a multiplier stage. If the base–emitter junction is good and there is sufficient input drive, you will find a negative voltage at the base of Q1. This is because the RF input is rectified by the junction and the current flows through R1. When the RF is rectified by the base–emitter junction, current pulses rich in harmonics are amplified and filtered by the tuned circuit in the collector of the transistor.

If a spectrum analyzer is available, the technician can verify that the stage is producing the proper multiple of the input frequency by loosely coupling the analyzer to the output coils. A two- or three-turn coil $\frac{1}{2}''$ in diameter connected to the analyzer will do.

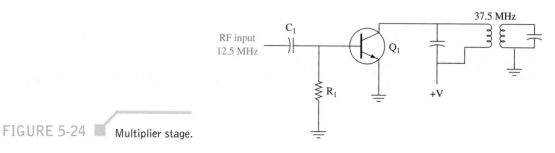

FIGURE 5-24   Multiplier stage.

Either the coils or the capacitors will be adjustable. You should be able to peak the output on the proper frequency with these adjustments. If not, check the capacitors and the inductors.

If you do not have a spectrum analyzer, simply measuring the bias voltage on the next stage may be sufficient. You can be reasonably sure that the multiplier is working if you can peak up the drive to the next stage with the adjustments. However, there is some possibility of tuning to the wrong harmonic. If the adjustments are all the way to one end, you may have done this or some component has failed. Another indication of improper tuning is that you will probably not be able to tune the next stage.

A typical transmitter will have three multiplier stages. At some point in the chain you will be able to find enough signal to run a frequency counter. Be careful, too much input to the counter will damage it. At the output of the transmitter, you can use a high-power attenuator.

2. INCORRECT FREQUENCY   Most of these transmitters will have trimmer capacitors to adjust the frequency, while some will have inductors. If the oscillator is off frequency, check the voltages first, then the capacitors. Intermittent capacitors are the hardest to find. Try cooling the capacitor with an aerosol spray sold for this purpose.

3. INCORRECT DEVIATION   There are usually two adjustments here, one for microphone gain and one on a limiter. The limiter prevents the user from overmodulating the transmitter. When adjusting deviation, make sure the limiter is adjusted so as not to affect the gain adjustment. Refer to Fig. 5-25.

Mobile radios are usually set for a peak deviation of 5 kHz. A good service shop will have a deviation meter. If you don't have one, a fair job can be done by simply comparing a known good transmitter with the unit under test by listening to both with any receiver.

If a spectrum analyzer is available, recall Carson's rule and speak into the microphone while adjusting the bandwidth of the transmitted signal to about 16 kHz.

One can also use the zero carrier amplitude method shown in Fig. 5-4. While viewing the transmitter output on the spectrum analyzer, apply a 2-kHz tone to the microphone input. Adjust the gain from zero up until the carrier null and you have 5-kHz peak deviation.

Reactance Modulator Circuit Operation

The reactance modulator is efficient and provides a large deviation. It is popular and used often in FM transmitters. Figure 5-26 illustrates a typical reactance modulator circuit. Refer to this figure throughout the following discussion. The circuit consists of the reactance circuit and the master oscillator. The reactance circuit operates on the master oscillator to cause its resonance frequency to shift up or shift down depending on the modulating signal being applied. The reactance circuit appears capacitive in nature to the

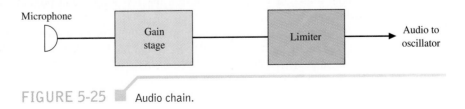

FIGURE 5-25 ▨ Audio chain.

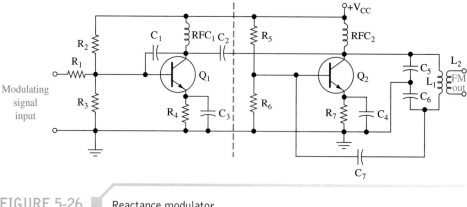

FIGURE 5-26 ■ Reactance modulator.

master oscillator. In this case, the reactance looks like a variable capacitor in the oscillator's tank circuit.

Transistor Q1 makes up the reactance modulator circuit. Resistors R2 and R3 establish a voltage divider network that biases Q1. Resistor R4 furnishes emitter feedback to thermally stabilize Q1. Capacitor C3 is a bypass component that prevents ac input signal degeneration. Capacitor C1 interacts with transistor Q1's inter-electrode capacitance to cause a varying capacitive reactance directly influenced by the input modulating signal.

The master oscillator is a Colpitts oscillator built around transistor Q2. Coil L1, capacitor C5, and capacitor C6 make up the resonant tank circuit. Capacitor C7 provides the required regenerative feedback to cause the circuit to oscillate. Q1 and Q2 are impedance coupled, and capacitor C2 effectively couples the changes at Q1's collector to the tank circuit of transistor Q2 while blocking dc voltages.

When a modulating signal is applied to the base of transistor Q1 via resistor R1, the reactance of the transistor changes in relation to that signal. If the modulating voltage goes up the capacitance of Q1 goes down, and if the modulating voltage goes down the reactance of Q1 goes up. This change in reactance is felt on Q1's collector and also at the tank circuit of the Colpitts oscillator transistor Q2. As capacitive reactance at Q1 goes up the resonant frequency of the master oscillator, Q2, decreases. Conversely, if Q1's capacitive reactance goes down the master oscillator resonant frequency increases.

Troubleshooting the Reactance Modulator

The FM output signal from coil L2 can be lost due to open bias resistors, open RF chokes RFC1 and RFC2, or an open winding at coil L1. In addition, a leaky coupling capacitor C2 may shift the collector voltage of Q2 of the master oscillator, causing a low FM output signal to be present at coil L2. Low collector voltages and weak transistors may produce a low FM signal output condition, as well as changes in bias resistors R2 and R3 in transistor Q1's circuit and resistors R5 and R6 in Q2's circuit. Changes in the emitter resistors of both transistor circuits will lessen the FM output and possibly shut the modulator down.

The master oscillator may operate without being influenced by the reactance circuit. The oscillator output signal at L2 would not be FM. This situation could occur if the modulating signal were missing from the base of transistor Q1. An open R1 would block the modulating signal from getting to the base of Q1. Without the modulating signal present, Q1's reactance would not change. A leaky or shorted C1 could also kill the reactance

TABLE 5-2

SYMPTOM	PROBLEM	PROBABLE CAUSE
No signal out from L2	No FM modulator output	Open bias resistors in Q1 and Q2 circuits; open RFC1 or RFC2; C2 open or leaky; feedback capacitor C7 open
No FM output at L2, master oscillator output only	No FM modulation taking place	Q1 not functioning, check C1, RFC1, and R4; resistor R1 may be open
Amplitude of FM output low	Low modulator output	Changes in bias resistor values, check R2, R3, R5, and R6; change in emitter resistors, check R4 and R7; Q2's gain has decreased

response of Q1. Transistor Q1 may still be operating perfectly but the reactance changes might not be passed to Q2's tank circuit due to an open C2.

The following table is a symptom guide to help you troubleshoot the reactance modulator circuit.

Check the resistors with your DMM for proper values. Resistors in the reactance modulator circuit will be precision types with close tolerances. For low FM signal outputs check capacitors C1, C2, C5, C6, and C7 with a capacitor checker. Any one of these capacitors could cause low output. If any of these capacitors open or become leaky, the FM output may cease altogether. Check coils for open windings using the DMM's continuity function. Look for cold solder joints by observing a dull appearance of the solder connection. The DMM will give a high resistance indication for a cold solder joint.

The Spectrum of a Wideband FM Signal

Figure 5-27 illustrates the frequency spectrum of perhaps the most common form of wideband FM signal being broadcast today. Note on the far left of the diagram the carrier of the transmitted signal. It is a station's carrier frequency to which we tune our radios. All frequencies in the figure are relative to that of the carrier. For example, the suppressed subcarrier at 38 kHz is 38 kHz above the carrier.

The signal has two major parts. The first part extends from the carrier up to 53 kHz. It consists of the three components making up the stereo (stereophonic) portion of the signal: the left-plus-right audio channel extending from 0.05 to 15 kHz, the pilot carrier at 19 kHz, and the left-minus-right audio channel from 23 to 53 kHz.

The second part of our FM spectrum is the SCA (Subsidiary Communications Authorization) signal extending from 60 to 74 kHz above the carrier. Because the station's voice ID (which by law must be broadcast at regular intervals) is carried on the stereo channel, the SCA channel can legally transmit, for example, weather announcements or perhaps only music. That music, however, is at such a high frequency (note it is centered around 67 kHz) it cannot be heard on regular radios. Instead, the radio station

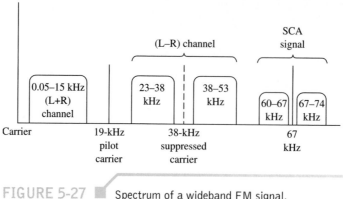

FIGURE 5-27 ■ Spectrum of a wideband FM signal.

leases special equipment that drops the frequencies to normal hearing range to supply background music to grocery stores and dentists' offices. Figure 5-28 shows the block diagram of a system that generates a stereo/SCA FM signal.

Troubleshooting the stereo/SCA generator is best done with a spectrum analyzer connected as shown in Fig. 5-29. Note first the (L + R) signals from 0.05 to 15 kHz in Fig. 5-27. They will be present whether the station is transmitting monophonic or stereo. The 19-kHz pilot carrier and (L − R) channels, however, are only present during stereo broadcasts. Should the oscillator generating the 19-kHz carrier fail, both that signal as well as the (L − R) channel from 23 to 53 kHz will disappear from the analyzer display. The oscillator will probably be contained within an IC that develops the entire stereo signal. If so, the chip must be replaced (assuming you've determined that V_{cc} and any other required inputs are present). Older units, however, might have a circuit constructed of separate components. In that case, use your regular troubleshooting techniques of checking for proper voltages, resistances, etc., to locate the problem.

The (L − R) channel is a DSBSC (double-sideband suppressed carrier) signal generated by a balanced modulator. The carrier it suppresses is the second harmonic (two

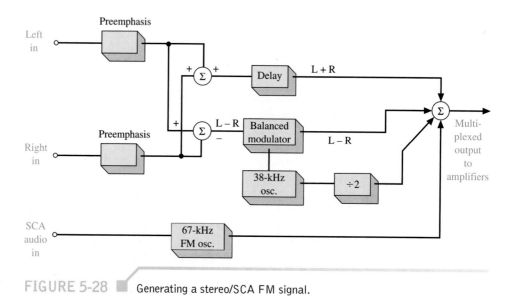

FIGURE 5-28 ■ Generating a stereo/SCA FM signal.

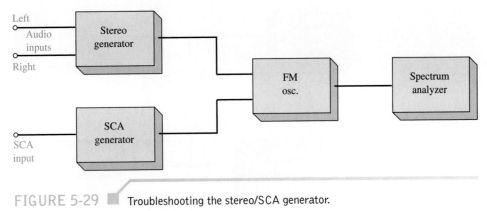

FIGURE 5-29 ■ Troubleshooting the stereo/SCA generator.

times) of the 19-kHz pilot carrier. Troubleshooting and balance adjustment of balanced modulators were discussed in Chapter 4.

Wideband FM Transmitters' Frequency and Deviation Test

Wideband FM transmissions are defined as those having a modulation index greater than 1. By FCC regulations, these signals are allowed a deviation no greater than ±75 kHz. Since they are modulated by audio signals from 30 Hz to 15 kHz, calculations tell us the modulation index can range from 5 to 2500.

We will discuss two important tests that are regularly made on FM transmitters. These are carrier frequency and deviation tests.

The block diagram of a wideband FM transmitter is seen in Fig. 5-30. Note the input to the carrier oscillator. It is the output of the stereo/SCA generator discussed above. We say the stereo/SCA generator frequency modulates the transmitter's carrier oscillator.

The FCC requires an FM broadcast station to hold its carrier frequency accurate to within ±2000 Hz. To achieve this stability, the transmitter employs an AFC (automatic frequency control) circuit. A reference crystal oscillator drives the AFC unit. Its frequency is measured very accurately. To hold this frequency stable against temperature changes, the entire unit is contained in a thermostatically controlled oven.

The AFC control compares the carrier oscillator frequency against that of the reference oscillator. Should there be an error between the two, in other words, should the car-

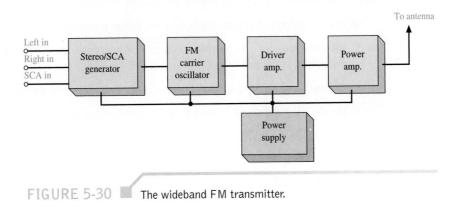

FIGURE 5-30 ■ The wideband FM transmitter.

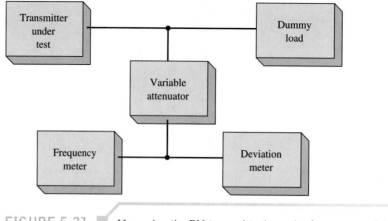

FIGURE 5-31 ■ Measuring the FM transmitter's carrier frequency and deviation.

rier oscillator change frequency or drift, the AFC circuitry shifts the carrier frequency to return the error to zero.

The test setup of Fig. 5-31 is used to measure a station's carrier frequency. Note the dummy load. It dissipates the energy developed by the transmitter as heat so no signal goes out "over the air" during testing and troubleshooting. The input impedance of the dummy load must "match" (be the same as) the impedance of the transmission line carrying the transmitter's output signal to the antenna.

The frequency meter or counter used to measure carrier frequency in Fig. 5-31 must be accurate, especially if the readings are being taken for FCC certification. Should the frequency be out of specs, check the reference oscillator for proper output with the same frequency meter. Look for problems in the AFC circuitry. The modulating signal should be turned off before making any carrier measurements.

Carrier deviation can also be measured using the deviation meter in the circuit of Fig. 5-31. As stated above, deviation must not exceed ±75 kHz for wideband FM broadcasting transmitters.

SUMMARY

In Chapter 5 we studied the concept of frequency modulation (FM) and learned the basics of FM transmitters. The major topics the student should now understand include:

- the definitions of angle, frequency, and phase modulation

- the generation of FM by using a capacitor microphone and the effects of changes in voice amplitude and frequency on the FM signal

- the analysis of FM using modulation index and Bessel functions

- the determination of FM deviation using the zero-carrier condition

- the analysis of noise suppression by limiter circuits and by using phasors and signal-to-noise ratio (S/N)

- the analysis of direct generation FM circuits, including varactor diodes, the reactance modulator, the LIC VCO, and the Crosby modulator

- the operation of an indirect FM system using the Armstrong modulator
- the generation of FM using a phase-locked loop (PLL)
- The changes made to a standard FM transmitter to enable broadcast stereo operation
- the advantage of FM versus SSB or AM

QUESTIONS AND PROBLEMS

SECTION 1

1. Define *angle modulation* and list its subcategories.
*2. What is the difference between frequency and phase modulation?
3. Even though PM is not actually transmitted, provide two reasons that make it important in the study of FM.

SECTION 2

4. Describe the effect of an intelligence signal's amplitude and frequency when it frequency-modulates a carrier.
5. Explain how a condenser microphone can very easily be used to generate FM.
6. In an FM transmitter, the output is changing between 90.001 and 89.999 MHz 1000 times a second. The intelligence signal amplitude is 3 V. Determine the carrier frequency and intelligence signal frequency. If the output deviation changes to between 90.0015 and 89.9985 MHz, calculate the intelligence signal amplitude. (90 MHz, 1 kHz, 4.5 V)
*7. What determines the rate of frequency swing for an FM broadcast transmitter?

SECTION 3

8. Define *modulation index* (m_f) as applied to an FM system.
*9. What characteristic(s) of an audio tone determines the percentage of modulation of an FM broadcast transmitter?
10. Explain the difference between modulation index for PM versus FM. How can a modulating signal be modified so that allowing it to phase-modulate a carrier results in FM?
11. Explain what happens to the carrier in FM as m_f goes from 0 up to 15.
12. Calculate the bandwidth of an FM system (using Table 5-1) when the maximum deviation (δ) is 15 kHz and $f_i = 3$ kHz. Repeat for $f_i = 2.5$ and 5 kHz. (48 kHz, 45 kHz, 60 kHz)
13. Does the maximum deviation directly determine the bandwidth of an FM system? Explain the relationship between bandwidth and deviation.
14. Explain the purpose of the *guard bands* for broadcast FM. How wide is an FM broadcast channel?
*15. What frequency swing is defined as 100% modulation for an FM broadcast station?
*16. What is the meaning of the term *center frequency* in reference to FM broadcasting?
*17. What is the meaning of the term *frequency swing* in reference to FM broadcast stations?
*18. What is the frequency swing of an FM broadcast transmitter when modulated 60%? (±45 kHz)

*19. An FM broadcast transmitter is modulated 40% by a 5-kHz test tone. When the percentage of modulation is doubled, what is the frequency swing of the transmitter?

*20. An FM broadcast transmitter is modulated 50% by a 7-kHz test tone. When the frequency of the test tone is changed to 5 kHz and the percentage of modulation is unchanged, what is the transmitter frequency swing?

*21. If the output current of an FM broadcast transmitter is 8.5 A without modulation, what is the output current when the percentage of modulation is 90%?

22. An FM transmitter delivers, to a 75-Ω antenna, a signal of $v = 1000 \sin (10^9 t + 4 \sin 10^4 t)$. Calculate the carrier and intelligence frequencies, power, modulation index, deviation, and bandwidth. (159 MHz, 1.59 kHz, 6.67 kW, 4, 6.37 kHz, ~ 16 kHz)

23. An FM transmitter puts out 1 kW of power. Determine the power in the carrier and all significant sidebands when $m_f = 2$. Verify that their sum is 1 kW.

*24. In an FM radio communications system, what is the meaning of *modulation index?* Of *deviation ratio?* What values of deviation ratio are used in an FM radio communications system?

25. Assuming that the 9.892-kW result of Ex. 5-6 is exactly correct, determine the total power in the J_2 sidebands and higher. (171 W)

SECTION 4

*26. What types of radio receivers do not respond to static interference?

*27. What is the purpose of a limiter stage in an FM broadcast receiver?

28. Explain why the limiter does *not* eliminate all noise effects in an FM system.

29. Calculate the amount of frequency deviation caused by a limited noise spike that still causes an undesired phase shift of 35° when f_i is 5 kHz. (3.05 kHz)

30. In a broadcast FM system, the input $S/N = 4$. Calculate the worst-case S/N at the output if the receiver's internal noise effect is negligible. (19.8:1)

31. Explain why narrowband FM systems have poorer noise performance than wideband systems.

32. Explain the *capture effect* in FM, and include the link between it and FM's inherent noise reduction capability.

*33. Why is narrowband FM rather than wideband FM used in radio communications systems?

*34. What is the purpose of preemphasis in an FM broadcast transmitter? Of deemphasis in an FM receiver? Draw a circuit diagram of a method of obtaining preemphasis.

*35. Discuss the following for frequency modulation systems:
(a) The production of sidebands.
(b) The relationship between the number of sidebands and the modulating frequency.
(c) The relationship between the number of sidebands and the amplitude of the modulating voltage.
(d) The relationship between percent modulation and the number of sidebands.
(e) The relationship between modulation index or deviation ratio and the number of sidebands.
(f) The relationship between the spacing of the sidebands and the modulating frequency.
(g) The relationship between the number of sidebands and the bandwidth of emissions.

(h) The criteria for determining bandwidth of emission.

(i) Reasons for preemphasis.

Section 5

36. Draw a schematic diagram of a varactor diode FM generator and explain its operation.

*37. Draw a schematic diagram of a frequency-modulated oscillator using a reactance modulator. Explain its principle of operation.

38. Using the specifications in Fig. 5-15, draw a schematic of an FM generator using the SE/NE 566 LIC function generator VCO. The center frequency is to be 500 kHz, and the output is to be a sine wave. Show all component values. How much center frequency drift can be expected from a temperature rise of 50°C?

39. Explain the principles of a Crosby-type modulator.

*40. How is good stability of a reactance modulator achieved?

*41. If an FM transmitter employs one doubler, one tripler, and one quadrupler, what is the carrier frequency swing when the oscillator frequency swing is 2 kHz? (48 kHz)

42. Draw a block diagram of a broadcast-band Crosby-type FM transmitter operating at 100 MHz, and label all frequencies in the diagram.

Section 6

*43. Draw a block diagram of an Armstrong-type FM broadcast transmitter complete from the microphone input to the antenna output. State the purpose of each stage, and explain briefly the overall operation of the transmitter.

44. Explain the difference in the amount of deviation when passing an FM signal through a mixer as compared to a multiplier.

Section 7

*45. Draw a block diagram of a stereo multiplex FM broadcast transmitter complete from the microphone inputs to the antenna output. State the purpose of each stage, and explain briefly the overall operation of the transmitter.

46. Explain how stereo FM is able to effectively transmit twice the information of a standard FM broadcast while still using the same bandwidth. How is the *S/N* at the receiver affected by a stereo transmission as opposed to monophonic?

Section 8

47. Explain the operation of the PLL FM transmitter shown in Fig. 5-20.

Section 9

*48. What are the merits of an FM communications system compared to an AM system?

*49. Why is FM undesirable in the standard AM broadcast band?

FREQUENCY
MODULATION:
RECEPTION

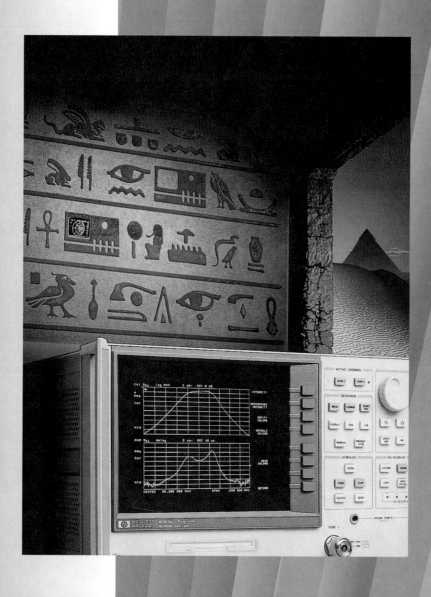

OBJECTIVES

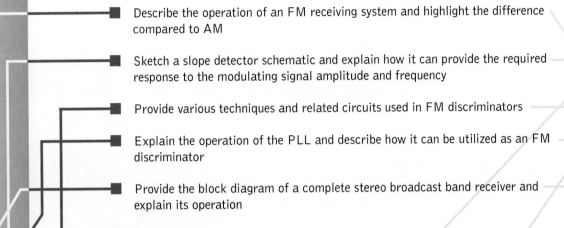

■ Describe the operation of an FM receiving system and highlight the difference compared to AM

■ Sketch a slope detector schematic and explain how it can provide the required response to the modulating signal amplitude and frequency

■ Provide various techniques and related circuits used in FM discriminators

■ Explain the operation of the PLL and describe how it can be utilized as an FM discriminator

■ Provide the block diagram of a complete stereo broadcast band receiver and explain its operation

■ Analyze the operation of a LIC used as a stereo decoder

■ Analyze and understand a complete FM receiver schematic

6-1 BLOCK DIAGRAM

The basic FM receiver uses the superheterodyne principle. In block diagram form, it has many similarities to the receivers covered in previous chapters. In Fig. 6-1, the only apparent differences are the use of the word *discriminator* in place of *detector,* the addition of a deemphasis network, and the fact that AGC may or may not be used as indicated by the dashed lines.

The *discriminator* extracts the intelligence from the high-frequency carrier and can also be called the detector, as in AM receivers. By definition, however, a discriminator is a device in which amplitude variations are derived in response to frequency or phase variations, and it is the preferred term for describing an FM demodulator.

The deemphasis network following demodulation is required to bring the high-frequency intelligence back to the proper amplitude relationship with the lower frequencies. Recall that the high frequencies were preemphasized at the transmitter to provide them with greater noise immunity, as explained in Sec. 5-4.

The fact that AGC is optional in an FM receiver may be surprising to you. From your understanding of AM receivers, you know that AGC is essential to their satisfactory operation. However, the use of limiters in FM receivers essentially provides an AGC

Photo: Spectrum/network analyzer provides response of a bandpass filter.
(Courtesy of Hewlett-Packard.)

223

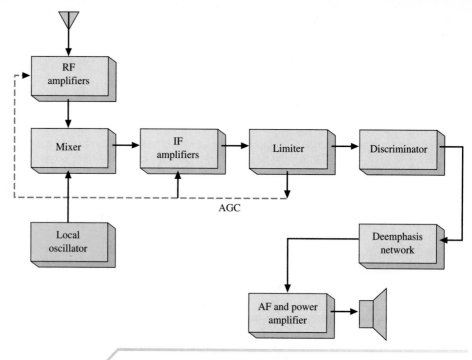

FIGURE 6-1 ■ FM receiver block diagram.

function, as will be explained in Sec. 6-3. Many older FM receivers also included an *automatic frequency control* (AFC) function. This is a circuit that provides a slight automatic control over the local oscillator circuit. It compensates for drift in LO frequency that would otherwise cause a station to become detuned. It was necessary because it had not yet been figured out how to make an economical *LC* oscillator at 100 MHz with sufficient frequency stability. The AFC system is not needed in new designs.

The mixer, local oscillator, and IF amplifiers are basically similar to those discussed for AM receivers and do not require further elaboration. It should be noted that higher frequencies are usually involved, however, because of the fact that FM systems generally function at higher frequencies. The universally standard IF frequency for FM is 10.7 MHz, as opposed to 455 kHz for AM. Because of significant differences in all the other portions of the block diagram shown in Fig. 6-1, they are discussed in the following sections.

6-2 RF AMPLIFIERS

Broadcast AM receivers normally operate quite satisfactorily without any RF amplifier. This is rarely the case with FM receivers, however, except for frequencies in excess of 1000 MHz (1 GHz), when it becomes preferable to omit it. The essence of the problem is that FM receivers can function with smaller received signals than AM or SSB receivers because of their inherent noise reduction capability. This means that FM receivers can function with a lower sensitivity, and are called upon to deal with input signals of 1 μV or less as compared with perhaps a 30-μV minimum input for AM. If a 1-μV signal is fed directly into a mixer, the inherently high noise factor of an active mixer stage destroys the intelligibility of the 1-μV signal. It is, therefore, necessary to amplify the 1-μV level in a RF stage to get the signal up to at least 10 to 20 μV before mixing occurs. The FM system can tolerate 1 μV of noise from a mixer on a 20-μV signal but obviously cannot cope with 1 μV of noise with a 1-μV signal.

This reasoning also explains the abandonment of RF stages for the ever-increasing FM systems at the 1-GHz-and-above region. At these frequencies, transistor noise is increasing while gain is decreasing. The frequency is reached where it is advantageous to feed the incoming FM signal directly into a diode mixer so as to immediately step it down to a lower frequency for subsequent amplification. Diode (passive) mixers are less noisy than active mixers.

Of course, the use of a RF amplifier reduces the image frequency problem, as explained in Chapter 3. Another benefit is the reduction in *local oscillator reradiation* effects. Without a RF amp, the local oscillator signal can more easily get coupled back into the receiving antenna and transmit interference.

FET RF Amplifiers

Virtually all RF amps used in quality FM receivers utilize FETs as the active element. You may think that this is done because of their high input impedance, but this is *not* the reason. In fact, their input impedance at the high frequency of FM signals is greatly reduced because of their input capacitance. The fact that FETs do not offer any significant impedance advantage over other devices at high frequencies is not a deterrent, however, since the impedance that a RF stage works from (the antenna) is only several hundred ohms or less anyway.

The major advantage is that FETs have an input/output square-law relationship while vacuum tubes have a $\frac{3}{2}$-power relationship and BJTs have a diode-type exponential characteristic. A square-law device has an output signal at the input frequency and a smaller distortion component at two times the input frequency, whereas the other devices mentioned have many more distortion components, with some of them occurring at frequencies close to the desired signal. The use of a FET at the critical small signal level in a receiver means that the device distortion components are easily filtered out by its tuned circuits, since the closest distortion component is two times the frequency of the desired signal. This becomes an extreme factor when you tune to a weak station that has a very strong adjacent signal. If the high-level adjacent signal gets through the input tuned circuit, even though greatly attenuated, it would probably generate distortion components at the desired signal frequency by a non-square-law device, and the result is audible noise in the speaker output. This form of receiver noise is called *cross-modulation*. This is similar to *intermodulation distortion*, which is characterized by the mixing of *two* undesired signals, resulting in an output component that is equal to the desired signal's frequency. The possibility of intermodulation distortion is also greatly minimized by use of FET RF amplifiers. Further discussion of intermodulation distortion is included in Sec. 8-4.

MOSFET RF Amplifiers

A dual-gate, common-source MOSFET RF amplifier is shown in Fig. 6-2. The use of a dual-gate device allows a convenient isolated input for an AGC level to control device gain. The MOSFETs also offer the advantage of increased *dynamic range* over JFETs. That is, a wider range of input signal can be tolerated by the MOSFET while still offering

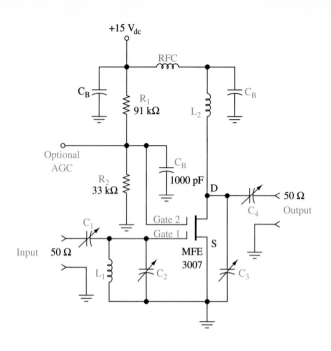

FIGURE 6-2 ▇ MOSFET RF amplifier.
(Courtesy of Motorola Semiconductor
Products, Inc.)

VHF Amplifier

The following component values are used
for the different frequencies:

Component Values	100 MHz	400 MHz
C_1	8.4 pF	4.5 pF
C_2	2.5 pF	1.5 pF
C_3	1.9 pF	2.8 pF
C_4	4.2 pF	1.2 pF
L_1	150 nH	16 nH
L_2	280 nH	22 nH
C_B	1000 pF	250 pF

the desired square-law input/output relationship. A similar arrangement is often utilized
in mixers, since the extra gate allows for a convenient injection point for the local oscillator signal. The accompanying chart in Fig. 6-2 provides component values for operation
at 100-MHz and 400-MHz center frequencies. The antenna input signal is coupled into
gate 1 via the coupling/tuning network comprised of C_1, L_1, and C_2. The output signal is
taken at the drain, which is coupled to the next stage by the L_2–C_3–C_4 combination. The
bypass capacitor C_B next to L_2 and the radio-frequency choke (RFC) ensure that the signal frequency is not applied to the dc power supply. The RFC acts as an open to the signal
while appearing as a short to dc, and the bypass capacitor acts in the inverse fashion.
These precautions are necessary to RF frequencies because while power supply impedance is very low at low frequencies and dc, it looks like a high impedance to RF and can
cause appreciable signal power loss. The bypass capacitor from gate 2 to ground provides

a short to any high-frequency signal that may get to that point. It is necessary to maintain the bias stability set up by R_1 and R_2. The MFE 3007 MOSFET used in this circuit provides a minimum power gain of 18 dB at 200 MHz.

6-3 LIMITERS

A limiter is a circuit whose output is a constant amplitude for all inputs above a critical value. Its function in an FM receiver is to remove any residual (unwanted) amplitude modulation and the amplitude variations due to noise. Both of these variations would have an undesirable effect if carried through to the speaker. In addition, the limiting function also provides AGC action, since signals from the critical minimum value up to some maximum value provide a constant input level to the detector. By definition, the discriminator (detector) ideally would not respond to amplitude variations anyway, since the information is contained in the amount of frequency deviation and the rate at which it deviates back and forth around its center frequency.

A transistor limiter is shown in Fig. 6-3. Notice the dropping resistor, R_c, which limits the dc collector supply voltage. This provides a low dc collector voltage, which makes this stage very easily overdriven. This is the desired result. As soon as the input is large enough to cause clipping at both extremes of collector current, the critical limiting voltage has been attained and limiting action has started.

The input/output characteristic for the limiter is shown in Fig. 6-4, and it shows the desired clipping action and the effects of feeding the limited (clipped) signal into an LC tank circuit tuned to the signal's center frequency. The natural flywheel effect of the tank removes all frequencies not near the center frequency and thus provides a sinusoidal output signal as shown. The omission of an LC circuit at the limiter output is desirable for some demodulator circuits. The quadrature detector (Sec. 6-4) uses the square-wave-like waveform that results.

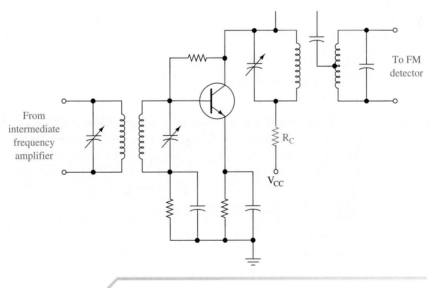

FIGURE 6-3 ■ Transistor limiting circuit.

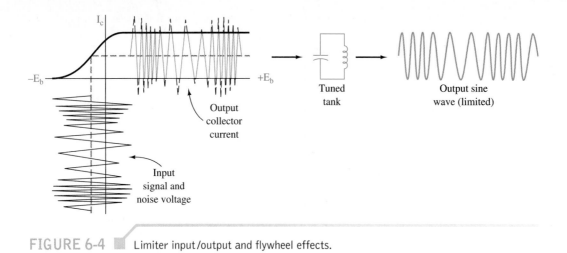

FIGURE 6-4 Limiter input/output and flywheel effects.

Limiting and Sensitivity

A limiter, such as shown in Fig. 6-3, requires about 1 V of signal to begin limiting. Much amplification of the received signal is therefore needed prior to limiting, which explains its position following the IF stages. When enough signal arrives at the receiver to start limiting action, the set *quiets,* which means that background noise disappears. The *sensitivity* of an FM receiver is defined in terms of how much input signal is required to produce a specific level of quieting, normally 30 dB. This means that a good-quality receiver with a rated 1.5-μV sensitivity will have background noise 30 dB down from the desired input signal that has a 1.5-μV level.

The minimum required voltage for limiting is called the *quieting, threshold,* or *limiting knee* voltage. The limiter then provides a constant-amplitude output up to some maximum value that prescribes the limiting range. Going above the maximum value results either in a reduced and/or distorted output. It is possible that a single-stage limiter will not allow for adequate range, thereby requiring a double limiter or the development of AGC control on the RF and IF amplifiers to minimize the possible limiter input range.

EXAMPLE 6-1

A certain FM receiver provides a voltage gain of 200,000 (106 dB) prior to its limiter. The limiter's quieting voltage is 200 mV. Determine the receiver's sensitivity.

SOLUTION

To reach quieting, the input must be

$$\frac{200 \ mV}{200,000} = 1 \ \mu V$$

The receiver's sensitivity is therefore 1 μV.

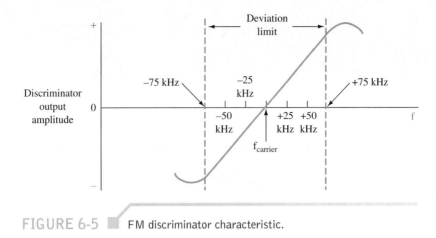

FIGURE 6-5 FM discriminator characteristic.

It is most common for today's FM receivers to use IC IF amplification. In these cases, the ICs have a built-in limiting action of very high quality (i.e., wide dynamic range). Section 6-7 provides an example of these ICs.

6-4 DISCRIMINATORS

The FM discriminator (detector) extracts the intelligence that has been modulated onto the carrier via frequency variations. It should provide an intelligence signal whose amplitude is dependent on instantaneous carrier frequency deviation and whose frequency is dependent on the carrier's rate of frequency deviation. A desired output amplitude versus input frequency characteristic for a broadcast FM discriminator is provided in Fig. 6-5. Notice that the response is linear in the allowed area of frequency deviation and that the output amplitude is directly proportional to carrier frequency deviation. Keep in mind, however, that FM detection takes place following the IF amplifiers, which means that the ±75-kHz deviation is intact but that carrier frequency translation (usually to 10.7 MHz) has occurred.

Slope Detector

The easiest FM *discriminator* to understand is the slope detector in Fig. 6-6. The *LC* tank circuit which follows the IF amplifiers and limiter is detuned from the carrier frequency so that f_c falls in the middle of the most linear region of the response curve. When the FM signal rises in frequency above f_c, the output amplitude increases while deviations below f_c cause a smaller output. The slope detector thereby changes FM into AM, and a simple diode detector then recovers the intelligence contained in the AM waveform's envelope. In an emergency, an AM receiver can be used to receive FM by detuning the tank circuit feeding the diode detector. Slope detection is not widely used in FM receivers because the slope characteristic of a tank circuit is not very linear, especially for the large-frequency deviations of wideband FM.

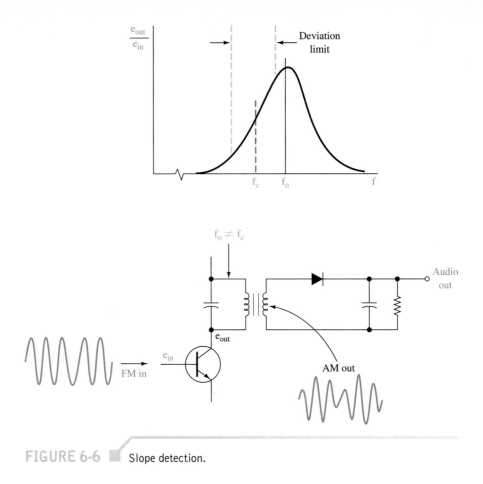

FIGURE 6-6   Slope detection.

Foster–Seely Discriminator

The two classical means of FM detection are the Foster–Seely discriminator and the ratio detector. While their once widespread use is now diminishing due to new techniques afforded by ICs, they remain a popular means of discrimination using a minimum of circuitry. A typical Foster–Seely discriminator circuit is shown in Fig. 6-7. In it, the two tank circuits [L_1C_1 and $(L_2 + L_3)C_2$] are tuned exactly to the carrier frequency. Capacitors C_c, C_4, and C_5 are shorts to the carrier frequency. The following analysis applies to an unmodulated carrier input:

1. The carrier voltage e_1 appears directly across L_4 because C_c and C_4 are shorts to the carrier frequency.

2. The voltage e_s across the transformer secondary (L_2 in series with L_3) is 180° out of phase with e_1 by transformer action, as shown in Fig. 6-8(a). The circulating $L_2L_3C_2$ tank current, i_s, is in phase with e_s since the tank is resonant.

3. The current i_s, flowing through inductance L_2L_3, produces a voltage drop that lags i_s by 90°. The individual components of this voltage, e_2 and e_3, are thus displaced by 90° from i_s, as shown in Fig. 6-8(a), and are 180° out of phase with each other because they are the voltage from the ends of a center-tapped winding.

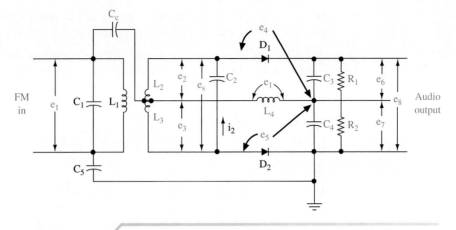

FIGURE 6-7 ■ Foster–Seely discriminator.

4. The voltage e_4 applied to the diode D_1, C_3, and R_1 network will be the vector sum of e_1 and e_2 [Fig. 6-8(a)]. Similarly, the voltage e_5 is the sum of e_1 and e_3. The magnitude of e_6 is proportional to e_4 while e_7 is proportional to e_5.

5. The output voltage, e_8, is equal to the sum of e_6 and e_7 and is zero since the diodes D_1 and D_2 will be conducting current equally (since $e_4 = e_5$) but in opposite directions through the R_1C_3 and R_2C_4 networks.

The discriminator output is zero with no modulation (zero frequency deviation), as is desired. The following discussion now considers circuit action at some instant when the input signal e_1 is above the carrier frequency. The phasor diagram of Fig. 6-8(b) is used to illustrate this condition:

1. Voltages e_1 and e_s are as before, but e_s now sees an inductive reactance, because the tank circuit is above resonance. Therefore, the circulating tank current, i_s, lags e_s.

2. The voltages e_2 and e_3 must remain 90° out of phase with i_s, as shown in Fig. 6-8(b). The new vector sums of $e_2 + e_1$ and $e_3 + e_1$ are no longer equal, so e_4 causes a heavier conduction of D_1 than exists for D_2.

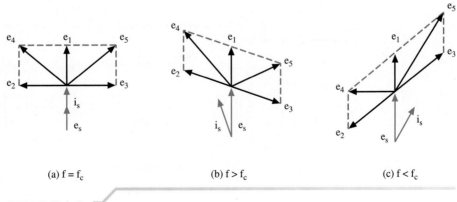

(a) $f = f_c$ (b) $f > f_c$ (c) $f < f_c$

FIGURE 6-8 ■ Discriminator phase relations.

3. The output, e_8, which is the sum of e_6 and e_7, will go positive since the current down through R_1C_3 is greater than the current up through R_2C_4 (e_4 is greater than e_5).

The output for frequencies above resonance (f_c) is therefore positive, while the phasor diagram in Fig. 6-8(c) shows that at frequencies below resonance the output goes negative. The amount of output is determined by the amount of frequency deviation, while the frequency of the output is determined by the rate at which the FM input signal varies around its carrier or center value.

Ratio Detector

While the Foster–Seely discriminator just described offers excellent linear response to wideband FM signals, it also responds to any undesired input amplitude variations. The *ratio detector* does not respond to input variations and minimizes the required limiting before detection.

The ratio detector, shown in Fig. 6-9, is a circuit designed to respond only to frequency changes of the input signal. Amplitude changes in the input have no effect upon the output. The input circuit of the ratio detector is identical to that of the Foster–Seely discriminator circuit. The most immediately obvious difference is the reversal of one of the diodes.

The ratio detector circuit operation is similar to the Foster–Seely. A detailed analysis will therefore not be given. Notice the large electrolytic capacitor, C_5, across the R_1–R_2 combination. This maintains a constant voltage that is equal to the peak voltage across the diode input. This feature eliminates variations in the FM signal, thus providing amplitude limiting. The sudden changes in the input signal's amplitude are suppressed by the large capacitor. The Foster–Seely discriminator does not provide amplitude limiting. The voltage E_s is

$$E_s = e_1 + e_2$$

and

$$e_0 = \frac{E_s}{2} - e_2 = \frac{e_1 + e_2}{2} - e_2$$

$$= \frac{e_1 - e_2}{2}$$

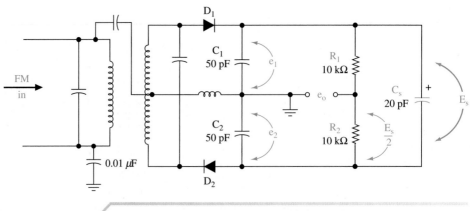

FIGURE 6-9 ▉ Ratio detector.

When $f_{in} = f_c$, $e_1 = e_2$ and hence the desired zero output occurs. When $f_{in} > f_c$, $e_1 > e_2$, and when $f_{in} < f_c$, $e_1 < e_2$. The desired frequency dependent output characteristic results.

The component values shown in Fig. 6-9 are typical for a 10.7-MHz IF FM input signal. The output level of the ratio detector is one-half that for the Foster–Seely circuit.

Quadrature Detector

The Foster–Seely and ratio detector circuits do not lend themselves to integration on a single chip due to the transformer required. This has led to increased usage of the quadrature detector and phase-locked loop (PLL). The PLL is introduced in the next section.

Quadrature detectors derive their name from use of the FM signal in phase and 90° out of phase. The two signals are said to be in *quadrature*—at a 90° angle. The circuit in Fig. 6-10 shows an FM quadrature detector using an Exclusive-OR gate. The limited IF output is applied directly to one input and the phase-shifted signal to the other. Notice that this circuit uses the limited signal that has not been changed back to a sine wave. The *L*, *C*, and *R* values used at the circuit's input are chosen to provide a 90° phase shift at the carrier frequency to the signal 2 input. The signal 2 input is a sine wave due to the *LC* circuit effects. The upward and downward frequency deviation of the FM signal results in a

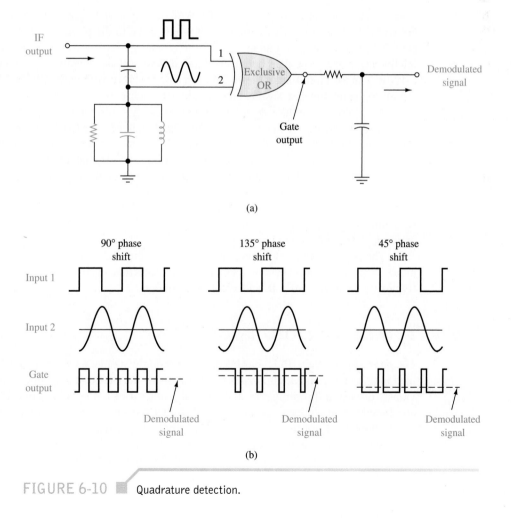

FIGURE 6-10 Quadrature detection.

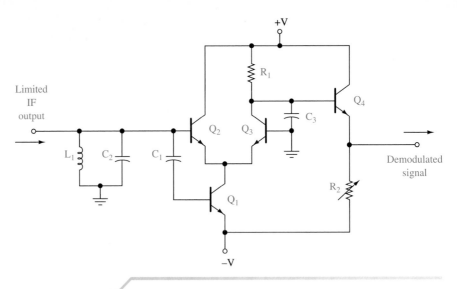

FIGURE 6-11 ▨ Analog quadrature detector.

corresponding higher or lower phase shift. With one input to the gate shifted, the gate output will be a series of pulses with a width proportional to the phase difference. The low-pass *RC* filter at the gate output "sums" the output, giving an average value that is the intelligence signal. The gate output for three different phase conditions is shown at Fig. 6-10(b). The *RC* circuit output level for each case is shown with dashed lines. This corresponds to the intelligence level at those particular conditions.

An analog quadrature detector is possible using a differential amplifier configuration, as shown in Fig. 6-11. A limited FM signal switches the transistor current source (Q_1) of the differential pair $Q_2 + Q_3$. L_1 and C_2 should be resonant at the IF frequency. The L_1–C_2–C_1 combination causes the desired frequency-dependent phase shift between the two signals applied to Q_2 and Q_1. The conduction through Q_3 depends on the coincident phase relationships of these two signals. The pulses generated at Q_3's collector are "summed" by the R_1–C_3 low-pass filter, and the resulting intelligence signal is taken at Q_4's emitter. R_2 is adjusted to yield the desired zero-volts output when an undeviating FM carrier is the circuit's input signal.

The popular 3089 LIC shown in Sec. 6-7 uses the analog quadrature detection technique. It provides an excellent total harmonic distortion (THD) specification of 0.1% (typically) for a 10.7-MHz IF and ±75-kHz deviation.

6-5 PHASE-LOCKED LOOP

The *phase-locked loop* (PLL) has become increasingly popular as a means of FM demodulation in recent years. It eliminates the need for the intricate coil adjustments of the previously discussed discriminators and has many other uses in the field of electronics. It is an example of an old idea, originated in 1932, that was given a new life by integrated circuit technology. Prior to its availability in a single IC package in 1970, its complexity in discrete circuitry form made it economically unfeasible for most applications.

The PLL is an electronic feedback control system as represented by the block diagram in Fig. 6-12. The input is to the *phase comparator* or *phase detector,* as it is also

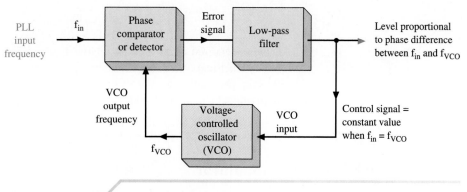

FIGURE 6-12 ■ PLL block diagram.

called. The VCO within the PLL generates the other signal applied to the comparator. The comparator "compares" the two signals and develops an output that is constant if the two input frequencies are identical. If the two are not identical, then the comparator's output, when passed through the low-pass filter, is a level that is applied to the VCO's input. This action "closes" the feedback loop since the level applied to the VCO input changes the VCO frequency in an attempt to make it exactly match the PLL input frequency. If the VCO frequency equals the input frequency, the PLL has achieved "lock" and the control voltage will be constant for as long as the PLL input frequency remains constant.

PLL Capture and Lock

If the VCO starts to change frequency, it is in the *capture* state. It then continues to change frequency until its output is the same frequency as the input. At that point, the PLL is *locked*. The PLL has three possible states of operation:

1. Free-running
2. Capture
3. Locked or tracking

If the input and VCO frequency are too far apart, the PLL free-runs at the nominal VCO frequency, which is determined by an external timing capacitor. This is not a normally used mode of operation. If the VCO and input frequency are close enough, the capture process begins and continues until the locked condition is reached. Once tracking (lock) begins, the VCO can remain locked over a wider input-frequency-range variation than was necessary to achieve capture. The tracking and capture ranges are a function of external resistors and/or capacitors selected by the user.

EXAMPLE 6-2

A PLL is set up such that its VCO free-runs at 10 MHz. The VCO does not change frequency until the input is within 50 kHz of 10 MHz. After that condition, the VCO follows the input to ±200 kHz of 10 MHz before the VCO starts to free-run again. Determine the lock and capture ranges of the PLL.

PLL FM Demodulator

If the PLL input is an FM signal, the low-pass filter output (VCO input) is the demodulated signal. The VCO input control signal (demodulated FM) causes the VCO output to match the FM signal applied to the PLL (comparator input). If the FM carrier (center) frequency drifts because of local oscillator drift, the PLL readjusts itself and no realignment is necessary. In a conventional FM discriminator, any shift in the FM carrier frequency results in a distorted output, since the *LC* detector circuits are then untuned. The PLL FM discriminator requires no tuned circuits nor their associated adjustments and "adjusts" itself to any carrier frequency drifts caused by LO or transmitted carrier drift. In addition, the PLL normally has large amounts of internal amplification, which allows the input signal to vary from the microvolt region up to several volts. Since the phase comparator responds only to phase changes and not to amplitudes, the PLL is seen to provide a limiting function of extremely wide range. The use of PLL FM detectors is widespread in current designs.

560 LIC PLL

The specifications for the 560 PLL IC are provided in Fig. 6-13. A look at the applications listed provides a clue to the versatility of the PLL. The 560 PLL provides operation for inputs up to at least 15 MHz and typically to 30 MHz at amplitudes from 100 μV to 1 V (a ratio of 10,000 to 1!). It is ideally suited for FM IF amplification and detection at the standard 10.7-MHz IF frequency. A typical FM demodulation circuit is shown in Fig. 2 of the specifications. The *tracking filter* shown in Fig. 3 has the ability to follow an input signal with a fixed bandpass, even as the signal's center frequency is varying. It can, therefore, limit the noise (by limiting bandwidth) while tracking a low-level varying-frequency signal.

Further analysis and detail of the FM detector shown in Fig. 2 of the 560 specifications will help in the understanding of PLL operation and application. The FM detector circuit shown indicates the "outboard" components required. Notice that no *LC* tuned circuits are necessary. That fact, combined with the superior linearity characteristics of PLL FM demodulators, explains their widespread use in many new designs. The outboard components R_1C_1 (used twice), C_0, and C_D are chosen to achieve the required VCO frequency, deemphasis response, lock range, threshold sensitivity, and bandwidth.

The formulas used for component calculation are provided in manufacturer's specification and applications sheets. The 560 PLL, one of the most widely used, is analyzed as follows:

1. *VCO center frequency:*

$$C_0 = \frac{3 \times 10^{-4}}{f_0}$$

DESCRIPTION

The NE560B Phase Locked Loop (PLL) is a monolithic signal conditioner, and demodulator system comprising a VCO, Phase Comparator, Amplifier and Low Pass Filter, interconnected as shown in the accompanying block diagram. The center frequency of the PLL is determined by the free running frequency (f_o) of the VCO. This VCO frequency is set by an external capacitor and can be fine tuned by an optional Potentiometer. The low pass filter, which determines the capture characteristics of the loop, is formed by the two capacitors and two resistors at the Phase Comparator output.

The PLL system has a set of self biased inputs which can be utilized in either a differential or single ended mode. The VCO output, in differential form, is available for signal conditioning frequency synchronization, multiplication and division applications. Terminals are provided for optional extended control of the tracking range, VCO frequency, and output DC level.

The monolithic signal conditioner-demodulator system is useful over a wide range of frequencies from less than 1 Hz to more than 15 MHz with an adjustable tracking range of $\pm 1\%$ to $\pm 15\%$.

FEATURES

- **FM DEMODULATION WITHOUT TUNED CIRCUITS**
- **NARROW BANDPASS - TO ± 1% ADJUSTABLE**
- **TRACKING RANGE**
- **EXACT FREQUENCY DUPLICATION IN HIGH**
- **NOISE ENVIRONMENT**
- **WIDE TRACKING RANGE ±15%**
- **HIGH LINEARITY - 1% DISTORTION MAX**
- **FREQUENCY MULTIPLICATION AND DIVISION**
- **THROUGH HARMONIC LOCKING**

APPLICATIONS

TONE DECODERS

FM IF STRIPS

TELEMETRY DECODERS

DATA SYNCHRONIZERS

SIGNAL RECONSTITUTION

SIGNAL GENERATORS

MODEMS

TRACKING FILTERS

SCA RECEIVERS

FSK RECEIVERS

WIDE BAND HIGH LINEARITY DETECTORS

ABSOLUTE MAXIMUM RATINGS

Maximum Operating Voltage	26V
Input Voltage	1V Rms
Storage Temperature	-65°C to 150°C
Operating Temperature	0°C to 70°C
Power Dissipation	300 mw

Limiting values above which serviceability may be impaired

PIN CONFIGURATION

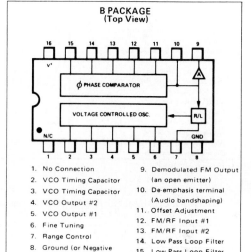

B PACKAGE
(Top View)

ORDER PART NO. NE560B

1. No Connection
2. VCO Timing Capacitor
3. VCO Timing Capacitor
4. VCO Output #2
5. VCO Output #1
6. Fine Tuning
7. Range Control
8. Ground (or Negative Power Supply)
9. Demodulated FM Output (an open emitter)
10. De-emphasis terminal (Audio bandshaping)
11. Offset Adjustment
12. FM/RF Input #1
13. FM/RF Input #2
14. Low Pass Loop Filter
15. Low Pass Loop Filter
16. Positive Power Supply

BLOCK DIAGRAM

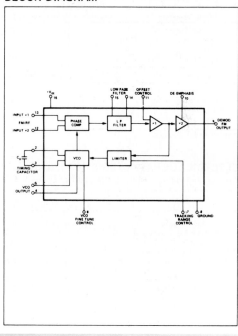

FIGURE 6-13 ■ 560 PLL Specifications. (Courtesy of Signetics.)

GENERAL ELECTRICAL CHARACTERISTICS

(15KΩ Pin 9 to GND, Input Pin 12 or Pin 13 AC Ground Unused Input, Optional Controls Not Connected, V+ = 18V Unless Otherwise Specified T_A = 25°C)

CHARACTERISTICS	LIMITS			UNITS	TEST CONDITIONS
	MIN	TYP	MAX		
Lowest Practical Operating Frequency		0.1		Hz	
Maximum Operating Frequency	15	30		MHz	
Supply Current	7	9	11	Ma	
Minimum Input Signal for Lock		100		μV	
Dynamic Range		60		dB	
VCO Temp Coefficient*		±0.06	±0.12	%/°C	Measured at 2 MHz, with both inputs AC grounded
VCO Supply Voltage Regulation		±0.3	±2	%/V	Measured at 2 MHz
Input Resistance		2		KΩ	
Input Capacitance		4		Pf	
Input DC Level		+4		V	
Output DC Level	+12	+14	+16	V	
Available Output Swing		4		Vp-p	Measured at Pin 9
AM Rejection*	30	40		dB	See Figure 1
De-emphasis Resistance		8		KΩ	

*ACC Test Sub Group C.

ELECTRICAL CHARACTERISTICS (For FM Applications, Figure 2) (15KΩ Pin 9 to GND, Input Pin 12 or 13, AC Ground Unused Input, Optional Controls Not Connected, V+ = 18V Unless Otherwise Specified T_A = 25°C)

CHARACTERISTICS	LIMITS			UNITS	TEST CONDITIONS
	MIN	TYP	MAX		
10.7 MHz Operation Deviation 75 kHz Source Impedance = 50Ω					
Detection Threshold		120	300	μV	V_{in} = 1 mv Rms Modulation Frequency 1 kHz
Demodulated Output Amplitude	30	60		mV	V_{in} = 1 mv Rms Modulation Frequency 1 kHz
Distortion*		.3	1	% T.H.D.	V_{in} = 1 mv Rms Modulation Frequency 1 kHz
Signal to Noise Ratio $\frac{S+N}{N}$		35		dB	V_{in} = 1 mv Rms Modulation Frequency 1 kHz
4.5 MHz Operation Deviation = 25 kHz, Source Impedance = 50Ω					
Detection Threshold		120	300	μV	V_{in} = 1 mv Rms Modulation Frequency 1 kHz
Demodulated Output Amplitude	30	60		mV	V_{in} = 1 mv Rms Modulation Frequency 1 kHz
Distortion		0.3	1.0	% T.H.D.	V_{in} = 1 mv Rms Modulation Frequency 1 kHz
Signal to Noise Ratio $\frac{S+N}{N}$		35		dB	V_{in} = 1 mv Rms Modulation Frequency 1 kHz
Wide Deviation ΔF/f_o = 5% Input = 4.5 MHz Deviation = 225 kHz @ 1 kHz Modulation Rate					
Detection Threshold		1	5	mV	
Demodulated Output	0.2	0.5		Vrms	V_{in} = 5 mv Rms
Distortion		0.8		% T.H.D.	V_{in} = 5 mv Rms
Signal to Noise Ratio $\frac{S+N}{N}$		50		dB	V_{in} = 5 mv Rms

*ACC Test Sub Group C.

ELECTRICAL CHARACTERISTICS (For Tracking Filter, Figure 3) (15KΩ Pin 9 to GND, Input Pin 12 or Pin 13 AC Ground Unused Input, Optional Controls Not Connected, V+ = 18V Unless Otherwise Specified T_A = 25°C)

CHARACTERISTICS	LIMITS			UNITS	TEST CONDITIONS
	MIN	TYP	MAX		
Tracking Range	±5	±15		% of f_o	V_{in} = 5 mv Rms
Minimum Signal to Sustain Lock 0°C to 70°C		0.8		mv Rms	Input 2 MHz - See Characteristic Curves
VCO Output Impedance		1		kΩ	
VCO Output Swing	0.4	0.6		Vp-p	Input 2 MHz Measured with high impedance Probe with less than 10 Pf Capacitance
VCO Output DC Level		+6.5		V	
Side Band Suppression		35		dB	Input 2 MHz with ±100 kHz Side Band Separation and 3 kHz Low Pass Filter Input 1 mv Peak for Carrier Each Side Band C_1 = 0.01 μF R_1 = 0

FIGURE 6-13 ▓ *(Continued)*

TYPICAL TEST CIRCUITS

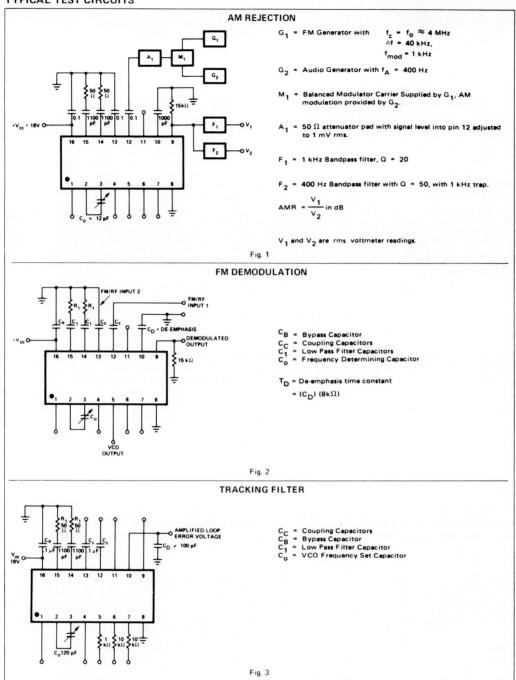

AM REJECTION

G_1 = FM Generator with $f_c = f_o \approx 4$ MHz
$\Delta f = 40$ kHz,
$f_{mod} = 1$ kHz

G_2 = Audio Generator with $f_A = 400$ Hz

M_1 = Balanced Modulator Carrier Supplied by G_1, AM modulation provided by G_2.

A_1 = 50 Ω attenuator pad with signal level into pin 12 adjusted to 1 mV rms.

F_1 = 1 kHz Bandpass filter, Q = 20

F_2 = 400 Hz Bandpass filter with Q = 50, with 1 kHz trap.

$$AMR = \frac{V_1}{V_2} \text{ in dB}$$

V_1 and V_2 are rms voltmeter readings.

Fig. 1

FM DEMODULATION

C_B = Bypass Capacitor
C_C = Coupling Capacitors
C_1 = Low Pass Filter Capacitors
C_o = Frequency Determining Capacitor

T_D = De-emphasis time constant
= (C_D) (8kΩ)

Fig. 2

TRACKING FILTER

C_C = Coupling Capacitors
C_B = Bypass Capacitor
C_1 = Low Pass Filter Capacitor
C_o = VCO Frequency Set Capacitor

Fig. 3

FIGURE 6-13 ■ *(Continued)*

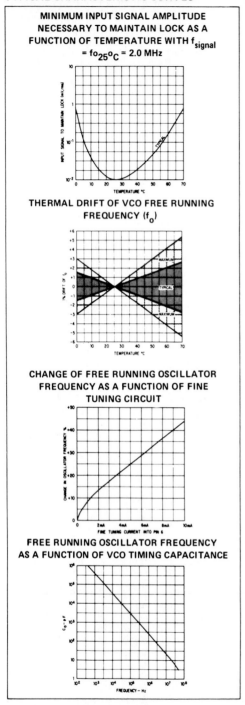

MINIMUM INPUT SIGNAL AMPLITUDE NECESSARY TO MAINTAIN LOCK AS A FUNCTION OF TEMPERATURE WITH f_{signal} = $f_{o_{25^oC}}$ = 2.0 MHz

THERMAL DRIFT OF VCO FREE RUNNING FREQUENCY (f_o)

CHANGE OF FREE RUNNING OSCILLATOR FREQUENCY AS A FUNCTION OF FINE TUNING CIRCUIT

FREE RUNNING OSCILLATOR FREQUENCY AS A FUNCTION OF VCO TIMING CAPACITANCE

AM REJECTION AS A FUNCTION OF INPUT SIGNAL LEVEL f_o = 10 MHz

TYPICAL TRACKING RANGE AS A FUNCTION OF INPUT SIGNAL

CHANGE OF FREE RUNNING OSCILLATOR FREQUENCY AS A FUNCTION OF RANGE CONTROL CURRENT

NORMALIZED TRACKING RANGE AS A FUNCTION OF RANGE CONTROL CURRENT

FIGURE 6-13 ▦ *(Continued)*

EXTERNAL CONTROLS

1. Loop Low Pass Filter (Pins 14 and 15)

The equivalent circuit for the loop low-pass filter can be represented as:

where RA (6K Ω) is the effective resistance seen looking into Pin #14 or Pin #15.

The corresponding filter transfer characteristics are:

$$\frac{V_2}{V_1}(S) = (S) = \frac{1 + S\,R_1\,C_1}{1 + S\,(R_1 + R_A)\,C_1}$$

where S is the complex frequency variable.

2. Loop Gain (Threshold) Control

The overall Phase Locked Loop gain can be reduced by connecting a feedback resistor, R_F, across the low-pass filter terminals, Pins #14 and #15. This causes the loop gain and the detection sensitivity to decrease by a factor α ($\alpha < 1$) where:

$$\alpha = \frac{R_F}{2\,R_A + R_F}$$

Reduction of loop gain may be desirable at high input signal levels ($V_{in} > 30$ mV) and at high frequencies ($f_o > 5$ MHz) where excessively high loop gain may cause instability.

3. Tracking Range Control (Pin 7)

Any bias current, I_p, injected into the tracking range control, reduces the tracking range of the PLL by decreasing the output of the limiter. The variation of the tracking range and the center frequency, as a function of I_p, are shown in the characteristic curves with I_p defined positive going into the tracking range control terminal. This terminal is normally at a DC level of +0.6 Volts and presents an impedance of 600 Ω.

4. External Fine Tuning (Pin 6)

Any bias current injected into the fine tuning terminal increases the frequency of oscillation, f_o, as shown in the characteristic curves. This current is defined Positive into the fine tuning terminal. This terminal is at a typical DC level of +1.3 Volts and has a dynamic impedance of 100Ω to ground.

5. Offset Adjustment (Pin 11)

Application of a bias voltage to the offset adjustment terminal modifies the current in the output amplifier setting the DC level at the output. The effect on the loop is to modify the relationship between the VCO free running frequency and the lock range, allowing the VCO free running frequency to be positioned at different points throughout the lock range.

Nominally this terminal is at +4V DC and has an input impedance of 3K Ω. The offset adjustment is optional. The characteristics specified correspond to operation of the circuit with this terminal open circuited.

6. De-emphasis Filter (Pin 10)

The de-emphasis terminal is normally used when the PLL is used to demodulate Frequency Modulated Audio signals. In this application, a capacitor from this terminal to ground provides the required de-emphasis. For other applications, this terminal may be used for band shaping the output signal. The 3 dB bandwidth of the output amplifier in the system block diagram (see Figure 2.) is related to the de-emphasis capacitor, C_D, as:

$$f_{3dB} = \frac{1}{2\ R_a\ C_D}$$

where R_D is the 8000 ohm resistance seen looking into the de-emphasis terminal.

When the PLL system is utilized for signal conditioning, and the loop error voltage is not utilized, de-emphasis terminal should be AC grounded.

FIGURE 6-13 ■ (Continued)

If the incoming signal is the FM IF frequency of 10.7 MHz, we have

$$C_0 = \frac{3 \times 10^{-4}}{10.7 \text{ MHz}} = 28 \text{ pF}$$

2. *Bandwidth (selectivity) of PLL:* The C_1 capacitors determine the low-pass filter characteristics of the 560 PLL if the desired bandwidth is the 15-kHz audio response of broadcast FM.

$$C_1 = \frac{13.3 \times 10^{-6}}{BW} = \frac{13.3 \times 10^{-6}}{15 \text{ kHz}} = 887 \text{ pF}$$

3. *The deemphasis function:* This is accomplished by C_D:

$$C_D = \frac{\text{desired time constant}}{8 \times 10^3}$$

For the standard 75-μs time constant,

$$C_D = \frac{75\ \mu s}{8 \times 10^3} = 9.38 \text{ nF}$$

The 8×10^3 factor comes from the pin 10 internal resistance of 8 kΩ.

4. *Lock range and threshold sensitivity:* The value of R_1 is used to control the lock range and threshold sensitivity of the VCO in the 560 PLL. These characteristics specify the extent for which the VCO can track the incoming RF signal (pin 12). The VCO control voltage is developed at the phase detector output—this control voltage is dependent on the phase difference and amplitude of the RF input signal as compared to the VCO output signal. The loop function requires a certain minimum signal level to achieve VCO tracking (phase lock). The minimum VCO input is termed *threshold sensitivity.* For a given RF input level, the frequency deviation range that provides control voltage levels greater than the threshold sensitivity is called its *lock* (or tracking) *range.* The 560 PLL is specified to have an optimum lock range of ±15% of f_0. Broadcast FM deviation of ±75 kHz represents about 1% of the 10.7-MHz IF (f_0) frequency. The lock range should therefore be reduced by increasing the value of R_1 and thereby reducing threshold sensitivity. The R_1 value is computed by

$$R_1 = \frac{12 \times 10^3}{\text{RF} - 1}$$

where RF is the lock range reduction factor (15% to 1%, or 15 in this case). Thus,

$$R_1 = \frac{12 \times 10^3}{15 - 1} = 857 \ \Omega$$

The PLL used in a frequency synthesizer is detailed in Chapter 8.

6-6 STEREO DEMODULATION

FM stereo receivers are identical to standard receivers up to the discriminator output. At this point, however, the discriminator output contains the 30-Hz to 15-kHz (L + R) signal *and* the 19-kHz subcarrier *and* the 23- to 53-kHz (L − R) signal. If a nonstereo receiver is tuned to a stereo station, its discriminator output may contain the additional frequencies, but even the 19-kHz subcarrier is above the normal audible range, and its audio amplifiers and speaker would probably not pass it anyway. Thus, the nonstereo receiver reproduces the 30-Hz to 15-kHz (L + R) signal (a full monophonic broadcast) and is not affected by the other frequencies. This effect is illustrated in Fig. 6-14.

The stereo receiver block diagram becomes more complex after the discriminator. At this point, the three signals are separated by filtering action. The (L + R) signal is obtained through a low-pass filter and given a delay so that it reaches the matrix network in step with the (L − R) signal. A 23- to 53-kHz bandpass filter "selects" the (L − R) double-side band signal. A 19-kHz bandpass filter takes the pilot carrier and is multiplied by 2 to 38 kHz, which is the precise carrier frequency of the DSB suppressed carrier 23- to 53-kHz (L − R) signal. Combining the 38-kHz and (L − R) signals through the nonlinear device of an AM detector generates sum and difference outputs of which the 30-Hz to 15-kHz (L − R) components are selected by a low-pass filter. The (L − R) signal is thereby retranslated back down to the audio range and it and the (L + R) signal are applied to the matrix network for further processing.

Figure 6-15 illustrates the matrix function and completes the stereo receiver block diagram of Fig. 6-14. The (L + R) and (L − R) signals are combined in an adder that cancels R since (L + R) + (L − R) = 2L. The (L − R) signal is also applied to an inverter, providing −(L − R) = (−L + R), which is subsequently applied to another adder along with

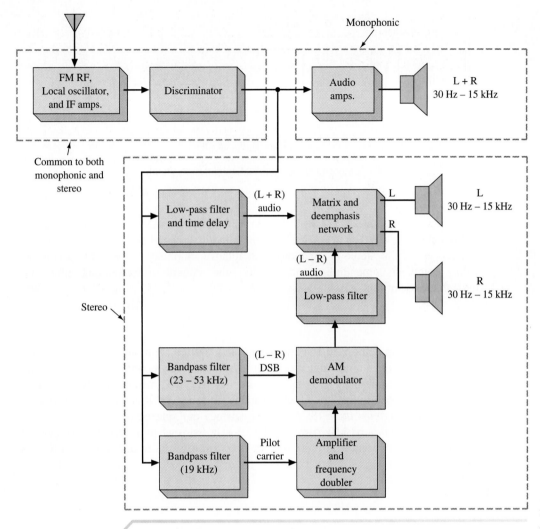

FIGURE 6-14 ▮ Monophonic and stereo receivers.

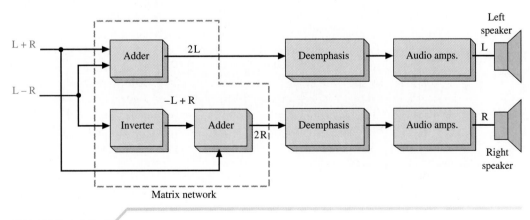

FIGURE 6-15 ▮ Stereo signal processing.

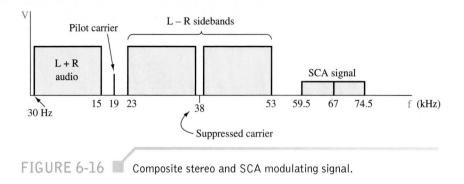

FIGURE 6-16 ◼ Composite stereo and SCA modulating signal.

$(L + R)$, which produces $(-L + R) + (L + R) = 2R$. The two individual signals for the right and left channels are then deemphasized and individually amplified to their own speaker. The process of FM stereo is ingenious in its relative simplicity and effectiveness in providing complete compatibility and doubling the amount of transmitted information through the use of multiplexing.

SCA Decoder

The FCC has also authorized FM stations to broadcast an additional signal on their carrier. It may be a voice communication or other signal for any nonbroadcast-type use. It is often used to transmit music programming that is usually commercial-free but paid for by subscription of department stores, supermarkets, and the like. It is termed the *subsidiary communication authorization* (SCA). It is frequency-multiplexed on the FM modulating signal, usually with a 67-kHz carrier and ±7.5-kHz (narrowband) deviation, as shown in Fig. 6-16. An SCA decoder circuit using the 565 PLL is provided in Fig. 6-17. A resistive

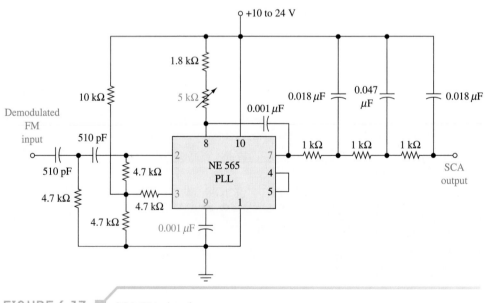

FIGURE 6-17 ◼ SCA PLL decoder.

voltage divider is used to establish a bias voltage for the input (pins 2 and 3). The demodulated FM signal is fed to the input through a two-stage high-pass filter (510 pF, 4.7 kΩ, 510 pF, 4.7 kΩ), both to allow capacitive coupling and to attenuate the stronger level of the stereo signals. The PLL is tuned to about 67 kHz, with the 0.001-μF capacitor from pin 9 to ground and the 5-kΩ potentiometer providing fine adjustment. The demodulated output at pin 7 is fed through a three-stage low-pass filter to provide deemphasis and attenuate the high-frequency noise that often accompanies SCA transmission.

LIC Stereo Decoder

The decoding of the stereo signals is normally accomplished via special function ICs. The RCA CA3090 is such a device with a functional block diagram provided in Fig. 6-18. The input signal from the detector is amplified by a low-distortion preamplifier and simultaneously applied to both the 19- and 38-kHz synchronous detectors (see Sec. 3-2). A 76-kHz signal, generated by a local voltage-controlled oscillator (VCO), is counted down by two frequency dividers to a 38-kHz signal and to two 19-kHz signals in phase quadrature. The 19-kHz pilot tone supplied by the FM detector is compared to the locally generated 19-kHz signal in the synchronous detector. The resultant signal controls the voltage-controlled oscillator so that it produces an output signal to phase-lock the stereo decoder with the pilot tone. A second synchronous detector compares the locally generated 19-kHz signal with the 19-kHz pilot tone. If the pilot tone exceeds an externally adjustable threshold voltage, a Schmitt trigger circuit is energized. The signal from the Schmitt trigger lights the stereo indicator, enables the 38-kHz synchronous detector, and

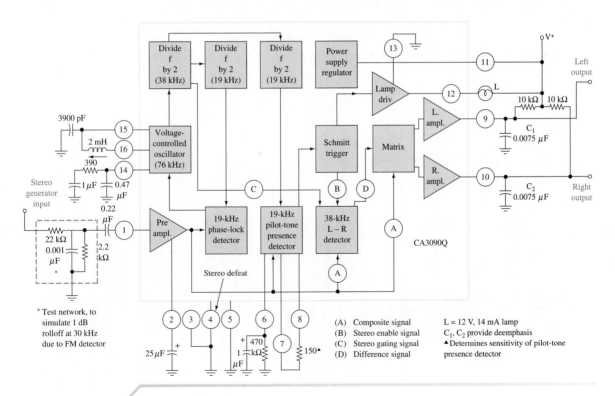

FIGURE 6-18 ▮ CA3090 stereo decoder. (Courtesy of RCA.)

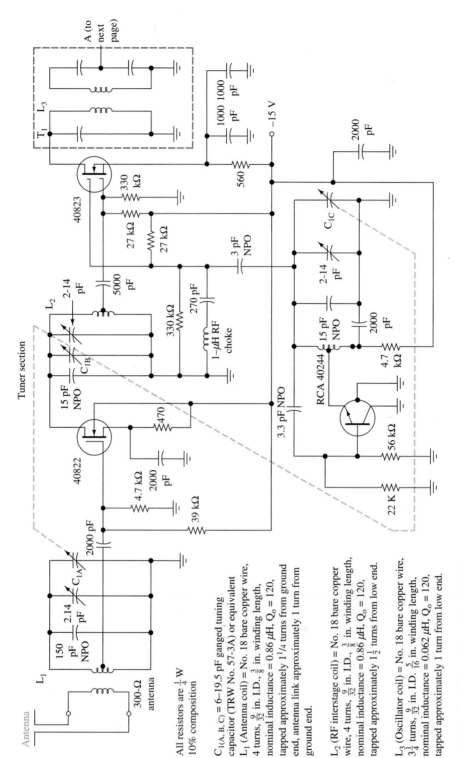

All resistors are $\frac{1}{4}$ W
10% composition

$C_{1(A, B, C)} = 6$–19.5 pF ganged tuning
capacitor (TRW No. 57-3A) or equivalent
L_1 (Antenna coil) = No. 18 bare copper wire,
4 turns, $\frac{9}{32}$ in. I.D., $\frac{3}{8}$ in. winding length,
nominal inductance = 0.86 μH, $Q_o = 120$,
tapped approximately $1\frac{1}{4}$ turns from ground
end, antenna link approximately 1 turn from
ground end.

L_2 (RF interstage coil) = No. 18 bare copper
wire, 4 turns, $\frac{9}{32}$ in. I.D., $\frac{3}{8}$ in. winding length,
nominal inductance = 0.86 μH, $Q_o = 120$,
tapped approximately $1\frac{1}{2}$ turns from low end.

L_3 (Oscillator coil) = No. 18 bare copper wire,
$3\frac{1}{4}$ turns, $\frac{9}{32}$ in. I.D. $\frac{5}{16}$ in. winding length,
nominal inductance = 0.062 μH, $Q_o = 120$,
tapped approximately 1 turn from low end.

T_1 (Input transformer to IF amplifier) = TRW
No. 21124-R2 (Design characteristics given in
RCA Application Note AN-3466)

FIGURE 6-19 ■ Complete 88- to 108-MHz stereo receiver.

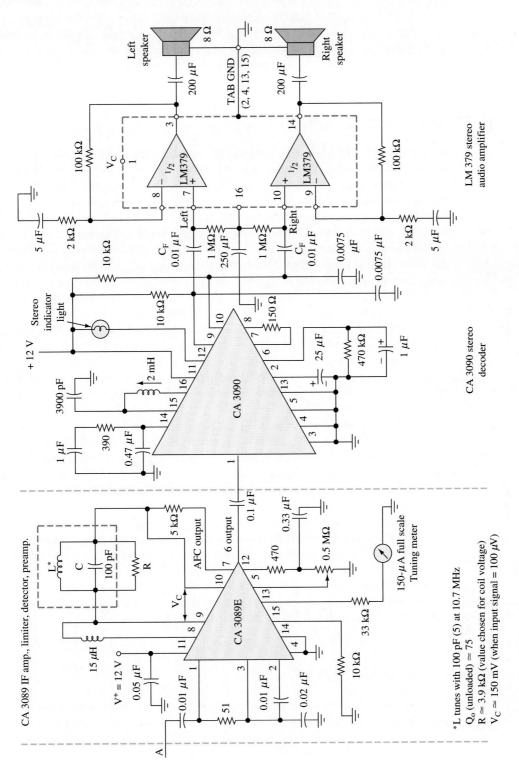

FIGURE 6-19 (Continued)

automatically switches the CA3090 from monaural to stereo operation. The output signal from the 38-kHz detector and the composite signal from the preamplifier are applied to a matrixing circuit, from which emerge the resultant left and right channel audio signals. These signals are applied to their respective left and right channel amplifiers for amplification to a level sufficient to drive most audio power amplifiers.

6-7 FM RECEIVERS

A typical quality FM receiver involves use of discrete MOSFET RF and mixer stages with a separately excited bipolar transistor local oscillator, as shown in Fig. 6-19. The antenna input signal is applied through the tuning circuit L_1, C_{1A} to the gate of the 40822 MOSFET RF amplifier. Its output at the drain is coupled to the lower gate of the 40823 mixer MOSFET through the C_{1B}–L_2 tuned circuit. The 40244 BJT oscillator signal is applied to the upper gate of the mixer stage. The local oscillator tuned circuit that includes C_{1C} uses a tapped inductor indicating a Hartley oscillator configuration. The tuning condensor, C_1, has three separate ganged capacitors that vary the tuning range of the RF amp and mixer tuned circuits from 88 to 108 MHz while varying the local oscillator frequency from 98.7 to 118.7 MHz to generate a 10.7-MHz IF signal at the output of the mixer. The mixer output is applied to the commercially available 10.7-MHz double-tuned circuit T_1.

MOSFET receiver front ends offer superior cross-modulation and intermodulation performance as compared to other types, as explained in Sec. 6-2. The *Institute of High Fidelity Manufacturers* (IHFM) sensitivity for this front end is about 1.75 μV. It is defined as the minimum 100% modulated input signal that reduces the total receiver noise and distortion to 30 dB below the output signal. In other words, a 1.75-μV input signal produces 30-dB *quieting*.

The front-end output through T_1 in Fig. 6-19 is applied to a CA3089 IC. The CA3089 provides three stages of IF amplification—limiting, demodulation, and audio preamplification. It provides demodulation with an analog quadrature detector circuit (Sec. 6-4). It also provides a signal to drive a tuning meter and an AFC output for direct control of a varactor tuner. Its audio output includes the 30-Hz to 15-kHz (L + R) signal, 19-kHz pilot carrier, and 23- to 53-kHz (L − R) signal, which are then applied to the FM stereo decoder IC, the CA3090. The CA3090 was explained in Sec. 6-6 and provides the separated left and right channel outputs as well as a signal to light a stereo indicator light.

The CA3090 audio outputs are then applied to a dual ganged volume control potentiometer (not shown) and then to a LM379 dual 6-W audio amplifier. It has two separate audio amplifiers in one 16-lead IC and has a minimum input impedance of 2 MΩ per channel. It typically provides a voltage gain of 34 dB, *total harmonic distortion* (THD) of 0.07% at 1-W output, and 70 dB of channel separation.

TROUBLESHOOTING

The basic approach to troubleshooting the FM receiver is no different from that of an AM receiver. The FM radio is a superheterodyne receiver like the AM receiver with a few differences. The methods you learn in this section will teach you how to isolate defects in the FM receiver. Upon completing this section you should be able to

- Identify defective stages in an FM receiver
- Describe the principal test setup for checking each receiver stage
- Troubleshoot a quadrature detector
- Test a semiconductor diode junction

The FM Receiver

Figure 6-20 represents a typical FM radio receiver. Troubleshooting begins at the limiter stage. Feed in a test signal at the input of the limiter stage, point A in Fig. 6-20. Based on the results of this test the signal injection point will move toward the antenna or toward the audio section. We will assume for the sake of this discussion that the trouble complaint is a dead broadcast band FM receiver and the power supply is good. The signal generator used for this procedure must be capable of producing a test signal at the operating frequencies of the FM receiver and at its IF. In addition, an audio signal is needed to modulate the FM test signal.

Locating a Defective Stage

Feed a modulated test signal at 10.7 MHz (the FM receiver's IF frequency) into the limiter stage. Use a 400- to 1000-Hz signal as the modulating signal. Set the output signal amplitude of the signal generator to approximately 4 V peak-to-peak (consult the service literature for the exact signal levels to inject for each stage being checked). Wobble the frequency control to each side of the IF center frequency of 10.7 MHz

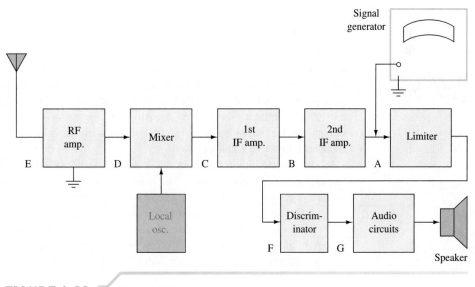

FIGURE 6-20 ■ Typical FM receiver.

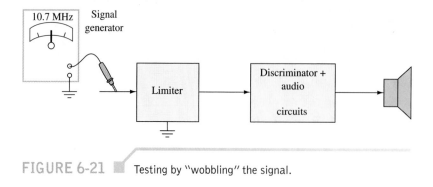

FIGURE 6-21 Testing by "wobbling" the signal.

(see Fig. 6-21). If the limiter, discriminator, and audio amplifier circuits are operating correctly then the wobbled signal will be heard at the speaker as a sound that goes from loud to low as the dial is changed on and off the IF frequency.

With the limiter, discriminator, and audio circuits working properly, move the injected signal to the input of the last IF amplifier, point B in Fig. 6-20. Lower the signal in amplitude and feed it into the base of the IF transistor circuit. For IF circuits composed of IC chips, connect the test signal to the proper input pin on the IC. Consult the service literature for proper signal input points and signal strength. Failure to hear the note indicates that the IF stage being tested is bad. If the note is heard at the speaker output, then move the test signal to point C. Continue in this fashion until the defective stage is located. The stage where the test tone is no longer heard is the defective circuit. If the local oscillator or the mixer is determined to be at fault, then troubleshoot these sections as described in the Chapter 3 Troubleshooting section.

Quadrature Detectors

Many FM receivers use quadrature detectors because no adjustment is required if the IF is aligned to the detector. If found in a communications receiver, they will usually be operated at 455 kHz and if in a television receiver, the frequency will be 4.5 MHz. The quadrature detector is usually included in an integrated circuit that combines the IF amplifier and the demodulator (see Fig. 6-22). Although the quadrature detector is usually a narrowband device, some home broadcast receivers use them at 10.7 MHz. In that case, C1 might actually be an inductor.

The crystal-like component X1 is not a quartz crystal. It will be either a ceramic resonator or an *LC* tuned circuit. The filter will probably be a ceramic type, too. Remember, in a communications receiver this type of IF system usually runs at 455 kHz, so an oscilloscope can be used to measure all voltages. The IF input will be less than 1 mV and will be difficult to see. Recovered audio should be around 0.25 V.

A modulated signal generator can be used to provide a test signal. These are narrowband circuits, so the modulating signal should be about 1 kHz and the deviation set to 5 kHz.

First look for the IF signal at the filter's input and output. A RF voltmeter such as the Boonton 91D (an old model but still around) may be necessary because the voltage will be in the order of millivolts. The signal at the filter output will be about one-half the

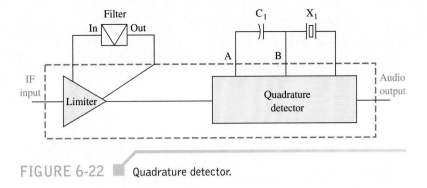

FIGURE 6-22 ▮ Quadrature detector.

input. Sweep the signal generator ±20 kHz from the center frequency and verify that the filter attenuates signals more than 7.5 kHz from the center frequency.

At the center frequency, the voltage at point B should be 90° out of phase with the voltage at point A. This can be observed with your oscilloscope. If you have a single channel oscilloscope, connect point A to the external sync. Note the position of the trace when looking at point A. The voltage at point B should be delayed from A by one-quarter of a sinewave.

If you do not have an oscilloscope, you can at least observe a peak in the voltage at point B when the signal generator is tuned to center frequency.

Typical problems are:

1. C1 is open or its value has changed. This can cause no output or distorted audio output.

2. Improper alignment. The center frequency of the IF amplifier must be the same as the tuned circuit, X1. One or both may be adjustable. Look for open coils or capacitors that have changed value.

3. Quadrature detectors sometimes have a resistor in the tuned circuit to lower the Q. If the resistor is missing or open, the Q will be too high, causing distorted audio.

Discriminators

Discriminators or ratio detectors (Figs. 6-7 and 6-9) are used in better broadcast receivers. These circuits may drift out of alignment over time and coils or capacitors may short or open. All cause distorted or no audio. Alignment is best done with a sweep generator, but in a pinch apply a fully modulated signal to the receiver and adjust the tuning for minimum distortion.

Stereo Demodulator

When troubleshooting a stereo demodulator (probably an IC), the first thing to look for is proper signal input level. An acceptable input is about 100-mV rms.

Make sure the 76-kHz oscillator is present and on frequency. Look for an open tuning capacitor or inductor if the oscillator is not running or cannot be properly adjusted. Check dc voltages on bypass capacitors. If these capacitors leak, the internal IC bias will be wrong and the IC will not work.

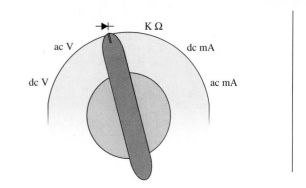

FIGURE 6-23 ■ Diode test range.

Testing Diodes and Transistors

To test a diode that you suspect is defective, set your DMM to the diode test function, as shown in Fig. 6-23. The procedure for testing diodes and transistors is illustrated in Fig. 6-24 (a) and (b). With the DMM in the diode test position, make a reading, then switch the test leads to the opposite ends of the diode and make another reading. Judge the diode based on the following criteria: if one reading shows a value and the other reading shows over range, then the device is good. If both readings result in over range, the device is open. If the two readings are zero or very low, the device is shorted. Refer to Table 6-1 for typical readings. The first and second readings may be in reverse order.

Testing the transistor is very similar to testing the diode if we think of the transistor as two diodes back-to-back. The transistor's collector–base junction is tested as if it were one diode, and the base–emitter junction is tested as another diode, shown in Fig. 6-24(b). Apply the DMM's test leads across the transistor's collector and base connections and make the first reading. Reverse the test leads and make the second reading. Compare your results to the readings in Table 6-1. Next, apply the DMM's test leads to the transistor's base and emitter connections. Make the first reading, then reverse the test leads and make the second reading. Compare the results to the readings in Table 6-1. Good readings should

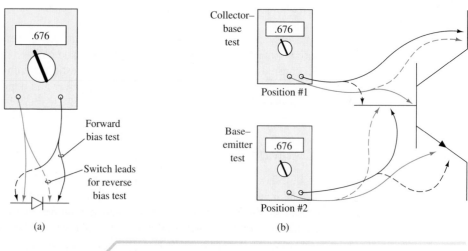

FIGURE 6-24 ■ Diode and transistor testing.

TABLE 6-1 *Displayed Values for a Diode Test*

First reading	Second reading	Conclusion
Approximately 0.4 V	Over range	Good germanium type
Approximately 0.6 V	Over range	Good silicon type
Over range	Over range	Device is open
Very small or zero	Very small or zero	Device is shorted

be fairly close to those in the table. For example, if your readings were 0.676 V and over range, this would indicate a good semiconductor junction. Consult the booklet that came with your DMM for the particular values of readings you may get. Remember, two over-range readings or two small-value readings indicate a defective semiconductor junction.

SUMMARY

In Chapter 6 we discussed the basis of an FM receiver and showed the similarities and differences compared to an AM receiver. The major topics the student should now understand include:

- the operation of an FM receiver using a block diagram as a guide, including complete descriptions of the discriminator, the deemphasis network, and the limiter functioning as AGC

- the benefits of RF amplifiers, including image frequency attenuation and local oscillator reradiation effects

- the detailed functioning of a transistor limiter circuit

- the description and comparison of slope detector, Foster–Seely discriminator, ratio detector, and quadrature detector circuits

- the description and operation of a phase-locked-loop (PLL) FM demodulator, including its three possible states

- the analysis of a stereo FM demodulation process using a block diagram

- the operation of the subsidiary communication authorization (SCA) decoder operation

- the operation of a complete 88–108-MHz stereo FM receiver by analysis of the schematic

QUESTIONS AND PROBLEMS

SECTION 1

*1. What is the purpose of a discriminator in an FM broadcast receiver?
2. Explain why the automatic frequency control (AFC) function is usually not necessary in today's FM receivers.

*3. Draw a block diagram of a superheterodyne receiver designed for reception of FM signals.
4. The local FM stereo rock station is at 96.5 MHz. Calculate the local oscillator frequency and the image frequency for a 10.7-MHz IF receiver. (107.2 MHz, 117.9 MHz)

Section 2

5. How does the noise reduction capability of a communications system affect its ultimate sensitivity rating?
6. Explain the desirability of a RF amplifier stage in FM receivers as compared to AM receivers. Why is this not generally true at frequencies over 1 GHz?
7. Describe the meaning of *local oscillator reradiation,* and explain how a RF stage helps to prevent it.
8. Why are FETs preferred over other devices as the active elements for RF amplifiers?
9. List two advantages of using a dual-gate MOSFET over a JFET in RF amplifiers.
10. Explain the need for the radio-frequency choke (RFC) in the RF amplifier shown in Fig. 6-2.

Section 3

*11. What is the purpose of a limiter stage in an FM broadcast receiver?
*12. Draw a diagram of a limiter stage in an FM broadcast receiver.
13. Explain fully the circuit operation of the limiter shown in Fig. 6-3.
14. Explain why a limiter minimizes or eliminates the need for the AGC function.
15. What is the relationship among limiting, sensitivity, and quieting for an FM receiver?
16. An FM receiver provides 100 dB of voltage gain prior to the limiter. Calculate the receiver's sensitivity if the limiter's quieting voltage is 300 mV. (3 μV)

Section 4

17. Draw a schematic of an FM slope detector and explain its operation. Why is this method not often used in practice?
18. Draw a schematic of a Foster–Seely discriminator, and provide a step-by-step explanation of what happens when the input frequency is below the carrier frequency. Include a phase diagram in your explanation.
*19. Draw a diagram of an FM broadcast receiver detector circuit.
*20. Draw a diagram of a ratio detector and explain its operation.
21. Explain the relative merits of the Foster–Seely and ratio detector circuits.
*22. Draw a schematic diagram of each of the following stages of a superheterodyne FM receiver:
(a) Mixer with injected oscillator frequency.
(b) IF amplifier.
(c) Limiter.
(d) Discriminator.
Explain the principles of operation. Label adjacent stages.
23. Describe the process of quadrature detection.

24. Draw a block diagram of a phase-locked loop (PLL) and briefly explain its operation.
25. Explain in detail how a PLL is used as an FM demodulator.
26. List the three possible states of operation for a PLL and explain each one.
27. A PLL's VCO free-runs at 7 MHz. The VCO does not change frequency until the input is within 20 kHz of 7 MHz. After that condition the VCO follows the input to ±150 kHz of 7 MHz before the VCO starts to free-run again. Determine the PLL's lock and capture ranges. (300 kHz, 40 kHz)
28. Draw a schematic of the 560 PLL of Fig. 6-13 used as an FM demodulator. Pick C_0 for operation at 10.7 MHz, C_D to provide a 75-μs deemphasis time constant, C_1 to provide a 15-kHz audio bandwidth, and R_1 to accommodate a ±75-kHz RF deviation. What is the typically required input signal, and how much output is to be expected? (28 pF, 3.13 nF, 2.66 nF, 190 Ω, 120 μV, 60 mV)

SECTION 6

29. Draw a block diagram of an FM stereo demodulator. Explain in detail the function of the AM demodulator and the matrix network. Make a circuit addition that simply energizes a light to indicate reception of a stereo station.
30. Explain how separate left and right channels are obtained from the (L + R) and (L − R) signals.
*31. What is SCA? What are some possible uses of SCA?
32. Determine the maximum reproduced audio signal frequency in a SCA system. Why does SCA cause less FM carrier deviation, and why is it thus less noise resistant than standard FM? (*Hint:* Refer to Fig. 6-16.) (7.5 kHz)
33. Explain the principle of operation for the CA3090 stereo decoder.

SECTION 7

34. The receiver front end in Fig. 6-19 is rated to have noise below the signal by 30 dB in the output with a 1.75-μV input. Calculate its output *S/N* ratio with a 1.75-μV input signal. (31.6 to 1)
35. The LIC dual audio amplifiers in Fig. 6-19 are rated to provide 70 dB of channel separation. If the left channel has 1 W of output power, calculate the wattage of the right channel that is included. (0.1 μW)

TELEVISION

OBJECTIVES

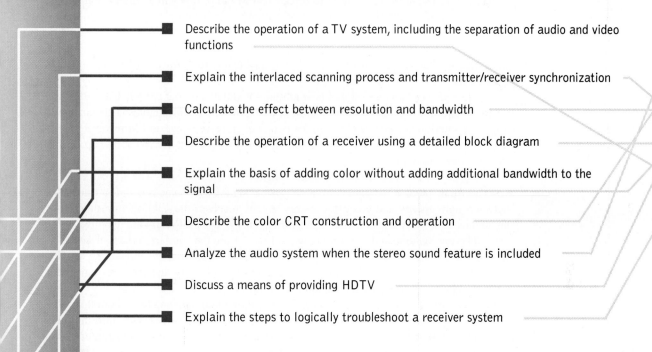

- Describe the operation of a TV system, including the separation of audio and video functions

- Explain the interlaced scanning process and transmitter/receiver synchronization

- Calculate the effect between resolution and bandwidth

- Describe the operation of a receiver using a detailed block diagram

- Explain the basis of adding color without adding additional bandwidth to the signal

- Describe the color CRT construction and operation

- Analyze the audio system when the stereo sound feature is included

- Discuss a means of providing HDTV

- Explain the steps to logically troubleshoot a receiver system

7-1 INTRODUCTION

Television is a field of electronic technology that has more direct effect on the people of our world than any other. It is a very specialized branch of technology that utilizes many of the principles already explained and many new ones.

The concept of television was developed in the 1920s, feasibility was shown in the 1930s, commercial broadcasting started in the 1940s, and the ensuing years have seen the mushrooming growth of an industry so far-reaching that some sociologists make the study of its effects their life's work. The technology, while still undergoing continued improvements, has reached a certain level of maturity. Today's 27-in color TV that can be purchased for $500 contains the complexity of industrial equipment selling for $5000. Mass-production techniques and specialized ICs utilized in TV assembly enable the average consumer to afford this truly sophisticated piece of electronics.

Photo: Flat-screen TV. (Courtesy of Fujitsu Microelectronics, Inc.)

7-2 TRANSMITTER PRINCIPLES

A TV transmitter is actually two separate transmitters. The *aural* or sound transmitter is actually an FM system very similar to broadcast FM radio. It is still a "high-fidelity" system since the same 30-Hz to 15-kHz audio range is transmitted. The major difference between broadcast FM and TV audio systems is that TV uses a ±25-kHz deviation, while you will recall broadcast FM uses a ±75-kHz deviation. Thus, the TV aural signal has the same fidelity but is less effective in cancelling the indirect noise effects explained in Chapter 5.

The *video* or "picture" signal is amplitude-modulated onto a carrier. Thus, the composite transmitted signal is a combination of both AM and FM principles. This is done to minimize interference effects between the two at the receiver since an FM receiver is relatively insensitive to amplitude modulation and an AM receiver has rejection capabilities to frequency modulation.

Figure 7-1 shows a simplified block diagram for a TV system. The TV camera converts a visual "picture" or scene into an electrical signal. The camera is thus a transducer between light energy and electrical energy. At the receiver, the CRT picture tube is the analogous transducer that converts the electrical energy back into light energy.

The microphone and speaker shown in Fig. 7-1 are the similarly related transducers for the sound transmission. There are actually two more transducers shown, the sending and receiving antennas. They convert between electrical energy and the electromagnetic energy required for transmission through the atmosphere.

The *diplexer* shown in Fig. 7-1 feeding the transmitter antenna feeds both the visual and aural signals to the antenna while not allowing either to be fed back into the other transmitter. Without the diplexer, the low-output impedance of either transmitter power amplifier would dissipate much of the output power of the other transmitter. The synchronizing signal block will be explained in the next section.

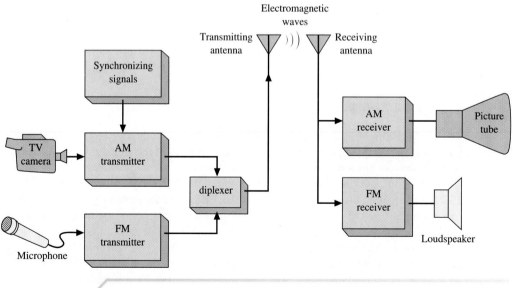

FIGURE 7-1 Simplified TV system.

TV Cameras

The TV camera is optically focused so that the scene to be transmitted appears on its light-sensitive area. The early TV cameras were the *iconoscope* and the *image orthicon*. The most widely used type today is the *vidicon*. It has comparable performance to its forerunners but is much less costly and more compact in size. Figure 7-2 shows a sketch of a typical vidicon. Notice the three very thin layers at the tube's front surface. These are the light-active areas. The first is a transparent conductive film. Then a semiconductor photoresistive layer is deposited on the conductive film. A photoconductive *mosaic* layer is then deposited on the semiconductor layer. The middle semiconductor layer exhibits very high resistance when dark, but it reduces greatly when struck by photons of light. The photoconductive mosaic has around a million individual separate areas that act as tiny capacitors, with the photoresistive layer acting as the dielectric and the conductive film as the other common electrode. The dielectric leakage is thus variable and dependent on the amount of light striking each area (capacitor).

The electron beam scans across (and slowly down) the mosaic areas so as to charge up each of the many tiny capacitors. Light on the mosaic areas discharges the capacitors through the load resistor R in Fig. 7-2. The scanning electron beam (developed by the cathode K and three grids) recharges the mosaic capacitors and produces a video signal voltage drop across R that is proportional to the light intensity at the individual areas being scanned. Thus, the video output is a signal whose instantaneous output is the result of scanning just one of the tiny capacitors at a time.

Less demanding camera applications are handled via charge-coupled devices (CCD). The unit pictured in Fig. 7-3 is designed for camcorder cameras but is also useful for security monitors, video phones, imaging, and video conferencing. Its color imaging capabilities include 410,000 pixels in a $\frac{1}{4}$-in area. Pixel pitch is 4.9 by 5.6 μm. Pixel density is 811 pixels on the horizontal axis and 507 on the vertical axis. Despite its high resolution, this CCD fits into a standard 14-pin DIP.

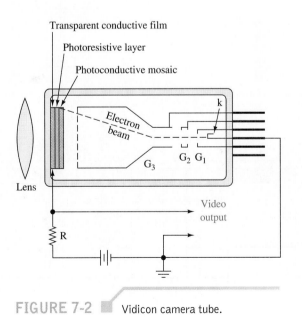

FIGURE 7-2 ▪ Vidicon camera tube.

FIGURE 7-3 ■ CCD imaging unit.
(Courtesy of Sharp Electronics Corp.)

Scanning

To understand how these tiny individual outputs can serve to represent an entire scene, refer to Fig. 7-4. In this simplified system the camera focuses the letter "T" onto the capacitors of the vidicon, but instead of a million capacitors this system has just 30, arranged in 6 rows with 5 capacitors per row. Each separate area is called a *pixel,* which is short for picture element. The greater the number of pixels, the greater can be the quality (or resolution) of the transmitted picture.

The letter "T" is focused on the light-sensitive area such that all of rows 1 and 6 are illuminated [Fig. 7-4(b)], while all of row 2 is dark and the centers of rows 3, 4, and 5 are dark. Now, if the electron beam is made to scan each row sequentially and if the *retrace*

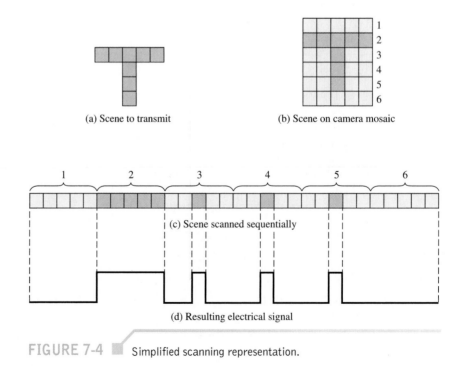

(a) Scene to transmit

(b) Scene on camera mosaic

(c) Scene scanned sequentially

(d) Resulting electrical signal

FIGURE 7-4 ■ Simplified scanning representation.

time is essentially zero, then Fig. 7-4(c) shows the sequential breakup of information. The *retrace* interval is the time it takes the moving electron beam to move from the end of one line back to the start of the next lower line. It is usually accomplished very rapidly. The variable light on the photoresistive capacitor dielectric causes a variable capacitive discharge, which results in a similar variable voltage being developed at the vidicon's output, as shown in Fig. 7-4(d). The visual scene has been converted to a video (electrical) signal and can now be suitably amplified and used to amplitude-modulate a carrier for broadcast.

The picture for broadcast TV has been standardized at a 4:3 ratio of the width to height. This is termed the *aspect ratio* and was selected as the most pleasing picture orientation to the human eye.

7-3 TRANSMITTER/RECEIVER SYNCHRONIZATION

When the video signal is detected at the receiver, some means of *synchronizing* the transmitter and receiver is necessary:

1. When the TV camera starts scanning line 1, the receiver must also start scanning line 1 on the CRT output display. You do not want the top of a scene appearing at the center of the TV screen.

2. The speed that the transmitter scans each line must be exactly duplicated by the receiver scanning process to avoid distortion in the receiver output.

3. The *horizontal retrace,* or time when the electron beam is returned back to the left-hand side to start tracing a new line, must occur coincidentally at both transmitter and receiver. You do not want the horizontal lines starting at the center of the TV screen.

4. When a complete set of horizontal lines has been scanned, moving the electron beam from the end of the bottom line to the start of the top line (vertical flyback or retrace) must occur simultaneously at both transmitter and receiver.

Visual transmissions are more complex than audio because of these synchronization requirements. At this point, voice transmission seems elementary because it can be sent on a continuous basis without synchronization. Thus, the other major function of the transmitter besides developing the video and audio signals is to generate synchronizing signals that can be used by the receiver so that it stays in step with the transmitter.

In the scanning process, the electron beam for both transmitter and receiver starts at the upper left-hand corner and sweeps horizontally to the right side. It then is rapidly returned to the left side, and this interval is termed *horizontal retrace.* An appropriate analogy to this process is the movement of your eye as you read this line and rapidly retrace to the left and drop slightly for the next line. When all the horizontal lines have been traced, the electron beam must move from the lower right-hand corner up to the upper left-hand corner for the next "picture." This *vertical retrace interval* is analogous to the time it takes the eye to move from the bottom of one page to the top of the next.

FCC regulations stipulate that U.S. TV broadcasts shall consist of 525 horizontal scanning lines. Of these, about 40 lines are lost as a result of the vertical retrace interval. This leaves 485 visible lines that you can actually see if a TV screen is viewed at close range. The number of visible lines does not depend on the TV screen size. Because this scanning occurs rapidly, persistence of vision and CRT phosphor persistence cause us to perceive these 485 lines as a complete image.

Interlaced Scanning

The *frame frequency* is the number of times per second that a complete set of 485 lines (complete picture) is traced. That rate for broadcast TV is 30 times per second. Stated another way, a complete scene (frame) is traced every $\frac{1}{30}$ s (second). Thirty frames per second is not enough to keep the human eye from perceiving *flicker* as a result of a non-continuous visual presentation. This flicker effect is observed when watching old-time movies. If the frame frequency were increased to 60 per second, the flicker would no longer be apparent, but the video signal bandwidth would have to be doubled. Instead of that solution, the process of *interlaced scanning* is used to "trick" the human eye into thinking it is seeing 60 pictures per second.

Figure 7-5 illustrates the process of interlaced scanning. The first set of lines (the first *field*) is traced in $\frac{1}{60}$ s, and then the second set of lines (the second *field*) that comprises a full scene (485 lines total) is interleaved between the first lines in the next $\frac{1}{60}$ s. Therefore, lines 2, 4, 6, etc., occur during the first field, with lines 1, 3, 5, etc., interleaved between the even-numbered lines. The field frequency is thus 60 Hz with a frame frequency of 30 Hz. This illusion is enough to convince the eye that 60 pictures per second occur when, in fact, there are only 30 full pictures per second.

The process of interlacing in TV is analogous to a trick used in motion picture projection to prevent flicker (noncontinuous motion). In motion pictures the goal is to conserve film rather than bandwidth, and this is accomplished by flashing each of the

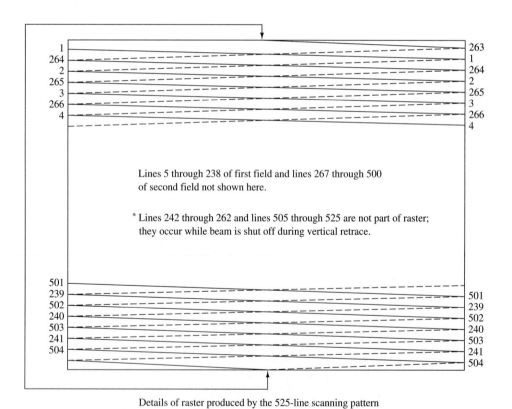

Details of raster produced by the 525-line scanning pattern

FIGURE 7-5 Interlaced scanning.

24 frames per second onto the screen twice so as to create the illusion of 48 pictures per second.

Horizontal Synchronization

To accommodate the 525 lines (485 visible) every $\frac{1}{30}$ s, the transmitter must send a synchronization (sync) pulse between every line of video signal so that perfect transmitter-receiver synchronization is maintained. The detail of these pulses is shown in Fig. 7-6. Three horizontal sync pulses are shown along with the video signal for two lines. The actual horizontal sync pulse rides on top of a so-called blanking pulse, as shown in the figure. The blanking pulse is a strong enough signal so that the electron beam retrace at the receiver will be "blacked" out and thus invisible to the viewer. The interval before the horizontal sync pulse appears on the blanking pulse is termed the *front porch,* while the interval after the end of the sync pulse, but before the end of the blanking pulse, is called the *back porch.* Notice in Fig. 7-6 that the back porch includes an eight-cycle sine-wave burst at 3,579,545 Hz. It is appropriately called the *color burst,* as it is used to calibrate the receiver color subcarrier generator. Further explanation on it will be provided in Sec. 7-10. Naturally enough, a black-and-white broadcast does not include the color burst.

The two lines of video picture signal shown in Fig. 7-6 can be described as follows:

Line 2: It starts out nearly full black at the left-hand side and gradually lightens to full white at the right-hand side.

Line 4: It starts out medium gray and stays there until one-third of the way over, when it gradually becomes black at the picture center. It suddenly shifts to white and gradually turns darker gray at the right-hand side.

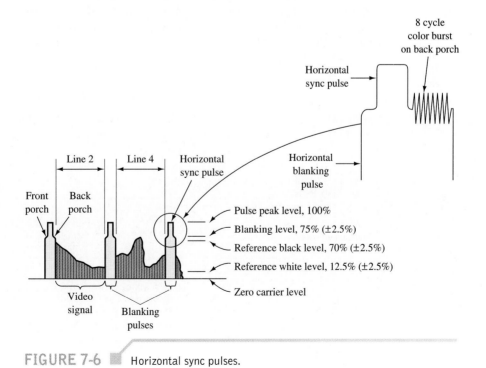

FIGURE 7-6 ▓ Horizontal sync pulses.

Since the horizontal sync pulses occur once for each of the 525 lines every $\frac{1}{30}$ s, the frequency of these pulses will be

$$525 \times 30 = 15.75 \text{ kHz}$$

Thus, both transmitter and receiver must contain 15.75-kHz *horizontal* oscillators to control horizontal electron beam movement.

Vertical Synchronization

The vertical retrace and thus vertical sync pulses must occur after each $\frac{1}{60}$ s since the two interlaced fields that make up one frame (picture) occur 60 times per second. The video signal just before, during, and after vertical retrace is shown in Fig. 7-7. Notice that two horizontal sync pulses and the last two lines of video information of a field are initially shown. These are followed in succession by

1. Equalizing pulses at a frequency double the horizontal sweep rate, or 15.75 kHz × 2 = 31.5 kHz. They each have a duration of about 2.7 μs with a period of 1/31.5 kHz, or 31.75 μs. They are used to keep the receiver horizontal oscillator in sync during the relatively long (830 to 1330 μs) vertical blanking period.

2. One vertical sync pulse with a 190-μs pulse width and five serrations having a duration of 4.4 μs at 27.3-μs intervals. These serrations are used to keep the horizontal oscillator synchronized during the vertical sync pulse interval.

3. More equalizing pulses.

4. Horizontal sync pulses until the entire vertical blanking period has elapsed.

Notice that the vertical blanking period is variable since the number of visible lines transmitted can vary between 482 and 495 at the discretion of the station. All other aspects of the pulses such as number, width, and rise and fall times are tightly specified by the FCC so that all receiver manufacturers will know precisely what type of signals their sets will have to process.

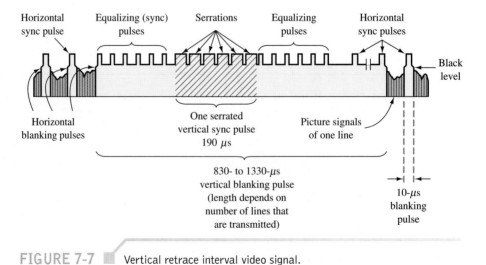

FIGURE 7-7 ▪ Vertical retrace interval video signal.

The vertical sync pulses occur at a frequency of 60 Hz, which is the same frequency as the ac line voltage in the United States. This allows for good stability of the vertical oscillator in the receiver. In Europe, where 50-Hz line voltage exists, a 50-Hz vertical oscillator system is used.

7-4 RESOLUTION

To provide adequate resolution, the video signal must include modulating frequency components from dc up to 4 MHz. This requires a truly wideband amplifier, and amplifiers that have bandpass characteristics from dc up into the MHz region have come to be known as *video amplifiers*.

Resolution is the ability to resolve detailed picture elements. We already have an idea as to resolution in the vertical direction. Since about 485 separate horizontal lines are traced per picture, it might seem that the vertical resolution would be 485 lines. *Vertical resolution* may be defined as the number of horizontal lines that can be resolved. However, the actual resolution turns out to be about 0.7 of the number of horizontal lines, or

$$0.7 \times 485 = 339$$

Thus, the vertical resolution of broadcast TV is about 339 lines.

Horizontal resolution is defined as the number of vertical lines that can be resolved. A little mathematical analysis will show this capability. The maximum modulating frequency has already been stated as 4 MHz. The more vertical lines to resolve, the higher will be the frequency of the resulting video signal. The horizontal trace occurs at a 15.75-kHz frequency, and thus each line is 63.5 μs (1/15.75 kHz) in duration. The horizontal blanking time is about 10 μs, leaving 53.5 μs. Since two consecutive lines can be converted into the highest rate video signal, the number of vertical lines resolvable is

$$4 \text{ MHz} \times 53.5 \mu s \times 2 = 428$$

Thus, the horizontal resolution is about 428 lines. Note that the 428 vertical lines conform nicely to the 339 horizontal lines when one remembers that a TV screen has a 4:3 width to height (aspect) ratio (428/339 $\simeq$ 4/3). Thus, equal resolution exists in both directions, as is desirable. The student should recognize that increased modulating signal rates above 4 MHz would allow for increased vertical or horizontal resolutions or some increase for both. This is shown in the following examples.

EXAMPLE 7-1

Calculate the increase in horizontal resolution possible if the video modulating signal bandwidth were increased to 5 MHz.

SOLUTION

The 53.5 μs allocated for each visible trace could now develop a maximum 5-MHz video signal. Thus, the total number of vertical lines resolvable is

$$53.5 \ \mu s \times 5 \text{ MHz} \times 2 = 535 \text{ lines}$$

EXAMPLE 7-2

Determine the possible increase in vertical resolution if the video frequency were allowed up to 5 MHz.

SOLUTION

The visible horizontal trace time can now be decreased if the horizontal resolution can stay at 428 lines. That new trace time is

$$\text{trace time} \times 5 \text{ MHz} \times 2 = 428$$

$$\text{trace time} = 42.8 \ \mu s$$

Once again, assuming that 10 μs is used for horizontal blanking, that means 52.8 μs total can be allocated for each horizontal trace. With $\frac{1}{30}$ s available for a full picture, that implies a total number of horizontal traces of

$$\frac{\frac{1}{30} \text{ s}}{52.8 \ \mu s} = 632 \text{ lines}$$

Allowing 32 lines for vertical retrace means a vertical resolution of

$$600 \times 0.7 = 420 \text{ lines}$$

Examples 7-1 and 7-2 are excellent proofs of Hartley's law (see Sec. 1-5). It is very plain to see that an increase in bandwidth led to the possibility of greater transmitted information (in the form of increased resolution).

7-5 THE TELEVISION SIGNAL

The maximum modulating rate for the video signal is 4 MHz. Since it is amplitude-modulated onto a carrier, a bandwidth of 8 MHz is implied. However, the FCC allows only a 6-MHz (*only* is a relative term here since 6 MHz is enough to contain 600 AM radio broadcast stations of 10 kHz each) bandwidth per TV station, and that must also include the FM audio signal. The TV signal that is transmitted is shown in Fig. 7-8.

The lower visual sideband extends only 1.25 MHz below its carrier with the remainder filtered out, but the upper sideband is transmitted in full. The audio carrier is 4.5 MHz above the picture carrier with FM sidebands as created by its ±25-kHz deviation. The 54- to 60-MHz limit shown in Fig. 7-8 is the allocation for channel 2. Table 7-1 shows the complete allocation for all the VHF and UHF channels. Notice the VHF channels are broken up into two bands—54 to 88 MHz and 174 to 216 MHz. The UHF band (channels 14 to 69) is continuous and eats up a tremendous chunk of the usable frequency spectrum, as can be seen.

The lower sideband is mostly removed by filters that occur near the transmitter output. While only one sideband is necessary, it would be impossible to filter out the entire lower sideband without affecting the amplitude and phase of the lower frequencies of the upper sideband and the carrier. Thus, part of the 6-MHz bandwidth is occupied by a "vestige" of the lower sideband (about 0.75 MHz out of 4 MHz). It is therefore commonly referred to as *vestigial-sideband* operation. It offers the added advantage that carrier rein-

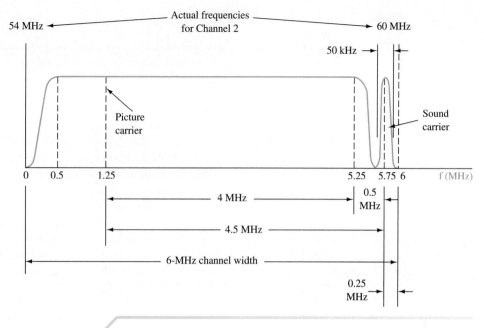

FIGURE 7-8 ■ Transmitted TV signal.

sertion at the receiver is not necessary as in SSB since the carrier is not attenuated in vestigial-sideband systems.

Once the entire TV signal is generated, it is amplified and driven into an antenna that converts the electrical energy into radio (electromagnetic) waves. These waves travel through the atmosphere to be intercepted by a TV receiving antenna and fed into the receiver once again as an electrical signal. That signal consists of the video, audio, and synchronizing signals. The synchronizing signals are contained in the video signal, as previously shown.

TABLE 7-1 *TV Channel Allocations*

LOWER VHF BAND		UPPER VHF BAND		UHF BAND	
CHANNEL	LOWEST FREQUENCY (MHz)	CHANNEL	LOWEST FREQUENCY (MHz)	CHANNEL	LOWEST FREQUENCY (MHz)
2	54	7	174	14	470
3	60	8	180	24	530
4	66	9	186	34	590
	(4 MHz skipped)	10	192	44	650
5	76	11	198	54	710
6	82	12	204	64	770
		13	210	69	800

7-6 TELEVISION RECEIVERS

A TV receiver utilizes the superheterodyne principle, as do virtually all other types of receivers. It does become a bit more complex than most others because it must handle video and synchronizing signals as well as the audio that previously studied receivers do. A block diagram for a typical TV receiver is shown in Fig. 7-9.

The incoming signal is selected and amplified by the RF amplifier and stepped down to the IF frequency by the mixer-local oscillator blocks. The IF amplifiers handle the composite TV signal, and then the video detector separates the sound and video signals. The sound signal detected out of the video detector is the FM signal that is sent into the sound channel block, which is a complete FM receiver system in itself. The other video detector output is the video (plus sync) signal. The actual video portion of the video signal is amplified in the video amplifier and subsequently controls the strength of the electron beam that is scanning the phosphor of the CRT. The sync separator separates the horizontal and vertical sync signals, which are then used to precisely and periodically calibrate the horizontal and vertical oscillators. The oscillator outputs are then amplified and used to precisely control the horizontal and vertical movement of the electron beam that is scanning the phosphor of the CRT. They are applied to a coil around the yoke of the CRT tube whose magnetic fields cause the electron beam to be deflected in the proper fashion. This coil around the tube yoke is commonly referred to as the *yoke*.

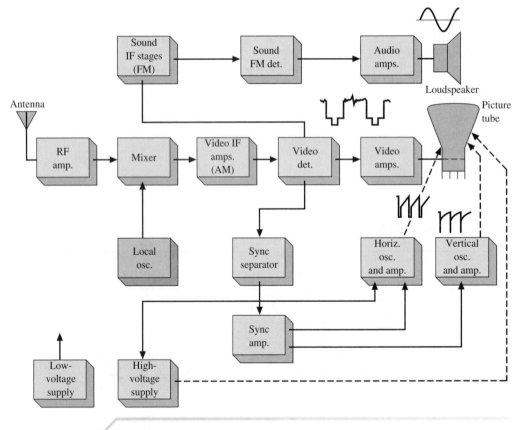

FIGURE 7-9 ■ TV receiver block diagram.

The low-voltage power supply shown in the receiver block diagram is used to power all the electronic circuitry. The high-voltage output is derived by stepping up the horizontal output signal (15.75 kHz) via transformer action. This transformer, usually termed the *flyback transformer,* has an output of 10 kV or more, which is required by the CRT anode to make the electron beam travel from its cathode to the phosphor. In the following sections we shall provide greater detail on the basic operation just presented.

7-7 THE FRONT END AND IF AMPLIFIERS

The front end of a TV receiver is also called the *tuner* and contains the RF amplifier, mixer, and local oscillator. Its output is fed into the first IF amplifier. It is the obvious function of the tuner to select the desired station and to reject all others, but these important functions are also performed:

1. It provides amplification.
2. It prevents the local oscillator signal from being driven into the antenna and thus radiating unwanted interference.
3. It steps the received RF signal down to the frequency required for the IF stages.
4. It provides proper impedance matching between the antenna–feed line combination into the tuner itself. This allows for the largest possible signal into the tuner and thus the largest possible signal-to-noise ratio.

Figure 7-10 provides a block diagram of a VHF/UHF tuner. There is no RF amplifier or transistor-type mixer, as you would probably expect. Instead, the UHF signal is immediately stepped down to the IF frequency and then goes to the VHF RF amplifier and mixer, which provide the gain that makes the UHF and VHF signals of about equal strength when they go into the IF amplifiers. This is done because RF amplifiers and transistor mixers (that supply gain) tend to be more costly and also offer relatively poor signal-to-noise ratios at the high UHF frequencies. The large majority of tuners are synthesized, which allows for the remote control feature that is found on most sets.

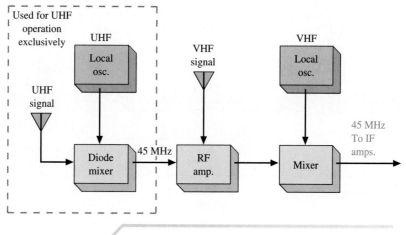

FIGURE 7-10 ▮ VHF/UHF tuner block diagram.

IF Amplifiers

The IF amplifier section is fed from the mixer output of the tuner. It is often referred to as the video IF even though it is also processing the sound signal. Sets that process the sound and video in the same IF stages are known as *intercarrier* systems. Very early sets used completely separate IF amps for the sound and video signals. The IF stages of intercarrier sets are often referred to as the video IF even though they also handle the sound signal because the sound signal is also processed by another IF stage after it has been extracted from the video signal. From now on the video IF will be referred to simply as the IF.

The major functions of the TV IF stage are the same as in a regular radio receiver: to provide the bulk of the set's selectivity and amplification. The standard IF frequencies are 45.75 MHz for the picture carrier and 41.25 MHz (45.75 MHz minus 4.5 MHz) for the sound carrier. Recall that mixer action causes a reversal in frequency when the IF amplifier accepts the difference between the higher local oscillator frequency and the incoming RF signal. Therefore, the sound carrier that is 4.5 MHz above the picture carrier in the RF signal ends up being 4.5 MHz below it in the mixer output into the first IF stage. The inversion effect of IF frequencies when receiving channel 5 is shown in Table 7-2. The IF frequencies are always equal to the difference between the local oscillator and RF frequencies.

Stagger Tuning

A major difference between radio and TV IF amplifiers is that most radio receivers require relatively high-Q tuned circuits since the desired bandwidth is often less than 10 kHz. A TV IF amp requires a passband of about 6 MHz because of the wide frequency range necessary for video signals. Hence, the problem here is not how to get a very narrow bandwidth with high-Q components, but instead how to get a wide enough bandwidth but still have relatively sharp falloff at the passband edges. Most TV IF amplifiers solve this problem through the use of *stagger* tuning. Stagger tuning is the technique of cascading a number of tuned circuits with slightly different resonant frequencies, as shown in Fig. 7-11. The response of three separate LC tuned circuits is used to obtain the total resultant passband shown with dashed lines. The use of a lower-Q tuned circuit in the middle helps provide a flatter overall response than would otherwise be possible.

Another interesting point illustrated in Fig. 7-11 is the attenuation given to the video side frequencies right around the picture carrier. This is done to reverse the

TABLE 7-2 *IF Signal Frequency Inversion*

CHANNEL 5 76–82 MHz	TRANSMITTED RF FREQUENCY (MHz)	LOCAL OSCILLATOR FREQUENCY (MHz)	IF FREQUENCY (MHz)
Upper channel frequency	82	123	41
Sound carrier	81.75	123	41.25
Picture carrier	77.25	123	45.75
Lower channel frequency	76	123	47

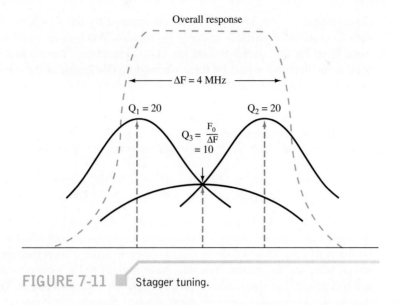

FIGURE 7-11 ■ Stagger tuning.

vestigial-sideband characteristic generated at the transmitter. Refer back to Fig. 7-8 to refresh your memory on the transmitted characteristic. If the receiver IF response were equal for all the video frequencies, the lower ones (up to 0.75 MHz) would have excessive output because they have both upper- and lower-sideband components.

SAW Filters

Color television receivers require a very complex IF alignment procedure due to the critical nature of their required bandpass characteristic. High-quality sets are now using *surface acoustic wave* (SAW) filters. They also find use in modern radar equipment since their characteristics can be matched to the reflected pulse from a target. This relatively new technology is now spreading into many new applications.

Recall that crystals rely on the effects in an entire solid piezoelectric material to develop a frequency sensitivity. SAW devices instead rely on the surface effects in a piezoelectric material such as quartz or lithium niobate. It is possible to cause mechanical vibrations (i.e., surface acoustic waves) that travel across the solid's surface at about 3000 m/s.

The process for setting up the surface wave is illustrated in Fig. 7-12. A pattern of interdigitated metal electrodes is deposited by the same photolithography process used to produce integrated circuits, and great precision is therefore possible. Since the frequency

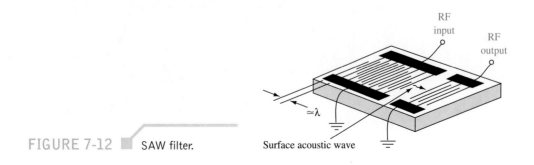

FIGURE 7-12 ■ SAW filter.

characteristics of the SAW device are determined by the geometry of the electrodes, a very accurate and repeatable response is provided. When an ac signal is applied, a surface wave is set up that travels toward the output electrodes. The surface wave is converted back to an electrical signal by these electrodes. The length of the input and output electrodes determines the strength of a transmitted signal. The spacing between the electrode "fingers" is approximately one wavelength for the center frequency of interest. The number of fingers and their configuration determines the bandwidth, shape of the response curve, and phase relationships.

IF Amplifier Response

The ideal overall IF response curve in Fig. 7-13 provides some interesting food for thought. The sound IF carrier and its narrow sidebands are amplified at only one-tenth the midband IF gain. This is done to minimize interference effects that the sound would otherwise have on the picture. You may have noticed a TV with normal picture when no audio is present but visual interference in step with the sound output. This is an indication that the sound signal in the IF is not attenuated enough, and it can often be remedied by adjustment of the set's fine-tuning control.

Wavetraps

To obtain the steep attenuation curve for the sound carrier shown in Fig. 7-13 it is necessary to incorporate a *wavetrap*, more simply termed *trap*, in the IF stage. A trap is a high-*Q* band*stop* circuit that attenuates a narrow band of frequencies. It can be a series resonant circuit that shorts a specific frequency to ground, as in Fig. 7-14(a), or a parallel resonant circuit that blocks a specific frequency, as in Fig. 7-14(b). Even greater attenuation to a specific frequency is obtained with the *bridged-T* trap in Fig. 7-14(c). Traps are

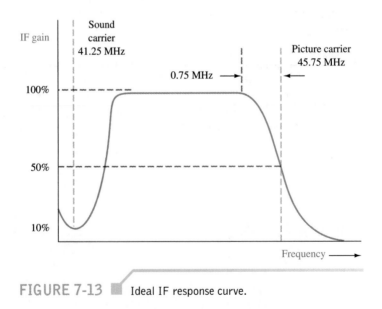

FIGURE 7-13 Ideal IF response curve.

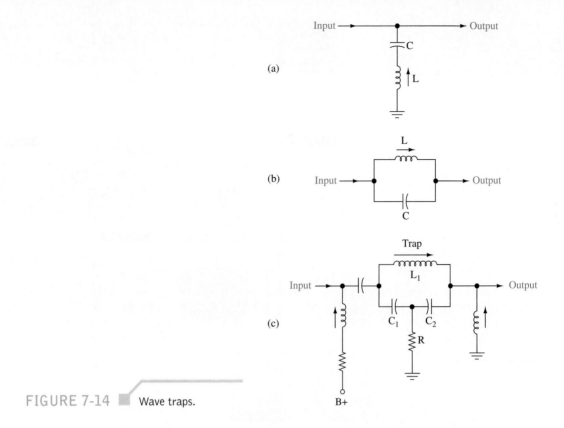

(a)

(b)

(c)

FIGURE 7-14 ▮ Wave traps.

B+

also employed in high-quality sets to eliminate carrier signals of adjacent channels. The carrier signal of an upper adjacent channel occurs at 39.75 MHz. While adjacent channels are not assigned in the same city, it is possible for a location midway between two adjacent channel stations to receive severe interference without a 39.75-MHz trap. A similar problem can exist with the sound carrier of a lower adjacent channel, which would occur at 47.25 MHz.

7-8 THE VIDEO SECTION

The function of the video section is outlined in the block diagram in Fig. 7-15. It takes the output of the video detector (0 to 4 MHz) and amplifies it to sufficient level to be applied to the picture-tube cathode. Once applied to the cathode, this signal varies or *modulates* the electron beam strength such that white and black spots of a scene are white and black spots on the CRT face. It, of course, causes electron beam strengths of "in-between" magnitudes to provide various shades of gray.

The contrast control block in Fig. 7-15 is analogous to the volume control of a radio receiver—it simply varies the amplitude of the signal applied to the CRT. The larger the difference in amplitude between maximum and minimum, the greater will be the picture contrast (difference between black and white).

The sync takeoff block in Fig. 7-15 is the point where the horizontal and vertical sync pulses are extracted from the video signal. Section 7-9 provides further elaboration

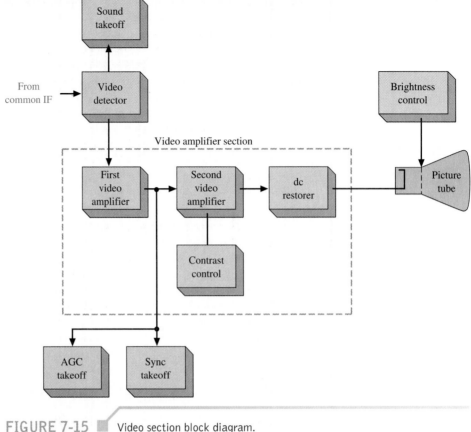

FIGURE 7-15 Video section block diagram.

on this subject. In some receivers the sync takeoff occurs after the final video amplifier stage. The sound takeoff may occur after several stages of video amplification rather than at the video detector, as shown in Fig. 7-15. In color sets, however, the sound takeoff must occur before the video detector.

The *dc restoration* block in Fig. 7-15 is not necessary if the video amplifiers use direct coupling. If, however, capacitive coupling is used, as is usually more economical, then the dc portion of the video signal is lost. Without dc restoration the picture background levels will be erroneous, and in color sets their color will be incorrect.

The brightness control (Fig. 7-15) is a user adjustment, just as is the contrast control. The brightness control simply varies the dc level applied to the control grid or cathode of the CRT. It is *not* connected to the video signal amplitude in any way. It controls the overall picture brightness and *not* the min–max video signal level, as does the contrast control.

The video section also provides a takeoff point for the AGC signal. That signal is used to control the gain of previous amplifying stages such as the RF amp, mixer, and IF stages. This is necessary so that both strong and weak stations end up supplying the CRT cathode with approximately the same signal level and thus provide the same picture illumination. If the received signal is too weak, however, the electrical noise predominates over the desired signal and results in a "snowy" picture.

7-9 SYNC AND DEFLECTION

The video section provides a takeoff point for the sync signals. Since the set needs both vertical (at 60 Hz) and horizontal (at 15.75 kHz) sync pulses, a means to separate one from the other is necessary. The *sync separator* is the circuit that performs this function. The key factor that enables separation is the fact that the vertical sync pulse is of long duration, while the horizontal sync pulse is of very short duration. In addition to separating one from the other, the sync separator *clips* the sync pulse off the video signal. This prevents the sweep instability that could occur because of false synchronization of the sweep oscillators by spurious video signals. Because of this, the sync separator is sometimes referred to as the *clipper.*

Once the sync separator has clipped the sync signals from the lower-level video signal, the two types of sync pulses are applied to both low- and high-pass filters. The output of the low-pass filter will be the lower-frequency vertical sync pulse at 60 Hz since it is a wide pulse rich in low-frequency components. The output from the high-pass filter will be the horizontal sync pulse at 15.75 kHz since it is a very narrow pulse that is rich in high-frequency content. A low-pass filter is also termed an *integrator,* while a high-pass filter is classified as a *differentiator.*

This entire process of clipping and separation is shown in Fig. 7-16. The low-pass filter can simply be a shunt capacitance that shorts high frequencies to ground. The high-pass filter includes a series capacitance that blocks low frequencies from reaching its output.

The vertical sync pulse is applied to the vertical oscillator. The vertical oscillator by itself generates a signal at *about* 60 Hz but must be at *precisely* the same frequency as the

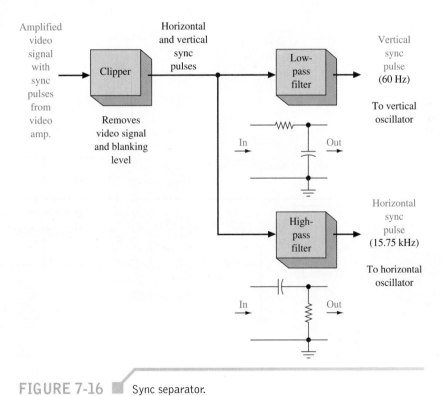

FIGURE 7-16 ■ Sync separator.

transmitter's vertical oscillator to prevent the picture from "rolling" in a vertical direction. The vertical adjustment control available to the set user adjusts the vertical oscillator's frequency to enable it to be brought into the range necessary so that it can "lock" onto the frequency of the sync pulse. The foregoing discussion also applies to the horizontal oscillator section except that the frequency is 15.75 kHz, and loss of horizontal sync results in heavy slanting streaks across the screen. This phenomenon can be witnessed by simply misadjusting the horizontal control on a TV set.

Horizontal Deflection and High Voltage

A typical block diagram for the horizontal deflection and high-voltage systems is shown in Fig. 7-17. The horizontal sync pulses are used to calibrate the horizontal oscillator, which is then amplified to a powerful level by the horizontal output amplifier and then applied to a high-voltage transformer commonly referred to as the *flyback transformer*. Its outputs drive the horizontal yoke windings and provide the high voltages for the CRT after rectification. A *damper* function is also provided, which will be subsequently explained. The horizontal system is seen to be a complex one, and because of this and the high voltages and powers involved, it is probably the most failure-prone section of a TV receiver.

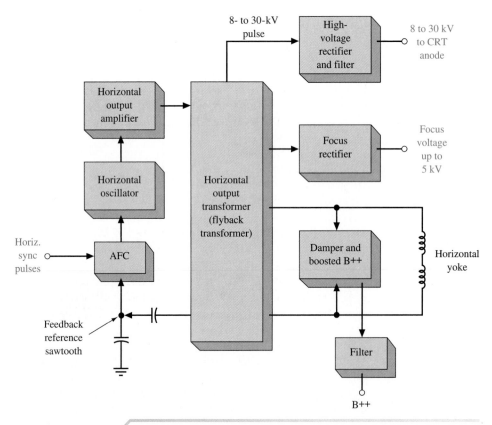

FIGURE 7-17 ■ Horizontal deflection block diagram.

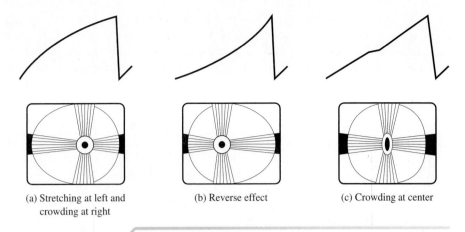

(a) Stretching at left and crowding at right

(b) Reverse effect

(c) Crowding at center

FIGURE 7-18 Nonlinear horizontal scanning. (From Bernard Grob, *Basic Television Principles and Servicing,* 4th ed., 1977; courtesy of McGraw-Hill, Inc., New York.)

As with vertical scanning, a linear sawtooth (current) waveform is required for linear horizontal deflection. If not, distortion in the picture results, as indicated in Fig. 7-18. A horizontal linearity control is sometimes provided at the rear of the set to correct for these conditions.

Horizontal Deflection Circuit

Figure 7-19 provides a schematic for a typical horizontal system. The horizontal oscillator frequency is held in sync by comparing the horizontal sync pulses to a signal fed back from the horizontal output in the phase detector diodes, D_1 and D_2. Any difference is detected as a phase difference and applied as a dc level to the base of the horizontal oscillator to correct its frequency. A PLL IC is used in many sets for this application. Note also the user-controlled variable inductor that adjusts the frequency of oscillation into the range that allows the sync pulses to exercise control. The horizontal oscillator signal is transformer-coupled into the base of the horizontal output transistor for amplification so that the signal has sufficient strength to drive the flyback transformer.

As the sawtooth level builds up on the horizontal amp base, its collector current builds up through the transformer primary and damper diode. When the sawtooth level suddenly changes (during retrace), the collector current drops to zero. The magnetic field around the horizontal yoke coils collapses, rapidly inducing a high-amplitude flyback EMF across the transformer secondary. This induces a pulse of current in the secondary of the transformer and a high-induced flyback EMF in the primary. The kilovolts of ac thus induced are rectified by the high-voltage rectifier and applied to the CRT as its required dc anode voltage.

During this flyback period, the energy of the horizontal yoke coils' collapsing magnetic field tends to produce damped oscillations that interfere with the start of the next sawtooth waveform. The *damper* diode serves as a short during this flyback interval so that the unwanted oscillations are very quickly damped. An auxiliary secondary winding on the flyback transformer provides a stepped *voltage boost* dc level of about 100 V for all the circuitry requiring more than the 12 to 20 V dc used elsewhere.

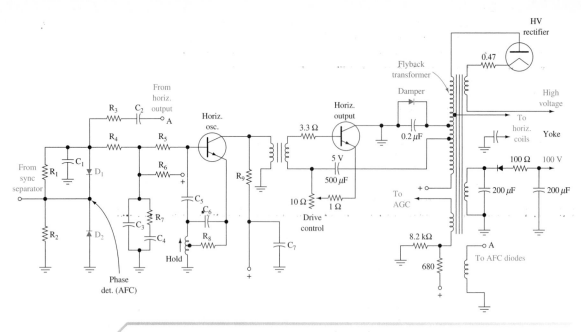

FIGURE 7-19 ▨ Horizontal system schematic.

7-10 PRINCIPLES OF COLOR TELEVISION

We have thus far been mainly concerned with black-and-white or *monochrome* television. While color TV presents a much greater degree of sophistication, the student who has mastered monochrome principles reasonably well can advance to the color set by adding a few more basic ideas.

Our system for color TV was instituted in 1953 and is termed *compatible.* That is, a color transmission can be reproduced in black-and-white shades by a monochrome receiver and a monochrome transmission is reproduced in black and white by a color receiver. To remain compatible, the same total 6-MHz bandwidth must be used, but more information (color) must be transmitted. This problem is overcome by a form of multiplexing, as when FM stereo was added to FM broadcasting. It turns out that the video signal information is clustered at 15.75-kHz intervals throughout its 4-MHz bandwidth. Midway between these 15.75-kHz clusters (harmonics) of information are unused frequencies, as indicated in Fig. 7-20. By generating the color information around just the right color subcarrier frequency (3.579545 MHz), it becomes centered in clusters exactly between the black-and-white signals. This is known as *interleaving.*

At the color TV transmitter the scene to be televised is actually scanned by three separate pickup tubes in the camera, each being sensitive to just one of the three primary colors, red, blue, and green. Since various combinations of these three colors can be mixed to form any color to which the human eye is sensitive, an electrical representation of a complete color scene is possible. The three color cameras scan the scene in unison, with the red, green, and blue color content separated into three different signals. At the receiver these three separate signals are made to properly illuminate groups of red, green, and blue phosphor dots (called *triads*), and the original scene is reproduced in color.

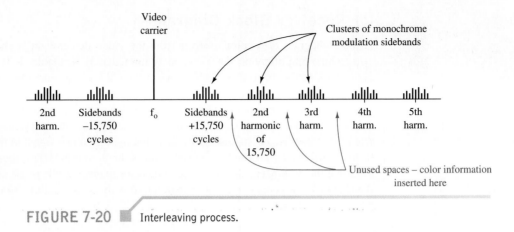

FIGURE 7-20 Interleaving process.

After generation these three separate color signals are fed into the transmitter signal processing circuits (*matrix*) and create the Y, or *luminance,* signal and the *chroma,* or color, signals I and Q. The Y signal contains just the right proportion of red, blue, and green such that it creates a normal black-and-white picture. It modulates the video carrier just as does the signal from a single black-and-white camera with a 4-MHz bandwidth. The chroma signals, I and Q, are used to phase-modulate the 3.58-MHz color subcarrier, which then *interleaves* their color information in the gaps left by luminance Y signal's sidebands. This modulation by the I and Q signals is accomplished in a balanced modulator, thus suppressing the 3.58-MHz subcarrier, as it would cause interference at the receiver. The composite transmitted signal is shown in Fig. 7-21.

At the receiver, a monochrome set will simply detect the Y signal and thus present a normal black-and-white rendition of a color picture. The chroma signals (I and Q) cannot be detected in a monochrome set because their 3.58-MHz subcarrier was suppressed and is not present in the received signal. Thus, a color set must have a means to generate and reinject the 3.58-MHz subcarrier to enable detection of the I and Q signals.

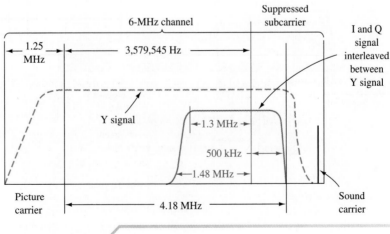

FIGURE 7-21 Composite color TV transmission.

Color Receiver Block Diagram

A block diagram of a color receiver from the video detector on is shown in Fig. 7-22. After video amplification the Y signal is immediately available. It is given a delay of about 1 μs as shown so that it will arrive at the CRT at the same time as the I and Q signals. This is necessary because the I and Q signals undergo considerably more processing, which takes about 1 μs. The chroma signals are amplified and then sent into a 2- to 4.2-MHz bandpass amplifier and then I and Q detectors. These detectors also have inputs from the 3.58-MHz crystal oscillator such that the difference signal in the I detector is the 0- to 1.5-MHz I signal and in the Q detector is the 0- to 0.5-MHz Q signal. Notice in Fig. 7-22 that the 3.58-MHz signal for the Q detector is given a 90° phase shift, which is how the Q signal was generated at the transmitter. It is this phase shift that makes them separable at the receiver.

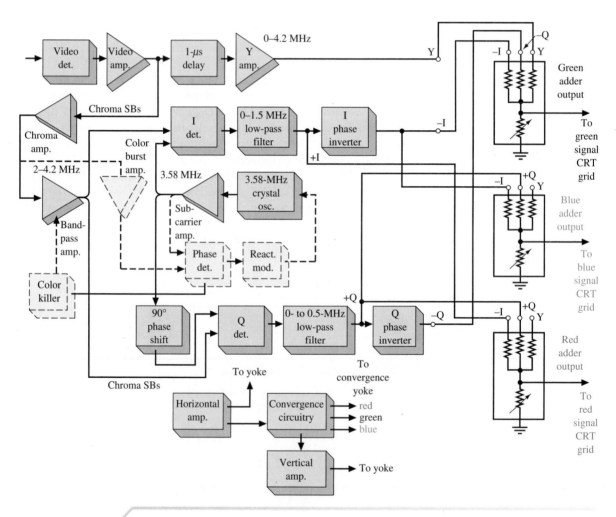

FIGURE 7-22 ■ Color receiver block diagram.

Once the I and Q signals are detected and passed through their respective low-pass filters, they are given a phase inversion that allows for both + and − chroma signals. This is necessary since

$$green = -I - Q + Y$$

$$blue = -I + Q + Y$$

$$red = +I + Q + Y$$

The I, Q, and Y signals are summed in the three-color adder circuits, with the resistor values providing the proper proportion of each signal. The output of each color adder is then applied to the appropriate CRT grid to control beam intensity. Notice the rheostat in each adder circuit. It allows for the intensity of each color signal to be varied in proportion to the other colors.

The color subcarrier crystal oscillator is not precise enough by itself to allow proper chroma signal detection. This is surprising since crystal oscillators are extremely stable and accurate. An accuracy of 1 part of 10^{12} is necessary to obtain the correct chroma signal. Recall that color transmissions eliminate this carrier from the video signal but do include a sample of it on the back porch of the horizontal blanking pulse, as shown in Fig. 7-23. The color burst amp shown in Fig. 7-22 is receptive to that portion of the overall video signal. Its frequency is compared with the 3.58-MHz crystals in the phase detector, and if they are not precisely equal, the phase detector applies a dc level to vary the reactance of the reactance modulator. It, in turn, causes the crystal's frequency to "pull" in the proper direction to bring it back into precise synchronization with the color burst frequency.

Notice that the phase detector in Fig. 7-22 also has an output that is applied to the *color killer.* The name is very descriptive since a monochrome broadcast has no color burst, and thus the phase detector has a large dc output that the color killer circuit uses to "kill" the 2- to 4.2-MHz bandpass amplifier. The purpose is to prevent any signals out of the chroma circuits during a monochrome broadcast. A defective color killer results in colored noise, called *confetti,* on the screen of a color receiver during a black-and-white transmission. The confetti looks like snow but with larger spots, in color.

The Color CRT and Convergence

Color receiver CRTs are a marvel of engineering precision. As previously mentioned, they are made up of triads of red, blue, and green phosphor dots. The trick is to get the proper electron beam to strike its respective colored phosphor dot. This is accomplished

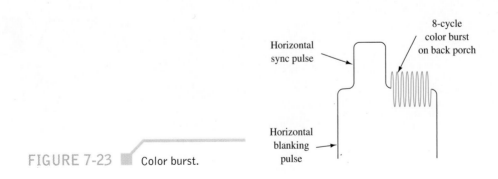

FIGURE 7-23 ▌ Color burst.

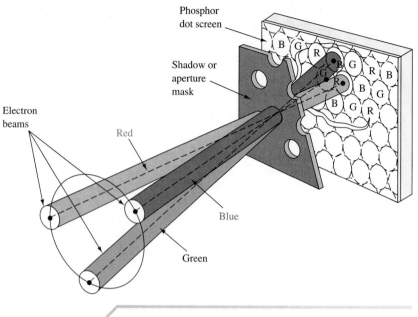

FIGURE 7-24 ■ Color CRT construction.

by passing the three beams through a single hole in the *shadow mask,* as shown in Fig. 7-24. The shadow mask prevents the "red" beam from spilling over onto an adjacent blue or green phosphor dot, which would certainly destroy the color rendition. A typical color CRT has over 200,000 holes in the shadow mask and triads of phosphor dots. To make the three beams properly "converge" on their color dot of phosphor throughout the face of the tube requires special modification to the horizontal and vertical deflection systems.

Static convergence refers to proper beam convergence at the center of the CRT's face. This adjustment is made by dc level changes in the horizontal and vertical amplifiers. Convergence away from the center becomes more of a problem and is referred to as *dynamic convergence.* It is necessary because the tube face away from the center is not a perfectly spherical shape (it is more nearly flat), and thus the beams tend to converge in

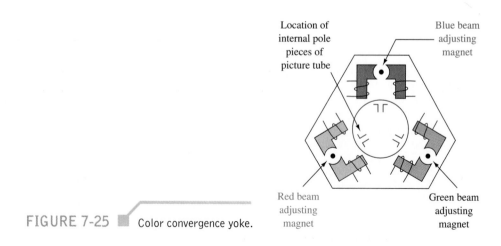

FIGURE 7-25 ■ Color convergence yoke.

front of the shadow mask away from the tube center. Special dynamic convergence voltages are derived from the horizontal and vertical amplifier signals and are applied to a special color convergence yoke placed around the tube yoke, as shown in Fig. 7-25. The dynamic convergence of a set involves the shown magnet adjustment and a number of adjustments on the convergence board, usually 12, that have interaction effects. The process is quite involved and time consuming.

7-11 SOUND AND PICTURE IMPROVEMENTS

Modern television systems have entered a period of rapid change and improvement. Low-power broadcast stations are planning to offer programming tailored to their community. Cable operators have wired cities and suburbs with cable systems that provide more than 70 channels of entertainment programming and data services. Satellite system operators offer direct broadcast systems (DBS) that broadcast television directly to individual subscribers from geostationary satellites. The use of videocassette recorders (VCRs) has become commonplace, providing the user with improved flexibility. With all of these changes there is also a move to improve the quality of TV audio and video. The audio improvement was accomplished when the FCC approved a new system in 1984. The video revolution is currently developing.

Enhanced Audio

The system approved by the FCC in 1984 is called Zenith/dbx. It has similarities to the FM stereo system but it provides a better signal. The multiplexing scheme used is illustrated in Fig. 7-26. Recall that in FM radio, a double-sideband carrier centered around 38 kHz is used. In the Zenith/dbx system, the new component, (L − R), is centered at two times the horizontal scan rate, f_H, about 2×15.7 kHz or ≈ 31.4 kHz. Notice also the separate audio program (SAP) and the voice/data channels centered at 5 and 6.5 times f_H. With SAP, a station can transmit a simultaneous foreign-language translation or an entirely unrelated service. The sub-carrier at $6.5f_H$, known as the professional subchannel, is intended for transmitting voice or data wholly unrelated to video programming. This could include radio reading services, market and financial data, paging and calling, and traffic-control signal switching.

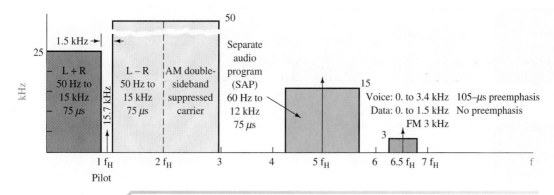

FIGURE 7-26 Zenith/dbx stereo system.

The stereo subcarrier, (L − R), centered at $2f_H$, deviates at ±50 kHz, or twice the regular monophonic signal (L + R). The composite modulating signal creates a bandwidth that exceeds the 200 kHz allowed for FM radio. This is allowable since the sound track is transmitted 4.5 MHz above the picture carrier and 250 kHz below the top of the channel. This increased bandwidth is one reason that TV stereo is a better signal than FM radio, but the main reason is the dbx companding system.

As explained in Chapter 4, "compand" is from "*com*press" and "ex*pand*," in which a variable-gain circuit at the transmitter increases its gain for low-level input signals to provide better noise performance. A complementary circuit at the receiver reverses the process. The monophonic audio channel (L + R) is not companded so as to maintain compatibility with nonstereo TVs.

The reduced level of the (L − R) signal compared to (L + R) in FM radio has been one of its major drawbacks. In TV, as shown in Fig. 7-26, it is actually increased in amplitude, which was possible due to the increased bandwidth TV has available. Moreover, as the left–right separation decreases, the difference channel amplitude declines with respect to the sum channel. The increased amplitude of the (L − R) signal in TV is therefore especially beneficial. In the dbx system, the gain control signal is a function of the rms audio signal and compression is 2:1. This means that for every 2-dB decrease in audio level below maximum, the transmitted audio is decreased 1 dB from the maximum. Although the effectiveness of companding has long been known, it has proven difficult to implement. Now that the complex circuitry required can be inexpensively fabricated on ICs, its use in mass-produced equipment is possible (see Sec. 4-4). In areas of good signal recep-

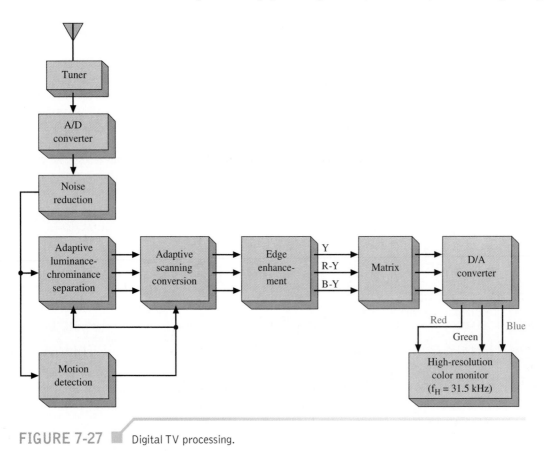

FIGURE 7-27 ▌ Digital TV processing.

tion, a receiver can provide a good signal with or without companding. The companded TV audio signal, however, remains noise-free well beyond the point at which a stereo FM radio signal is noticeably degraded.

Enhanced Video

At this time, much work is taking place worldwide to develop standards for high-definition TV (HDTV) systems. The goal is to offer a TV picture that will rival the definition of motion picture film presentations. Principal barriers to implementation of HDTV include the need for wide transmission bandwidths, the lack of worldwide standards for both production and transmission of the signals, and the difficulty of achieving compatibility with existing TV receivers. It is generally felt that HDTV receivers will use digital signal processing that converts the received signal from analog to digital, as shown in Fig. 7-27. The VSLI chips to accomplish this have been developed but are still being refined. Besides the higher overall picture quality, they promise such features as freeze-frame, zoom focusing, and picture-within-a-picture.

A proposal for a system advanced by the New York Institute of Technology is shown in Fig. 7-28. It uses the two-signal approach. One camera produces a standard

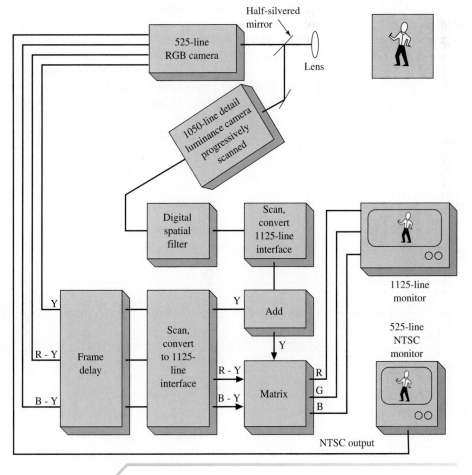

FIGURE 7-28 ■ HDTV scheme.

525-line output; a second produces a narrowband detail output. When the two are transmitted, standard receivers display the 525-line program, while the HDTV receivers can display the 1125-line resolution of the combined signals. Notice that a different aspect ratio (16:9) is used for the HDTV picture. By the time you are reading this, the technical and political challenges to perfect a working system may have been accomplished. The NTSC (National Television Standards Committee) output shown in Fig. 7-28 is the 525-line system of TV used in the United States and Japan.

Video for Multimedia Computing

Multimedia is basically data, voice, and imaging working around a personal computer (PC) or workstation. This rapidly growing field has major implications for television; in fact, the televisions of the future will likely be some kind of TV/computer combination.

The rather complex circuitry that allows the marriage of the digital computer and analog audio/video has become available in the mid 1990s. The basics of one such system are shown in Fig. 7-29. The TDA8708 provides the analog-to-digital conversion that outputs to the heart of this system—the digital video decoder and scaler (DESC). The SAA7194 DESC integrates a multistandard color decoder with a resizer circuit, bidirectional bus, and clock circuit. Packaged in a 120-lead quad flat pack, the DESC chip converts digitized composite video signals into brightness, color, and timing data, then downscales the image for display on the computer monitor. Video inputs can be NTSC (National Television Standards Committee) color signals used in the U.S., PAL (phases-alternate line) used in most of Europe, and SECAM (sequential color and memory) system color signals used in France and Russia. It supports all major video compression standards in conjunction with the shown video compression codec (coder/decoder). After data for matting to the various output formats, the scaled video is buffered in a 32-bit-by-16-word FIFO whose output links with the VRAM (shown in Fig. 7-29).

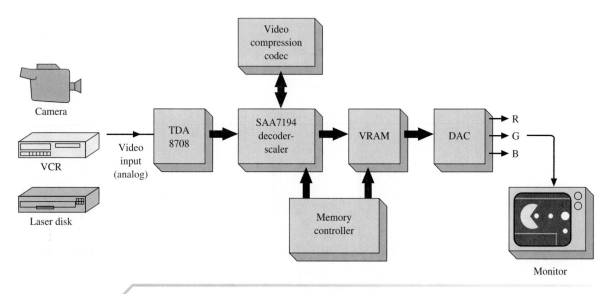

FIGURE 7-29 ■ Video for multimedia computing.

The decoding done via the system shown in Fig. 7-29 can be done in software to a very limited degree. Some of the very high-powered microprocessors can accomplish the decoding if real-time video is not required. Decoding in real time still requires significant hardware support like the system shown in Fig. 7-29.

TROUBLESHOOTING

As you recall from Sec. 7-2, the TV set is basically an AM receiver for picture information and an FM receiver for sound reception. The basic approach for troubleshooting a TV set is to proceed as if it were two radios. In this section you will learn techniques that will help you identify and isolate faulty sections within the TV receiver. These techniques may be used when troubleshooting any kind of TV set.

Upon completing this section you should be able to

- Identify defective stages in a TV receiver

- Describe faulty sections within the TV based on viewed symptoms

Looking into the back of a TV set for the first time can be a fearful sight. Inside the TV set are many very complex circuits and hundreds of components. To be an effective service technician, you must have a thorough knowledge and understanding of how a TV works. It becomes quite apparent that service literature is needed when servicing a TV. Make sure you get the service literature pertaining to the model and chassis that you are repairing. The TV setup should be as close as you can get it to normal viewing. Organize your thinking in a logical pattern before ever making any circuit measurements. Observe and classify any abnormality such as sound, raster, video, or color problems. Study the service literature and identify functional sections in the TV set. For example, identify the location of the horizontal output section, vertical section, video section, and others. Try to localize the problem to a specific section from the symptoms being observed.

The TV set stands out from other communications receivers in that circuit defects often show up on the screen. From these visual symptoms faulty sections can be singled out before ever opening the back of the TV set. For example, symptoms like no video, no sound, and good raster would lead us to check the IF section of the TV since both picture information and sound information are amplified there.

Study the problem. Make a decision where to start troubleshooting based on the viewed symptoms. Try to isolate the defective stage from the symptoms on the screen. Is a picture present? Does the picture roll up and down or left to right? Does the picture have snow in it? Is the sound present or not? By observing these signposts you quickly determine the defective section within the TV. Table 7-3 gives symptoms, causes, and the area in which to look. It is not all-inclusive of the problems that might occur in a TV set.

Consult manufacturer's service literature for diagnostic charts or other troubleshooting guidelines. Often, when these troubleshooting aids are available, they relate to common failures incurred for that model TV receiver or manufacturing defects that may exist in the set.

Pull the back off the set, and with the power off, look for obvious problems. Look for loose wires or connectors, burned components, broken or burned PCB traces, and cold solder joints. With the power on, listen for unusual sounds like hissing (normally associated with horizontal output transformers that have developed high-voltage leaks), arcing,

TABLE 7-3

Symptom	Cause	Stage/Area of Trouble
Set is dead, no sound, no video, no raster	No power to circuits	Check main power supply, start-up circuits, main fuses, and line cord
Set blows fuses	A short circuit exists in main power supply or horizontal output	Check for shorted diodes, shorted regulator transistors, and shorted filter capacitors in main power supply, or shorted horizontal output transistor and shorted horizontal output transformer
Sound normal, no video, no raster	No high voltage	Check the horizontal output circuit/high-voltage section
Normal raster, no video, no sound	Video and sound signal missing	Check antenna, tuner, and IF amplifiers
Raster and video normal	Sound signal missing	Check sound IF amplifiers, detector section, audio amplifiers, and speaker
Raster, video, and sound normal, no color	Color signal missing	Check color killer and color processing circuits
Picture has snow, noise heard in sound	Signal-to-noise ratio high	Check RF amplifier in tuner
Vertical roll (up and down)	Vertical sync missing	Check sync separator, vertical oscillator
Horizontal white line across screen	Vertical output signal missing	Check the vertical output circuit, vertical oscillator, or yoke
Horizontal roll (left and right)	Horizontal sync missing	Check sync separator circuit, horizontal AFC, horizontal oscillator
Vertical white line on screen	Horizontal output signal missing	Check horizontal output circuit, horizontal oscillator, yoke

and high-pitched sequels from the horizontal oscillator. Do you smell anything burning? Look for brown areas on the PCB indicating overheating components.

If the preliminary inspection fails to localize a defective component, continue troubleshooting by doing voltage and resistance measurements on the suspected stage in the TV set. Compare the results with specified values from the service literature. Use the oscilloscope to check for proper waveforms in the defective section and associated sections. Schematic diagrams usually giv°e pictures indicating what the correct waveform looks like for all the common signals in the set. Base your troubleshooting on an organized approach and not a disorganized one. Having an organized strategy will save you valuable time and enhance your troubleshooting.

SUMMARY

In Chapter 7 the complete TV system was introduced, including the signal generation, transmission, and reception. The video (AM) and audio (FM) are combined to make up the composite TV signal. The major topics the student should now understand include:

- the description of a complete but simplified TV system

- the operation of the vidicon and charge-coupled device (CCD) in the TV camera

- the definition of scanning, pixel, horizontal/vertical retrace interval, and aspect ratio

- the explanation of interlaced scanning with frame definition and timing

- the calculation and definition of horizontal and vertical resolution

- the explanation of TV receiver operation using a block diagram

- the changes necessary to the black-and-white signal to allow inclusion of color information in a compatible fashion

- the description of audio and video improvements, including stereo audio and HDTV

QUESTIONS AND PROBLEMS

SECTION 2

*1. Does the sound transmitter at a television broadcast station employ frequency or amplitude modulation?

2. In what way is TV audio equivalent to broadcast FM radio, and in what way is it inferior?

*3. Does the video transmitter at a television broadcast station employ frequency or amplitude modulation?

4. Explain the major benefit of combining AM and FM techniques in television broadcasting.

5. List and explain the functions of the six transducers used in a complete TV system.

*6. Why is a diplexer a necessary stage of most TV transmitters?

7. Describe the operation of a vidicon camera tube.

*8. What is a mosaic plate in a television camera?

9. Sketch an electrical video signal as would result from scanning the letter "E" in the setup shown in Fig. 7-4.

*10. In television broadcasting, what is the meaning of the term *aspect ratio?*

*11. Numerically, what is the aspect ratio of a picture as transmitted by a television broadcast station?

SECTION 3

*12. What is the purpose of synchronizing pulses in a television broadcast signal?

13. Provide an analogy between horizontal and vertical retrace as compared to reading a book.

*14. If the cathode-ray tube in a television receiver is replaced by a larger tube such that the size of the picture is changed from 6 by 8 in to 12 by 16 in, what change, if any, is made in the number of scanning lines per frame?

*15. How many frames per second do television broadcast stations transmit?

*16. Why is a scanning technique known as *interlacing* used in television broadcasting?

*17. What are synchronizing pulses in a television broadcast and receiving system?

*18. What are blanking pulses in a television broadcasting and receiving system?

19. Calculate the frequency required for the horizontal sync pulses. (15.8 kHz)

*20. What is the field frequency of a television broadcast transmitter?

*21. In television broadcasting, why is the field frequency made equal to the frequency of the commercial (ac) power source?

*22. Besides the camera signal, what other signals and pulses are included in a complete television broadcast signal?

Section 4

23. Describe the characteristics of a video amplifier.

24. Define *resolution, vertical resolution,* and *horizontal resolution.*

25. Explain why vertical resolution is less than the number (about 0.7) of horizontal lines. (*Hint:* Consider what might happen if a pattern of 495 alternate black-and-white horizontal lines were scanned by a TV camera such that each scan saw half of a white-and-black line.)

26. Calculate the horizontal resolution of a broadcast TV picture. ($\approx$428 lines)

27. Calculate the decrease in horizontal resolution if the video signal bandwidth were reduced from 4 to 3.5 MHz. (from 428 to 375 lines)

28. Calculate the vertical resolution if the video signal bandwidth were reduced from 4 to 3.5 MHz, assuming that the horizontal resolution was not to change. (307 lines)

Section 5

*29. If a television broadcast station transmits the video signals on channel 6 (82 to 88 MHz), what is the center frequency of the aural transmitter?

*30. What is meant by 100% modulation of the aural transmitter at a television broadcast station?

31. What TV channel is most likely to be heard on an FM broadcast receiver? Explain why.

*32. What is *vestigial-sideband transmission* of a television broadcast station?

Section 6

33. Draw a TV receiver block diagram, and briefly explain the function of each block.

Section 7

34. State what a TV *front end* consists of and the important functions it performs.

35. Show how the VHF tuner is used in conjunction with VHF reception. Why is the VHF signal stepped down in frequency before it is given any amplification?

36. Calculate the sound and picture carrier frequency for channel 10 before and after frequency translation to the IF frequency. What is the required local oscillator frequency? (41 to 47 MHz, 197.75 MHz, 193.25 MHz, 41.25 MHz, 45.75 MHz, 239 MHz)

37. Explain *stagger tuning* and explain why it is often used in TV IF amplifiers.

38. Calculate the approximate "finger" spacing for a SAW filter operating at 44 MHz. (0.0682 mm)

39. Why is the sound carrier and its sidebands only given one-tenth the amplification of the video by the IF response curve? Explain why part of the video signal is given less amplification also.

40. Discuss the function of a wavetrap and the need for such traps in TV receivers.

SECTION 8

41. If an amplifier stage of the video section shown in Fig. 7-15 became inoperative, would the receiver's sound be affected?

42. In detail, explain the difference between adjustment of the brightness and contrast controls.

43. What is the function of dc restoration, and what kind of video sections require it?

SECTION 9

44. What is the function of the sync separator? How is it able to differentiate between the horizontal and vertical sync pulses? What types of circuits are used for each?

45. Explain the relationship between the horizontal deflection system and the CRT anode high-voltage supply. Why is this a failure-prone area in a TV receiver?

46. What are the possible effects of a nonlinear deflection waveform?

47. Explain the operation of the horizontal system schematic shown in Fig. 7-19.

48. Explain the function of the damper system and flyback transformer.

SECTION 10

49. Describe the process of interleaving, and explain its role in making color TV broadcast possible on the same bandwidth used in the monochrome system.

*50. Describe the scanning process employed in connection with color TV broadcast transmission.

51. Describe the important features of the Y, I, and Q signals in a color TV broadcast.

52. Define the meaning of *compatibility* with respect to color and monochrome TV. How does a monochrome set properly display a color transmission?

*53. Describe the composition of the chrominance subcarrier used in the authorized system of color television.

54. Explain how the Y, I, and Q signals are processed by a color TV receiver.

55. Why is extreme accuracy required of the color subcarrier oscillator within a color TV receiver? Explain how this accuracy is obtained in the receiver.

56. Explain the operation of the color killer. Describe the effect of a defective color killer.

57. Describe the important characteristics and construction of the color CRT. Include the need for convergence and how it is accomplished in this discussion.

SECTION 11

58. Describe the stereo audio system used for TV and explain its superior performance compared to broadcast FM stereo.

59. The audio signal for TV stereo reaches a maximum level of 20 dBm at the transmitter. Calculate the companded audio level for 4 dBm and 15 dBm. (12 dBm, 17.5 dBm)

60. What is HDTV? Give a reason why it has an increased aspect ratio over regular TV.

COMMUNICATIONS TECHNIQUES

OBJECTIVES

- Describe double conversion and up-conversion and explain their advantages

- Analyze the advantages of delayed AGC and auxiliary AGC

- Explain the features and their operation that a high-quality receiver may include as compared to a basic receiver

- Analyze and explain the relationships between noise, receiver sensitivity, dynamic range, and the third-order intercept

- Troubleshoot an amplifier suspected of excessive IMD

- Explain the operation of a frequency synthesizer

- Describe the operation of a DDS system and provide advantages and drawbacks compared to analog synthesizers

- Provide the basics of spread-spectrum systems and describe their advantages compared to standard communication systems

8-1 INTRODUCTION

Communications equipment may be loosely defined as that which is *not* used for entertainment. Since much of this equipment is of a vital nature, it is not surprising that communications equipment contains more sophistication than a standard broadcast receiver. In addition, since communications tends to require two-way capabilities, we find the use of transceivers prevalent. A *transceiver* is simply a transmitter *and* receiver in a single package.

8-2 FREQUENCY CONVERSION

Double Conversion

One of the most likely areas of change from broadcast receivers to communications receivers is in the mixing process. The two major differences are the widespread use of *double conversion* and the increasing popularity of *up-conversion* in communications equipment. Both of these refinements have as a major goal the minimization of image frequency problems (refer to Chapter 3 for a review of these phenomena).

Photo: Spread-spectrum RF front-end chip set designed for telemetry and wireless local area network (LAN) applications. (Courtesy of M/A-COM, Inc., Lowell, MA.)

Double conversion is the process of stepping down the RF signal to a first, relatively high IF frequency and then mixing down again to a second, lower, final IF frequency. Figure 8-1 provides a block diagram for a typical double-conversion system. Notice that the first local oscillator is variable so as to allow a constant 10-MHz frequency for the first IF amplifier. Now the input into the second mixer is a constant 10 MHz, which allows the second local oscillator to be a fixed 11-MHz crystal oscillator. The difference component (11 MHz − 10 MHz = 1 MHz) out of the second mixer is accepted by the second IF amplifier, which is operating at the comfortable low frequency of 1 MHz. The following example will serve to illustrate the ability of double conversion to eliminate image frequency problems.

EXAMPLE 8-1

Determine the image frequency for the receiver illustrated in Fig. 8-1.

SOLUTION

The image frequency is the one that when mixed with the 30-MHz first local oscillator signal will produce a first mixer output frequency of 10 MHz. The desired frequency of 20 MHz mixed with 30 MHz yields a 10-MHz component, of course, but what *other* frequency provides a 10-MHz output? A little thought shows that if a 40-MHz input signal mixes with a 30-MHz local oscillator signal, an output of 40 MHz − 30 MHz = 10 MHz is also produced. Thus, the image frequency is 40 MHz.

Example 8-1 shows that in this case the image frequency is double the desired signal (40 MHz versus 20 MHz), and even the relatively broadband tuned circuits of the RF and mixer stages will almost totally suppress the image frequency. On the other hand, if this receiver uses a single conversion directly to the final 1-MHz IF frequency, it is found that the image frequency will not be fully suppressed.

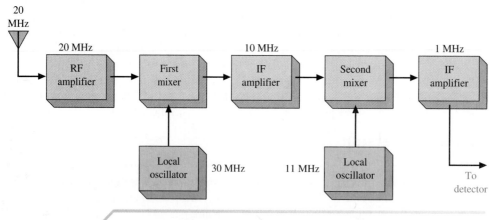

FIGURE 8-1 Double-conversion block diagram.

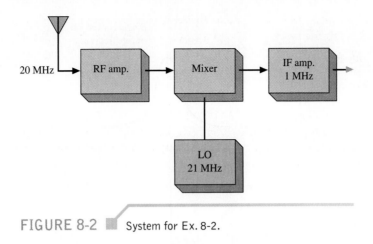

FIGURE 8-2 ▇ System for Ex. 8-2.

EXAMPLE 8-2

Determine the image frequency for the receiver illustrated in Fig. 8-2.

SOLUTION

If a 22-MHz signal mixes with the 21-MHz local oscillator, a difference component of 1 MHz is produced just as when the desired 20-MHz signal mixes with 21 MHz. Thus, the image frequency is 22 MHz.

The 22-MHz image frequency of Ex. 8-2 is very close to the desired 20-MHz signal. While the RF and mixer tuned circuits will certainly provide attenuation to the 22-MHz image, if it is a strong signal it will certainly get into the IF stages and will not be removed from that point on. The graph of RF and mixer tuned circuit response in Fig. 8-3 serves to illustrate the tremendous image frequency response rejection provided by the double-conversion scheme.

Image frequencies are not a major problem for low-frequency carriers, say below 4 MHz. For example, a single-conversion setup for a 4-MHz carrier and a 1-MHz IF means that a 5-MHz local oscillator will be used. The image frequency is 6 MHz, which is far enough away from the 4-MHz carrier so as not to present a problem. At higher frequencies, where images are a problem, the situation is aggravated by the enormous number of transmissions taking place in our crowded communications bands.

EXAMPLE 8-3

Why do you suppose that images tend to be somewhat less of a problem in FM versus AM or SSB communications?

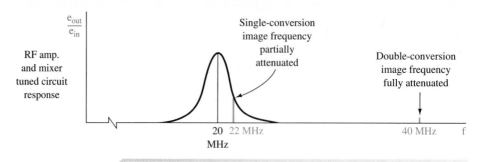

FIGURE 8-3 ■ Image frequency rejection.

SOLUTION

Recall the concept of the *capture* effect in FM systems (Chapter 5). It was shown that if a desired and undesired station are picked up simultaneously, the stronger one tends to be "captured" by inherent suppression of the weaker signal. Thus, a 2:1 signal-to-undesired-signal ratio may result in a 10:1 ratio at the output. This contrasts with AM systems (SSB included) where the 2:1 ratio is carried through to the output.

Up-Conversion

Until recently, the double-conversion scheme, with the lowest IF frequency (often the familiar 455 kHz) doing most of the receiver's selectivity, has been standard practice because the components available made it easiest to achieve the necessary selectivity at low IF frequencies. However, now that VHF crystal filters (30 to 120 MHz) are available for IF circuitry, conversion to a higher IF than RF frequency is popular in sophisticated communications receivers. As an example, consider a receiver tuned to a 30-MHz station and using a 40-MHz IF frequency, as illustrated in Fig. 8-4. This represents an *up-conversion* system in that the IF is a higher frequency than the received signal. The 70-MHz local oscillator mixes with the 30-MHz signal to produce the desired 40-MHz IF. Sufficient IF selectivity at 40 MHz is possible with a crystal filter.

EXAMPLE 8-4

Determine the image frequency for the system of Fig. 8-4.

SOLUTION

If a 110-MHz signal mixes with the 70-MHz local oscillator, a 40-MHz output component results. The image frequency is therefore 110 MHz.

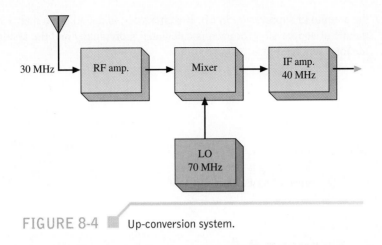

FIGURE 8-4
Up-conversion system.

Example 8-4 shows the superiority of up-conversion. It is highly unlikely that the 110-MHz image could get through the RF amplifier tuned to 30 MHz. There is no need for double conversion and all its necessary extra circuitry. The only disadvantage to up-conversion is the need for a higher-Q IF filter and better high-frequency response IF transistors. The current state-of-the-art in these areas now makes up-conversion economically attractive. Additional advantages over double conversion include better image suppression and less tuning range requirements for the oscillator. The smaller tuning range for up-conversion is illustrated in the following example and minimizes the tracking difficulties of a widely variable local oscillator.

EXAMPLE 8-5

Determine the local oscillator tuning range for the systems illustrated in Figs. 8-1 and 8-4 if the receivers must tune from 20 to 30 MHz.

SOLUTION

The double-conversion local oscillator in Fig. 8-1 is at 30 MHz for a received 20-MHz signal. To provide the same 10-MHz IF frequency for a 30-MHz signal means that the local oscillator must be at 40 MHz. Its tuning range is from 30 to 40 MHz or 40 MHz/30 MHz = 1.33. The up-conversion scheme of Fig. 8-4 has a 70-MHz local oscillator for a 30-MHz input and requires a 60-MHz oscillator for a 20-MHz input. Its tuning ratio is then 70 MHz/60 MHz, or a very low 1.17.

The tuned circuit(s) prior to the mixer is often referred to as the *preselector*. The preselector is responsible for the image frequency rejection characteristics of the receiver. If an image frequency rejection is the result of a single tuned circuit of known Q, the amount of image frequency rejection can be calculated. The following equation predicts the amount of suppression in dB. If suppression is due to more than a single tuned circuit, the dB of suppression for each is calculated individually and the results added to provide the total suppression:

$$\text{image rejection (dB)} \cong 20 \log\left[\left(\frac{f_i}{f_s} - \frac{f_s}{f_i}\right)Q\right]$$ (8-1)

where f_i = image frequency
f_s = desired signal frequency
Q = tuned circuit's Q

EXAMPLE 8-6

An AM broadcast receiver has two identical tuned circuits prior to the IF stage. The Q of these circuits is 60 and the IF frequency is 455 kHz, and the receiver is tuned to a station at 680 kHz. Calculate the amount of image frequency rejection.

SOLUTION

Using Eq. (8-1), the amount of image frequency rejection per stage is calculated:

$$\text{image rejection (dB)} \cong 20 \log\left[\left(\frac{f_i}{f_s} - \frac{f_s}{f_i}\right)Q\right]$$ (8-1)

The image frequency is 680 kHz + (2 × 455 kHz) = 1590 kHz. Thus,

$$20 \log\left[\left(\frac{1590 \text{ kHz}}{680 \text{ kHz}} - \frac{680 \text{ kHz}}{1590 \text{ kHz}}\right)60\right]$$

$$= 20 \log 114.6$$

$$= 41 \text{ dB}$$

Thus, the total suppression is 41 dB plus 41 dB, or 82 dB. This is more than enough to provide excellent image frequency rejection.

8-3 SPECIAL TECHNIQUES

Delayed AGC

The simple automatic gain control (AGC) discussed in Chapter 3 has a minor disadvantage. It provides some gain reduction even to very weak signals. This is illustrated in Fig. 8-5. As soon as even a weak received signal is tuned, simple AGC provides some gain

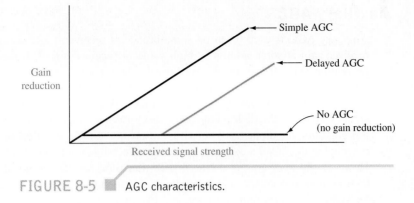

FIGURE 8-5 ◾ AGC characteristics.

reduction. Since communications equipment is often dealing with marginal (weak) signals, it is usually advantageous to add some additional circuitry to provide a *delayed* AGC, that is, an AGC that does not provide any gain reduction until some arbitrary signal level is attained and therefore has no gain reduction for weak signals. This characteristic is also shown in Fig. 8-5.

A simple means of providing delayed AGC is shown in Fig. 8-6. A reverse bias is applied to the cathode of D_1. Thus, the diode looks like an open circuit to the ac signal from the last IF amplifier unless that signal reaches some predetermined instantaneous level. For small IF outputs when D_1 is open, the capacitor C_1 sees a pure ac signal, and thus no dc AGC level is sent back to previous stages to reduce their gain. If the IF output increases, eventually a point is reached where D_1 will conduct on its peak positive levels. This will effectively *short* out the positive peaks of IF output, and C_1 will therefore see a more negative than positive signal and filter it into a relatively constant negative level used to reduce the gain of previous stages. The amplitude of IF output required to start feedback of the "delayed" AGC signal is adjustable by the delayed AGC control potentiometer of Fig. 8-6. This may be an external control so that the user can adjust the amount of delay to suit conditions. For instance, if mostly weak signals are being received, the control might be set so that no AGC signal is developed except for very strong stations. This means the delay interval shown in Fig. 8-5 is increased.

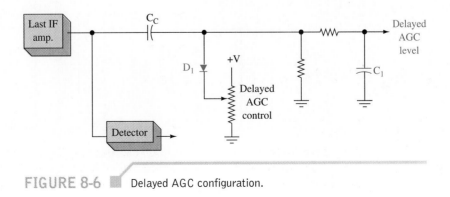

FIGURE 8-6 ◾ Delayed AGC configuration.

Auxiliary AGC

Auxiliary AGC is used (even on some broadcast receivers) to cause a step reduction in receiver gain at some arbitrarily high value of received signal. It then has the effect of preventing very strong signals from overloading a receiver. A simple means of accomplishing the auxiliary AGC function is illustrated in Fig. 8-7. Notice the auxiliary AGC diode connected between the collectors of the mixer and first IF transistors. Under normal signal conditions the dc level at each collector is such that the diode is reverse-biased. In this condition the diode has a very high resistance and has no effect on circuit action. The potential at the mixer's collector is constant since it is not controlled by the normal AGC. However, the AGC control on the first IF transistor, for very strong signals, causes its dc base current to decrease, and hence the collector current also decreases. Thus, its collector voltage becomes more positive, and the diode starts to conduct. The diode resistance goes low, and it loads down the mixer tank (L_1C_1) and thereby produces a step reduction of the signal coupled into the first IF stage. The dynamic AGC range has thereby been substantially increased.

Variable Sensitivity

Despite the increased dynamic range provided by delayed AGC and auxiliary AGC, it is often advantageous for a receiver also to include a variable sensitivity control. This is a manual AGC control in that the user controls the receiver gain (and thus sensitivity) to suit the requirement. A communications receiver may be called upon to deal with signals over a 100,000:1 ratio, and even the most advanced AGC system does not afford that amount of range. Receivers that are designed to provide high sensitivity and that also are able to handle very large input signals incorporate a manual sensitivity control.

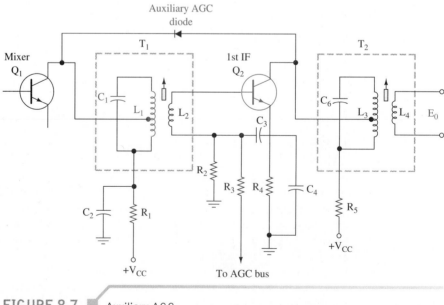

FIGURE 8-7 Auxiliary AGC.

Variable Selectivity

Many communications receivers provide detection to more than one kind of transmission. They may detect code transmissions, SSB, AM, and FM all in one receiver. The required bandwidth to avoid picking up adjacent channels may well vary from 1 kHz for code up to 30 kHz for narrowband FM. Variable selectivity may be accomplished by having a potentiometer across the primary and/or secondary of the last IF stage. The variable-resistance controls the Q and hence BW of the tuned circuit. This control can also be advantageous in situations where a received signal with a 10-kHz bandwidth is being adversely affected by noise. By reducing the receiver's selectivity to only 5 kHz, the external noise should be roughly cut in half, making a marginal reception acceptable. This action will, however, cut off some of the received sidebands and therefore decrease some of the intelligibility. The reduced selectivity therefore is a compromise adjustment in this situation.

The use of variable notch filters (wavetraps) as described in Chapter 7 for TV receivers is also encountered in communications receivers. Recall that a *notch filter* provides high attenuation to a specifically unwanted frequency. If the notch filter's rejection frequency is controllable by the user, you can tune out a specific disturbing signal that may occasionally ruin reception.

Noise Limiter

Man-made sources of external noise are extremely troublesome to highly sensitive communications receivers as well as to any other electronics equipment that is dealing with signals in the microvolt region or less. The interference created by these man-made sources, such as ignition systems, motor communication systems, and switching of high current loads, is a form of *electromagnetic interference* (*EMI*). Reception of EMI by a receiver creates undesired amplitude modulation, sometimes of such large magnitude as to adversely affect FM reception, to say nothing of the complete havoc created in AM systems. While these noise impulses are usually of short duration, it is not uncommon for them to have amplitudes up to 1000 times that of the desired signal. A noise limiter circuit is employed to silence the receiver for the duration of a noise pulse, which is preferable to a very loud crash from the speaker. These circuits are sometimes referred to as automatic noise limiter (ANL) circuits.

A common type of circuit for providing noise limiting is shown in Fig. 8-8. It uses a diode, D_2, that conducts the detected signal to the audio amplifier as long as it is not greater than some prescribed limit. Greater amplitudes cause D_2 to stop conducting until the noise impulse has decreased or ended. The varying audio signal from the diode detector, D_1, is developed across the two 100-kΩ resistors. If the received carrier is producing a -10-V level at the AGC takeoff, the anode of D_2 is at -5 V and the cathode is at -10 V. The diode is "on" and conducts the audio into the audio amplifier. Impulse noise will cause the AGC takeoff voltage to instantaneously increase, which means the anode of D_2 also does. However, its cathode potential does not change instantaneously since the voltage from cathode to ground is across a 0.001-μF capacitor. Remember that the voltage across a capacitance cannot instantaneously change. Therefore, the cathode stays at -10 V, and as the anode approaches -10 V, D_2 turns "off" and the detected audio is blocked from entering the audio amplifier. The receiver is silenced for the duration of the noise pulse.

The switch across D_2 allows the noise limiter action to be disabled by the user. This may be necessary when a marginal (noisy) signal is being received and the set output is being turned off excessively by the ANL.

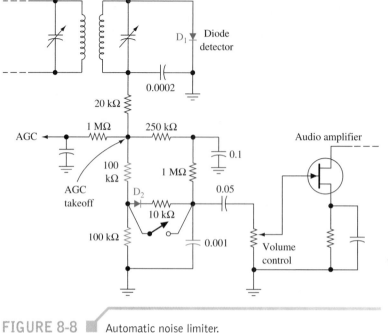

FIGURE 8-8 ▨ Automatic noise limiter.

Metering

Many communications receivers are equipped with a meter that provides a visual indication of received signal strength. It is known as the *S meter* and often is found in the emitter leg of an AGC controlled amplifier stage (RF or IF). It reads dc current, which is usually inversely proportional to received signal strength. With no received signal, there is no AGC bias level, which causes maximum dc emitter current flow and therefore maximum stage voltage gain. As the AGC level increases, indicating an increasing received signal, the dc emitter current goes down, which thereby reduces gain. The S meter can thus be used as an aid to accurate tuning as well as providing a relative guide to signal strength.

In some receivers, the S meter can be electrically switched into different areas so as to allow its use in troubleshooting a malfunction. In those cases, the operator's manual provides a troubleshooting guide on the basis of meter readings obtained in different areas of the receiver.

Squelch

When a sensitive receiver receives no carrier, the AGC action causes maximum system gain, which results in a high degree of noise output. This occurs in many communications applications where the user is constantly monitoring a transmission, such as in police service, but there is no transmission most of the time. Without a squelch system to cut off the receiver's output during transmission lulls, the noise output would cause the user severe aggravation. Squelch circuitry is also useful in minimizing noise output that occurs when tuning between stations. In fact, even better-quality broadcast FM receivers provide

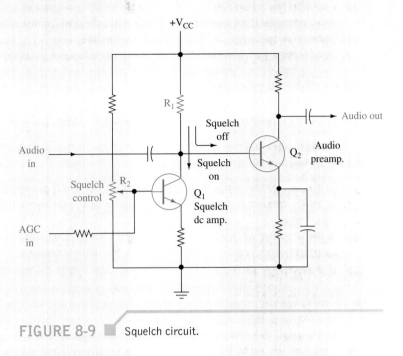

+V_CC

R_1

Squelch off

Audio out

Audio in

Squelch on

Q_2 Audio preamp.

R_2

Squelch control

Q_1
Squelch dc amp.

AGC in

FIGURE 8-9 ■ Squelch circuit.

a squelch capability, but in these applications it is usually termed *muting*. Squelch circuitry is also referred to as the *quieting* or simply the Q circuit.

Figure 8-9 shows a squelch configuration that causes the audio amplifier stage to be cut off whenever no carrier (or an extremely weak station) is being received. In that case the AGC level is zero, which causes the dc amplifier stage, Q_1, to be on. This draws the current being supplied by R_1 into Q_1's collector away from Q_2's base, which means the audio preamp transistor, Q_2, is cut off. Thus, no ac signal is available from Q_2's collector for subsequent power amplification to the speaker. When a signal is picked up by the receiver, the AGC level goes to some negative dc value, which causes Q_1 to turn off. This allows the dc current being supplied by R_1 to enter Q_2's base, which biases it on so that audio preamplification can take place.

User adjustment of the squelch control (R_2) allows the cut-in point of quieting to be varied. This is necessary so that a very weak station, one that generates a small AGC level, will not cause the receiver's output to be squelched.

8-4 RECEIVER NOISE, SENSITIVITY, AND DYNAMIC RANGE RELATIONSHIPS

Now that you have become more knowledgeable about receivers, it is appropriate to expand on the noise considerations provided in Chapter 1. As will be seen, there are various trade-offs and relationships between noise figure, sensitivity, and dynamic range when dealing with high-quality receiver systems.

To understand fully these relationships for a receiver, it is first necessary to recognize the factors limiting sensitivity. In one word, the factor most directly limiting sensitivity is "noise." Without noise it would only be necessary to provide enough amplification to receive any signal, no matter how small. Unfortunately, noise is always present and must be understood and controlled as much as possible.

As explained in Chapter 1, there are many sources of noise. The overwhelming effect in a receiver is thermal noise caused by electron activity in a resistance. From Chapter 1 the noise power is

$$P_n = kT\Delta f \tag{1-2}$$

For a 1-Hz bandwidth (Δf) and at 290 K,

$$P = 1.38 \times 10^{-23} \text{ J/K} \times 290 \text{ K} \times 1 \text{ Hz}$$

$$= 4 \times 10^{-21} \text{ W} = -174 \text{ dBm}$$

For a 1-Hz, 1-K system,

$$P = 1.38 \times 10^{-23} \text{ W} = -198 \text{ dBm}$$

The preceding shows the temperature variable is of interest since it is possible to lower the circuit temperature and decrease noise without changing other system parameters. At 0 K there is no noise generated. Unfortunately, it is very expensive and difficult to operate systems even near 0 K. Most receiving systems are operated at ambient temperature. The other possible means to lower thermal noise is to lower the bandwidth. However, the designer has limited capability in this regard.

Noise and Receiver Sensitivity

What is the sensitivity of a receiver? This question cannot be answered directly without making certain assumptions or knowing certain facts that will have an effect on the result. Examination of the following formula illustrates the dependent factors in determining sensitivity.

$$S = \text{sensitivity} = -174 \text{ dBm} + \text{NF} + 10 \log_{10} \Delta f + \text{desired } S/N \tag{8-2}$$

where -174 dBm is the thermal noise power at room temperature (290 K) in a 1-Hz bandwidth. It is the performance obtainable at room temperature if no other degrading factors are involved. The $10 \log_{10} \Delta f$ factor in Eq. (8-2) represents the change in noise power due to change above a 1-Hz bandwidth. The wider the bandwidth, the greater the noise power and the higher the noise floor. S/N is the desired signal-to-noise ratio in dB. It can be determined for the signal level, which is barely detectable, or it may be considered to be the level allowing an output at various ratings of fidelity. Often, a 0-dB S/N is used, which means that the signal and noise power at the output are equal. The signal can therefore also be said to be equal to the *noise floor* of the receiver. The receiver noise floor and the receiver output noise are one and the same thing.

Consider a receiver that has a 1-MHz bandwidth and a 20-dB noise figure. If a S/N of 10 dB is desired, the sensitivity (S) is

$$S = -174 + 20 + 10 \log(1,000,000) + 10$$

$$= -84 \text{ dBm}$$

It can be seen from this that if a lower *S/N* is required, better receiver sensitivity is necessary. If a 0-dB *S/N* is used, the sensitivity would become −94 dBm. The −94-dBm figure is the level at which the signal power equals noise power in the receiver's bandwidth. If the bandwidth were reduced to 100 kHz while maintaining the same input signal level, the output *S/N* would be increased to 10 dB due to noise power reduction.

Dynamic Range

The *dynamic range* of an amplifier or receiver is the input power range over which it provides a useful output. It should be stressed that a receiver's dynamic and AGC ranges are usually two different quantities. The low-power limit is essentially the sensitivity specification discussed in the preceding paragraphs. It is a function of the noise. The upper limit has to do with the point at which the system no longer provides the same linear increase as related to the input increase. It also has to do with certain distortion components and their degree of effect.

When testing a receiver (or amplifier) for the upper dynamic range limit, it is common to apply a single test frequency and determine the *1-dB compression point*. As shown in Fig. 8-10, this is the point on the input/output relationship where the output has just reached a level where it is 1 dB down from the ideal linear response. The input power at that point is then specified as the upper power limit determination of dynamic range.

In practice, certain distortion characteristics that affect the normally encountered multifrequency signals are often a major factor. When two frequencies (f_1 and f_2) are amplified, the second-order distortion products are generally out of the system passband and are therefore not a problem. They occur at $2f_1$, $2f_2$, $f_1 + f_2$, and $f_1 - f_2$. Unfortunately, the third-order products at $2f_1 + f_2$, $2f_1 - f_2$, $2f_2 - f_1$, and $2f_2 + f_1$ usually have components in the system bandwidth. The distortion thereby introduced, *intermodulation distortion* (IMD), was mentioned in Chapter 6. Recall that the use of MOSFETs at the critical RF and mixer stages is helpful in minimizing these third-order effects. Intermodulation effects have such a major influence on the upper dynamic range of a receiver (or amplifier) that they are often specified via the *third-order intercept point* (or *input intercept*).

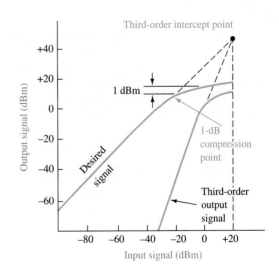

FIGURE 8-10 ▮ Third-order intercept and compression point illustration.

This is illustrated in Fig. 8-10. It is the input power at the point where straight-line extensions of desired and third-order input/output relationships meet. It is about 20 dBm in Fig. 8-10. It is used only as a figure of merit. The better a system is with respect to intermodulation distortion, the higher will be its input intercept.

The dynamic range of a system is usually approximated as

$$\text{dynamic range (dB)} \cong \tfrac{2}{3}\left(\text{input intercept} - \text{noise floor}\right) \tag{8-3}$$

Poor dynamic range causes problems, such as undesired interference and distortion, when a strong signal is received. The current state of the art is a dynamic range of about 100 dB.

EXAMPLE 8-7

A receiver has a 20-dB noise figure (NF), a 1-MHz bandwidth, a +5-dBm third-order intercept point, and a 0-dB S/N. Determine its sensitivity and dynamic range.

SOLUTION

$$S = -174 \text{ dBm} + \text{NF} + 10 \log_{10} \Delta f + \frac{S}{N} \tag{8-2}$$

$$= -174 \text{ dBm} + 20 \text{ dB} + 10 \log_{10} 10^6 + 0 = -94 \text{ dBm}$$

$$\text{dynamic range} \cong \tfrac{2}{3}\left(\text{input intercept} - \text{noise floor}\right) \tag{8-3}$$

$$= \tfrac{2}{3}\left[5 \text{ dBm} - (-94 \text{ dBm})\right]$$

$$= 66 \text{ dB}$$

EXAMPLE 8-8

The receiver from Ex. 8-7 has a preamplifier put at its input. The preamp has a 24-dB gain and a 5-dB NF. Calculate the new sensitivity and dynamic range.

SOLUTION

The first step is to determine the overall system noise ratio (NR). Recall from Chapter 1 that

$$\text{NR} = \log^{-1} \frac{\text{NF}}{10}$$

Letting NR_1 represent the preamp and NR_2 the receiver, we have

$$NR_1 = \log^{-1} \frac{5 \text{ dB}}{10} = 3.16$$

$$NR_2 = \log^{-1} \frac{20 \text{ dB}}{10} = 100$$

The overall NR is

$$NR = NR_1 + \frac{NR_2 - 1}{P_{G_1}} \qquad (1\text{-}9)$$

and

$$P_{G_1} = \log^{-1} \frac{24 \text{ dB}}{10} = 251$$

$$NR = 3.16 + \frac{100 - 1}{251} = 3.55$$

$$NF = 10 \log_{10} 3.55 = 5.5 \text{ dB}$$

$$= \text{total system NF}$$

$$S = -174 \text{ dBm} + 5.5 \text{ dB} + 60 \text{ dB} = -108.5 \text{ dBm}$$

The third-order intercept point of the receiver alone had been +5 dBm but is now preceded by the preamp with 24-dB gain. Assuming that the preamp can deliver 5 dBm to the receiver without any appreciable intermodulation distortion, the system's third-order intercept point is +5 dBm − 24 dB = −19 dBm. Thus,

$$\text{dynamic range} \cong \tfrac{2}{3}\left[-19 \text{ dBm} - (-108.5 \text{ dBm})\right]$$

$$= 59.7 \text{ dB}$$

EXAMPLE 8-9

The 24-dB gain preamp in Ex. 8-8 is replaced with a 10-dB gain preamp with the same 5-dB NF. What is the system's sensitivity and dynamic range?

SOLUTION

$$NR = 3.16 + \frac{100 - 1}{10} = 13.1$$

$$NF = 10 \log_{10} 13.1 = 11.2 \text{ dB}$$

$$S = -174 \text{ dBm} + 11.2 \text{ dB} + 60 \text{ dB} = -102.8 \text{ dBm}$$

$$\text{dynamic range} \cong \tfrac{2}{3}\left[-5 \text{ dBm} - (-102.8 \text{ dB})\right] = 65.2 \text{ dB}$$

The results of Exs. 8-7 to 8-9 are summarized as follows:

	RECEIVER ONLY	RECEIVER AND 10-dB PREAMP	RECEIVER AND 24-dB PREAMP
NF (dB)	20	11.2	5.5
Sensitivity (dBm)	−94	−102.8	−108.5
Third-order intercept point (dBm)	+5	−5	−19
Dynamic range (dB)	66	65.2	59.7

An analysis of the examples and these data shows that the greatest sensitivity can be realized by using a preamplifier with the lowest noise figure and highest available gain in order to mask the higher NF of the receiver. It must be remembered that as gain increases, so does the chance of spurious signals and intermodulation distortion components operating up into the nonlinear region. A preamplifier used prior to a receiver input has the effect of decreasing the third-order intercept proportionally to the gain of the amplifier while the increase in sensitivity is less than the gain of the amplifier. Therefore, to maintain a high dynamic range it is best only to use the amplification needed to obtain the desired noise figure. It is not helpful in an overall sense to use excessive gain. The data in the chart show that adding the 10-dB gain preamplifier improved sensitivity by 8.8 dB and decreased dynamic range by only 0.8 dB. The 24-dB gain preamp improved sensitivity by 14.5 dB but decreased the dynamic range by 6.3 dB.

Intermodulation Distortion Testing

It is common to test an amplifier for its intermodulation distortion (IMD) by comparing two test frequencies to the level of a specific IMD product. As previously mentioned, the second-order products are usually outside the frequency range of concern. This is generally true for all the even-order products and is illustrated in Fig. 8-11. It shows some second-order products that would be outside the bandwidth of interest for most systems.

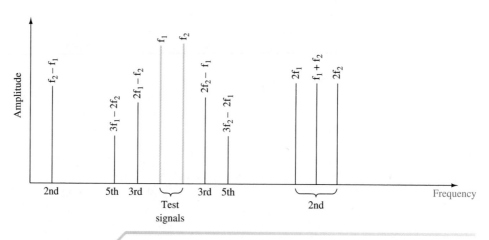

FIGURE 8-11 IMD products (second-, third-, and fifth-order for two test signals).

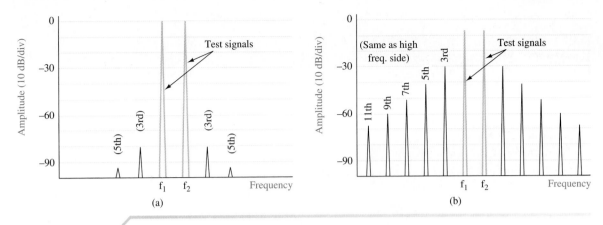

FIGURE 8-12 ░ IMD testing: (a) mixer; (b) Class AB linear power amplifier.

The odd products are of interest since some of them can be quite close to the test frequencies f_1 and f_2 shown in Fig. 8-11. The third-order products shown ($2f_2 - f_1$ and $2f_1 - f_2$) have the most effect, but even the fifth-order products ($3f_2 - 2f_1$ and $3f_1 - 2f_2$) can be troublesome. Figure 8-12(a) shows a typical spectrum analyzer display when two test signals are applied to a mixer or small-signal amplifier. Notice that the third-order products are shown 80 dB down from the test signals while the fifth-order products are more than 90 dB down.

Figure 8-12(b) shows the IMD testing result when applying two frequencies to a typical Class AB linear power amplifier. The higher odd-order products (up to the eleventh in this case) are significant for the power amplifier. Fortunately, these effects are less critical in power amplifiers than in the sensitive front-end of a radio receiver.

8-5 FREQUENCY SYNTHESIS

Most transceiver designs use frequency synthesizers to generate the highly accurate frequencies used for the transmitter carrier and receiver local oscillator. They are also widely used in signal generators and instrumentation systems such as spectrum analyzers and modulation analyzers. The concept of frequency synthesis has been around since the 1930s, but the cost of the circuitry necessary was prohibitive for most designs until integrated circuit technology started offering the phased-locked loop (PLL) in a single, low-cost chip. Recall that basic PLL theory was provided in Chapter 6.

A basic frequency synthesizer is shown in Fig. 8-13. Besides the PLL, the synthesizer includes a very stable crystal oscillator and the divide-by-N programmable divider. The output frequency of the voltage-controlled oscillator (VCO) is a function of the applied control voltage.

The output of the phase comparator is a voltage termed the *error voltage,* which is proportional to the phase difference between the signals at its two inputs. This output controls the frequency of the VCO so that the phase comparator input from the VCO via the variable divider ($\div N$) remains at a constant phase difference with the reference input, f_R, so that the frequencies are equal. The VCO frequency is thus maintained at Nf_R. Such a synthesizer will produce a number of frequencies separated by f_R and is the most basic form of phase-locked synthesizer. Its stability is directly governed by the stability of the

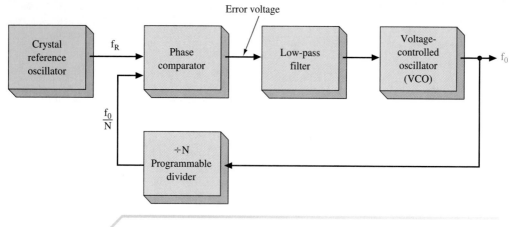

FIGURE 8-13 ▦ Basic frequency synthesizer.

reference input f_R, although it is also related to noise in the phase comparator, noise in any dc amplifier between the phase comparator and the VCO, and the characteristics of the low-pass filter usually placed between the phase comparator and the VCO.

Consider the case where the programmable divider in Fig. 8-13 can divide by integers from N equals 1 to 10. If the reference frequency is 100 kHz and $N = 1$, then the output should be 100 kHz. If $N = 2$, f_0 must equal 200 kHz to provide a constant phase difference for the phase comparator. Similarly, for $N = 5$, $f_0 = 500$ kHz. The pattern should be apparent to you now. A synthesizer with outputs of 100 kHz, 200 kHz, 300 kHz, and so on, is not useful for most applications. Much smaller spacing between output frequencies is necessary, and means to attain that condition are discussed next.

The design of frequency synthesizers using the principle described above involves the design of various subsystems, including the VCO, the phase comparator, any low-pass filters in the feedback path, and the programmable dividers.

Programmable Division

A typical programmable divider is shown in Fig. 8-14. It consists of three stages with division ratios K_1, K_2, and K_3, which may be programmed by inputs P_1, P_2, and P_3, respectively. Each stage divides by K_n except during the first cycle after the program

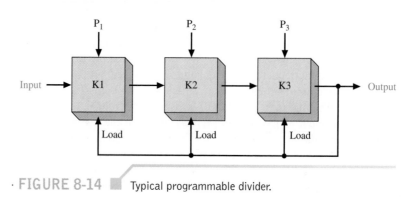

· FIGURE 8-14 ▦ Typical programmable divider.

input P_n is loaded, when it divides by P (which may have any integral value from n to K). Hence the counter illustrated divides by $P_3 \times (K_1K_2) + P_2K_1 + P_1$, and when an output pulse occurs the program inputs are reloaded. The counter will divide by any integer between 1 and $(K_1K_2K_3 - 1)$.

The most common programmable dividers are either decades or divide-by-16 counters. These are readily available in various logic families, including CMOS and TTL. It is possible to buy CMOS quad decades in a single package. Using such a package one can program a value of N from about 3 to 9999. The theoretical minimum count of 1 is not possible because of the effects of circuit propagation delays. The use of such counters permits the design of frequency synthesizers, which are programmed with decimal thumbwheel switches and use a minimum of components. If a synthesizer is required with an output of nonconsecutive frequencies and steps, a custom programmable counter may be made using some custom logic family, such as PMOS, NMOS, CMOS, or I²L. This is the case for the citizen's band synthesizer discussed later in this section.

The maximum input frequency of a programmable divider is limited by the speed of the logic used, and more particularly by the time taken to load the programmed count. Few programmable dividers of the type discussed will operate with test frequencies much above 20 MHz. The faster types, operating at perhaps 50 MHz, use Schottky TTL, which consumes considerable power and has a tendency to inject HF and VHF noise (EMI) into supply lines. The output frequency of the simple synthesizer in Fig. 8-13 is of course limited to the maximum frequency of the programmable divider.

There are many ways of overcoming this limitation on synthesizer frequency. The VCO output may be mixed with the output of a crystal oscillator and the resulting difference frequency fed to the programmable divider, or the VCO output may be multiplied from a low value in the operating range of the programmable divider to the required high output frequency. Alternatively, a fixed ratio divider capable of operating at a high frequency may be interposed between the VCO and the programmable divider. These methods are shown in Fig. 8-15(a), (b), and (c), respectively.

All the methods discussed above have their problems, although all have been used and will doubtless continue to be used in some applications. Method (a) is the most useful technique since it allows narrower channel spacing or high reference frequencies (hence faster lock times and less loop-generated jitter) than the other two, but it has the drawback that since the crystal oscillator and the mixer are within the loop, any crystal oscillator noise or mixer noise appears in the synthesizer output. Nevertheless, this technique has much to recommend it.

The other two techniques are less useful. Frequency multiplication introduces noise and both techniques must either use a very low reference or rather wide channel spacing. What is needed is a programmable divider that operates at the VCO frequency—one can then discard the techniques described above and synthesize directly at whatever frequency is required.

Two-Modulus Dividers

Considerations of speed and power make it impractical to design programmable counters of the type described above, even using ECL, at frequencies much into the VHF band (30 to 300 MHz) or above. A different technique exists, however, using two-modulus dividers; that is, in one mode it divides by N and in the other mode by $N + 1$.

Figure 8-16 shows a divider using a two-modulus prescaler. The system is similar to the one shown in Fig. 8-13(c), but in this case the prescaler divides by either N or $N +$

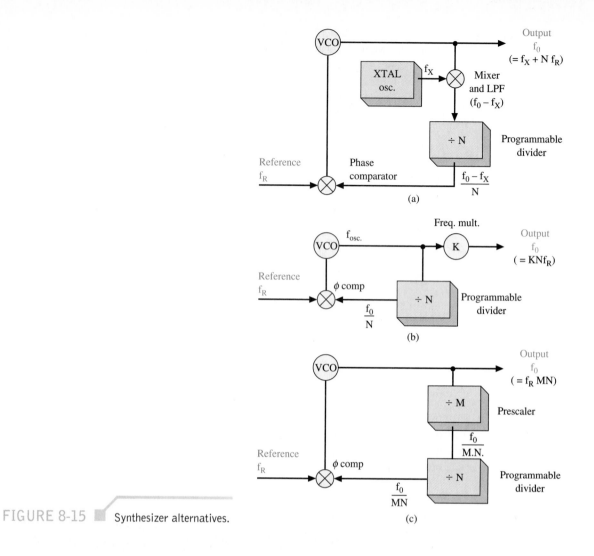

FIGURE 8-15 ■ Synthesizer alternatives.

1, depending on the logic state of the control input. The output of the prescaler feeds two normal programmable counters. Counter 1 controls the two-modulus prescaler and has a division ratio A. Counter 2, which drives the output, has a division ratio M. In operation the $N/(N + 1)$ prescaler (Fig. 8-16) divides by $N + 1$ until the count in programmable counter 1 reaches A and then divides by N until the count in programmable counter 2 reaches M, when both counters are reloaded and a pulse passes to output and the cycle restarts. The division ratio of the entire system is $A(N + 1) + N(M - A)$, which equals $NM + A$. There is only one constraint on the system—since the two-modulus prescaler does not change modulus until counter 1 reaches A, the count in counter 2 (M) must never be less than A. This limits the minimum count the system may reach to $A(N + 1)$, where A is the maximum possible value of count in counter 1.

The use of this system entirely overcomes the problems of high-speed programmable division mentioned earlier. A number of ÷ 10/11 counters working at frequencies of up to 500 MHz and also ÷ 5/6, ÷ 6/7, and ÷ 8/9 counters working up to 500 MHz are now readily available. There is also a pair of circuits intended to allow ÷ 10/11 counters to be used in ÷ 40/41 and ÷ 80/81 counters in 25-kHz and 12.5-kHz channel VHF synthesizers. It is not necessary for two-modulus prescalers to divide by $N/(N + 1)$. The same principles

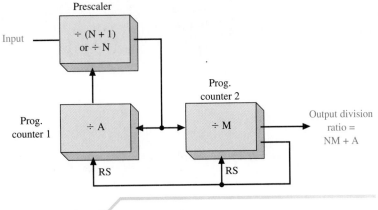

Prescaler

Input → $\div (N+1)$ or $\div N$

Prog. counter 1 → $\div A$

Prog. counter 2 → $\div M$

Output division ratio = $NM + A$

RS   RS

FIGURE 8-16 ▪ Divider system with two-modulus prescaler.

apply to $\div N/(N + Q)$ counters (where Q is any integer), but $\div N/(N + 1)$ tends to be most useful.

CITIZEN'S BAND SYNTHESIZER   The availability of low-cost PLL and programmable divider ICs has led to the use of synthesizers in virtually all channelized transceivers. This is true even for the very low-cost systems used on the 40-channel citizen's band. A typical CB transceiver is shown in Fig. 8-17.

The synthesizer circuit shown at Fig. 8-18 allows all necessary frequencies for a CB transceiver to be generated by using a single-crystal oscillator. The 40 channels are spaced at 10-kHz intervals (with some gaps) between 26.965 and 27.405 MHz. Local oscillator frequencies for the reception of these channels with intermediate frequencies of 455 kHz, 10.240 MHz, and 10.700 MHz are also synthesized. Table 8-1 shows the rela-

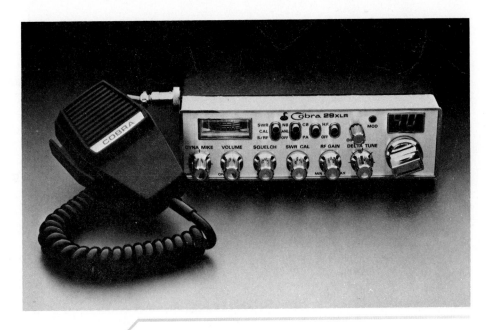

FIGURE 8-17 ▪ CB transceiver. (Courtesy of Dynascan Corp.)

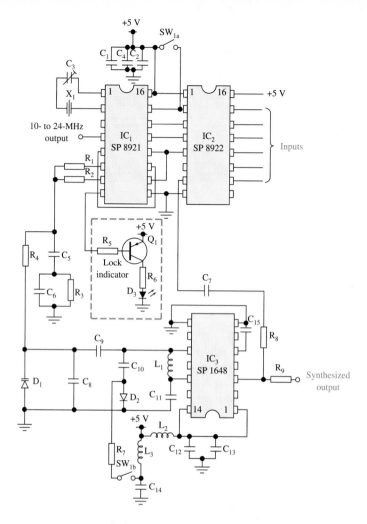

All resistors are $\frac{1}{8}$ W ± 10% unless otherwise stated. Capacitor values are in microfarads unless otherwise stated.

IC_1	SP8921	R_7	1 kΩ
IC_2	SP8922/SP8923	R_8	1 kΩ
IC_3	SP1648	R_9	470 (adjust for required output level)
Q_1	2N 3906	C_1	0.1
D_1	ZC822, Ferranti varactor diode	C_2	100, 10 V solid tantalum
D_2	1N4148 Silicon diode	C_3	2-22 pF variable
D_3	LED lock indicator	C_4	100 10 V solid tantalum
X_1	10.240-MHz crystal, series mode	C_5	10 V solid tantalum
L_1	11 turns 30-gauge cotton-covered wire	C_6	0.1
	on Neosid A7 assembly	C_7	1000 pF
L_2	100-μH RF choke	C_8	22 pF +10%
L_3	100-μH RF choke	C_9	0.01
R_1	1.0 kΩ ± 5%	C_{10}	100 pF ± 10%
R_2	1.0 kΩ ± 5%	C_{11}	0.01
R_3	8.2 kΩ ± 5%	C_{12}	0.1
R_4	33 kΩ	C_{13}	10 solid tantalum
R_5	10 kΩ	C_{14}	0.1
R_6	150 (adjust for LED brightness)	C_{15}	1000 pF
		SW_1	2-pole, 1-way switch, (receive/transmit)

FIGURE 8-18 ▧ CB synthesizer circuit.

TABLE 8-1 *Input Code for CB Synthesizer of Fig. 8-18*

CHANNEL NUMBER	INPUT CODE						OUTPUT FREQUENCY WITH $R/T = 0$ (MHz)
	F	E	D	C	B	A	
1	0	0	0	1	1	1	26.965
2	0	0	1	0	0	0	26.975
3	0	0	1	0	0	1	26.985
4	0	0	1	0	1	1	27.005
5	0	0	1	1	0	0	27.015
6	0	0	1	1	0	1	27.025
7	0	0	1	1	1	0	27.035
8	0	1	0	0	0	0	27.055
9	0	1	0	0	0	1	27.065
10	0	1	0	0	1	0	27.075
11	0	1	0	0	1	1	27.085
12	0	1	0	1	0	1	27.105
13	0	1	0	1	1	0	27.115
14	0	1	0	1	1	1	27.125
15	0	1	1	0	0	0	27.135
16	0	1	1	0	1	0	27.155
17	0	1	1	0	1	1	27.165
18	0	1	1	1	0	0	27.175
19	0	1	1	1	0	1	27.185
20	0	1	1	1	1	1	27.205
21	1	0	0	0	0	0	27.215
22	1	0	0	0	0	1	27.225
23	1	0	0	1	0	0	27.255
24	1	0	0	0	1	0	27.235
25	1	0	0	0	1	1	27.245
26	1	0	0	1	0	1	27.265
27	1	0	0	1	1	0	27.275
28	1	0	0	1	1	1	27.285
29	1	0	1	0	0	0	27.295
30	1	0	1	0	0	1	27.305
31	1	0	1	0	1	0	27.315
32	1	0	1	0	1	1	27.325
33	1	0	1	1	0	0	27.335
34	1	0	1	1	0	1	27.345
35	1	0	1	1	1	0	27.355
36	1	0	1	1	1	1	27.365
37	1	1	0	0	0	0	27.375
38	1	1	0	0	0	1	27.385
39	1	1	0	0	1	0	27.395
40	1	1	0	0	1	1	27.405

tionship between the program input and the channel selected. By using a program other than one of the 40 given, other frequencies may be selected—in fact, there are 64 channels at 10-kHz separation available from 26.895 to 27.525 MHz and programming starts at all zeros on input A through F for 26.895, and each increase of one bit to the binary number on these inputs increases the channel frequency by 10 kHz until all 1s give 27.535 MHz. The A input is the least significant bit, F the most significant. The programming input on pin 16 of the SP8922 IC in Fig. 8-18 is normally kept high, but making it low increases the programmed frequency by 5 kHz. Table 8-2 shows the programming required to obtain various offsets. This synthesizer is intended for use in double-conversion receivers with IFs of 10.695 MHz and 455 kHz and generates either the frequency programmed or the frequency programmed less 10.695 MHz.

If other offsets are programmed in connections to pin 15 of the SP8921 IC and pin 2 of the SP8922 IC, they must be altered according to Table 8-2. The synthesizer consists of the SP8921 and the SP8922 plus an SP1648 voltage-controlled oscillator. The programming inputs to the SP8922 are as shown in Table 8-1. Logic 1 is +3 V or more; logic 0 is either ground or an open circuit.

The crystal oscillator in the SP8921 (Fig. 8-18) is trimmed by a small variable capacitor, C_3, which must be set up during alignment of the synthesizer so that the output frequency on pin 4 is 10.240000 MHz. The only other adjustment is to set the core of L_1 so that the varicap control voltage (D_1) is 2.85 V when the synthesizer is set to channel 30 transmit. Since the difference between transmit and receive frequencies is over 10 MHz, it is not possible to tune both with the same tuned circuit, and an extra capacitor is switched by means of a diode during reception.

The phase/frequency comparator of the SP8921 can have an output swing from 0.5 V to 3.8 V, but it is better to work in the range 1.5 to 3.0 V, as the phase-error output voltage is more linear in this region. The ZC822 tuning diode specified for this synthesizer may be replaced by any other tuning diode provided that it will tune the VCO over the required range, or a little more, as the control voltage goes from 1.5 V to 3.0 V. With slight coil changes the MV2105 has been used successfully in this synthesizer.

The low-pass filter of the PLL consists of C_5, C_6, and R_3. If faster lock (at the expense of larger noise and reference sidebands) is required, the filter may be redesigned. If the synthesizer is used in a scanning receiver, a switched filter should be used to give fast lock during scanning but a slower lock and cleaner signal during normal operation. A scanning receiver automatically "searches" a number of channels until it finds a good received signal. It then stays tuned to that channel until "nudged" by the operator and/or the reception is lost. The lock output on pin 8 of the SP8921 is used to light an indicator when the loop is not locked and should also be used, in a transmitter or transceiver, to prevent transmission when the loop is unlocked.

TABLE 8-2 Offset Codes for Synthesizer of Fig. 8-18

Offset	SP8921	SP8922
0	0	0
455 kHz	0	1
10.240 MHz	1	0
10.695 MHz	1	1

Figure 8-19 shows the circuit board layout and component location of this synthesizer. It requires a single +5-V supply and draws about 60 mA. The performance is improved if a double-sided board is used with a ground plane on one side. A small further improvement would come from the use of a grounded screening can over the whole system to prevent stray noise pickup.

The synthesizer has reference frequency sidebands 50 dB down at 1.25 kHz from the carrier. All output over 5 kHz from the carrier is over 70 dB down. Lock time for a change from channel 0 to channel 40 (a frequency change of 440 kHz) is around 35 ms. Stepping from transmit to receive, or vice versa, takes somewhat longer because of the much larger change of frequency but is generally complete within 75 ms.

LOW-FREQUENCY SYNTHESIZER   The generation of low frequencies is generally accomplished with *RC* oscillators. The techniques recently developed for frequency synthesis can now be used conveniently when very precise low frequencies are required.

A low-frequency sine-wave synthesizer will demonstrate accuracy and stability characteristics that can be measured in parts per million (even at 0.01 Hz) if it is crystal-controlled. The synthesizer in Fig. 8-20 is based on an 8-bit digital-to-analog converter (AD7523), which presents a precise triangle wave to the timing of a function-generator chip (ICL8038). The converter receives an 8-bit code from an up/down counter (CD4040B), which operates at a frequency 2^9 times greater than the desired sine-wave output. The crystal frequency required is

$$f_{crystal} = 2^9 \times N \times f_{out}$$

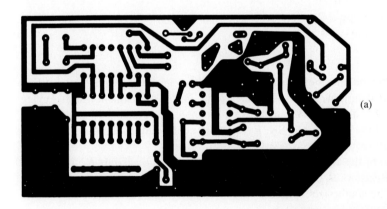

(a)

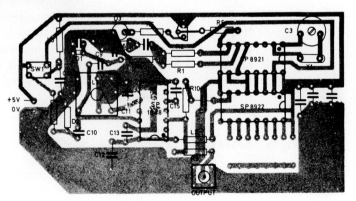

(b)

FIGURE 8-19 ▮ Printed circuit board details: (a) printed circuit layout for CB synthesizer; (b) component layout for CB synthesizer.

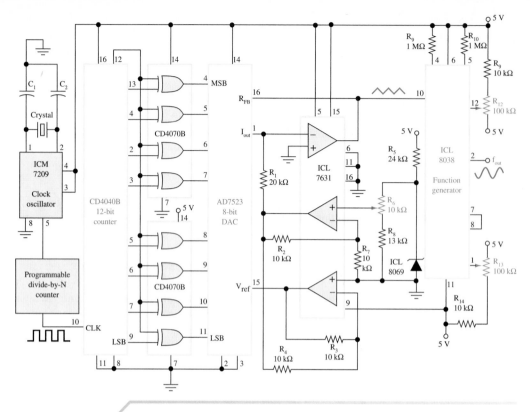

FIGURE 8-20 ■ Low-frequency synthesizer. (Reprinted with permission from *Electronic Design,* Vol. 29, No. 9; copyright Hayden Publishing Co., Inc., 1981.)

The phase of the output sine wave can be precisely controlled by using a presettable up/down counter to determine the cycle start point. For easy frequency-shift keying (FSK), a programmable divide-by-N counter is placed after the crystal oscillator. Chapter 10 provides detail on the concept of FSK. By digitally controlling N, the user modulates the sine-wave output frequency.

The amplitude of the converter's triangle-wave output is adjusted by R_6 to approximately 3.3 V p-p. Resistors R_{12} and R_{13} are adjusted to give the lowest distortion to the output sine wave. The synthesizer consumes about 100 mW and can be used in portable equipment.

The crystal control has two advantages over other methods for generating low-frequency sinusoids. First, the circuit's temperature stability is superior because the crystal determines these characteristics; RC sine-wave oscillators depend on stable and accurate resistors and capacitors to get good temperature performance. Second, unlike RC oscillators that require tuning, the synthesized, crystal-controlled approach requires no trimming to get accurate frequencies.

8-6 DIRECT DIGITAL SYNTHESIS

Direct digital synthesis (DDS) systems became economically feasible in the late 1980s. They offer some advantages over the analog synthesizers discussed in the previous section but generally tend to be somewhat more complex and expensive. They are, however, useful for some applications as will be discussed. The digital logic used can improve on the repeatability and drift problems of analog units that often require select-by-test components. These advantages also apply to digital filters that have replaced some standard analog ones in recent years. The disadvantages of DDS (and digital filters) are the relatively limited maximum output frequency and greater complexity/cost considerations.

A block diagram for a basic DDS system is provided in Fig. 8-21. The numerically controlled oscillator (NCO) contains the phase accumulator and read-only memory (ROM) look-up table. The NCO provides the updated information to the digital-to-analog converter (DAC) to generate the RF output.

The phase accumulator generates a phase increment of the output waveform based upon its input (Δ phase in Fig. 8-21). The input (Δ phase) is a digital word that, in conjunction with the reference oscillator (f_{CLK}), determines the frequency of the output waveform. The output of the phase accumulator serves as a variable-frequency oscillator generating a digital ramp. The frequency of the signal is defined by Δ phase as

$$f_{out} = \frac{(\Delta \ phase)f_{CLK}}{2^N} \tag{8-4}$$

for a N-bit phase accumulator.

Translating phase information from the phase accumulator into amplitude data is accomplished by means of the look-up table stored in memory. Its digital output (amplitude data) is converted into an analog signal by the DAC. The low-pass filter provides a spectrally pure sine-wave output.

The final output frequency is typically limited to about 40% of f_{CLK}. The phase accumulator size is chosen based upon the desired frequency resolution, which is equal to $f_{CLK} \div 2^N$ for a N-bit accumulator. The physical layout for a DDS IC is shown in Fig. 8-22. The Analog Devices AD9955 can operate with clock frequencies as high as 100 MHz. It contains a 32-bit phase accumulator and a 15-bit phase to 12-bit sine amplitude converter. Recall that the phase-to-amplitude transformation is performed by the read-only memory look-up table.

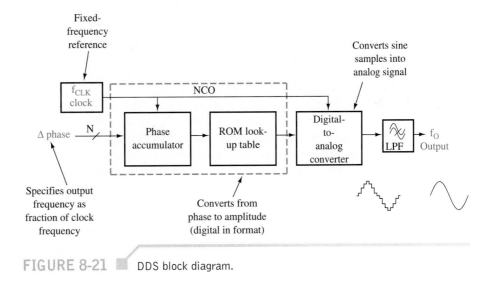

FIGURE 8-21 ▮ DDS block diagram.

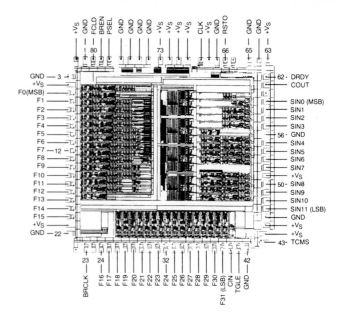

FIGURE 8-22 AD9955 chip layout.
(Courtesy of Analog Devices.)

EXAMPLE 8-10

Calculate the maximum output frequency and frequency resolution for the AD9955 DDS when operated at $f_{\text{CLK MAX}}$.

SOLUTION

The maximum output frequency is approximately 40% of $f_{\text{CLK MAX}}$

$$= 0.40 \times 100 \text{ MHz}$$

$$= 40 \text{ MHz}$$

The frequency resolution is given by $f_{\text{CLK}} \div 2^N$

$$= \frac{100 \text{ MHz}}{2^{32}} \cong 0.023 \text{ Hz}$$

The preceding example shows that DDS offers the possibility for extremely small frequency increments. This is one of the advantages offered by DDS over the analog synthesizers described in Sec. 8-5. Another DDS advantage is the ability to shift frequencies very quickly. This characteristic is very useful for the spread-spectrum systems described in Sec. 8-8.

The disadvantages of DDS include the limit on maximum output frequency and higher *phase noise*. Spurious changes in phase of the synthesizer's output result in energy at frequencies other than the desired one. This phase noise is often specified for all types of oscillators and synthesizers. It is usually specified in dB/$\sqrt{\text{Hz}}$ at a particular offset from center frequency. A specification of -90 dB/$\sqrt{\text{Hz}}$ at a 10-kHz offset means that noise energy at a 1-Hz bandwidth 10 kHz away from the center frequency should be 90 dB lower than the center frequency output. In a sensitive receiver, phase noise will mask out a weak signal that would otherwise be detected.

8-7 FM COMMUNICATIONS TRANSCEIVERS

A large amount of two-way communication takes place in the 150- to 174-MHz band via frequency modulation. In fact, this band is so heavily utilized by business and industry that the overflow is going to allocations at 450 or 470 MHz and around 900 MHz. The radio described in this section is a basic design selected to show you, in detailed fashion, complete circuit and troubleshooting information.

The unit shown in Fig. 8-23 is the General Electric Phoenix model. It is a basic FM transceiver available at one specific frequency of operation in the range 150 to 174 or 450 to 470 MHz. We shall detail a 150- to 174-MHz version, but it is very similar to the higher-frequency model. These units provide very reliable communication up to about 40 miles using transmitter output powers of 25 W (150 to 174 MHz) or 20 W (450 to 470 MHz).

The only major feature this transceiver has that has not previously been described is the *channel guard* function. It causes a very specific audio frequency to be encoded onto the carrier together with the audio. If the intended receiver does not detect this tone, it inhibits reception of the call. This feature minimizes unwanted reception and allows for a certain degree of privacy in these extremely crowded channels.

The detailed block diagram for this transceiver is shown in Fig. 8-24. The detailed circuit analysis to follow will be easier to understand if you take a few minutes now to familiarize yourself with the functional block operation.

Circuit Analysis

TRANSMITTER (FIG. 8-24) The transmitter utilizes a crystal-controlled, frequency-modulated exciter for two-frequency operation in the 150- to 174-MHz frequency band. The transmitter consists of audio processor U_{101}, oscillator Q_{151}, buffer Q_{153},

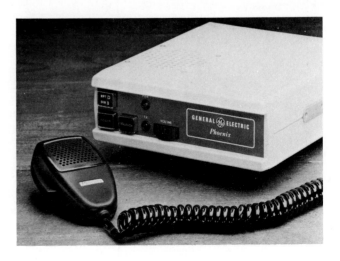

FIGURE 8-23 ■ Basic communications transceiver. (Courtesy of General Electric Company.)

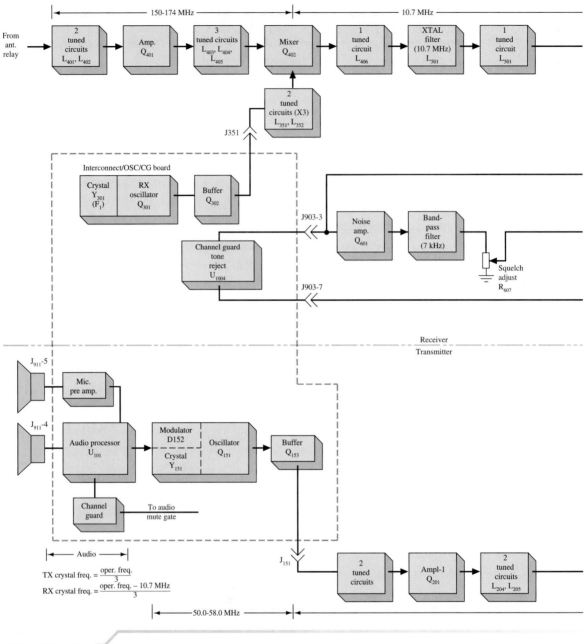

$\qquad$ 150-174 MHz $\qquad$ | $\qquad$ 10.7 MHz $\qquad$

From ant. relay → | 2 tuned circuits L_{401}, L_{402} | Amp. Q_{401} | 3 tuned circuits L_{403}, L_{404}, L_{405} | Mixer Q_{402} | 1 tuned circuit L_{406} | XTAL filter (10.7 MHz) L_{501} | 1 tuned circuit L_{501}

2 tuned circuits (X3) L_{351}, L_{352}

J351

Interconnect/OSC/CG board

Crystal Y_{301} (F_1) | RX oscillator Q_{301} | Buffer Q_{302}

Channel guard tone reject U_{1004}

J903-3 | Noise amp. Q_{601} | Band-pass filter (7 kHz)

Squelch adjust R_{607}

J903-7

Receiver
Transmitter

J_{911}-5 | Mic. pre amp.

J_{911}-4 | Audio processor U_{101} | Modulator D152 / Crystal Y_{151} | Oscillator Q_{151} | Buffer Q_{153}

Channel guard | To audio mute gate

|← Audio →|

J_{151}

2 tuned circuits | Ampl-1 Q_{201} | 2 tuned circuits L_{204}, L_{205}

TX crystal freq. = $\dfrac{\text{oper. freq.}}{3}$

RX crystal freq. = $\dfrac{\text{oper. freq.} - 10.7 \text{ MHz}}{3}$

|← $\qquad$ 50.0-58.0 MHz $\qquad$ →|

FIGURE 8-24 ■ Block diagram. (Courtesy of General Electric Company.)

exciter stages Q_{201} through Q_{203}, PA amplifier Q_{251} and Q_{252}, and power control circuit Q_{254} through Q_{257}. The exciter provides approximately 300 to 350 mW of modulated RF to the power amplifier, which provides rated output power. Figure 8-24 is a block diagram of the radio showing both the transmitter and receiver.

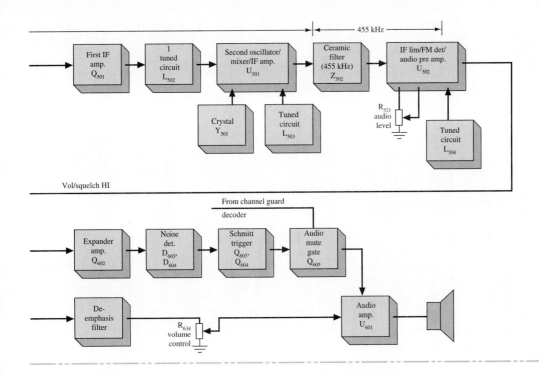

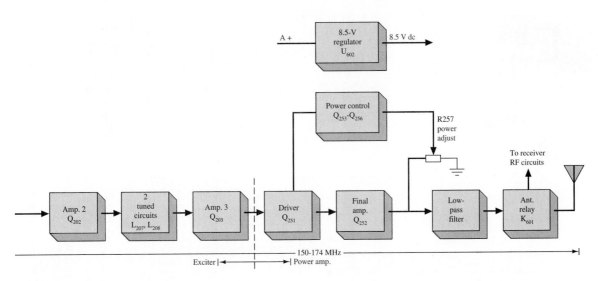

FIGURE 8-24 ▥ (Continued)

MICROPHONE PREAMPLIFIER (FIG. 8-25)   A preamplifier stage (Q_{901} and associated circuitry) is provided for the standard electret microphone without a built-in preamplifier. The preamplifier circuit is located on the IOC board.

With this microphone, MIC HI is coupled through J_{911-5} to the preamplifier stage. The J_{911-5} refers to the J_{911} connector (on the IOC board in Fig. 8-25) and indicates pin 5 of that connector. The amplified output is coupled through R_{908} and C_{905} to the audio

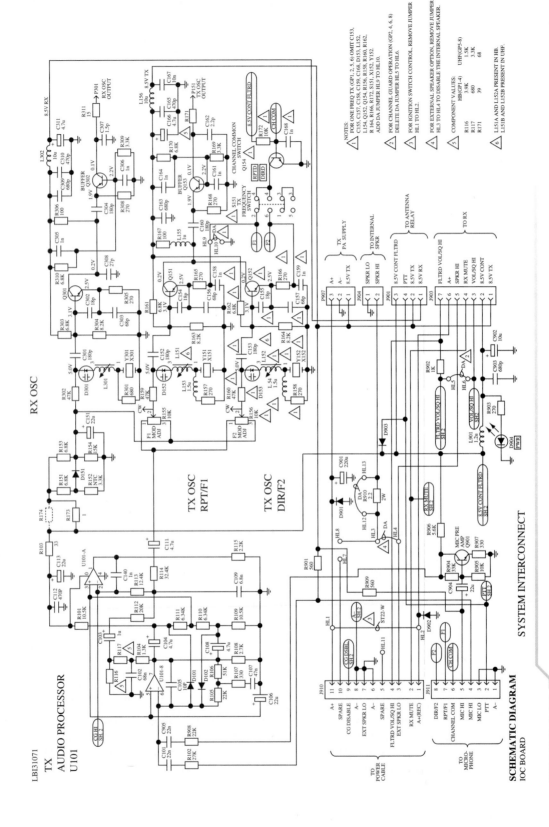

SCHEMATIC DIAGRAM
IOC BOARD

FIGURE 8-25 ■ Audio, IOC, and oscillator schematic. (Courtesy of General Electric Company.)

processor. For optional microphones with a built-in preamplifier, audio is coupled through J_{911-4}, bypassing MIC preamp Q_{901}.

AUDIO PROCESSOR U_{101} (FIG. 8-25)   The audio processor provides audio preemphasis with amplitude limiting and postlimiter filtering. A total gain of approximately 24 dB is realized through the audio processor. Twenty dB is provided by U_{101B} and 4 dB by U_{101A}.

The 8.5-V regulator powers the audio processor and applies regulated +8.5 V through P_{903-2} to a voltage divider consisting of R_{101}, R_{111}, R_{110}, and R_{109}. The +4.25-V output from the voltage divider establishes the operating reference point for both operational amplifiers. C_{106} provides an ac ground at the summing input of both operational amplifiers (U_{101A} and U_{101B}).

Resistors R_{109}, R_{110}, and R_{111} and diodes D_{101} and D_{102} provide limiting for U_{101B}. Diodes D_{101} and D_{102} are reverse-biased at +1.7 V dc. Voltage-divider network R_{109}, R_{110}, and R_{111} provides +5.9 V dc at the cathode of D_{101} and +2.6 V dc at the anode of D_{102}. The voltage at the junction of D_{101} and D_{102} is 4.25 V dc. C_{104} and C_{108} permit a dc-level change between U_{101B-7} and the voltage-divider network for diode biasing.

When the input signal to U_{101B-6} is of a magnitude such that the amplifier output at U_{101B-7} does not exceed 4 V p-p, the amplifier provides a nominal 20-dB gain. When the audio signal level at U_{101B-7} exceeds 4 V p-p, diodes D_{101} and D_{102} conduct on the positive and negative half-cycles, providing 100% negative feedback to reduce the amplifier gain to 1. This limits the audio amplitude at U_{101B-7} to 5 V p-p.

Resistors R_{105}, R_{106}, and R_{107} and C_{107} comprise the audio preemphasis network that enhances the signal-to-noise ratio (see Sec. 5-4). R_{107} and C_{107} control the preemphasis curve below limiting. R_{106} and C_{107} control the cutoff point for high-frequency preemphasis. As high frequencies are attenuated, the gain of U_{101} is increased.

Audio from the preamplifier or microphone is coupled to the input of operational amplifier U_{101B-6}. The amplified output of U_{101B} is coupled through R_{114}, R_{112}, R_{104}, and R_{117} to a second operational amplifier, U_{101A}.

The channel guard (CG) tone input is applied to U_{101A-2}. The CG tone is then combined with the microphone audio.

A postlimiter filter consisting of U_{101A}, R_{112}–R_{114}, C_{108}, and C_{110} provides 12 dB/octave roll-off. R_{104} and C_{102} provide an additional 6 dB/octave roll-off, for a total of 18 dB.

The output of the postlimiter filter is coupled through C_{110} to the temperature-compensated transmitter oscillators Q_{151} and Q_{152}.

TRANSMITTER OSCILLATOR (FIG. 8-25)   A temperature-compensating network consisting of R_{151}, R_{152}, R_{153}, R_{154}, D_{151}, and C_{151} maintains oscillator frequency over a temperature range of −30 to +60° C. The temperature-compensating dc voltage and audio are applied to FM modulator D_{152} through MOD ADJ control R_{155} and R_{159}. Modulator varactor D_{151} varies the transmit frequency at the audio rate applied from the audio processor.

Q_{151}, Y_{151}, and associated circuitry comprise a Colpitts oscillator that generates a frequency that is one-third of the desired carrier. The transmit oscillator is adjusted to the assigned operating frequency by L_{151}. The oscillator output is applied to buffer Q_{153} and then coupled through P_{151} to the exciter circuitry on the transmitter/receiver (Tx/Rx) board.

EXCITER (FIG. 8-26)   The exciter consists of three amplifier stages that provide a minimum of 300 mW to the PA section of the transmitter. In addition to providing approximately 22 dB of gain, the exciter contains filters that determine the bandwidth and spurious characteristics of the transmitter.

The output of the oscillator and buffer stages is coupled to the input of the exciter via J_{151}, which is connected to a tap on L_{201}. This tap also supplies voltage to buffer transistor Q_{153} (Fig. 8-25). L_{201} and L_{202} select the third harmonic (150 MHz), which is present at J_{151}.

C_{206} and C_{207} match the output of the two-pole filter to the base of Q_{201}. Q_{201} provides approximately 10 dB of gain. C_{213} and C_{214} match the collector of Q_{201} to the input of a second two-pole filter, consisting of L_{204} and L_{205}. The emitter voltage on Q_{201} (AMPL-1) can be monitored at TP_{201} and is typically +0.3 V.

The collector of Q_{201} is matched to the second two-pole filter, consisting of L_{204} and L_{205}. The output of this filter (approximately 10 to 15 mW) is matched to the base of amplifier Q_{202}. The collector of this transistor in turn provides the signal to the third two-pole filter, consisting of L_{207} and L_{208}. The emitter voltage on Q_{202} is monitored at TP_{202} and is typically +0.5 V.

The output of the third two-pole filter (approximately 60 to 80 mW) is applied to amplifier Q_{203}, which supplies 300 to 350 mW (typically) to the PA driver. An "L" network (L_{209}, L_{210}, and C_{232}) matches the output impedance of Q_{203} to 50 Ω. A 50-Ω microstrip (W_{251}) applies the 300- to 350-mW signal to the PA circuitry. (See Chapter 15 for an explanation of microstrip circuit connections.) An additional metering point is available on this microstrip to monitor power at the exciter-PA interface. TP_{251} consists of two pins that will accept a special RF detector probe that can be used with any multimeter.

POWER AMPLIFIER (FIG. 8-26)   The two-stage power amplifier consists of driver Q_{251}, power amplifier Q_{252}, and associated circuitry. Collector voltage for driver Q_{251} is applied from A+ through pass transistor Q_{256} and L_{253}, L_{252}, and R_{252}. The collector voltage for Q_{251} is a result of the output power setting and voltage variations at any given time. The output of driver Q_{251} is coupled to the base of power amplifier Q_{252} through an impedance-matching network consisting of C_{258}, C_{259}, L_{254}–L_{256}, C_{261}, C_{262}, and R_{253}. Collector voltage for Q_{252} is provided from A+ through L_{258}, L_{257}, and R_{255}.

The 25-W output of the power amplifier is connected to the low-pass filter by W_{252}–W_{254} and then to antenna relay K_{601}. The output power adjust circuit allows the transmitter to be set to rated output power. The power adjustment is attained by controlling the dc collector voltage to drive Q_{251} through pass transistor Q_{256}. The pass transistor is controlled by a feedback loop consisting of Q_{253} through Q_{256}. The power is set by potentiometer R_{257}.

A change in output power is sensed by D_{251}, causing the base voltage of Q_{253} to change accordingly. For example, if the output power increases, the base of Q_{253} goes more positive, causing it to increase conduction, which lowers its collector voltage. Q_{253} controls Q_{254}; therefore, as Q_{253} increases conduction Q_{254} decreases conduction. This raises the voltage applied to the base of Q_{255}. The conduction of Q_{255} decreases proportionally, lowering the base voltage of pass transistor Q_{256}. The resulting decrease in conduction of Q_{256} lowers the collector voltage of driver Q_{251}, thereby lowering the output power in proportion to the excessive power originally sensed by the base circuit of Q_{253}.

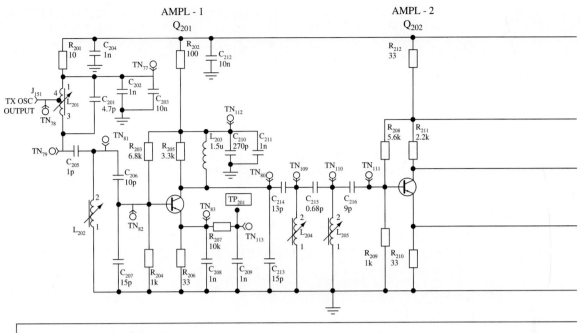

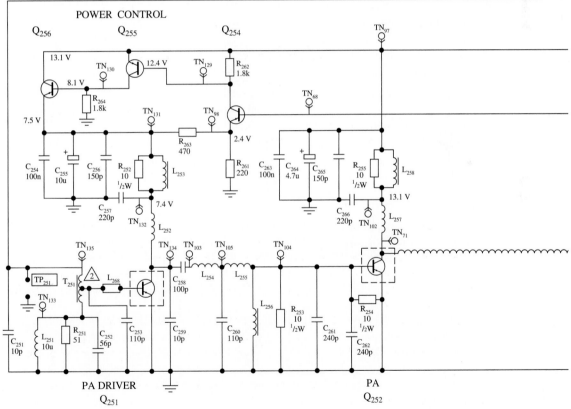

FIGURE 8-26 ▪ Transmitter schematic. (Courtesy of General Electric Company.)

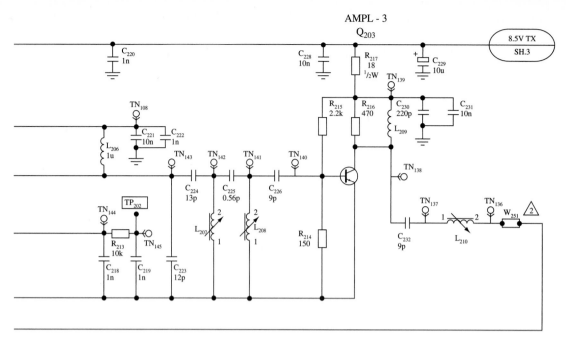

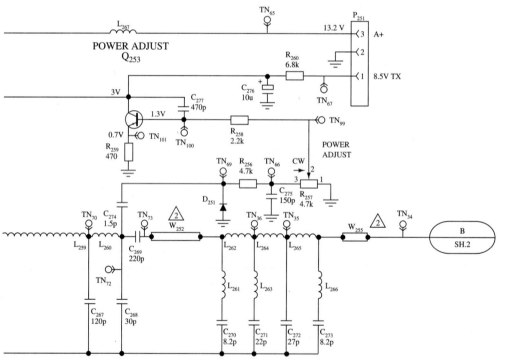

FIGURE 8-26 ■ (Continued)

RECEIVER (FIG. 8-24)   The receiver is dual-conversion, superheterodyne FM designed for one-frequency operation in the frequency range 150 to 174 MHz. A regulated 8.5 V is used for all receiver stages except for the audio PA IC, which operates from the A+ supply.

The receiver has intermediate frequencies of 10.7 MHz and 455 kHz. Adjacent channel selectivity is obtained by using two bandpass filters: a 10.7-MHz crystal filter and a 455-kHz ceramic filter.

All of the receiver circuitry except the oscillator is mounted on the transmitter/receiver (Tx/Rx) board (Fig. 8-27). The receiver consists of:

Receiver front end

10.7-MHz first IF circuitry

First and second oscillators

455-kHz second IF circuitry with FM detector

Audio PA circuitry

Squelch circuitry

RECEIVER FRONT END (FIG. 8-27)   A RF signal from the antenna is coupled through antenna relay K_{601} and two tuned circuits (L_{401}, C_{401}, C_{402}; and L_{402}, C_{404}, C_{405}) to the emitter of RF amplifier Q_{401}. The output of Q_{401} is coupled through three additional tuned circuits (L_{403}, C_{408}; L_{404}, C_{411}; and L_{405}, C_{413}) to the gate of first mixer Q_{402}. The front-end selectivity is provided by these five tuned circuits.

OSCILLATOR AND MULTIPLIER (FIG. 8-25) Q_{301}, Y_{301}, and associated circuitry make up a Colpitts oscillator. The frequency is controlled by a third-mode crystal operated at one-third of the required injection frequency. Voltage-variable capacitor D_{301}, L_{301}, and Y_{301} are connected in series to provide compensation capability. The compensation voltage used to control the transmitter oscillators is applied to D_{301} to maintain stability. L_{301} is adjustable to set the oscillator frequency. R_{301} is in parallel with Y_{301} to ensure operation on the third overtone of the crystal.

The output of Q_{301} is coupled through C_{304} to the emitter of buffer Q_{302}. The output of Q_{302} is coupled through P_{301} to two tuned circuits (L_{351} and L_{352}) on the Tx/Rx board. L_{351} and L_{352} are tuned to the third harmonic of the oscillator frequency, which is applied to the source input of mixer Q_{402}.

The RF frequency from the oscillator/multiplier chain and input level to the mixer can be measured at TP_{401}. The meter reading at TP_{401} is typically 2 to 4 V.

FIRST MIXER (FIG. 8-27)   The first mixer uses a junction FET (Q_{402}) as the active device. The FET mixer provides a high-input impedance, high-power gain, and an output relatively free of intermodulation products.

In the mixer stage, RF from the front-end filter is applied to the gate of the mixer. Injection voltage from the multiplier stages is applied to the source of the mixer. The 10.7-MHz mixer first IF output signal is coupled from the drain of Q_{402} through an impedance-matching network (L_{406}, C_{415} and C_{416}) to crystal filter Z_{501}.

The highly selective crystal filter provides the first portion of the receiver IF selectivity. The output of the filter is coupled through impedance-matching network L_{501} to the first IF amplifier.

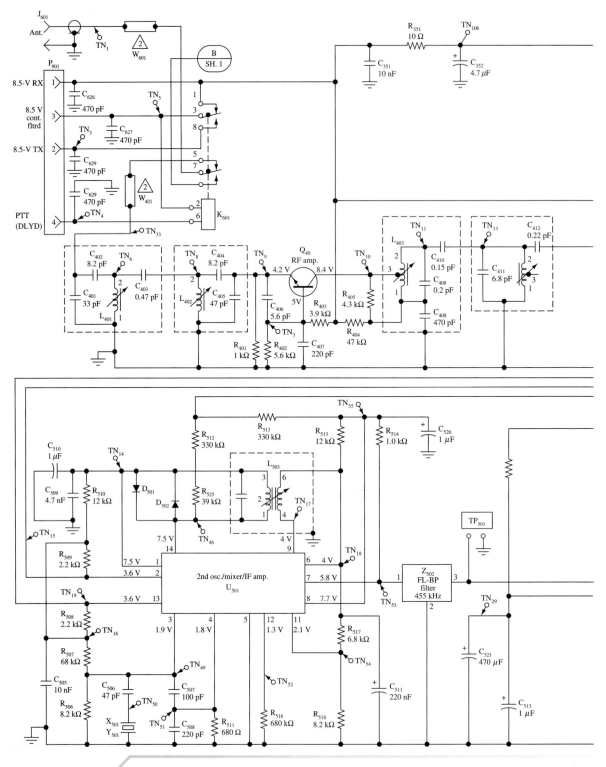

FIGURE 8-27 ▌ Receiver schematic. (Courtesy of General Electric Company.)

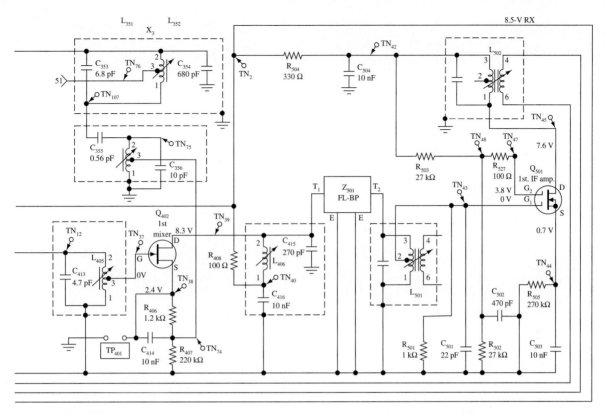

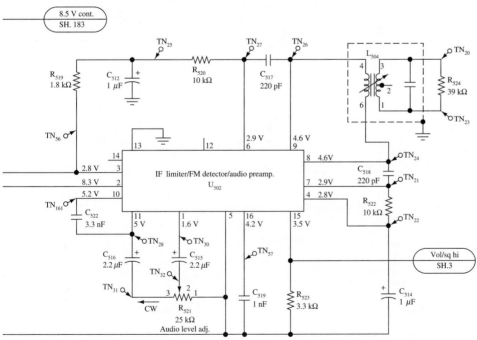

FIGURE 8-27 ■ (Continued)

Notes:

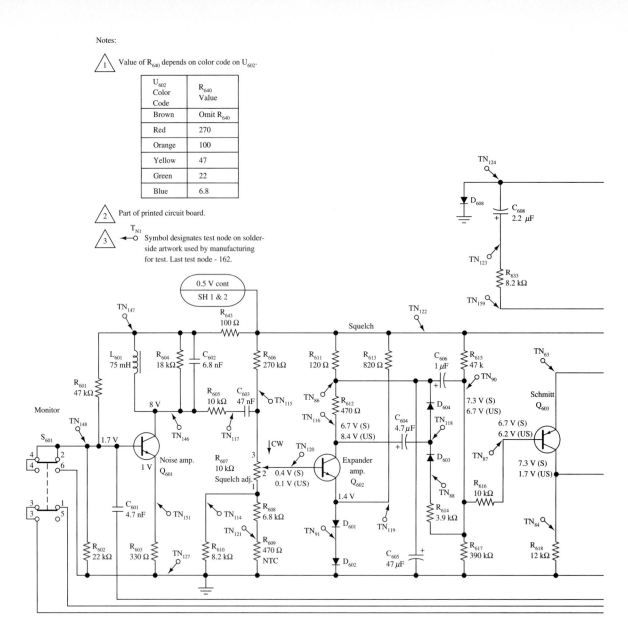

1. Value of R_{640} depends on color code on U_{602}.

U_{602} Color Code	R_{640} Value
Brown	Omit R_{640}
Red	270
Orange	100
Yellow	47
Green	22
Blue	6.8

2. Part of printed circuit board.

3. T_{N1} — Symbol designates test node on solder-side artwork used by manufacturing for test. Last test node - 162.

All Resistors are 1/4 W unless otherwise specified.

Voltage Readings

Voltage readings are typical readings measured to system negative with a 20,000-Ω/V dc voltmeter under the following conditions:
1. No signal input
2. Volume control (R_{634}) set to minimum
3. Monitor switch (S_{601}) in out position
4. Unsquelched (US)-squelch adjust (R_{607}) set to minimum (CW)
5. Squelched (S)-squelch adjust (R_{607}) set to maximum (CCW)

FIGURE 8-27 ■ (Continued)

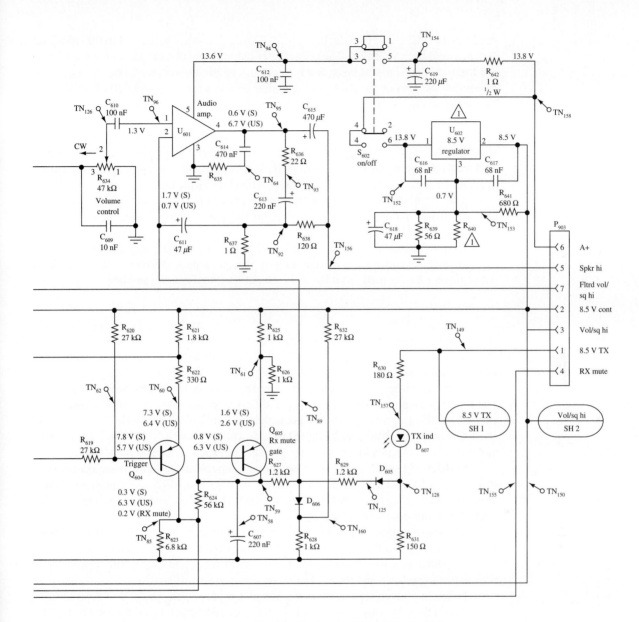

Model No.	Description
19D900599G1	TX/RX BD assembly
19D900602G1	Chassis assembly

FIGURE 8-27 ▦ *(Continued)*

First IF amplifier Q_{501} is a dual-gate MOSFET. The crystal filter output is applied to gate 1 of the amplifier, and the amplified signal is taken from the drain. The biasing on gate 2 and the drain load determines the gain of the stage. The amplifier provides approximately 20 dB of IF gain. The output of Q_{501} is coupled through an impedance-matching network (L_{502}) that matches the amplifier output to the input of IC U_{501}.

U_{501} and associated circuitry consist of the second oscillator, mixer, and second IF amplifier. The crystal for the oscillator is Y_{501}, and the oscillator operates at 10.245 MHz. This frequency is mixed with the 10.7-MHz input. The output of the mixer is limited by D_{501} and D_{502}. L_{503} is tuned for the 455-kHz second IF frequency.

The output of U_{501} is coupled through ceramic filter Z_{502}, which provides the 455-kHz selectivity, and applied to U_{502}. Test point TP_{501} is used in aligning the receiver and can be used to check the output of U_{501}.

U_{502} and associated circuitry consist of a 455-kHz limiter, a quadrature-type FM detector, and an audio preamplifier. L_{504} is the quadrature detector coil. Audio-level potentiometer R_{521} is used to set the audio output level to the audio amplifier.

Audio and squelch circuits (Fig. 8-27)   **Audio.** Audio is applied to the channel guard tone reject filter through $P_{903\text{-}3}$ and back to the deemphasis network through $P_{903\text{-}7}$. The audio passes through the deemphasis network (R_{633}, C_{608} and C_{609}) to volume control R_{634}.

Audio amplifier IC U_{601} drives the speaker at the desired audio level (up to 3 W). The feedback loop containing R_{637}, R_{638}, and C_{611} determines the amplifier's closed-loop gain. R_{636} and C_{613} provide the audio high-frequency roll-off above 6 kHz.

The audio amplifier can be muted by a dc voltage from 8.5 V Tx, or from receiver mute gate Q_{605}. The two logic inputs for Q_{605} are a squelch signal and Rx MUTE.

Squelch (Fig. 8-27). The squelch circuit operates on the noise components contained in the FM detector output. The output of U_{502} is applied to frequency-selective noise amplifier Q_{601}, which has a resonant circuit (L_{601}, R_{604}, and C_{602}) as the collector load. The output is noise in a band around 7 kHz.

The noise output is coupled through squelch control R_{607} to expander amplifier Q_{602}, which improves the level discrimination characteristics of the circuit. The output of Q_{602} is applied to a passive voltage-doubler circuit (D_{603} and D_{604}). This circuit has a high source impedance and operates as an average-value rectifier.

Following the voltage doubler is a Schmidt trigger (Q_{603}–Q_{604}). The Schmidt trigger provides the necessary hysteresis and a well-defined output signal for Rx MUTE gate Q_{605}.

With no RF signal present, the detected noise at the voltage-doubler output turns on Q_{603}, turning off Q_{604}. This causes Q_{605} to turn on, applying +1 V to pin 2 of audio amplifier U_{601}. This voltage turns off U_{601} and mutes the receiver.

When a RF signal is received, the noise at the output of Q_{601} decreases and drive to Q_{603} is removed. This turns off Q_{603} and allows Q_{604} to turn on. With Q_{604} turned on, Rx MUTE gate Q_{605} turns off. This turns on U_{601} so that audio is heard at the speaker.

The squelch sensitivity is adjusted by R_{607} in the base circuit of expander amplifier Q_{602}.

Pressing in the MONITOR pushbutton on the front of the radio opens the Rx MUTE to disable the channel guard. It also grounds the base of Q_{601} and disables the squelch function.

CHANNEL GUARD (FIG. 8-28)   The channel guard is a continuous-tone encoder/decoder for operation on tone frequencies in the range 71.9 to 210.7 Hz. The encoder provides tone-coded modulation to the transmitter. The decoder operates in conjunction with the receiver to inhibit all calls that are not tone-coded with the proper channel guard frequency.

The channel guard circuitry consists of discrete components for the encode disable, PTT switch, and receiver mute switch; and four thick-film integrated circuit modules consisting of decode module U_{1001}, encode module U_{1002}, frequency-switchable selective amplifier (FSSA) U_{1003}, plug-in Versatone network Z_{1001}, and IC U_{1004} in the tone reject filter.

For a functional diagram of the channel guard encoder/decoder, refer to the troubleshooting procedures.

FSSA (FIG. 8-28)   Frequency-switchable selective amplifier (FSSA) U_{1003} is a highly stable active bandpass filter for the frequency range 71.9 to 210.7 Hz. The selectivity of the filter is shifted across the bandpass frequency range by switching Versatone networks in the filter circuit.

The gain of the FSSA is a function of the tone frequency (Table 8-3). The tone frequency is determined by the tone network connected in the FSSA circuit. Versatone network Z_{1001} is a precision resistor network.

ENCODE (FIG. 8-28)   When the PTT switch is operated, the channel guard encode tone is generated by coupling the output or FSSA bandpass filter U_{1003} back to its input through a phase-inverting amplifier circuit and a limiter circuit. The output of the FSSA is coupled from U_{1003} to the input of the phase-inverting amplifier at $U_{1002\text{-}9}$.

An amplifier provides 180° phase shift of the tone frequency at the output. The output of the phase-inverting amplifier circuit is coupled from $U_{1002\text{-}6}$ to the input of the limiter circuit at $U_{1002\text{-}5}$. A limiting network sets the tone output coupled from $U_{1002\text{-}4}$ to the input of the FSSA ($U_{1003\text{-}12}$) at 53 mV p-p.

The limiter circuit is also used as an encode switch. Keying the transmitter applies +5.4 V to $U_{1002\text{-}2}$. This starts the circuit oscillating. The tone frequency is determined by the tone network connected in the FSSA circuit.

TABLE 8-3   *Standard Tone Frequencies (Hz)*

71.9	88.5	107.2	131.8	162.2
74.4	91.5	110.9	136.5	167.9
77.0	94.8	114.8	141.3	173.8
79.7	97.4	118.8	146.2	179.9
82.5	100.0	123.0	151.4	186.2
85.4	103.5	127.3	156.7	192.8
			203.5	
			210.7	

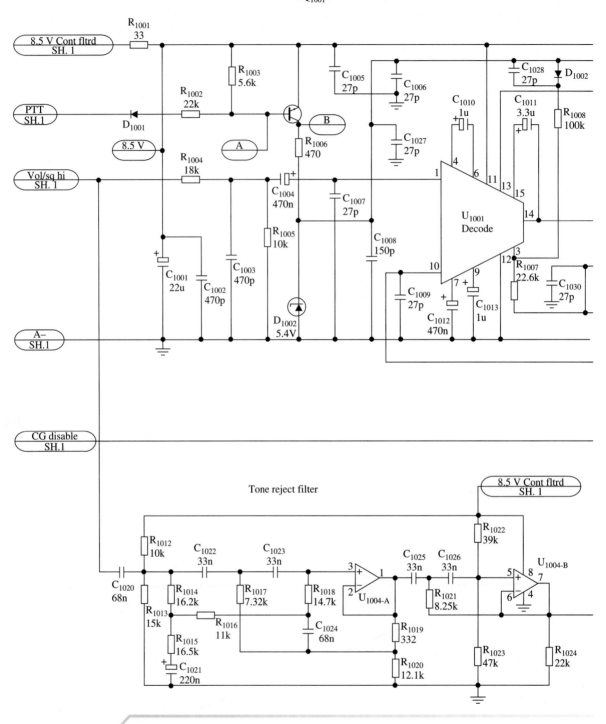

FIGURE 8-28 Channel guard schematic. (Courtesy of General Electric Company.)

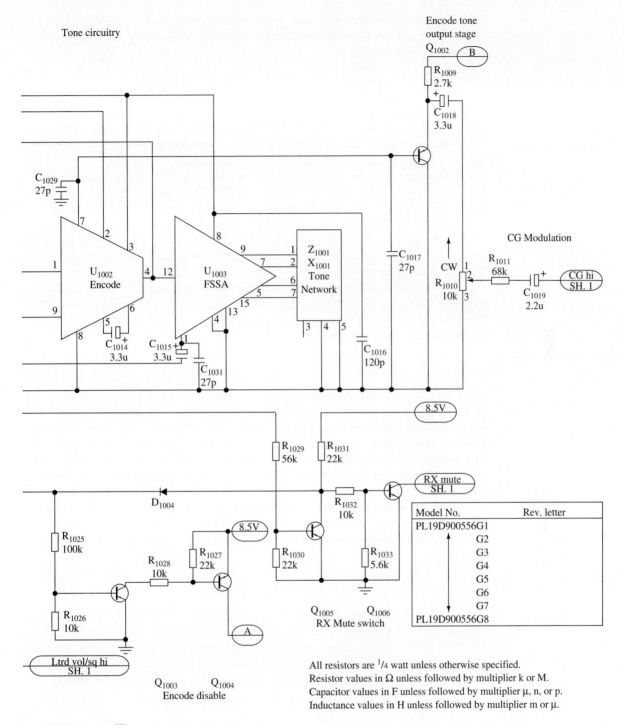

Tone circuitry

Encode tone output stage

CG Modulation

All resistors are $1/4$ watt unless otherwise specified.
Resistor values in Ω unless followed by multiplier k or M.
Capacitor values in F unless followed by multiplier μ, n, or p.
Inductance values in H unless followed by multiplier m or μ.

FIGURE 8-28 ■ (Continued)

The tone output of the encoder circuit is taken from $U_{1002\text{-}7}$ and coupled through tone output amplifier Q_{1002} and modulation adjustment R_{1010} to the audio processor on the transmitter/receiver board.

Audio from volume/squelch high that contains the correct frequency is coupled to pin 1 of decode module U_{1001}. Pin 1 of U_{1001} is the input of an active, three-stage, low-pass filter. The low-pass filter attenuates frequencies over 210.7 Hz. The output of the low-pass filter at $U_{1001\text{-}15}$ is applied to $U_{1001\text{-}14}$. $U_{1001\text{-}14}$ is the input of a limiter circuit, limiting the output at $U_{1001\text{-}13}$ to 55 mV p-p.

The output from the limiter is coupled to pin 12 of FSSA U_{1003}. Since the tone is the proper frequency, the FSSA will allow it to pass. The output of the FSSA is coupled to $U_{1001\text{-}3}$. $U_{1001\text{-}3}$ is the input to an amplifier circuit. The output of the amplifier at $U_{1001\text{-}4}$ is coupled to the input of a threshold detector at $U_{1001\text{-}6}$.

In the mute mode, when the tone decoder in U_{1001} detects the channel guard frequency, Q_{1005} turns Q_{1006} off. This unmutes the receiver audio. In the squelch mode, Q_{1006} is operating, grounding the Rx MUTE lead and muting the receiver audio.

Audio from VOL/SQ HI is applied to the tone reject filter. The tone reject filter is an active filter consisting of U_{1004} and associated circuitry. All frequencies from 80 to 210.7 Hz are rejected by the filter while passing all other audio frequencies back to the receiver audio circuits (filtered VOL/SQ HI).

The encode disable circuit consists of Q_{1003} and Q_{1004}. To disable the encode circuit, a positive voltage (+8.5 to 14 V dc) is applied to connector $P_{910\text{-}9}$ at the rear of the radio. This is accomplished by temporarily jumpering $P_{910\text{-}9}$ (GC DISABLE) to $P_{910\text{-}11}$ (A+). This positive voltage is applied to the base of Q_{1003}, turning on both Q_{1003} and Q_{1004}. When turned on, Q_{1004} applies +8.5 V dc to the base of the PTT switch Q_{1001}, forcing it off. With Q_{1001} off, the operating voltage for the encoder IC U_{1002} and encode tone output stage transistor Q_{1002} is removed, preventing any tone output.

The encode disable circuit has been incorporated as a maintenance aid for the service technician. This circuit disables the channel guard encode circuit and allows for transmitter distortion and modulation checks without removing the cover from the radio.

Troubleshooting (Figs. 8-29 and 8-30)

Troubleshooting charts are provided in Figs. 8-29, pages 339 and 340, and 8-30 for all major areas of the transceiver. The reader is now referred to these self-explanatory guides. Note that the receiver troubleshooting information (Fig. 8-30) takes the form of a flow-chart path to solution. The transmitter information (Fig. 8-29) takes the more conventional approach of listing symptoms, probable causes, and troubleshooting procedures.

8-8 SPREAD-SPECTRUM TECHNIQUES

The first spread-spectrum systems were utilized by the U.S. government toward the end of World War II. These techniques allow transmissions that cannot be jammed (i.e., rendered useless by a counter "noise" signal at the same frequency) nor detected by the

Symptom	Typical Voltage/Signal	Probable Cause	Troubleshooting Procedure
No power	0 V at TP_{203}	No exciter output	Repair exciter.
	0.65 V at TP_{203} but 0 V at TP_{251}	C_{251} stage inoperative	Check for A+ at collector of Q_{251} and check associated circuitry.
	0.8 V at TP_{251} but 0 V on collector Q_{252}	Power control circuitry faulty	Repair power control circuit (see circuit analysis for operation).
	Radio draws 4 to 5 A	Open circuit after Q_{253} output	Check U_{256}, filter, relay, etc., for open circuit.
	Radio current less than 2 A but 0.8 V on TP_{251} and 12 V collector Q_{251}	Q_{252} or Q_{253} stages inoperative	If Q_{252} draws around 0.75 A problem is Q_{253} or associated circuitry. If Q_{253} draws zero current, problem is Q_{253} or associated circuitry.
Low power	Significantly less than 0.65 V at TP_{203}	Less exciter output	Repair exciter.
	Significantly less than 0.8 V at TP_{251}	Low output Q_{251} stage	Problem is Q_{251} or associated circuitry.
	Less than 11 V on collector Q_{252}	Power control	Adjust, then if necessary, repair power control.
	12 V on collector Q_{252}, radio draws approx. 5A	Antenna filter	Tune filter.
	Radio current less than 4 A	Q_{252} or Q_{253} stages faulty	If Q_{252} draws around 0.75 A, problem is Q_{253} or associated circuitry. If Q_{252} draws significantly less than 0.75 A, problem is Q_{252} or associated circuitry.

Troubleshooting

Symptom	Procedure
Unit does not decode	1. Place squelch switch in the unsquelched position and check for proper receiver operation.
	2. If the receiver operates properly, set squelch to the "out" position. Apply the proper channel guard tone to the radio and check for 8.5 V dc at position $U_{1001-10}$.
	3. If reading is not correct, check voltage readings on connections between the tone network FL_{1001} and U_{1005}.
	4. If the readings between the tone network and U_{1003} are incorrect, ensure good contact between the tone network and the network socket.
	5. If readings are correct, check voltage readings at all other points identified.
Unit does not encode	1. Check for 3.1 V dc at U_{1002-7}.
	2. If reading is correct, check mod. adj. R_{1001}, then check the transmitter oscillator module.
	3. If reading is not correct, check voltage readings on connections between the tone network FL_{1001} and U_{1003}.
	4. If the readings between the tone network and U_{1003} are incorrect, ensure good contact between the tone network and the network socket.
	5. If readings are correct, check voltage readings at all other points identified.

FIGURE 8-29 Transmitter troubleshooting. (Courtesy of General Electric Company.)

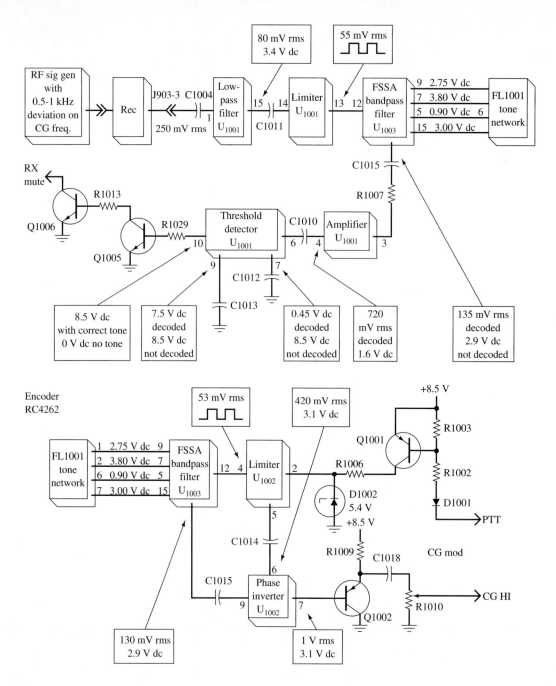

FIGURE 8-29 ■ (Continued)

enemy. Military applications are still widespread, but as explained subsequently, commercial use is increasing due to performance advantages.

Spread spectrum started to gain serious commercial attention in 1985 when the Federal Communications Commission (FCC) opened the industrial, scientific, and military (ISM) band for unlicensed operation of devices under FCC technical regulations 15.247. The FCC permits spread-spectrum modulation at a maximum transmitter power of 1 W in three bands—902–928 MHz, 2400–2483.5 MHz, and 5725–5850 MHz.

Spread spectrum is becoming the technology of choice in many wireless applications, including wireless local-area networks (WLANs), automated data collection systems using portable scanners of UPC codes, and cellular telephones. Other applications include remote heart monitoring, industrial security systems, and Very Small Aperture Satellite Terminals (VSAT). Multiple spread-spectrum users can coexist in the same bandwidth if each user is assigned a different "spreading code," which will subsequently be explained. Systems utilizing this technique are called *Code Division Multiple Access* (CDMA) systems.

As we have seen, regular modulation schemes tend to fully utilize a single band of frequencies. Noise in that band will obviously degrade the signal, and therefore it is vulnerable to jamming. This single-frequency band also allows detection by undesired recipients who may track down the signal source via direction-finding (DF) techniques. An antenna commonly used in DF applications is described in Chapter 14. The spread-spectrum solution to these problems takes one of three forms described below.

Frequency Hopping

The transmitted signal is contained by a carrier that is switched in a pseudorandom fashion. *Pseudorandom* implies a sequence that can be recreated (e.g., at the receiver) but has the properties of randomness. The time of each carrier block is termed the *dwell time*. Dwell times are usually less than 10 ms. The receiver knows the order of frequency switching, picks up the successive blocks, and assembles them into the original message.

Time Hopping

This is similar to frequency hopping except that the hopping occurs in the time domain. The transmitter emits its signal during very short time intervals. These intervals are controlled via a random sequence that is known to the receiver. The receiver thus knows when it can find the signal. This is the least popular of the three basic spread-spectrum systems.

Direct Sequence

This is used in the transmission of digital bits. Its pseudorandom sequence uses pulses shorter than the message bits, called *chips*. The chips successively modulate fractions of the bits that typically phase shift the carrier. The receiver multiplies the incoming signal by the same chip signal sequence to recover the original digital signal.

All three of these spread-spectrum techniques utilize a far wider bandwidth than that required by conventional modulation schemes. Besides the two advantages for the military already described, other advantages exist that have led to nonmilitary applications. Spread spectrum permits many transmitters to operate over the same channel with minimal interference. This is true since the different pseudorandom sequences will only rarely coincide. This coincidence, termed a *hit*, adds only a low-level noise to overall reception. The net result is a more efficient channel utilization (more transmissions), especially under conditions where continuous transmissions are not being made. In fact, spread-spectrum communications are now being added to bands "fully" used by conventional communication techniques with minimal interference. Another advantage of spread spectrum is the reduction in fading it affords compared to conventional (narrowband) systems. Different parts of the frequency spectrum tend to fade at variable rates. Thus each

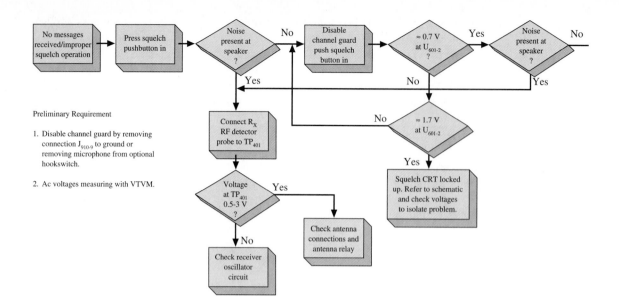

Preliminary Requirement

1. Disable channel guard by removing connection J_{910-9} to ground or removing microphone from optional hookswitch.

2. Ac voltages measuring with VTVM.

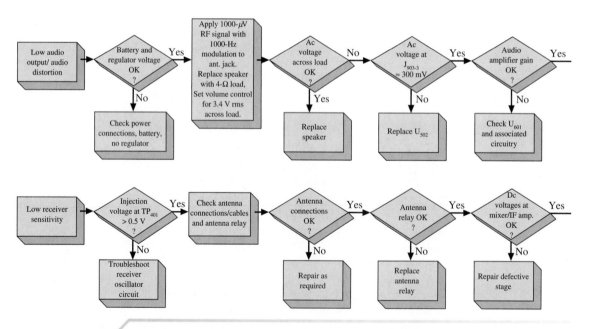

FIGURE 8-30 ■ Receiver troubleshooting. (Courtesy of General Electric Company.)

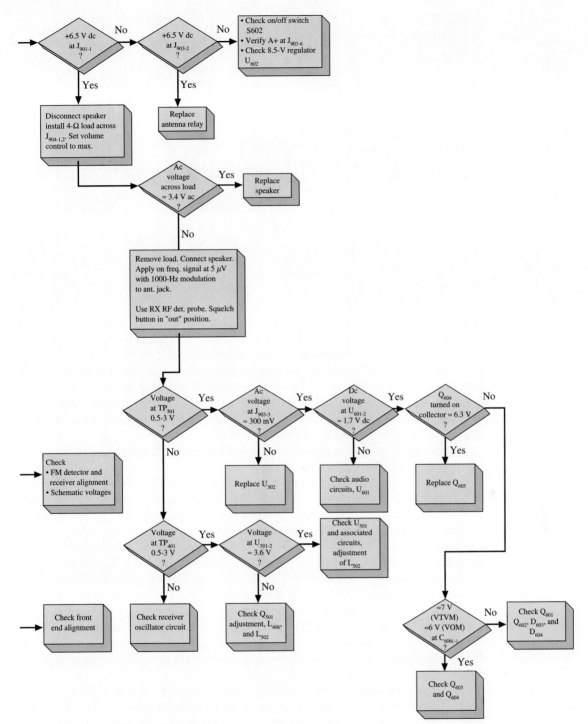

FIGURE 8-30   (Continued)

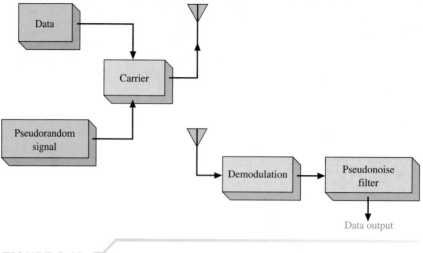

FIGURE 8-31 ▓ Direct-sequence spread spectrum.

hop of a frequency-hopping signal has an independent fading characteristic. The ratio of maximum to minimum received signal strength (fading) is typically 2 to 3 dB compared with 20 to 30 dB for conventional transmissions.

Spread-spectrum techniques are also widely used in radar systems (Chapter 15). The spread carrier, as determined by the pseudorandom sequence, allows the receiver to determine very accurately the time the transmitter sent its energy with minimal effects from noise, and so on. The distance to a transmitter (ranging) or object via a reflection can thereby be accurately made with high reliability.

Figure 8-31 shows the basic format of a direct-sequence spread-spectrum system. These systems are also referred to as pseudonoise (PN) spread-spectrum signaling systems. The spreading signal is a carrier modulated with a binary sequence having a pseudorandom (pseudonoise or PN) nature generated by a shift register with feedback connections. The sequence generated is called the *signature sequence*. The clock rate of the shift register is very high compared with the data rate to be transmitted. One period of the shift-register sequence contains thousands of binary transitions called chips, and they are modulated with one bit of binary data. The receiver uses a shift register identically programmed (via logic connections) to the transmitters to recognize the signal being received. The position of the two shift registers along the pseudorandom sequence is synchronized by analysis of the received waveform, with a decision circuit picking the signal with the highest correlation level. Other users in the band may have similar transmitters and receivers but use different logic connections in their shift registers so as to present a different "signature." Receivers do not recognize other signatures; they see them as background noise.

A frequency-hopping system block diagram is provided in Fig. 8-32. Essentially identical programmable-frequency synthesizers and hopping-sequence generators are the basis of these systems. The receiver must synchronize itself to the transmitter's hopping sequence. This is done via the synchronization logic. Spread-spectrum systems obviously have an extra degree of complexity compared to conventional systems. This is increasingly being found economically feasible due to the advantages offered. Of course, this has always been true in many military applications.

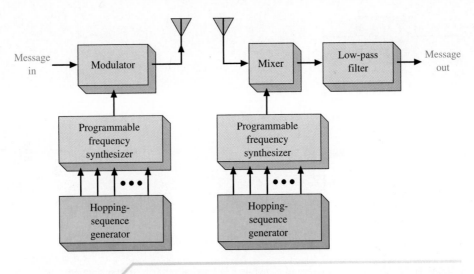

FIGURE 8-32 ■ Frequency-hopping spread spectrum.

TROUBLESHOOTING

Transceivers, or two-way radios, are found in many commercial applications. In this section we are going to look at troubleshooting the transmitter portion of a mobile transceiver. General troubleshooting techniques are presented in this section. You should always consult the service manual before disassembling a transceiver and making any adjustments or repairs on it.

Today's communication equipment usually includes digital logic circuits to control various functions. We will learn to troubleshoot some basic logic circuits. We'll also consider troubleshooting a frequency synthesizer.

After completing this section you should be able to

- Describe the signal flow in a mobile FM transmitter circuit
- Describe common mobile transmitter failures
- Troubleshoot basic logic circuits
- Troubleshoot a frequency synthesizer

Transceiver Transmitter

The block diagram in Fig. 8-33 depicts the transmitter portion of a mobile transceiver. Mobile transmitters may differ somewhat in design. For example, this particular transmitter uses several frequency multiplier circuits in the exciter stage to step-up the frequency to the necessary operating frequency. A press-to-talk microphone feeds the voice signal into an audio amplifier. The voice signal is amplified and sent to the phase modulator. The phase modulator is also fed by a crystal-controlled oscillator. The signal driving the power amplifier is FM from the phase modulator that has been amplified and multiplied in frequency by the exciter stage. The power amplifier delivers a specified output power to the antenna via the harmonic filter and the antenna-switching relay. Typical output power ratings are 20 to 25 W.

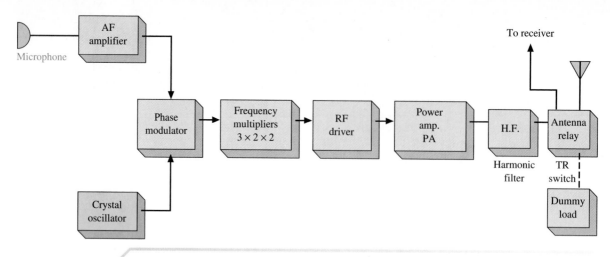

FIGURE 8-33 Block diagram of a mobile FM transceiver, transmitter portion.

Transmitter Troubles

Transmitter troubles fall into several categories: low-power output, exciter troubles, oscillator troubles, power amplifier troubles, transmit-receive (TR) antenna-switching relay problems, modulator troubles, and microphone-associated problems. We will look at some of these common transmitter troubles in the following discussion.

1. MICROPHONE AND AUDIO TROUBLES   Microphone failures are more often the problem in mobile radios than audio amplifiers. This is true because of the extensive use the microphones get. In addition, these radios are exposed to temperature extremes and moisture. Being subject to these weather conditions will cause the microphone's cords to become stiff and brittle. Flexing the cord may damage the insulation or cause the cord to break at its plug. Solder connections can become poor conductors over a period of time and weathering. Press-to-talk switches become contaminated and corroded from dirt and moisture. These switches get a tremendous amount of use, and often the internal contacts fail to make good connection. For inoperative microphones, replace them and the cords.

2. MODULATOR TROUBLES   A common modulator problem is low modulation. This low modulation reduces the transmitter's operational range. Leaky coupling capacitors, open bypass capacitors, and off-tolerance resistors are likely candidates in the modulator circuits that can reduce the modulation level. Check bias resistors in the transistor circuits of the modulator. Also, review the Chapter 5 Troubleshooting section for the reactance modulator.

3. TR SWITCHING TROUBLES   The TR switch is responsible for switching the antenna between the transmitter and the receiver in the transceiver radio. By this switching action a common antenna is shared by both the receiver and the transmitter. Absence of

RF power at the antenna could be due to contact trouble in the antenna-switching relay. Figure 8-29 shows a typical antenna-switching relay, K801. Notice the contact switching positions. A defective relay can usually be checked by pressing on the contacts using a plastic wand while keying the transmitter. If the relay works with pressure applied, then it should be replaced. Cleaning the contacts may work if the relay points are not pitted.

4. PA TROUBLES   The presence of a strong exciter output, but the absence of a signal or a weak signal at the power-amplifier output, indicates power-amplifier troubles. A very common cause for this missing output is blown output transistors. Output transistors can blow from not being terminated and from impedance mismatches with the antenna or dummy load. Never key a transmitter unless it is connected to the antenna or the dummy load. For no- or low-output power check the transistors. If the transistors are not bad, look for passive components that might be defective and for changes in resistor values.

5. OSCILLATOR TROUBLES   Improper tuning of the oscillator stage may cause unstable transmitter operation. Sometimes a tank inductance is varied in the oscillator circuit. In other circuits, the tank is fixed and a capacitor may be used to trim the frequency. Improper tuning in either case can put the oscillator at the edge of stable or unstable operation. Weak crystals will cause the oscillator's output to decrease. The weak crystal may even shut the oscillator down completely. Verify the oscillator's operation by monitoring the transmitter output or by measuring the voltage at a frequency multiplier test point. Note any changes in frequency at the transmitter's output or any voltage variations at the test point. Consult the service manual for specific steps to repair, adjust, and replace crystals.

Logic Problems

Today's receivers and transmitters usually utilize a good bit of digital logic. The NAND gate shown in Fig. 8-34 is the basic building block of all logic functions. Four of these gates are found in one IC, the 7400. This particular example is called transistor-transistor logic, or TTL for short.

When troubleshooting any digital circuit, the technician will first look to see if minimum logic levels are being met. We will assume that the parts are operating from a +5-V supply. The part is guaranteed to output at least 2.4 V (logic one) and less than 0.4 V for a logic zero. Current ratings go with the voltages that guarantee the number of loads a gate can drive.

The part is guaranteed to accept any voltage more than 2.0 V as a logic one on its input, point A or B. The gate is guaranteed to accept any input voltage less than 0.8 V as a logic zero.

Note that the part is guaranteed to give more than it requires, 2.4 V vs. 2.0 V. This is done so that there is a guaranteed margin to allow for age and noise. It sounds simple.

Now the technician has to understand what the gate is supposed to do. All gates have a truth table to tell you what the relationship of the input to the output is. Table 8-4 shows what a NAND gate will do. Now, what can go wrong?

LOGIC ZERO INCORRECT   To place a logic zero on the TTL gate's input, we must draw a small amount of current from the gate. If the part driving the gate cannot do this,

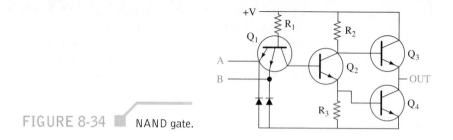

FIGURE 8-34 ▮ NAND gate.

the voltage will not fall in the guaranteed region and the gate will not know what to do. For example, if Q4 has failed and will not sink current, it cannot place a zero on the next gate.

It is also possible that the gate input has shorted to the supply. In this case it is impossible for any output stage to drive the bad input. You will have to change one or the other or both. Which one? See if other gates in that IC are working. Is one hot? It might be the bad one.

LOGIC ONE INCORRECT   This is exactly the opposite of the above problem. Q3 must be able to supply current. TTL is guaranteed to supply 40 μa. If one of the diodes in the following gate is shorted, Q3 will not be able to output the 2 V the next gate requires to function. Again, it is difficult to tell which gate is malfunctioning, input or output. You will have to make an educated guess and change one.

Synthesizer Problems

Figure 8-13 is a block diagram of a synthesizer found in many communications receivers. The output of the VCO is connected to the first mixer to set the receiver frequency. Here are some typical troubles.

1.   *Small frequency error:* Check the frequency of the oscillator. Remember, error at the oscillator is multiplied by N.

TABLE 8-4   *7400 NAND Gate*

INPUTS		OUTPUTS
A	B	
0	0	1
1	0	1
0	1	1
1	1	0

2. *Large frequency error:* Is the loop locked? Probably not. Look at the output of the phase comparator. If the loop is not locked, you will see a waveform that is the difference in the divider output and the reference oscillator. Check components in the VCO and look at logic levels in the divider. Check to see that the division ratio is correct.

3. *No output at all:* A failure in the system may drive the varactor in the VCO to some condition that will not allow oscillations. You might isolate the VCO from the low-pass filter. Check all VCO components.

SUMMARY

In Chapter 8 we described various improvements to receiver design and discussed some of the more complicated specifications used in high-quality receivers. The concept of spread-spectrum communications was also introduced. The major topics the student should now understand include:

- the analysis of advanced techniques for image frequency reduction, including double conversion and up-conversion

- the description and explanation of special techniques for improving receiver operation, including delayed AGC, auxiliary AGC, manual sensitivity control, variable notch filters, ANL circuits, and squelch control

- the analysis of the relationship between noise, sensitivity, and dynamic range in a high-quality receiver

- the analysis of intermodulation distortion (IMD) testing

- the description and analysis of various frequency synthesizers

- the method used to obtain direct digital synthesis (DDS) systems

- the analysis of spread-spectrum techniques, including description of CDMA, frequency hopping, time hopping, and direct sequence

QUESTIONS AND PROBLEMS

SECTION 2

1. Explain the difference between an FM stereo receiver and a communications transceiver.
2. Draw a block diagram for a double-conversion receiver when tuned to a 27-MHz broadcast using a 10.7-MHz first IF and 1-MHz second IF. List all pertinent frequencies for each block. Explain the superior image frequency characteristics as compared to a single-conversion receiver with a 1-MHz IF, and provide the image frequency in both cases.
3. Describe the process of up-conversion. Explain its advantages and disadvantages compared to double conversion.

4. Draw block diagrams and label pertinent frequencies for a double-conversion *and* up-conversion system for receiving a 40-MHz signal. Discuss the economic merits of each system and the effectiveness of image frequency rejection.
5. A receiver tunes the HF band (3 to 30 MHz), utilizes up-conversion with an intermediate frequency of 40.525 MHz, and uses high-side injection. Calculate the required range of local oscillator frequencies. (43.5 to 70.5 MHz)
6. An AM broadcast receiver's preselector has a total effective Q of 90 to a received signal at 1180 kHz and uses an IF of 455 kHz. Calculate the image frequency and its dB of suppression. (2090 kHz, 40.7 dB)

Section 3

7. Discuss the advantages of delayed AGC over normal AGC and explain how it may be attained.
8. Explain the function of auxiliary AGC and give a means of providing it.
9. Explain the need for variable sensitivity and show with a schematic how it could be provided.
10. What is the need for a noise limiter circuit? Explain the circuit operation of the noise limiter shown in Fig. 8-8.
11. List some possible applications for *metering* on a communications transceiver.
*12. What is the purpose of a squelch circuit in a radio communications receiver?
13. List two other names for a squelch circuit. Provide a schematic of a squelch circuit and explain its operation.
14. Describe the operation of an automatic noise limiter (ANL).

Section 4

15. It is desired to operate a receiver with NF = 8 dB at S/N = 15 dB over a 200-kHz bandwidth at ambient temperature. Calculate the receiver's sensitivity. (−98 dBm)
16. Define receiver dynamic range. What are the factors that determine the upper and lower limits of dynamic range?
17. Explain the significance of a receiver's 1-dB compression point. For the receiver represented in Fig. 8-10, determine the 1-dB compression point. ($\cong$ 10 dBm)
18. Define intermodulation distortion and explain how the third-order intercept-point specification provides a figure of merit for a receiver's ability to handle this distortion.
19. Determine the third-order intercept for the receiver illustrated in Fig. 8-10, ($\cong$ +20 dBm)
20. The receiver described in Problem 15 has the input/output relationship shown in Fig. 8-10. Calculate its dynamic range. (78.7 dB)
21. The receiver in Problem 20 has a 6-dB NF preamp (gain = 20 dB) added to its input. Calculate the system's sensitivity and dynamic range. (−99.5 dBm, 66.3 dB)

Section 5

22. Explain the operation of a basic frequency synthesizer as illustrated in Fig. 8-13. Calculate f_0 if f_R = 1 MHz and N = 61. (61 MHz)

23. Discuss the relative merits of the synthesizers shown in Fig. 8-15(a), (b), and (c) as compared to the one in Fig. 8-13.
24. Describe the operation of the synthesizer divider in Fig. 8-16. What basic problem does it overcome with respect to the varieties shown in Figs. 8-13 and 8-15?
25. Calculate the output frequency of a synthesizer using the divider technique shown in Fig. 8-16 when the reference frequency is 1 MHz, $A = 26$, $M = 28$, and $N = 4$. (138 MHz)
26. Determine the output frequency for the synthesizer of Fig. 8-18 when the input code is 100011. (27.245 MHz)
27. Explain the operation of the low-frequency synthesizer of Fig. 8-20. What advantages does it provide over conventional low-frequency sine-wave oscillators?

Section 6

28. Briefly explain DDS operation based upon the block diagram shown in Fig. 8-21.
29. A DDS system has a $f_{CLK\ MAX} = 60$ MHz and a 28-bit phase accumulator. Calculate its approximate maximum output frequency and frequency resolution when operated at $f_{CLK\ MAX}$. (24 MHz, 0.223 Hz)

Section 7

30. Describe the function of the channel guard function in a communication transceiver.
31. Analyze the system block diagram in Fig. 8-26. Redraw it more simply so as to explain its operation to technicians who are not experts in the field of electronic communication.
32. The relay (K601) in Fig. 8-29 is malfunctioning. The contacts are not releasing when the coil is deenergized. Describe the effect this would have on the transceiver's performance.

Section 8

33. Briefly explain spread-spectrum systems, including origination, basic types, and advantages over conventional modulation systems.
34. After doing some research at a library, provide a report concerning some specific aspect of spread-spectrum systems.
35. Describe a signature sequence for direct-sequence systems.

DIGITAL
COMMUNICATIONS:
CODING
TECHNIQUES

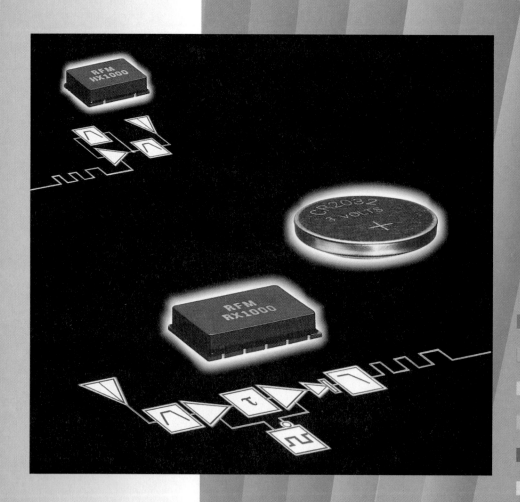

OBJECTIVES

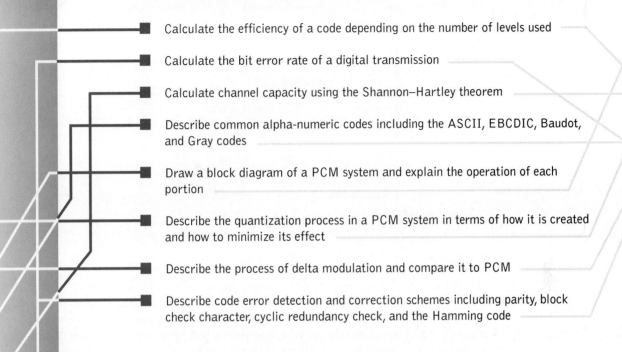

- Calculate the efficiency of a code depending on the number of levels used

- Calculate the bit error rate of a digital transmission

- Calculate channel capacity using the Shannon–Hartley theorem

- Describe common alpha-numeric codes including the ASCII, EBCDIC, Baudot, and Gray codes

- Draw a block diagram of a PCM system and explain the operation of each portion

- Describe the quantization process in a PCM system in terms of how it is created and how to minimize its effect

- Describe the process of delta modulation and compare it to PCM

- Describe code error detection and correction schemes including parity, block check character, cyclic redundancy check, and the Hamming code

9-1 INTRODUCTION

The field of digital and data communications has experienced explosive growth in recent years. In general, this field includes the transfer of analog signals using digital techniques and the transfer of digital data using digital and/or analog techniques. It is difficult to separate the two topics totally due to their interrelationships.

Digital communications is the transfer of information in digital form. As shown in Fig. 9-1, if the information is analog (voice in this case), it is converted to digital for transmission. At the receiver it is reconverted to analog. Figure 9-1 also shows a digital computer signal transmitted to another computer. Notice that this is shown to represent both digital and data communications. The third system in Fig. 9-1 shows a computer's digital signal converted to analog for transmission and then reconverted to digital by the modem. We look at modems in detail in Chapter 11.

The reason that various techniques are used boils down to performance and cost. This will be apparent as we take a close look at the systems involved.

The move to digital and/or data communications is due to a number of factors. It is occurring despite the increased complexity and bandwidth necessary for transmission.

Photo: Chip set for a digital receiver. (Courtesy of RF Monolithics, Inc.)

353

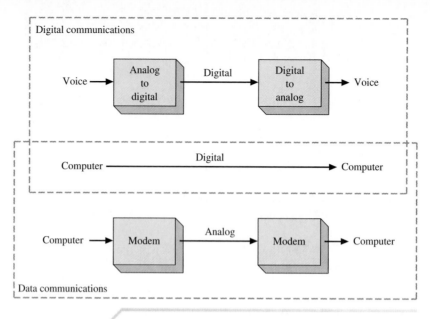

FIGURE 9-1 Digital/data communications.

Noise performance is one of two major advantages. Consider an analog signal with an instantaneous received value of 1 mV. If at the same time an instantaneous 0.1-mV noise spike changes the received value to 1.1 mV, there is normally no way of knowing the correct value of the signal. In a digital system, however, the received signal may ultimately be changed to either a 0- or 1-V level. Now the received noise of 0.1 mV may still be there but certainly would not cause an error.

The digital system is able to recreate the original signal by having circuits that change any signal up to 0.5 V into the 0-V level and any signal above 0.5 V into the 1-V level. This ability to restore a noise-corrupted signal to its original value is called *regeneration*. Obviously, if the noise is so great as to cause the 0-V level to be seen as a 0.6-V level, an error will occur.

Another advantage of using a digital format involves the ability to process the signal at the transmitter (pre-processing) and/or the receiver (post-processing). Both of these operations are termed *digital signal processing*. Signals in digital format can be stored in computer memory and be easily manipulated by *algorithms*—a plan or set of instructions followed to achieve a specific goal. They can be implemented by digital circuitry, which is a *hardware* solution. Increasingly, they are implemented by *software* instructions (via a *computer program*) that instruct a microprocessor how to perform specific manipulations. Many communications systems use *microcontrollers* that are microprocessors programmed to do one basic task only.

9-2 CODING

Some of the earliest forms of electrical communications used coding to send messages rather than direct transmission of *voice*. The telegraph demonstration by Samuel Morse in 1843 is an example. It is ironic that digital communications (as telegraphy is) now promise to help ease the problems of overcrowded voice transmission facilities. The future will certainly bring more and more coded speech, transmitted in digital format because of the following advantages:

1. Less sensitive to noise
2. Less crosstalk (cochannel interference)
3. Lower distortion levels
4. Faded signals more easily recreated
5. Greater transmission efficiency

What at first seemed a barrier to digital transmission—that is, the need to encode (convert) an analog signal into digital form—is proving to be an advantage. Coding now allows speech to be compressed to its minimum essential content and therefore permits the greatest possible efficiency in transmission.

Coding may be defined as the process of transforming messages or signals in accordance with a definite set of rules. There are many different codes available for use, but one thing they universally share is the use of two levels. We can then refer to this as a binary system. In such a system the next signal will either be *high* or *low* and should have an equal (50:50) chance of being one or the other. A *bit* is a unit of information required to allow proper selection of one out of two equally probable events. For example, assuming that heads or tails when flipping a coin are equally probable, let a high condition represent heads and a low condition represent tails. It is required to have 1 bit of information to predict the result of the coin toss. If that 1 bit of information is a high condition (usually termed the 1 level), then one can correctly predict that heads came up.

The high and low conditions are referred to as 1 and 0 or *mark* and *space,* respectively. With respect to an electrical signal used to represent these conditions, the relationship is shown in Fig. 9-2. In this figure, 7 bits of information are provided with the following sequence: 1 0 1 1 0 0 1. With 7 bits available it is possible to select the correct result out of 128 equiprobable events since the 7 bits can be arranged in 128 different configurations, and each one may be allowed to represent 1 of 128 equiprobable events in a code. Since 2 raised to the seventh power equals 128, we can infer the following relationship:

$$B = \log_2 N \qquad (9\text{-}1)$$

where B is the number of bits of information required to predict one out of N equiprobable events.

If it were required to use a binary code to represent the 26 letters in the alphabet, Eq. (9-1) shows that

$$B = \log_2 26 = 4.7$$

Since the number of bits required must always be an integer, 5 bits are necessary. Since 5 bits means that a choice out of 32 different events ($2^5 = 32$) is possible, not all the capability of a 5-bit system is utilized. The efficiency is

$$\mu = \frac{4.7}{5} \times 100\% = 94\%$$

It is not absolutely necessary to use a binary system for coding, but it is easy to show why it is used almost exclusively. A decimal system or 10 different levels is sometimes

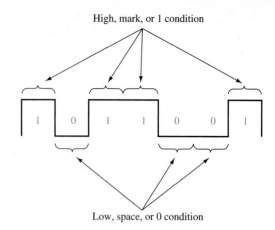

High, mark, or 1 condition

FIGURE 9-2 ▇ Binary information.

Low, space, or 0 condition

utilized. In it, the *dit* (decimal digit) is the basic unit of information. If it were used to represent our alphabet, the number of required dits would be $\log_{10} 26 = 1.415$ dits. Thus, 2 dits would be required to represent the 26 different letters, which means the efficiency of the decimal system in this case would be

$$\mu = \frac{1.415}{2} \times 100\% = 71\%$$

It is a fact that binary coding is more efficient than other coding except in isolated instances. It is also simpler to implement than other coding schemes.

EXAMPLE 9-1

Determine the number of bits required for a binary code to represent 110 different possibilities, and compare its efficiency with a decimal system to accomplish the same goal.

SOLUTION

In a binary system,

$$B = \log_2 N$$
$$= \log_2 110 = 6.78 \tag{9-1}$$

(Solution of the equation above can be accomplished by taking the natural log of 110 divided by the natural log of 2.

$$\frac{\ln 110}{\ln 2} = 6.78$$

and therefore $2^{6.78} = 110$.)

Thus, 7 bits are required, and the efficiency is

$$\mu = \frac{6.78}{7} \times 100\% = 97\%$$

In a decimal system the number of dits required is $\log_{10} 110 = 2.04$, or a total of 3 dits. The efficiency is

$$\mu = \frac{2.04}{3} \times 100\% = 68\%$$

Code Noise Immunity

We have seen that binary coding systems are generally more efficient than other systems. Perhaps an even greater advantage is their superior noise immunity. Consider a binary system where 0 V represents 0 and 9 V represents 1. In that system it would take a noise level of roughly 5 V or more to cause an error in output intelligibility. In a decimally coded system of the same total output power, 0 is represented by 0 V, 1 by 1 V, 2 by 2 V, etc., up to 9 by 9 V. Thus, the 10 discrete levels are 1 V apart, and a 0.5-V noise level can impair intelligibility. If the output levels in this decimal system were 0, 10 V, 20 V, . . . , 90 V, then the 9-V binary and 90-V decimal system would have comparable noise immunities. Since power is proportional to the square of voltage, the decimal system requires 10^2 or 100 times the power of the binary system to offer the same noise immunity. By now in your study of communications you may have come to the correct conclusion that noise is the single most important consideration. The reasoning just concluded regarding efficiency and noise immunity is an elementary example of the work of an information theory specialist. Recall from Chapter 1 that *information theory* is the branch of learning concerned with optimization of transmitted information.

Errors occur as a result of noise. There are many sources of noise, as explained in Chapter 1. The *error probability* in a digital system is the number of errors per total number of bits received. For instance, if 1 error bit per 100,000 bits occurs, the error probability is 1/100,000, or 10^{-5}. The acceptable error probability in communications systems ranges from about 10^{-5} up to 10^{-12} in more demanding applications. The average number of errors in a transmission m bits long can be calculated as

$$\text{average number of errors} = m \times \text{error probability} \qquad (9\text{-}2)$$

The most common method of referring to the quality of a digital communications system is its *bit error rate* (BER). The BER is simply the average number of correct bits before an error bit occurs. The BER is the reciprocal of the error probability. Therefore, the acceptable range of 10^{-5} to 10^{-12} for error probability translates into a BER of 10^5 to 10^{12}.

EXAMPLE 9-2

A digital transmission has an error probability of 10^{-4} and is 10^9 bits long. Calculate the expected number of error bits and the BER.

$$\text{Average number of errors} = m \times \text{error probability}$$

$$= 10^9 \times 10^{-4} \tag{9-2}$$

$$= 10^5$$

BER is $1/10^{-4}$ or 10^4.

It was previously shown that the transmitted power must be raised significantly to allow a constant signal-to-noise ratio when the number of coding levels is raised. The Shannon–Hartley theorem relates the capacity of a channel when its bandwidth and noise are known.

$$C = \text{BW} \log_2\left(1 + \frac{S}{N}\right) \tag{9-3}$$

where C = channel capacity (b/s)
  BW = bandwidth (Hz)
  S/N = signal-to-noise power ratio

EXAMPLE 9-3

Calculate the capacity of a telephone channel that has a S/N of 1023.

SOLUTION

The telephone channel has a bandwidth of about 3 kHz. Thus,

$$C = \text{BW} \log_2(1 + S/N) \tag{9-3}$$

$$= 3 \times 10^3 \log_2(1 + 1023)$$

$$= 3 \times 10^3 \log_2(1024)$$

$$= 3 \times 10^3 \times 10$$

$$= 30{,}000 \text{ bits per second}$$

Example 9-3 shows that a telephone channel could theoretically handle 30,000 b/s. In practice, speeds of 9600 b/s are not normally exceeded. The Shannon–Hartley theorem represents a fundamental limitation. The only consequence of exceeding it is a very high error rate. Generally, an acceptable bit error rate (BER) of 10^5 or better requires significant reductions from the Shannon–Hartley theorem prediction.

Two of the most common alpha-numeric coding schemes for binary data are *ASCII,* the American Standard Code for Information Interchange; and *EBCDIC,* the Extended Binary-Coded Decimal Interchange Code. Each of these codes can be found in many digital systems, with the ASCII code being the most prevalent.

The ASCII Code

ASCII is a 7-bit code used for representing alpha-numeric symbols with a distinctive code word. The ASCII code was developed by a committee of the American National Standards Institute (ANSI) for the purpose of coding binary data. ASCII-77 is the adopted international standard. Figure 9-3 provides a list of the codes.

There are 2^7 (128) possible 7-bit code words available with an ASCII system. The binary codes are ordered sequentially, which simplifies the grouping and sorting of the characters. The 7-bit words are ordered with the least significant bit (LSB) given as bit 1

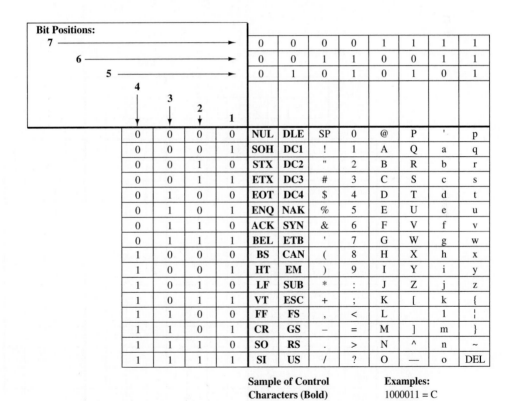

Sample of Control
Characters (Bold)

STX = Start of text
EOT = End of transmission
CR = Carriage return
HT = Horizontal tabulation

Examples:

1000011 = C
0110011 = 3
1010000 = P
0110000 = 0 (Zero)
0100000 = SP (space)

FIGURE 9-3 ▮ American Standard Code for Information Interchange (ASCII).

(b_1), while the most significant bit (MSB) is bit 7 (b_7). Notice that a binary value is not specified by the code for bit 8 (b_8). Usually the bit 8 (b_8) position is used for parity checking. *Parity* is an error detection scheme that identifies whether an even or odd number of logical ones are present in the code word. This concept is discussed in greater detail in Sec. 9-8. For ASCII data used in a serial transmission system b_1, the LSB bit, is transmitted first.

The ASCII system is based on the binary-coded-decimal (BCD) code in the last 4 bits. The first 3 bits indicate whether a number, letter, or character is being specified. Notice that 0110001 represents "1," while 1000001 represents "A" and 1100001 represents "a." It uses the standard binary progression (i.e., 0110010 represents "2"), and this makes mathematical operations possible. Since the letters are also represented with the binary progression, alphabetizing is also achieved via binary mathematical procedures. The student should also be aware that analog waveform coding is accomplished simply by using the BCD code for PCM systems covered in Sec. 9-4.

In some systems the actual transmission of these codes includes an extra pulse at the beginning (start) and ending (stop) for each character. When start/stop pulses are used in the coding of signals, it is called an *asynchronous* (nonsynchronous) transmission. A synchronous transmission (without start/stop pulses) allows more characters to be trans-

EBCDIC CODES

Bit Positions 4,5,6,7	Second Hex Digit	0	1	2	3	4	5	6	7	8	9	A	B	C	D	E	F
0000	0	NUL	DLE	DS		SP	&	-						(	)	\	0
0001	1	SOH	DC1	SOS		RSP		/		a	j	~		A	J	NSP	1
0010	2	STX	DC2	FS	SYN					b	k	s		B	K	S	2
0011	3	ETX	DC3	WUS	IR					c	l	t		C	L	T	3
0100	4	SEL	RES/ENP	BYP/INP	PP					d	m	u		D	M	U	4
0101	5	HT	NL	LF	TRN					e	n	v		E	N	V	5
0110	6	RNL	BS	ETB	NBS					f	o	w		F	O	W	6
0111	7	DEL	POC	ESC	BOT					g	p	x		G	P	X	7
1000	8	GE	CAN	SA	SBS					h	q	y		H	Q	Y	8
1001	9	SPS	EM	SPE	IT				▲	i	r	z		I	R	Z	9
1010	A	RPT	UBS	SM/SW	RFF	¢	!	|	:						SHY		
1011	B	VT	CU1	CSP	CU3	.	$	,	#								
1100	C	FF	IFS	MFA	DC4	<	*	%	@								
1101	D	CR	IGS	ENQ	NAK	(	)	_	▲								
1110	E	SO	IRS	ACK		+	;	>	=								
1111	F	SI	IUS/ITB	BEL	SUB	¦	¬	?	"								BO

Bit Positions 0,1: 00 (cols 0–3), 01 (cols 4–7), 10 (cols 8–B), 11 (cols C–F); Bit Positions 2,3: 00, 01, 10, 11; First Hexadecimal Digit.

FIGURE 9-4 ■ The Extended Binary-Coded Decimal Interchange Code.

mitted within a given sequence of bits. The transmission of information between various computer installations may require the less efficient asynchronous transmitting mode depending on computer characteristics.

The EBCDIC Code

The Extended Binary-Coded Decimal Interchange Code (*EBCDIC*) is an 8-bit alpha-numeric code. The term "binary-coded decimal" is used in the name because of the structure present in the coding scheme, which uses only the 0–9 positions. A list of the code words for the EBCDIC system is given in Fig. 9-4, and the acronyms for the control characters are listed in Table 9-1.

The Baudot Code

Another interesting code is the *Baudot* code. The Baudot code was developed in the days of teletype machines such as the ASR-33 Teletype terminal. Baudot is an alpha-numeric code based on five binary values. The Baudot code is not very powerful, but it does have

TABLE 9-1 *The EBCDIC Code—List of Acronyms*

ACK	Acknowledge	ETB	End of Transmission	RFF	Required Form Feed
BEL	Bell	ETX	End of Text	RNL	Required New Line
BS	Backspace	FF	Form Feed	RPT	Repeat
BYP/	Bypass/Inhibit	FS	Field Separator	SA	Set Attribute
INP	Presentation	GE	Graphic Escape	SBS	Subscript
CAN	Cancel	HT	Horizontal Tab	SEL	Select
CR	Carriage Return	IFS	Interchange File Sep.	SFE	Start Field Extend
CSP	Control Sequence Prefix	IGS	Interchange Group Sep.	SI	Shift In
CU1	Customer Use 1	IR	Index Return	SM/ SW	Set Mode/Switch
CU3	Customer Use 3	IRS	Interchange Record Sep.	SO	Shift Out
DC1	Device Control 1	IT	Indent Tab	SOH	Start of Heading
DC2	Device Control 2	IUS/ ITB	Interchange Unit Sep./ Intermediate Text Block		
DC3	Device Control 3	LF	Line Feed	SOS	Start of Significance
DC4	Device Control 4	MFA	Modify Field Attribute	SPS	Superscript
DEL	Delete	NAK	Negative Acknowledge	STX	Start of Text
DLE	Data Link Escape	NBS	Numeric Backspace	SUB	Substitute
DS	Digit Select	NL	New Line	SYN	Synchronous Idle
EM	End of Medium	NUL	Null	TRN	Transparent
ENQ	Enquiry	POC	Program-Operator Comm.	UBS	Unit Backspace
EO	Eight Ones	PP	Presentation Position	VT	Vertical Tab
EOT	End of Transmission	RES/ NEP	Restore/Enable Presentation	WUS	Word Underscore
ESC	Escape				

Character Shift		Binary Code
		BIT
Letter	Figure	4 3 2 1 0
A	–	1 1 0 0 0
B	?	1 0 0 1 1
C	:	0 1 1 1 0
D	$	1 0 0 1 0
E	3	1 0 0 0 0
F	!	1 0 1 1 0
G	&	0 1 0 1 1
H	#	0 0 1 0 1
I	8	0 1 1 0 0
J	'	1 1 0 1 0
K	(	1 1 1 1 0
L	)	0 1 0 0 1
M	.	0 0 1 1 1
N	,	0 0 1 1 0
O	9	0 0 0 1 1
P	0	0 1 1 0 1
Q	1	1 1 1 0 1
R	4	0 1 0 1 0
S	BEL	1 0 1 0 0
T	5	0 0 0 0 1
U	7	1 1 1 0 0
V	;	0 1 1 1 1
W	2	1 1 0 0 1
X	/	1 0 1 1 1
Y	6	1 0 1 0 1
Z	"	1 0 0 0 1
Figure Shift		1 1 1 1 1
Letter Shift		1 1 0 1 1
Space		0 0 1 0 0
Line Feed		0 1 0 0 0
Null		0 0 0 0 0

FIGURE 9-5 ▓ The Baudot code.

its place in communications history and still can be of use in limited circumstances. The Baudot code is provided in Fig. 9-5.

The alphabet has 26 letters, and there is an almost equal number of commonly used symbols and numbers. The 5-bit Baudot code is capable of handling these possibilities. A 5-bit code can have only 2^5 or 32 bits of information but actually provides 26×2 bits by transmitting a 11111 to indicate all following items are "letters" until a 11011 transmission occurs, indicating "figures." Notice that no provision for lowercase letters is provided.

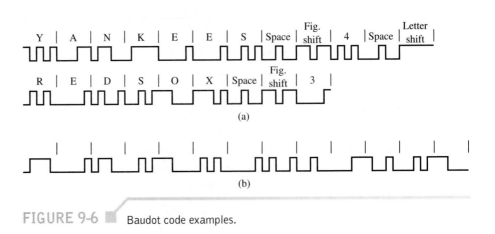

FIGURE 9-6 ▓ Baudot code examples.

Figure 9-6(a) shows an example of the Baudot code to transmit "YANKEES 4 REDSOX 3." Be sure to work out the code in Fig. 9-6(b) on your own; it is the only "X-rated" part of this book that I was allowed to include.

The Gray Code

The last alpha-numeric code we will look at is the *Gray* code. The Gray code is a numeric code for representing the decimal values from 0 to 9. It is based on the relationship that only one bit in a binary word changes for each binary step. For example, the code for 7 is 0010 while the code for 8 is 0011. Notice that only one binary bit changes when the decimal value changes from 7 to 8. This is true for all of the numbers (0–9). The Gray code is shown in Fig. 9-7.

The Gray code is used most commonly in telemetry systems that have slow changing data. This coding scheme works well for detecting errors in slowly changing outputs, such as data from a temperature sensor (thermocouple). If more than one change is detected when words are decoded, then the receiving circuitry assumes that an error is present. More information on telemetry systems is provided in Chapter 10.

9-4 PULSE-CODE MODULATION

Pulse-code modulation (PCM) is the most common technique used today in digital communications for representing an analog signal by a digital word. PCM is used in many applications, such as your telephone system, digital audio recording (DAT or Digital Audio Tape), CD laser disks, digitized video special effects, voice mail, and many other applications. PCM techniques and applications are a primary building block for many of today's advanced communications systems.

Our discussion of PCM systems begins with a look at the analog-to-digital conversion process—ADC or A/D converter. A simple example of the A/D process is to think in terms of our voice and the mechanism required to convert it into a digital data format suitable for inputting to a computer. This requires that the analog signal, which is a continuous-time signal, be converted into a series of quantized values that then represent the original analog signal in digital form. This digitized signal can then be digitally processed by the interface circuitry to the computer or by the computer itself. The data can then be held in the computer until accessed by the DAC, the digital-to-analog converter.

Pulse-code modulation is a technique for converting the analog signals into a digital representation. The PCM architecture consists of a sample-and-hold (S/H) circuit and a

Binary	#
0000	0
1000	1
1100	2
0100	3
0110	4
1110	5
1010	6
0010	7
0011	8
1011	9

1 bit change for each step value.

FIGURE 9-7 ▪ The Gray code.

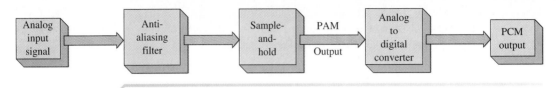

FIGURE 9-8 ■ Block diagram of the PCM process.

system for converting the sampled signal into a representative binary format. First, the analog signal is inputted into a sample-and-hold circuit. At fixed time intervals, the analog signal is sampled and held at a fixed voltage level until the circuitry inside the A/D converter has time to complete the conversion process of generating a binary value. A block diagram of the process is shown in Fig. 9-8. We'll discuss the anti-aliasing filter shown in coming paragraphs.

The Sample-and-Hold Circuit

Most A/D integrated circuits come with S/H circuits integrated into the system, but it is still necessary for the user to have a good understanding of its operation, including both the capabilities and the limitations of the S/H circuit.

A typical S/H circuit is shown in Fig. 9-9. The analog signal is typically inputted into a buffer circuit. The purpose of the buffer circuit is to isolate the input signal from the S/H circuit and to provide proper impedance matching as well as drive capability to the hold circuit. Many times the buffer circuit is also used as a current source to charge the hold capacitor. The output of the buffer is fed to an analog switch, which is typically the drain of a JFET, junction field-effect transistor, or a MOSFET, metal-oxide semiconductor field-effect transistor. The JFET or MOSFET is wired as an analog switch and controlled at the gate by a sample pulse generated by the sample clock. When the JFET's or MOSFET's gate is asserted, the switch will short the analog signal from drain to source. This connects the buffered input signal to a hold capacitor. The capacitor begins to charge to the input voltage level at a time constant determined by the hold capacitor's capacitance and the analog switch's and buffer circuit's "on" channel resistance.

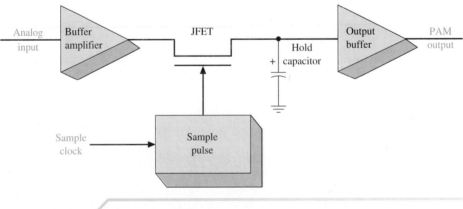

FIGURE 9-9 ■ A sample-and-hold circuit.

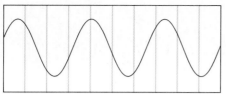

(a) Sample intervals for an input sinusoid.

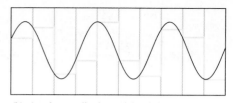

(b) A pulse-amplitude-modulated signal (PAM).

FIGURE 9-10 ■ Generation of PAM.

When the analog switch is turned off, the sampled analog signal voltage level is held by the hold capacitor. Figure 9-10(a) shows a picture of a sinusoid on the input of the S/H circuit. The sample times are indicated by the green vertical lines. In Fig. 9-10(b) the sinusoid is redrawn as a sampled signal. Note that the sampled signal maintains a fixed voltage level between samples. The region where the voltage level remains relatively constant is called the hold time. The resulting waveform, shown in Fig. 9-10(b), is called a pulse-amplitude-modulated (PAM) signal. The S/H circuit is designed so that the sampled signal is held long enough to be converted by the A/D circuitry into a binary representation.

The time required for a S/H circuit to complete a sample is based partly on the acquisition and aperture times. The *acquisition* time is the amount of time it takes for the hold circuit to reach its final value (during this time the analog switch connects the input signal to the hold capacitor). The acquisition time is controlled by the sample pulse. The *aperture* time is the time that the S/H circuit must hold the sampled voltage. The aperture and acquisition times limit the maximum frequency at which the S/H circuit can accurately process the analog signal.

To provide a good-quality S/H circuit a couple of design considerations must be made. The analog switch "on" resistance must be small. The output impedance of the input buffer must also be small. By keeping the input resistance minimal the overall time constant for sampling the analog signal can be controlled by the selection of an appropriate hold capacitor. Ideally, a low capacitance should be selected so that a fast charging time is possible, but a small capacitor will have trouble holding a charge for a very long period. A 1-nano-farad hold capacitor is a popular choice for many circuit designers. It is also important that the hold capacitor be of high quality. High-quality capacitors have dielectrics of polyethylene, polycarbonate, or teflon. These types of dielectrics minimize voltage variations due to capacitor characteristics.

Pulse-Amplitude Modulation

The concept of pulse-amplitude modulation (PAM) has been introduced in this section, but there are a few specifics regarding the creation of a pulse-amplitude-modulated signal at the output of a sample-and-hold circuit that necessitate discussion.

There are two basic sampling techniques used to create a PAM signal. The first is called *natural sampling*. Natural sampling is when the tops of the sampled waveform (the sampled analog input signal) retain their natural shape. An example of natural sampling is shown in Fig. 9-11(a). Notice that the one side of the analog switch is connected to ground. When the gate is asserted the JFET will short the signal to ground, but it will pass

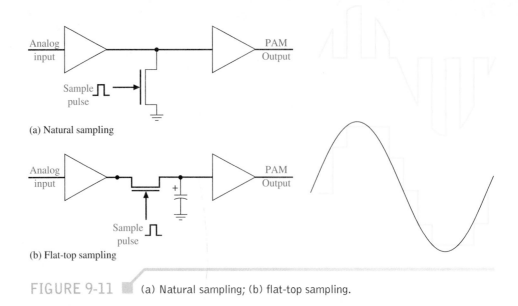

(a) Natural sampling

(b) Flat-top sampling

FIGURE 9-11   (a) Natural sampling; (b) flat-top sampling.

the unaltered signal to the output when the gate is not asserted. Note too that there is not a hold capacitor present in the circuit.

Probably the most popular type of sampling used in PAM systems is called *flat-top sampling*. In flat-top sampling the sample signal voltage is held constant between samples. The method of sampling creates a staircase that tracks the changing input signal. This method is popular because it provides a constant voltage during a window of time for the binary conversion of the input signal to be completed. An example of flat-top sampling is shown in Fig. 9-11(b). Please note that this is the same type of waveform as shown in Fig. 9-9(b). With flat-top sampling the analog switch connects the input signal to the hold capacitor.

The Sample Frequency

One of the most critical specifications in a PCM system is the selection of the sample frequency. The sample frequency is governed by the *Nyquist* rate. The Nyquist rate states that the sample frequency (f_s) must be at least twice the highest input frequency (f_a).

$$f_s \geq 2f_a \qquad (9\text{-}4)$$

EXAMPLE 9-4

A CD audio laser-disk system has a frequency bandwidth of 20 Hz to 20 kHz. What is the minimum sample rate required to satisfy the Nyquist sampling rate?

$$f_s \geq 2f_a$$

$$f_s \geq 2 \times 20 \text{ kHz} \qquad (9\text{-}4)$$

$$f_s \geq 40 \text{ kHz}$$

If the sampling criterion is not met, the original analog signal frequency is lost and an *alias* frequency is produced instead. The frequency of the alias signal is given by

$$f_{\text{alias}} = f_{\text{in}} - f_s \qquad (9\text{-}5)$$

This is shown in Fig. 9-12, where the sampling rate is $\frac{2}{3}$ times the 1-kHz signal rate. The output signal in Fig. 9-12 is at 1000 − 667 Hz, or 333 Hz, and will bear no resemblance to the original signal.

The fact that an erroneous alias signal can occur leads to the use of *antialiasing* filters in PCM systems. A block diagram for a basic PCM system is provided in Fig. 9-13. The antialiasing filter shown is a sharp-cutoff low-pass filter used to ensure that no frequencies above twice the sampling rate (clock in Fig. 9-13) reach the analog-to-digital converter (ADC). These frequencies could be the unnecessary high-frequency portions of an analog signal and/or noise that has corrupted the signal. In either event, it is absolutely necessary to prevent the generation of alias frequencies, making the filter a necessity.

Notice the parallel-to-serial conversion stage in the transmitter section of Fig. 9-13. This process is normally used so that multiple transmitting mediums are not required. An exception to this practice sometimes occurs when the distance between transmitter and receiver is very short, such as from a computer to a printer. In that instance, an eight-wire ribbon cable is not too much of a problem.

The clock signal (at the sampling rate) is also shown in Fig. 9-13. This is necessary so that the digital-to-analog converter (DAC) knows when all the input bits are valid as a group. Erroneous outputs would otherwise occur since small differences in the physical length or momentary shifts (*jitter*) caused by circuitry action cause bits to arrive at differ-

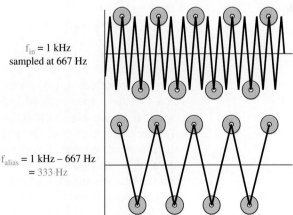

$f_{\text{in}} = 1 \text{ kHz}$
sampled at 667 Hz

$f_{\text{alias}} = 1 \text{ kHz} - 667 \text{ Hz}$
$= 333 \text{ Hz}$

FIGURE 9-12   Generation of alias frequency.

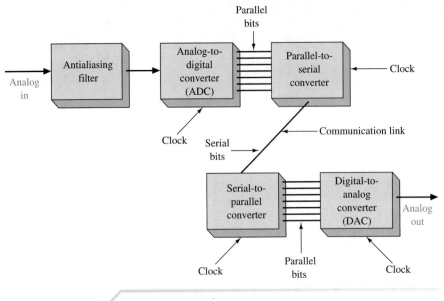

FIGURE 9-13 ■ PCM communication system.

ent times. Without the clock, the DAC will respond to the digital pattern at any instant, which may include bits from two different transmitter conversions. This results in false outputs called *glitches*.

Codec

The A/D circuitry in PCM systems is often referred to as the encoder. The D/A circuitry at the receiver is correspondingly termed the decoder. These functions are often combined in a single LSI chip termed a *codec*. The block diagram for a typical codec is provided in Fig. 9-14. These devices are widely used in the telephone industry to allow voice transmission to be accomplished in digital form.

PCM/TDM Repeaters

PCM is the most noise-resistant transmission system available. It also lends itself to time-division multiplexing (TDM) and because of its true digital nature is very adaptable to microprocessor-controlled communications equipment and to the transmission of digital data. For these reasons, PCM is the fastest-growing method of getting information from one place to another. Its popularity in telephone systems has already been mentioned. Another example of its use is in the distribution of live TV from its point of origination to all parts of the country. To accomplish this, microwave towers at 20- to 30-mi intervals are used to *relay* the TV signals throughout the country. Each tower contains a receiver that receives the signal, reamplifies it, and transmits it to the next tower. The towers are termed *repeaters* and make up for the huge power loss incurred from each transmitter to receiver. Since each reamplification of signal and noise can only aggravate the *S/N* ratio, it is critical to use the most noise-immune system of modulation possible—PCM. It can be theoretically shown that a signal with *S/N* ratio of 21 dB or better can be "repeated" indefinitely without any degradation using PCM.

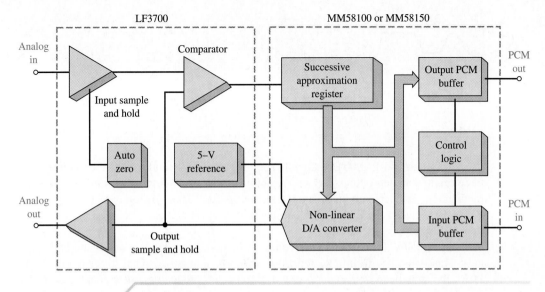

FIGURE 9-14 ■ Codec block diagram.

A PCM/TDM repeater system is shown in Fig. 9-15. It is interesting to notice that the TDM sampler initially creates a PAM signal that is then quantized into an *n*-bit PCM/TDM signal. It is then modulated onto a carrier and transmitted to the first repeater. While only two repeaters are shown, the number can be as large as necessary, and in fact hundreds exist for TV signal distribution. At the ultimate signal destination the signal is applied to a demodulator, which outputs the binary-coded pulses to the decoder. The

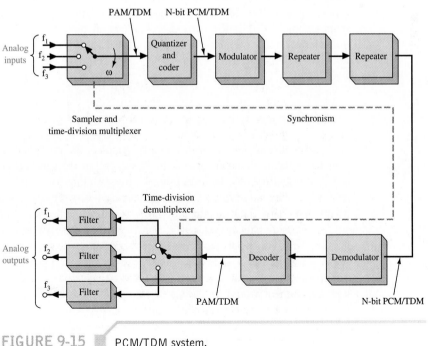

FIGURE 9-15 ■ PCM/TDM system.

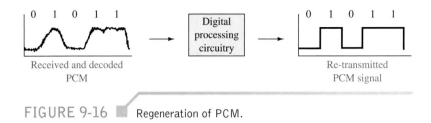

FIGURE 9-16 ▓ Regeneration of PCM.

decoder output is a PAM/TDM signal. The time-division *de*multiplexer then delivers the three original analog outputs, which are applied to low-pass filters that remove unwanted modulation products produced by the demultiplexer. The synchronism link shown in Fig. 9-15 between the multiplexer and demultiplexer is necessary to keep the three distinct signals properly separated.

In actuality, the repeaters are complete demodulators/modulators of the PCM signals. This is the key to infinite repetition without cumulative signal degeneration. With PCM, the signal can be regenerated at each repeater, producing a new signal essentially free from noise. A reference to the pulses shown in Fig. 9-16 serves to illustrate this point. The received PCM signal has been affected by noise but not to the point where it isn't recognizable as a 01011 signal. The signal's amplitude, pulse width, and pulse position have been altered by noise and a finite bandwidth system that causes pulse rise and fall times to have visible slope. Applying this signal to appropriate digital processing circuitry allows the signal to be "cleaned" up to its original condition for retransmission. This assumes the noise is not so great as to make it impossible to distinguish between a high or low on a reliable basis. This process is not possible with standard AM or FM. The noise errors could not be cleaned up, and cumulative degradation occurs. PCM relies solely on the presence or absence of pulses, not absolute amplitude, duration, or position.

PCM should be given consideration for applications involving TDM, minimum transmitted power, combinations of analog and digital messages, computer control, or many repeater stations. Since long-distance telephone transmission has all of these requirements, PCM is the current technology in modern telephony. In routine applications, however, the high cost of hardware for PCM makes it prohibitive compared to analog modulation systems.

9-5 PCM QUANTIZATION EFFECTS

Once an analog signal has been properly sampled, the process of converting the sampled signal to a binary value can begin. In PCM systems the sampled signal is segmented into different voltage levels, with each level corresponding to a different binary number. This process is called *quantization.* The quantization levels also determine the resolution of the digitizing system. Each quantization level step-size is called a *quantile* or *quantile interval.*

Analog signals are quantized to the closest binary value provided in the digitizing system. This is an approximation process. For example, if our numbering system is the set of whole numbers 1, 2, 3, . . . etc., and the number 1.4 must be converted (rounded-off) to the closest whole number, then 1.4 is translated to 1. If the input number were 1.6, then the number would be translated to a 2. If the number were 1.5, then we have the same error if the number is rounded off to a 1 or 2.

In PCM, the electrical representation of voice is converted from analog form to digital form. This process of encoding is shown in Fig. 9-17. There are a set of amplitude

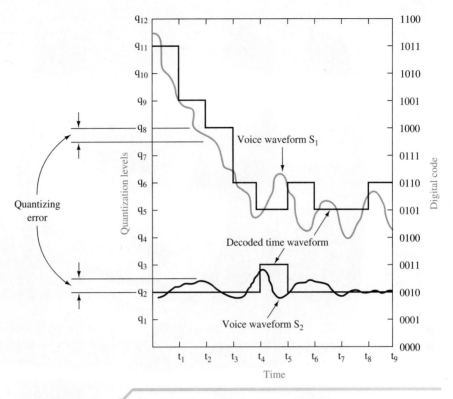

FIGURE 9-17 ■ PCM encoding. (From the November 1972 issue of the *Electronic Engineer,* with the permission of the publisher.)

levels and sampling times. The amplitude levels are termed *quantizing levels,* and 12 such levels are shown. At each sampling interval the analog amplitude is quantized into the closest available quantization level, and the analog-to-digital converter (ADC) puts out a series of pulses representing that level in the binary code.

For example, at time t_2 in Fig. 9-17, voice waveform S_1 is closest to level q_8, and thus the coded output at that time is the binary code 1000, which represents 8 in binary code. Note that the quantizing process resulted in an error, which is termed the *quantizing error* or *quantizing noise.* Voice waveform S_2 provides a 0010 code at time t_2, and its quantizing error is also shown in Fig. 9-17. The amount of this error can be minimized by increasing the number of quantizing levels, which of course lessens the space between each one. The 4-bit code shown in Fig. 9-17 allows for a maximum of 16 levels since $2^4 = 16$. The use of a higher-bit code decreases the error at the expense of transmission time and/or bandwidth since, for example, a 5-bit code (32 levels) means transmitting 5 high or low pulses instead of 4 for each sampled point. The sampling rate is also critical and must be greater than twice the highest significant frequency, as previously described. It should be noted that the sampling rate in Fig. 9-17 is lower than the highest-frequency component of the information. This is not a practical situation but was done for illustrative purposes.

While a 4- or 5-bit code may be adequate for voice transmission, it is not adequate for transmission of television signals. Figure 9-18 provides an example of TV pictures for 5-bit and 8-bit (256 levels) PCM transmissions, each with 10-MHz sampling rates. In the

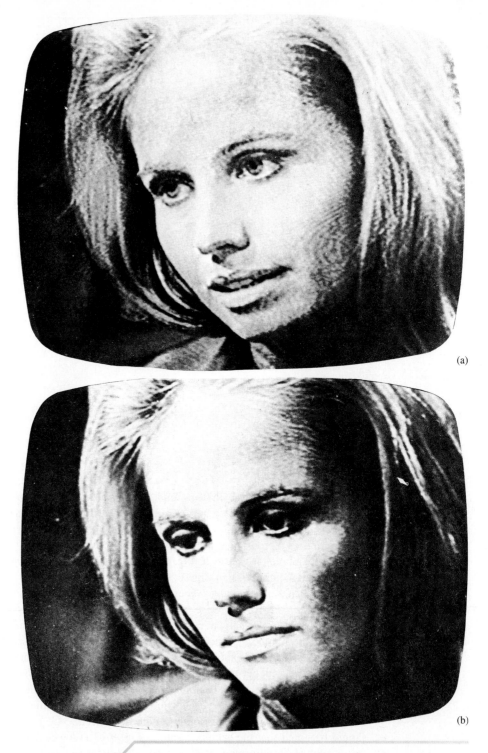

(a)

(b)

FIGURE 9-18 ■ PCM TV transmission: (a) 5-bit resolution; (b) 8-bit resolution.

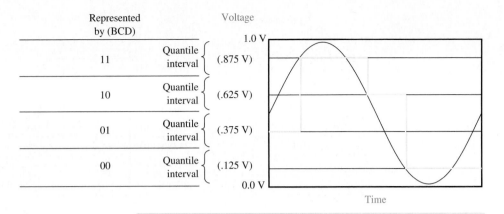

Represented by (BCD)		Voltage
11	Quantile interval	(.875 V)
10	Quantile interval	(.625 V)
01	Quantile interval	(.375 V)
00	Quantile interval	(.125 V)

FIGURE 9-19 ■ Voltage levels for a quantized signal.

first (5-bit) picture, contouring in the forehead and cheek areas is very pronounced. The 8-bit resolution results in an excellent-fidelity TV signal that is not discernibly different from a standard continuous modulation transmission.

Notice in Fig. 9-19 that at the sample intervals, the closest quantization level is selected for representing the sine-wave signal. The resulting waveform has poor resolution with respect to the sine-wave input. *Resolution* with respect to a digitizing system refers to the accuracy of the digitizing system in representing a sampled signal. For example, the analog input to our PCM system has a minimum voltage of 0.0 V and a maximum of 1.0 V. If a 2-bit system is used for quantizing a signal, then 2^2, or four, quantized levels are used. Referring to Fig. 9-19 we see that the quantized levels (quantile intervals) are each .25 V in magnitude. Typically it is stated that this system has 2-bit resolution.

To increase the resolution of a digitizing system requires that the number of quantization levels be increased. To increase the number of quantization levels requires that the number of binary bits representing each voltage level be increased. If the resolution of the example in Fig. 9-19 is increased to 3 bits, then the input signal will be converted to 1 of 8 possible values. The 3-bit example with improved resolution is shown in Fig. 9-20.

Represented by (BCD)		Voltage
111	Quantile interval	.9375 V
110	Quantile interval	.8125 V
101	Quantile interval	.6875 V
100	Quantile interval	.5625 V
011	Quantile interval	.4375 V
010	Quantile interval	.3125 V
001	Quantile interval	.1875 V
000	Quantile interval	.0625 V

FIGURE 9-20 ■ An example of 3-bit quantization.

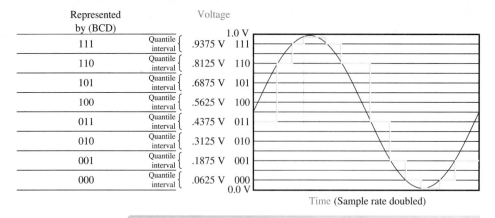

Represented by (BCD)		Voltage	
		1.0 V	
111	Quantile interval	.9375 V	111
110	Quantile interval	.8125 V	110
101	Quantile interval	.6875 V	101
100	Quantile interval	.5625 V	100
011	Quantile interval	.4375 V	011
010	Quantile interval	.3125 V	010
001	Quantile interval	.1875 V	001
000	Quantile interval	.0625 V	000
		0.0 V	

Time (Sample rate doubled)

FIGURE 9-21 ■ An example of 3-bit quantization with increased sample rate.

Another way of improving the accuracy of the quantized signal is to increase the sample rate. Figure 9-21 shows the sample rate doubled but still using a 3-bit system. The resultant signal is dramatically improved by this change in sampling rate.

Dynamic Range and Signal-to-Noise Calculations

Dynamic range (DR) for a PCM system is defined as the ratio of the maximum input or output voltage level to the smallest voltage level that can be quantized and/or reproduced by the converters. This value is expressed as:

$$\text{Dynamic range (DR)} = \frac{V_{MAX}}{\frac{V_{LSB}}{2}} \qquad (9\text{-}6)$$

Dynamic range is typically expressed in terms of decibels. For a binary system, each bit can have two logic levels, either a logical low or logical high. Therefore the dynamic range for a single-bit binary system can be expressed logarithmically, in terms of dB, by the expression:

$$\text{DR}_{dB} = 20 \log_{10}(2) \qquad (9\text{-}7)$$

$$\text{DR} = 6.02 \text{ dB/bit}$$

The value "2" is used in Eq. (9-7) since there are 2 states in a binary system. Therefore, the dynamic range for a binary system is expressed as 6.02 dB/bit or 6.02 × *n*, where *n* represents the number of quantizing bits. To calculate the dynamic range for a multiple-bit system, simply multiply the number of quantizing bits (*n*) times 6.02 dB per bit. For example, an 8-bit system will have a dynamic range (expressed in dB) of:

$$(8 \text{ bits})(6.02 \text{ dB/bit}) = 48.16 \text{ dB}$$

The *signal-to-noise ratio* (*S/N*) for a digitizing system is written as:

$$S/N = [1.76 + 6.02n]$$ (9-8)

where n = the number of bits used for quantizing the signal
S/N = signal-to-noise ratio in dB

Example 9-5 shows how Eqs. (9-7) and (9-8) can be used to obtain the number of quantizing bits required to satisfy a specified dynamic range and determine the signal-to-noise ratio for a digitizing system.

EXAMPLE 9-5

A digitizing system specifies 55 dB of dynamic range. How many bits are required to satisfy the dynamic range specification? What is the signal-to-noise ratio for the system?

SOLUTION

First solve for the dynamic range (DR).

$$DR = 6.02 \text{ dB/bit} (n)$$ (9-7)

$$55 \text{ dB} = 6.02 \text{ dB/bit} (n)$$

$$n = \frac{55}{6.02} = 9.136$$

Therefore, ten bits are required to achieve 55 dB of dynamic range. Nine bits will only provide a dynamic range of 54.18 dB. The tenth bit is required to meet the 55 dB of required dynamic range. Ten bits provides a dynamic range of 60.2 dB. To determine the signal-to-noise (*S/N*) ratio for the digitizing system:

$$S/N = [1.76 + 6.02n]\text{dB}$$ (9-8)

$$S/N = [1.76 + (6.02)10]\text{dB}$$

$$S/N = 61.96 \text{ dB}$$

Therefore, the system will have a signal-to-noise ratio of 61.96 dB.

Companding

Up to this point our discussion and analysis of PCM systems have been developed around *uniform* or *linear* quantization levels. In linear (uniform) quantization systems each quantile interval is the same step-size. An alternative to linear PCM systems is *non-linear* or *non-uniform* coding in which each quantile interval step-size may vary in magnitude.

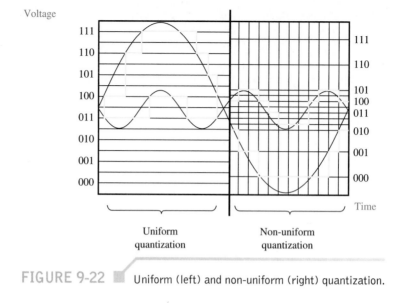

Voltage

111
110
101
100
011
010
001
000

111
110
101
100
011
010
001
000

Time

Uniform
quantization

Non-uniform
quantization

FIGURE 9-22 ▓ Uniform (left) and non-uniform (right) quantization.

It is quite possible for the amplitude of an analog signal to vary throughout its full range. In fact, this is expected for systems exhibiting a wide dynamic range. The signal will change from a very strong signal (maximum amplitude) to a weak signal (minimum amplitude—V_{lsb} for quantized systems). For the system to exhibit good signal-to-noise characteristics, either the input amplitude must be increased with reference to the quantizing error or the quantizing error must be reduced.

The justification for the use of a non-uniform quantization system has been presented, but let's discuss some general considerations before proceeding. How can the quantization error be modified in a non-uniform PCM system in such a way that the result is improved *S/N*? The answer can be obtained by first examining a waveform that has uniform quantile intervals as shown in Fig. 9-22. Notice that poor resolution is present in the weak signal regions, yet the strong signal regions exhibit a reasonable facsimile of the original signal. Figure 9-22 shows how the quantile intervals can be changed to provide smaller step-sizes within the area of the weak signal. This will result in an improved *S/N* ratio for the weak signal.

What is the price paid for incorporating a change such as this in a PCM system? The answer is that the large amplitude signals will have a slightly degraded *S/N*, but this is an acceptable situation if the goal is improving the weak signal's *S/N*.

Amplitude Companding

The other form of companding is called *amplitude companding*. Amplitude companding involves the process of volume compression before transmission and expansion after detection. This is illustrated in Fig. 9-23. Notice how the weak portion of the input is made nearly equal to the strong portion by the compressor but restored to proper level by the expander. Companding is essential to quality transmission using PCM and the delta modulation technique introduced in the next section.

The use of time-division-multiplexed (TDM) PCM transmission for telephone transmissions has proven its ability to cram more messages into short-haul cables than

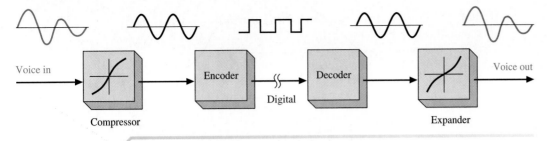

FIGURE 9-23 Companding process.

frequency-division-multiplexed (FDM) analog transmission. The TDM PCM methods were started by Bell Telephone in 1962 and are now the only method used in new designs except for delta modulation schemes. Once digitized, these voice signals can be electronically switched and restored without degradation. The standard PCM system in U.S. telephony uses *μ-law* companding. The companded signal is predicted by

$$V_{out} = \frac{V_{max} \times \ln\left(1 + \mu \, V_{in}/V_{max}\right)}{\ln\left(1 + \mu\right)} \qquad (9-9)$$

The prevalent systems have $\mu = 100$. These systems multiplex 24 voice channels onto a single twisted pair. Each channel is coded into 8 bits and the sampling frequency is 8 kHz. The 8 bits multiplied by 24 channels equals 192 bits. An additional bit is added for synchronizing, which makes up a *frame* period of 193 bits. The signaling rate is therefore

$$193 \text{ bits} \times 8 \text{ kHz} = 1.544 \text{ Mb/s}$$

This defines the standard T1 telephone line (see Sec. 11-2). This assumes that each bit (square wave) can be adequately resolved by its fundamental (sine-wave) component. This is, in fact, true in actual practice.

9-6 DELTA MODULATION

Delta modulation (DM), sometimes called *slope modulation,* is another truly digital system (as is PCM). It transmits information to only indicate whether the analog signal it encodes is to "go up" or "go down." This process is shown in Fig. 9-24. Note that the encoder outputs are highs or lows that "instruct" whether to go up or go down, respectively. The relative simplicity of this system is readily apparent as compared to PCM. Delta modulation takes advantage of the fact that voice signals do not change abruptly and there is generally only a small change in level from one sample to the next. On the other hand, PCM is able to respond to very abrupt level changes between samples, such as analog signals that have been time-domain multiplexed before encoding.

A schematic for a delta modulator is shown in Fig. 9-25. A demodulator would consist of an integrator (just like the one in Fig. 9-25) followed by a sharp-cutoff low-pass filter. The integrator output would look like waveform B in Fig. 9-25. The filter smooths it out to provide the final analog signal. The student is asked to fully describe/explain the modulator's operation in a question at the end of the chapter.

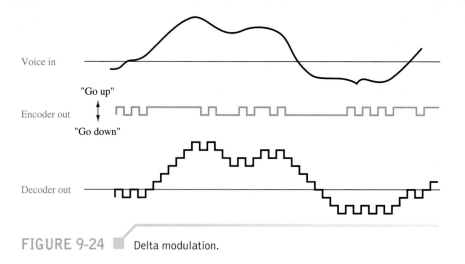

FIGURE 9-24 ■ Delta modulation.

The major advantage of delta modulation is simplicity. The delta modulator is called a *tracking* ADC because it follows the contours of the input and provides output that shows input changes rather than exact values. It does not require the synchronization of PCM systems and inherently provides a serial stream of bits, so that there is no need for a parallel-to-serial converter.

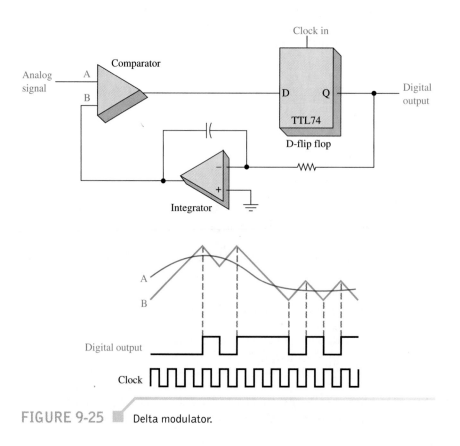

FIGURE 9-25 ■ Delta modulator.

A difficulty faced by these systems is *slope overload.* When the analog signal has a high rate of change, the delta modulator can "fall behind" and a distorted output results. An increased sample rate could be used, but this necessitates a higher bandwidth for transmission of the signal. In systems where this is a problem, a technique called *continuously variable slope delta* (CVSD) modulation is used. A typical CVSD scheme is to increase the step size whenever there is a longer run than three successive 1's or 0's. When the modulator catches up (as indicated by a change from 1 to 0 or 0 to 1), the step size returns to normal.

CVSD systems are typically implemented using a single LSI chip for both the transmitter and receiver sections. Similarly, the algorithm used for transmitting is used to reverse the process for reception. The net overall effect of CVSD can be compared to companding, except that CVSD depends on past values and adapts while companding is a fixed algorithm.

PCM/Delta Modulation Comparison

The choice between PCM and delta modulation (with or without CVSD) involves technical trade-offs, as you might expect. The inherent simplicity of delta modulation comes with the price of increased bandwidth (bit rate), as compared to PCM by a factor of 2 or 3. In applications where increased bandwidth is not a problem, delta modulation would typically be used. Additionally, delta modulation (with or without CVSD) exhibits less serious degradation in the presence of digital noise interference than with PCM. In PCM, a noice-induced error in one of the more significant bits in a digital word can give the listener a painful audio burst. In CVSD, an error in any bit produces only a minor transient noise. It has been demonstrated that CVSD transmissions are quite intelligible even with 10% random bit errors introduced.

9-7 DIGITAL SIGNAL ENCODING FORMATS

Transmission of digital data using a binary format (+5 V—hi, 0.0 V—low) is usually limited to short distances such as a computer to a printer interface. Typically the binary data are transmitted serially over a single wire, fiber, or RF link. This requires that the binary data be encoded in such a way so that highs and lows can easily be detected. The transmission systems typically use a serial transmission system that is either asynchronous or synchronous. This requires the addition of clocking information in the data for synchronous systems.

The digital signal encoding formats presented in this section are the most commonly used PCM waveforms. (Note that we are identifying the encoding format as a pulse-code-modulated waveform.) The waveforms are classified as one of four encoding groups:

1. NRZ—nonreturn-to-zero
2. RZ—return-to-zero
3. Phase-encoded and delay modulation
4. Multilevel binary

The encoding formats described are of the *baseband* type. A baseband signal is one that is not modulated. These waveforms are still in a binary or pseudo-binary format, so therefore they are classified as baseband.

The NRZ Group

The NRZ group is a popular method for encoding binary data. NRZ codes are also one of the easiest to implement. NRZ codes get their name from the fact that the data signal does not return to zero during an interval. In other words, NRZ codes remain constant during an interval. Because of this the code has a dc component in the waveform. For example, a data stream containing a chain of 1's or 0's will appear as a dc signal at the receive side. Look at the waveform for the NRZ-L code that is shown in Fig. 9-26. Notice that the code remains constant for several clock cycles for a series of zeros or ones.

Another important factor to consider is that the NRZ codes do not contain any self-synchronizing capability. NRZ codes will require the use of start bits or some kind of synchronizing data pattern to keep the transmitted binary data synchronized. There are three coding schemes in the NRZ group, NRZ-L (level), NRZ-M (mark), and NRZ-S (space). The waveforms for these formats are provided in Fig. 9-26. The NRZ code descriptions are provided in Table 9-2.

The RZ Codes

The *RZ-unipolar* code shown in Fig. 9-26 has the same limitations and disadvantages as the NRZ group. A dc level appears on the data stream for a series of 1's or 0's. Synchronizing capabilities are also limited. These deficiencies are overcome by modifications in the coding scheme, which include using bi-polar signals and alternating pulses. The *RZ-bipolar* code provides a transition at each clock cycle, and a bipolar pulse technique is used to minimize the dc component. Another RZ code is *RZ-AMI*. The alternate-mark-inversion code provides alternating pulses for the 1's. This technique virtually removes the dc component from the data stream, but since a data value of 0 is 0 V the system can have poor synchronizing capabilities if a series of 0's is transmitted. This deficiency can also be overcome by transmission of the appropriate start, synchronizing, and stop bits. Table 9-3 provides descriptions of the RZ codes.

Bi-phase and Miller Codes

The popular names for phase-encoded and delay-modulated codes are *bi-phase* and *Miller* codes. Bi-phase codes are very popular for use in optical systems, satellite telemetry links, and magnetic recording systems. *Bi-phase M* is used for encoding SMPTE

TABLE 9-2 *NRZ Codes*

NRZ-L (Nonreturn to zero—level)
    1 (Hi)—high level
    0 (Low)—low level

NRZ-M (Nonreturn to zero—mark)
    1 (Hi)—transition at the beginning of the interval
    0 (Low)—no transition

NRZ-S (Nonreturn to zero—space)
    1 (Hi)—no transition
    0 (Low)—transition at the beginning of the interval

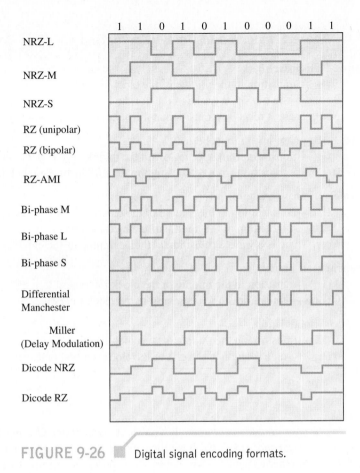

FIGURE 9-26 ▮ Digital signal encoding formats.

(Society of Motion Picture and Television Engineers) time-code data for recording on videotapes. The bi-phase code is an excellent choice for this type of media because the code does not have a dc component to it. Another important benefit of the code is *self-synchronizing*, or *self-clocking*. This feature allows the data stream speed to vary (tape shuttle in fast and slow search modes) while still providing the receiver with clocking information.

TABLE 9-3 *RZ Codes*

RZ (unipolar) (Return-to-zero)
  1 (Hi)—transition at the beginning of the interval
  0 (Low)—no transition

RZ (bipolar) (Return-to-zero)
  1 (Hi)—positive transition in the first half of the clock interval
  0 (Low)—negative transition in the first half of the clock interval

RZ-AMI (Return-to-zero—alternate-mark inversion)
  1 (Hi)—transition within the clock interval alternating in direction
  0 (Low)—no transition

TABLE 9-4 *Phase-Encoded and Delay-Modulation (Miller) Codes*

Bi-phase M (Bi-phase-mark)
    1 (Hi)—transition in the middle of the clock interval
    0 (Low)—no transition in the middle of the clock interval
    Note: There is always a transition at the beginning of the clock interval.

Bi-phase L (Bi-phase-level / manchester)
    1 (Hi)—transition from high-to-low in the middle of the clock interval
    0 (Low)—transition from low-to-high in the middle of the clock interval

Bi-phase S (Bi-phase-space)
    1 (Hi)—no transition in the middle of the clock interval
    0 (Low)—transition in the middle of the clock interval
    Note: There is always a transition at the beginning of the clock interval.

Differential Manchester
    1 (Hi)—transition in the middle of the clock interval
    0 (Low)—transition at the beginning of the clock interval

Miller / delay modulation
    1 (Hi)—transition in the middle of the clock interval
    0 (Low)—no transition at the end of the clock interval unless followed by a zero

The *bi-phase L* code is commonly known as *Manchester Coding*. This code is used on the *ethernet* standard IEEE 802.3 for local area networks (LANs). Chapter 11 provides more detail on ethernet and LANs. Figure 9-26 provides examples of these codes and Table 9-4 summarizes their characteristics.

Multilevel Binary Codes

Codes that have more than two levels representing the data are called *Multilevel binary* codes. In many cases the codes will have three levels. We have already examined two of these codes in the RZ group, RZ (bipolar) and RZ-AMI. Also included in this group are *Dicode NRZ* and *Dicode RZ*. Table 9-5 summarizes the multilevel binary codes.

TABLE 9-5 *Multilevel Binary Codes*

Dicode NRZ
    One-to-zero and zero-to-one data transitions change the signal polarity.
    If the data remain constant, then a zero-level is output.

Dicode RZ
    One-to-zero and zero-to-one data transitions change the signal polarity in half-step
    voltage increments. If the data don't change, then a zero-voltage level is output.

9-8 CODE ERROR DETECTION AND CORRECTION

Codes and raw digital data are being transmitted with increasing volume every year. Unless some means of error detection is used, it is not possible to know when errors have occurred. These errors are caused by noise and transmission system impairments. In contrast, it is obvious when a voice transmission has been impaired by noise or equipment problems.

Redundancy is used as the means of error detection when codes and digital data are transmitted. A basic redundancy system is to transmit everything twice and to make sure that exact correlation exists. Fortunately, schemes have been developed that do not require such a high degree of redundancy.

Parity

The most common method of error detection is the use of parity. A single bit called the *parity bit* is added to each code representation. If it makes the total number of 1's even, it is termed *even parity,* and an odd number of 1's is *odd parity.* For example, if the ASCII code for A is to be generated, the code is P1000001 and P is the parity bit. Odd parity would be 11000001 since the number of 1's is now 3. The receiver checks for parity. If an even number of 1's occurs in a character grouping (word), an error is indicated and the receiver usually requests a retransmission. Unfortunately, if two errors (an even number) occur, parity systems will not indicate an error. In many systems a burst of noise causes two or more errors, so that more elaborate error-detection schemes may be required.

There are many circuits used as parity generators and/or checkers. A simple technique is shown in Fig. 9-27. If there are *n* bits per word, *n* − 1 Exclusive-OR (XOR) gates are needed. The first two bits are applied to the first gate and the remaining individual bits to each subsequent gate. The output of this circuit will always be a 1 if there is an odd number of 1's and a 0 output for an even number of 1's. If odd parity is desired, the output is fed through an inverter. When used as a parity checker, the word and parity bit is applied to the inputs. If no errors are detected, the output is low for even parity and high for odd parity.

When an error is detected there are two basic system alternatives:

1. An automatic request for retransmission
2. Display of an unused symbol for the character with a parity error (called *symbol substitution*)

Most systems use a request for repeat system. A block of data is transmitted—if no error is detected, a positive acknowledgment (ACK) is sent back to the transmitter. If a parity error is detected, a negative acknowledgment (NAK) is made and the transmitter repeats that block of data. Exactly what happens after error detection falls under the subject of *protocols* introduced in Chapter 11.

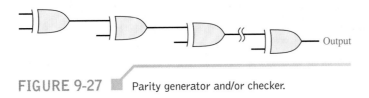

Output

FIGURE 9-27 Parity generator and/or checker.

Block Check Character

A more sophisticated method of error detection than simple parity is needed in high-data-rate systems (typically, 2400 b/s and above). At these speeds, telephone data transmission is usually synchronous and blocked. A block is a group of characters transmitted with no time gap between them. It is followed by an *end of message* (EOM) indicator and then a *block check character* (BCC). A block size is typically 256 characters. The transmitter uses a predefined algorithm to compute the BCC. The same algorithm is used at the receiver based on the block of data received. The two BCCs are compared, and if identical, the next block of data is transmitted.

There are many algorithms used to generate a BCC. The most elementary one is an extension of parity into two dimensions, called the *longitudinal redundancy check* (LRC). This method is illustrated with the help of Fig. 9-28. Shown is a small block of 4-bit characters using odd parity. The BCC is formed as an odd-parity bit for each vertical column. Now suppose that a double error occurred in character 2 as shown in Fig. 9-28(b). As described previously, simple parity would not pick up this error. With the BCC, however, the error is detected. If a single error occurs [Fig. 9-28(c)], the erroneous bit can be pinpointed as the intersection of the row and column containing the error.

The error location process just described is not usually utilized. Rather, the receiver checks for character and LRC errors and if either (or both) occur, a retransmission is requested. Occasionally, an error will occur in the BCC. This is unavoidable, but the only negative consequence is an occasional unnecessary retransmission.

Cyclic Redundancy Check

A more sophisticated set of algorithms than parity schemes for BCC calculation is the *cyclic redundancy check* (CRC). It is easily the most powerful error-detection technique in common use. The entire message block is treated as a long binary number. Binary division is performed by some binary constant and the remainders are subsequently compared at the receiver. A difference between the remainder transmitted by the transmitter and generated at the receiver causes that message block to be retransmitted. The division process usually involves special 12- or 16-bit shift registers at the transmitter and receiver. The CRC technique does not burden the transmitter with the continuous sending of parity bits and thus offers more efficient data transmission. The CRC pattern is only 16 bits or less, and it need only be sent at the end of blocks containing several thousand bits.

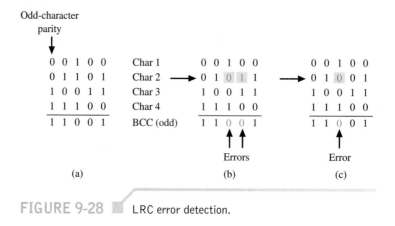

FIGURE 9-28 LRC error detection.

Hamming Code

The error-detection schemes thus far presented require retransmission if errors occur. Techniques that allow correction at the receiver are called *forward error-correcting* (FEC) codes. The basic requirement of such codes is for sufficient redundancy so as to allow error correction without further input from the transmitter. The *Hamming code* is an example of an FEC code named for R. W. Hamming, an early developer of error-detection/correction systems.

If m represents the number of bits in a data string and n represents the number of bits in the Hamming code, n must be the smallest number such that

$$2^n \geq m + n + 1 \qquad (9\text{-}10)$$

Consider a 4-bit data word 1101. The minimum number of parity bits to be used is 3 when Eq. (9-6) is referenced. A possible setup, then, is

$$
\begin{array}{ccccccc}
P_1 & P_2 & 1 & P_3 & 1 & 0 & 1 \\
1 & 2 & 3 & 4 & 5 & 6 & 7 \quad \text{bit location}
\end{array}
$$

We'll let the first parity bit, P_1, provide even parity for bit locations 3, 5, and 7. P_2 does the same for 3, 6, and 7, while P_3 checks 5, 6, and 7. The resulting word, then, is

$$
\begin{array}{ccccccc}
1 & 0 & 1 & 0 & 1 & 0 & 1 \\
1 & 2 & 3 & 4 & 5 & 6 & 7 \quad \text{bit location} \\
P_1 & P_2 & & P_3 & & &
\end{array}
$$

When checked, a 1 is assigned to incorrect parity bits, while a 0 represents a correct parity bit. If an error occurs so that bit location 5 becomes a 0, the following process takes place. P_1 is a 1 and indicates an error. It is given a value of 1 at the receiver. P_2 is not concerned with bit location 5 and is correct and therefore given a value of 0. P_3 is incorrect and is therefore assigned a value of 1. These three values result in the binary word 101. Its decimal value is 5, and this means that bit location 5 has the wrong value and the receiver has pinpointed the error without a retransmission. It then changes the value of bit location 5 and transmission continues. The Hamming code is not able to detect multiple errors in a single data block. More complex (and more redundant) codes are available if necessary.

TROUBLESHOOTING

Digital communications provide vital links to transfer information in today's world. Digital data are used in every aspect of electronics in one form or another. In this section we are going to look at digital pulses and the effects that noise, impedance, and frequency have on them. Digital communications troubleshooting requires that the technician be able to recognize digital pulse distortion and what causes it.

After completing this section you should be able to

- Identify a good pulse waveform

- Identify frequency distortion

- Describe effects of incorrect impedance on the square wave

- Identify noise on a digital waveform

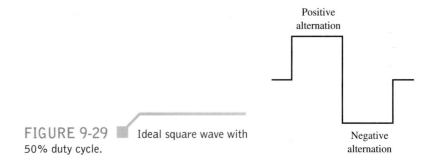

FIGURE 9-29 ▨ Ideal square wave with 50% duty cycle.

Positive alternation

Negative alternation

The Digital Waveform

A square wave signal is a digital waveform and is illustrated in Figure 9-29. The square wave shown is a periodic wave that continually repeats itself. It is made up of a positive alternation and a negative alternation. The ideal square wave will have sides that are vertical. These sides represent the high-frequency components. The flat-top and bottom lines represent the low-frequency components. A square wave is composed of a fundamental frequency and an infinite number of odd harmonics, as described in Chapter 1 in the Fourier analysis section.

Figure 9-30 shows this same square wave stretched out to illustrate a more true representation of it. Notice the sides are not ideally vertical but have a slight slope to them. The edges are rounded off since transition time is required for the low pulse to go high and back to low again. Figure 9-30 shows the positive alternation as a positive pulse and the negative alternation as a negative pulse. Rise time refers to the pulse's low-to-high transition and is normally measured from the 10% point to the 90% point on the waveform. Fall time represents the high-to-low transition and is measured from the 90% and 10% points. From the pulse's maximum low point to its maximum high point is the amplitude measured in volts. By observing the pulse waveform in response to circuit conditions that it may encounter, much can be determined about the circuit.

Effects of Noise on the Pulse

From previous discussions about noise you learned that noise has an additive effect on a signal. A signal's amplitude is changed by noise adding to it or subtracting from it. This concept is depicted in Figure 9-31. The positive pulse and the negative pulse have

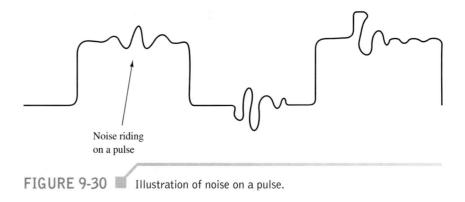

Noise riding on a pulse

FIGURE 9-30 ▨ Illustration of noise on a pulse.

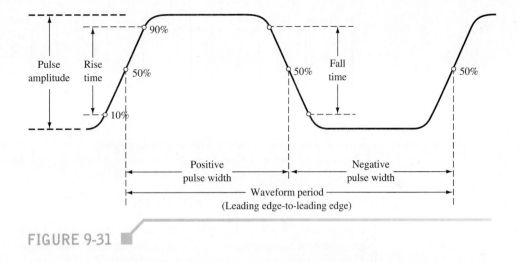

changed from the ideal as a direct result of encountering noise. The noise has changed the true amplitudes of the pulses. If the noise becomes too severe, the positive and negative noise excursions might be mistaken by logic circuits as high and low pulses. Proper noise compensating techniques, shielding, and proper grounds help to reduce noise. If the digital waveforms under test show signs of deterioration due to noise, troubleshoot by checking compensation circuits, shielding, and ensuring proper grounds are made.

Effects of Impedance on the Pulse

The square wave pulse can show the effects of impedance mismatches, as seen in Figure 9-32. The pulse in Figure 9-32(a) is severely distorted by an impedance that is below that required. For example, if RG-58/AU coaxial cable, carrying data pulses, became shorted or if one of its connectors developed a low-resistance leakage path to ground, then the data pulses would suffer low-impedance distortion. Figure 9-32(b) shows a type of distortion on the top and bottom of the square wave called ringing. Ringing is the result of the effects of high impedance on the pulse. If a transmission line is improperly terminated or develops a high resistance for some reason, then ringing can occur. A tank circuit with too high of a Q will cause ringing. By adjusting the tank's Q or by adding proper termination to a transmission line, the waveform can be brought back to normal, as shown in Figure 9-32(c). When working with data lines, ensure that proper terminations are made. The effect of impedance loading is reduced in communications equipment like transmitters, receivers, and data handling circuits when proper repair and alignment maintenance techniques are used.

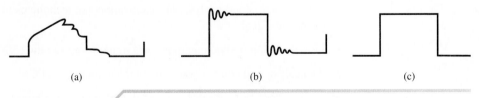

(a)   (b)   (c)

FIGURE 9-32   Effects of impedance mismatches: (a) Impedance is too low. (b) Impedance is too high (ringing). (c) Impedance is matched.

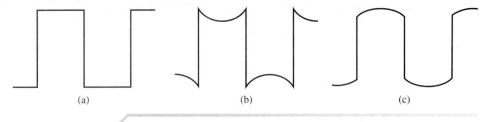

(a)   (b)   (c)

FIGURE 9-33 ■ Effects of frequency on data pulses:
(a) Good pulses. (b) Low-frequency attenuation. (c) High-frequency attenuation.

Effects of Frequency on the Pulse

Digital pulses will not be distorted when passing through an amplifier with a sufficient bandwidth or a transmission line with a sufficient bandwidth. Figure 9-33(a) shows that the pulse does not become distorted in any fashion. The dip at the top and bottom of the waveform in Figure 9-33(b) represents low-frequency attenuation. The high-frequency harmonic components pass without being distorted, but the low-frequency components are attenuated. Coupling capacitors in communication circuits can cause low-frequency distortion. Low-frequency compensating network malfunctions will also attenuate the low-frequency components of the pulse waveform.

High-frequency distortion occurs when the high-frequency harmonic components of a digital square wave are lost. The edges become rounded off, as seen in Figure 9-33(c). The high harmonic frequencies are lost when the media bandwidth becomes too narrow to pass the pulse in its entirety. A circuit's bandwidth can change when the value of circuit components such as coils, capacitors, and resistors increases or decreases. Open and shorted capacitors found in tank circuits can cause high-frequency losses as well.

SUMMARY

In Chapter 9 we studied the coding techniques commonly used in digital communications. The operation and advantages of pulse-code modulation (PCM) systems were explained. The major topics the student should now understand include:

- the definition and efficiency of coding systems

- the calculation of bit error rate (BER) and channel capacity

- the description of the following numeric and alpha-numeric codes—ASCII, EBCDIC, Baudot, and Gray

- the description of pulse-code modulation (PCM), including operation of the sample-and-hold (S/H) circuit and antialiasing filter

- the description and analysis of quantization and techniques to improve accuracy of quantized signals

- the calculation of dynamic range and signal-to-noise ratio (S/N) in PCM systems

- the analysis and comparison of delta modulation with PCM

- the description of various baseband digital signal encoding formats

- the explanation of various code error detection and correction methods, including parity, block check character, cyclic redundancy check (CRC), and the Hamming code

SECTION 1

1. With the assistance of Fig. 9-1, describe a transmission that is
   (a) Digital but not data.
   (b) Both digital and data.
   (c) Data but not digital.
2. Describe the process of regeneration as it applies to digital/data communication.
3. Define *digital signal processing*. Provide an example that is not described in this book.

SECTION 2

4. In what ways are coded voice transmission advantageous over direct transmission? What are some possible disadvantages?
5. Define *coding* and *bit*.
6. Sketch a voltage waveform that could be used to represent the binary number 11010100100.
7. Determine the number of bits required to encode a system of 50 equiprobable events with a binary code. Calculate the efficiency of this code. Calculate the efficiency of a decimal code to accomplish the same goal. (94%, 85%)
8. Explain the noise immunity advantages of the binary code over any other code.
9. Calculate the error probability and BER in a system that produces 7 error bits out of 5,700,000 total bits. (1.23×10^{-6}, 8.14×10^{5})
10. Calculate the channel capacity (bits per second) of a standard phone line that has a *S/N* of 511. (27 kb/s)

SECTION 3

11. What do the acronyms ASCII and EBCDIC stand for?
12. Provide the ASCII code for 5, a, A, and STX.
13. Provide the EBCDIC code for 5, a, A, and STX.
14. Describe the Gray code and provide a common application.

SECTION 4

15. Provide a block diagram of a system to generate PCM and explain the function of each block.
16. Describe the operation of a sample-and-hold circuit and explain its importance in the analog-to-digital conversion process.
17. Explain PAM as it relates to the generation of PCM.
18. Describe methods to generate PAM via natural and flat-top sampling.
19. Calculate the Nyquist rate for a telephone line that carries signals from 300 Hz to 4 kHz.
20. What is an alias frequency? Explain how to prevent the generation of an alias frequency.
21. A 2-kHz sine wave is being sampled at a 1-kHz rate. Is an alias frequency developed, and if so, what is its frequency?

22. Explain the need for the clock signal at a PCM receiver. Include the concepts of jitter and glitches in your explanation.
23. Describe the function of a codec with reference to Fig. 9-14.
24. List the advantages of PCM systems. Explain why PCM is adaptable to systems requiring many repeaters and TDM. The information shown in Fig. 9-16 should be used in this explanation.

SECTION 5

25. Describe the meaning and importance of quantizing error in a PCM system. Why does a PCM TV transmission require more quantizing levels than a PCM voice transmission?
26. Calculate the resolution of a 3-bit system that is sampling a 0- to 10-V signal.
27. Define the resolution of a PCM system. Provide two ways resolution can be improved.
28. Calculate the dynamic range of the system described in Problem 26.
29. Calculate the number of bits required to satisfy a dynamic range of 48 dB. Calculate the signal-to-noise ratio of this system.
30. Explain the process of companding and the benefit it provides.
31. A μ-law companding system with $\mu = 100$ is used to compand a 0- to 10-V signal. Calculate the systems output for inputs of 0, 0.1, 1, 2.5, 5, 7.5, and 10 V. (0, 1.5, 5.2, 7.06, 8.52, 9.38, and 10 V)

SECTION 6

32. Explain the process of delta modulation.
33. Provide a detailed explanation of the delta modulator shown in Fig. 9-25.
34. Why is delta modulation a truly digital system? Contrast its difference from PCM and major advantage over PCM.
35. Draw a sketch of the encoder and decoder output for a pure sine wave in a DM system.
36. Explain how slope overload occurs in DM and how the CVSD feature helps to overcome it.
37. Compare CVSD to companding.
38. Describe the trade-offs involved in choosing PCM versus DM.

SECTION 7

39. Describe the characteristics of the four basic encoding groups—NRZ, RZ, phase-encoded and delay modulation, and multilevel binary.
40. Sketch the data waveforms for 11010 using NRZ-L, bi-phase M, differential Manchester, and dicode RZ.

SECTION 8

41. Describe the use of parity for error-detection systems.
42. What are the alternatives when parity systems detect an error?

43. Explain the error-detection process known as the block check character. Describe why it is advantageous compared to simple parity systems.
44. Describe the cyclic redundancy check (CRC) for error detection.
45. Calculate the CRC for 101001000 if a 1011 divisor is being used. (100)
46. Explain the Hamming code and the advantage it provides compared to the CRC technique.
47. Calculate the number of Hamming bits required for a 16-bit data string.
48. The data word 0101 is to be transmitted using the Hamming code system described in the text. A burst of noise causes the receiver to read it as 0100. Determine the parity bits transmitted (P_1, P_2, P_3) and the values assigned by the receiver to pinpoint the error.
49. In the following blocks of data, the left column is odd character parity and the bottom row is an odd LRC. Determine if any errors are present, and if so, indicate the control bit that provides this error information.

(a) 1 0 0 1 1
   0 0 1 0 0
   1 1 1 0 0
   0 1 0 1 1
   ‾‾‾‾‾‾‾‾‾
   1 1 0 0 1

(b) 1 0 0 1 1
   0 1 1 0 0
   1 0 1 0 0
   0 1 1 0 1
   ‾‾‾‾‾‾‾‾‾
   1 1 0 0 1

DIGITAL COMMUNICATIONS: TRANSMISSION

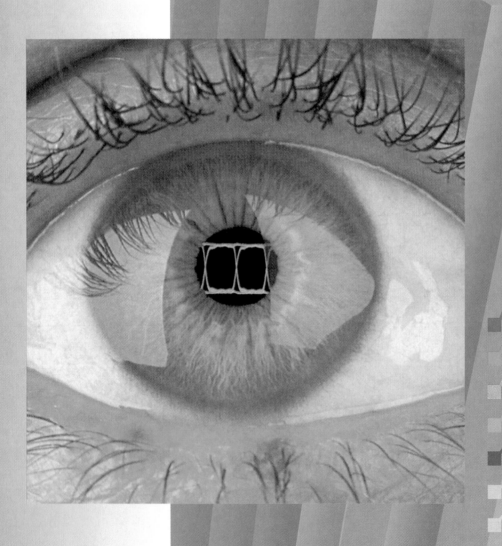

OBJECTIVES

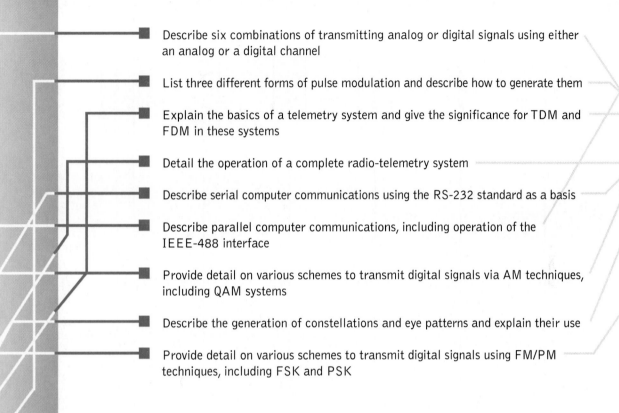

■ Describe six combinations of transmitting analog or digital signals using either an analog or a digital channel

■ List three different forms of pulse modulation and describe how to generate them

■ Explain the basics of a telemetry system and give the significance for TDM and FDM in these systems

■ Detail the operation of a complete radio-telemetry system

■ Describe serial computer communications using the RS-232 standard as a basis

■ Describe parallel computer communications, including operation of the IEEE-488 interface

■ Provide detail on various schemes to transmit digital signals via AM techniques, including QAM systems

■ Describe the generation of constellations and eye patterns and explain their use

■ Provide detail on various schemes to transmit digital signals using FM/PM techniques, including FSK and PSK

10-1 INTRODUCTION

Chapter 9 provided the coding techniques utilized in digital communications. In this chapter we will consider the important aspects of the transmission of signals as they relate to digital communications. Before we do that it would be helpful to consider different transmission schemes for digital and analog signals. This will provide perspective to the various methods presented in this chapter and help you to compare them with some other transmission systems.

Six possible methods for transmitting information are provided in Fig. 10-1. In Fig. 10-1(a) the analog signal is transmitted directly. An example of this is a basic intercom system where a microphone creates an electrical analog of your voice, which is transmitted directly via a pair of wires. Of course that signal is usually amplified at both the input and output sides of the channel (pair of wires). Notice that the channel is labeled an analog baseband channel in Fig. 10-1(a). *Baseband* means the signal is transmitted at its base frequencies, and no modulation to another frequency range has occurred.

Eye patterns are used to analyze data signals. (Courtesy of W. L. Gore & Associates, Inc.)

393

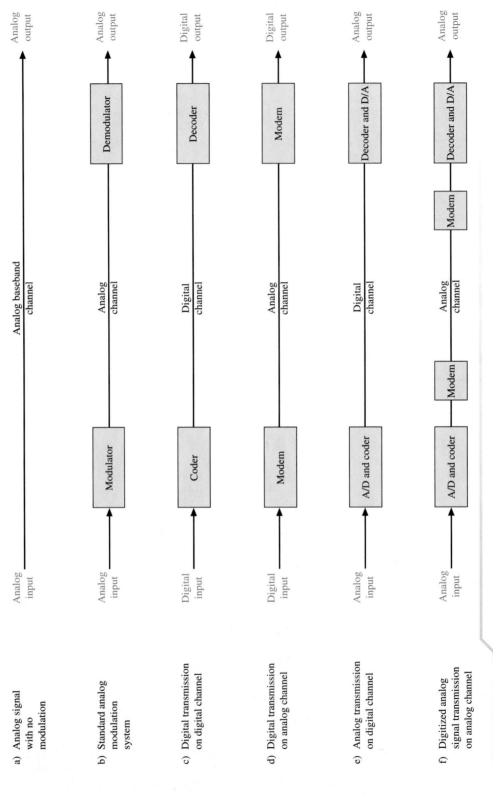

FIGURE 10-1 Transmission schemes for analog and digital signals.

a) Analog signal with no modulation

b) Standard analog modulation system

c) Digital transmission on digital channel

d) Digital transmission on analog channel

e) Analog transmission on digital channel

f) Digitized analog signal transmission on analog channel

Figure 10-1(b) shows a standard analog modulation scheme. Broadcast radio and television are examples of this, and we spent Chapters 2–8 studying these systems.

When a computer transmits digital data to another computer or a computer peripheral such as a printer, it may transmit that signal directly. This situation is illustrated in Fig. 10-1(c). Notice that a coder and decoder are shown in this system. We'll consider these types of communications in Secs. 5 and 6 of this chapter.

The transmission of digital signals via an analog channel is shown in Fig. 10-1(d). A modem is shown to allow the digital signal to be transmitted on an analog channel. This is the scheme used by most personal computers to transmit their information over telephone lines. We'll look at this in Chapter 11.

You may wonder, why go to the complexities shown in Fig. 10-1(e) and (f) just to transmit an analog signal? In Fig. 10-1(e) the signal is converted to digital form by the analog-to-digital converter and transmitted as digital pulses on the digital channel. In Fig. 10-1(f) the same processing occurs only this channel cannot carry the digital pulses, so the modem is used to allow an analog channel. As will be shown in Secs. 2, 3, and 4 of this chapter, the digitization of analog signals offers advantages when multiplexing more than one signal in a transmission. There can also be advantages when noise is a significant problem.

10-2 PULSE MODULATION

You have undoubtedly drawn graphs of *continuous* curves many times during your education. To do that, you took data at some finite number of discrete points, plotted each point, and then drew the curve. Drawing the curve may have resulted in a very accurate replica of the desired function even though you did not look at every possible point. In effect, you took *samples* and guessed where the curve went in between the samples. If the samples had sufficiently close spacing, the result is adequately described. It is possible to apply this line of thought to the transmission of an electrical signal, that is, to transmit only the samples and let the receiver reconstruct the total signal with a high degree of accuracy. This is termed *pulse modulation.*

The key distinction between pulse modulation and normal AM or FM is that in AM or FM some parameter of the modulated wave varies continuously with the message, whereas in pulse modulation some parameter of a sample pulse is varied by each sample value of the message. The pulses are usually of very short duration so that a pulse modulated wave is "off" most of the time. This factor is the main reason for using pulse modulation since it allows

1. Transmitters to operate on a very low duty cycle ("off" more than "on"), as is desirable for certain microwave devices and lasers
2. The time intervals between pulses to be filled with samples of other messages

The latter reason conveniently allows a number of different messages to be transmitted on the same channel. This is the form of multiplexing known as *time-division multiplexing* (TDM). It is analogous to computer time sharing, where a number of users simultaneously utilize a computer.

It was shown in Chapter 9 that a signal sampled at twice the rate of its highest significant frequency component can be fully reconstructed at the receiver to a high degree of accuracy. Stated inversely, a given bandwidth can carry pulse signals of half its high-frequency cutoff. This is known as the *Nyquist rate.* In the case of voice transmission the standard sampling rate is 8 kHz, it being just slightly more than twice the highest significant frequency component. This implies a pulse rate of 8 kHz or 125-μs period. Since a pulse duration of 1 μs may be adequate, it is easy to see that a number of different messages could be multiplexed (TDM) on the channel, or alternatively it would allow a high peak transmitted power with a much lower (1/125) average power. The high peak power can provide a very high signal-to-noise ratio or a greater transmission range.

The student should be careful to realize that a price must be paid for system gains obtained by pulse modulation schemes. More important than the greater equipment complexity is the requirement for greater channel (bandwidth) size. If a maximum 3-kHz signal directly amplitude-modulates a carrier, a 6-kHz bandwidth is required. If a 1-μs pulse does the modulating, just allowing its fundamental component of 1/1 μs or 1 MHz do the modulating means a 2-MHz bandwidth is required in AM. In spite of the large bandwidth required, TDM is still preferable (if not the only possible way) to using 100 different transmitters, antennas or transmission lines, and receivers in cases where large numbers of messages must be conveyed simultaneously.

In its strictest sense, pulse modulation is not modulation but rather a *message-processing* technique. The message to be transmitted is sampled by the pulse, and the pulse is subsequently used to either amplitude- or frequency-modulate the carrier. The three basic forms of pulse modulation are illustrated in Fig. 10-2. There are numerous varieties of pulse modulation, and no standard terminology has yet evolved. The three types we shall consider here are usually termed *pulse-amplitude modulation* (PAM), *pulse-width modulation* (PWM), and *pulse-position modulation* (PPM). For the sake of clarity, the illustration of these modulation schemes has greatly exaggerated the pulse widths. Since a major application of pulse modulation occurs when TDM is to be used, shorter pulse durations, leaving room for more multiplexed signals, are obviously desirable. As shown in Fig. 10-2, the pulse parameter that is varied in step with the analog signal is varied in direct step with the signal's value at each sampling interval. Notice that the pulse amplitude in PAM and pulse width in PWM is not zero when the signal is minimum. This is done to allow a constant pulse rate and is important in maintaining synchronization in TDM systems.

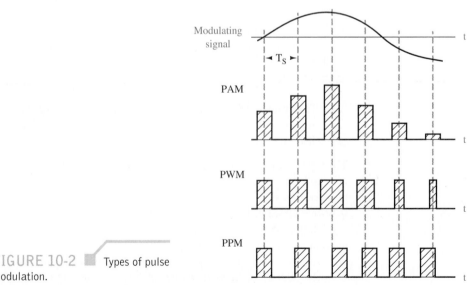

FIGURE 10-2 ■ Types of pulse modulation.

Pulse-Amplitude Modulation

Pulse-amplitude modulation (PAM) was introduced in Chapter 9, where it was described as an intermediate step in the development of pulse-code modulation (PCM) systems. That is the most important application for PAM, but we'll show how it can be used directly to modulate a carrier for transmission of information.

In pulse-amplitude modulation (PAM), the pulse amplitude is made proportional to the modulating signal's amplitude. This is the simplest pulse modulation to create in that a simple sampling of the modulating signal at a periodic rate can be used to generate the pulses which are subsequently used to modulate a high-frequency carrier. An eight-channel TDM PAM system is illustrated in Fig. 10-3. At the transmitter, the eight signals to be transmitted are periodically sampled. The sampler illustrated is a rotating machine making periodic brush contact with each signal. A similar rotating machine at the receiver is used to distribute the eight separate signals, and it must be synchronized to the transmitter. A mechanical sampling system such as this may be suitable for low sampling rates such as encountered in some telemetry systems but would not be adequate for the 8-kHz rate required for voice transmissions. In that case an electronic switching system would be incorporated as was described in Sec. 9-4.

At the transmitter, the variable amplitude pulses are used to frequency-modulate a carrier. A rather standard FM receiver recreates the pulses, which are then applied to the electromechanical *distributor* going to the eight individual channels. This distributor is virtually analogous to the distributor in a car that delivers high voltage to eight spark

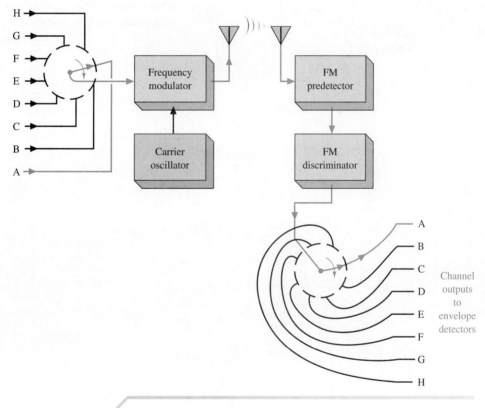

FIGURE 10-3 ■ Eight-channel TDM PAM system.

plugs in a periodic fashion. The pulses applied to each line go into an envelope detector that serves to recreate the original signal. This can be a simple low-pass *RC* filter such as is used following the detection diode in a standard AM receiver.

While PAM finds some use due to its simplicity, PWM and PPM use constant amplitude pulses and provide superior noise performance. The PWM and PPM systems fall into a general category termed *pulse-time modulation* (PTM) since their timing, and not amplitude, is the varied parameter.

Pulse-Width Modulation

Pulse-width modulation (PWM), a form of PTM, is also known as *pulse-duration modulation* (PDM) and *pulse-length modulation* (PLM). A simple means of PWM generation is provided in Fig. 10-4 using a 565 PLL. It actually creates PPM at the VCO output (pin 4), but by applying it and the input pulses to an Exclusive-OR gate, PWM is also created. For the phase-locked loop (PLL) to remain locked, its VCO input (pin 7) must remain constant. The presence of an external modulating signal upsets the equilibrium. This causes the phase detector output to go up or down to maintain the VCO input (control) voltage. However, a change in phase detector output also means a change in phase difference between the input signal and the VCO signal. Thus, the VCO output has a phase

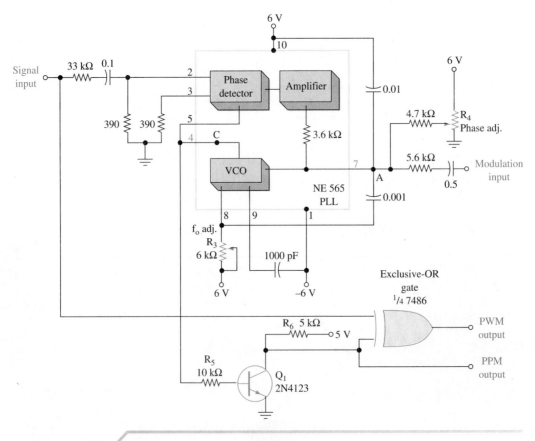

FIGURE 10-4 ■ PLL generation of PWM and PPM.

shift proportional to the modulating signal amplitude. This PPM output is amplified by Q_1 in Fig. 10-4 just prior to the output. The Exclusive-OR circuit provides a "high" output only when just one of its two inputs is "high." Any other input condition produces a "low" output. By comparing the PPM signal and the original pulse input signal as inputs to the Exclusive-OR circuit, the output is a PWM signal at twice the frequency of the original input pulses.

Adjustment of R_3 varies the center frequency of the VCO. The R_4 potentiometer may be adjusted to set up the quiescent PWM duty cycle. The outputs (PPM or PWM) of this circuit may then be used to modulate a carrier for subsequent transmission.

CLASS D AMPLIFIER AND PWM GENERATOR   PWM forms the basis for a very efficient form of power amplification. The circuit in Fig. 10-5 is a so-called class D amplifier since the actual power amplification is provided to the PWM signal, and since it is of constant amplitude, the transistors used can function between cutoff and saturation. This allows for maximum efficiency (in excess of 90%) and is the reason for the increasing popularity of class D amplifiers as a means of amplifying any analog signal.

The circuit of Fig. 10-5 illustrates another common method for generation of PWM and also illustrates class D amplification. The Q_6 transistor generates a constant current to provide a linear charging rate to capacitor C_2. The unijunction transistor, Q_5, discharges C_2 when its voltage reaches Q_5's firing voltage. At this time C_2 starts to charge again. Thus, the signal applied to Q_7's base is a linear sawtooth as shown at A in Fig. 10-6. That sawtooth following amplification by the Q_7 emitter-follower in Fig. 10-5 is applied to the op amp's inverting input. The modulating signal or signal to be amplified is applied to its noninverting input, which causes the op amp to act as a comparator. When the sawtooth waveform at A in Fig. 10-6 is less than the modulating signal B, the comparator's output (C) is high. At the instant A becomes greater than B, C goes low. The comparator (op amp) output is therefore a PWM signal. It is applied to a push-pull amplifier (Q_1, Q_2, Q_3,

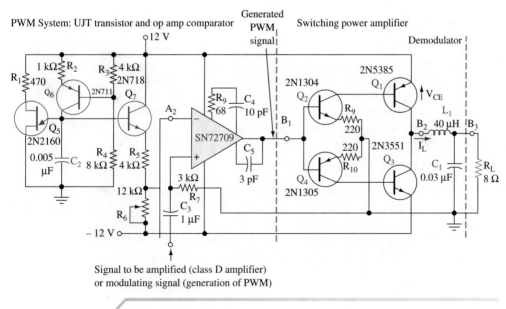

FIGURE 10-5   PWM generator and class D power amplifier.

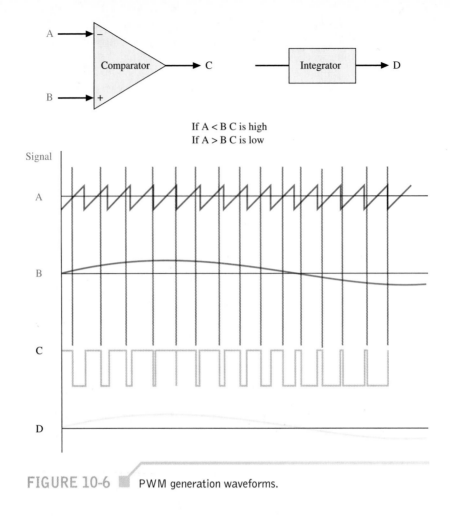

If A < B C is high
If A > B C is low

FIGURE 10-6 ■ PWM generation waveforms.

Q_4) in Fig. 10-5, which is a highly efficient switching amplifier. The output of this power amp is then applied to a low-pass LC circuit (L_1, C_1) that converts back to the original signal (B) by integrating the PWM signal at C, as shown at D in Fig. 10-6. The output of the op amp in Fig. 10-5 would be used to modulate a carrier in a communications system, while a simple integrating filter would be used at the receiver as the detector to convert from pulses to the original analog modulating signal.

Pulse-Position Modulation

PWM and pulse-position modulation (PPM) are very similar, a fact that is underscored in Fig. 10-7, which shows PPM being generated from PWM. Since PPM has superior noise characteristics, it turns out that the major use for PWM is to generate PPM. By inverting the PWM pulses in Fig. 10-7 and then differentiating them, the positive and negative spikes shown are created. By applying them to a Schmitt trigger sensitive to only positive levels, a constant amplitude and constant pulse-width signal is formed. However, the position of these pulses is variable and now proportional to the original modulating signal, and the desired PPM signal has been generated. The information content is *not* contained in either the pulse amplitude or width as in PAM and PWM, which means the signal now has a greater resistance to any error caused by noise. In addition, when PPM

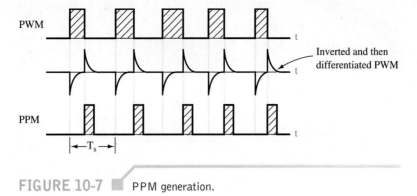

FIGURE 10-7 ■ PPM generation.

modulation is used to amplitude-modulate a carrier, a power savings results since the pulse width can be made very small.

At the receiver, the detected PPM pulses are usually converted to PWM first and then converted to the original analog signal by integrating as previously described. Conversion from PPM to PWM can be accomplished by feeding the PPM signal into the base of one transistor in a flip-flop. The other base is fed from synchronizing pulses at the original (transmitter) sampling rate. The period of time that the PPM-fed transistor's collector is low depends on the difference in the two inputs, and it is therefore the desired PWM signal.

This detection process illustrates the one disadvantage PPM has compared to PAM and PWM. It requires a pulse generator synchronized from the transmitter. However, its improved noise and transmitted power characteristics make it the most desirable pulse modulation scheme.

10-3 TELEMETRY

Telemetry may be defined as remote metering. It is the process of gathering data on some particular phenomenon without the presence of human monitors. The gathered data may be recorded on chart recorders, tape recorders, or computer memory and then picked up at some convenient time. If the data are transmitted as a radio wave, it is called *radio telemetry.* This process started during World War II when telemetry systems were developed to obtain flight data from aircraft and missiles. It offered an alternative to having human observers on board when that was impractical or considered too dangerous. Besides the military applications many new commerical uses are being developed. Today's telemetry markets range from remote reading of gas and electric meters to credit card validation, security monitoring, token-free highway toll systems, and remote inventory control.

Since situations to be remotely metered invariably involve more than one measurement, the different signals are always multiplexed. This allows the use of a single transmitter/receiver and in fact was the first major use of multiplexing techniques. It was also the beginning of pulse modulation techniques, described in the previous section, because of the ease with which they can be multiplexed.

Telemetry Block Diagram

A radio telemetry system block diagram is shown in Fig. 10-8. The process begins with the system to be monitored. Transducers (sensors) convert from the entity to be measured to an electrical signal. Five transducer outputs are shown in Fig. 10-8, but sophisticated

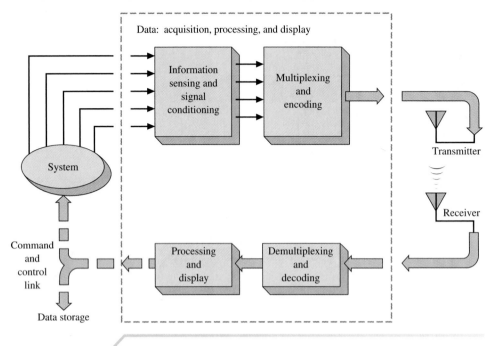

FIGURE 10-8   Radio-telemetry block diagram.

systems, such as a Mars exploration probe, may include hundreds of transmitted measurements. In any measurement system, stimuli such as temperature, pressure, movement, or acceleration must be converted to a form that can be processed electrically. While a mercury thermometer produces a good visual output of temperature, a thermistor or thermocouple transducer converts temperature to an electrical signal usable by a telemetry system.

Following the sensing, Fig. 10-8 shows a signal conditioning function. The outputs from the transducers will be electrical in nature but may not have much else in common. The outputs may be variable resistance, capacitance, inductance, voltage, or current of many different magnitudes. The conditioning circuits turn the raw data into uniform digestible information. Following conditioning, the signals are applied to the multiplexing and encoding block. They are then transmitted to the receiver for demultiplexing and decoding. It is followed by the processing and display function. Complex telemetry systems incorporate a computer for maximum efficiency in data processing. From this block a command and control link is shown in Fig. 10-8 for those systems that send control signals back to the system under measurement. This may be done to change the performance to a desired level. This "completes the loop" except for the path to data storage shown in Fig. 10-8.

Typical Telemetry System

Telemetry systems may use FDM or TDM and in some cases both, as shown in Fig. 10-9. In this instance, three wideband channels and six narrowband channels are provided. Many times there are some conditions that must be monitored more often than others.

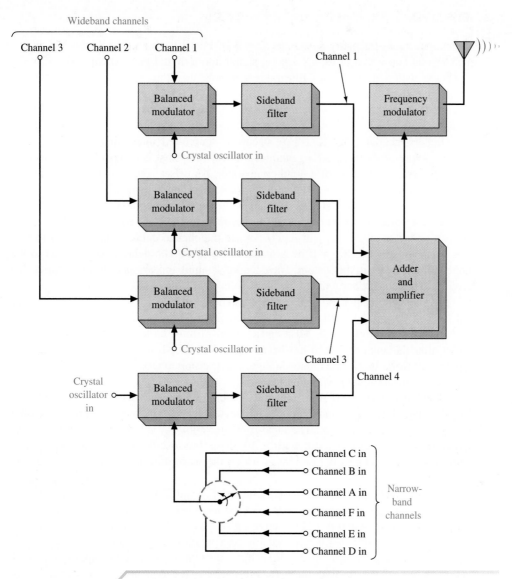

TDM and FDM telemetry transmitter system.

They are represented by channels 1, 2, and 3 in Fig. 10-9. The six conditions that do not require rapid monitoring (subchannels *A* through *F*) are time-division-multiplexed (TDM) onto channel 4. They are referred to as subchannels since 6 of them make up one of the frequency-division-multiplexed (FDM) main channels. There are countless modulation combinations possible in radio telemetry. In the system illustrated in Fig. 10-9, the sub-channels are PCM-encoded to amplitude-modulate a subcarrier that is converted into SSB, which is used to frequency-modulate the main carrier. The three wideband channels use SSB directly to frequency-modulate the main carrier. This is termed a PCM/SSB/FM system. However, any other method is possible (and has probably been used), such as PAM/FM/FM, PWM/AM/FM, PPM/AM/AM, and so on.

A complete radio-telemetry system is shown in Fig. 10-10. Part (a) shows the analog encoder that converts a 0- to 5-V analog signal into a digital pulse stream. The LM 325 IC shown within dashed lines illustrates a simple way this system can be used in a temperature-monitoring telemetry system. The LM 325 IC contains the temperature transducer and circuitry to provide the linear 0- to 5-V signal representing a range of temperatures.

In the encoder [Fig. 10-10(a)], a current-source-fed integrating capacitor at the inverting input of A_1 (0.22 μF) generates a linear ramp that is compared with the input voltage by A_1, one section of a quad comparator. An offset voltage is summed with the input so that the circuit will generate a transmittable, nonzero signal for a ground-level input to A_2 and A_3. Comparator A_3 discharges the 0.22 μF integrating capacitor while A_2 provides a separate buffered output. When the ramp is reset, A_1 goes high. Since A_1 has an open-collector output, A_2 and A_3 do not see the change until C_2 has charged to 6 V. This delay assures that C_1 will be discharged on reset. The delay is short so as to not induce timing errors in the system. The encoder's output appears as a string of pulses that have cycle periods proportional to the input voltage (PWM).

The transmitter portion of the system is shown in Fig. 10-10(b). It is based on the LM 1871 IC designed specifically for radio telemetry. The analog encoder's output signal modulates the transmitter's 49-MHz carrier. Coil T_1 must be trimmed for maximum RF output after the antenna length has been selected. A length of 2 ft is sufficient; shorter lengths will work, but with a reduced range. The transmit range is about 150 m but can be extended with additional power amplification.

The receiver [Fig. 10-10(c)] contains a local oscillator, a mixer, a 455-kHz IF section, and digital detection and decoding circuitry. Receiver alignment is simple and does not require special equipment. The local-oscillator coil (L_1) is tuned while that section's output signal (pin 2) is viewed with a 10-pF or less oscilloscope probe. The coil is trimmed to the point just before that at which the waveform amplitude peaks and then disappears.

The antenna input transformer (T_3), mixer (T_1), and IF transformer (T_2) are adjusted by means of the companion transmitter and the intended receiving antenna. The internal automatic-frequency-control (AFC) circuitry is disabled by grounding pin 16. The 455-kHz IF output (pin 15) is adjusted for peak amplitude. The amplitude must remain below 400 mV p-p throughout the adjustment procedures to prevent clipping of the IF waveform. To control the amplitude, you must remove the transmitter's antenna and separate the transmitter and receiver to obtain a usable IF output level.

Once the first adjustment is complete, the scope should be connected to the unused T_2 secondary to prevent detuning. Transformers T_1, T_2, and T_3 should be trimmed for maximum waveform amplitude. This LM 1871 and 1872 system can handle more than one channel of telemetry data, as explained in the manufacturer's specifications and application notes.

In the analog decoder [Fig. 10-10(d)] a reference-current source combines with an operational amplifier/integrator (LF356) to generate a second linear ramp. This signal is put into a sample-and-hold (S/H) circuit (LM398), which "looks" for a time specified by the incoming pulse string. A negative offset is also summed into the S/H input to correct for the voltage added during coding. Output pulses from the receiver trigger a dual one-shot (74LS123), which generates two successive pulses of approximately 5-μs duration. The first pulse triggers the S/H, and the second resets the integrator. The LM 398 output is therefore a signal that is proportional to the pulse width from the receiver, as is desired.

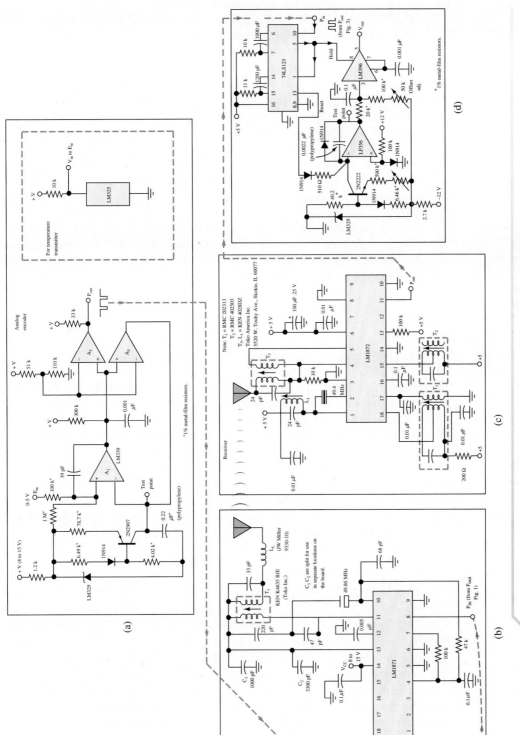

FIGURE 10-10 Complete radio-telemetry system: (a) encoder; (b) transmitter; (c) receiver; (d) decoder. (Reprinted with permission from *Electronic Design*, Vol. 29, No. 10; copyright Hayden Publishing Co., Inc., 1981.)

The gain and offset adjustments are somewhat interactive. With a two-point calibration, however, no more than three iterations should be needed.

With the addition of a 10-kΩ resistor and a two-terminal temperature sensor, the circuit can serve as a remote temperature transmitter, with the voltage output equaling 10 mV/K. The most significant contributor to circuit inaccuracy is the temperature coefficient of the integrating capacitors in the encoder and decoder circuitry. The circuits in Fig. 10-10(a) and (d) use polypropylene capacitors with a temperature coefficient of approximately −150 ppm/°C. By paralleling capacitors with opposite temperature coefficients, the designer can reduce the temperature sensitivity of the circuit by a factor of 5. A combination of silver–mica and polystyrene in a 3:1 ratio would have such an effect.

For analog data acquisition, RF telemetry presents an attractive alternative to the usual hard-wired approaches. The most common complications in analog data acquisition—source inaccessibility, the lack of practical line routes, and the need for nonstationary sensors—can be sidestepped. In addition, the RF route takes care of problems such as ground loops and wire losses.

A wide variety of uses with this system is possible, including the transmission of physiological data from human and animal subjects without need for their confinement; collection of data from rotating or otherwise moving machines without the need for brushes or slip rings; and wireless outdoor-to-indoor links for weather and other information.

10-5 COMPUTER COMMUNICATION—SERIAL

The data communication that takes place between computers and peripheral equipment is of two basic types—serial and parallel. Both Baudot and ASCII codes are used, but ASCII is most often used. In addition, the data that are sent in serial form (i.e., one bit after another on a single pair of wires) may be classified as either synchronous or asynchronous.

In an *asynchronous* system, the transmit and receive clocks free-run at approximately the same speed. Each computer "word" is preceded by a *start* bit and followed by at least one *stop* bit to "frame" the word. In a *synchronous* system both sender and receiver are exactly synchronized to the same clock frequency. This is most often accomplished by having the receiver derive its clock signal from the received data stream.

RS-232 Standard

The majority of serial data communications follow a standard called *RS-232*. More correctly this is called RS-232 C. Usually everyone is referring to the "C" version of RS-232 since that is what is currently in use, but oftentimes the "C" is omitted. Be aware that even though we may refer to the standard as RS-232, we really mean RS-232 C. The RS-232 C standard is set by the Electronics Industry Association (EIA).

In addition to setting a standard of voltages, timing, etc., standard connectors have also been developed. This normally consists of what is called a DB-25 connector. This is a connector with two rows of pins, arranged so that there are 13 in one row and 12 in the other. A diagram of this connector is provided in Fig. 10-11.

It should be noted that even though the DB-25 connector is usually used, the actual RS-232 C standard specifications do not define the actual connector. In the last ten years

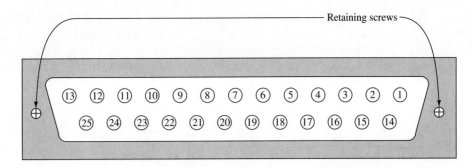

RS-232 "D-type" connector—front view.

Pin	Name	Abbreviation
1	Frame ground	FG
2	Transmit data	TD
3	Receive data	RD
4	Request to send	RTS
5	Clear to send	CTS
6	Data set ready	DSR
7	Signal ground	SG
8	Data carrier detect	DCD
20	Data terminal ready	DTR

Most popular pins implemented in RS-232 connections.

FIGURE 10-11 ■ DB-25 connector.

or so, another connector has also been used for RS-232. This is the DB-9 connector. This has become a sort of "quasi-standard" for use on IBM compatible personal computers. See Fig. 10-12 for a diagram of the DB-9 connector. You may ask, "How can 25 pins from a DB-25 connector all fit into the 9 pins of a DB-9 connector?" As we will see, all 25 pins of the DB-25 connector are not used, and in fact 9 pins are enough to do the job.

The original purpose of RS-232 was to provide a means of interfacing a computer with a modem. The computer in this case was likely a mainframe computer, since personal computers, at least as we know them today, had not yet been developed. Modems were always external, so it was necessary that some means of connection between the modem and the computer be made. It would be even nicer if this connection would be made standard, so that all computers, and all modems, could be connected interchangeably. This was the original purpose of RS-232. However, it has "evolved" into being many other things.

Today an RS-232 interface is used to interface a mouse to a personal computer, to interface a printer to a personal computer, and probably to interface about as many other things as one can think of to a personal computer. In many cases it is used to interface instrumentation to a PC. This means that the standard has "evolved," and in the process of this evolution has changed in terms of the "real world."

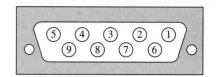

Front view

Pin	1	DCD	(8)
	2	RD	(3)
	3	TD	(2)
	4	DTR	(20)
	5	SG	(7)
	6	DSR	(6)
	7	RTS	(4)
	8	RTS	(5)
	9	RI	(22)

FIGURE 10-12 ▪ DB-9 connector.

The standard did not define the connector but did define signal levels and different lines that could be used. The signal levels that are defined are very broad. The voltage levels are to be between 3 and 25 V, and a minus voltage indicates a "1" and a plus voltage indicates a "0." Although the definition covers 3 to 25 V, the real signal levels are usually a nominal 12 to 15 V. Many chips available today will not respond to the 3-V levels, so from a real-world practical point of view, the signal level is between 5 and 15 V. This is still a very broad range and can obviously allow for a lot of loss in a cable.

In addition to the signal levels, the RS-232 standard specifies that the maximum distance for a cable is 50 ft, and the capacitance of the cable can not exceed 2500 pF. In reality, it is the capacitance of the cable that limits the distance. In fact, distances that far exceed 50 ft are commonly used today for serial transmission.

The RS-232 standard also contains another interesting statement. It says that if any two pins are shorted together, the equipment should not be damaged. This obviously requires a good buffer. This buffering is normally provided. It should be noted that the standard only says the equipment will not be damaged. It does not say that the equipment will work in that condition. In other words, if you short the pins on an RS-232 connector, there should be no smoke, but it might not work!

Perhaps the most important part of the standard from a technical point of view is that it defines the way the computer should "talk" to the modem; the timing involved, including the sequence of signals; and how each is to respond.

RS-232 Line Descriptions

Now that we know a little about what RS-232 is designed to do, it is time to look at the actual signal lines involved and see what they do. As explained earlier, the DB-25 connector definition is not a part of the original standard, but since it has become a *de facto* standard, we will use it in our discussion. Refer to Fig. 10-11 as a reference in this description. The complete signal description chart in Fig. 10-13 will also be helpful.

1. Ground pins: Actually, there are two ground pins in RS-232. They are *not* the same, however, and serve very different purposes.

   Pin 1 is the *protective ground (GND)*. It is connected to the chassis ground and is there simply to make certain that no potential difference exists between the

PIN NO.	EIA CKT.	CCITT CKT.	Signal description	Common abbrev.	From DCE	To DCE
1	AA	101	Protective (chassis) ground	GND		
2	BA	103	Transmitted data	TD		X
3	BB	104	Received data	RD	X	
4	CA	105	Request to send	RTS		X
5	CB	106	Clear to send	CTS	X	
6	CC	107	Data set ready	DSR	X	
7	AB	102	Signal ground/common return	SG	X	X
8	CF	109	Received line signal detector	DCD	X	
9			Reserved			
10			Reserved			
11			Unassigned			
12	SCF	122	Secondary received line signal detector		X	
13	SCB	121	Secondary clear to send		X	
14	SBA	118	Secondary transmitted data			X
15	DB	114	Transmitter signal element timing (DCE)		X	
16	SBB	119	Secondary received data		X	
17	DD	115	Receiver signal element timing		X	
18			Unassigned			
19	SCA	120	Secondary request to send			X
20	CD	108/2	Data terminal ready	DTR		X
21	CG	110	Signal quality detector	SQ	X	
22	CE	125	Ring indicator	RI	X	
23	CH	111	Data signal rate selector (DTE)			X
23	CI	112	Data signal rate selector (DCE)		X	
24	DA	113	Transmitter signal element timing (DTE)			X
25			Unassigned			

FIGURE 10-13 ▓ Signal description for DB-25.

chassis of the computer and the chassis of the peripheral equipment. This is *not* the signal ground. The protective ground functions much the same as the third prong of a 115-V ac "3-prong" outlet. The circuit will work without a connection to this pin, but the operator will also lose protection. In other words, make certain this pin is connected.

Pin 7 is the *signal ground (SG)*. This is the pin that is used for the ground return of all of the other signal lines. Look at the location of the signal ground, pin 7. One of the problems often associated with RS-232 and especially DB-25 connectors is to know if you are looking from the front or back and which end you should start counting from. Many connectors have the pin numbers printed on them, but this printing is usually so small that you cannot read it. Note that regardless of which end you count from, pin 7 always comes out the same place!

2. Data signal pins: Data can be sent from both the computer and peripheral equipment. Therefore a bi-directional path is necessary.

Pin 2 is *transmit data (TD)*. In theory, this pin will contain the actual data flowing from the computer to the peripheral equipment.

Pin 3 is *receive data (RD)*. In theory, this pin has the actual data flowing from the peripheral equipment to the computer. Note that it is the same as pin 2 listed above, but in the opposite direction.

The problem is that theory and reality are not always the same. What if you want to link two computers together? Which one is sending data, and which one is receiving data? It would seem that there should be an easy answer to this, but such is not the case. Which end of the cable are you looking at? In this instance, when one is transmitting data, those same data become receive data to the other computer. What this really means is that it is not easy to define which is receive and which is transmit.

Because of this problem, what is usually called a "null modem" cable has been developed. It has pins 2 and 3 "crisscrossed," i.e., pin 2 at one end is connected to pin 3 at the other end, and vice versa. Of course, this problem does not exist if we were to use RS-232 as orginally intended, i.e., for the computer to talk to a modem.

3. Handshaking pins: Pins 4 and 5 are used for "handshaking" or flow control. More correctly, pin 4 is called the *request to send (RTS)* pin, while pin 5 is the *clear to send (CTS)*. These two lines work together to determine that everything is fine for data to flow. Originally, this was used to turn on the modem's carrier, but today it is more often used to check for buffer overflow. Almost all modern modems (and other serial devices) have some sort of buffer that is used when receiving and sending data. It would not be satisfactory for that buffer to be sending more information than it could actually hold. Pins 4 and 5 may be used to handle this problem. It should be noted that if one computer is talking to another one via a "null modem" cable, these pins need to be crisscrossed just like pins 2 and 3.

    In many cases, software flow control is used, and pins 4 and 5 do not serve any useful purpose at all. In this case, pins 4 and 5 need to be jumpered at the DB-25 connector. You may wonder why it would be necessary to jumper these pins if they are not actually used. This is because many serial ports will check these pins before they will send any data. Many serial interface cards are configured so that this signal is required for any data transfer to take place.

4. Equipment ready pins: Pins 6 and 20 are complementary pairs much as pins 4 and 5 are. Actually, pin 6 is called *data set ready (DSR)*, while pin 20 is called *data terminal ready (DTR)*. The original purpose of these pins was simply to make certain that the power of the external modem was turned on and that the modem was "ready to go to work." In some instances, these pins were also used to indicate if the phone was "off hook" or "on hook."

    Today, these pins may be used for any number of purposes, including such things as "paper out" indicators on printers. The two pins work together and normally should be jumpered on the DB-25 connector like pins 4 and 5 if their use is not expected.

    It should be noted that up to this point, the complementary pairs of pins have been adjacent to one another. Obviously, pins 6 and 20 are not in consecutive order. An inspection of the DB-25 connector, however, will reveal that pins 6 and 20 are almost directly above and below each other in their respective rows.

5. Signal detect pin: Pin 8 is usually called the *data carrier detect (DCD)* or sometimes simply the *carrier detect (CD)* line. In an actual modem, it is used to indicate that a carrier (or signal) is present. It may also indicate that the signal-to-noise ratio is such that data transmission may take place.

    Many computer interface cards will require that this signal be present before they will communicate. If the RS-232 connection is not to a modem requiring this signal, it is often tied together with pins 6 and 20 (DSR and DTR). As has been the case with other pins, this pin is probably not used for its original purpose. What it really is used for is highly dependent on the device in use.

6. Ring indicator pin: Pin 22 is known as the *ring indicator (RI)*. Its original purpose was to do just what its name indicates—to let the computer know when the phone was ringing. Most modems are equipped with "automatic answer" capabilities. Obviously, the modem needed to tell the computer that someone with data to send was calling!

    Pin 22 is only forced true when the ringing voltage is present. This means that this signal will go on and off in rhythm with the actual telephone ring. As is true with other signals, this pin may not be used. Often it is tied together with pins 6, 20, and 8. In other cases, it is simply ignored. Again, the actual use of this pin today will vary widely with the equipment connected.

7. Other pins: Up to this point we have discussed the 10 pins most often used. In reality, the chassis ground is usually accomplished by grounding each piece of equipment separately and is not necessary with battery-powered computer equipment. These considerations have led to the 9-pin connector usage shown in Fig. 10-12. When the DB-25 connector is used, the remaining pins are occasionally used for special situations as warranted by the equipment connected.

    It should be noted that the terms "computer," "modem," and "peripheral equipment" have been used in the discussion thus far. Two more technically correct terms may be used in literature, texts, and diagrams. The term *data terminal equipment (DTE)* is used to indicate a computer, computer terminal, personal computer, etc. The term *data communications equipment (DCE)* is used to indicate the peripheral equipment or the modem (such as printer, mouse, etc.). Notice that the DCE label is used in Fig. 10-12.

10-6 COMPUTER COMMUNICATION—PARALLEL

In the previous section, we discussed serial data communications. In this section, we will discuss parallel data communications, along with serial-to-parallel data conversion. This section is divided into three major parts. The first part will deal with serial/parallel conversion. The second part will investigate the parallel data port that is common on most personal computers today. The last part of the section will deal with the IEEE-488 parallel interface that is often used in modern instrumentation.

Figure 10-14 shows the difference between serial and parallel data transfer. It is very obvious that parallel data transfer requires more wires than serial. This also means that parallel data transfer is usually used for shorter distances than is serial data transfer. The two major places that parallel data transfer is used today are on the internal bus of a computer (or PC) and between the computer and a printer. We are also seeing a fair amount of usage in a third place—between a personal computer and "automatic" instrumentation.

Series-Parallel Conversion

In the modern computers that we use today, it is often necessary to convert data from parallel to serial form, or vice versa. At one time, this was a very difficult process, involving gates, shift registers, etc. Today a device called a *UART* is used. UART stands for *universal asynchronous receiver transmitter*. There are several different types of UARTs. Each

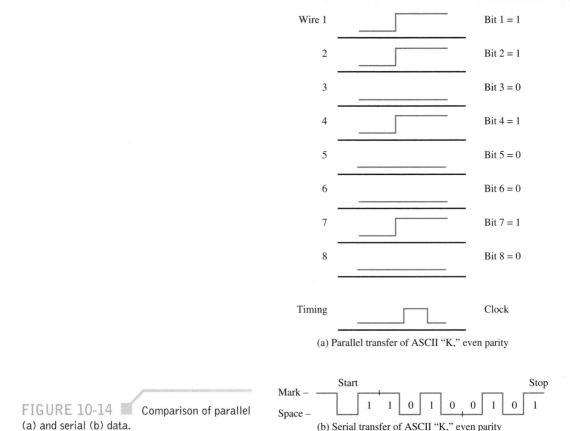

Wire 1   Bit 1 = 1

2   Bit 2 = 1

3   Bit 3 = 0

4   Bit 4 = 1

5   Bit 5 = 0

6   Bit 6 = 0

7   Bit 7 = 1

8   Bit 8 = 0

Timing   Clock

(a) Parallel transfer of ASCII "K," even parity

Start   Stop

Mark –

Space –   1   1   0   1   0   0   1   0   1

FIGURE 10-14   Comparison of parallel (a) and serial (b) data.

(b) Serial transfer of ASCII "K," even parity

of them is really a small microprocessor in itself. The UART can be programmed just like a microprocessor. In some cases it may be "hard wired" for specific purposes, but usually it is programmed by software, just like a microprocessor.

Most UARTs are used to interface a personal computer to the "outside world." The 8250 was used in the original IBM personal computer. The 8250 is used primarily in the early 8-bit personal computers (the PCs and XTs) and not in the newer 16- or 32-bit machines (286, 386, 486, and pentiums, called "AT class" machines). Since most modems today operate at 14.4 K baud or higher, the 16550A is the current standard.

As newer UARTs with higher capabilities become available, it is necessary that they be "downward compatible." This means they will run the software that was written for earlier UARTs. It means that after reset, all registers are the same as earlier UARTs. Obviously, this chip must have registers. That in itself makes it sound like a microprocessor. It has a FIFO (first in, first out) buffer in both transmit and receive modes. It has shift registers and holding registers. In addition, it has a programmable baud rate generator. The earlier UARTs required jumpers to set the baud rate and were only usable at certain standard baud rates. The 16550A is capable of operating at almost any baud rate one would choose.

As with any UART, the 16550A will add or subtract start bits, stop bits, or parity bits. It can recognize 5-, 6-, 7-, or 8-bit words and be set for no parity or odd or even parity. Although 1 stop bit is used most of the time today, it is capable of 1, 1.5, or 2 stop bits. It will also operate up to 256 K baud—a rate that is above what is commonly used today.

It is also interesting to note that the receive portion and the transmit portion can be set up with separate parameters. Note also that this UART is capable of implementing modem control by using the CTS, RTS, DSR, DTR, RI, and DCD lines that were discussed in the previous section on serial communications.

You should also be aware that most UARTs (including the 16550A) operate on TTL levels (0 and +5 V). A level converter to the plus and minus 12 V for RS-232 levels is provided by other chips, such as the 1488 and 1489.

Personal Computer Parallel Printer Port

Most IBM-compatible personal computers use a printer that has a parallel interface. A few may use a serial interface, but that is the exception rather than the rule. In addition, most printers, at the printer end of the connection, use a 37-pin connector and are "Centronics compatible."

At the computer end, the common interface is a DB-25 connector. This should sound familiar, since it is the same connector as discussed in serial data communications in the previous section. You may wonder why the same connector would be used for both serial and parallel data communications. The answer is that it is not really the same connector. They are both DB-25 connectors but of opposite gender. On the back of the PC, the male connector is the serial port, while the female connector is the parallel printer port. Obviously the cable has the opposite gender to mate with it.

In Fig. 10-15 the DB-25 connector, along with a list of what each of the pins are used for, is shown. Note that the parallel connector uses pins 18 through 25 as a ground.

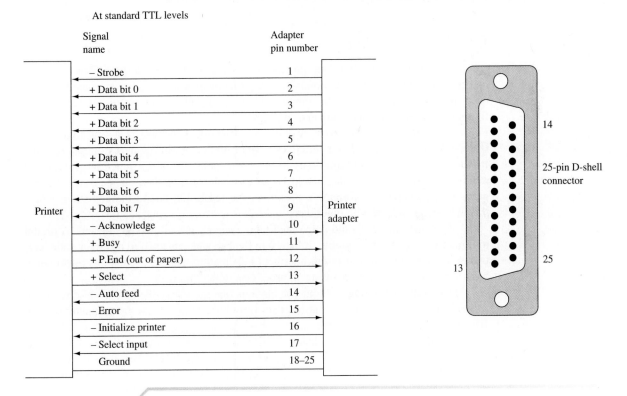

FIGURE 10-15 ■ DB-25 connector for parallel data printer connection.

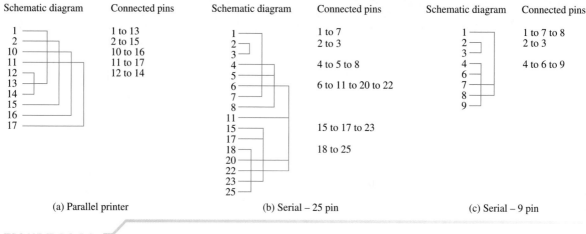

Schematic diagram	Connected pins	Schematic diagram	Connected pins	Schematic diagram	Connected pins
1	1 to 13	1	1 to 7	1	1 to 7 to 8
2	2 to 15	2	2 to 3	2	2 to 3
10	10 to 16	3		3	
11	11 to 17	4	4 to 5 to 8	4	4 to 6 to 9
12	12 to 14	5		6	
13		6	6 to 11 to 20 to 22	7	
14		7		8	
15		8		9	
16		11			
17		15	15 to 17 to 23		
		17			
		18	18 to 25		
		20			
		22			
		23			
		25			
(a) Parallel printer		(b) Serial – 25 pin		(c) Serial – 9 pin	

FIGURE 10-16 Wrap plugs.

This means there is a total of eight grounds—one for each of the data bit lines. To test the parallel printer port, it is necessary to return some of the signals being sent to the printer back to the PC. To do this a "wrap plug" is normally used. Figure 10-16(a) shows how a wrap plug would be wired. The connections for wrap plugs for both 25-pin and 9-pin serial ports, discussed in the last section, are included in Fig. 10-16(b) and (c). Almost any diagnostic program requires wrap plugs in the serial or parallel port for proper testing. These connections should work for most of these programs.

IEEE-488 Interface

Another example of parallel data transmission is the IEEE-488 interface. This is a standard set by IEEE, but it is also often called *GPIB* or general purpose interface bus. It is used mostly for instrumentation/computer interfacing. Different instruments can be configured either as "listeners," "talkers," or both.

An example of a "listener" would be a function generator that was told by the IEEE-488 interface to output a certain type wave and a certain frequency, amplitude, etc. It listens to what it is told and does it, but no response from the function generator is required. A frequency counter would be an example of a "talker." It measures the frequency and sends this information back over the IEEE-488 bus to the computer, where this data are processed with other data.

Most instruments would come under the category of "talker and listener." A digital multimeter would be an example. It listens to the bus and sets itself to a certain scale, certain type wave, etc. Once it is set, it gathers information and then sends this information back to the computer, where it is again processed with other data.

From this description, it should be obvious that the IEEE-488 bus is a bi-directional parallel interface. This means that the data flow both to and from the computer. Figure 10-17 shows the general concept.

In order for the IEEE-488 bus to work, each instrument that will use the bus must have an address. This address is usually set by hardware. It may be set permanently by the manufacturer of the instrument, may be set by means of a switch on the instrument, or in some cases is set by jumpers on the control card inside the instrument. It should be

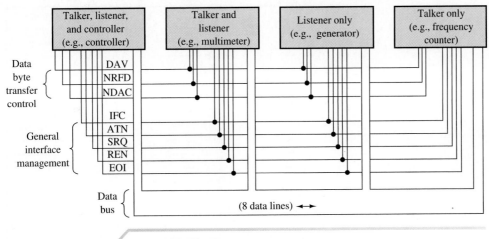

Data byte transfer control

General interface management

DAV
NRFD
NDAC

IFC
ATN
SRQ
REN
EOI

Talker, listener, and controller (e.g., controller)

Talker and listener (e.g., multimeter)

Listener only (e.g., generator)

Talker only (e.g., frequency counter)

Data bus

(8 data lines) ◄─►

FIGURE 10-17 ■ IEEE-488 (GPIB) interface.

noted that special instruments with an IEEE-488 interface built in are necessary to work with this standard. Most recently manufactured test equipment include this feature.

Special software is available to use a standard PC as a controller on the IEEE-488 bus. Some of this software permits very sophisticated interfacing and often includes analysis software as part of the package. By using this bus, completely automated test procedures can be set up. An example application is a high-volume production line of communication transceivers that could be individually tested for a number of specifications.

10-7 DATA TRANSMISSION VIA AM

The earliest form of data transmission using amplitude modulation occurred in the very first days of radio about 100 years ago. The International Morse code was transmitted by simply turning a carrier on and off.

The Morse code is not a true binary code in that it not only includes marks and spaces but also differentiates between the duration of these conditions. The Morse code is still used in amateur radio-telegraphic communications. A human skilled at code reception can provide highly accurate decoding. The International Morse code is shown in Fig. 10-18. It consists of dots (short mark), dashes (long mark), and spaces. A *dot* is made by pressing the telegraph key down and allowing it to spring back rapidly. The length of a dot is one basic time unit. The *dash* is made by holding the key down (keying) for three basic time units. The spacing between dots and dashes in one letter is one basic time unit and between letters is three units. The spacing between words is seven units.

The most elementary form of transmitting highs and lows is to simply key a transmitter's carrier on and off. Figure 10-19(a) shows a dot, dash, dot waveform, while Fig. 10-19(b) shows the resulting transmitter output if the mark allows the carrier to be transmitted and space cuts off transmission. Thus, the carrier is conveying intelligence by simply turning it on or off according to a prearranged code. This type of transmission is called *continuous wave* (CW); however, since the wave is periodically interrupted, it

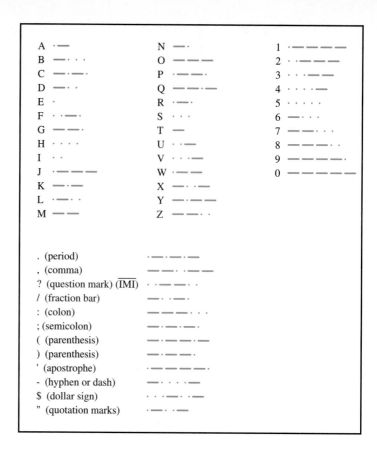

A	·—	N	—·	1	·—————	
B	—···	O	———	2	··———	
C	—·—·	P	·——·	3	···——	
D	—··	Q	——·—	4	····—	
E	·	R	·—·	5	·····	
F	··—·	S	···	6	—····	
G	——·	T	—	7	——···	
H	····	U	··—	8	———··	
I	··	V	···—	9	————·	
J	·———	W	·——	0	—————	
K	—·—	X	—··—			
L	·—··	Y	—·——			
M	——	Z	——··			

. (period)	·—·—·—
, (comma)	——··——
? (question mark) (IMI)	··——··
/ (fraction bar)	—··—·
: (colon)	———···
; (semicolon)	—·—·—·
((parenthesis)	—·——·
) (parenthesis)	—·——·—
' (apostrophe)	·————·
- (hyphen or dash)	—····—
$ (dollar sign)	···—··—
" (quotation marks)	·—··—·

FIGURE 10-18 International Morse code.

might more appropriately be called an interrupted continuous wave. As a concession to the CW misnomer, it is sometimes called *interrupted continuous wave* (ICW).

Whether the CW shown in Fig. 10-19(b) is created by a hand-operated key, a remote-controlled relay, or an automatic system such as punched tape, the rapid rise and fall of the carrier presents a problem. The steep sides of the waveform are rich in harmonic content, which means the channel bandwidth for transmission would have to be extremely wide or else adjacent channel interference would occur. This is a severe problem in that a major advantage of coded transmission versus direct voice transmission is narrow bandwidth channels. The situation is remedied by use of a *LC* filter, as shown in Fig. 10-20. The inductor L_3 slows down the rise time of the carrier, while the capacitor C_2 slows down the decay. This filter is known as a *keying filter* and is also effective in block-

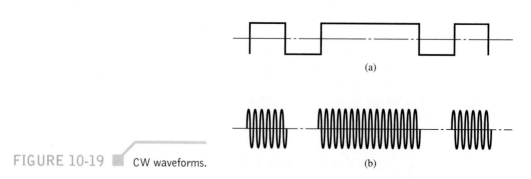

(a)

(b)

FIGURE 10-19 CW waveforms.

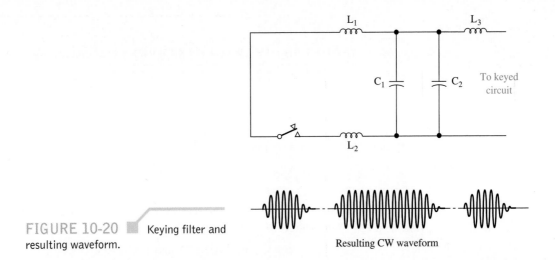

FIGURE 10-20 ▇ Keying filter and resulting waveform.

Resulting CW waveform

ing the RFI (radio frequency interference), created by arcing of the key contacts, from being transmitted. This is accomplished by the L_1, L_2 RF chokes and capacitor C_1 that form a low-pass filter.

CW is a form of AM and therefore suffers from noise to a much greater extent than FM systems. The space condition (no carrier) is also troublesome to a receiver since at that time the receiver's gain is increased by AGC action so as to make received noise very troublesome. Manual receiver gain control helps but not if the received signal is fading between high and low levels, as is often the case. Its simplicity and narrow bandwidth make it attractive to radio amateurs, but its major value is to show the historical development of data transmission.

Machine Code Transmission

In time, telegraphic transmission by hand gave way to machines. This was done through use of a *teleprinter.* It is a transceiver that not only handles the radio-related functions but also has a typewriter-like keyboard. It is commonly referred to as a *Teletype* machine, which is a trade name of the major manufacturer of this equipment. When a key is depressed, the machine mechanically creates the appropriate code for the desired letter or symbol, which is subsequently processed for transmission either to an antenna or over standard telephone lines. Another Teletype machine receiving the signal decodes it and provides a typewriter-style copy. The transmission medium can be via antennas (radioteletype) or by standard telephone lines.

The maximum rate at which information can be transmitted is limited by operator typing speed and/or the bandwidth being used. The faster an operator types, the shorter must be the duration of the pulses in the transmitted binary code. Of course, shorter pulses mean higher-frequency content in the pulses since frequency is inversely proportional to pulse width. As a result of this, a measure of telegraph speed is expressed as the reciprocal of time for the shortest transmitted pulse. This specifies how fast the signal states are changing and is measured in symbols per second or *baud.* For binary (two-level) transmission, the bit rate and the baud rate are the same.

The standard shortest-duration pulse in telegraphy is 22 ms, and its reciprocal, $\frac{1}{22}$ ms or 45 baud, is the system's speed. In addition, each letter (coded) is preceded by a

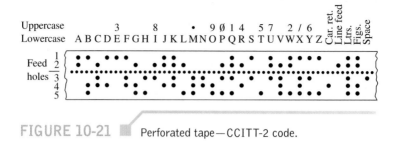

Uppercase   3   8   •   9 0 1 4   5 7   2 / 6

Lowercase A B C D E F G H I J K L M N O P Q R S T U V W X Y Z   Car. ret. Line feed Ltrs. Figs. Space

Feed holes 1 2 3 4 5

FIGURE 10-21 ■ Perforated tape—CCITT-2 code.

22-ms space and followed by a 33-ms mark. Since the average word has six letters, the time for each word using a standard 5-bit code is

$$6 \times \left[22 \ \text{ms} + (5 \times 22 \ \text{ms}) + 33 \ \text{ms} \right] \approx 1 \ \text{s}$$

This implies a transmission speed of 60 words/min because each word, on the average, takes 1 s.

To ensure that a channel is used at its maximum capacity, most Teletypes have a perforated tape unit. The operator can then type the message at any convenient speed without transmitting. The message is stored on a perforated tape and then can be transmitted at a convenient time at whatever maximum speed the system has. This provides maximum channel usage efficiency. Figure 10-21 illustrates a commonly used code, the CCITT-2, as it appears on perforated tape. This code is an adaptation of the Baudot code shown in Fig. 9-5. Each hole (perforation) in the tape represents a mark for the particular bit. The lack of a hole represents a space condition.

Two-Tone Modulation

Two-tone modulation is a form of AM, but in it the carrier is always transmitted. Instead of simply turning the carrier on and off, the carrier is amplitude-modulated by two different frequencies representing either mark or space. The two frequencies are usually separated by 170 Hz. An example of such a telegraphy system is provided in Fig. 10-22. When the transmitter is keyed, the carrier is modulated by a 470-Hz signal (mark condition), while it is modulated by a 300-Hz signal for the space condition. At the receiver, after detection, either 300- or 470-Hz signals are present. A 470-Hz bandpass filter provides an output for the mark condition that makes the output high whenever 470 Hz is present and low otherwise.

EXAMPLE 10-1

The two-tone modulation system shown in Fig. 10-22 operates with a 10-MHz carrier. Determine all possible transmitted frequencies and the required bandwidth for this system.

SOLUTION

This is an amplitude modulation system, and therefore when the carrier is modulated by 300 Hz the output frequencies will be 10 MHz and 10 MHz ±

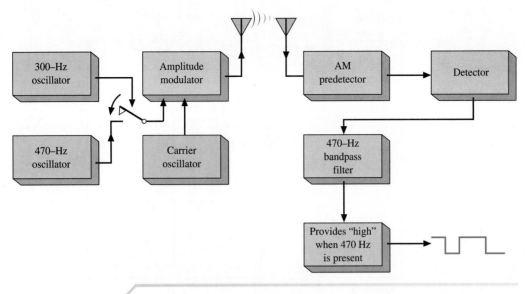

Two-tone modulation system (AM).

300 Hz. Similarly, when modulated by 470 Hz the output frequencies will be 10 MHz and 10 MHz ± 470 Hz. Those are all possible outputs for this system. The bandwidth required is therefore 470 Hz × 2 = 940 Hz, which means that a 1-kHz channel would be adequate.

Example 10-1 shows that two-tone modulation systems are very effective with respect to bandwidth utilized. One hundred 1-kHz channels could be sandwiched in the frequency spectrum from 10 MHz to 10.1 MHz. The fact that a carrier is always transmitted eliminates the receiver gain control problems previously mentioned, and the fact that three different frequencies (a carrier and two side frequencies) are always being transmitted is another advantage over CW systems. In CW either one frequency, the carrier, or none is transmitted. Single-frequency transmissions are much more subject to ionospheric fading conditions than multifrequency transmissions. This phenomenon will be elaborated on in Chapter 13.

Quadrature Amplitude Modulation

A number of special modulation techniques are used, over and above those previously described, to transmit digital signals. The most popular system to achieve high data rates in limited bandwidth channels is called *quadrature amplitude modulation* (QAM).

The block diagram in Fig. 10-23 shows a QAM transmitter. The binary data are first fed to a ÷ data block that essentially produces two data signals at half the original bit rate by feeding every other data bit to its two outputs. These two-level outputs are then converted to four-level baseband streams. The resultant four-level symbol streams of the I and Q channels are then applied to the modulators in Fig. 10-23. Notice the carrier for the

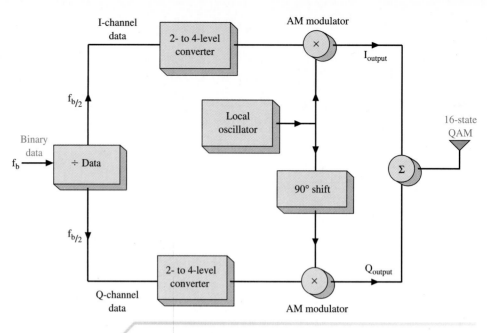

FIGURE 10-23 ▮ 16-QAM transmitter.

Q-channel modulator. It is shifted 90° from the I channel and is said to be in quadrature. This explains its Q designation, while the I channel is so named since it is the in-phase channel. The net result of this is the ability to transmit large amounts of digital data through limited bandwidth channels.

The QAM demodulator reverses the modulator process, thereby providing the original binary data signal. Some insight into QAM systems is provided by feeding the demodulated I signal into the horizontal input and the Q signal into the vertical input of an oscilloscope. The result (shown in Fig. 10-24) is called a *constellation pattern* because of its resemblance to stars. The gain and position of each channel must be properly adjusted and a signal must also be applied to the scope's Z-axis input to kill the scope intensity during the digital state transition times.

The constellation pattern can be used to analyze the systems' linearity and noise performance. Since QAM involves AM, linearity of the transmitter's power amplifiers can be a cause of system error. Linearity problems are indicated by unequal spacing in the constellation pattern. Noise problems are indicated by excessive blurring and spreading out of the points.

QAM digital transmission systems have become dominant in recent years. Besides the 4×4 system introduced here, 2×2 and 8×8 systems are also commonly used.

Loopbacks

Many digital modulation systems include a *loopback* capability. The receiver takes the received data and sends them back to the transmitter. The data are then compared with the originally transmitted data to provide an indication of system performance. Since bit errors can occur both in the original transmission and in the loopback, the bit error rate cannot be pinpointed since it will not be known where the error occurred. Nonetheless, the loopback test is very helpful in diagnosing the basic system performance.

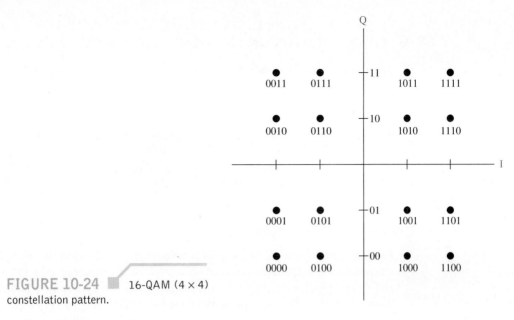

FIGURE 10-24 ■ 16-QAM (4 × 4) constellation pattern.

Eye Patterns

Another technique that is extremely helpful in diagnosing the performance of a digital modulation system is the generation of *eye patterns*. They are generated by "overlaying" on an oscilloscope all the digital bit signals received. Ideally, this would result in a rectangular display because of the persistence of the CRT phosphor. The effects of transmission cause various rounding effects, which result in a display resembling an eye.

Refer to Fig. 10-25 for the various possible patterns. The opening of the eye represents the voltage difference between a 1 or a 0. A large variation indicates noise problems, while nonsymmetrical shapes indicate system distortion. Jitter and undesired phase shifting are also discernible on the eye pattern.

The eye pattern can be viewed while making adjustments to the system. In that way you can immediately see the effects of filter or circuit or antenna adjustments as they are made.

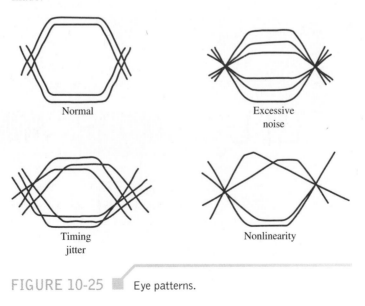

FIGURE 10-25 ■ Eye patterns.

10-8 DATA TRANSMISSION VIA FM/PM

The transmission of digital data via frequency or phase modulation offers some of the same advantages over amplitude modulation as occur in standard analog systems. In this section we'll examine the common FM/PM schemes used to transmit digital data. Keep in mind that these schemes are used for transmission of any type of binary code. Recall that the most common codes used were described in Chapter 9. These include the ASCII, EBCDIC, Baudot, Gray, and PCM codes. Remember that all of these except PCM are typically used to code computer information, while PCM is used to convert analog signals into digital form. This also applies to the AM systems explained in the previous section.

Frequency Shift Keying

Frequency shift keying (FSK) is a form of frequency modulation in which the modulating wave shifts the output between two predetermined frequencies—usually termed the mark and space frequencies. It may be considered as an FM system in which the carrier frequency is midway between the mark and space frequencies and is modulated by a rectangular wave, as shown in Fig. 10-26. The mark condition causes the carrier frequency to increase by 42.5 Hz, while the space condition results in a 42.5-Hz downward shift. Thus, the transmitter frequency is constantly changing by 85 Hz as it is keyed. This 85-Hz shift is the standard for narrowband FSK, while an 850-Hz shift is the standard for wideband FSK systems.

EXAMPLE 10-2

Determine the channel bandwidth required for the narrowband and wideband FSK systems.

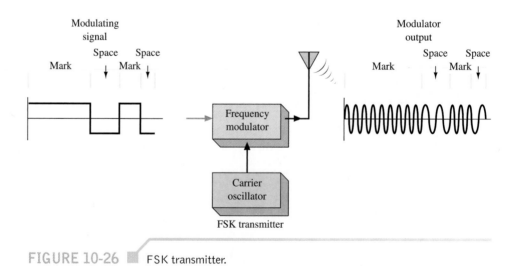

FIGURE 10-26 FSK transmitter.

The fact that narrowband FSK shifts a total of 85 Hz does *not* mean the bandwidth is 85 Hz. While shifting 85 Hz it creates an infinite number of sidebands, with the extent of significant sidebands determined by the modulation index. If this is difficult for you to accept, it would be wise to review the basics of FM in Chapter 5.

In practice, most narrowband FSK systems utilize a channel of several kilohertz, while wideband FSK uses 10 to 20 kHz. Because of the narrow bandwidths involved, FSK systems offer only slightly improved noise performance over the AM two-tone modulation scheme. However, the greater number of sidebands transmitted in FSK allows better ionospheric fading characteristics than two-tone AM modulation schemes.

FSK Generation

The generation of FSK can be easily accomplished by switching an additional capacitor into the tank circuit of an oscillator when the transmitter is keyed. In narrowband FSK it is often possible to get the required frequency shift by shunting the capacitance directly across a crystal, especially if frequency multipliers follow the oscillator, as is usually the case. FSK can also be generated by applying the rectangular wave modulating signal to a VCO. Such a system is shown in Fig. 10-27. The VCO output is the desired FSK signal,

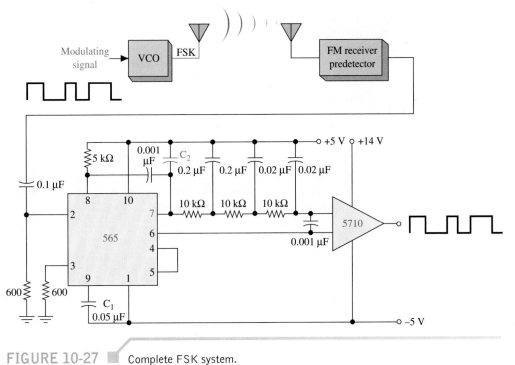

FIGURE 10-27 Complete FSK system.

which is then transmitted to an FM receiver. The receiver is a standard unit up through the IF amps. At that point a 565 PLL is used for detecting the original modulating signal. As the IF output signal appears at the PLL input, the loop locks to the input frequency and tracks it between the two frequencies with a corresponding dc shift at its output, pin 7. The loop filter capacitor C_2 is chosen to set the proper overshoot on the output, and the three-stage ladder filter is used to remove the sum frequency component. The PLL output signal is a rounded-off version of the original binary modulating signal and is therefore applied to the comparator circuit (the 5710 in Fig. 10-27) to make it logic compatible.

Phase Shift Keying

One of the most efficient methods for data modulation is *phase shift keying (PSK)*. PSK systems provide a low probability of error. The incoming data cause the phase of the carrier to phase-shift a defined amount. This relationship is expressed as:

$$V_o(t) = V \sin\left[\omega_c(t) + \frac{2\pi(i-1)}{M}\right]$$

(10-1)

where: $i = 1, 2, \ldots, M$
$M = 2^N$, number of allowable phase states
N = the number of data bits needed to specify the phase state
ω_c = angular velocity of carrier

There are many versions of the PSK signal. Three common versions are shown in Table 10-1.

With M (the number of allowable phase states) greater than four the systems are referred to as *M-ary* systems, and the output signal is called a *constellation*. In a BPSK signal the phase of the carrier will shift by 180° (i.e., +/– sin $\omega_c t$). For a QPSK signal the phase changes 90° for each possible state. A diagram showing both a BPSK and QPSK constellation is provided in Fig. 10-27.

Binary Phase Shift Keying

It has been defined that for a BPSK signal, M = 2 and N = 1. The +sin($\omega_c t$) vector provides the logical "1" and the −sin($w_c t$) vector provides the logical "0" as shown in Fig. 10-28(a). The BPSK signal does not require that the frequency of the carrier be shifted as with the FSK system. Instead, the carrier is directly phase modulated, meaning that the phase of the carrier is shifted by the incoming binary data. This relationship is shown in Fig. 10-29 for an alternating pattern of 1's and 0's.

TABLE 10-1 *Common PSK Systems*

Binary Phase Shift Keying—BPSK	M = 2	N = 1
Quadrature Phase Shift Keying—QPSK	M = 4	N = 2
8PSK	M = 8	N = 3

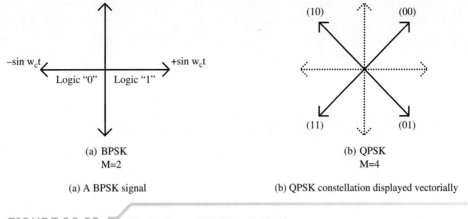

(a) BPSK
M=2

(b) QPSK
M=4

(a) A BPSK signal

(b) QPSK constellation displayed vectorially

FIGURE 10-28 ■ Illustrations of BPSK and QPSK.

Generation of the BPSK signal can be accomplished in many ways. A block diagram of a simple method is shown in Fig. 10-30. The carrier frequency [+sin($\omega_c t$)] is phase shifted 180°. The + and − values are then fed to a 1 of 2 selector circuit, which is driven by the binary data. If the binary data is a 1, then the output is +sin($\omega_c t$). If the binary-input data is a 0, then the −sin($\omega_c t$) signal is selected for the output. The actual devices selected for performing this operation are dependent on the binary input data rate and the transmit carrier frequency.

A BPSK receiver requires that the phase shift be detected. One possible way of constructing a BPSK receiver is by using a mixer circuit. The received BPSK signal is fed into the mixer circuit. The other input to the mixer circuit is driven by a reference oscillator synchronized to sin($\omega_c t$). This is referred to as *coherent carrier recovery*. The recovered carrier frequency is mixed with the BPSK input signal to provide the demodulated binary output data. A block diagram of the receive circuit is provided in Fig. 10-31. Mathematically, the BPSK receiver shown in Fig. 10-31 can provide a 1 and a 0 as shown:

$$\text{"1" Output} = \big[\sin(\omega_c t)\big]\big[\sin(\omega_c t)\big] = \sin^2(\omega_c t)$$

and by trig identity $\sin^2 A = \frac{1}{2}[1 - \cos 2A]$. Therefore

$$\text{"1" Output} = \tfrac{1}{2} - \tfrac{1}{2}\big[\cos(2\omega_c t)\big]$$

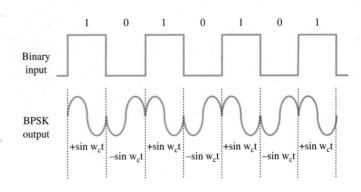

FIGURE 10-29 ■ The output of a BPSK modulator circuit for a 1010101 input.

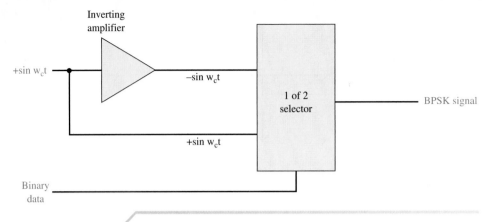

FIGURE 10-30 ■ A circuit for generating a BPSK signal.

The $\frac{1}{2}$ [cos $(2\omega_c t)$] term is filtered out by the low-pass filter shown in Fig. 10-31. This leaves

$$\text{“1” Output} = \tfrac{1}{2}$$

Similar analysis will show

$$\text{“0” Output} = \big[-\sin(\omega_c t)\big]\big[\sin(\omega_c t)\big]$$

$$= -\sin^2(\omega_c t)$$

$$= -\tfrac{1}{2}$$

The +/− 1/2 represent dc values that correspond to the 1 and 0 binary values. The +/− 1/2 values can be conditioned for the appropriate input level for the receive digital system.

The QPSK System

The QPSK constellation provides four vectors for representing the binary data. The savings by transmitting this way are in the reduced bandwidth requirement. The QPSK system uses two data channels. These are identified as the I and Q channels. Each channel contributes to the direction of the vector within the phase constellation. The four possible

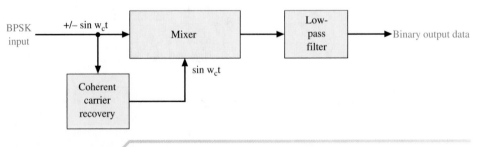

FIGURE 10-31 ■ A BPSK receiver using coherent carrier recovery and a mixer circuit.

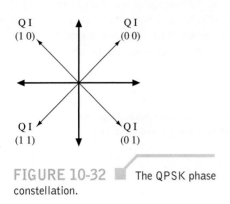

FIGURE 10-32 ■ The QPSK phase constellation.

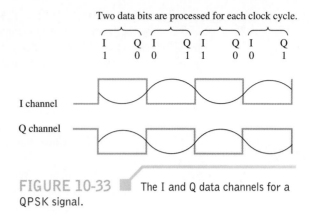

FIGURE 10-33 ■ The I and Q data channels for a QPSK signal.

values for I and Q are shown in Fig. 10-32. The BPSK signal requires a $BW_{min} = f_b$, where the QPSK signal will only require $f_b/2$ for each channel. The quantity f_b refers to the frequency of each of the original bits in the data. With QPSK transmission, a *data bandwidth compression* is realized. This means that more data are being compressed into the same available bandwidth. This relationship is shown pictorially in Fig. 10-33.

TROUBLESHOOTING

Telemetry systems have become big business in recent years. In addition to the military applications, many commercial uses are being developed. Today's telemetry markets range from remote reading of gas and electric meters to credit card validation, security monitoring, token-free highway toll systems, and remote inventory control.

After completing this section you should be able to:

- Troubleshoot a radio-telemetry receiver
- Troubleshoot a radio-telemetry transmitter

Receiver Problems

Let's start by looking at some possible receiver problems in Fig. 10-10. Since this receiver has a low-frequency IF amplifier, an oscilloscope can be used to check signal levels and adjust the tuned circuits. However, a typical oscilloscope probe presents a capacitive load to the circuit you connect it to. The best probe for this work will be a X10 type. These probes divide the signal to the oscilloscope by 10 but only shunt the circuit under test with 8 to 10 pF, whereas a X1 probe would have a shunt capacitance of 100 pF. Sometimes even 10 pF is significant, but it is usually not a problem if the technician is aware of the effect.

No OUTPUT: Assume you know that a good transmitted signal is present. Check the easy things first: battery or power supply and the antenna connections. Refer to Fig. 10-10 in this discussion.

1. Disconnect the receiver from the decoder. It is possible the decoder is shorted, therefore killing the receiver.

2. Check the local oscillator. You can check the LO signal at pin 2 with an oscilloscope by using the low-capacitance probe. If you don't have one, just use a small capacitor, 10 pF or less in series with your probe. The capacitor forms a voltage divider with the probe capacitance. Assuming the capacitance of the added capacitor is small compared to that of the probe, the loss is approximately the ratio of the capacitances.

   If the oscillator is not running, try adjusting the inductor, L1. After the oscilloscope runs, you should be able to find a peak amplitude followed by a point where it quits. Adjust the coil away from the quitting point and just below the output peak.

   If adjustment is fruitless, check the crystal as described in Chapter 1 or substitute another. Don't forget the inductor and its tuning capacitor. Integrated circuits are usually more reliable than the rest of the parts and harder to unsolder.

3. Check for signal in the IF amplifier stages. Since there are no means of adjusting the oscillator frequency, the IF must be aligned to the transmitter's frequency. The manufacturer recommends killing the AGC by grounding pin 16. The secondary of T3 is a good test point because it isolates the IF amplifier from the oscilloscope probe. The signal amplitude will be down by a factor of 8 according to the data sheet, but we are only looking for a peak. Adjust T1, T2, and T3 for maximum output. After some signal is obtained, you may find that one coil will not tune properly. Check the inductance and its associated capacitor.

Transmitter Problems

In this case, you will need some means of monitoring the transmitter's RF output without actually making a physical connection to the transmitter. A spectrum analyzer is nice, but almost any receiver capable of operating on the proper frequency will do.

1. No output. Check for modulation at the input, pin 8; and at the internal modulator's output, pin 13. The modulator simply turns the oscillator's power supply on and off. The off-on rate will be in the low audio frequency region. If these are present and there is no RF output, it's time to check the crystal and the tuned circuits.

2. Low output. This can be caused by low modulator output. Check the modulation voltage peak amplitude at pin 13. For the LM1871, you should have 4.5 V.

   Monitor the signal strength with a receiver or the spectrum analyzer while adjusting the oscillator transformer, T1. Look for a definite peak. If none can be found, check the associated 220-pF and 47-pF capacitors.

Encoder/Decoder Problems

The current source is the heart of both the encoder and the decoder. Both circuits also depend on a high-quality capacitor to integrate the current provided by the current source.

   First, a short discussion of the current source. The LM329 is an active 6.9-V reference. This voltage is divided by the 6.49- and the 4.02-kΩ resistors to provide a fixed base–emitter voltage on the 2N2907. Since this voltage is fixed, the voltage across the emitter resistor is fixed. Therefore, the collector current is constant. Now, let's look at some typical problems.

1. Assume the 2N2907 has a base-to-collector short. Now the current is only limited by the emitter resistor. The capacitor will charge rapidly. Therefore, a lower signal

voltage will cause the comparator (A1) to trip at a lower voltage, and the gain will appear to be too high. Since the voltage across the capacitor is no longer linear, the system output will also be nonlinear.

2. Assume the integrating capacitor has developed a high-resistance leak. Now part of the current is not building charge on the capacitor. The gain will appear to be lower and some nonlinearity may appear. This applies to both the encoder and decoder.

3. Assume a failure of the 74LS123 in the decoder. It must provide the proper logic levels to reset the integrator and trigger the sample-and-hold IC. Incorrect logic levels result in no output.

4. Assume the sample-and-hold capacitor (0.001 μFd) has developed a leak. The operational amplifier may have enough drive to charge the capacitor, but voltage will immediately begin to fall. Looking at the output with an oscilloscope, you will see a sawtooth waveform, and the output voltage will be too low.

SUMMARY

In Chapter 10 we examined various aspects of digital communications transmission. These included pulse modulation, telemetry, serial and parallel communication, and data transmission using AM techniques. The major topics the student should now understand include:

- the different schemes possible for transmission of analog and digital signals

- the description and implementation of pulse modulation, including pulse-amplitude (PAM), pulse-width (PWM), and pulse-position (PPM)

- The basic concept of telemetry and description of a complete radio-telemetry system

- the analysis and description of asynchronous and synchronous serial computer communication with emphasis on the RS-232 communication standard

- the analysis of parallel computer communication, including serial-to-parallel conversion, PC parallel port description, and IEEE-488 interface description

- the description of various data transmission methods using amplitude modulation

- the analysis of quadrature amplitude modulation (QAM), including constellation patterns, loopback, and eye tests

- the analysis of data transmission techniques using FM/PM, including frequency and phase shift keying

QUESTIONS AND PROBLEMS

SECTION 1

1. Explain what is meant by a baseband transmission.
2. With reference to Fig. 10-1, which transmission scheme of the six shown is used by a broadcast FM radio station?

3. Provide some possible reasons why an analog signal is digitized when an analog output is desired. A block diagram of such a system is shown in Fig. 10-1(f).

SECTION 2

4. Describe the key distinction between pulse modulation and amplitude or frequency modulation.
5. List two advantages of pulse modulation and explain their significance.
6. A signal that varies from 20 Hz to 5 kHz is to be processed via a pulse modulation scheme. Determine the minimum sampling rate that will still allow adequate reproduction at the receiver. Calculate the number of different time-division-multiplexed signals that could be transmitted if each sample takes 10 μs. (10 kHz, 10)
7. With a sketch similar to Fig. 10-2, explain the basics of PAM, PWM, and PPM.
8. Describe a means of generating and detecting PWM.
9. Describe a means of generating and detecting PPM.

SECTION 3

10. What is a telemetry system? Explain why telemetry invariably involves multiplexing and pulse modulation techniques.
11. List the basic functions performed by a complete telemetry system and explain them briefly.
12. Why does the telemetry transmitting system shown in Fig. 10-9 have six subchannels feeding channel 4, while channels 1, 2, and 3 do not? Why is this arrangement better than using nine different channels—one for each condition being monitored?

SECTION 4

13. Briefly describe the operation of the radio-telemetry system shown in Fig. 10-10.
14. Troubleshoot the following failure modes of the radio-telemetry system of Fig. 10-10. You should provide at least one, and preferably two, *specific* possible failure causes.
    (a) The decoder output is not a linear representation of the temperature being sensed.
    (b) The transmitter's output power is significantly below normal.
    (c) The receiver's output (LM 1872, pin 11) is near zero even though the received signal is normal.
    (d) The decoder's output is not changing at the proper intervals.
15. Explain in detail how the decoder in Fig. 10-10 is able to convert the PWM received signal back to the original analog signal.

SECTION 5

16. Serial data are classified as either asynchronous or synchronous. Explain the difference between them.
17. Describe the origin of the RS-232C standard and discuss its current status as a standard.
18. What two connections are commonly used with RS-232C? List the signal lines they specify.
19. Explain the meaning of the acronyms DTE and DCE.

20. Describe the difference between serial and parallel data.
21. Explain the function of a UART and why it should be "downward compatible."
22. What is the purpose of a wrap plug? Suppose the pin 10 to pin 16 connection has opened and you're unaware of this. Explain some possible effects this could have on your troubleshooting.
23. What is the purpose of the IEEE-488 interface? Explain what a talker and a listener are in this system.
24. Your employer is producing a high volume of *LC* bandpass filters. Describe a means to use the IEEE-488 system shown in Fig. 10-17 to automatically test the upper and lower cut-off frequency.

SECTION 7

25. Define *continuous-wave transmission*. In what way is this an inappropriate name? Explain the role that a keying filter plays in a CW transmission.
26. Explain the AGC difficulties encountered in the reception of CW. What is two-tone modulation, and how does it remedy the receiver AGC problem of CW?
27. Calculate all possible transmitted frequencies for a two-tone modulation system using a 21-MHz carrier with 300- and 470-Hz modulating signals to represent mark and space. Calculate the channel bandwidth required.
28. Explain the functions of a teleprinter. Include in this explanation the meaning of a *baud* and how it relates to the standard transmission rate of 60 words/min.
29. Describe the advantages of using a perforated tape unit in conjunction with a teleprinter.
30. Provide a brief description of a QAM system and explain why it is an efficient user of the frequency spectrum.
31. Describe how to generate a constellation pattern and how to interpret it.
32. Explain the necessary change to the 16 QAM transmitter in Fig. 10-23 to allow it to function as a 64 QAM (8 × 8) system.
33. What is a loopback? Describe several applications.
34. Explain how to generate an eye pattern and how it may be utilized.

SECTION 8

35. What is a frequency shift keying system? Describe several methods of generating FSK.
*36. Explain briefly the principles involved in frequency shift keying (FSK). How is this signal detected?
37. Describe the PSK process and include the significance of M and N shown in Table 10-1.
38. Explain a method to generate BPSK using Fig. 10-30 as a basis.
39. Describe a method of detecting the binary output data from a BPSK signal, using Fig. 10-31 as a basis.
40. What is coherent carrier recovery?
41. Describe a QPSK system with the assistance of Fig. 10-32.
42. How does QPSK achieve a data bandwidth compression with respect to BPSK?

NETWORK COMMUNICATIONS

OBJECTIVES

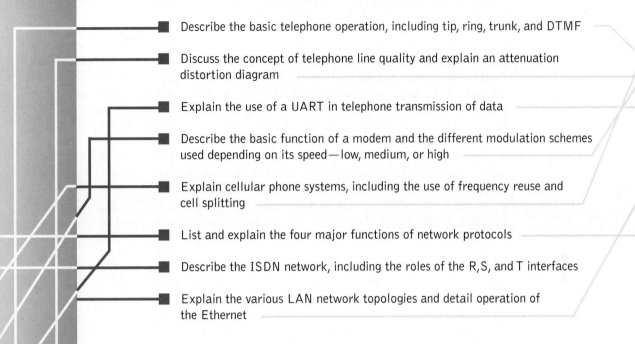

■ Describe the basic telephone operation, including tip, ring, trunk, and DTMF

■ Discuss the concept of telephone line quality and explain an attenuation distortion diagram

■ Explain the use of a UART in telephone transmission of data

■ Describe the basic function of a modem and the different modulation schemes used depending on its speed—low, medium, or high

■ Explain cellular phone systems, including the use of frequency reuse and cell splitting

■ List and explain the four major functions of network protocols

■ Describe the ISDN network, including the roles of the R, S, and T interfaces

■ Explain the various LAN network topologies and detail operation of the Ethernet

11-1 INTRODUCTION

A network is an interconnection of users that allows communication with each other. The most extensive existing network is the worldwide telephone grid. It allows direct connection between two users simply by dialing an access code. Behind that apparent simplicity is an extremely complex system that we look at in Sec. 11-2. Since the telephone network is so convenient and inexpensive, it is often used to allow one computer to "speak" with another. The techniques that allow transmission of computer bits over telephone lines are the topics of Secs. 11-3 and 11-4. The mushrooming cellular phone network is explored in Sec. 11-5.

Computers often need to communicate with more than one other computer or terminal device. In fact, the proliferation of low-cost computer systems has spawned the growth of networks that include hundreds of computers. In them, one computer may wish to send data to all the others or to just a few specified units. Networks that allow this form the basis for the rest of this chapter.

11-2 BASIC TELEPHONE OPERATION

The Greek word *tele* means "far" and *phone* means "sound." The telephone system represents a worldwide grid of connections that enables point-to-point communications between the many subscribers. The early systems used mechanical switches to provide routing of a call. Initially, the *Strowger stepping switch* was used, and subsequently *crossbar switching* was used. Today's systems utilize solid-state electronics for switching under computer control to determine and select the best routing possibilities. The possible routes for local calls include hard-wired paths or fiber-optic systems. Long-distance calls are routed using these same paths, but additionally use radio transmission via satellite or microwave transmission paths. It is possible for a call to use all of these paths in getting from its source to destination. These multipath transmissions are especially tough on digital transmissions, as we will later see.

The telephone company (telco) provides two-wire service to each subscriber. One wire is designated the *tip* and the other the *ring*. The telco provides −48 V dc on the ring and grounds the tip, as shown in Fig. 11-1. The telephone circuitry must work with three signal levels: the received voice signal, which could be as low as a few millivolts; the transmitted voice signal of 1 to 2 V rms; and an incoming ringing signal of 90 V rms. They also accept the dc power of −48 V at 15 to 80 mA. Until the phone is removed from the hook, only the ring circuits are connected to the line. The subscriber's telephone line is usually No. 22 twisted-pair wire, which handles the 300-Hz to 3-kHz audio voice signal but in reality can work up to several megahertz. The twisted-pair line, or *trunk* or *local loop* as it is often called, runs up to a few miles to a central phone office, or in a business setting to a PBX (private business exchange). When a subscriber lifts the handset, a switch is closed that completes a dc short between the tip and the ring line through the phone's microphone. The handset earpiece is transformer or electronically coupled into this circuit also. The telco senses the off-hook condition and responds with an audi-

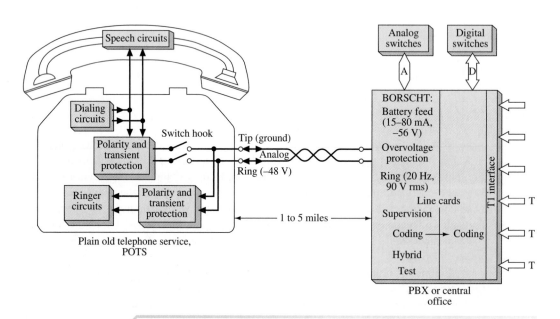

FIGURE 11-1 Telephone representation.

697	1	2	3
770	4	5	6
852	7	8	9
941	*	0	#
Frequency (Hz)	1209	1336	1477

FIGURE 11-2 ■ Touch-tone dialing.

ble dial tone. The central office/PBX function shown in Fig. 11-1 will be detailed later in this section.

At this point the subscriber dials or keys in the desired number. Dial pulsing is the interruption of the circuit according to the number dialed. A 2 interrupts the circuit twice and an 8 interrupts it eight times. Tone-dialing systems utilize a dual-tone multifrequency (DTMF) electronic oscillator to provide this information. Figure 11-2 shows the arrangement of this system. When selecting the digit 8, a dual-frequency tone of 852 Hz and 1336 Hz is transmitted. When the telco receives the entire number selected, its central computer makes a path selection. It then either sends the destination a 90-V ac ringer signal or sends the originator a busy signal if the destination is already in use.

The connection paths for telephone service were all initially designed for voice transmission. As such, the band of frequencies of interest were about 300 to 3000 Hz. To meet the increasing demands of data transmission, the telco now provides special dedicated lines with enhanced performance and bandwidths up to 30 MHz for high-speed applications. The characteristics of these lines have a major effect on their usefulness for data transmission applications. Line-quality considerations are explored later in this section.

Telephone Converter Circuit

A circuit that can be used to power circuits added to the basic phone is shown in Fig. 11-3. This may be more economical/convenient than adding a separate supply from the ac power line or using battery power. It takes the −48 V provided by the telco and efficiently converts it to a well-regulated 5 V.

This converter circuit does not require the bulk and expense of a transformer, which is usually used in these designs. It takes the −48 V (it can actually function well with inputs of −35 to −75 V) and provides up to 150 mA at 5 V. The switching regulator (IC_1) in Fig. 11-3 operates in the boost configuration powered by a zener-regulated 6.2 V with reference to the −48 V. It therefore acts like it was converting 48 V to 53 V. A feedback signal is shifted in level from 5 V to the IC's feedback input (pin 10) by the current source transistor, Q_2. Transistor Q_3 is included to compensate for the V_{be} temperature drift in Q_2. Transistor Q_3 can be omitted if about −2mV/°C voltage variations are acceptable.

The switching FET (Q_1) exhibits a typical $R_{DS(on)}$ of 1Ω when operating with the 6-V gate drive provided by the circuit. The efficiency of this circuit is about 75%, and the 1-Ω sense resistor (R_2) limits peak current to about 200 mA.

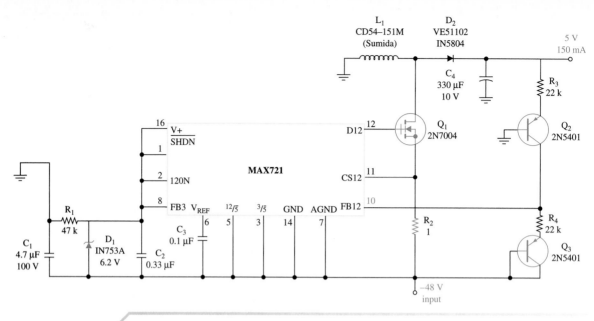

FIGURE 11-3 ▮ −48-V to 5-V converter circuit.

Telephone Systems

A complete telephone system block diagram is shown in Fig. 11-4. On the left, three subscribers are shown. The top one is an office with three phones and a terminal/computer (T/C). The digital signal of the T/C is converted to analog by the modem. The office system is internally tied together via its PBX. The PBX also connects it to the outside world, as shown. The next subscriber in Fig. 11-4 is a home with two phones, while the third subscriber is a home with a phone and a T/C. Notice the manual switch used for voice/data communications.

The primary function of the PBX and central office is the same: switching one telephone line to another. In addition, most central offices multiplex many conversations onto one channel or line. The multiplexing may be based on time or frequency division, and the transmitted signals may be analog or digital.

Before switching at the PBX or central office, circuitry residing on the analog line cards (digital in the future) handles the so-called BORSCHT function for their line. These interface circuits are also called the *subscriber loop interface circuit* (SLIC). BORSCHT is an acronym generated as follows:

Battery feeding: supplying the −48 V at 15- to 80-mA power required by the telephone service. Batteries under continuous charge at the central office allow phone service even during power failures.

Overvoltage protection: guarding against induced lightning strike transients and other electrical pickups.

Ringing: producing the 90-V rms signal shared by all lines; a relay or high-voltage solid-state switch connects the ring generator to the line.

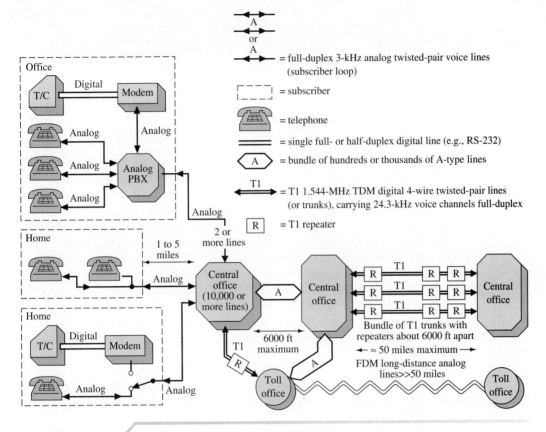

FIGURE 11-4 ■ Telephone system block diagram. (Reprinted with permission from *Electronic Design,* Vol. 33, No. 1; copyright Hayden Publishing Co., Inc., 1985.)

Supervision: alerting the central office to on- and off-hook conditions (dial tone, ringing, operator requests, busy signals); also the office's way of auditing the line for billing.

Coding: if the central office uses digital switching or is connected to TDM/PCM digital lines, there are codes and their associated filters on the line card.

Hybrid: separating the two-wire subscriber loop into two-wire pairs for transmitting and receiving. The phone system is obviously a full-duplex system. Other than the local loop, four wires are used by the phone company, two for transmitting and two for receiving. This arrangement is shown in Fig. 11-5. The hybrid was a transformer-based circuit, but today an electronic circuit contained within an IC provides the BORSCHT functions.

Testing: enabling the central office to test the subscriber's line.

Today, a portion of central-office switching and that handled by PBXs is analog. Analog switches range from stepping switches and banks of relays to crossbar switches and crosspoint arrays of solid-state switches. They are slow, have limited bandwidth, eat up space, consume lots of power, and are difficult to control with microprocessors or

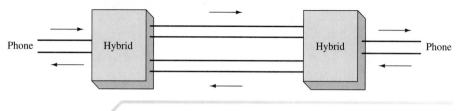

FIGURE 11-5 ▮ Two- to four-wire conversion.

even CPUs. To overcome the limitations inherent in analog switching and to enable central offices to use TDM/PCM [see Chapter 9 for a discussion of pulse-code modulation (PCM)] links in urban areas, telephone companies have been converting to digital transmission using 8-bit words. The conversions will be virtually complete in the late 1990s.

Bundles of twisted-pair wires running between line cards and switches have been replaced with a few serial time-division-multiplexed PCM lines to the switches. Those switches are now complex logic circuits controlled by CPUs and microprocessors. A given voice channel occupies just one time slot in the TDM/PCM signal.

Digitally switched digital signals may flow between two analog lines coming into the central office. In that case, they are returned to a line card, reconverted into analog form, and sent over an analog line either directly to a subscriber or in a bundle of lines to a central office.

On the other hand, if the digital words are multiplexed onto a T1 carrier (see Fig. 11-4) en route to another office, the serial bit stream is converted to the equivalent of 24 voice channels plus supervisory, signaling, and framing bits. Frames are transmitted every 125 μs (an 8-kHz rate).

Bundles of T1 lines carry most voice channels between central offices in densely populated areas. If the lines stretch more than 6000 ft or so, a repeater amplifies the signal and regenerates its timing. Glass fibers have replaced most of the copper wire used in telephone systems, especially in urban areas. Refer to Chapter 17 for details on these fiber-optic systems.

Line Quality Considerations

An ideal telephone line would transmit a perfect replica of a signal to the receiver. Ideally, this would be true for a basic analog voice signal, an analog version of a digital signal, or a pure digital signal. Unfortunately, as I'm sure you would expect, the ideal does not occur. We now look at the various reasons that a signal received via telephone lines is less than perfect (i.e., is distorted).

Attenuation Distortion

The local loop for virtually all telephone transmissions is a two-wire twisted-pair cable. This rudimentary transmission line is usually made up of copper conductors surrounded by polyethylene plastic for insulation. The transmission characteristics of this line are dependent on the wire diameter, conductor spacing, and dielectric constant of the insulation. The resistance of the copper causes signal attenuation. This attenuation is essentially constant for all frequencies. Unfortunately, as explained in Chapter 12, transmission lines

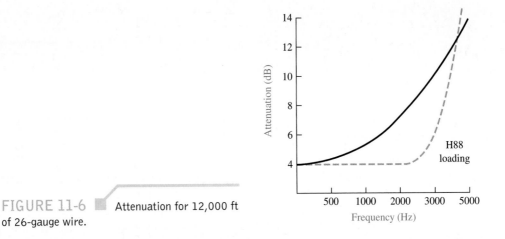

FIGURE 11-6 ■ Attenuation for 12,000 ft of 26-gauge wire.

have inductance and capacitance that have a frequency-dependent effect. A frequency versus attenuation curve for a typical twisted pair cable is shown in Fig. 11-6. The higher frequencies are obviously attenuated much more than the lower ones. This distortion can be very troublesome to digital signals since the pulses become very rounded and data errors (is it a 1 or a 0?) occur. This distortion is also troublesome to analog signals.

This higher-frequency attenuation can be greatly curtailed by adding inductance in series with the cable. The dashed curve in Fig. 11-6 is a typical frequency versus attenuation response for a cable that has had 88 mH added in series every 6000 ft. Notice that the response is nearly flat below 2 kHz. This type of cable, termed *loaded,* is universally used by the phone company. Loaded cable is denoted by the letters H, D, or B to indicate added inductance every 6000, 4500, or 3000 ft, respectively. Standard values of added inductance are 44, 88, or 135 mH. A twisted-pair cable with a 26 D 88 label indicates a 26-gauge wire with 88 mH added every 4500 ft.

Attenuation distortion is limited by requirements of the Federal Communications Commission (FCC). Their definition of *attenuation distortion* is the difference in gain at some frequency with respect to a reference tone of 1004 Hz. The basic telephone line (called a 3002 channel) specification is illustrated graphically in Fig. 11-7. As can be seen, from 500 to 2500 Hz, the signal cannot be more than 2 dB above the 1004-Hz level or 8 dB below the 1004-Hz level. From 300 to 500 Hz and 2500 to 3000 Hz the allowable limits are +3 dB and −12 dB. A subscriber can sometimes lease a better line if necessary. A commonly encountered "improved" line is designated as a C2 line. It has limits of +1 dB and −3 dB between 500 and 2800 Hz. From 300 to 500 Hz and 2800 to 3000 Hz the C2 limits are +2 dB and −6 dB.

Delay Distortion

A signal traveling down a transmission line experiences some delay from input to output. That is not normally a problem. Unfortunately, not all frequencies experience the same amount of delay. The *delay distortion* can be quite troublesome to data transmissions. The basic 3002 channel is given an *envelope delay* specification by the FCC of 1750 μs between 800 and 2600 Hz. This specifies that the delay between any two frequencies cannot exceed 1750 μs. The improved C2 channel is specified to be better than 500 μs from

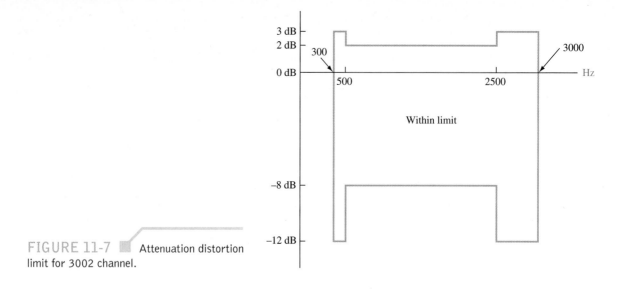

FIGURE 11-7 ■ Attenuation distortion limit for 3002 channel.

1000 to 2600 Hz, 1500 μs from 600 to 1000 Hz, and 3000 μs from 500 to 600 and 2600 to 2800 Hz.

The delay versus frequency characteristic for a typical phone line is shown with dashed lines in Fig. 11-8, while the characteristics after delay equalization are also provided. The *delay equalizer* is a complex *LC* filter that provides increased delay to those frequencies least delayed by the phone line, so that all frequencies arrive at nearly the same time.

11-3 THE UART

There is a gradual transition taking place in types of communications traffic. Formerly, it was mostly analog in nature, but we now are experiencing a more digital than analog situation. This transition has been caused by the increasing use of digital coding (i.e., PCM) for transmission of analog signals and the surging need for computers to talk between themselves and to remote terminals. The links used in these communications systems are

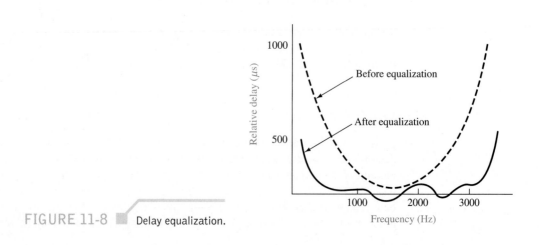

FIGURE 11-8 ■ Delay equalization.

often the standard voice-grade telephone lines or antenna to antenna systems, usually at microwave carrier frequencies.

Designers of the required equipment have been adapting rapidly to this change in traffic mix and the tremendous upsurge in volume with the following changes:

1. The use of computer control systems to find the unused portion of a multiplexed system to maximize use of available channels.
2. The application of digital switching theory to increase channel capacity.
3. Sharing of communications links by voice and data signals.

The use of microcomputers is occurring more frequently in even relatively small, portable-type communications equipment. These computers serve as network controllers; are used for frequency synthesis, diagnostic tools, digital filters, storage devices, and signal processors; and can be used automatically to make many of the adjustments normally required of an operator.

Telephone Line Transmission

The largest communications network in the world, by far, is the telephone system. It also turns out to be the most economical link for transmitting digital data when a large volume of data traffic between two locations does not exist. The use of telephone lines to carry digital data presents several problems, however, and therefore requires the use of complex processing circuitry. The problems arise for the following reasons:

1. The bandwidth of individual telephone lines is barely 3000 Hz, they do not pass dc, and they introduce considerable phase distortion and noise.
2. The data to be transmitted are often parallel data from a number of lines in a computer and therefore cannot be transmitted directly over a two-wire transmission line.

Because of limited bandwidth, the steep-sided, flat-topped pulses cannot be directly transmitted. The pulses must be used to modulate a carrier that the 3000-Hz bandwidth lines can handle. In addition, the parallel data format must be converted to serial form for transmission over two-wire lines.

LSI UART

The process for handling digital data for transmission is shown in block diagram form in Fig. 11-9. The computer outputs are accepted by the *universal asynchronous receiver/transmitter (UART)*. The UART, in simplified terms, converts the parallel computer data into the required serial data format. In Chapter 10 we considered UARTs as used in personal computer communications. In this section we'll look at the more general case in more detail. The UART's output is applied to a *modem,* which is an acronym for "modulator/demodulator." The serial pulses entering the modem are outputted as different frequency tones that are coupled to the telephone line either directly or by acoustic coupling with the telephone handset. At the receive end, the process is reversed. The system shown in Fig. 11-9 connects two computers, but many other schemes are possible. It may connect a computer to a computer terminal, two terminals, or a digital system of remote environmental sensors to a central monitoring computer, as examples.

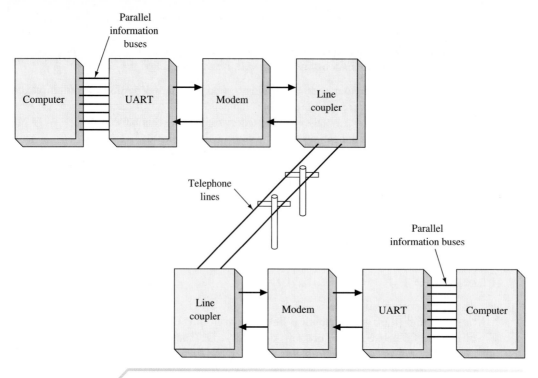

Computer data transmission system.

UART Operation

UARTs are available, in chip form, for around $10 and up depending on system complexity. A glance at the transmit section of a typical unit shown in Fig. 11-10 provides an indication of the complexities involved. In simplified fashion, the transmit function is achieved as follows:

1. The logic mode controls (pins 34 to 39) are applied either manually or by computer:
   a. *Pin 36:* This is the stop bit control. A logic 0 causes one stop bit to follow each serialized set of data bits in the UART output (pin 25). A logic 1 (+ 5 V) causes two stop bits to follow the data.
   b. *Pin 39:* This is the parity select control. *Parity* pertains to the use of a self-checking code in which the total number of 1's (or 0's) in each code expression is always odd or even. A 0 at pin 39 checks for odd parity, while a 1 checks even parity. If the number of bits per code expression is in error, a failure indication is provided at the receive UART.
   c. *Pin 35:* A 1 applied to this pin disables transmission of the parity bit and the receiver parity check and forces pin 13 to go to 0.
   d. *Pins 37 and 38:* These select the desired character length of 5, 6, 7, or 8 bits/character.
   e. *Pin 34:* A logic 1 causes the above-listed controls to be entered into the holding register.
2. After the above-mentioned controls are set up, the parallel data bits may be entered into pins 26 to 33. The least significant bit (LSB) is entered into pin 26 up to the most significant bit (MSB) at pin 33 if all 8 bits are used.

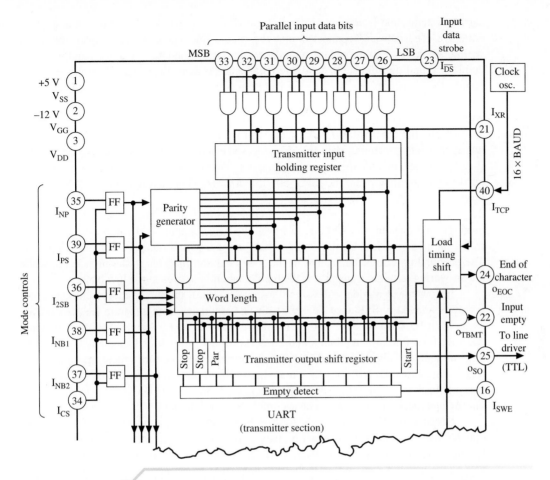

FIGURE 11-10 ■ UART transmit section.

3. After data entry of step 2, a negative-going pulse is sent to pin 23 by the computer after each character entry. It enters the parallel data into the input holding register.

4. The clock oscillator input at pin 40 allows the serial data output to occur at one data bit for every 16 clock pulses. In this way the transmission rate is controlled by the operator to suit the capabilities of the following modem and telephone lines.

5. The parallel data in the holding register are transferred to the transmitter output shift register, which actually performs the parallel-to-serial conversion.

The receiving section of the UART is similar to the transmitting section but functions in reverse order. The UART handles many complex operations on a single monolithic chip. It can be termed a microprocessor dedicated to a specific function—data handling.

11-4 MODEMS

The voice-frequency channels of the general switched telephone network are used extensively for the transmission of digital data. To use these channels, the data must be put in a form that can be sent over a limited bandwidth line. In voice-grade lines, transformers, carrier systems, and loading considerations attenuate all signals below 300 Hz and above 3400 Hz. The data are connected into the phone line via acoustic couplers or direct wiring. An *acoustic coupler* uses a cradle to support a standard telephone handpiece and employs sound transducers to send and receive audio tones through the telephone. These couplers offer portability and electric isolation from the telephone line but do provide a somewhat lower baud-rate capability than that of the more common practice of direct wiring into the phone line.

While the bandwidth from 300 to 3400 Hz is suitable for voice transmission, it is not appropriate for digital data transmission since pulse signals have many components (i.e., harmonics) outside this range. To transmit data via phone lines requires a conversion into a signal totally within the range 300 to 3400 Hz. This conversion is performed by a modem.

In full-duplex data transmission, frequency-division multiplexing (FDM) can be used for data rates up to 2400 b/s. In FDM, the voice channel is divided into two bands, called the high and low bands. One is used for sending and the other for receiving data. The originating terminal transmits in the low band and receives in the high band, while the answering terminal transmits in the high band and receives in the low band.

In low-speed modems (300-b/s transmission rate), the modulation technique commonly employed is FSK (see Sec. 9-7). In FSK modems, four separate frequencies are used: 1070 Hz for a 0 (space) in the low band, 1270 Hz for a 1 (mark) in the low band, 2025 Hz for a 0 in the high band, and 2225 Hz for a 1 in the high band. The transmitting modem takes the digital 1's and 0's from the terminal and converts them into the proper tones, which are then sent over the phone lines. The receiving modem takes the tones and converts back to 1's and 0's. Since four frequencies are used (FDM), simultaneous transmitting and receiving of data can be accomplished (full duplex).

Because of the phone line's limited bandwidth, FSK modems only work up to 600 b/s for full-duplex transmission. This is due to the frequency spectrum generated by the FSK. The faster the transmission, the wider the spectrum. There are 1200-b/s FSK modems, but they are only half-duplex—they can only send or receive data at that rate. Higher-speed modems use other modulation techniques. We will study a low- and a medium-speed modem and discuss high-speed modems in this section. The student should consult a data communication textbook for more detail on high-speed modems.

Modem Interfacing

The Electronic Industries Association (EIA) has developed a number of standards so that data communications equipment from all the manufacturers can interface with one another.* Similar international standards are provided by CCITT. The most commonly encountered standard is the RS (recommended standard) 232C specifications. These specifications identify the electrical, mechanical, and functional description of the interface. The RS-232C specifies that a 25-wire cable/connector be used for the transmission of digital data and control signals. The drivers for these signals are to be +5 to +15 V for mark and −5 to −15 V for space. The maximum cable length is usually 50 ft. Greater detail on this specification was provided in Chapter 10.

---

* These standards are all available from Electronic Industries Association, 2001 Eye Street, Washington, DC 20006.

Low-Speed Modem

The system shown in Fig. 11-11 is a direct-wired two-IC modem operating in full-duplex mode. Notice the *line-hybrid* transformer shown between the phone line and circuitry. It allows full-duplex operation by providing isolation between the transmit and receive legs of the system.

The transmit leg of the modem uses the XR-2206 IC. For optimum operation of the FSK transmitter it offers the following features:

1. Frequency stability of less than 50 ppm/°C.
2. A continuous phase output as the frequency keys from mark to space and back. This simplifies the required demodulator circuitry and prevents switching transients from being transmitted into telephone lines.
3. A rapid keying response such that the output switches frequencies within a half-cycle of the transmitting frequency.

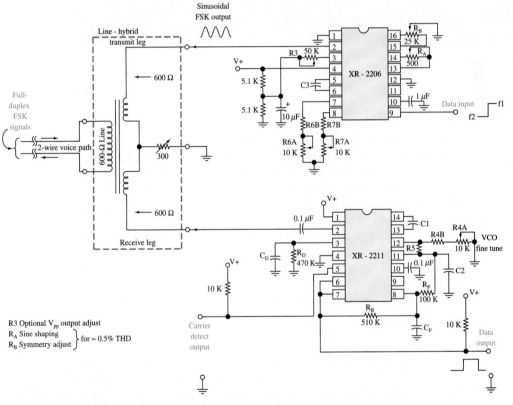

FSK band			XR-2206						XR-2211					
Baud rate	f_L	f_H	R_{6A}	R_{6B}	R_{7A}	R_{7B}	C_3	R_{4A}	R_{4B}	R_5	C_1	C_2	C_F	C_D
300	1070	1270	10	18	10	20	.039	10	18	100	0.039	0.01	0.005	0.05
300	2025	2225	10	16	10	18	.022	10	18	200	0.022	0.0047	0.005	0.05
1200	1200	2200	10	16	20	30	.022	10	18	30	0.027	0.01	.0022	0.01

Units: Frequency—Hz; Resistors—kΩ; Capacitors—μF

FIGURE 11-11 Full-duplex modem. (Reprinted with permission from *Electronic Design,* Vol. 27, No. 8; copyright Hayden Publishing Co., Inc., 1979.)

4. Independent mark and space frequency adjustments so that the output frequencies are adjustable via separate controls.

The frequency adjust for the XR-2206 is provided by the R_{6A} and R_{7A} potentiometers. The mark and space frequencies are determined by their values in conjunction with the C_3 capacitor value. The complete circuit gives sine-wave outputs with lower than 1% total harmonic distortion.

The FSK demodulator shown in Fig. 11-11 is the XR-2211. It provides independent bandwidth and center-frequency adjustment for operation over a wide range of transmission rates, channels, and bandwidths. Its output is logic-compatible, which eliminates the need to shape up the signal before application to the receiving digital system. The XR-2211 consists of a PLL that tracks the input signal within the pass band and quadrature-phase detector that operates on the carrier to tell when the unit is locked. It provides a carrier detect output logic level (pin 5) that can be used to indicate a detect problem or shut the system down.

The FSK XR-2211 demodulator provides an acceptable output with input signal levels from 2 mV to 3 V. To set the VCO center frequency, R_{4A} is adjusted in conjunction with C_1. It is normally set to fall midway between the mark and space frequencies of the FSK signal to be detected. The C_F and C_D capacitors are parts of the frequency shift keying and lock detection filters.

Medium-Speed Modem

In medium-speed, full-duplex modems (1200-b/s transmission rate), a different modulation technique is usually employed. Phase shift keying (PSK) uses one carrier frequency for the high band—2400 Hz—and one for the low band—1200 Hz—for sending and receiving data. For each carrier frequency (one for transmitting and one for receiving), one of four phase angles is used: 0, 90, 180, and 270°. The data are sent two bits at a time, or in *dibits*. Since there are four ways to send two bits at a time—00, 01, 10, or 11—each of the four phases represents one unique dibit. While the data rate is 1200 b/s, the baud rate (the rate at which information packets are sent) is 600 because two bits (a dibit) are sent in each packet. Again, 600 packets per second (600 baud and in this case, 1200 b/s) is the limit for transmitting full-duplex data over the general switched telephone network using frequency-division-multiplexing techniques with either FSK or PSK. More complex techniques allow high-speed (up to 28,800 b/s) modem operation, as discussed in the following section on high-speed modems.

Modems have often been stand-alone, add-on devices for most data communications equipment. The recent integration of the complex functions a modem performs into a single chip has allowed manufacturers to economically build the modem right into the microcomputer or data terminal equipment. An example of such a modem in block diagram form, the SC11004, is shown in Fig. 11-12. All of the signal processing functions needed for a full-duplex, 300/1200-b/s modem, including the FSK and PSK modulators and demodulators, are integrated on a single chip. It includes capabilities for call progress monitoring and for generating all required guard tones. It also includes analog loopback and remote digital loopback functions for self-testing (see Sec. 9-8). It operates in a synchronous or asynchronous mode and handles 8-, 9-, 10-, or 11-bit words.

Like all modems, the SC11004 needs a controller to determine the mode of operation, initiate the call to the remote modem (either pulse or DTMF tone dialing), set up the handshaking sequence with the remote modem, monitor the call progress tones on the line (ringing, busy, answer tone, and voice), and switch into the data mode. A simple

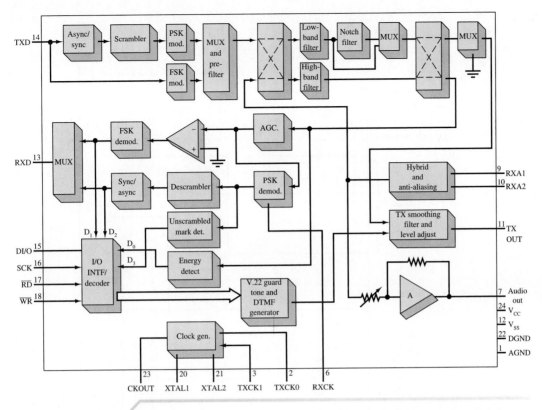

FIGURE 11-12 ▧ Single-chip-modem block diagram. (Courtesy of Sierra Semiconductor.)

four-line serial data interface is used to enable it to work with just about any 8-bit micro-controller or microprocessor. The control lines are Data Input/Output, Shift, Clock, Read, and Write. A complete modem/controller package is shown in Fig. 11-13, page 448. The SC11004 modem is paired with the SC11007 controller. The SC11007 handles all the modem control functions, as well as the interface to a system bus. Besides including an 8-bit microprocessor, 8K by 8 bytes of RAM, it contains the functionality of an 8250B UART. This greatly simplifies the interface to a parallel system bus such as used in IBM's PC. All of the communication software written for the PC will work with the system shown in Fig. 11-13.

Another version of the controller, the SC11008, is intended for RS-232C applications. It contains the same processor, memory, and UART as the SC11007 and has the same interface to the modem chip. The difference is that the UART is turned around so that serial data from the RS-232C port are converted to parallel data handled by the internal processor.

High-Speed Modems

High-speed modems usually use the complex M-ary phase shift keying schemes described in Sec. 10-8. As was described, the higher number phase-state schemes can cram more data into a given analog bandwidth. Some high-speed modems use the quadrature AM (QAM) system described in Sec. 10-8. As higher and higher data rates are

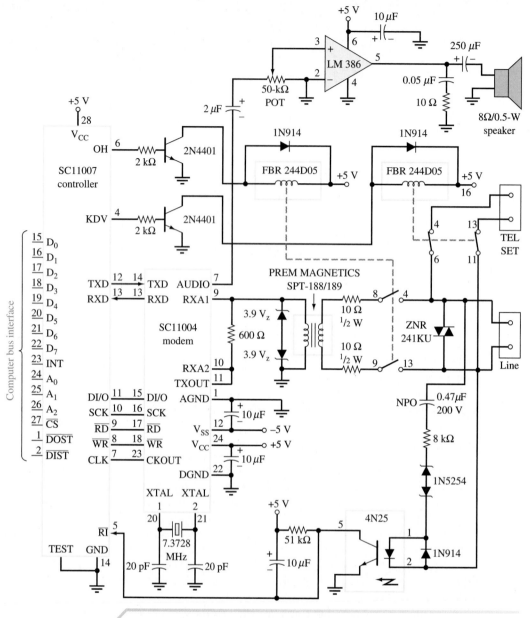

FIGURE 11-13 ▓ Modem and controller schematic. (Courtesy of Sierra Semiconductor.)

achieved, the possibility of error is increased. The error problems are particularly acute when using cellular phone lines (Sec. 11-5). All high-speed modems must use complex software systems for error detection and control, and the ones intended for cellular applications especially so.

The fastest modem speed that's widely supported by commercial and corporate networks is 14,400 bps, usually called 14.4, fourteen-four, or fourteen-dot-four. Modems have one speed for data and a lower speed for faxing. The 14.4 modems usually have a faxing speed of 9,600 bps.

The industry is starting to provide 28,800-bps modems, but at this point it is tough to find things that can communicate with you at that rate. Additionally, none of the on-line services can handle data at this speed, and when they can it's likely they'll charge a premium for faster service. By the time you read this 28,800 bps may be the norm, but as I write (May 1995) the changeover has not occurred.

Facsimile

Facsimile is the process whereby fixed graphic material such as pictures, drawings, or written material is scanned and converted to an electrical signal, transmitted, and after reception used to produce a likeness (facsimile) of the subject copy. It is roughly comparable to the transmission of a single TV frame except the output is reproduced on paper rather than on the face of a CRT. Facsimile has been used for years for the rapid transmission of photos to local newspapers by news services (e.g., AP Wirephoto) and aboard ships for reception of up-to-date weather charts or maps.

In recent years facsimile (*fax*) equipment has been designed to appeal to the industrial world. This equipment generally uses standard telephone lines as the communica-

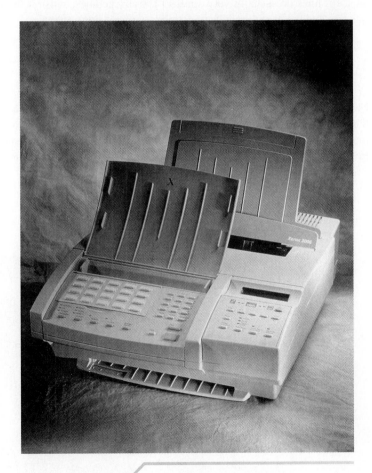

FIGURE 11-14 ▆ Facsimile unit. (Courtesy of The Document Company, Xerox Corp.)

tions link and is electrically or acoustically coupled by simply placing a telephone handset into a special acoustic coupler. This enables industry to rapidly send important business papers across town or around the world using standard telephone lines. As the cost and problems concerned with standard mail delivery escalate, it is not far-fetched to envision facsimile postal service.

A facsimile unit is shown in Fig. 11-14. The copy to be transmitted is scanned by a laser light source. Light reflected from the light and dark elements on each page is gathered by a solid-state photoreceptor. This signal is then used to frequency-modulate a low-frequency carrier before it is applied to the phone line by using a modem.

Standard phone lines have a very limited bandwidth of less than 3 kHz. Recall Hartley's law (Chapter 1), which states that the information transmitted is proportional to bandwidth times the time of transmission. Today's fax machines use *compression* techniques to shorten document transmission time. The receiver is programmed to reverse the data compression scheme. The basic compression technique uses the concept that documents often contain a large amount of blank (white) area. The fax transmits information that indicates how many consecutive "white" areas occur in the scanned line, thereby greatly reducing the time for transmission. This involves usage of computer memory at both transmitter and receiver. More advanced (faster) fax machines rely on the fact that adjacent lines are usually very similar. These machines transmit the difference between one line and the next to reduce transmission time even further.

11-5 CELLULAR TELEPHONE

Mobile telephone service originated in the late 1940s. It was never a widely used system, due to its limited frequency spectrum allocation and the high cost of the required equipment. Additional frequency spectrum was afforded by the FCC's reallocation of the 800- to 900-MHz band away from UHF TV in the mid-1970s. This allowed for the hundreds

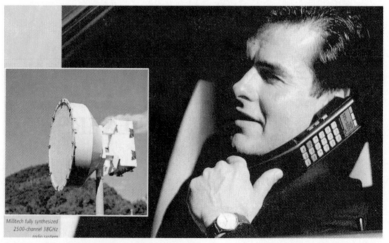

Cellular telephone system. (Courtesy of © 1994 by Millitech Corporation.)

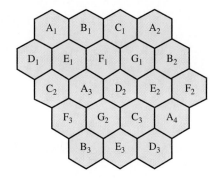

FIGURE 11-15 ▦ Cellular phone system layout.

of voice channels required for large-scale mobile service, while semiconductor advances and ingenious system design were able to bring costs under control.

The Bell Telephone Company developed the current system called Advanced Mobile Phone Service (AMPS). It is commonly referred to as cellular mobile radio for reasons that will soon be obvious. The cellular concept involves an essentially regular array of transmitter–receiver stations called *cell sites*. Figure 11-15 shows a cellular system with 21 "cells" being served by seven different channel groups. The cells cover the entire geographic area to be served by the system. The hexagon shape was chosen after detailed studies showed it led to the most cost-efficient and easily managed system. The two major concepts of cellular systems are *frequency reuse* and *cell splitting*.

Frequency Reuse

Frequency reuse is the process of using the same carrier frequency (channel) in different cells that are geographically separated. Power levels are kept low enough so that cochannel interference is not objectionable. Thus cell sites A_1 and A_2 in Fig. 11-15 use channels at the same frequency but have enough separation so as to not interfere with each other. Actually, this process is used in most radio services but not on the shrunken geographic scale of cellular telephone. Instead of covering an entire metropolitan area from a single transmitter site with high-power transceivers as the previous mobile phone systems had, the service provider distributes moderate-power systems at each cell site. Through frequency reuse, a cellular system can handle a number of simultaneous calls, greatly exceeding the number of allocated channels. The multiplier by which the system capacity exceeds allocated channels depends primarily on the number of cell sites.

Cell Splitting

A number of frequency channels are assigned to each cell in the system. This is called a channel set. Obviously, one channel is required for each phone call taking place at any one instant of time. If all traffic in a cell increases beyond reasonable capacity, a process called *cell splitting* is utilized. Figure 11-16 illustrates this process. An area from Fig. 11-15 is split into a number of cells that perhaps corresponds to the city's downtown area where phone traffic is heaviest. Successive stages of cell splitting would further increase the available call traffic, if necessary. The techniques of frequency reuse and cell splitting permit service to a large and growing area while using a relatively small frequency spectrum allocation.

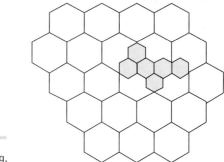

FIGURE 11-16 ▨ Cell splitting.

System Operation

The AMPS system uses frequency modulation with a peak deviation of 12-kHz and 30-kHz channel spacing. A duplex phone conversation requires a 30-kHz channel for transmitting and one for receiving. A typical metropolitan system with 666 duplex channels thereby requires a 40-MHz (30 kHz × 2 × 666) spectrum allocation.

The system includes a Mobile Telephone Switching Office (MTSO), the call sites, and the mobile units. The MTSO central processor controls the switching equipment needed to interconnect mobile users with the land telephone network. It also controls cell-site actions and many of the mobile unit actions through commands relayed to them by the cell sites. An 11-cell system is shown in Fig. 11-17, with the MTSO providing a link with two mobile users.

The MTSO is linked to each cell site with land telephone connections over which they exchange information necessary for processing calls. Each cell site contains one transceiver for each of its assigned voice channels and the transmitting and receiving antennas for those channels. The cell site also includes signal-level monitoring equipment and a "setup" radio, as explained subsequently.

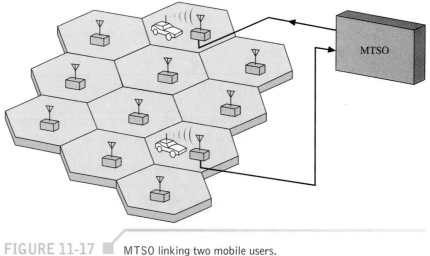

FIGURE 11-17 ▨ MTSO linking two mobile users.

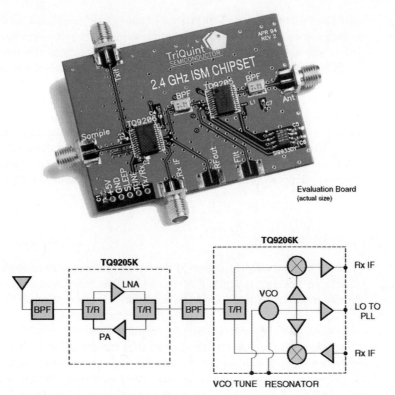

Chip set for cellular phone system. (Courtesy of TriQuint Semiconductor. Ad produced by Victor Imaharan, Marketing To Go.)

The mobile equipment includes a control unit, logic unit, transceiver, and antenna. The control unit contains the user interfaces—handset, dialer, and indicator lights. The transceiver includes a synthesizer to tune all allocated channels. The logic unit interprets customer actions and various system commands. Subsequently, it controls the transceiver and control units. One of the antennas is used for transmission, and both antennas are used for reception in a space-diversity configuration, as explained later in this section.

A few of the radio channels are used for "setup" and allow exchange of information needed to establish (setup) calls. Whenever a mobile unit is turned on but not in use, it is monitoring the setup channels. The mobile unit samples all received "setup" channels and selects the strongest one. Its current cell location has thereby been determined. It then synchronizes with the data stream being transmitted and interprets it. The setup channel data include identification numbers of mobile units to which calls are currently being directed.

When a mobile unit determines that it is being called, it samples the signal strength of all received "setup" channels again and responds through the cell site offering the strongest signal. It transmits its choice to that cell site, and that cell transmits back the voice channel assignment. The mobile unit tunes to the assigned channel and then receives a command to alert the mobile user (i.e., the telephone "rings"). The sequence of events is similar but reversed when the mobile user originates a call.

The system examines the call being received every few seconds at the cell site. When necessary, the system "looks" for another site to serve the call. If need for another site is determined, the system sends a command to the user to retune to a channel allocated to the new cell site. This process of changing channels is called "handoff." This procedure causes only a brief interruption of the conversation—typically, 50 ms. This handoff is not noticed by the user. A command signal from the base station causes the frequency synthesizer (under microprocessor control) in the phone to switch automatically to the carrier frequency of the new cell.

The power output is from 0.7 to 3 W. Its level is controlled by a power up/down signal from the base station in seven 1-dB steps. This is done to reduce interference with other phones and to minimize overloading of the base station receiver.

The means of speech transmission is standard (analog) narrowband FM using a 12.5- to 30-kHz bandwidth. Frequency-division multiple access separates the signals of different terminals. The second-generation cellular systems will use digital speech transmission.

Rayleigh Fading

The FM capture effect (Chapter 5) is very helpful in minimizing cochannel interference effects in cellular systems. Unfortunately, this benefit is considerably degraded by *Rayleigh fading*—a rapid variation in signal strength received by mobile units in urban environments. To maintain adequate signal strength during the fades, transmitter power must be increased by the "fading margin" of up to 20 dB.

The signal received by the mobile user can take many paths in an urban area. The signal reflects off buildings and many other obstructions. This multipath reception causes the signal to contain components from many different path lengths. The fading results because in some relative positions, phases of the signals arriving from the various paths interfere constructively, while in other positions the phases add destructively. The received signal in cellular systems varies by as much as 30 dB (Rayleigh fading) at a very rapid rate since the signal can go from one extreme to the other during a half-wavelength movement by the receiving vehicle. This does not take long for operation in the range 800 to 900 MHz, where a wavelength is about one-third of a meter!

The second-generation cellular system is currently under development to increase capacity and reduce interference. The current analog system suffers from congestion in major cities. Transferring to a digital system will at least triple capacity while creating cleaner channels with greater security and privacy.

It is unclear which channel accessing technique will become standard. One contender is time-division multiple access (TDMA), where *n* users share a single carrier by using the *n*th time slot for their information. The other system being considered is code-division multiple access (CDMA), where each phone uses a different binary sequence to modulate the carrier, spreading the spectrum of the waveform (spread spectrum). The signals are separated at the receiver by using a "correlator" that accepts only the signal from the selected binary sequence. Since there are currently strong proponents of TDMA and CDMA systems, they may both be used (in different geographic areas) as a means to make a final decision on the best system.

A related development is the *personal communication network* (PCN). Unlike cellular systems, the PCN is designed to operate independently as well as interface with the standard wired telephone network. The PCN will operate in the 1.9-GHz region. The radios are intended to be very small (shirt pocket or wrist). Transmitted power will be low, and therefore more base stations will be required than for cellular systems. PCNs are therefore termed *micro-cellular systems*. Each person will have a personal identification number (PIN), eliminating the need for separate home and work numbers.

Protocols

The vast number of data facilities now in existence require complex networks to allow equipment to "talk" to one another. To maintain order during the interchange of data, rules to control the process are necessary. Initially, procedures allowing orderly interchange between just a central computer and remote sites were needed. These rules and procedures were called *handshaking.* As the complexity of data communication systems increased, the need for something more than a "handshake" was necessitated. Thus developed sets of rules and regulations that are called *protocols.* A protocol may be defined as a set of rules designed to force the devices sharing a channel to observe orderly communication procedures.

Protocols have four major functions:

1. *Framing:* Data are normally transmitted in blocks, as indicated in the Chapter 9 section on error detection and control. The *framing* function deals with the separation of blocks into the information (text) and control sections. A maximum block size is dictated by the protocol. Each block will normally contain control information such as an *address field* to indicate the intended recipient(s) of the data and the block check character (BCC) for error detection. The protocol also prescribes the process for error correction when one is detected.

2. *Line control: Line control* is the procedure used to decide which device has permission to transmit at any given time. In a simple full-duplex system with just two devices, line control is obviously not necessary. However, systems with three or more devices (*multipoint circuits*) require line control.

3. *Flow control:* Often there is a limit on the rate with which a receiving device can accept data. A computer printer is a prime example of this condition. *Flow control* is the protocol process used to monitor and control these rates.

4. *Sequence control:* This is necessary for complex systems where a message must pass through numerous links before it reaches its final destination. *Sequence control* keeps message blocks from being lost or duplicated and ensures that they are received in the proper sequence. This has become an especially important consideration in packet-switching systems. They are introduced later in this section.

Protocols are responsible for integration of control characters within the data stream. Control characters are indicated by specific bit patterns that can occur in the data stream. To circumvent this problem, the protocol can use a process termed *character insertion.* This procedure is also called *character stuffing* or *bit stuffing.* If a control character sequence is detected in the data, the protocol causes an insertion of a bit or character that allows the receiving device to view the preceding sequence as valid data. This method of control character recognition is called *transparency.*

Protocols are classified according to their framing technique. Character-oriented protocols (COP) use specific binary characters to separate segments of the transmitted information frame. The most common COP is the binary synchronous communications (BISYNC) protocol introduced by IBM in 1968. Bit-oriented protocols (BOP) use frames made up of well-defined fields between 8-bit start and stop flags. A flag is fixed in both length and pattern: 01111110. The BOPs are variations of high-level data link control

(HDLC). They all have similar frame structures. They are therefore independent of codes, line configurations, and peripheral devices.

There are many protocol variations. They are not difficult to learn, but the magnitude of detail involved does not justify further study at this time. When you become involved with a specific protocol application, that will be the appropriate time to get into the specifics.

Network Organization

A data communications network includes the devices to be connected (often termed nodes) and the arrangement of facilities that interconnect them. A network may be simply the connection between two devices or be so complex as to include the millions of interconnections that make up the public telephone network. There are two basic network configurations. A *centralized network* is shown in Fig. 11-18. The other configuration is the *distributed network,* which is an interconnection of more than one centralized network. Notice that the centralized network has a single processing station. It controls all transmissions between point-to-point and multipoint lines. The point-to-point lines are dedicated to transmission between just two devices. The multipoint lines are shared by more than two devices under control of the central processing station. The central control may use time- or frequency-division multiplexing to allow multipoint operation.

The switching of digital data can be handled in one of three basic modes:

1. *Circuit switching:* This refers to a dedicated line for the entire communication. This is analogous to maintaining a voice telephone connection even during periods when no conversation is taking place. Once a call has been placed, the line is dedicated until the users agree to terminate it.

2. *Message switching:* This technique sends a message to a switching center, where it is forwarded and/or stored. The possibility of buffering (temporary storage) means that immediate connection is not required. Thereby, the data get forwarded when

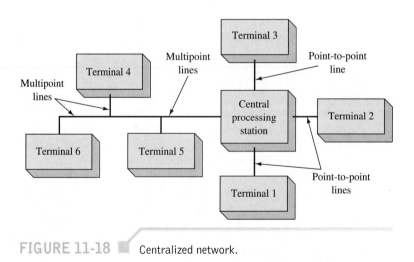

FIGURE 11-18 ▧ Centralized network.

the next circuit becomes available. This allows facilities to be shared by a number of users on a per message basis and provides more efficient network usage. The obvious disadvantage of message switching compared to circuit switching is the time delay associated with message storage. Real-time computer communication requires the dedicated line afforded by circuit switching or packet switching.

3. *Packet switching:* In this relatively recent development, data are divided up into small segments called *packets.* A typical packet size is 1000 bits. The individual packets of an overall message do not necessarily take the same path to a destination. They are held for very short periods of time at switching centers and are therefore transmitted in near-real time. The processors at the switching centers continuously monitor the packets from the standpoint of source, destination, and priority. The processors then direct each packet so as to use the network most efficiently. The process is termed *statistical concentration.* The price paid for the very high efficiency afforded and near-real-time transmission is the need for very complex protocols and switching arrangements. As packet switching technology progresses, the user may not need to choose in advance between packet and circuit switching; the techniques may merge. This could be the result of "fast" packet switches. The development of packet-switch systems operating at millions of packets per second could eliminate the need for circuit switching.

The ISDN

A worldwide digital network is now evolving. Telephone companies throughout the world are developing standards for this network, called the integrated-services digital network (ISDN). A key role in establishing international ISDN standards is being played by the International Telegraph and Telephone Consultive Committee (CCITT), a part of a specialized treaty agency of the United Nations. The goal is to provide all the required digital communication links for both voice and data through a single set of standardized interfaces.

For business, the primary attractions will be increased capability, flexibility, and decreasing cost. If one type of service—say, facsimile—is required in the morning and another in the afternoon—perhaps teleconferencing or computer links—it can easily shift back and forth. At present, the hookup for a given service might take a substantial amount of time to complete. With ISDN, new capacities will be available just by asking for them through a terminal.

The ISDN contains four major interface points as shown in Fig. 11-19. The R, S, T, and U interface partitions allow for a variety of equipment to be connected into the system. Type 1, or TE1, equipment includes digital telephones and terminals that comply with ISDN recommendations. Type 2, or TE2, gear is not compatible with ISDN specifications. It needs a terminal adapter to change the data to the ISDN's 64-kb/s B channel rate. The TE2 equipment interfaces the network via the R reference point.

The ISDN standards also define two network termination (NT) points. The NT_1 represents the telephone companies' network termination as viewed by the customer. NT_2 represents the termination of such things as local area networks (see Sec. 11-7) and private branch exchanges (PBXs).

The customer ties in with the ISDN's NT_1 point with the S interface. If an NT_2 termination also exists, an additional T reference point linking both NT_2 and NT_1 termina-

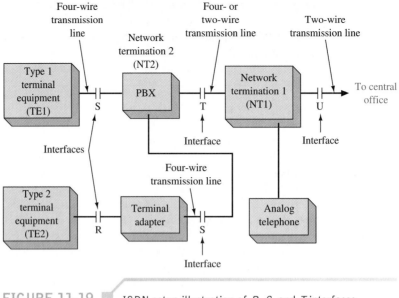

FIGURE 11-19 ■ ISDN setup illustration of *R, S,* and *T* interfaces.

tions will act as an interface. Otherwise, the *S* and *T* reference points are identical. The CCITT recommendations call for both *S* and *T* reference points to be four-wire synchronous interfaces that operate at a basic access rate of 192 kb/s. They are called the local loop. Reference point *U* links NT$_1$ points on either side of a pair of users over a two-wire 192-kb/s span. The two termination points are essentially the central office switches.

The ISDN specifications spell out a basic system as two *B* channels and one *D* channel (2*B* + *D*). The two *B* channels operate at 64 kb/s each while the *D* channel is at 16 kb/s for a total of 144 kb/s. The 48-kb/s difference between the basic 192-kb/s access rate and the 2*B* + *D* rate of 144 kb/s is mainly for containment of protocol signaling. The 2*B* + *D* channels are what the *S* and *T* four-wire reference points see. The *B* channels carry voice and data while the *D* channel handles signaling, low-rate packet data, and low-speed telemetry transmissions. Your power company may be doing its "meter reading" via the *D* channel in the near future.

The CCITT defines two types of communication channels from the ISDN central office to the user. They are shown in Fig. 11-20. The basic-access service is the 192-kb/s channel already discussed and will serve small installations. The primary-access channel will have a total overall data rate of 1.544 Mb/s and will serve installations with large data rates. This channel contains 23 64-kb/s *B* channels plus a 64-kb/s *D* channel. From any angle the potential for ISDN is enormous. The capabilities of worldwide communications are now taking a quantum leap forward.

11-7 LOCAL AREA NETWORKS

The dramatic decrease in computer system cost and increase in availability have led to an explosion in their usage. Organizations such as corporations, colleges, and government agencies have acquired large numbers of single-user computer systems. They may be

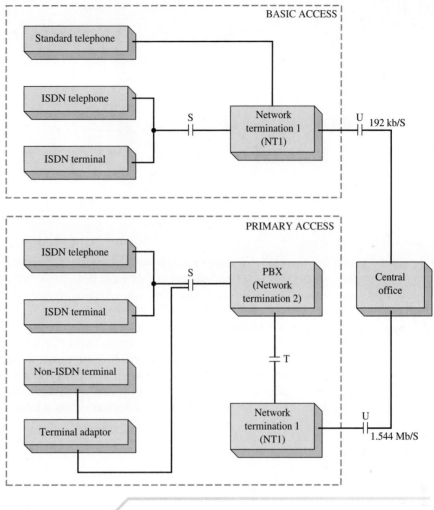

FIGURE 11-20 ■ Basic and primary access ISDN system.

dedicated to word processing, scientific computation, process control, etc. A means to interconnect these locally distributed computer networks soon became apparent. Interconnection allows the users to send messages to the other network members. It also allows resource sharing of expensive equipment such as high-quality graphics printers or access to a large mainframe computer to run programs too complicated for the local computer. The local computer is usually a personal micro-computer-type system. The network used to accomplish this is called a *local area network* (LAN). LANs are typically limited to separations of a mile or two and to several hundred users, but are usually smaller in scope.

One possible solution to the networking requirement is to tie each system through a modem into the organization's telephone system (PBX). This allows one system to talk to a number (or all) of the other systems simultaneously. It also limits the data rate to that of a phone line. Another approach is to connect each system to all others with cables. This

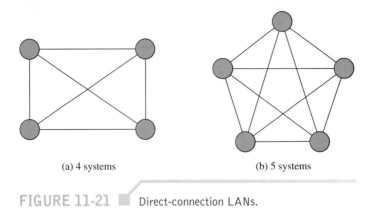

(a) 4 systems   (b) 5 systems

FIGURE 11-21 ■ Direct-connection LANs.

may be acceptable for small organizations. Four systems connected together are shown in Fig. 11-21. This is not so bad, as only six cable connections are necessary for the interconnections. The addition of just one more system is also shown, and now 10 interconnections are required. Mathematically, it can be shown that every time you add the nth system, $n - 1$ additional interconnections become necessary. This obviously becomes impractical when computer system size becomes larger than a few users.

The most common architectures for LANs are shown in Fig. 11-22. Another term used for architecture is the *topology* of the network. The star network is used when many users need access to a dominant central mainframe computer/controller. It is also the configuration for most PBX systems. The development of most LANs has been evolving into

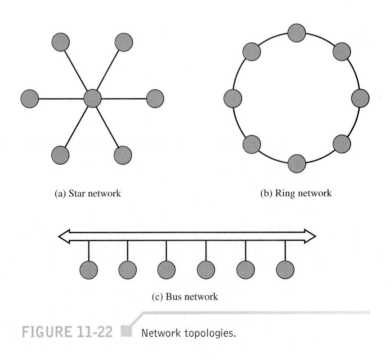

(a) Star network   (b) Ring network

(c) Bus network

FIGURE 11-22 ■ Network topologies.

the use of either the ring or bus networks shown in Fig. 11-22. They are decentralized systems not relying on a central computer to maintain operation. Most LANs use these topologies. The cables used are either 50-Ω coaxial or fiber optic links.

The transmission of information on these systems is via baseband or via a modulation scheme. *Baseband* is the system where the digital signals are placed directly on the coaxial cable. Transmission of voice at the *base* frequency (audio in this case) is another example of baseband operation. If the baseband is used to modulate a carrier, it is referred to as a *broadband system.* This permits multiplexing different frequency channels (FDM) and allows digital data and voice/video signals to be transmitted simultaneously. These broadband LANs are obviously more complicated than the baseband systems.

Another important characteristic of a LAN is the type of *channel access* used. Channel access defines how a user gets control of the channel so as to allow transmission. The channel access methods in common use are *carrier sense, multiple access with collision detection* (CSMA/CD), and *token passing.*

CSMA/CD is used by most baseband systems. In it a user that wishes to transmit "listens" to see if anyone else is using the line. If no one is, transmission begins. This is fine unless another station tries to begin transmission at the same time. In that case a "collision" occurs. Both stations then stop transmitting for a random period of time before attempting another transmission.

The token-passing technique is well suited to the ring network topology. An electrical "token" is placed in the channel and circulates around the ring. If a user wishes to transmit, the station must wait until possession of the token exists. Each station is assured access for transmission of its messages. This is an advantage over CSMA/CD systems, where heavy traffic conditions can actually cause throughput to decrease due to excessive collisions and the resulting delays. A disadvantage of this system is that if an error changes the token pattern, it causes the token to stop circulating. Additionally, ring networks rely on each system to relay all data to the next user. A failed station causes data traffic to cease. Solutions to these reliability problems have been developed, but the complexities involved have led to the greatest popularity for the bus topology.

There are an extensive number of LAN systems currently available. Many are applicable to specific manufacturer's equipment only. The Institute of Electrical and Electronic Engineers (IEEE) standards board in 1983 approved LAN standards. These IEEE 802 standards provide impetus for different manufacturers to use the same codes, signal levels, etc. The most widely used LAN is now presented to provide an operational view of LANs.

Ethernet LAN

Ethernet is a baseband, bus topology, CSMA/CD local area network system. It originated in 1972, and the full specification was provided via a joint effort between Xerox, Digital Equipment Corporation, and Intel in 1980. It is a very high data rate system. The time for each bit in this baseband system is 100 ns, which translates into a 10-Mb/s data rate. The 0's and 1's use the format called *Manchester encoding,* introduced in Sect. 9-7. The 100-ns period always has a state transition in the center, as shown in Fig. 11-23. All 1's are indicated by a positive and 0's by a negative, center transition. The end of each interval goes to whatever level is necessary to allow for the correct transition at the next mid-interval point. This encoding process guarantees a transition during each bit interval. Each data element therefore provides timing (clock) information, and a 100-ns interval without a transition indicates that transmission has ended or an error has occurred.

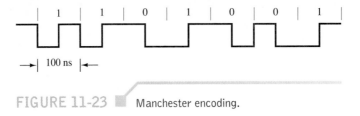

FIGURE 11-23 ■ Manchester encoding.

A simple Ethernet system is shown in Fig. 11-24. The 50-Ω coaxial cable has color bands every 2.5 m that indicate an acceptable point to tap into the cable. These taps are specially constructed and consist of two probes. One contacts the center conductor, and the other probe contacts the shield (ground). These taps do not require cable cutting when adding a new station.

An Ethernet system is allowed to have three segments. The system in Fig. 11-24 has two segments. Each segment can be up to 500 m long and can contain a maximum of 100 stations. The segments are connected via amplifier repeaters. No more than two repeaters can exist between two stations. Notice the use of 50-Ω terminators to prevent reflections on the 50-Ω coaxial cable.

The controller/transceiver block between each station and tap in Fig. 11-24 performs a number of functions. Figure 11-25 provides additional detail on this function. Two LSI circuits are usually used: the local area network controller for Ethernet

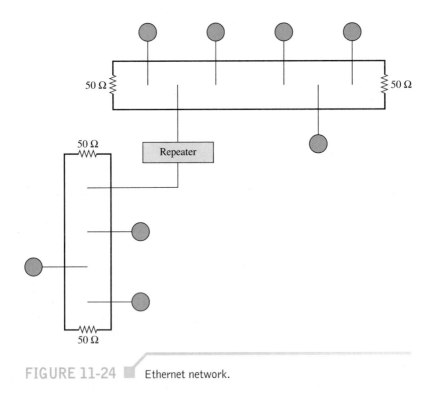

FIGURE 11-24 ■ Ethernet network.

(LANCE) and a serial interface adapter (SIA). The LANCE is a 48-pin, microprogrammed intelligent device that performs parallel-to-serial and serial-to-parallel data conversions for transmission and reception, respectively. It also handles buffer linking and management, interrupt generation, several types of frame checking and error detection, and two types of looping self-testing. Using its full 24-bit linear address space allows it to interface with most popular 16-bit processors that may be used in the user station's computer. The CSMA/CD channel management and data packetization are thereby handled by the LANCE.

The SIA mentioned in the previous paragraph encodes and decodes between Manchester format and 1's/0's. Additionally, the SIA's 20-MHz crystal-controlled oscillator supplies a 10-MHz clock signal to synchronize data transfers.

IEEE-488

The IEEE-488 standard can be loosely considered to be a LAN. It is also referred to as the General-Purpose Interface Bus (GPIB). It was created for computerized test stations that gather large amounts of data from many sources. That application was introduced in Sec. 10-6. It can also be used to connect computers for information sharing as with Ethernet, but in limited fashion. Up to 15 stations may be interconnected on one line, which may be up to 20 m long.

An illustrative IEEE-488 system is shown in Fig. 11-25. Data are carried asynchronously by 16 lines in bit-parallel, byte-serial form, bidirectionally. Eight parallel lines carry data at up to 1 Mb/s. Five control wires and three handshaking wires make up the other eight lines. The three handshaking lines allow an asynchronous transfer of each 8-bit character, with each listener acknowledging receipt before the next is sent. This allows devices of different speed capabilities to be used in the IEEE-488 system. The five control wires contain status signals sent between the controller and all other devices on the line to indicate when a transmission is needed and possible.

Three classes of equipment are shown in Fig. 11-25: talkers, listeners, and controllers. A frequency counter is a good example of a talker. It reads the frequency of a signal and transmits the information to the network. An example of a listener is a digitally controlled frequency synthesizer. It gets instructions from the network and generates signals of the requested frequency and power output. Modulated outputs are possible if so requested. A digital multimeter functions as a talker–listener. It listens for instructions such as dc volts, 20-V scale, and negative polarity. It then talks back with the reading it makes. The controller makes the decision as to what is to be done based upon human input via a program.

The SN series ICs shown in Fig. 11-25 perform the transceiver interface function. The SN75160A is used for all data bus interconnections between the data bus and each instrument. It transmits in either direction under the controller's direction. The SN75161A is used to implement the eight-line control bus (five control, three handshaking) functions. The SN75162 implements control bus operation for controller instruments.

The control and handshaking functions required to allow IEEE-488 system operation are rather detailed. Careful study of the lengthy standard is necessary for complete understanding. Many instruments from many manufacturers are available with the IEEE-488 interface built in. The ability to design and implement complex test stations is thereby greatly facilitated.

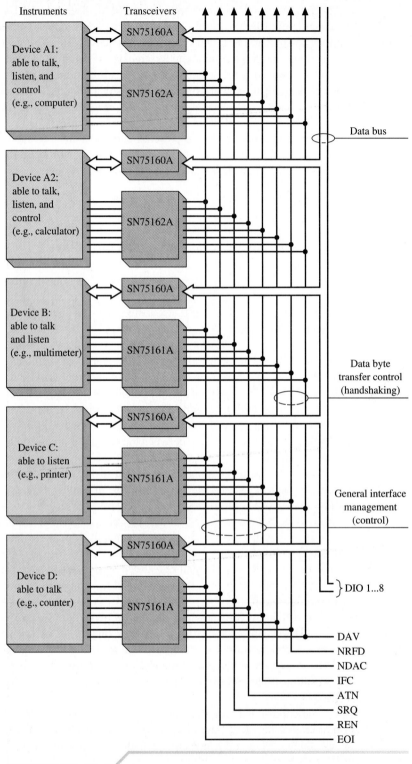

FIGURE 11-25 IEEE-488 network. (Courtesy of *Microwaves and RF,* February 1984.)

11-8 LAN INTERCONNECTION

The utility of LANs led to the desire to connect two (or more) together. For instance, a large corporation may have had separate networks for its research and engineering and for its manufacturing units. Typically, these two systems used totally different technologies, but it was deemed necessary to "tie" them together. This led to *metropolitan area networks* (MAN)—two or more LANs linked together within a limited geographical area. Once the techniques were in place to do this, it was decided that it would be helpful to link the MAN with the marketing division on the other side of the country. Now two or more LANs were linked together over a wide geographical area, resulting in a *wide area network* (WAN).

To allow different types of networks to be linked together, an *open systems interconnection* (OSI) reference model was developed by the International Organization for Standardization. It contains seven layers, as shown in Fig. 11-26. It provides for everything from the actual physical network interface to software applications interfaces.

1. *Physical layer:* provides the electrical connection to the network. It doesn't speak to the modulation or physical medium used.
2. *Data link layer:* handles error recovery, flow control (synchronization), and sequencing (which terminals are sending and which are receiving). It is considered the "media access control layer."
3. *Network layer:* accepts outgoing messages and combines messages or segments into packets, adding a header that includes routing information. It acts as the network controller.
4. *Transport layer:* is concerned with message integrity between the source and destination. It also segments/reassembles (the packets) and handles flow control.
5. *Session layer:* provides the control functions necessary to establish, manage, and terminate the connections as required to satisfy the user request.
6. *Presentation layer:* accepts and structures the messages for the application. It translates the message from one code to another if necessary.
7. *Application layer:* logs the message in, interprets the request, and determines what information is needed to support the request.

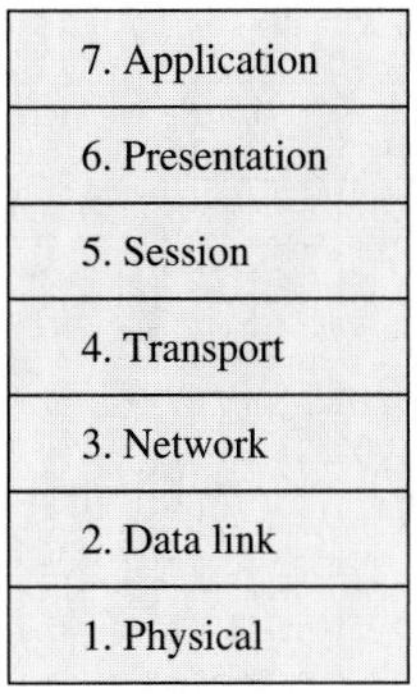

7. Application
6. Presentation
5. Session
4. Transport
3. Network
2. Data link
1. Physical

FIGURE 11-26 ▓ OSI reference model.

Interconnecting LANs

The interconnection of two or more LANs (into a MAN or WAN) is accomplished in a number of ways, depending on the LAN similarities.

Bridges: Bridges use only the bottom two OSI layers to link LANs that usually have identical protocols at the physical and data link layers. This is illustrated in Fig. 11-27.

Routers: Routers interconnect LANs running identical internetwork protocols by using the bottom three OSI layers, as shown in Fig. 11-28. They manage traffic congestion by employing a flow control mechanism to direct traffic to alternative paths.

Gateways: When connected, LANs employ different high-level protocols, a gateway is used, as shown in Fig. 11-29. A gateway encompasses all seven OSI layers. It interconnects two networks that use different protocols and formats.

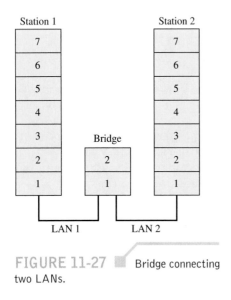

FIGURE 11-27 ▓ Bridge connecting two LANs.

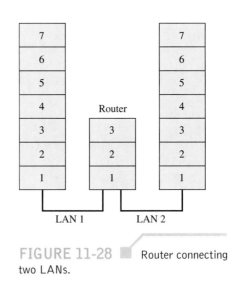

FIGURE 11-28 ▓ Router connecting two LANs.

FIGURE 11-29 ▓ Gateway connecting two LANs.

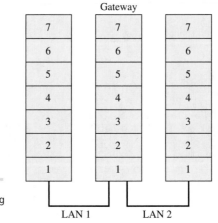

TROUBLESHOOTING

Local area networks (LANs) are finding their way into every kind of business at an ever-increasing rate. From small offices to very large government agencies, LANs are becoming an indispensable part of business communications. Computer data and audio and video information are shared on LANs every day. In this section, you will be introduced to a typical LAN configuration and to some common LAN problems. Opportunities abound in LAN technology to those who are willing to prepare with specialized training. LAN seminars, community college classes, and hands-on training will help prepare you for this technology.

After completing this section you should be able to

- Describe a typical LAN configuration
- Define near end crosstalk interference
- Describe two common problems when using twisted pair
- Name the two types of modular eight connectors and their proper use

LAN Configuration

A typical LAN configuration today is an IBM Token ring network. As illustrated in Fig. 11-30, the main backbone uses fiber-optic cable (see Chapter 17), and the hub-to-workstation connections use unshielded twisted pair (UTP). Network interface cards (NICs) are installed in each workstation. The workstation is usually a personal computer (PC) and has specialized software installed on it to be able to communicate over the LAN. The

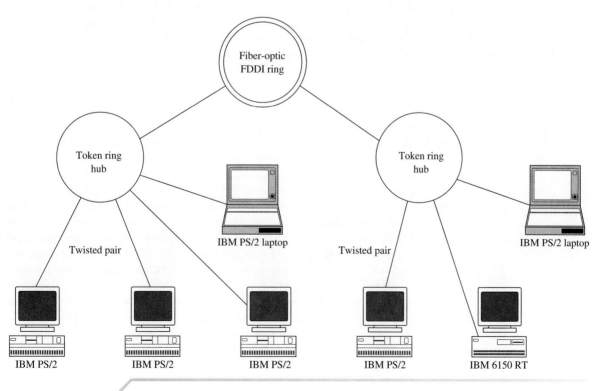

FIGURE 11-30 ■ Typical LAN configuration.

peer-to-peer LAN consists of a few users; the client-server LAN can have hundreds of users. Small or large, LANs suffer from similar troubles. Since the crux of the network is cabling, in one form or another, this is the weak area for most LANs. As a matter of fact, a major percentage of cataloged LAN failures are related to the media—the wiring.

Troubleshooting a LAN

The most common maintenance situation for larger LAN installations is to set up a help desk. LAN users who experience problems call the help desk. Usually a technician is dispatched if the problem can't be resolved over the phone. Let's take a look at some of the problems the technician may encounter when dispatched.

Some preliminary checks should be made first. Ensure that the workstation is plugged into electrical power. Is it turned on? Is the CRT brightness turned down? These are obvious things that are easily overlooked and warrant a look. The network interface card should be checked for proper installation in the PC. Check the hub-to-workstation connection. Is the user's account set up properly on the server? Sometimes passwords and accounts get deleted by accident. Check the user's LAN connection software for proper boot-up. This software can become corrupted and may need to be reinstalled on the workstation.

TROUBLESHOOTING CABLE-BASED NETWORKS   Networks that use coaxial cable as the installed base can cause unique problems. Usually this type of network exists as a bus topology, and a fault anywhere along the bus can affect all the workstations on it. Some common faults associated with coaxial LANs include broken or missing terminators, improperly disconnected T junctions, and damaged cables. A missing terminator from a coax link will shut down all the workstations on that link. A properly seated T junction on the network interface card connector will guarantee a good LAN connection. Coaxial cables often get pinched, crimped, and stretched, any of which can cause trouble for the workstation and possibly all the workstations on that link.

TROUBLESHOOTING UNSHIELDED TWISTED-PAIR NETWORKS   Twisted-pair networks represent a different challenge to the troubleshooter than coax. Two common problems that occur with unshielded twisted-pair wiring are that the pairs become crossed or split up. Both conditions produce data signal degeneration. Near end crosstalk (NEXT) is generated from split pairs. NEXT stems from interference between the twisted pairs. Let's use the example of a signal being transmitted from the workstation to the hub. The signal is smallest (maximum attenuation) at the hub. A transmitted signal originating at the hub—a strong signal—will feed over into the attenuated weak signal. Preventing crossed and split pairs is the best insurance against NEXT. Crossed pairs are not difficult to find but usually require a certain amount of wire tracing and continuity checking. Split pairs are more difficult to find, and special test instruments should be used to track the splits down. A LAN cable meter will have several specialized test functions, and the mis-wired feature will be one of them.

Another common problem that happens with unshielded twisted-pair wiring is that the wrong kind of connector is used. Stranded copper conductors need a piercing-type modular eight connector, and solid core conductors use a connector that straddles the wire. Since both types of connectors look alike they can easily be mixed up and often are. When the wrong connector is used, the result is an open or an intermittent connection. Use care in replacing connectors. Keep these connectors in separate parts in clearly labeled bins.

SOME CABLING TIPS   With a little extra precaution, many LAN problems can be eliminated. Be sure to keep wiring links short without stretching the wire tight. Coax or twisted pair should never be placed near ac power lines or other noise sources. Install your cable base carefully. Wiring runs should be dry. Moisture causes corrosion over a period of time. Use only good-quality connectors. Never untwist more twists than necessary when making twisted-pair connections. Finally, keep good-quality wiring diagrams of the network installation.

SUMMARY

In Chapter 11 we studied the various networks encountered in digital and analog communications. These include the telephone and cellular networks and local area networks. The major topics the student should now understand include:

- the basis of telephone operation, including definitions of tip, ring, trunk, PBX, DTMF, BORSCHT function, and T1 line

- the line quality considerations, including the effects of attenuation and signal delay distortion

- the application of a UART in telephone data transmission

- the analysis of various speed modems with description of modulation methods

- the analysis of cellular telephone systems, including frequency reuse, cell splitting, and Rayleigh fading

- the description of network protocol functions, including framing, line control, flow control, and sequence control

- the explanation of switching modes for digital data, including circuit switching, message switching, and packet switching

- the explanation of the integrated-services digital network (ISDN)

- the description of local area network (LAN) topologies

- the description of the OSI seven-layer reference model and definitions of internetworking devices such as bridges, routers, and gateways

QUESTIONS AND PROBLEMS

SECTION 1

1. Describe the basic limitation of using the telephone system in computer communication.

SECTION 2

2. List the three signal levels that telephone circuitry must work with.
3. Describe the sequence of events taking place when a telephone call is initiated through to its completion.

4. Transcribe your phone number into the two possible electrical signals commonly used in the phone system.
5. What is a PBX and what is its function?
6. List the BORSCHT functions.
7. Define *attenuation distortion* and explain its causes.
8. Define *loaded cable.*
9. What is a C2 line? List its specification.
10. Define *delay distortion* and explain its causes.
11. Describe the envelope delay specification for a 3002 channel.
12. Explain the need for delay equalization when phone lines are used for signal transmission. If uncorrected, what would be the result of unequal delays to the different frequency components of a received signal?

SECTION 3

13. Explain some possible applications for microprocessors in communications transceivers.
14. Why are telephone lines widely used for transmission of digital data? Explain the problems involved with their use.
15. Describe the function of each block of the computer data transmission system shown in Fig. 11-9. This description should include the functions performed by modems, UARTs, and line couplers.

SECTION 4

16. Describe the difference between a half-duplex and full-duplex communications system.
17. Describe the function of an acoustic coupler and an alternative to its use.
18. Explain the general function of a modem and the function of recommended standards for interfacing them.
19. Provide a brief operational description of the full-duplex modem in Fig. 11-11.
20. Discuss the operation of a typical medium-speed modem. Include PSK and dibits in your discussion.
21. Briefly describe the function/operation of the modem represented in Figs. 11-12 and 11-13.
22. Research high-speed modems in your library and write a paper on the current state-of-the-art. Include information on maximum speeds and modulation techniques.
23. In general terms, explain the facsimile process.
24. A standard TV broadcast transmits a full picture in $\frac{1}{30}$ s. Why does a typical facsimile transmission require about half of a minute?

SECTION 5

25. Redesign the cellular system in Fig. 11-15 such that only six different channel groups (*A, B, C, D, E, F*) are used instead of the seven shown.
26. Describe the concepts of frequency reuse and cell splitting.
27. Design the split-cell system in Fig. 11-16 so that the minimum number of channel groups are used.
28. Describe the sequence of events when a mobile user makes a call to a land-based phone. Include in your description the handoff once the call has been made.

29. The cellular system in Fig. 11-17 requires splitting of the two cells surrounded by other cells. They need to be split into five cells (from the original two). Determine the minimum number of channel sets that can serve the original and new systems.

30. Describe Rayleigh fading and explain how to minimize its effects.

31. A cellular system operates at 840 MHz. Calculate the Rayleigh fading rate for a mobile user traveling at 40 mph (100 fades per second).

32. A mobile user is transmitting two steps above minimum power. Calculate its power output. (1.11 W)

33. Explain TDMA and CDMA as related to future cellular systems.

34. Describe the proposed personal communication network (PCN).

SECTION 6

35. Can handshaking be called a protocol? What is the differentiation between these two concepts?

36. List and briefly describe the four major functions of a protocol.

37. Explain the integration of control characters within the data stream by a protocol. Include character insertion, bit stuffing, and transparency in your explanation.

38. Draw a block diagram for a system that has a distributed network organization. Explain the difference between centralized and distributed networks.

39. List the three basic techniques for switching digital data and provide brief explanations of each.

40. Explain the objective of the ISDN and briefly explain its organization with the help of Figs. 11-19 and 11-20.

SECTION 7

41. Provide a general description of a local area network.

42. A direct connection system LAN for 11 computers is planned. Calculate the number of required connections.

43. List the basic topologies available for LANs and explain them.

44. What is the difference between baseband and modulated LANs? Provide an advantage of each compared to the other.

45. What is channel access? Describe the two basic methods for channel access.

46. Provide a brief introduction to the Ethernet LAN. Describe the Manchester coding system it uses.

47. Determine the maximum possible number of stations in a basic Ethernet system and the maximum distance between two stations.

48. Provide a brief introduction to the IEEE-488 system. How does its basic objective differ from the Ethernet LAN?

49. List the three basic classes of equipment connected by the IEEE-488 system.

50. Draw a block diagram of an IEEE-488 LAN that would be useful for some of your laboratory work.

SECTION 8

51. Describe the differences between LANs, MANs, and WANs.

52. Provide a brief description of the functions addressed by the OSI reference model.

53. Explain the function of bridges, routers, and gateways and indicate how they differ.

TRANSMISSION LINES

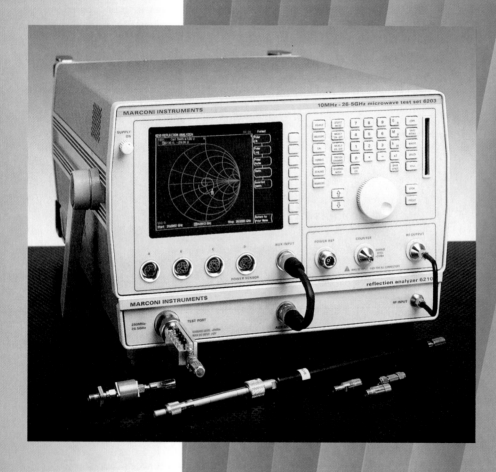

OBJECTIVES

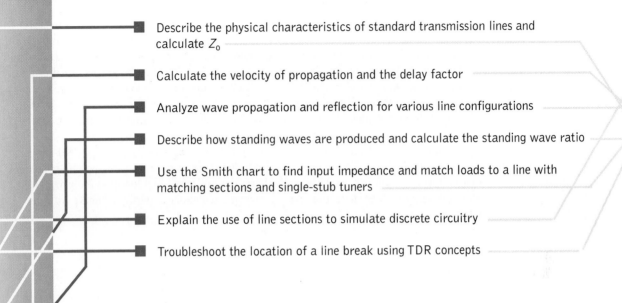

- Describe the physical characteristics of standard transmission lines and calculate Z_0

- Calculate the velocity of propagation and the delay factor

- Analyze wave propagation and reflection for various line configurations

- Describe how standing waves are produced and calculate the standing wave ratio

- Use the Smith chart to find input impedance and match loads to a line with matching sections and single-stub tuners

- Explain the use of line sections to simulate discrete circuitry

- Troubleshoot the location of a line break using TDR concepts

12-1 INTRODUCTION

In previous chapters we have been concerned with the generation and reception of communications signals. In Chapters 12 to 15 we shall learn of the methods of getting these signals from transmitter to receiver. Transmission may take place via transmission lines, antennas, waveguides, or optical fibers. Sometimes a combination such as transmission line from transmitter to its antenna, to receiving antenna, to transmission line, and to receiver is used. A *transmission line* may be defined as the conductive connections between system elements that carry signal power. You may be thinking that if the wire connection between two points is a transmission line, why is a whole chapter of study required? It turns out that at very high frequencies even simple wire connections start behaving in a peculiar fashion. What appears to be a short circuit may no longer be one. Or, energy sent down the wire is reflected back. It is these phenomena and others that form the basis of this chapter. A good understanding of the material presented in this chapter is a necessary prerequisite for the antenna and waveguide chapters to follow.

Test set for time-domain reflectometry. (Courtesy of Marconi Instruments, Inc.)

Two-Wire Open Line

One type of parallel line is the two-wire open line illustrated in Fig. 12-1. This line consists of two wires that are generally spaced from $\frac{1}{4}$ to 6 in apart. It is sometimes used as a transmission line between antenna and transmitter or antenna and receiver. An advantage of this type of line is its simple construction. The principal disadvantages are the high radiation losses and noise pickup due to the lack of shielding. Radiation losses are caused by the changing fields that are produced by the changing currents and voltages in each conductor. Some of these lines of force will be radiated from the transmission line in much the same manner that energy is radiated from an antenna.

Another type of parallel line is the twin lead or two-wire ribbon type. This line is illustrated in Fig. 12-2. This line is essentially the same as the two-wire open line, except that uniform spacing is assured by embedding the two wires in a low-loss dielectric, usually polyethylene. The dielectric space between conductors is partly air and partly polyethylene.

Twisted Pair

The twisted-pair transmission line is illustrated in Fig. 12-3. As the name implies, the line consists of two insulated wires twisted to form a flexible line without the use of spacers. It is not used for high frequencies due to the high losses that occur in the rubber insulation. When the line is wet, the losses increase greatly. Local area networks (LANs) are often wired using twisted pair.

Shielded Pair

The shielded pair, shown in Fig. 12-4, consists of parallel conductors separated from each other, and surrounded by, a solid dielectric. The conductors are contained within a copper braid tubing that acts as a shield. The assembly is covered with a rubber or flexible composition coating to protect the line from moisture or mechanical damage.

The principal advantage of the shielded pair is that the conductors are balanced to ground; that is, the capacitance between the cables is uniform throughout the length of the line. This balance is due to the grounded shield that surrounds the conductors with a

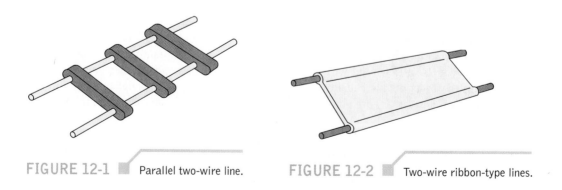

FIGURE 12-1 ▌ Parallel two-wire line.   FIGURE 12-2 ▌ Two-wire ribbon-type lines.

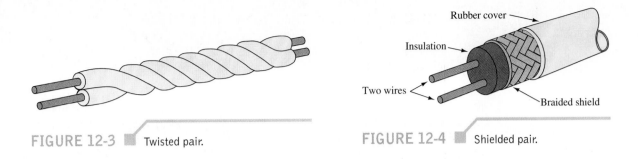

FIGURE 12-3 ▌ Twisted pair.

FIGURE 12-4 ▌ Shielded pair.

uniform spacing along their entire length. The copper braid shield isolates the conductors from external noise pickup. It also prevents the signal on the cable from radiating to and interfering with other systems.

Coaxial Lines

There are two types of coaxial lines: the rigid or air coaxial line and the flexible or solid coaxial line. The electrical configuration of both types is the same; each contains two concentric conductors.

The rigid air coaxial line consists of a wire mounted inside of, and coaxially with, a tubular outer conductor. This line is shown in Fig. 12-5. In some applications the inner conductor is also tubular. The inner conductor is insulated from the outer conductor by insulating spacers, or beads, at regular intervals. The spacers are made of Pyrex, polystyrene, or some other material possessing good insulating characteristics and low loss at high frequencies.

The chief advantage of this type of line is its ability to minimize radiation losses. The electric and magnetic fields in the two-wire parallel line extend into space for relatively great distances, and radiation losses occur. No electric or magnetic fields extend outside of the outer (grounded) conductor in a coaxial line. The fields are confined to the space between the two conductors; thus, the coaxial line is a perfectly shielded line. Noise pickup from other lines is also prevented.

This line has several disadvantages: It is expensive to construct, it must be kept dry to prevent excessive leakage between the two conductors, and although high-frequency losses are somewhat less than in previously mentioned lines, they are still excessive enough to limit the practical length of the line.

FIGURE 12-5 ▌ Air coaxial: cable with washer insulator.

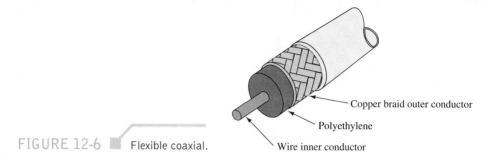

FIGURE 12-6 Flexible coaxial.

Copper braid outer conductor

Polyethylene

Wire inner conductor

The condensation of moisture is prevented in some applications by the use of an inert gas, such as nitrogen, helium, or argon, pumped into the line at a pressure of from 3 to 35 psi. The inert gas is used to dry the line when it is first installed, and a pressure is maintained to ensure that no moisture enters the line.

Concentric cables are also made, with the inner conductor consisting of flexible wire insulated from the outer conductor by a solid, continuous insulating material. Flexibility may be gained if the outer conductor is made of braided wire. Early attempts at obtaining flexibility employed the use of rubber insulators between the two conductors. The use of rubber insulators caused excessive losses at high frequencies, while the bead arrangement allowed moisture-carrying air to enter the line, resulting in high leakage current and arc-over when high voltages were applied. These problems were solved by the development of polyethylene plastic, a solid substance that remains flexible over a wide range of temperatures. A coaxial line with a polyethylene spacer is shown in Fig. 12-6. Polyethylene is unaffected by seawater, gasoline, oils, and liquids. High-frequency losses due to the use of polyethylene, although greater than the losses would be if air were used, are lower than the losses resulting from the use of most other practical solid dielectric material. Solid flexible coaxial transmission lines are the most frequently used type of transmission line.

12-3 ELECTRICAL CHARACTERISTICS OF TRANSMISSION LINES

Two-Wire Transmission Line

The end of a two-wire transmission line that is connected to a source is ordinarily called the *generator end* or *input end*. The other end of the line, if connected to a load, is called the *load end* or *receiving end*.

The electrical characteristics of the two-wire transmission line are dependent primarily on the construction of the line. Since the two-wire line can be viewed as a long capacitor, the change of its capacitive reactance will be noticeable as the frequency applied to it is changed. Since the long conductors will have a magnetic field about them when electrical energy is being passed through them, the properties of inductance will also be observed. The values of the inductance and capacitance present are dependent on various physical factors, and the effects of the line's associated reactances will be dependent on the frequency applied. Since no dielectric is perfect (electrons will manage to move from one conductor to the other through the dielectric), there will be a conductance

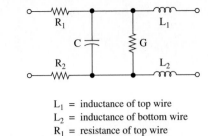

FIGURE 12-7 ▮ Equivalent circuit for a
two-wire transmission line.

L_1 = inductance of top wire
L_2 = inductance of bottom wire
R_1 = resistance of top wire
R_2 = resistance of bottom wire
G = conductance between wires
C = capacitance between wires

value for each type of two-wire transmission line. This conductance value will represent the value of current flow that may be expected through the insulation. If the line is uniform (all values equal at each unit length), one small section of the line may be represented as shown in Fig. 12-7. Such a diagram may represent several feet of line.

In many applications, the values of conductance and resistance are insignificant and may be neglected. If they are neglected, the circuit will appear as shown in Fig. 12-8. Notice that this network is *terminated* with a resistance that represents the impedance of the infinite number of sections exactly like the section of line under consideration. The termination is considered to be a load connected to the line.

Characteristic Impedance

A line infinitely long can be represented by an infinite number of inductors and capacitors. If a voltage is applied to the input terminals of the line, current would begin to flow. Since there are an infinite number of these sections of line, the current would flow indefinitely. If the infinite line were uniform, the impedance of each section would be the same as the impedance offered to the circuit by any other section of line of the same unit length. Therefore, the current would be of some finite value. If the current flowing in the line and the voltage applied across it are known, the impedance of the infinite line could be determined by using Ohm's law. This impedance is called the *characteristic impedance* of the line. The symbol used to represent the characteristic impedance is Z_0. If by some means the characteristic impedance of the line were measured at any point on the line, it would be found that it would be the same. The characteristic impedance is sometimes called the *surge impedance.*

In Fig. 12-8 the distributed inductance of the line is divided equally into two parts in the horizontal arms of the T. The distributed capacitance is lumped and shown connected in the central leg of the T. The line is terminated in a resistance equal to that of the characteristic impedance of the line as seen from terminals *AB*. The reasons for using this value of resistive termination will be fully explained in Sec. 12-6. Since the circuit in Fig. 12-8 is nothing more than a series–parallel *LCR* circuit, the impedance of the network may be determined by the formula that will now be developed.

The impedance, Z_0, looking into terminals *AB* of Fig. 12-8 is

$$Z_0 = \frac{Z_1}{2} + \frac{Z_2\left[(Z_1/2) + Z_0\right]}{Z_2 + (Z_1/2) + Z_0} \qquad (12\text{-}1)$$

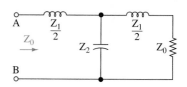

FIGURE 12-8 ■ Simplified circuit
terminated with its characteristic impedance.

Simplifying yields

$$Z_0 = \frac{Z_1}{2} + \frac{(Z_1 Z_2/2) + Z_0 Z_2}{Z_2 + (Z_1/2) + Z_0} \qquad (12\text{-}2)$$

Expressing the right-hand member in terms of the least common denominator, we obtain

$$Z_0 = \frac{Z_1 Z_2 + (Z_1^2/2) + Z_1 Z_0 + (2Z_1 Z_2/2) + 2Z_0 Z_2}{2[Z_2 + (Z_1/2) + Z_0]} \qquad (12\text{-}3)$$

If both sides of this equation are multiplied by the denominator of the right-hand member, the result is

$$2Z_2 Z_0 + \frac{2Z_1 Z_0}{2} + 2Z_0^2 = Z_1 Z_2 + \frac{Z_1^2}{2} + Z_1 Z_0 + \frac{2Z_1 Z_2}{2} + 2Z_0 Z_2 \qquad (12\text{-}4)$$

Simplifying gives us

$$2Z_0^2 = 2Z_1 Z_2 + \frac{Z_1^2}{2} \qquad (12\text{-}5)$$

or

$$Z_0^2 = Z_1 Z_2 + \left(\frac{Z_1}{2}\right)^2 \qquad (12\text{-}6)$$

If the transmission is to be accurately represented by an equivalent network, the T-network section of Fig. 12-8 must be replaced by an infinite number of similar sections. Thus, the distributed inductance in the line will be divided into n sections, instead of the number (2) as indicated in the last term of Eq. (12-6). As the number of sections approaches infinity, the last term Z_1/n will approach zero. Therefore,

$$Z_0 = \sqrt{Z_1 Z_2} \qquad (12\text{-}7)$$

Since the term Z_1 represents the inductive reactance and the term Z_2 represents the capacitive reactance,

$$Z_0 = \sqrt{2\pi f L \times \frac{1}{2\pi f C}}$$

and

$$Z_0 = \sqrt{\frac{L}{C}} \qquad (12\text{-}8)$$

The derivation resulting in Eq. (12-8) shows that a line's characteristic impedance is dependent on its inductance and capacitance.

EXAMPLE 12-1

A commonly used coaxial cable, RG-8A/U, has a capacitance of 29.5 pF/ft and inductance of 73.75 nH/ft. Determine its characteristic impedance for a 1-ft section and for a length of 1 mi.

SOLUTION

For the 1-ft section,

$$Z_0 = \sqrt{\frac{L}{C}} \tag{12-8}$$

$$= \sqrt{\frac{73.75 \times 10^{-9}}{29.5 \times 10^{-12}}} = \sqrt{2500} = 50 \ \Omega$$

For the 1-mi section,

$$Z_0 = \sqrt{\frac{5280 \times 73.75 \times 10^{-9}}{5280 \times 29.5 \times 10^{-12}}} = \sqrt{\frac{5280}{5280} \times 2500} = 50 \ \Omega$$

Example 12-1 shows that the line's characteristic impedance is independent of length and is in fact a *characteristic* of the line. The value of Z_0 depends on the ratio of the distributed inductance and the capacitance in the line. An increase in the separation of the wires increases the inductance and decreases the capacitance. This effect takes place because the effective inductance is proportional to the flux that may be established between the two wires. If the two wires carrying current in opposite directions are placed farther apart, more magnetic flux is included between them (they cannot cancel their magnetic effects as completely as if the wires were closer together), and the distributed inductance is increased. The capacitance is lowered if the plates of the capacitor (the plates are the two conducting wires) are more widely spaced.

Thus, the effect of increasing the spacing of the two wires is to increase the characteristic impedance, because the L/C ratio is increased. Similarly, a reduction in the diameter of the wires also increases the characteristic impedance. The reduction in the size of the wire affects the capacitance more than the inductance, for the effect is equivalent to decreasing the size of the plates of a capacitor in order to decrease the capacitance. Any change in the dielectric material between the two wires also changes the characteristic impedance. If a change in the dielectric material increases the capacitance between the wires, the characteristic impedance, by Eq. (12-8), is reduced.

The characteristic impedance of a two-wire line with air as the dielectric may be obtained from the formula

$$Z_0 \simeq 276 \ \log_{10} \frac{2D}{d} \tag{12-9}$$

where D = spacing between the wires (center to center)
   d = diameter of one of the conductors
   ϵ = dielectric constant of the insulating material relative to air

The characteristic impedance of a concentric or coaxial line also varies with L and C. However, because the difference in construction of the two lines causes L and C to vary in a slightly different manner, the following formula must be used to determine the characteristic impedance of the coaxial line:

$$Z_0 = \frac{138}{\sqrt{\epsilon}} \log_{10} \frac{D}{d} \qquad (12\text{-}10)$$

where D = inner diameter of the outer conductor
$\quad\;\; d$ = outer diameter of the inner conductor
$\quad\;\; \epsilon$ = dielectric constant of the insulating material relative to air

The relative dielectric constant of air is 1, while polyethylene has a value of 2.3.

EXAMPLE 12-2

Determine the characteristic impedance of
(a) A parallel wire line with D/d = 2 with air dielectric.
(b) An air dielectric coaxial line with D/d = 2.35.
(c) RG-8A/U coaxial cable with D = 0.285 in and d = 0.08 in. It uses a
*   polyethylene dielectric.*

SOLUTION

(a)
$$Z_0 = \frac{276}{\sqrt{\epsilon}} \log_{10} \frac{2D}{d} \qquad (12\text{-}9)$$

$$= \frac{276}{1} \log_{10} 4$$

$$= 166 \ \Omega$$

(b)
$$Z_0 = \frac{138}{\sqrt{\epsilon}} \log_{10} \frac{D}{d} \qquad (12\text{-}10)$$

$$= \frac{138}{1} \log_{10} 2.35$$

$$= 51.2 \ \Omega$$

(c)
$$Z_0 = \frac{138}{\sqrt{2.3}} \log_{10} \frac{0.285}{0.08}$$

$$= 50 \ \Omega$$

Transmission Line Losses

Whenever the electrical characteristics of lines are explained, the lines are often thought of as being loss-free. Although this allows for simple and more readily understood explanations, the losses in practical lines cannot be ignored. There are three major losses that occur in transmission lines: copper losses, dielectric losses, and radiation or induction losses.

The resistance of any conductor is never zero. When current flows through a transmission line, energy is dissipated in the form of I^2R losses. A reduction in resistance will minimize the power loss in the line. The resistance is indirectly proportional to the cross-sectional area. Keeping the line as short as possible will decrease the resistance and the I^2R loss. The use of a wire with a large cross-sectional area is also desirable; however, this method has its limitations due to the resulting cost increases.

At high frequencies the I^2R loss is mainly due to the *skin effect*. When a dc current flows through a conductor, the movement of electrons through its cross section is uniform. The situation is somewhat different when ac is applied. The expanding and collapsing fields about each electron encircle other electrons. This phenomenon, called *self-induction,* retards the movement of the encircled electrons. The flux density at the center is so great that electron movement at this point is reduced. As frequency is increased, the opposition to the flow of current in the center of the wire increases. Current flow in the center of the wire is reduced, and most of the electron flow is on the wire surface. When the frequency applied is 100 MHz or higher, the electron movement in the center is so small that the center of the wire could be removed with virtually no effect.

Dielectric losses are proportional to the voltage across the dielectric. They increase with frequency and coupled with skin effect losses limit most practical operation to a maximum frequency of about 18 GHz. These losses are lowest when air dielectric lines are used. In many cases the use of a solid dielectric is required, for example, in the flexible coaxial cable, and if losses are to be minimized, an insulation with a low dielectric constant is used. Polyethylene allows the construction of a flexible cable whose dielectric losses, though higher than air, are still much lower than the losses that occur with other types of low-cost dielectrics. Since I^2R losses and dielectric losses are proportional to length, they are usually lumped together and expressed in decibels of loss per meter. The loss versus frequency effects for some common lines are shown in Fig. 12-9.

The electrostatic and electromagnetic fields that surround a conductor also cause losses in transmission lines. The action of the electrostatic fields is to charge neighboring objects, while the changing magnetic field induces an EMF in nearby conductors. In either case, energy is lost.

Radiation and induction losses may be greatly reduced by terminating the line with a resistive load equal to the line's characteristic impedance and by proper shielding of the line. Proper shielding can be accomplished by the use of coaxial cables with the outer conductor grounded. The problem of radiation loss is of consequence, therefore, only for parallel wire transmission lines.

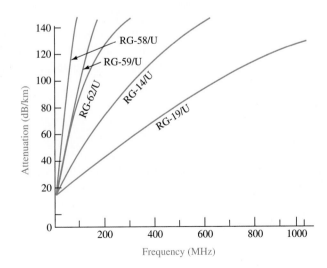

FIGURE 12-9 ▮ Line attenuation characteristics.

Physical Explanation of Propagation

To better understand the characteristics of a transmission line with an ac voltage applied, the infinitely long transmission line will first be analyzed with a dc voltage applied. This will be accomplished using a circuit illustrated in Fig. 12-10. In this circuit the resistance of the line is not shown. The line will be assumed to be loss-free.

Considering only the capacitor C_1 and the inductor L_1 as a series circuit, when voltage is applied to the network, capacitor C_1 will have the ability to charge through inductor L_1. It is characteristic of an inductor that at the first instant of time when voltage is applied a maximum voltage is developed across it and minimum current is permitted to pass through it. At the same time, the capacitor will have a minimum of voltage across it and pass a maximum current. The maximum current is not permitted to flow at the first instant because of the action of the inductor, which is in the charge path of the capacitor. At this time the voltage across points c and d is zero. Since the remaining portion of the line is connected to points c and d, there will be 0 V developed across it at the first instant of time. The voltage across the rest of the line is dependent on the charging action of the capacitor, C_1. It will require some finite amount of time for capacitor C_1 to charge through inductor L_1. As capacitor C_1 is charging, the ammeter records the changing current. When C_1 charges to a voltage that is near the value of the applied voltage, capacitor C_2 will begin to charge through inductors L_1 and L_2. The charging of capacitor C_2 will again require time. In fact, the time required for the voltage to reach points e and f from points c and d will be the same time as was necessary for the original voltage to reach points c and d. This is true because the line is uniform, and the values of the reactive components are the same throughout its entire length. This action will continue in the same manner until all of the capacitors in the line are charged. Since the number of capacitors in an infinite line is infinite, the time required to charge the entire line would be an infinite amount of time. It is important to note that current is flowing continuously in the line and that it has some finite value.

Velocity of Propagation

When a current is moving down the line, its associated electric and magnetic fields are said to be *propagated* down the line. Time is required to charge each unit section of the line, and if the line were infinitely long, it would require an infinitely long time to charge. The time for a field to be propagated from one point on a line to another may be computed, for if the time and the length of the line are known, the *velocity of propagation* may

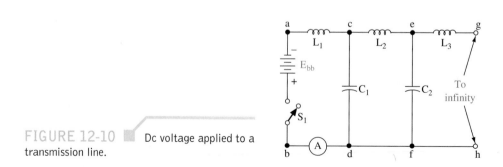

FIGURE 12-10 Dc voltage applied to a transmission line.

be determined. The network shown in Fig. 12-11 is the circuit that will be used to compute the time required for the voltage wavefront to pass a section of line of specified length. The total charge (Q) in coulombs on capacitor C_1 is determined by the relationship

$$Q = Ce \qquad (12\text{-}11)$$

Since the charge on the capacitor in the line had its source at the battery, the total amount of charge removed from the battery will be equal to

$$Q = it \qquad (12\text{-}12)$$

Because these charges are equal, they may be equated:

$$Ce = it \qquad (12\text{-}13)$$

As the capacitor C_1 charges, capacitor C_2 contains a zero charge. Since capacitor C_1's voltage is distributed across C_2 and L_2, at the same time the charge on C_2 is practically zero, the voltage across C_1 (points c and d) must be, by Kirchhoff's law, entirely across L_2. The value of the voltage across the inductor is given by

$$e = L\frac{\Delta i}{\Delta t} \qquad (12\text{-}14)$$

Since current and time start at zero, the change in time and the change in current are equal to the final current and the final time. Equation (12-14) becomes

$$et = Li \qquad (12\text{-}15)$$

Solving the equation for i, we have

$$i = \frac{et}{L} \qquad (12\text{-}16)$$

Solving the equation that was a statement of the equivalency of the charges for current [Eq. (12-13)], we obtain

$$i = \frac{Ce}{t} \qquad (12\text{-}17)$$

Equating both of these expressions yields

$$\frac{et}{L} = \frac{Ce}{t} \qquad (12\text{-}18)$$

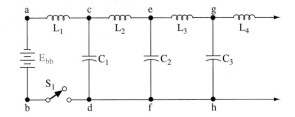

FIGURE 12-11   Circuit for computing time of travel.

Solving the equation for t gives us

$$t^2 = LC$$

or

$$t = \sqrt{LC} \tag{12-19}$$

Since velocity is a function of both time and distance ($V = dt$), the formula for computing propagation velocity is

$$V_P = \frac{d}{\sqrt{LC}} \tag{12-20}$$

where V_P = velocity of propagation

d = distance of travel

$\sqrt{LC}$ = time (t)

It should again be noted that the time required for a wave to traverse a transmission line segment will depend on the value of L and C and that these values will be different, depending on the type of the transmission line considered.

Delay line. (Courtesy of Damaskos, Inc.)

Delay Line

The velocity of electric waves through a vacuum is the speed of light, or 3×10^8 m/s. It is just slightly reduced for travel through air. It has just been shown that a transmission line decreases this velocity due to its inductance and capacitance. This property is put to practical use when it is desired to delay a signal by some specific amount of time. A transmission line used for this purpose is called a *delay line*.

EXAMPLE 12-3

Determine the amount of delay and the velocity of propagation introduced by a 1-ft section of RG-8A/U coaxial cable used as a delay line.

SOLUTION

From Ex. 12-1 we know that this cable has a capacitance of 29.5 pF/ft and inductance of 73.75 nH/ft. The delay introduced by 1 ft of this line is

$$t = \sqrt{LC}$$

$$= \sqrt{73.75 \times 10^{-9} \times 29.5 \times 10^{-12}} = 1.475 \times 10^{-9} \text{ s} \qquad (12\text{-}19)$$

$$\text{or} \quad 1.475 \text{ ns}$$

The velocity of propagation is

$$V_P = \frac{d}{\sqrt{LC}} \qquad (12\text{-}20)$$

$$= \frac{1 \text{ ft}}{1.475 \text{ ns}} = 6.78 \times 10^8 \text{ ft/s} \quad \text{or} \quad 2.07 \times 10^8 \text{ m/s}$$

Example 12-3 showed that the energy velocity for RG-8A/U cable is roughly two-thirds the velocity of light. This ratio of actual velocity to the velocity in free space is termed the *velocity constant* or *velocity factor* of a line. It can range from about 0.55 up to 0.97, depending on the type of line, the D/d ratio, and the type of dielectric. As an approximation for non-air-dielectric coaxial lines, the velocity factor, v_f, is

$$v_f \simeq \frac{1}{\sqrt{\epsilon}} \qquad (12\text{-}21)$$

where v_f = velocity factor
  ϵ = relative dielectric constant

EXAMPLE 12-4

Determine the velocity factor for RG-8A/U cable by using the results of Ex. 12-3 and also by using Eq. (12-21).

SOLUTION

From Ex. 12-3, the velocity was 2.07×10^8 m/s. Therefore,

$$v_f = \frac{2.07 \times 10^8 \text{ m/s}}{3 \times 10^8 \text{ m/s}} = 0.69$$

Using Eq. (12-21), we obtain

$$v_f \simeq \frac{1}{\sqrt{\epsilon}} = \frac{1}{\sqrt{2.3}} = 0.66$$

Wavelength

A wave that is radiated through space travels at about the speed of light, or 186,000 mi/s (3×10^8 m/s). The velocity of this wave is constant regardless of frequency, so that the distance traveled by the wave during a period of one cycle (called one *wavelength*) can be found by the formula

$$\lambda = \frac{v}{f} \tag{12-22}$$

where λ (the Greek lowercase letter lambda, used to symbolize wavelength) is the distance in meters from the crest of one wave to the crest of the next, f is the frequency, and v is the velocity of the radio wave in meters per second. It should be noted that the wavefront travels more slowly on a wire than it does in free space.

EXAMPLE 12-5

Determine the wavelength (λ) of a 100-MHz signal in free space and while traveling through an RG-8A/U coaxial cable.

SOLUTION

In free space the wave's velocity is 3×10^8 m/s. Therefore,

$$\lambda = \frac{v}{f} \tag{12-22}$$

$$= \frac{3 \times 10^8 \text{ m/s}}{1 \times 10^8 \text{ Hz}} = 3 \text{ m}$$

In RG-8A/U cable we found that the velocity of propagation is 2.07×10^8 m/s in Ex. 12-3. Therefore,

$$\lambda = \frac{v}{f} \tag{12-22}$$

$$= \frac{2.07 \times 10^8 \text{ m/s}}{1 \times 10^8 \text{ Hz}} = 2.07 \text{ m}$$

Example 12-5 shows that line wavelength is less than free-space wavelength for any given frequency signal.

12-5 NONRESONANT LINE

Traveling dc Waves

A *nonresonant line* is defined as one of infinite length or one that is terminated with a resistive load equal in ohmic value to the characteristic impedance of the line. In a nonresonant line, all of the energy transferred down the line is absorbed by the load resistance and any inherent resistance in the line. The voltage and current waves are called *traveling waves* and move in phase with one another from the source to the load.

Since the nonresonant line may be either an infinite line or one terminated in its characteristic impedance, the physical length is not critical. In the resonant line that will be discussed shortly, the physical length of the line is quite important.

The circuit in Fig. 12-12 shows a line terminated with a resistance equal to its characteristic impedance. The charging process and the ultimate development of a voltage across the load resistance will now be described. At the instant switch S_1 is closed, the total applied voltage is felt across inductor L_1. After a very short time has elapsed, capacitor C_1 begins to assume a charge. C_2 cannot charge at this time because all the voltage felt between points c and d is developed across inductor L_2 in the same way the initial voltage was developed across inductor L_1. Capacitor C_2 is unable to charge until the charge on C_1 approaches the amplitude of the supply voltage. When this happens, the voltage charge on capacitor C_2 begins to rise. The voltage across capacitor C_2 will be felt between points e and f. Since the load resistor is also effectively connected between points e and f, the voltage across the resistor will be equal to the voltage appearing across C_2. The voltage input has been transferred from the input to the load resistor. While the capacitors were charging, the ammeter recorded a current flow. After all of the capacitors are charged, the ammeter will continue to indicate the load current that will be flowing through the dc resistance of the inductors and the load resistor. The current will continue to flow as long as switch S_1 is closed. When it is opened, the capacitors will discharge through the load resistor in much the same way as filter capacitors discharge through a bleeder resistor.

Traveling ac Waves

There is little difference between the charging of the line when an ac voltage is applied to it. The charging sequence of the line with an ac voltage applied will now be discussed. Refer to the circuit and waveform diagrams in Fig. 12-13.

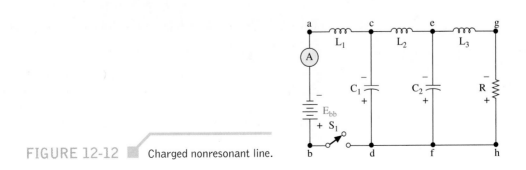

FIGURE 12-12 ▪ Charged nonresonant line.

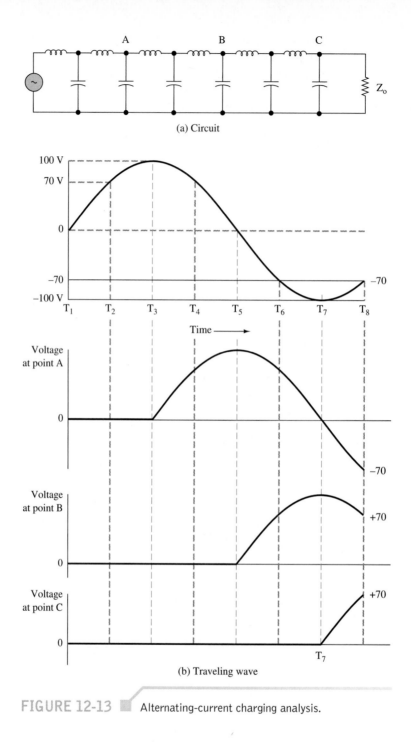

(a) Circuit

(b) Traveling wave

FIGURE 12-13 ■ Alternating-current charging analysis.

As the applied voltage begins to go positive, the voltage wave begins traveling down the line. At time t_3, the first small change in voltage arrives at point A, and the voltage at that point starts increasing in the positive direction. At time t_5, the same voltage rise arrives at point B, and at time t_7, the same voltage rise arrives at the end of the line. The waveform has moved down the line as a wavefront. The time required for the voltage changes to move down the line is the same as the time required for the dc voltage to move down the line. The time for both of these waves to move down the line for a specified length may be computed by using Eq. (12-19). The following general remarks concerning the ac charging of the line may now be made. All of the instantaneous voltages produced by the generator travel down the line in the order in which they were produced. If the voltage waveform is plotted at any point along the line, the resulting waveform will be a duplicate of the generator waveform. Since the line is terminated with its characteristic impedance, all of the energy produced by the source will be absorbed by the load impedance.

12-6 RESONANT TRANSMISSION LINE

A *resonant line* is defined as a transmission line that is terminated with an impedance that is *not* equal to its characteristic impedance. Unlike the nonresonant line, the length of the resonant line is critical. In some applications, the resonant line may be terminated in either an open or a short. When this occurs, some very interesting effects may be observed.

DC Applied to an Open-Circuited Line

A transmission line of finite length terminated in an open circuit is illustrated in Fig. 12-14. The characteristic impedance of the line may be assumed to be equal to the internal impedance of the source. Since the impedance of the source is equal to the impedance of the line, the applied voltage will be divided equally between the impedance of the source and the impedance of the line. When switch S_1 is closed, current begins to flow as the capacitors begin to charge through the inductors. As each capacitor charges in turn, the voltage will move down the line. As the last capacitor is charged to the same voltage as every other capacitor, there will be no difference in potential between points e and g. This is true because the capacitors will possess exactly the same charge. The inductor L_3 is also connected between points e and g. Since there is no difference in potential between points e and g, there can no longer be current flow through the inductor. This means that the magnetic field about the inductor will no longer be sustained. The magnetic field must collapse. It is characteristic of the field about an inductor to tend to keep current flowing in the same direction when the magnetic field collapses. This additional current must flow into the capacitive circuit of the open circuit, C_3. Since the energy stored in the magnetic field is equal to that stored in the capacitor, the charge on capacitor C_3 will double. The voltage on capacitor C_3 will be equal to the value of the applied voltage. Since there is no

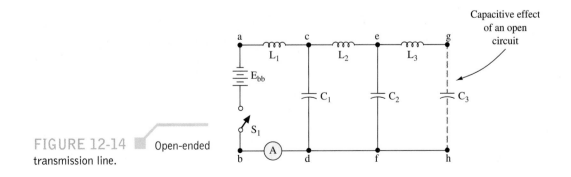

FIGURE 12-14 Open-ended transmission line.

difference of potential between points c and e, the magnetic field about inductor L_2 will collapse, forcing the charge on capacitor C_2 to double its value. The field about inductor L_1 will also collapse, doubling the voltage on C_1. The combined effect of the collapsing magnetic field about each inductor in turn causes a voltage twice the value of the original to apparently move back toward the source. This voltage movement in the opposite direction caused by the conditions just described is called *reflection*. The reflection of voltage occurred in the same polarity as the original charge. Therefore, it is said of a transmission line terminated in an open that the reflected voltage wave will always be of the same polarity and amplitude as the incident voltage wave. When this reflected voltage reaches the source, the action stops because of the cancellation of the voltages. The current, however, is reflected back with an opposite polarity because when the field about the inductor collapsed, the current dropped to zero. As each capacitor is charged, causing the reflection, the current flow in the inductor that caused the additional charge drops to zero. When capacitor C_1 is charged, current flow in the circuit stops, and the line is charged. It may also be said that the line now "sees" that the impedance at the receiving end is an open.

Incident and Reflected Waves

The situation for a 100-V battery with a 50-Ω source resistance is illustrated in Fig. 12-15. The battery is applied to an open-circuited 50-Ω characteristic impedance line at time $t = 0$. Initially, a 50-V level propagates down the line, and 1 A is drawn from the battery. Notice that 50 V is dropped across R_s at this point in time. The reflected voltage from the open circuit is also 50 V so that the resultant voltage along the line is 50 V + 50 V, or

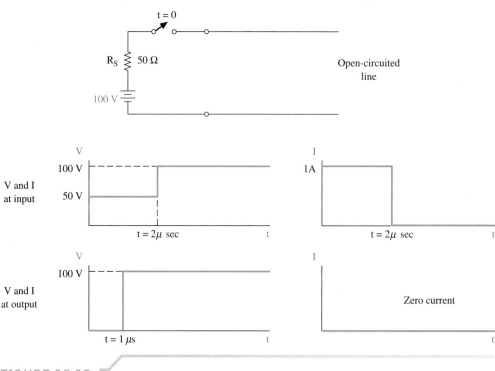

FIGURE 12-15   Direct current applied to open-circuited line.

100 V, once the reflection gets back to the battery. If it takes 1 μs for energy to travel the length of this transmission line, the voltage at the line's input will initially be 50 V until the reflection gets back to the battery at $t = 2$ μs. Remember, it takes 1 μs for the energy to get to the end of the line and then 1 μs additional for the reflection to get back to the battery. Thus, the voltage versus time condition shown in Fig. 12-15 for the line's input is 50 V until $t = 2$ μs, when it changes to 100 V. The voltage at the load is zero until $t = 1$ μs, as shown, when it jumps to 100 V due to the 50 V just reaching it and the resulting 50-V reflection.

The current conditions on the line are also shown in Fig. 12-15. The incident current is 1 A, which is 100 V $\div$ ($R_s + Z_0$). The reflected current for an open-circuited line is out of phase and, therefore, is -1 A with a resultant of 0 A. The current at the line's input is 1 A until $t = 2$ μs, when the reflected current reaches the source to cancel the incident current. After $t = 2$ μs the current is therefore zero. The current at the load is always zero since when the incident 1 A reaches the load it is immediately cancelled by the reflected -1 A.

It is thus seen that after 1 μs at the load and after 2 μs at the source, the results are predicted by standard analysis; that is, the voltage on an open-circuited wire is equal to the source voltage, and the current is zero. Unfortunately, when ac signals are applied to a transmission line, the results are not so simple since the incident and reflected signals occur on a continuous repetitive basis.

DC Applied to a Short-Circuited Line

The condition of applying a 100-V, 50-Ω source resistance battery to a shorted 50-Ω transmission line is illustrated in Fig. 12-16. Once again, assume that it takes 1 μs for

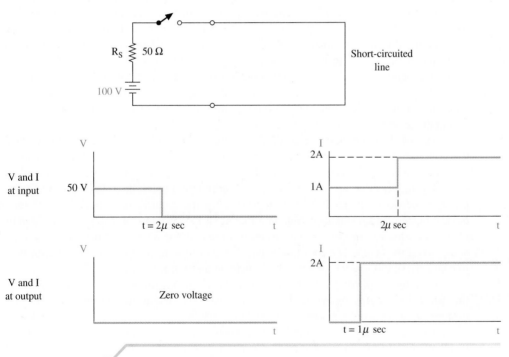

FIGURE 12-16 ▓ Direct current applied to shorted line.

energy to travel the length of the line. A complete analysis as presented with the open-circuited line would now be repetitive. Instead, only the differences and final results will be provided.

When the incident current of 1 A [100 V ÷ $(R_s + Z_0)$] reaches the short-circuited load, the reflected current is 1 A and in phase so that the load current becomes 1 A + 1 A = 2 A. The incident voltage (+50 V) is reflected back out of phase (−50 V) so that the resultant voltage at the short circuit is zero, as it must be across a short. The essential differences here from the open-circuited line are:

1. The voltage reflection from an open circuit is in phase, while from a short circuit it is out of phase.
2. The current reflection from an open circuit is out of phase, while from a short circuit it is in phase.

The resultant load voltage is therefore always zero, as it must be across a short circuit, and is shown in Fig. 12-16. However, the voltage at the line's input is initially +50 V until the reflected out-of-phase level (−50 V) gets back in 2 μs, which causes the resultant to be zero. The load current is zero until $t = 1$ μs, when the incident and reflected currents of +1 A combine to cause 2 A of current to flow. The current at the line's input is initially 1 A until the reflected current of 1 A arrives at $t = 2$ μs to cause the total current to be 2 A.

Standing Waves: Open Line

The conditions of an open or shorted transmission line are extraordinary conditions. Reflected waves are, on the whole, highly undesirable. It is known that when the impedance of the source is equal to the characteristic impedance of the line and that line is terminated in its characteristic impedance, there is a maximum transfer of power, complete absorption of energy by the load, and no reflected waves. If the line is not terminated in its characteristic impedance, there will be reflected waves present on the line. The type and amount of reflected waves are dependent on the type and amount of mismatch. When a mismatch occurs, there is an interaction between the incident and reflected waves. When the applied signal is ac, this interaction results in the creation of a new kind of wave called a *standing wave*. This name is given to these waveforms because they apparently remain in one position, varying only in amplitude. These waves, and the variations in amplitude, are illustrated in Fig. 12-17.

The left-hand column in Fig. 12-17 represents the in-phase reflection of the voltage wave on the open line or current wave on a shorted line. Note the dashed line extending past the load in the top left-hand diagram. It is an extension of the incident traveling wave. By simply "folding" it back across the termination point, the reflected wave for in-phase reflection is obtained. All the diagrams showing in-phase reflection show the incident, reflected, and resultant waves on a line at various instants of time. They are graphs of a wave versus *position* at some instant of time and *not* waves versus time, as you are accustomed to seeing. The resultant waves for each instant of time are shown with blue line and are simply the vector sum of the incident and reflected waves.

Note that at positions d_1 and d_3 in Fig. 12-17 the resultant voltage (current) on the line (shown in blue) is always zero. If you stationed yourself at these points with an oscilloscope, no wave would ever be seen once the very first incident wave and then reflected wave had passed by. On the other hand, at position d_2, the generator and, at the load, the resultants have a maximum value. The resultant is truly a *standing* wave. If readings of the rms voltage and current waves for an open line were taken along the line, the

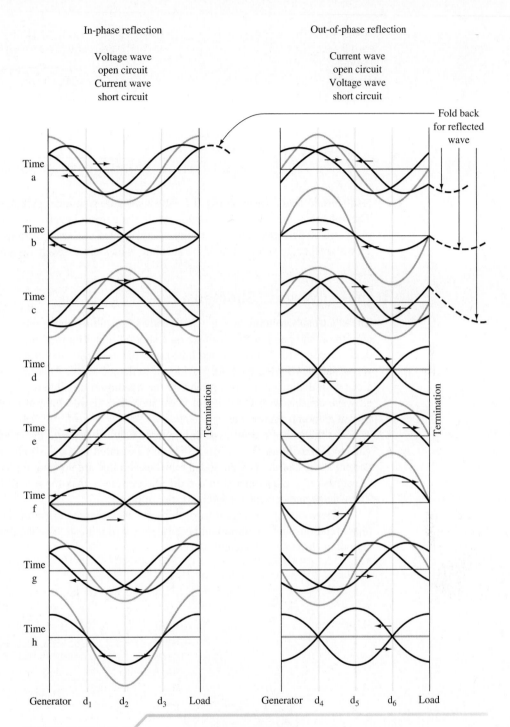

In-phase reflection

Voltage wave
open circuit
Current wave
short circuit

Out-of-phase reflection

Current wave
open circuit
Voltage wave
short circuit

Fold back
for reflected
wave

Time a

Time b

Time c

Time d

Time e

Time f

Time g

Time h

Termination

Termination

Generator   d₁   d₂   d₃   Load

Generator   d₄   d₅   d₆   Load

FIGURE 12-17 Development of standing waves.

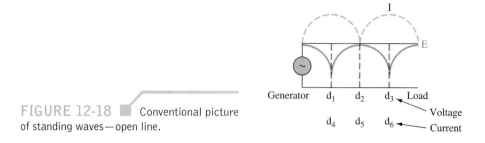

FIGURE 12-18 ■ Conventional picture of standing waves—open line.

result would be as shown in Fig. 12-18. This is a conventional picture of standing waves. The voltage at the open is maximum, while the current is zero, as it must be through an open circuit. The positions d_1 through d_6 correspond to those positions shown in Fig. 12-17 with respect to summation of the absolute values of the resultants shown in Fig. 12-17.

Standing Waves: Shorted Line

The right-hand column of Fig. 12-17 shows the out-of-phase reflection that occurs for current on an open line and voltage on a shorted line. The first graph shows that the incident wave is extended (in dashed lines) past the load *180° out of phase* from the incident wave and then folded back to provide the reflected wave. At *b* the reflected wave coincides with the incident wave, providing the maximum possible resultant. At *d* the reflected wave cancels with the incident wave such that the resultant at that instant of time is zero at all points on the line.

At the end of a transmission line terminated in an open, the current is zero and the voltage is maximum. This relationship may be stated in terms of phase. The voltage and current at the end of an open-ended transmission line are 90° out of phase. At the end of a transmission line terminated in a short, the current is maximum and the voltage is zero. The voltage and current are again 90° out of phase.

These current–voltage relationships are shown in the diagrams in Fig. 12-19. These phase relationships are important because they will indicate how the line will act at different points throughout its length.

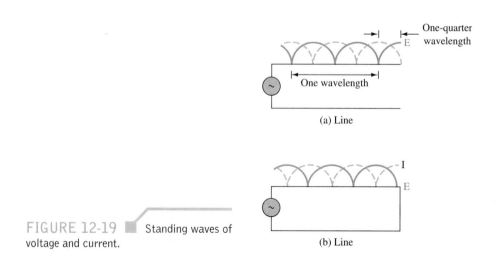

FIGURE 12-19 ■ Standing waves of voltage and current.

A transmission line will have points of maximum and minimum voltage as well as points of maximum and minimum current. The position of these points can be accurately predicted if the applied frequency and type of line termination are known.

EXAMPLE 12-6

An open-circuited line is 1.5λ long. Sketch the incident, reflected, and resultant voltage and current waves for this line at the instant the generator is at its peak positive value.

SOLUTION

Recall that to obtain the reflected voltage from an open circuit the incident wave should be continued on past the open (in your mind) and then folded back toward the generator. This process is shown in Fig. 12-20(a). Notice that the reflected wave coincides with the incident wave, making the resultant wave double the amplitude of the generator voltage. The current wave at an open circuit should be continued on 180° out of phase, as shown in Fig. 12-20(b), and then folded back to provide the reflected wave. In this case, the incident and reflected current waves cancel one another, leaving a zero resultant all along the line at this instant of time.

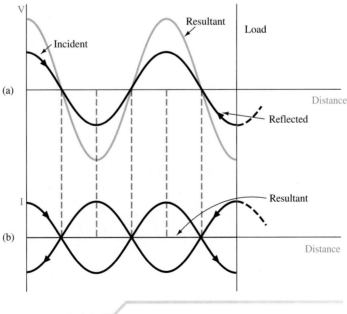

FIGURE 12-20 ▓ Diagram for Ex. 12-6.

12-7 STANDING WAVE RATIO

A standing wave is the result of an incident and reflected wave. The ratio of reflected voltage to incident voltage is called the reflection coefficient, Γ. That is,

$$\Gamma = \frac{E_r}{E_i} \tag{12-23}$$

where Γ is the reflection coefficient, E_r is the reflected voltage, and E_i is the incident voltage. When a line is terminated with a short circuit, open circuit, or purely reactive load, no energy can be absorbed by the load so that total reflection takes place. The reflected wave equals the incident wave so that $|\Gamma| = 1$. When a line is terminated with a resistance equal to the line's Z_0, no reflection occurs, and therefore $\Gamma = 0$. For all other cases of termination the load will absorb some of the incident energy (but not all), and the reflection coefficient will be between 0 and 1.

The reflection coefficient may also be expressed in terms of the load impedance:

$$\Gamma = \frac{Z_L - Z_0}{Z_L + Z_0} \tag{12-24}$$

where Z_L is the load impedance.

On a lossless line the voltage maximums and minimums of a standing wave have a constant amplitude. The ratio of the maximum voltage to minimum on a line is called the *voltage standing wave ratio* (VSWR):

$$VSWR = \frac{E_{max}}{E_{min}} \tag{12-25}$$

In a more general sense it is sometimes referred to as simply the *standing wave ratio* (SWR) since it is also equal to the ratio of maximum current to minimum current. Therefore,

$$SWR = VSWR = \frac{E_{max}}{E_{min}} = \frac{I_{max}}{I_{min}} \tag{12-26}$$

The VSWR can also be expressed in terms of Γ:

$$VSWR = \frac{1 + |\Gamma|}{1 - |\Gamma|} \tag{12-27}$$

The VSWR has an infinite value when total reflection occurs since E_{min} is 0 and has a value of 1 when no reflection occurs. No reflection means no standing wave so that the rms voltage along the line is always the same (neglecting losses), so that E_{max} equals E_{min} and the VSWR is 1.

Effect of Mismatch

This perfect condition of no reflection occurs only when the load is purely resistive and equal to Z_0. Such a condition is called a *flat* line and indicates a VSWR of 1. It is highly desirable since all the generator power capability is getting to the load. Also, if reflection occurs, the voltage maximums along the line may exceed the cable's dielectric strength, causing a breakdown. In addition, the existence of a reflected wave means greater I^2R (power) losses, which can be a severe problem when the line is physically long. The energy contained in the reflected wave is absorbed by the generator (except for the I^2R losses) and not "lost" unless the generator is not perfectly matched to the line, in which case a rereflection of the reflected wave occurs at the generator. In addition, a high VSWR also tends to accentuate noise problems and causes "ghosts" to be transmitted with video or data signals. Refer to Chapter 13 for a discussion of ghost problems.

To summarize, then, the disadvantages of not having a perfectly matched (flat line) system are as follows:

1. The full generator power does not reach the load.
2. The cable dielectric may break down as a result of high-value standing waves of voltage.
3. The existence of reflections (and rereflections) increases the power loss in the form of I^2R heating.
4. Noise problems are increased by mismatches.
5. "Ghost" signals are created.

The VSWR can be determined quite easily, if the load is a known value of pure resistance, by the following equation:

$$\text{VSWR} = \frac{Z_0}{R_L} \quad \text{or} \quad \frac{R_L}{Z_0} \quad \text{(whichever is larger)} \tag{12-28}$$

where R_L is the load resistance. Whenever R_L is greater than Z_0, R_L is used in the numerator so that the VSWR will be greater than 1. For instance, on a 100-Ω line ($Z_0 = 100\ \Omega$), a 200-Ω or 50-Ω R_L results in the same VSWR of 2. They both create the same degree of mismatch.

The higher the VSWR, the greater is the mismatch on the line. Thus, a low VSWR is the goal in a transmission line system except when the line is being used to simulate a capacitance, inductance, or tuned circuit. These effects are considered in following sections.

EXAMPLE 12-7

A citizen's band transmitter operating at 27 MHz with 4-W output is connected via 10 m of RG-8A/U cable to an antenna that has an input resistance of 300 Ω.
Determine
(a) The reflection coefficient.
(b) The electrical length of the cable in wavelengths (λ).
(c) The VSWR.
(d) The amount of the transmitter's 4-W output absorbed by the antenna.
(e) How to increase the power absorbed by the antenna.

(a)
$$\Gamma = \frac{Z_L - Z_0}{Z_L + Z_0} \qquad (12\text{-}24)$$

$$= \frac{300~\Omega - 50~\Omega}{300~\Omega + 50~\Omega} = \frac{5}{7} = 0.71$$

(b)

$$\lambda = \frac{v}{f} \qquad (12\text{-}22)$$

$$= \frac{2.07 \times 10^8~\text{m/s}}{27 \times 10^6~\text{Hz}} = 7.67~\text{m}$$

Since the cable is 10 m long, its electrical length is

$$\frac{10~\text{m}}{7.67~\text{m/wavelength}} = 1.3\lambda$$

(c) Since the load is resistive,

$$\text{VSWR} = \frac{R_L}{Z_0} \qquad (12\text{-}28)$$

$$= \frac{300~\Omega}{50~\Omega} = 6$$

or, an alternative solution, since Γ is known,

$$\text{VSWR} = \frac{1 + \Gamma}{1 - \Gamma} \qquad (12\text{-}27)$$

$$= \frac{1 + \frac{5}{7}}{1 - \frac{5}{7}} = \frac{\frac{12}{7}}{\frac{2}{7}} = 6$$

(d) The reflected voltage is Γ times the incident voltage. Since power is proportional to the square of voltage, the reflected power is $(5/7)^2 \times 4~\text{W} = 2.04~\text{W}$, and the power to the load is

$$P_{\text{load}} = 4~\text{W} - P_{\text{refl}}$$

$$= 4~\text{W} - 2.04~\text{W} = 1.96~\text{W}$$

(e) You do not have enough theory yet to answer this question. Read on in this chapter and in Chapter 14 on antennas.

Example 12-7 shows the effect of mismatch between line and antenna. The transmitted power of only 1.96 W instead of 4 W would seriously impair the effective range of the transmitter. It is little wonder that great pains are taken to get the VSWR down to as close to 1 as possible.

Quarter-Wavelength Transformer

One simple way to match a line to a resistive load is by use of a *quarter-wavelength matching transformer.* It is not physically a transformer but does offer the property of impedance transformation. To match a resistive load, R_L, to a line with characteristic

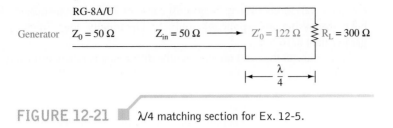

FIGURE 12-21 ▧ λ/4 matching section for Ex. 12-5.

impedance Z_0, a λ/4 section of line with characteristic impedance Z_0' should be placed between them. In equation form the value of Z_0' is

$$Z_0' = \sqrt{Z_0 R_L} \tag{12-29}$$

The required λ/4 section to match the 50-Ω line to the 300-Ω resistive load of Ex. 12-5 would have $Z_0' = \sqrt{50\ \Omega \times 300\ \Omega} = 122\ \Omega$. This solution is shown pictorially in Fig. 12-21. The input impedance looking into the matching section is 50 Ω. This is the termination on the 50-Ω cable, and thus the line is flat, as desired, from there on back to the generator. Remember that λ/4 matching sections are only effective working into a resistive load. The principle involved is that there will now be two reflected signals, equal but separated by λ/4. Since one of them travels λ/2 farther than the other, this 180° phase difference causes cancellation of the reflections.

Electrical Length

The *electrical length* of a transmission line is of importance to this overall discussion. Recall that when reflections occur, the voltage maximums occur at λ/2 intervals. If the transmission line is only λ/16 long, for example, the reflection still occurs, but the line is so short that virtually no voltage variation along the line exists. This situation is illustrated in Fig. 12-22.

Notice that the λ/16 length line has virtually no "standing" voltage variations along the line because of its short electrical length, while the line that is one λ long has two

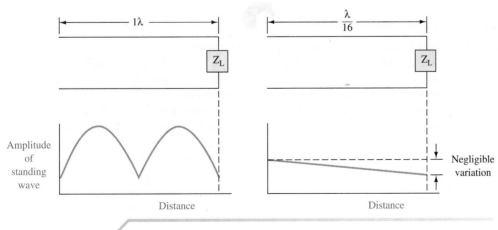

FIGURE 12-22 ▧ Effect of line electrical length.

complete variations over its length. Because of this, the transmission line effects we have been discussing are applicable only to electrically long lines—generally only those greater than $\lambda/16$ long.

Be careful that you understand the difference between electrical length and physical length. A line can be miles in length and still be electrically short if the frequency it carries is low. For example, a telephone line carrying a 300-Hz signal has a λ of 621 mi. On the other hand, a 10-GHz signal has a λ of 3 cm so that even extremely small circuit interconnections require application of transmission line theory.

12-8 THE SMITH CHART

Transmission Line Impedance

Transmission line calculations are very cumbersome from a purely mathematical stand-point. It is often necessary to know the input impedance of a line of a certain length with a given load. The impedance at any point on a line with standing waves is repetitive every half wavelength since the voltage and current waves are similarly repetitive. The impedance is therefore constantly changing along the line and is equal to the ratio of voltage to current at any given point. This impedance can be solved via the following equation for a lossless line:

$$Z_s = Z_0 \frac{Z_L + jZ_0 \tan \beta s}{Z_0 + jZ_L \tan \beta s} \qquad (12\text{-}30)$$

where Z_s = line impedance at a given point
$\quad Z_L$ = load impedance
$\quad Z_0$ = line's characteristic impedance
$\quad \beta s$ = distance from the load to the point where it is desired to know the line impedance (electrical degrees)

Since the tangent function is repetitive every 180°, the result will similarly be repetitive, as we know it should be. If the line is $\frac{3}{4}\lambda$ long (270 electrical degrees), its impedance will be the same as at a point $\lambda/4$ from the load.

Smith Chart Introduction

Equation (12-30) can be solved, but the solution using the *Smith chart* is far more convenient and versatile. This impedance chart was developed by P. H. Smith in 1938 and is still widely used for line, antenna, and waveguide calculations in spite of the widespread availability of computers and programmable calculators.

Figure 12-23 illustrates the Smith chart. It contains two sets of lines. The lines representing constant resistance are circular and are all tangent to each other at the right-hand end of the horizontal line through the center of the chart. The value of resistance along any one of these circles is constant, and its value is indicated just above the horizontal line through the center of the chart.

The second set of lines represents arcs of constant reactance. These arcs are also tangent to one another at the right-hand side of the chart. The values of reactance for each arc are labeled at the circumference of the chart and are positive on the top half and negative on the bottom half.

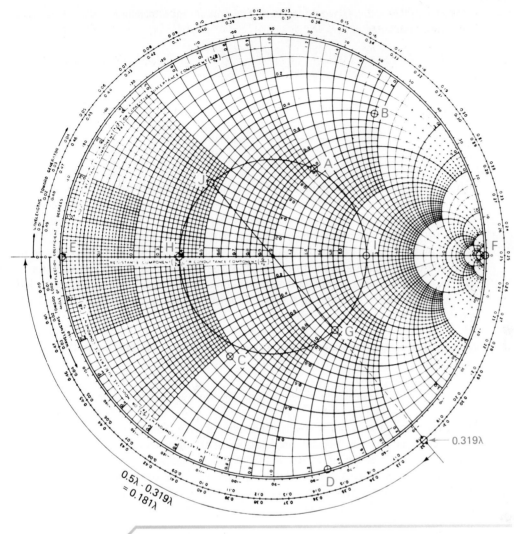

FIGURE 12-23 ■ Smith chart.

Several computer aided design (CAD) programs are available that can accurately predict high-frequency circuit performance. It may therefore seem strange that the pencil and paper (Smith chart) are still utilized. The intuitive and graphical nature of Smith chart design is still desirable, however, since the effect of each design step is clearly apparent. This contrasts with the "feed in the data" and print out the answer situation with most computer programs. A recent program for use on the Macintosh computer recognizes this deficiency and effectively offers a computerized Smith chart. This allows the user to interactively progress through a problem with the various "Smith chart" moves displayed and gives the designer a sense of the effect of each operation.

Using the Smith Chart

So that the chart can be used for a wide range of impedance and admittance values, a process of *normalizing* is employed. This involves dividing all impedances by the line's Z_0 so that if the impedance $100 + j50$ were to be plotted for a 50-Ω line, the value

$100/50 + j50/50 = 2 + j1$ would be used. The normalized impedance is then represented by a lowercase letter,

$$z = \frac{Z}{Z_0}$$ (12-31)

where z is the normalized impedance. When working with admittances,

$$y = \frac{Y}{Y_0}$$ (12-32)

where y is the normalized admittance and Y_0 is the line's characteristic admittance.

The point A in Fig. 12-23 is $z = 1 + j1$ or could be $y = 1 + j1$. It is the intersection of the one resistance circle and one reactive arc. Point B is $0.5 + j1.9$, while point C is $0.45 - j0.55$. Be sure that you now understand how to plot points on the Smith chart.

Point D (toward the bottom of the chart) in Fig. 12-23 is $0 - j1.3$. All points on the circumference of the Smith chart, such as point D, represent pure reactance except for the points at either extreme of the horizontal line through the center of the chart. At the left-hand side, point E, a short circuit ($z = 0 + j0$) is represented, while at the right-hand side, point F, an open circuit is represented ($z = \infty$).

Points G and J in Fig. 12-23 illustrate a practical application of the Smith chart. Assume that a 50-Ω transmission line has a load $Z_L = 65 - j55$ Ω. That load when normalized is

$$z_L = \frac{Z_L}{Z_0} = \frac{65}{50} - j\frac{55}{50} = 1.3 - j1.1$$

Point G is plotted as $z_L = 1.3 - j1.1$. By drawing a circle through point G and using the chart center as the circle's center, the locations of all impedances along the transmission line are described by the circle. Recall that the impedance along a line varies but repeats every half wavelength. A full revolution around the circle corresponds to a half-wavelength (180°) movement on a line. A clockwise (CW) rotation on the chart means you are moving toward the generator, while a counterclockwise (CCW) rotation indicates movement toward the load. The scales on the outer circumference of the chart indicate the amount of movement in wavelengths. For example, moving from the load (point G) toward the generator to point H brings us to a point on the line where the impedance is purely resistive and equal to $z = 0.4$. As shown in Fig. 12-23, that movement is equal to $0.5\lambda - 0.319\lambda = 0.181\lambda$. In other words, at a point 0.181λ from the load, the impedance on the line is purely resistive and has a normalized value of $z = 0.4$. In terms of actual impedance, the impedance is $Z = z \times Z_0 = 0.4 \times 50$ $\Omega = 20$ Ω.

Moving another $\lambda/4$ or halfway around the circle from point H in Fig. 12-23 brings you to another point of pure resistance, point I. At point I on the line the impedance is $z = 2.6$. That also is the VSWR on the line. Wherever the circle drawn through z_L for a transmission line crosses the right-hand horizontal line through the chart center, that point is the VSWR that exists on the line. For this reason, the circle drawn through a line's load impedance is often called its VSWR circle.

The Smith chart allows for a very simple conversion of impedance to admittance, and vice versa. The value of admittance corresponding to any impedance is always diagonally opposite and the same distance from the center. For example, if an impedance is equal to $1.3 - j1.1$ (point G), then the admittance y is at point J and is equal to $0.45 + j0.38$ as read from the chart. Since $y = 1/z$, you can mathematically show that

$$\frac{1}{1.3 - j1.1} = 0.45 + j0.38$$

but the Smith chart solution is much less tedious. Recall that mathematical z to y and y to z transformations require first a rectangular to polar conversion, taking the inverse, and then a polar to rectangular conversion.

Other applications of the Smith chart are most easily explained by solution of actual examples.

EXAMPLE 12-8

Find the input impedance and VSWR of a transmission line 4.3λ long when $Z_0 = 100 \ \Omega$ and $Z_L = 200 - j150 \ \Omega$.

SOLUTION

First normalize the load impedance:

$$z_L = \frac{Z_L}{Z_0}$$

$$= \frac{200 - j150 \ \Omega}{100 \ \Omega} = 2 - j1.5$$

(12-31)

That point is then plotted on the Smith chart in Fig. 12-24, and its corresponding VSWR circle is also shown. That circle intersects the right-hand half of the horizontal line at 3.3 as shown, and therefore the VSWR = 3.3. Now, to find the line's input impedance it is required to move from the load toward the generator (CW rotation) 4.3λ. Since each full revolution on the chart represents λ/2, it means that eight full rotations plus 0.3λ is necessary. Thus, just moving 0.3λ from the load in a CW direction will provide z_{in}. The radius extended through z_L intersects the outer wavelength scale at 0.292λ. Moving from there to 0.5λ provides a movement of 0.208λ. That means that further movement of 0.092λ (0.3λ − 0.208λ) brings us to the generator and provides z_{in} for a 4.3λ line (or 0.3λ, 0.8λ, 1.3λ, 1.8λ, etc., length line). The line's z_{in} is read as

$$z_{in} = 0.4 + j0.57$$

from the chart, which in ohms is

$$Z_{in} = z_{in} \times Z_0 = (0.4 + j0.57) \times 100 \ \Omega = 40 \ \Omega + j57 \ \Omega$$

Corrections for Transmission Loss

Example 12-8 assumed the ideal condition of a lossless line. If line attenuation cannot be neglected, the incident wave gets weaker as it travels toward the load, and the reflected wave gets weaker as it travels back toward the generator. This causes the VSWR to get weaker as we approach the source end of the line. A true standing wave representation on

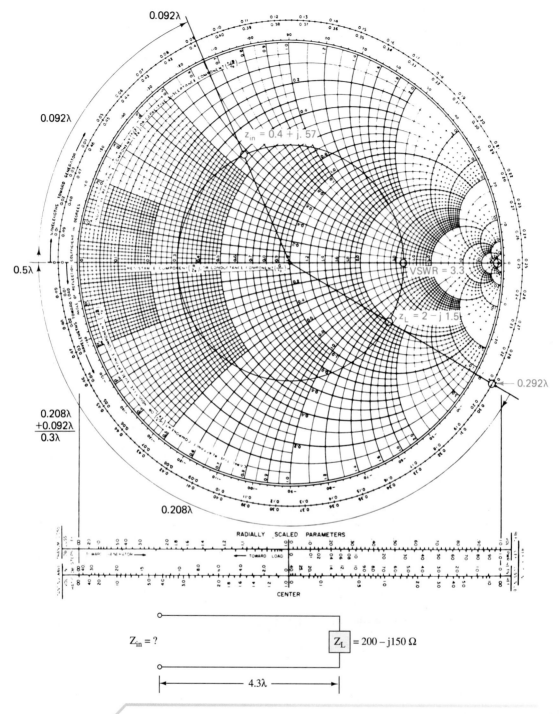

0.092λ

0.092λ

0.5λ

$z_{in} = 0.4 + j.57$

VSWR = 3.3

$z_L = 2 - j 1.5$

0.292λ

$\dfrac{\begin{array}{r} 0.208\lambda \\ +0.092\lambda \end{array}}{0.3\lambda}$

0.208λ

RADIALLY SCALED PARAMETERS

TOWARD GENERATOR →   ← TOWARD LOAD

CENTER

Z_{in} = ?

$\boxed{Z_L}$ = 200 − j150 Ω

4.3λ

FIGURE 12-24   Smith chart for Ex. 12-8.

the Smith chart would be a spiral. The correction for this condition is made using the "transm. loss, 1-dB steps" scale at the bottom left-hand side of the Smith chart. Other scales at the bottom of the chart can be used to provide the VSWR (which can be taken directly off the chart as previously shown), the voltage and power reflection coefficient, and decibel loss information.

Matching Using the Smith Chart

Many of the calculations involved with transmission lines pertain to matching a load to the line in order to keep the VSWR as low as possible. The following examples illustrate some of these situations.

EXAMPLE 12-9

A load of 75 Ω + j50 Ω is to be matched to a 50-Ω transmission line using a λ/4 matching section. Determine the proper location and characteristic impedance of the matching section.

SOLUTION

1. Normalize the load impedance:

$$z_L = \frac{Z_L}{Z_0} = \frac{75\ \Omega + j50\ \Omega}{50\ \Omega} = 1.5 + j1$$

2. Plot z_L on the Smith chart and draw the VSWR circle. This is shown in Fig. 12-25.

3. From z_L move toward the generator (CW) until a point is reached where the line is purely resistive. Recall that the λ/4 matching section only works between a pure resistance and the line.

4. Point A or B in Fig. 12-25 could be used as the point where the matching section is inserted. We shall select point A since it is closest to the load. It is 0.058λ from the load. At that point the line should be cut and the λ/4 section inserted as shown in Fig. 12-25.

5. The normalized impedance at point A is purely resistive and equal to $z = 2.4$. That also is the VSWR on the line with no matching. The actual resistance is 2.4 × 50 Ω = 120 Ω. Therefore, the characteristic impedance of the matching section is

$$Z_0' = \sqrt{Z_0 \times R_L}$$
$$= \sqrt{50\ \Omega \times 120\ \Omega} = 77.5\ \Omega$$

$$(12\text{-}29)$$

Stub Tuners

The use of short-circuited stubs is prevalent in matching problems. A *single-stub tuner* is illustrated in Fig. 12-26(a). In it the stub's distance from the load and location of its short

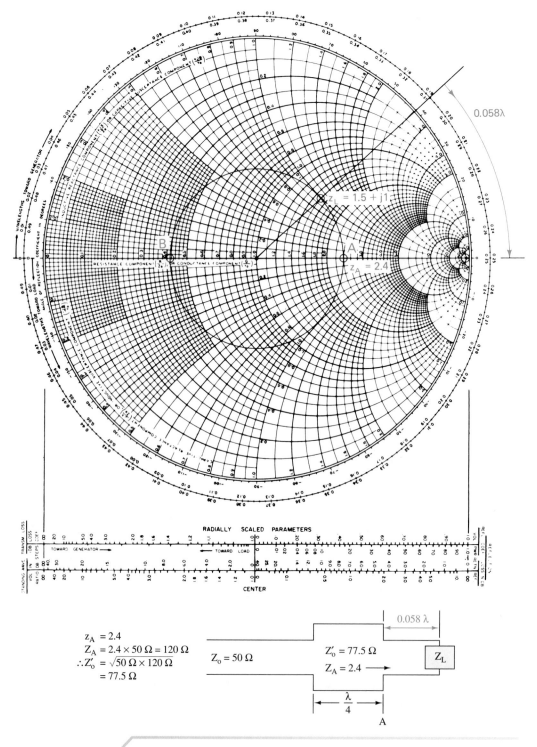

$$z_L = 1.5 + j1$$
$$z_A = 2.4$$

0.058λ

RADIALLY SCALED PARAMETERS

TOWARD GENERATOR → ← TOWARD LOAD

CENTER

$z_A = 2.4$
$Z_A = 2.4 \times 50\,\Omega = 120\,\Omega$
$\therefore Z_o' = \sqrt{50\,\Omega \times 120\,\Omega}$
$\quad = 77.5\,\Omega$

$Z_o = 50\,\Omega$

$0.058\,\lambda$

$Z_o' = 77.5\,\Omega$
$Z_A = 2.4 \rightarrow$

Z_L

$\dfrac{\lambda}{4}$

A

FIGURE 12-25 Smith chart for Ex. 12-9.

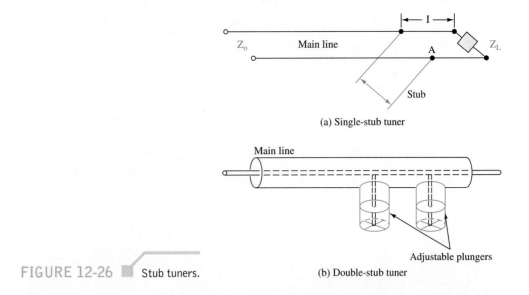

FIGURE 12-26 ■ Stub tuners.

(a) Single-stub tuner

Adjustable plungers

(b) Double-stub tuner

circuit are adjustable. The *double-stub* tuner shown in Fig. 12-26(b) has fixed stub locations, but the position of both short circuits is adjustable. The following example illustrates the procedure for matching with a single-stub tuner.

EXAMPLE 12-10

The antenna load on a 75-Ω line has an impedance of $50 - j100\ \Omega$. Determine the length and position of a short-circuited stub necessary to provide a match.

SOLUTION

1. $$z_L = \frac{Z_L}{Z_0} = \frac{50 - j100\ \Omega}{75\ \Omega} = 0.67 - j1.33$$

2. Plot z_L on the Smith chart (Fig. 12-27), and draw the VSWR circle.

3. Convert z_L to y_L by going to the diagonally opposite side of the VSWR circle from z_L. Read $y_L = 0.27 + j0.59$.

4. Move from y_L to the point where the admittance is $1 \pm$ whatever j term results. That is point A in Fig. 12-27. Read point A as $y_A = 1 + j1.75$ and note that point A is 0.093λ from the load. The short-circuited stub should be located 0.093λ from the load.

5. Now the $+j1.75$ term at point A must be canceled out by the stub admittance. If the stub admittance is $y_s = -j1.75$, the imaginary terms cancel out since the total admittance at point A with the parallel stub is $(1 + j1.75 - j1.75) = 1$. Recall that parallel admittances are directly additive.

6. The load admittance of the short-circuited stub is infinite. That is plotted on the Smith chart as point B. From point B move toward the gen-

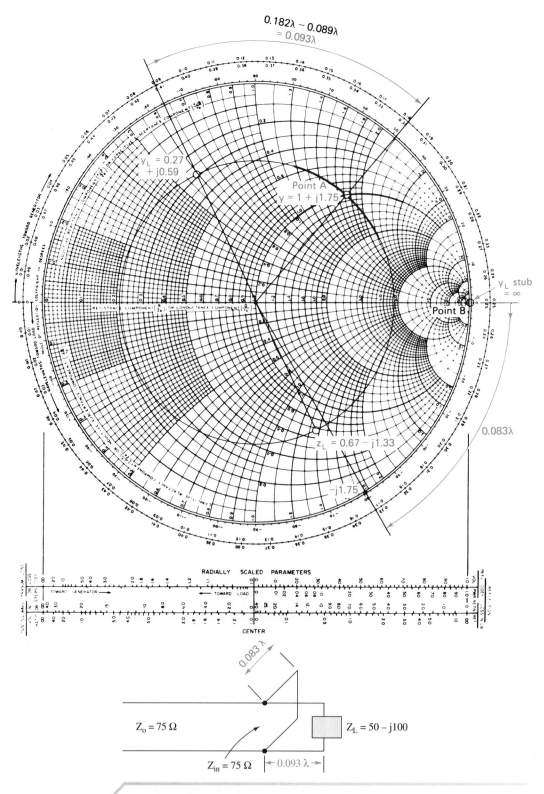

FIGURE 12-27 ▮ Smith chart for Ex. 12-10.

erator until the stub admittance is $-j1.75$. That corresponds to a distance of 0.083λ and is the required length for the short-circuited stub.

7. A match is now accomplished since the total admittance at the stub location is 1. This means that

$$z = \frac{1}{y} = \frac{1}{1} = 1$$

or $Z = 1 \times 75\Omega = 75\ \Omega$, which matches the line to the point where the stub is connected back toward the generator.

The key to matching with the single-stub tuner of the previous example is moving back from the load until the admittance takes on a normalized real term of 1. Whatever j term (imaginary) results can then be canceled out with a short-circuited stub made to look like the same j term with opposite polarity. While the Smith chart may at first be perplexing to work with, a bit of practice allows mastery in a short period of time.

12-9 TRANSMISSION LINE APPLICATIONS

Discrete Circuit Simulation

Transmission line sections can be used to simulate inductance, capacitance, and LC resonance. Figure 12-28 tabulates these effects for open and shorted sections with lengths equal to, less than, and greater than a quarter wavelength. A shorted quarter-wavelength section looks like a parallel LC circuit or ideally therefore like an open circuit. A shorted section less than $\lambda/4$ looks like a pure inductance and a section greater than $\lambda/4$ like a pure capacitance. These effects can be verified on a Smith chart by plotting z_L of 0 for a short-circuited line at the left center of the chart. Moving CW toward the generator less than $\lambda/4$ puts you on the top half of the chart, indicating a $+j$ term and thus inductance. Moving exactly $\lambda/4$ puts you at the center-right point on the chart, indicating the infinite impedance of an ideal tank circuit, and moving past that point provides the $-j$ impedance of capacitance.

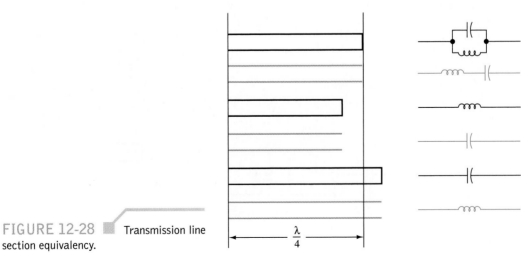

FIGURE 12-28 Transmission line section equivalency.

While open-circuit sections would seem to provide similar effects, they are seldom used. The open-circuited line tends to radiate a fair amount of energy off the end of the line so that total reflection does not take place. This causes the simulated circuit element to take on a resistive term, which greatly reduces its quality. These losses do not occur with shorted sections, and the simulated circuit has better quality than is possible using discrete inductors and/or capacitors. Short-circuited $\lambda/4$ sections offer Q's of about 10,000 as compared to a maximum of about 1000 using a very high-grade inductor and capacitor.

For this reason, it is normal practice to use transmission line sections to replace inductors and/or capacitors at frequencies above 500 MHz where the line section becomes short enough to be practically used. They are commonly found in the oscillator of UHF tuners for television, which operate from about 500 to 800 MHz.

Baluns

The parallel wire line generally carries two equal but 180° out-of-phase signals with respect to ground. Such a line is called a *balanced* line. In a coaxial line the outer conductor is usually grounded so that the two conductors do not carry signals with the same relationship to ground. The coaxial line is therefore termed an *unbalanced* line.

It is sometimes necessary to change from an unbalanced to a balanced condition such as when a coaxial line feeds a balanced load like a dipole antenna. They cannot be connected directly together, as the antenna lead connected to the cable's grounded conductor would ground that point of the antenna and impair operation.

This situation is solved through use of an unbalanced-to-balanced transformer, which is usually called a *balun*. It can be a transformer as shown in Fig. 12-29(a) or a special transmission line configuration as shown at (b). The use of the standard transformer is limited due to excessive losses at high frequencies. The balun in Fig. 12-29(b) does not suffer in that respect. The inner conductor of the coaxial line is tapped at 180° ($\lambda/2$) from the end. The tap and the end of the inner conductor provide two equal but 180° out-of-phase signals, neither of which is grounded. Thus, they supply the required signals for the balanced line. These baluns are reversible in that they function equally well in going from a balanced to an unbalanced condition.

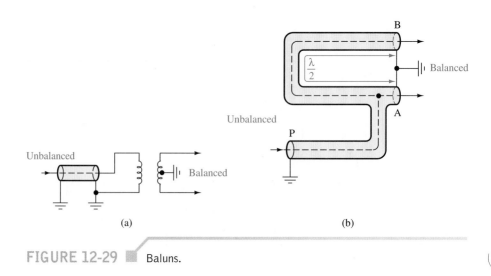

FIGURE 12-29 ▦ Baluns.

Transmission Lines as Filters

A quarter-wave section of transmission line can be used as an efficient filter or suppressor of even harmonics. Other types of filters may be used to filter out the odd harmonics. In fact, filters may be designed to eliminate the radiation of an entire single sideband of a modulated carrier.

Suppose that a transmitter is operating on a frequency of 5 MHz, and it is found that the transmitter is causing interference at 10 and 20 MHz. A shorted quarter-wave line section may be used to eliminate these undesirable harmonics. A quarter-wave line shorted at one end offers a high impedance at the unshorted end to the fundamental frequency. At a frequency twice the fundamental, such a line is a half-wave line, and at a frequency 4 times the fundamental, the line becomes a full-wave line. A half-wave or full-wave line offers zero impedance when its output is terminated in a short. Therefore, the radiation of even harmonics from the transmitter antenna can be eliminated almost completely by the circuit shown in Fig. 12-30.

The resonant filter line, *AB,* is a quarter wave in length at 5 MHz and offers almost infinite impedance at this frequency. At the second harmonic, 10 MHz, the line *AB* is a half-wave line and offers zero impedance at the antenna, thereby shorting this frequency to ground. The quarter-wave filter may be inserted anywhere along the nonresonant transmission line with a similar effect.

Slotted Lines

One of the simplest and yet most useful measuring instruments at very high frequencies is the *slotted line.* As its name implies, it is a section of coaxial line with a lengthwise slot cut in the outer conductor. A pickup probe is inserted into the slot, and the magnitude of signal picked up is proportional to the voltage between the conductors at the point of insertion. The probe rides in a carriage along a calibrated scale so that data can be obtained to plot the standing wave pattern as a function of distance. From these data, the following information can be determined:

1. VSWR
2. Generator frequency
3. Unknown load impedance

Time-Domain Reflectometry

Time-domain reflectometry (TDR) is a system whereby a short-duration pulse is transmitted into a line. While monitoring with an oscilloscope, the reflection of that pulse provides much information regarding the line. Of course, no reflection at all indicates an

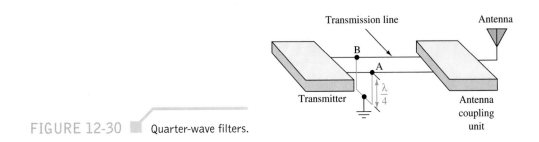

FIGURE 12-30 ■ Quarter-wave filters.

infinitely long line or (more likely) a line with a perfectly matched load and no disconti-nuities.

A very common problem with transmission line communications involves cable failure between communications terminals. These problems are usually due to chemical erosion or a mechanical break. Location of these problems is done by using a time-domain reflectometer. These instruments can pinpoint a fault within several feet at a distance of 10 mi. This is accomplished by simply measuring the time taken for the return pulse and then calculating distance based on the cable's propagation velocity.

TDR is also often used in laboratory testing. In these cases the transmitted signal is sometimes a fast-rise-time step signal. The setup for this condition is shown in Fig. 12-31(a). In Fig. 12-31(b) the case for an open-circuited line is shown. Assuming a lossless line, the reflected wave is in phase and effectively results in a doubling of the incident step voltage. The time T shown is the time of incidence plus reflection and therefore should be divided by 2 to calculate the distance to the open circuit. Thus, the distance is propagation velocity (V_p) times $T/2$.

The case for a shorted line is shown in Fig. 12-31(c). In this case the reflection is out of phase and results in cancellation. The situation of Z_L greater than Z_0 is shown in

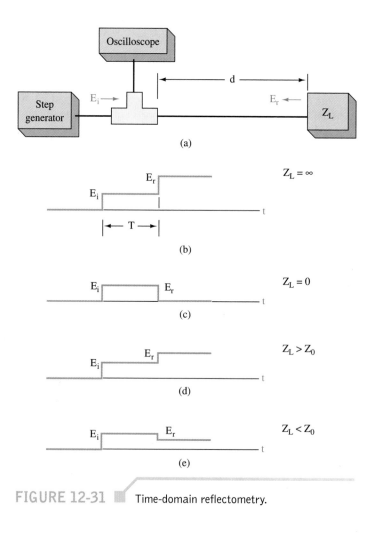

FIGURE 12-31 ▮ Time-domain reflectometry.

Fig. 12-31(d), while (e) shows the case for Z_L less than Z_0. The magnitudes shown in the scope display can be used to determine Z_L or the reflection coefficient Γ since

$$\Gamma = \frac{E_r}{E_i} \tag{12-23}$$

and

$$\Gamma = \frac{Z_L - Z_0}{Z_L + Z_0} \tag{12-24}$$

and the final voltage, E_F, is

$$E_F = E_i + E_r$$
$$= E_i(1 + \Gamma)$$
$$= E_i\left(1 + \frac{Z_L - Z_0}{Z_L + Z_0}\right)$$

TROUBLESHOOTING

Vital links of communication data flow over transmission lines daily. Phone conversations, computer data, and audio and video information all depend on the media's ability to pass these signals wholly without any loss of quality. In this section, we will look at a popular application for transmission lines. We will also look at problems encountered in the cabling application. Many opportunities exist today for the technician who can effectively track down and repair cabling troubles.

After completing this section you should be able to

- Identify popular types of cabling available for LANs and other uses
- Describe crosstalk interference and give an example of it
- Describe simple methods for testing cable resistance, insulation, and TDR
- Troubleshoot television antenna lines

Common Application

A very common application for miles and miles of wire today is the local area network (LAN). The most popular kinds of wiring used to connect these systems are shielded and unshielded twisted pair (STP and UTP, respectively) and coaxial cable. STP and UTP represent about 50% to 70% of the installed base for LANs and coax about 30%. Twisted pair and coax are used mainly to connect the workstation to the hub, or punch-down block. The vitality of a local area network depends on the cabling base—the transmission media. As a matter of fact, 75% of LAN failures are wiring related. Therefore, technicians must learn techniques to effectively track cable problems and repair them.

Losses on Transmission Lines

From previous discussions in this chapter, transmission line losses can be summed up as energy losses, magnetic field losses, and electric field losses. Energy losses are a result of wire heating and leakage currents in the dielectric material used as the insulator. Heat is produced when current flows in the wire that has resistance. The heat is radiated away

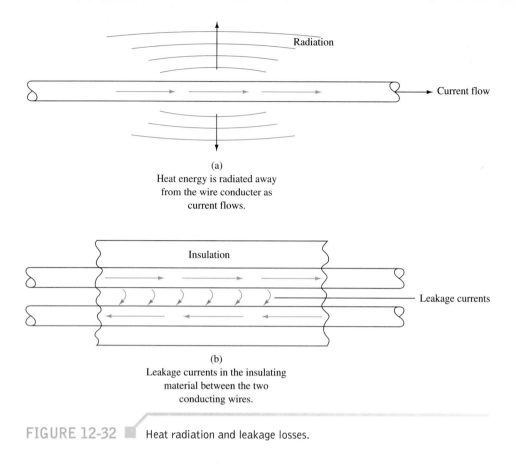

(a)
Heat energy is radiated away
from the wire conducter as
current flows.

(b)
Leakage currents in the insulating
material between the two
conducting wires.

FIGURE 12-32 ▮ Heat radiation and leakage losses.

from the wire as shown in Fig. 12-32(a). Figure 12-32(b) illustrates the effect of dielectric losses. Voltage potentials develop between the conductors that in turn cause minute currents to flow through the dielectric material. Heating and dielectric losses increase as signal frequency increases.

Magnetic field losses occur when currents are induced in a nearby conductor from the influence of another conductor's magnetic fields. Likewise, electrostatic fields tend to build a charge in a nearby conductor, causing electric field loss. Another effect of magnetic field and electrostatic field is induced interference, called crosstalk.

Interference on Transmission Lines

When UTP is used for the wire base in a LAN, there will be a transfer of energy between the wires, causing interference to be induced on wires. This unwanted coupling is caused by overlapping electric and magnetic fields and is called *crosstalk*. All of us have experience hearing another party's conversation while having our own on the telephone. This is an example of crosstalk. Crosstalk can be reduced by careful attention to wire separation and twists. Tighter twists create better coupling, which cancels the effects of magnetic and electrostatic fields, thus reducing crosstalk interference. Don't untwist twisted pair, even at the connection points. Crosstalk can be measured by special hand-held instruments that directly compare the crosstalk reading against set standards. Crosstalk is also reduced by using STP or coaxial cable, but expense goes up and must be weighed against the system's requirements.

Logic circuits can be fooled into creating data from noise. Noise can come from a variety of sources. Electric lights, electric generators, and switching devices generate noise that can become part of the signal if proper precautions are not observed. Do not run sensitive cabling near ac power lines or motors. Avoid running cable near overhead fluorescent lighting. In troubleshooting a system that is bothered by interference, check the wiring; someone may have ignored the above warnings. Shielding and proper cable terminations are also safeguards against noise interference.

Cable Testing

Test instruments have been developed to do sophisticated testing on cable links. Discussing these instruments in detail is beyond the scope of this material. However, some instruments will be mentioned. Simple testing techniques can be used to discover faulty cabling. Pin-to-pin continuity testing done from one end of the wire to the opposite end is a simple effective means to detect miswiring and shorted and open links. Resistance measurements can be made on lengths of single wire and compared to the cable manufacturer's specified resistance value. At the far end the wires to be tested would be shorted together, and the resistance would be measured at the near end. The resistance value would represent the resistance of two wires and must be taken under consideration. Resistance readings above or below the expected value indicates opens or shorts in the wires. Insulation testers or the megohm meter (megometer) give a readout or audible indication of the insulation test results. Normally the megometer is connected to the wires at the near end, and the wires are disconnected at the opposite end. The readout represents the condition of the insulation. Breakdowns of the insulation at any point along the wiring path are readily displayed.

Television Antenna Line Repair

Transmission lines seldom totally fail without human intervention; however, they do age. Sunlight is hard on plastic dielectrics used in ribbon-type lines, sometimes called twin-lead. Where twin-lead is used near salt water, it deteriorates more rapidly. Moisture and salt collect on the dielectric, causing the loss to rise to the point that the twin-lead is not usable.

Coaxial cables are relatively immune to aging due to weather. Older types suffered from contamination due to plasticizer in the outer cover. Water can enter the cable at connectors and breaks in the outer covering. Water has a very high dielectric constant and will change the characteristic impedance and increase the loss. Sometimes coaxial cable is installed with very tight turns around corners. Over time, the center conductor will walk through the dielectric and eventually short the cable.

Suppose you are called to service a television receiver with "snow" almost overwhelming the picture. TV twin-lead is being used and is too cheap to justify much testing. Simply inspect it, looking for cracks in the plastic and discoloration of the plastic. You might take the time to short the far end and check resistance with an ohmmeter; there should be almost no resistance. If the far end is open, the resistance should be infinite. When in doubt, change the twin-lead.

SUMMARY

In Chapter 12 we introduced basic transmission line theory. It was discovered that a basic wire doesn't just act like a strip of copper when the frequency is high enough. The major topics the student should now understand include:

- the definition and types of transmission lines, including two-wire open line, twisted pair, shielded pair, and coaxial cable

- the discussion of transmission line electrical characteristics, emphasizing characteristic impedance and the various line losses

- the explanation of dc voltage propagation, including velocity, delay, and reflections from both shorted and open-circuited lines

- the analysis of standing waves for open and shorted lines

- the definition of voltage standing wave ratio (VSWR), electrical length of a line, effects of a mismatched load, and use of a quarter-wavelength matching transformer

- the description of a Smith chart and its use in matching a load to a line

- the application of transmission lines as discrete circuit components, unbalanced-to-balanced transformers (baluns), and filters

- the testing of transmission lines using slotted lines and time-domain reflectometers

QUESTIONS AND PROBLEMS

SECTION 2

1. Define *transmission line*. If a simple wire connection can be a transmission line, why is an entire chapter of study devoted to it?
2. In general terms, discuss the various types of transmission lines. Include the advantages and disadvantages of each type.
*3. What would be the considerations in choosing a solid dielectric cable over a hollow pressurized cable for use as a transmission line?
*4. Why is an inert gas sometimes placed within concentric radio-frequency transmission cables?

SECTION 3

5. Draw an equivalent circuit for a transmission line, and explain the physical significance of each element.
6. Provide a physical explanation for the meaning of a line's characteristic impedance (Z_0).
7. Calculate Z_0 for a line that exhibits an inductance of 4 nH/m and 1.5 pF/m. (51.6 Ω)
8. Calculate the capacitance per meter of a 50-Ω cable that has an inductance of 55 nH/m. (22 pF/m)
9. In detail, explain how an impedance bridge could be used to determine Z_0 for a piece of transmission line.

*10. If the spacing of the conductors in a two-wire radio-frequency transmission line is doubled, what change takes place in the surge impedance (Z_0) of the line?

*11. If the conductors in a two-wire radio-frequency transmission are replaced by larger conductors, how is the surge impedance affected, assuming no change in the center-to-center spacing of the conductor?

*12. What determines the surge impedance of a two-wire radio-frequency transmission line?

13. Determine Z_0 for the following transmission lines:
   (a) Parallel wire, air dielectric with $D/d = 3$. (215 Ω)
   (b) Coaxial line, air dielectric with $D/d = 1.5$. (24.3 Ω)
   (c) Coaxial line, polyethylene dielectric with $D/d = 2.5$. (36.2 Ω)

14. List and explain the various types of transmission line losses.

*15. A long transmission line delivers 10 kW into an antenna; at the transmitter end, the line current is 5 A, and at the coupling house (load) it is 4.8 A. Assuming the line to be properly terminated and the losses in the coupling system negligible, what is the power lost in the line? (850 W)

SECTION 4

16. With the help of Fig. 12-10, provide a step-by-step explanation of how a dc voltage propagates through a transmission line.

17. Derive the equation for the time required for energy to propagate through a transmission line [Eq. (12-19)].

*18. What is the velocity of propagation for radio-frequency waves in space?

19. A delay line using RG-8A/U cable is to exhibit a 5-ns delay. Calculate the required length of this cable. (3.39 ft)

20. Explain the significance of the velocity factor for a transmission line.

SECTION 5

21. Define a *nonresonant* transmission line, and explain what its traveling waves are and how they behave.

*22. An antenna is being fed by a properly terminated two-wire transmission line. The current in the line at the input end is 3 A. The surge impedance of the line is 500 Ω. How much power is being supplied to the line? (4.5 kW)

23. With the help of Fig. 12-13, provide a charging analysis of a nonresonant line with an ac signal applied.

SECTION 6

24. Explain the properties of a resonant transmission line. What happens to the energy reaching the end of a resonant line? Are reflections a generally desired result?

25. A dc voltage from a 20-V battery with $R_s = 75$ Ω is applied to a 75-Ω transmission line at $t = 0$. It takes the battery's energy 10 μs to reach the load, which is an open circuit. Sketch current and voltage waveforms at the line's input and load.

26. Repeat Problem 25 for a short-circuited load.

27. What are *standing waves, standing wave ratio* (SWR), and *characteristic imped-*

ance, as referred to transmission lines? How can standing waves be minimized?

28. With the help of Fig. 12-17, explain how standing waves develop on a resonant line.

29. An open-circuited line is 1.75λ long. Sketch the incident, reflected, and resultant waveforms for both voltage and current at the instant the generator is at its peak negative value.

30. Repeat Problem 29 for a short-circuited line.

*31. If the period of one complete cycle of a radio wave is 0.000001 s, what is the wavelength? (300 m)

*32. If the two towers of a 950-kHz antenna are separated by 120 electrical degrees, what is the tower separation in feet? (345 ft)

SECTION 7

33. Define *reflection coefficient,* Γ, in terms of incident and reflected voltage and also in terms of a line's load and characteristic impedances.

34. Express SWR in terms of
    (a) Voltage maximums and minimums.
    (b) Current maximums and minimums.
    (c) The reflection coefficient.
    (d) The line's load resistance and Z_0.

35. Explain the disadvantages of a mismatched transmission line.

*36. What is the primary reason for terminating a transmission line in an impedance equal to the characteristic impedance of the line?

*37. What is the ratio between the currents at the opposite ends of a transmission line one-quarter wavelength long and terminated in an impedance equal to its surge impedance?

38. A SSB transmitter at 2.27 MHz and 200 W output is connected to an antenna (R_{in} = 150 Ω) via 75 ft of RG-8A/U cable. Determine
    (a) The reflection coefficient.
    (b) The electrical cable length in wavelengths.
    (c) The SWR.
    (d) The amount of power absorbed by the antenna.

*39. What should be the approximate surge impedance of a quarter-wavelength matching line used to match a 600-Ω feeder to a 70-Ω (resistive) antenna?

40. Explain why lines less than about λ/16 in length may be considered free of transmission line effects.

SECTION 8

41. Calculate the impedance of a line 675 electrical degrees long. Z_0 = 75 Ω and Z_L = 50 Ω + j75 Ω. Use the Smith chart *and* Eq. (12-30) as separate solutions, and compare the results.

42. Convert an impedance, 62.5 Ω – j90 Ω, to admittance mathematically *and* with the Smith chart. Compare the results.

43. Find the input impedance of a 100-Ω line, 5.35λ long, and with Z_L = 200 Ω + j300 Ω.

44. Match a load of 25 Ω + j75 Ω to a 50-Ω line using a quarter-wavelength matching

section. Determine the proper location and characteristic impedance of the matching section.

45. Repeat Problem 44 for a $Z_L = 110\ \Omega - j50\ \Omega$ load. Provide *two* separate solutions.

*46. Why is the impedance of a transmission line an important factor with respect to matching "out of a transmitter" into an antenna?

*47. What is *stub tuning*?

48. The antenna load on a 150-Ω transmission line is $225\ \Omega - j300\ \Omega$. Determine the length and position of a short-circuited stub necessary to provide a match.

49. Repeat Problem 48 for a 50-Ω line and an antenna of $25\ \Omega + j75\ \Omega$.

SECTION 9

50. Calculate the length of a short-circuited 50-Ω line necessary to simulate an inductance of 2 nH at 1 GHz.

51. Calculate the length of a short-circuited 50-Ω line necessary to simulate a capacitance of 50 pF at 500 MHz.

52. Describe two types of baluns, and explain their function.

*53. How may harmonic radiation of a transmitter be prevented?

*54. Describe three methods for reducing harmonic emissions of a transmitter.

*55. Draw a simple schematic diagram showing a method of coupling the radio-frequency output of the final power amplifier stage of a transmitter to a two-wire transmission line, with a method of suppression of second and third harmonic energy.

56. Explain the construction of a slotted line and some of its uses.

57. Explain the principle of TDR and some uses for this technique.

58. A pulse is sent down a transmission line that is not functioning properly. It has a propagation velocity of 2.1×10^8 m/s, and an inverted reflected pulse (equal in magnitude to the incident pulse) is returned in 0.731 ms. What is wrong with the line, and how far from the generator does the fault exist?

59. A fast-rise-time 10-V step voltage is applied to a 50-Ω line terminated with an 80-Ω resistive load. Determine Γ, E_F, and E_r. (0.231, 12.3 V, 2.3 V)

WAVE PROPAGATION

OBJECTIVES

- Discuss the makeup of an electromagnetic wave and the characteristics of an isotropic point source

- Explain the processes of wave reflection, refraction, and diffraction

- Describe ground- and space-wave propagation and calculate the ghosting effect in TV reception

- Calculate the approximate radio horizon based on antenna height

- Discuss the effects of the ionosphere on sky-wave propagation

- Define the critical angle and skip zone for sky-wave propagation

- Describe the important aspects of satellite communication

- Calculate the power received by a satellite based upon a SATCOM power budget analysis

13-1 ELECTRICAL TO ELECTROMAGNETIC CONVERSION

Early radios were often referred to as the "wireless." This new machine could speak without being "wired" to the source like the telegraph and telephone. The transmitter's output is coupled to its surrounding atmosphere and then intercepted by the receiver. We know that the atmosphere is *not* a conductor of electrons like a copper wire—air is, in fact, a very good insulator. Thus, the electrical energy fed into a transmitting antenna must be converted to another form of energy for transmission. In this chapter we shall study the effects of the transformed energy and its propagation.

The transmitting antenna converts its input electrical energy into electromagnetic energy. The antenna can thus be thought of as a *transducer*—a device that converts from one form of energy into another. In that respect, a light bulb is very similar to an antenna. The light bulb also converts electrical energy into electromagnetic energy—light. The only difference between light and the radio waves we shall be concerned with is their frequency. Light is an electromagnetic wave at about 5×10^{14} Hz, while the usable radio waves extend from about 1.5×10^4 Hz up to 3×10^{11} Hz. The human eye is responsive

Photo: Wave propogation using satellites. (Courtesy of W. L. Gore & Associates, Inc.)

521

(able to perceive) to the very narrow range of light frequencies, and consequently we are blind to the radio waves. Actually, that is a good thing since the great number of radio waves surrounding our earth would otherwise paint a very chaotic picture.

The receiving antenna intercepts the transmitted wave and converts it back into electrical energy. An analogous transducer for it is the photovoltaic cell that also converts a wave (light) into electrical energy. Since a basic knowledge of waves is necessary to your understanding of antennas and radio communications, the following section is presented prior to your further study of wave propagation.

13-2 ELECTROMAGNETIC WAVES

Electricity and electromagnetic waves are interrelated. An electromagnetic field consists of an electric field and a magnetic field. These fields exist with all electric circuits since any current-carrying conductor creates a magnetic field around the conductor, and any two points in the circuit with a potential difference (voltage) between them create an electric field. These two fields contain energy, but in circuits this field energy is usually returned to the circuit when the field collapses. If the field does not fully return its energy to the circuit, it means the wave has been at least partially *radiated,* or set free, from the circuit. This radiated energy is undesired, as it may cause interference with other electronic equipment in the vicinity. It is termed *radio-frequency interference* (RFI) if it is undesired radiation from a radio transmitter, and if from another source, it is termed *electromagnetic interference* (EMI) or, more simply, noise.

In the case of a radio transmitter, it is hoped that the antenna efficiently causes the wave energy to be set free. The antenna is designed so as to *not* allow the electromagnetic wave energy to collapse back into the circuit.

An electromagnetic wave is pictured in Fig. 13-1. In it $1\frac{1}{2}$ wavelengths of the electric field (E) and the magnetic field (H) are shown. The direction of propagation is shown to be perpendicular to both fields, which are also mutually perpendicular to each other. The wave is said to be *transverse* since the oscillations are perpendicular to the direction of propagation. The *polarization* of an electromagnetic wave is determined by the direction of its E field component. In Fig. 13-1 the E field is vertical (y direction), and the wave is therefore said to be vertically polarized. As will subsequently be shown, the antenna's orientation determines polarization. A vertical antenna results in a vertically polarized wave.

Wavefronts

If an electromagnetic wave were radiated equally in all directions from a point source in free space, a spherical wavefront would result. Such a source is termed an *isotropic point source.* A *wavefront* may be defined as a plane joining all points of equal phase. Two wavefronts are shown in Fig. 13-2. An *isotropic* source radiates equally in all directions. The wave travels at the speed of light so that at some point in time the energy will have reached the area indicated by wavefront 1 in Fig. 13-2. The power density, $\mathcal{P}$ (in watts per square meter), at wavefront 1 is inversely proportional to the square of its distance, r (in meters), from its source, with respect to the originally transmitted power, P_t. Stated mathematically,

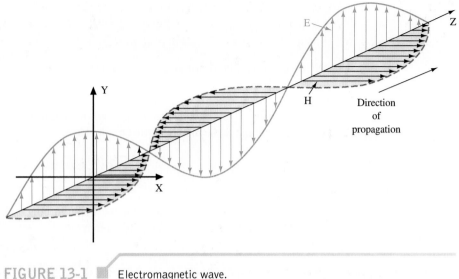

FIGURE 13-1 ▦ Electromagnetic wave.

$$\mathcal{P} = \frac{P_t}{4\pi r^2} \tag{13-1}$$

If wavefront 2 in Fig. 13-2 is twice the distance of wavefront 1 from the source, then its power density in watts per unit area is just one-fourth that of wavefront 1. Any section of a wavefront is curved in shape. However, at appreciable distances from the source, small sections are nearly flat. These sections can then be considered as *plane wavefronts,* which simplifies the treatment of their optical properties provided in Sec. 13-3.

Characteristic Impedance of Free Space

The strength of the electric field, $\mathcal{E}$ (in volts per meter), at a distance r from a point source is given by

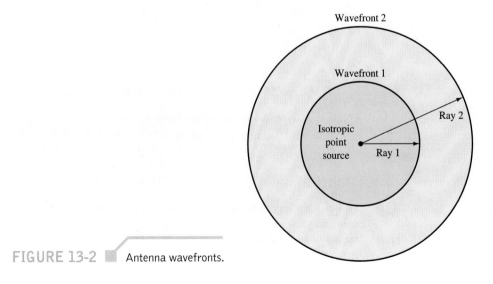

FIGURE 13-2 ▦ Antenna wavefronts.

$$\mathcal{E} = \frac{\sqrt{30P_t}}{r} \tag{13-2}$$

where P_t is the originally transmitted power in watts. This is one of Maxwell's equations, which were finalized in 1873 and allowed mathematical analysis of electromagnetic wave phenomena.

Power density $\mathcal{P}$ and the electric field $\mathcal{E}$ are related to impedance in the same way that power and voltage relate in an electric circuit. Thus,

$$\mathcal{P} = \frac{\mathcal{E}^2}{\mathcal{Z}} \tag{13-3}$$

where $\mathcal{Z}$ is the characteristic impedance of the medium conducting the wave. For free space, Eqs. (13-1) and (13-2) can be substituted into (13-3) to give

$$\mathcal{Z} = \frac{\mathcal{E}^2}{\mathcal{P}} = \frac{30P_t}{r^2} \div \frac{P_t}{4\pi r^2} = 120\pi = 377\ \Omega \tag{13-4}$$

Thus, it is seen that free space has a characteristic impedance just as does a transmission line.

The characteristic impedance of any electromagnetic wave-conducting medium is provided by

$$\mathcal{Z} = \sqrt{\frac{\mu}{\varepsilon}} \tag{13-5}$$

where μ is the medium's permeability and ϵ is the medium's permittivity.

For free space, $\mu = 1.26 \times 10^{-6}$ H/m and $\epsilon = 8.85 \times 10^{-12}$ F/m. Substituting in Eq. (13-5) yields

$$\mathcal{Z} = \sqrt{\frac{\mu}{\varepsilon}} = \sqrt{\frac{1.26 \times 10^{-6}}{8.85 \times 10^{-12}}} = 377\ \Omega$$

which agrees with the result from Eq. (13-4).

13-3 WAVES NOT IN FREE SPACE

Until now, we have discussed the behavior of waves in free space, which is a vacuum or complete void. We now consider the effects of our environment on wave propagation.

Reflection

Just as light waves are reflected by a mirror, radio waves are reflected by any conductive medium such as metal surfaces or the earth's surface. The angle of incidence is equal to the angle of reflection, as shown in Fig. 13-3. Note that there is a change in phase of the incident and reflected waves, as seen by the difference in the direction of polarization. The incident and reflected waves are 180° out of phase.

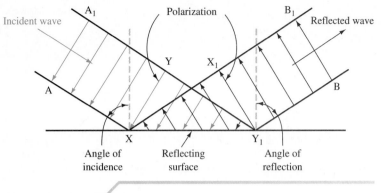

FIGURE 13-3 ■ Reflection of a wavefront.

Complete reflection occurs only for a theoretically perfect conductor and when the electric field is perpendicular to the reflecting element. For it the *coefficient of reflection* ρ is 1 and is defined as the ratio of the reflected electric field intensity divided by the incident intensity. It is less than 1 in practical situations due to the absorption of energy by the nonperfect conductor and also because some of the energy will actually propagate right through it.

The previous discussion is valid when the electric field is *not* normal to the reflecting surface. If it is fully parallel to the reflecting (conductive) surface, the electric field is *shorted* out, and all of the electromagnetic energy is dissipated in the form of generated surface currents in the conductor. If the electric field is partially parallel to the surface, it will be partially shorted out.

If the reflecting surface is curved, as in a parabolic antenna, the wave may be analyzed using the appropriate optical laws with regard to focusing the energy, etc. This is especially true with respect to microwave frequencies treated in Chapter 15.

Refraction

Refraction of electromagnetic radio waves occurs in a manner akin to the refraction of light. Refraction occurs when waves pass from one density medium to another.

An example of refraction is the apparent bending of a spoon when it is immersed in water. The bending seems to take place at the water's surface, or exactly at the point where there is a change of density. Obviously, the spoon does not bend from the pressure of the water. The light forming the image of the spoon is bent as it passes from the water, a medium of high density, to the air, a medium of comparatively low density.

The bending (refraction) of an electromagnetic wave (light or radio wave) is shown in Fig. 13-4. Also shown is the reflected wave. Obviously, the coefficient of reflection is less than 1 here since a fair amount of the incident wave's energy is propagated through the water—after refraction has occurred.

The angle of incidence, θ_1, and the angle of refraction, θ_2, are related by the following expression, which is Snell's law:

$$n_1 \sin \theta_1 = n_2 \sin \theta_2 \tag{13-6}$$

where n_1 is the refractive index of the incident medium and n_2 is the refractive index of the refractive medium.

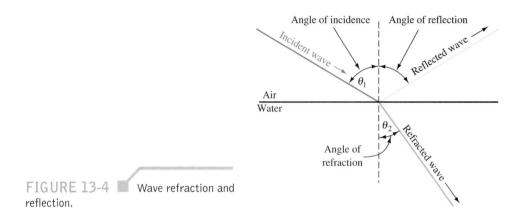

FIGURE 13-4 ▦ Wave refraction and reflection.

Recall that the refractive index for a vacuum is exactly 1 and approximately 1 for the atmosphere, while glass is about 1.5 and water is 1.33.

Diffraction

Diffraction is the phenomenon whereby waves traveling in straight paths bend around an obstacle. This effect is the result of Huygens' principle, advanced by the Dutch astronomer Christian Huygens in 1690. The principle states that each point on a spherical wavefront may be considered as the source of a secondary spherical wavefront. This concept is important to us since it explains radio reception behind a mountain or tall building. Figure 13-5 shows the diffraction process allowing reception beyond a mountain in all but a small area, which is called the *shadow zone*. The figure shows that electromagnetic waves are diffracted over the top and around the sides of an obstruction. The direct wavefronts that just get by the obstruction become new sources of wavefronts that start filling in the void, making the shadow zone a finite entity. The lower the frequency of the wave, the quicker is this process of diffraction (i.e., smaller shadow zone).

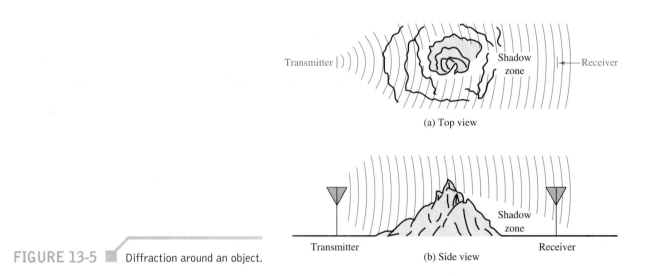

FIGURE 13-5 ▦ Diffraction around an object.

13-4 GROUND- AND SPACE-WAVE PROPAGATION

There are four basic modes of getting a radio wave from the transmitting to receiving antenna:

1. Ground wave
2. Space wave
3. Sky wave
4. Satellite communications

As will be seen in the following discussions, the frequency of the radio wave is of primary importance in considering the performance of each type of propagation.

Ground-Wave Propagation

A *ground wave* is a radio wave that travels along the earth's surface. It is sometimes referred to as a *surface wave*. The ground wave must be vertically polarized (electric field vertical) since the earth would short out the electric field if horizontally polarized. Changes in terrain have a strong effect on ground waves. Attenuation of ground waves is directly related to the surface impedance of the earth. This impedance is a function of conductivity and frequency. If the earth's surface is highly conductive, the absorption of wave energy, and thus its attenuation, will be reduced. Ground-wave propagation is much better over water (especially salt water) than say a very dry (poor conductivity) desert terrain.

The ground losses increase rapidly with increasing frequency. For this reason ground waves are not very effective at frequencies above 2 MHz. Ground waves are, however, a very reliable communications link. Reception is not affected by daily or seasonal changes such as with sky-wave propagation.

Ground-wave propagation is the only way to communicate into the ocean with submarines. To minimize the attenuation of seawater, extremely low frequency (ELF) propagation is utilized. ELF waves encompass the range 30 to 300 Hz. At a typically used frequency of 100 Hz, the attenuation is about 0.3 dB/m. This attenuation increases steadily with frequency such that at 1 GHz a 1000-dB/m loss is sustained!

Space-Wave Propagation

The two types of space waves are shown in Fig. 13-6. They are the direct wave and ground reflected wave. Do not confuse these with the ground wave just discussed. The direct wave is by far the most widely used mode of antenna communications. The propagated wave is direct from transmitting to receiving antenna and does not travel along the ground. The earth's surface, therefore, does not attenuate it.

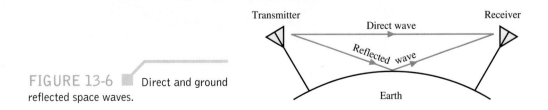

FIGURE 13-6 ▓ Direct and ground reflected space waves.

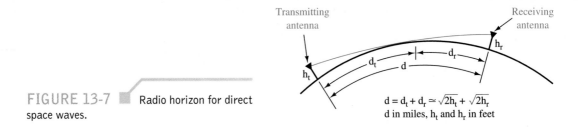

FIGURE 13-7 ■ Radio horizon for direct space waves.

$$d = d_t + d_r \approx \sqrt{2h_t} + \sqrt{2h_r}$$

d in miles, h_t and h_r in feet

The direct space wave does have one severe limitation—it is basically limited to so-called *line-of-sight* transmission distances. Thus, the antenna height and the curvature of the earth are the limiting factors. The actual *radio horizon* is about $\frac{4}{3}$ greater than the geometric line of sight due to diffraction effects and is empirically predicted by the following approximation:

$$d \approx \sqrt{2h_t} + \sqrt{2h_r} \tag{13-7}$$

where d = radio horizon (mi)
h_t = transmitting antenna height (ft)
h_r = receiving antenna height (ft)

The diffraction effects cause the slight wave curvature, as shown in Fig. 13-7. If the transmitting antenna is 1000 ft above ground level and the receiving antenna is 20 ft high, a radio horizon of about 50 mi results. This explains the coverage that typical broadcast FM and TV stations provide since they are utilizing direct space-wave propagation.

The reflected wave shown in Fig. 13-6 can cause reception problems. If the phase of these two received components is not the same, some degree of signal fading and/or distortion will occur. This can also result when both a direct and ground wave are received or when any two or more signal paths exist. A special case involving TV reception is presented next.

GHOSTING IN TV RECEPTION   Any tall or massive objects obstruct space waves. This results in diffraction (and subsequent shadow zones) and reflections. Reflections pose a specific problem since, for example, reception of a TV signal may be the combined result of a direct space wave and a reflected space wave, as shown in Fig. 13-8. This condition results in *ghosting,* which manifests itself in the form of a double-image distortion. This is due to the two signals arriving at the receiver at two different times—the reflected signal has a farther distance to travel. The reflected signal is weaker than the

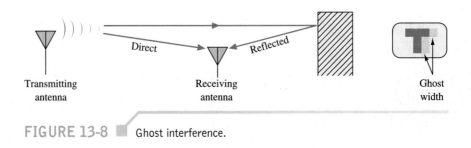

FIGURE 13-8 ■ Ghost interference.

EXAMPLE 13-1

Determine the ghost width on a TV screen 15 in wide when a reflected wave results from an object $\frac{1}{2}$ mi "behind" a receiver.

SOLUTION

The reflected wave travels 1 mi farther than the direct wave (2 × 0.5 mi). Each horizontal line on the receiver is 53.5 μs in duration (Chapter 7). Assuming the wave travels at the speed of light, the time of delay between the direct and reflected signal is

$$t = \frac{d}{v} = \frac{1 \text{ mi}}{186,000 \text{ mi/s}} = 5.38 \ \mu s$$

The ghost width will therefore be

$$\frac{5.38 \ \mu s}{53.5 \ \mu s/\text{trace}} \times 15 \text{ in} = 1.51 \text{ in}$$

direct signal because of the inverse square-law relationship of signal strength to distance [Eq. (13-1)] and because of losses incurred during reflection.

A possible solution to the ghosting problem shown in Ex. 13-1 is to detune the receiving antenna orientation so that the reflected wave is too weak to be displayed. Of course, the direct wave must exceed the receiver's sensitivity limit, as it will also be reduced in level when the antenna is detuned. It should be noted that ghosting can also be caused by transmission line reflections between antenna and set.

Ghosting is seen in the image on left. (Courtesy of Philips Consumer Electronics Company.)

13-5 SKY-WAVE PROPAGATION

One of the most frequently used methods of long-distance transmission is by the use of the *sky wave*. Sky waves are those waves radiated from the transmitting antenna in a direction that produces a large angle with reference to the earth. The sky wave has the ability to strike the ionosphere, be refracted from it to the ground, strike the ground, be reflected back toward the ionosphere, and so on. The refracting and reflecting action of the ionosphere and the ground is called *skipping*. An illustration of this skipping effect is shown in Fig. 13-9.

The transmitted wave leaves the antenna at point *A,* is refracted from the ionosphere at point *B,* is reflected from the ground at point *C,* is again refracted from the ionosphere at point *D,* and arrives at the receiving antenna *E.* The critical nature of the sky waves and the requirements for refraction will be discussed thoroughly in this section.

To understand the process of refraction, the composition of the atmosphere and the factors that affect it must be considered. Insofar as electromagnetic radiation is concerned, there are only three layers of the atmosphere: the troposphere, the stratosphere, and the ionosphere. The troposphere extends from the surface of the earth up to approximately 6.5 mi. The next layer, the stratosphere, extends from the upper limit of the troposphere to an approximate elevation of 23 mi. From the upper limit of the stratosphere to a distance of approximately 250 mi lies the region known as the ionosphere. Beyond the ionosphere is free space. The temperature in the stratosphere is considered to be a constant unfluctuating value. Therefore, it is not subject to temperature inversions, nor can it cause significant refractions. The constant temperature stratosphere is also called the *isothermal region.*

The ionosphere is appropriately titled because it is composed primarily of ionized particles. The density at the upper extremities of the ionosphere is very low and becomes progressively higher as it extends downward toward the earth. The upper region of the ionosphere is subjected to severe radiation from the sun. Ultraviolet radiation from the sun causes ionization of the air into free electrons, positive ions, and negative ions. Even though the density of the air molecules in the upper ionosphere is small, the radiation particles from space are of such high energy at that point that they cause wide-scale ionization of the air molecules that are present. This ionization extends down through the ionosphere with diminishing intensity. Therefore, the highest degree of ionization occurs at the upper extremities of the ionosphere, while the lowest degree occurs in the lower portion of the ionosphere.

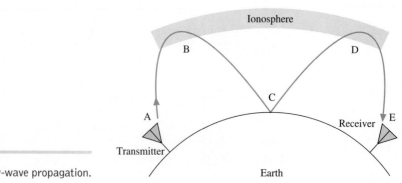

FIGURE 13-9 ■ Sky-wave propagation.

Ionospheric Layers

The ionosphere is composed of three layers designated, respectively, from lowest level to highest level as *D, E,* and *F*. The *F* layer is further divided into two layers designated F_1 (the lower layer) and F_2 (the higher layer). The presence or absence of these layers in the ionosphere and their height above the earth vary with the position of the sun. At high noon, radiation from the sun in the ionosphere directly above a given point is greatest, while at night it is minimum. When the radiation is removed, many of the ions that were ionized recombine. The interval of time between these conditions finds the position and number of the ionized layers within the ionosphere changing. Since the position of the sun varies with respect to a specified point on earth daily, monthly, and yearly, the exact characteristics of the layers are extremely difficult to predict. However, the following general statements can be made:

1. The *D* layer ranges from about 25 to 55 mi. Ionization in the *D* layer is low because it is the lowest region of the ionosphere (farthest from the sun). This layer has the ability to refract signals of low frequencies. High frequencies pass right through it but are partially attenuated in so doing. After sunset the *D* layer disappears because of the rapid recombination of its ions.

2. The *E* layer limits are from approximately 55 to 90 mi high. This layer is also known as the Kennelly–Heaviside layer because these two men were the first to propose its existence. The rate of ionic recombination in this layer is rather rapid after sunset and is almost complete by midnight. This layer has the ability to refract signals of a higher frequency than were refracted by the *D* layer. In fact, the *E* layer can refract signals with frequencies as high as 20 MHz.

3. The *F* layer exists from about 90 to 250 mi. During the daylight hours, the *F* layer separates into two layers, the F_1 and F_2 layers. The ionization level in these layers is quite high and varies widely during the course of a day. At noon, this portion of the atmosphere is closest to the sun, and the degree of ionization is maximum. Since the atmosphere is rarefied at these heights, the recombination of the ions occurs slowly after sunset. Therefore, a fairly constant ionized layer is present at all times. The *F* layers are responsible for high-frequency, long-distance transmission due to refraction for frequencies up to 30 MHz.

The relative distribution of the ionospheric layers is shown in Fig. 13-10. With the disappearance of the *D* and *E* layers at night, signals normally refracted by these layers are refracted by the much higher layer, resulting in greater skip distances at night. The layers that form the ionosphere undergo considerable variations in altitude, density, and thickness, due primarily to varying degrees of solar activity. The F_2 layer undergoes the greatest variation due to solar disturbances (sunspot activity). There is a greater concentration of solar radiation in the earth's atmosphere during peak sunspot activity, which recurs in 11-year cycles, as discussed in Chapter 1. During periods of maximum sunspot activity, the *F* layer is more dense and occurs at a higher altitude. During periods of minimum sunspot activity, the lower altitude of the *F* layer returns the sky waves (dashed lines) to points relatively close to the transmitter compared with the higher altitude *F* layer occurring during maximum sunspot activity. Consequently, skip distance is affected by the degree of solar disturbance.

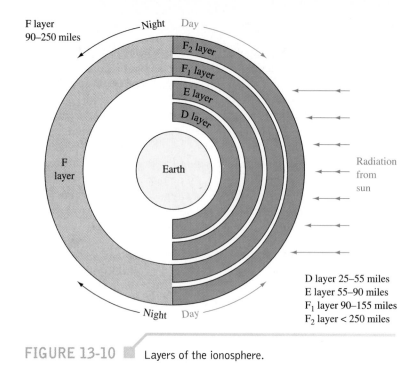

F layer
90–250 miles

Night   Day

F₂ layer
F₁ layer
E layer
D layer

F
layer

Earth

Radiation
from
sun

D layer 25–55 miles
E layer 55–90 miles
F₁ layer 90–155 miles
F₂ layer < 250 miles

Night   Day

FIGURE 13-10   Layers of the ionosphere.

Effects of the Ionosphere on the Sky Wave

The ability of the ionosphere to return a radio wave to the earth depends on the ion density, the frequency of the radio wave, and the angle of transmission. The refractive ability of the ionosphere increases with the degree of ionization. The degree of ionization is greater in summer than in winter and is also greater during the day than at night. As mentioned previously, abnormally high densities occur during times of peak sunspot activity.

CRITICAL FREQUENCY   If the frequency of a radio wave being transmitted vertically is gradually increased, a point will be reached where the wave will not be refracted sufficiently to curve its path back to earth. Instead, these waves continue upward to the next layer, where refraction continues. If the frequency is sufficiently high, the wave will penetrate all layers of the ionosphere and continue on out into space. The highest frequency that will be returned to earth when transmitted vertically under given ionospheric conditions is called the *critical frequency*.

CRITICAL ANGLE   In general, the lower the frequency, the more easily the signal is refracted; conversely, the higher the frequency, the more difficult is the refracting or bending process. Figure 13-11 illustrates this point. The angle of radiation plays an important part in determining whether a particular frequency will be returned to earth by refraction from the ionosphere. Above a certain frequency, waves transmitted vertically continue on into space. However, if the angle of propagation is lowered (from the vertical), a portion of the high-frequency waves below the critical frequency will be returned to earth. The highest angle at which a wave of a specific frequency can be propagated and still be returned (refracted) from the ionosphere is called the *critical angle* for that particular frequency. The critical angle is the angle that the wavefront path makes with a line extended to the center of the earth. Refer to Fig. 13-11, which shows the critical angle for

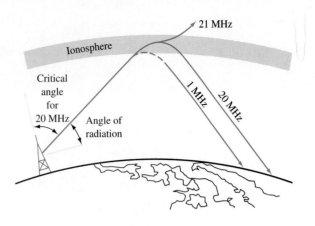

FIGURE 13-11 ■ Relationship of
frequency to refraction by the ionosphere.

20 MHz. Any wave above 20 MHz (e.g., the 21-MHz wave shown) is not refracted back
to earth but goes through the ionosphere out into space.

MAXIMUM USABLE FREQUENCY (MUF)   There is a best frequency for optimum
communication between any two points at any specific condition of the ionosphere. As
can be seen in Fig. 13-12, the distance between the transmitting antenna and the point at
which the wave returns to earth depends on the angle of propagation, which in turn is limited by the frequency. The highest frequency that is returned to earth at a given distance is
called the *maximum usable frequency* (MUF) and has an average monthly value for any
given time of the year. The *optimum working frequency* is the one which provides the
most consistent communication and is therefore the best one to use. For transmission
using the F_2 layer, the optimum working frequency is about 85% of the MUF, while propagation via the E layer will be consistent, in most cases, if a frequency near the MUF is
used. Since ionospheric attenuation of radio waves is inversely proportional to frequency,
using the MUF results in maximum signal strength.

Because of this variation in the critical frequency, nomograms and frequency tables
are issued that predict the maximum usable frequency for every hour of the day for every
locality in which transmissions are made. This information is prepared from data obtained
experimentally from stations scattered all over the world. All this information is pooled,
and the results are tabulated in the form of long-range predictions that remove most of the
guess-work from this type of radio communications.

SKIP ZONE   Between the point where the ground wave is completely dissipated and
the point where the first sky wave returns, *no* signal will be heard. This area is called the

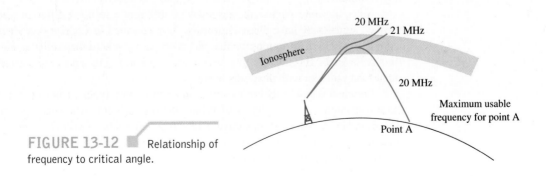

FIGURE 13-12 ■ Relationship of
frequency to critical angle.

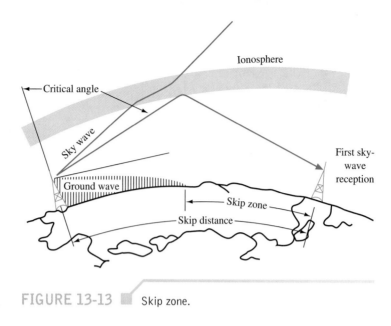

FIGURE 13-13 ■ Skip zone.

quiet or *skip zone* and is shown in Fig. 13-13. It is thus seen that the skip zone occurs for a given frequency, when propagated at its critical angle. The skip zone is the distance from the end of ground-wave reception to the point of the first sky-wave reception. This occurs for the energy propagated at the critical angle. Similarly, the skip distance is the minimum distance from transmitter to where the sky wave can be returned to earth and also occurs for energy propagated at the critical angle.

FADING   *Fading* is a term used to describe variations in signal strength that occur at a receiver during the time a signal is being received. Fading may occur at any point where both the ground wave and the sky wave are received, as shown in Fig. 13-14(a). The two waves may arrive out of phase, thus producing a cancellation of the usable signal. This type of fading is encountered in long-range communications over bodies of water where ground-wave propagation extends for a relatively long distance. In areas where sky-wave propagation is prevalent, fading may be caused by two sky waves traveling different distances, thereby arriving at the same point out of phase, as shown in Fig. 13-14(b). Such a condition may be caused by a portion of the transmitted wave being refracted by the *E* layer while another portion of the wave is refracted by the *F* layer. A complete cancellation of the signal would occur if the two waves arrived 180° out of phase with equal amplitudes. Usually, one signal is weaker than the other, and therefore a usable signal may be obtained.

Since the ionosphere causes somewhat different effects on different frequencies, a received signal may have phase distortion. As mentioned in Chapter 4, SSB is least susceptible to phase distortion problems. FM is so susceptible to these effects that it is rarely used below 30 MHz (where sky waves are possible). The greater the bandwidth, the greater the problem with phase distortion.

Frequency blackouts are closely related to certain types of fading, some of which are severe enough to completely blank out the transmission. The changing conditions in the ionosphere shortly before sunrise and shortly after sunset may cause complete blackouts at certain frequencies. The higher-frequency signals may then pass through the ionosphere, while the lower-frequency signals are absorbed.

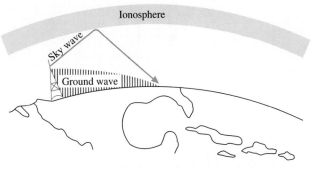

(a) Fading caused by arrival of ground wave and sky wave at the same point out of phase

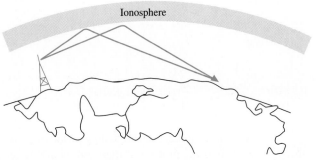

(b) Fading caused by arrival of two sky waves at the same point out of phase

FIGURE 13-14 ■ Fading.

Ionospheric storms (turbulent conditions in the ionosphere) often cause radio communications to become erratic. Some frequencies will be completely blacked out, while others may be reinforced. Sometimes these storms develop in a few minutes, and at other times they require as much as several hours to develop. A storm may last several days.

Tropospheric Scatter

Tropospheric scatter transmission can be considered as a special case of sky-wave propagation. Instead of aiming the signal toward the ionosphere, however, it is aimed at the troposphere. The troposphere ends just 6.5 mi above the earth's surface. Frequencies from about 350 MHz to 10 GHz are commonly used with reliable communications paths of up to 400 mi.

The scattering process is illustrated in Fig. 13-15. As shown, two directional antennas are pointed so that their beams intersect in the troposphere. The great majority of the transmitted energy travels straight up into space. However, by a little-understood process a small amount of energy is *scattered* in the forward direction. As shown in Fig. 13-15, some energy is also scattered in undesired directions. The best and most widely used frequencies are around 0.9, 2, and 5 GHz. Even then, however, the received signal is only one-millionth to one-billionth of the transmitted power. There is an obvious need for high-powered transmitters and extremely sensitive receivers. In addition, the scattering process is subject to two forms of fading. The first is due to multipath transmissions within the scattering path, with the effect occurring as quickly as several times per minute. Atmospheric changes provide a second, but slower, change in the received signal strength.

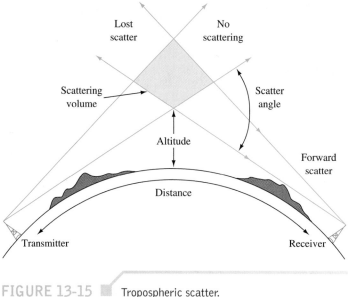

FIGURE 13-15 ■ Tropospheric scatter.

To accommodate these severe fading problems, some form of *diversity reception* is always used. This is the process of transmitting and/or receiving a number of signals and then either adding them all together at the receiver or selecting the best one at any given instant. The types of diversity reception utilized include one or combinations of the following:

Space diversity: comprised of two or more receiving antennas separated by 50 wavelengths or more. The best received signal at any instant is selected as input for the receiver.

Frequency diversity: transmission of the same information on slightly different frequencies. The different frequencies fade independently even when transmitted and received through the same antennas.

Angle diversity: transmission of information at two or more slightly different angles. This results in two or more paths based on illuminating different scattering volumes in the troposphere.

In spite of the high-power and diversity requirements and the more recent satellite communications, the use of tropospheric scatter continues since its first use in 1955. It provides reliable long-distance communications links in areas such as deserts and mountain regions and between islands. It is used for voice and data links by the military and commercial users.

13-6 SATELLITE COMMUNICATIONS

A final category of wave propagation is satellite communications (SATCOM). In these systems, a communications satellite is placed into *synchronous* orbit—which means its position remains fixed with respect to the earth's rotation. This is accomplished when the satellite is stationed approximately 22,000 mi above the earth's surface.

The transmitter sends a signal via a highly directional antenna, through the ionosphere, to the satellite's receiving antenna. It is then reamplified within the satellite and transmitted back to earth. The satellite is powered by a bank of batteries whose charge is maintained by panels of solar cells. The service life of communications satellites is from 5 to 10 years. The unit is designed to be extremely reliable, as service calls are tough to accomplish!

Satellite communication allows transoceanic links, and wide bandwidths are utilized to allow the multiplexing of a number of different signals. Frequencies used are in excess of 1 GHz. At these high frequencies, the effects of ionospheric refraction and attenuation are negligible. The frequencies used range from about 1 GHz up to 40 GHz. The signals received and subsequently retransmitted by the satellite are at different carrier frequencies. For example, the Intelsat III satellite shown in Fig. 13-16 receives signals (the *uplink*) at 5.93 to 6.42 GHz, amplifies, translates down to 3.705 to 4.195 GHz, and then reamplifies via a TWT output stage to a 7-W level for transmission back to earth (the *downlink*). The frequency translation is used to prevent interference between the two signals both at the ground station and satellite. The frequency bands commonly utilized and their designations are shown in Table 13-1. The Intelsat III would be classified as a C/S band satellite since its uplink is nominally in the C-band and downlink in the S-band.

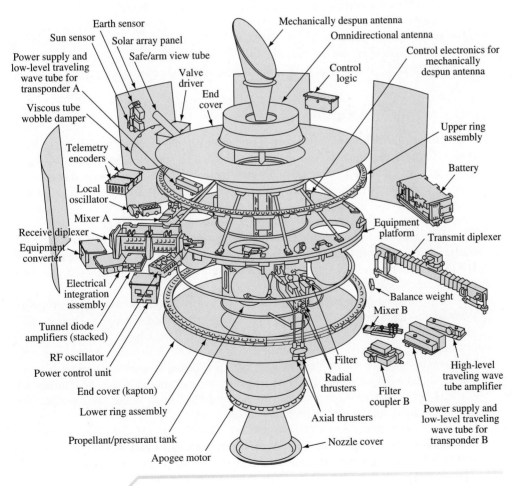

FIGURE 13-16   Intelsat III satellite. (Courtesy of TRW Systems.)

TABLE 13-1 Satellite Frequency Ranges and Band Designators

BAND DESIGNATOR	FREQUENCY (GHz)
L	1–2
S	2–4
C	4–8
X	8–12
Ku	12–18
K	18–27
Ka	27–40

An electronic system performing the reception, frequency translation, and retransmission is called a *transponder*. The total power consumption for satellite operation is about 150 W. The capacity of this satellite is for 1200 duplex voice channels or 4 TV broadcasts or any combination thereof. The Intelsat VII launched in 1992 handles 90,000 voice circuits.

The round-trip distance for a satellite relay is typically 90,000 km. The total transmission time is about 300 ms. Thus, in a transoceanic telephone conversation, a 600-ms delay occurs before you hear a reply. Because of this, care is exercised in the routing of international calls to ensure that no more than a single satellite hop is utilized. Additionally, special circuitry is incorporated to reduce delayed echo to reasonable levels.

Multiplexing Techniques

A single satellite will typically allow simultaneous communications between multiple users. Consider the situation shown in Fig. 13-17. The satellite shown has a *footprint* (coverage area) as shown. Some satellites use highly directional antennas so that the footprint may include two specific areas. For example, it may be desirable to utilize a satellite between Hawaii and the west coast of the United States. In that case there is no sense in wasting downlink signal power over a large portion of the Pacific Ocean.

In Fig. 13-17 communication between five earth stations is taking place simultaneously. Station A is transmitting to station B on path 1. Station C is transmitting to stations D and E on path 2. Control signals included with the original transmitted signals are used to allow reception at the appropriate receiver(s).

Two different multiplexing methods are commonly used to allow multiple transmissions with a single satellite. The early satellite systems all used *Frequency-Division Multiple-Access* (FDMA). In these systems the satellite is a wideband receiver/transmitter that includes a number of frequency channels, much as broadcast radio contains a number of channels. An earth station that sends a signal indicating a desire to transmit is sent a control signal telling it which available frequency to transmit on. When the transmission is complete the channel is released back to the "available" pool. In this fashion a *M*ultiple *A*ccess capability for the earth stations is provided—FD*MA*.

Most of the newer SAT COM systems use *Time-Division Multiple-Access* (TDMA) as a means to allow a single satellite to service multiple earth stations simultaneously. In TDMA all stations use the same carrier frequency, but they transmit one or more traffic bursts in nonoverlapping time frames. This is illustrated in Fig. 13-18, where three earth stations are transmitting simultaneously but never at the same instant of time. The traffic

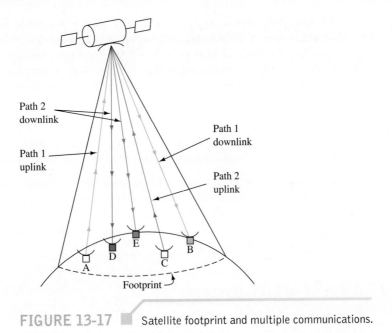

FIGURE 13-17 ■ Satellite footprint and multiple communications.

bursts are amplified by the satellite transponder and retransmitted in a downlink beam that is received by the desired station(s). The computer control of these systems is rather elaborate, as you can well imagine.

TDMA offers the following advantages over FDMA systems:

1. A single carrier for the transponder to operate on is a major advantage. Its TWT power amplifier is much less subject to intermodulation problems and can operate at a higher power output when dealing with a smaller range of frequency.

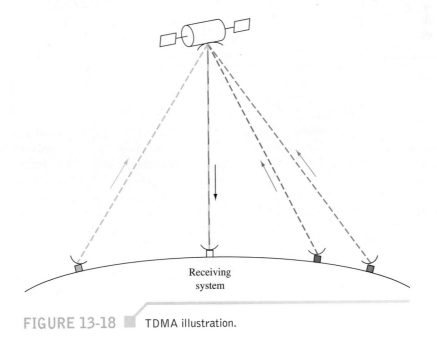

FIGURE 13-18 ■ TDMA illustration.

2. The use of the time domain rather than frequency domain to achieve selectivity is advantageous. In FDMA the earth station must transmit and receive on a multiplicity of frequencies and must provide a large number of frequency-selective up-conversion and down-conversion chains. In TDMA the selectivity is accomplished in time rather than frequency. This is much simpler and less expensive to accomplish.

3. TDMA is ideally suited to digital communications since they are naturally suited to the storage, rate conversions, and time-domain processing used in TDMA implementation. Additionally, TDMA is ideally suited to demand-assigned operation, in which the traffic burst durations are adjusted to accomodate demand.

It should be mentioned that a third multiplexing technique is now receiving some attention. *Code-Division Multiple-Access* (CDMA) also allows the use of just one carrier. In it each station uses a different binary sequence to modulate the carrier. The control computer uses a "correlator" that can separate and "distribute" the various signals to the appropriate downlink station.

VSAT and MSAT Systems

In recent years the two areas of satellite communications showing the greatest growth are (1) very small aperture terminal (VSAT) fixed satellite communication systems, and (2) ultrasmall aperture terminal mobile satellite (MSAT) systems. Technological advances

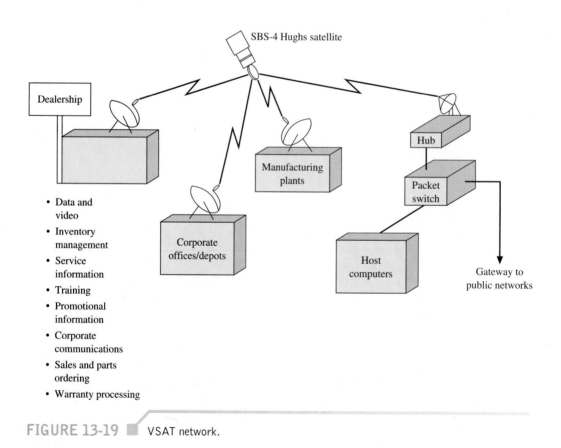

FIGURE 13-19 ■ VSAT network.

and market demand have driven the development of these new markets. MSAT terminals, which can be called "VSATs on wheels," have several features in common with VSATs. While VSATs take telecommunication services directly to fixed users, MSAT terminals take them to moving vehicles.

VSAT systems allow multiple inexpensive stations to be linked to a large central installation. For example, K-Mart recently installed small aperture antenna systems (VSATs) at over 2000 stores and links them with its central mainframe computer in Michigan. This allows them to verify checks and credit cards quickly and to convey data, such as what customers are buying and how much inventory is on hand. This allows them to supply each individual store with the items their customers are buying and speeds up the checkout process.

The cost of a complete VSAT terminal is about $10,000. The dish antenna is typically 1 km in diameter, and a transmitter power of just 2 to 3 W is sufficient. This market is expected to be immune to optical fiber competition for another 20 years or more, until fiber replaces copper cable. When that takes place, the economics may favor the terrestrial fiber transmission systems.

Figure 13-19 provides a pictorial representation of Chrysler's VSAT network. It connects the automaker's headquarters with more than 6000 dealerships and corporate facilities in North America. It is used to assist mechanics with repair and allows salespeople to order and confirm delivery dates for cars from a showroom computer. It will also help maintain proper inventories of automobiles and spare parts.

The first application of MSAT systems includes large national trucking firms. The system allows the dispatching center to maintain continuous communication with each of its trucks.

SATCOM Power Budget

The following equation relates transmitted and received power of any space wave, whether it be point to point on earth or between earth and a satellite.

$$\left(\frac{P_R}{P_T}\right)_{dB} \simeq (G_T)_{dB} + (G_R)_{dB} - \left[32.5 + 20\log_{10} d + 20\log_{10} f\right] dB \qquad (13\text{-}8)$$

where P_R = received power
P_T = transmitted power
G_T = gain of transmitting antenna (see Chapter 14)
G_R = gain of receiving antenna
d = distance (km) between antennas
f = frequency (MHz)

EXAMPLE 13-2

Calculate the power received at a satellite given the following conditions:

1. *The power gain of the transmitting parabolic dish antenna (see Chapter 16) is 30,000.*

2. *The transmitter drives 2 kW of power into the antenna at a carrier frequency of 6.21 GHz.*
3. *The satellite receiving antenna has a power gain of 30.*
4. *The transmission path is 45,000 km.*

SOLUTION

Using Eq. (13-8),

$$\left(\frac{P_R}{2\ kW}\right)_{dB} \simeq 10\log_{10}30,000 + 10\log_{10}30$$

$$- \left(32.5 + 20\log_{10}45,000 + 20\log_{10}6210\right)\ dB$$

$$\left(\frac{P_R}{2\ kW}\right)_{dB} \simeq 44.8\ dB + 14.8\ dB - (32.5 + 93.1 + 75.9)\ dB$$

$$= 59.6\ dB - 201.5\ dB$$

$$= -141.9\ dB$$

$$\frac{P_R}{2\ kW} = \frac{1}{antilog\ 14.19\ dB} = \frac{1}{1.55 \times 10^{14}}$$

$$P_R = \frac{2\ kW}{1.55 \times 10^{14}} = 1.29 \times 10^{-11}\ W$$

$$= 12.9\ pW!$$

Example 13-2 shows the great amount of energy lost in long-distance transmissions. The return link suffers the same type of attenuation, and this whole exercise helps in understanding the need for extremely sophisticated, low-noise, and costly equipment in satellite systems.

TROUBLESHOOTING

A radio communication system transmits a radio frequency signal and depends on a substantial amount of that signal being received at the receiver. In earlier chapters you learned about the kinds of noise and interference that could affect the transmitted signal. Now we are going to take a closer look at the problems interference causes in TV and FM radio systems since we all can identify with these. We will also discuss some methods used to resolve interference problems.

After completing this section you will be able to:

- Identify different types of interference
- Describe three methods to reduce interference
- Troubleshoot various antenna installation problems

Radio Interference

This discussion will be limited to TV and FM reception since we are most familiar with these. Sources of unwanted signals (noise and interference) happen naturally or are man-made. Review Chapter 1 for sources of noise. Good circuit and antenna design will reduce noise to negligible levels. Man-made sources generate most of the interference that usually disturbs the quality of the signal at the receiver site. There are various solutions to remedy interference problems at the receiver. The first step in eliminating an unwanted signal is to pinpoint the source of it. After finding the source of the interference that is causing the disturbance, remove it if possible. When it is not possible to remove the interference source, try increasing the distance of the undesirable source from the receiver. This will usually reduce the effects of interference on the receiver and may clear the problem. Using filters is another practical approach for removing unwanted signals. One more method is to protect the receiver by shielding the antenna, the ac input power line, or the whole receiver from the interference signal. The following paragraphs talk about the kinds of interference the communication technician may encounter and practical methods to resolve the problems.

CAPTURE AND CO-CHANNEL INTERFERENCE EFFECTS   In areas of congested radio, TV, and communication channels, such as metropolitan areas, receivers are subject to the capture effect and co-channel interference. From the discussion in Chapter 5, the capture effect causes a stronger station to overpower and replace a weaker station at the receiver. The weak station is usually lost completely. Co-channel interference takes the form of two or more broadcasting stations bleeding into each other at the receiver. To the listener, this bleeding-over effect turns into bothersome noise. The best solution for these kinds of interference is to rotate the antenna or obtain a more directional antenna.

EMI AND RFI   Section 13-2 introduced EMI and RFI. Electromagnetic interference (EMI) shows up on TV as vertical bands of dots moving on the screen. On FM, EMI causes distortion to the audio. Radio frequency interference (RFI) displays itself as several bars or wavy lines on the TV screen. Strong RFI will cause complete loss of the TV's picture. FM audio is affected by RFI to produce garbled sound, often causing it to be pure gibberish.

The automobile ignition or spark plugs and kitchen appliances like the blender or microwave oven are sources of EMI. In addition, computers and electric motors produce EMI. EMI can enter the receiver through the antenna, lead-in wire, or power line. To decide which is bringing in the EMI, disconnect the lead-in wire from the receiver and short the receiver antenna terminals together. If the interference disappears, then the source was the antenna or lead-in wire. If the interference continues, then the unwanted EMI is coming through the power line. When it is not possible to remove the EMI source or relocate it, use shielding or filtering to remove the interference. When the interference is entering the receiver by way of the antenna, it may be necessary to relocate the antenna.

Ham radio and CB radio transmitters are a common cause of RFI. If the source can be found (towering antennas in the neighborhood are usually a dead giveaway), attempt to contact the owner and let them know the problem exists. Install a high-pass filter between the antenna and receiver input on the lead-in wire to eliminate RFI. However, the best way to eliminate the interference is to remove the source of it.

FADING   Fading is one of the most troublesome hindrances in communications. Fading is the result of the signal arriving at the receiver from two different paths—a direct path and the skyway path. A typical example of this type of problem occurs when an airplane flies over an area where outside TV antennas are used. The airplane causes reflected signals to mix with the direct signal, and a fluttering results in the picture on the TV receiver. Using a high-gain directional antenna will often resolve this kind of interference.

REFLECTIONS   Figure 13-6 shows a problem that exists between any type of broadcast station and its receiver. In practice, the reflected wave is quite strong, almost as strong as the direct wave, but the path taken by the reflected wave is longer than that of the direct wave. The important thing to remember is that even though the wavelength may be only 1 m and the path is several miles, every time the path difference is equivalent to $\frac{1}{2}$ wavelength or 180°, there will be a null in the signal. Conversely, when the path difference is a multiple of wavelengths the signals add, potentially doubling the signal strength. Equation (13-9) will enable you to determine where a peak in signal strength might be found.

$$\theta = \frac{1.385 \times 10^{-4} \times H_t \times H_r \times f}{D} \tag{13-9}$$

Remember, every time θ is an odd multiple of 180° you are in a null. H_t is the transmitter height in feet, H_r is the receiver height in feet, D is the distance in miles, and f is the frequency in MHz.

The point is, when you need more signal, move the antenna. Intuition would tell you to increase the height, but you may actually be able to lower the antenna and find more signal.

Diffraction: Diffraction is much more complicated than reflection, but the solution is the same. As you move away from the mountain shown in Fig. 13-5 you will find hot spots. The technician who finds the hot spot can save a great deal of money on an antenna installation.

Ghosting in TV Reception: It may be possible to fight the ghost problem described in Fig. 13-8 with knowledge of your antenna patterns. Most TV antennas will have a fairly broad main beam and several null and sidelobes (see Fig. 13-20). Further information on antenna patterns is provided in Chapter 14.

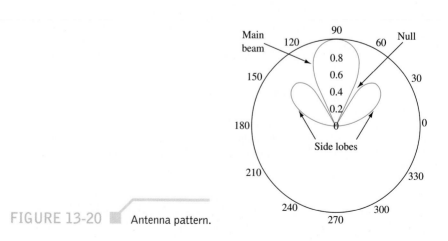

FIGURE 13-20 Antenna pattern.

The technician should try orienting the antenna to place the ghost signal in a null and the desired signal somewhere on the main beam. If this isn't possible, try a single-frequency antenna such as a corner reflector. Single-frequency antennas have a high rejection of side lobes.

Sky-Wave Propagation

Commercial use of the shortwave frequencies is steadily declining, but it is still the cheapest way to communicate with remote areas of the world. The technician or engineer will need some knowledge of system planning. People interested in shortwave propagation or forecasting should obtain a copy of a computer program called "Ioncap." It was developed by the National Bureau of Standards, now called National Institute of Standards and Technology, and can be purchased from the U.S. government printing office. Several commercial programs are available that use Ioncap as a core and are easier to use.

Satellite Communications

When servicing or installing a satellite system, your chief problem will be aligning the antenna with the satellite. Beams are no more than 2° wide and polarity might be unknown. All are usually adjustable and will often need adjustment when performance is not satisfactory.

SUMMARY

In Chapter 13 we studied various considerations of wave propagation. It was discovered that electrical energy can be converted to wave energy with many properties in common with light wave propagation. The major topics the student should now understand include:

- the definition of an electromagnetic wave, isotropic point source, wavefront, and characteristic impedance of free space

- the understanding of environmental effects on wave propagation, including reflection, refraction, and diffraction

- the explanation of ground- and space-wave propagation

- the description of ionospheric layers and their effects on sky-wave propagation

- the definitions of skipping, critical frequency, critical angle, maximum usable frequency (MUF), skip zone, fading, and tropospheric scatter

- the description and usage of satellite communications

- the explanations of multiplexing techniques used in satellite communications, including FDMA, TDMA, and CDMA

- the description of very small aperture terminal (VSAT) and ultrasmall aperture terminal (MSAT) satellite communication

- the power-loss calculations used in satellite communications analysis

QUESTIONS AND PROBLEMS

SECTION 1

1. Explain why an antenna can be thought of as a transducer.
2. List the similarities and dissimilarities between light waves and radio waves.

SECTION 2

3. What are the two components of an electromagnetic wave? How are they created? Explain the two possible things that can happen to the energy in an electromagnetic wave near a conductor.
4. Explain the difference between electromagnetic interference (EMI) and radio-frequency interference (RFI).
*5. What is *horizontal and vertical polarization* of a radio wave?
*6. What kinds of fields emanate from a transmitting antenna, and what relationships do they have to each other?
7. Define *wavefront*.
8. Calculate the power density in watts per square meter (on earth) from a 10-W satellite source that is 22,000 mi from earth. (6.35×10^{-16} W/m^2)
9. Calculate the power received from a 20-W transmitter, 22,000 mi from earth, if the receiving antenna has an effective area of 1600 m^2. (2.03×10^{-12} W)
10. Calculate the electric field intensity, in volts per meter, 20 km from a 1-kW source. How many decibels down will that field intensity be an additional 30 km from the source? (8.66 mV/m, 7.96 dB)
11. Calculate the characteristic impedance of free space using two different methods.
*12. How does the field strength of a standard broadcast station vary with distance from the antenna?

SECTION 3

13. In detail, explain the process of reflection for an electromagnetic wave.
14. With the aid of Snell's law, fully explain the process of refraction for an electromagnetic wave.
15. What is *diffraction* of electromagnetic waves? Explain the significance of the shadow zone and how it is created.

SECTION 4

16. List the three basic modes whereby an electromagnetic wave propagates from a transmitting to receiving antenna.
17. Describe ground-wave propagation in detail.
18. Explain why ground-wave propagation is more effective over sea water than desert terrain.
*19. What is the relationship between operating frequency and ground-wave coverage?
*20. What are the lowest frequencies useful in radio communications?
21. Fully explain space-wave propagation. Explain the difference between a direct and reflected wave.
22. Calculate the radio horizon for a 500-ft transmitting antenna and receiving antenna of 20 ft. Calculate the required height increase for the receiving antenna if a 10% increase in radio horizon were required. (37.9 mi, 31.2 ft)

23. Explain the phenomenon of *ghosting* in TV reception. What would be the effect if this occurred with a voice transmission?

24. Calculate the ghost width for a 17-in.-wide TV screen when a reflected wave results from an object $\frac{3}{8}$ mi "behind" a receiver. How could this effect be minimized? (1.28 in)

SECTION 5

25. List the course of events in the process of sky-wave propagation.

26. Provide a detailed discussion of the ionosphere—its makeup, its layers, its variations, and its effect on radio waves.

*27. What effects do sunspots and the aurora borealis have on radio communications?

28. Define and describe *critical frequency, critical angle,* and *maximum usable frequency* (MUF). Explain their importance to sky-wave communications.

29. What is the optimum working frequency, and what is its relationship to the MUF?

30. In the strictest sense, define *skip distance* and *skip zone.*

31. Explain the various ways in which fading occurs when sky waves are being received.

32. What frequencies have substantially straight-line propagation characteristics analogous to those of light waves and are unaffected by the ionosphere?

33. What radio frequencies are useful for long-distance communications requiring continuous operation?

34. In radio transmissions, what bearings do the angle of radiation, density of the ionosphere, and frequency of emission have on the length of the skip zone?

35. Why is it possible for a sky wave to "meet" a ground wave 180° out of phase?

36. What is the process of tropospheric scatter? Explain under what conditions it might be used.

*37. What is the purpose of a diversity antenna receiving system?

38. List and explain three types of diversity reception schemes.

SECTION 6

39. What are *satellite communications?* List reasons for their increasing popularity.

40. Describe a typical VSAT installation. How does it differ from a MSAT system?

41. The signal received by the satellite in Ex. 13-2 is amplified to a 7-W level at 4 GHz and retransmitted to earth via the same antennas. Calculate the power received by the earth station. (0.108 pW)

42. Explain the methods of multiplexing in SATCOM systems, and provide the advantages of TDMA over FDMA.

CHAPTER

14

ANTENNAS

RF MICRO·DEVICES

RF IC components in 3 process technologies:
Optimum Technology Matching™
only from RF Micro Devices

OBJECTIVES

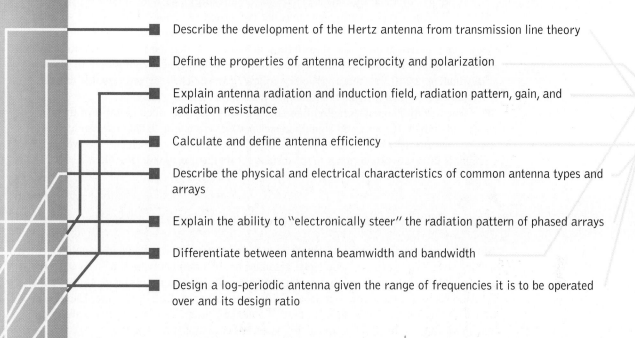

- Describe the development of the Hertz antenna from transmission line theory

- Define the properties of antenna reciprocity and polarization

- Explain antenna radiation and induction field, radiation pattern, gain, and radiation resistance

- Calculate and define antenna efficiency

- Describe the physical and electrical characteristics of common antenna types and arrays

- Explain the ability to "electronically steer" the radiation pattern of phased arrays

- Differentiate between antenna beamwidth and bandwidth

- Design a log-periodic antenna given the range of frequencies it is to be operated over and its design ratio

14-1 BASIC ANTENNA THEORY

In this chapter we introduce the fundamentals of antennas and describe the most commonly encountered types. Antennas for use at microwave frequencies are described in Chapter 16.

Antennas serve either or both of the following two functions: the generation and the collection of electromagnetic energy. In a transmitting system, a radio-frequency signal is developed, amplified, modulated, and applied to the antenna. The RF currents flowing through the antenna produce electromagnetic waves that radiate into the atmosphere. In a receiving system, electromagnetic waves "cutting" through the antenna induce alternating currents for use by the receiver.

To have adequate signal strength at the receiver, either the power transmitted must be extremely high or the efficiency of the transmitting and receiving antennas must be high because of the high losses in wave travel between the transmitter and the receiver.

Any receiving antenna transfers energy from the atmosphere to its terminals with the same efficiency with which it transfers energy from the transmitter into the atmosphere. This property of interchangeability for transmitting and receiving operations is known as antenna *reciprocity*. Antenna reciprocity occurs because antenna characteristics are essentially the same regardless of whether an antenna is sending or receiving electromagnetic energy.

Because of reciprocity, we will generally treat antennas from the viewpoint of the transmitting antenna, with the understanding that the same principles apply equally well when the antenna is used for receiving electromagnetic energy.

Antennas produce or collect electromagnetic energy and they should do so in an efficient manner. Consequently, antennas are composed of conductors arranged so as to permit efficient operation. Efficient operation also requires that the receiving antenna be of the same polarization as the transmitting antenna. *Polarization* is the direction of the electric field and is, therefore, the same as the antenna's physical configuration. Thus, a vertical antenna will transmit a vertically polarized wave. The received signal is theoretically zero if a vertical E field cuts through a horizontal receiving antenna.

The received signal strength of an antenna is usually described in terms of the electric field strength. If a received signal induces a 10-μV signal in an antenna 2 m long, the field strength is 10 μV/2 m, or 5 μV/m. Recall from Chapter 13 that the received field strength is inversely proportional to the distance from the transmitter [Eq. (13-2)].

14-2 HERTZ ANTENNA

Any antenna having a physical length that is one-half wavelength of the applied frequency is called a *Hertz* antenna. Hertz antennas are predominately used with frequencies above 2 MHz. It is unlikely that a Hertz antenna will be found in applications below 2 MHz because at these low frequencies this antenna is physically too large. Usually, at frequencies below 2 MHz a *Marconi* type of antenna is used. This is a quarter-wavelength antenna with the ground reflection acting as the other quarter wavelength.

Development of the Hertz Antenna

When the open two-wire transmission line was discussed in Chapter 12, it was found that one of its disadvantages was excessive radiation at high frequencies. Radiation from a transmission line is undesirable since the perfect transmission line would be one that possessed no losses. Although the two-wire transmission line was considered to be an adequate transmission medium at extremely high frequencies, it can become an effective antenna. For this reason, an analysis of the open-ended, quarter-wave transmission line will furnish an excellent introduction for understanding basic antenna theory. The open-ended quarter-wave transmission line segment is shown in Fig. 14-1.

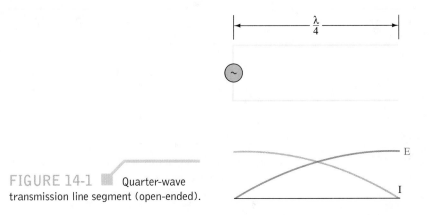

FIGURE 14-1   Quarter-wave transmission line segment (open-ended).

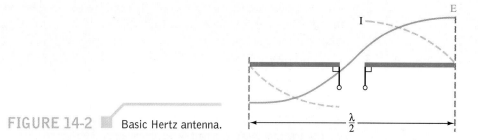

FIGURE 14-2 Basic Hertz antenna.

The characteristics of the open-ended line are such that the voltage at the end of the line is maximum, and the current at the end is zero. This is true of an open-ended line regardless of the wavelength of the line. On either the open or shorted line, standing waves will be produced. Since the voltage applied to the line is sinusoidal, the line will constantly be charging and discharging. Current will be flowing in the line continuously. Since the current at the ends of the line is minimum, a quarter wave back (at the source) the current must be maximum. The impedance at the sending end is low, and the imped-ance at the open circuit is high. The standing waves of current and voltage are shown on the quarter-wave section in Fig. 14-1.

It is desirable to have maximum radiation from an antenna. Under such conditions all energy applied to the antenna would be converted to electromagnetic waves and radiat-ed. This maximum radiation is not possible with the two-wire transmission line because the magnetic field surrounding each conductor of the line is in a direction that opposes the lines of force about the other conductor. Under these conditions, the quarter-wave transmission line proves to be an unsatisfactory antenna; however, with only a slight physical modification, this section of transmission line can be transformed into a relative-ly efficient antenna. This transformation is accomplished by bending each line outward 90° to form a *half-wave* or *Hertz antenna* or a λ/2 *dipole,* as shown in Fig. 14-2.

The antenna shown in Fig. 14-2 is composed of two quarter-wave sections. The electrical distance from the end of one to the end of the other is a half wavelength. If volt-age is applied to the line, the current will be maximum at the input and minimum at the ends. The voltage will be maximum between the ends and minimum between the input terminals.

Hertz Antenna Impedance

An impedance value may be specified for a half-wave antenna thus constructed. General-ly, the impedance at the ends is maximum, while that at the input is minimum. Conse-quently, the impedance value varies from a minimum value at the generator to a maximum value at the open ends. An impedance curve for the half-wave antenna is shown in Fig. 14-3. Notice that the line has different impedance values for different points along its length. The impedance values for half-wave antennas vary from about 2500 Ω at the open ends to 73 Ω at the source ends.

Radiation and Induction Field

Feeding the Hertz antenna at the center results in an input impedance that is purely resis-tive and equal to 73 Ω. Recall that with an open-circuited λ/4 transmission line the input impedance was 0 Ω, and it could therefore not absorb power. By spreading the open λ/4

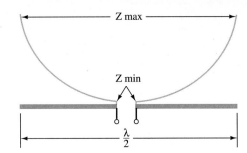

FIGURE 14-3 ▉ Impedance along a half-wave antenna.

transmission line out into a Hertz antenna its input impedance has taken on a finite resistive value. It can now absorb power, but the question is, how? The answer is that it can now efficiently accept electrical energy and radiate it into space as electromagnetic waves. While the mechanisms of launching a wave from a current-carrying wire are not fully understood, the fields surrounding the antenna do not collapse their energy back into the antenna but rather radiate it out into space. This radiated field is appropriately termed the *radiation field*. Antennas also have an *induction field* associated with them. It is the portion of field energy that *does* collapse back into the antenna and is therefore limited to the zone immediately surrounding the antenna. Its effect becomes negligible at a distance more than about one-half wavelength from the antenna.

Radiation Pattern

The radiation pattern for the $\lambda/2$ Hertz antenna is shown in Fig. 14-4(a). A *radiation pattern* is an indication of radiated field strength around the antenna. The pattern shown in Fig. 14-4(a) shows that maximum field strength for the $\lambda/2$ dipole occurs at right angles to the antenna, while virtually zero energy is launched "off the ends." Recall from Chapter 13 that we considered an isotropic source of waves. Its radiation pattern is spherical, or as shown in two dimensions [Fig. 14-4(b)], it is circular or *omnidirectional*. The Hertz antenna is termed *directional* in that it concentrates energy in certain directions at the expense of lower energy in other directions.

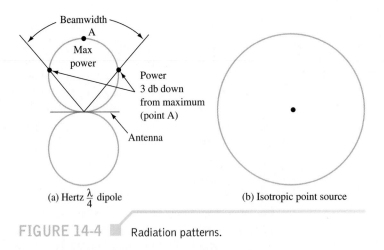

(a) Hertz $\frac{\lambda}{4}$ dipole   (b) Isotropic point source

FIGURE 14-4 ▉ Radiation patterns.

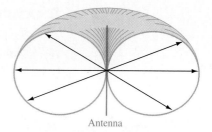

FIGURE 14-5 ▊ Three-dimensional radiation pattern for a λ/2 dipole.

Antenna

Another important concept is an antenna's *beamwidth*. It is the angular separation between the half-power points on its radiation pattern. It is shown for the λ/2 dipole in Fig. 14-4(a). A three-dimensional radiation pattern cross section for a vertically polarized λ/2 dipole is shown in Fig. 14-5. As can be seen, it is a doughnut-shaped pattern. If the antenna were mounted close to ground, the pattern would be altered somewhat by the effects of ground reflected waves.

Antenna Gain

The Hertz antenna has *gain* with respect to the theoretical isotropic radiator. *Antenna gain* is different from amplifier gain since feeding 50 W into a dipole does not result in more than 50 W of radiated field energy. It is, instead, a gain relative to a reference antenna. The dipole, therefore, has a gain relative to the isotropic radiator in a certain direction. The Hertz antenna has a 2.15-dB gain (at right angles to the antenna) as compared to an isotropic radiator. However, since a perfect isotropic radiator cannot be practically realized, the λ/2 dipole antenna is sometimes taken as the standard reference to which all other antennas are compared with respect to their *gain*. When the gain of an antenna is multiplied by its power input, the result is termed its effective radiated power (ERP). For instance, an antenna with a gain of 7 and fed with 1 kW has an ERP of 7 kW.

An antenna whose gain is provided with respect to an isotropic radiator is often expressed as *dBi*. In other words, the Hertz (dipole) antenna's gain can be expressed as 2.15 dBi. If an antenna's gain is given in decibels with respect to a dipole, it is expressed as *dBd*. This occurs somewhat less often than dBi in antenna literature. The gain of an antenna in dBi is 2.15 dB more than when expressed in dBd. Thus, an antenna with a gain of 3 dBd has a gain of 5.15 dBi (3 dB + 2.15 dB).

The amount of power received by an antenna through free space can be predicted by the following:

$$P_r = \frac{P_t G_t G_r \lambda^2}{16\pi^2 d^2}$$
(14-1)

where P_r = power received (W)
P_t = power transmitted (W)
G_t = transmitting antenna gain (ratio, not dB) compared to isotropic radiator
G_r = receiving antenna gain (ratio, not dB) compared to isotropic radiator
λ = wavelength (m)
d = distance between antennas (m)

EXAMPLE 14-1

Two λ/2 dipoles are separated by 50 km. They are "aligned" for optimum reception. The transmitter feeds its antenna with 10 W at 144 MHz. Calculate the power received.

SOLUTION

The two dipoles have a gain of 2.15 dB. That translates into a gain ratio of $\log^{-1} 2.15$ dB = 1.64.

$$P_r = \frac{P_t G_t G_r \lambda^2}{16\pi^2 d^2}$$

$$= \frac{10 \text{ W} \times 1.64 \times 1.64 \times \left(\dfrac{3 \times 10^8 \text{ m/s}}{144 \times 10^6}\right)^2}{16\pi^2 \times \left(50 \times 10^3 \text{ m}\right)^2} \qquad (14\text{-}1)$$

$$= 2.96 \times 10^{-10} \text{ W}$$

The received signal in Ex. 14-1 would provide a voltage of 147 μV into a matched 73-Ω receiver system [$(P = V^2/R, V = (2.96 \times 10^{-10} \text{ W} \times 73 \text{ }\Omega) = 147 \text{ }\mu$V]. This is a relatively strong signal since receivers can often provide a usable output with less than a 1-μV signal.

14-3 RADIATION RESISTANCE

The portion of an antenna's input impedance that is the result of power radiated into space is called the *radiation resistance*, R_r. It should be noted that R_r is not the resistance of the conductors that form the antenna. It is simply an effective resistance that is related to the power radiated by the antenna. Since a relationship exists between the power radiated by the antenna and the antenna current, radiation resistance can be mathematically defined as the ratio of total power radiated to the square of the effective value of antenna current, or

$$R_r = \frac{P}{I^2} \qquad (14\text{-}2)$$

where R_r = radiation resistance (Ω)
 I = effective rms value of antenna current at the feed point (A)
 P = total power radiated from the antenna

It should be mentioned at this point that not all of the energy absorbed by the antenna is radiated. Power may be dissipated in the actual antenna conductor by high-powered transmitters, by losses in imperfect dielectrics near the antenna, by eddy currents induced in metallic objects within the antenna's induction field, and by arcing effects in high-powered transmitters. These arcing effects are termed *corona discharge*. If these losses are represented by one lumped value of resistance, R_d, and the sum of R_d and R_r is called the antenna's total resistance, R_T, the antenna's efficiency can be expressed as

$$\eta = \frac{P_{\text{transmitted}}}{P_{\text{input}}} = \frac{R_r}{R_r + R_d} = \frac{R_r}{R_T} \tag{14-3}$$

Effects of Antenna Length

The radiation resistance varies with antenna length, as shown in Fig. 14-6. For a half-wave antenna, the radiation resistance measured at the current maximum (center of the antenna) is approximately 73 Ω. For a quarter-wave antenna, the radiation resistance measured at its current maximum is approximately 36.6 Ω. These are free-space values, that is, the values of radiation resistance that would exist if the antenna were completely isolated so that its radiation pattern would not be affected by ground or other reflections.

Ground Effects

For practical antenna installations, the height of the antenna above ground affects radiation resistance. Changes in radiation resistance occur because of ground reflections that intercept the antenna and alter the amount of antenna current flowing. Depending on their phase, the reflected waves may increase antenna current or decrease it. The phase of the reflected waves arriving at the antenna, in turn, is a function of antenna height and orientation.

At some antenna heights, it is possible for a reflected wave to induce antenna currents in phase with transmitter current so that total antenna current increases. At other antenna heights, the two currents may be 180° out of phase so that total antenna current is less than if no ground reflection occurred.

With a given input power, if antenna current increases, the effect is as if radiation resistance decreases. Similarly, if the antenna height is such that the total antenna current decreases, the radiation resistance is increased. The actual change in radiation resistance

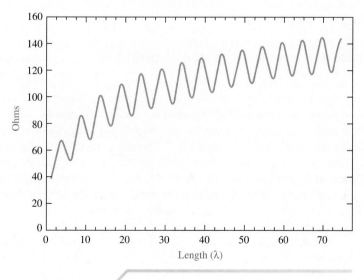

FIGURE 14-6 Radiation resistance of antennas in free space plotted against length.

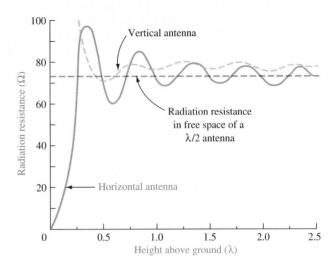

of a half-wave antenna at various heights above ground is shown in Fig. 14-7. The radiation resistance of the horizontal antenna rises steadily to a maximum value of 90 Ω at a height of about three-eighths wavelength. The resistance then continues to rise and fall around an average value of 73 Ω, which is the free-space value. As the height is increased, the amount of variation keeps decreasing.

The variation in radiation resistance of a vertical antenna is much less than that of the horizontal antenna. The radiation resistance (dashed line in Fig. 14-7) is a maximum value of 100 Ω when the center of the antenna is a quarter wavelength above ground. The value falls steadily to a minimum value of 70 Ω at a height of a half wavelength above ground. The value then rises and falls by several ohms about an average value slightly above the free-space value of a horizontal half-wave antenna.

Since antenna current is affected by antenna height, the field intensity produced by a given antenna also changes. In general, as the radiation resistance is reduced, the field intensity increases, whereas an increase in radiation resistance produces a drop in radiated field intensity.

Electrical versus Physical Length

If an antenna is constructed of very thin wire and is isolated in space, its electrical length corresponds closely to its physical length. In practice, however, an antenna is never isolated completely from surrounding objects. For example, the antenna will be supported by insulators with a dielectric constant greater than 1. The dielectric constant of air is arbitrarily assigned a numerical value equal to 1. Therefore, the velocity of a wave along a conductor is always slightly less than the velocity of the same wave in free space, and the physical length of the antenna is less (by about 5%) than the corresponding wavelength in space. The physical length can be approximated as about 95% of the calculated electrical length.

EXAMPLE 14-2

It is desired to build a λ/2 dipole to receive a 100-MHz broadcast. Determine the optimum length of the dipole.

Solution

At 100 MHz,

$$\lambda = \frac{c}{f} = \frac{3 \times 10^8 \text{ m/s}}{100 \times 10^6 \text{ Hz}} = 3 \text{ m}$$

Therefore, its electrical length is $\lambda/2$, or 1.5 m. Applying the 95% correction factor, the actual optimum physical length of the antenna is

$$0.95 \times 1.5 \text{ m} = 1.43 \text{ m}$$

Effects of Nonideal Length

The 95% correction factor is an approximation. If ideal results are desired, a trial-and-error procedure is used to find the exact length for optimum antenna performance. If the antenna length is not the optimum value, its input impedance will look like a capacitive circuit or an inductive circuit depending on whether the antenna is shorter or longer than the specified wavelength. A Hertz antenna slightly longer than a half wavelength will act like an inductive circuit, and an antenna slightly shorter than a half wavelength will appear to the source as a capacitive circuit. Compensation for additional length can be made by cutting the antenna down to proper length or by tuning out the inductive reactance by adding a capacitance in series. This added X_c will completely cancel the inductive reactance, and the source will then see a pure resistance, provided the proper size capacitor is used. If an antenna is shorter than the required length, the source end of the line will appear capacitive. This condition may be corrected by adding inductance in series with the antenna input.

14-4 ANTENNA FEED LINES

If energy is applied at the geometrical center of an antenna, the antenna is said to be *center-fed*. If energy is applied to the end of an antenna, it is known as an *end-fed* antenna. Although energy may be fed to an antenna in various ways, most antennas are either *voltage-fed* or *current-fed*. When energy is applied to the antenna at a point of high circulating current, the antenna is current-fed. When the generator energy is applied to a point of high voltage on the antenna, the antenna is voltage-fed. Both of these types of feed are shown in Fig. 14-8.

It is seldom possible to connect a generator directly to an antenna. It is usually necessary to transfer energy from the generator (transmitter) to the antenna by use of a transmission line (also called an antenna *feed line*). Such lines may be resonant, nonresonant, or a combination of both types.

Resonant Feed Line

The resonant transmission line is not widely used as an antenna feed method because it tends to be inefficient and is very critical with respect to its length for a particular operating frequency. However, in certain high-frequency applications, resonant feeders sometimes prove convenient.

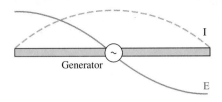

(a) Generator at current maximum means current feed

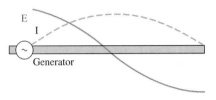

FIGURE 14-8 ■ Current and voltage feed.   (b) Generator at voltage maximum means voltage feed

In the current-fed antenna with a resonant line, as shown in Fig. 14-9, the transmission line is connected to the center of the antenna. This antenna has a low impedance at the center and, like the voltage-feeding transmission line, has standing waves on it. Constructing it to be exactly a half wavelength causes the impedance at the sending end to be low. A series resonant circuit is used to develop the high currents needed to excite the line. Adjusting the capacitors at the input compensates for slight irregularities in line and antenna length.

Although this example of an antenna feed system is a simple one, the principles described apply to antennas and to lines of any length provided both are resonant. The line connected to the antenna may be either a two-wire or coaxial line. In high-frequency applications, the coaxial line is preferred due to its lower radiation loss.

One advantage of connecting a resonant transmission line to an antenna is that it makes impedance matching unnecessary. In addition, it makes it possible to compensate for any irregularities in either the line or the antenna by providing the appropriate resonant circuit at the input. Its disadvantages are increased power losses in the line due to high standing waves of current, increased probability of arc-over because of high standing waves of voltage, very critical length, and production of radiation fields by the line due to the standing waves on it.

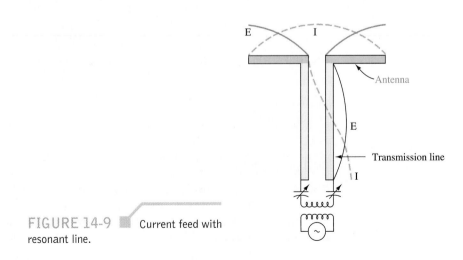

FIGURE 14-9 ■ Current feed with resonant line.

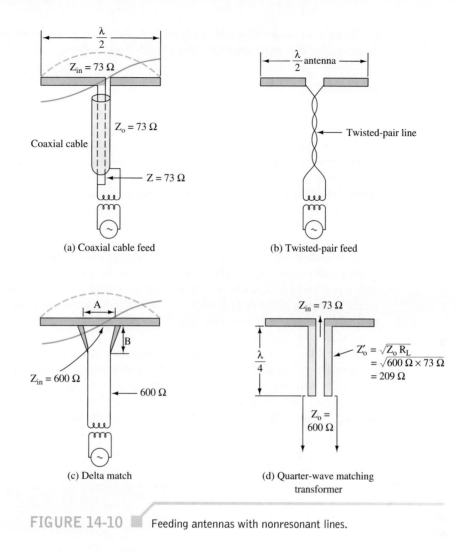

FIGURE 14-10 ▉ Feeding antennas with nonresonant lines.

Nonresonant Feed Line

The nonresonant feed line is the more widely used technique. The open-wire line, the shielded pair, the coaxial line, and the twisted pair may be used as nonresonant lines. This type of line has negligible standing waves if it is properly terminated in its characteristic impedance at the antenna end. It has a great advantage over the resonant line in that its operation is practically independent of its length.

The illustrations in Fig. 14-10 show the excitation of a half-wave antenna by non-resonant lines. If the input to the center of the antenna in Fig. 14-10(a) is 73 Ω and if the coaxial line has a characteristic impedance of 73 Ω, a common method of feeding this antenna is accomplished by connecting directly to the center of the antenna. This method of connection produces no standing waves on the line when the line is matched to a generator. Coupling to a generator is often made through a simple untuned transformer secondary.

Another method of transferring energy to the antenna is through the use of a twisted-pair line, as shown in Fig. 14-10(b). It is used as an untuned line for low frequencies. Due to excessive losses occurring in the insulation, the twisted pair is not used at higher frequencies. The characteristic impedance of such lines is about 70 Ω.

Delta Match

When a line does not match the impedance of the antenna, it is necessary to use special impedance matching techniques such as discussed with Smith chart applications in Chapter 12. An example of an additional type of impedance matching device is the *delta* match, shown in Fig. 14-10(c). Due to inherent characteristics, the open, two-wire transmission line does not have a characteristic impedance (Z_0) sufficiently low to match a center-fed dipole with $Z_{in} = 73$ Ω. Practical values of Z_0 for such lines lie in the range 300 to 700 Ω. To provide the required impedance match, a delta section [shown in Fig. 14-10(c)] is used. This match is obtained by spreading the transmission line as it approaches the antenna. In the example given the characteristic impedance of the line is 600 Ω, and the center impedance of the antenna is 73 Ω. As the end of the transmission line is spread, its characteristic impedance increases. Proceeding from the center of the antenna to either end, a point will be reached where the antenna impedance equals the impedance at the output terminals of the delta section. Recall that the antenna impedance increases as you move from its center to the ends. The delta section is then connected at this distance to either side of the antenna center.

The delta section becomes part of the antenna and, consequently, introduces radiation loss (one of its disadvantages). Another disadvantage is that trial-and-error methods are usually required to determine the dimensions of the A and B sections for optimum performance. Since both the distance between the delta output terminals (its width) and the length of the delta section are variable, adjustment of the delta match is difficult.

Antenna transmission line
measurement system.
(Courtesy of Wiltron.)

Quarter-Wave Matching

Still another impedance matching device is the quarter-wave transformer, or matching transformer, as shown in Fig. 14-10(d). This device is used to match the low impedance of the antenna to the line of higher impedance. Recall from Chapter 12 that the quarter-wave matching section is effective only between a line and purely resistive loads.

To determine the characteristic impedance (Z'_0) of the quarter-wave section, the following formula from Chapter 12 is used.

$$Z'_0 = \sqrt{Z_0 R_L} \tag{12-29}$$

where Z'_0 = characteristic impedance of the matching line
Z_0 = impedance of the feed line
R_L = resistive impedance of the radiating element

For the example shown, Z'_0 has a value slightly over 209 Ω. With this matching device, standing waves will exist on the λ/4 section but not on the 600-Ω line. Recall from Chapter 12 the use of stub matching techniques as another alternative.

This matching technique is useful for narrowband operation, while the delta section is more broadband in operation.

14-5 MARCONI ANTENNA

It was mentioned earlier that the Marconi antenna is used primarily with frequencies below 2 MHz. The difference between the Marconi antenna and the Hertz antenna is that the Marconi type requires a conducting path to ground, and the Hertz type does not. The Marconi antenna is usually a quarter-wave grounded antenna or any odd multiple of a quarter wavelength.

Effects of Ground Reflection

A Marconi antenna used as a transmitting element is shown in Fig. 14-11. The transmitter is connected between the antenna and ground. The actual length of the antenna is one-quarter wavelength. However, this type of antenna, by virtue of its connection to ground, uses the ground as the other quarter wavelength, making the antenna electrically a half wavelength. This is so because the earth is considered to be a good conductor. In fact, there is a reflection from the earth that is equivalent to the radiation that would be realized if another quarter-wave section were used. The reflection from the ground looks like it is coming from a λ/4 section beneath the ground. This is known as the *image antenna* and is shown in Fig. 14-11. By use of the Marconi antenna, which is a quarter wave in

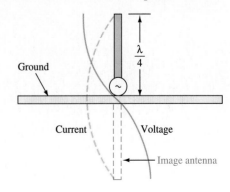

FIGURE 14-11 Grounded Marconi antenna.

actual physical length, half-wave operation may be obtained. All of the voltage, current, and impedance relationships characteristic of a half-wave antenna will also exist in this antenna. The only exception is the input impedance, which is approximately 36.6 Ω at the base. The effective current in the Marconi grounded antenna is maximum at the base and minimum at the top, while voltage is minimum at the bottom and maximum at the top.

When the conductivity of the soil in which the Marconi antenna is supported is very low, the reflected wave from the ground may be greatly attenuated. A great attenuation of the reflected signal is highly undesirable. To overcome this disadvantage, the site location could be moved to a location where the soil possesses a high conductivity. If it is impractical to move the site, provisions must be made to improve the reflecting characteristics of the ground by installing a buried ground screen.

The Counterpoise

When an actual ground connection cannot be used because of the high resistance of the soil or a large buried ground screen is impractical, a *counterpoise* may replace the usual direct ground connection. This is required for Marconi antennas mounted at the top of tall buildings. The counterpoise consists of a structure made of wire erected a short distance above the ground and *insulated from the ground.* The size of the counterpoise should be at least equal to, and preferably larger than, the size of the antenna.

The counterpoise and the surface of the ground form a large capacitor. Due to this capacitance, antenna current is collected in the form of charge and discharge currents. The end of the antenna normally connected to ground is connected through the large capacitance formed by the counterpoise. If the counterpoise is not well insulated from ground, the effect is much the same as that of a leaky capacitor, with a resultant loss greater than if no counterpoise were used.

Although the shape and size of the counterpoise are not particularly critical, it should extend for equal distances in all directions. When the antenna is mounted vertically, the counterpoise may have any simple geometric pattern, such as is shown in Fig. 14-12. The counterpoise is constructed so that it is nonresonant at the operating frequency. The operation realized by use of either the well-grounded Marconi antenna or the Marconi antenna using a counterpoise is the same as that of the half-wave antenna of the same polarization.

Radiation Pattern

The radiation pattern for a Marconi antenna is shown in Fig. 14-13(a). It is omnidirectional in the ground plane but falls to zero off the antenna's top. Thus, a large amount of energy is launched as a ground wave, but appreciable sky-wave energy also exists. By

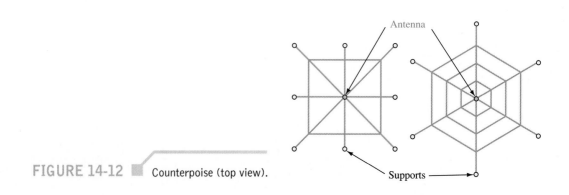

FIGURE 14-12 ▪ Counterpoise (top view).

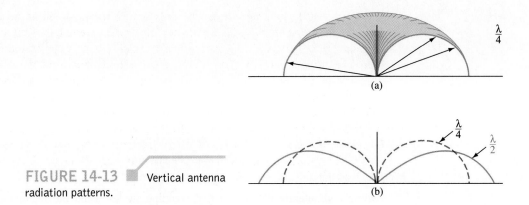

$\dfrac{\lambda}{4}$

(a)

$\dfrac{\lambda}{4}$

$\dfrac{\lambda}{2}$

(b)

FIGURE 14-13 ▓ Vertical antenna radiation patterns.

increasing the vertical height to $\lambda/2$, the ground-wave strength is increased, as shown in Fig. 14-13(b). The maximum ground-wave strength is obtained by using a length slightly less than $5/8\lambda$. Any greater length produces high-angle radiation of increasing strength, and horizontal radiation is reduced. At a height of 1λ there is no ground wave.

Loaded Antennas

In many low-frequency applications it is not practical to use an antenna that is a full quarter wavelength. This is especially true for mobile transceiver applications. Marconi antennas less than a quarter wavelength have an input impedance that is highly capacitive, and they become inefficient radiators. The reason for this is that a highly reactive load cannot accept energy from the transmitter. It will be reflected and will set up high standing waves on the feeder transmission line. An example of this is a $\lambda/8$ Marconi antenna, which exhibits an input impedance of about $8\ \Omega - j500\ \Omega$ at its base.

To remedy this situation, the *effective* height of the antenna should be $\lambda/4$, and this can be accomplished with several different techniques. Figure 14-14 shows a series inductance that is termed a *loading coil*. It is used to tune out the capacitive appearance of the antenna. The coil–antenna combination can thus be made to appear resonant (resistive) so that it can absorb the full transmitter power. The inductor can be variable to allow adjustment for optimum operation over a range of transmitter frequencies. Notice the standing wave of current shown in Fig. 14-14. It has maximum amplitude at the loading coil and thus does not add to the radiated power. This results in heavy I^2R losses in the coil instead of this energy being radiated. However, the transmission line feeding the

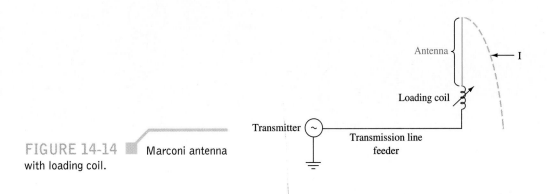

FIGURE 14-14 ▓ Marconi antenna with loading coil.

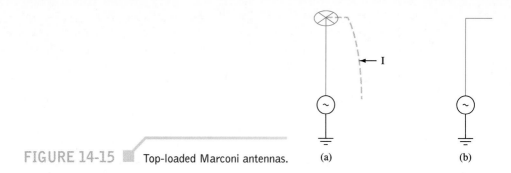

FIGURE 14-15 Top-loaded Marconi antennas.   (a)   (b)

loading coil/antenna will be free of standing waves when the loading coil is properly tuned.

A more efficient solution is the use of top loading, as shown in Fig. 14-15(a). Notice the high-current standing wave now exists at the base of the antenna so that maximum possible radiation now occurs. The metallic *spoked wheel* at the top adds shunt capacitance to ground. This additional capacitance reduces the antenna's capacitive reactance since C and X_c are inversely related. The antenna can, therefore, be made nearly resonant with the proper amount of top loading. This does not allow for convenient variable frequency operation as with the loading coil, but it is a more efficient radiator. The *inverted L* antenna in Fig. 14-15(b) accomplishes the same goal as the top-loaded antenna but is usually less convenient to physically construct.

14-6 ANTENNA ARRAYS

Hertz Antenna with Parasitic Element

The most elementary antenna array is shown in Fig. 14-16. It consists of a simple Hertz half-wave dipole and a nondriven (not electrically connected) half-wave element located a quarter wavelength behind the dipole. The non-driven element is also termed a *parasitic* element since it is not electrically connected.

The dipole radiates electromagnetic waves with the usual bidirectional pattern. However, the energy traveling toward the parasitic element, upon reaching it, induces voltages and currents but incurs a 180° phase shift in the process. These voltages and currents cause the parasitic element to also radiate a bidirectional wave pattern. However,

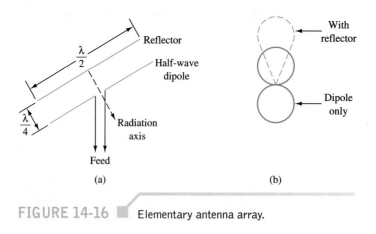

FIGURE 14-16 Elementary antenna array.

due to the 180° phase shift, the energy traveling away from the driven element cancels with that from the driven element. The energy from the parasitic element traveling toward the driven element reaches it in phase and causes a doubling of energy propagated in that direction. This effect is shown by the radiation pattern in Fig. 14-16(b). The parasitic element is also termed a *reflector* since it effectively "reflects" energy from the driven element. Notice that this simple array has resulted in a more directive antenna and thus exhibits gain with respect to a standard half-wavelength Hertz antenna.

Let us consider *why* the energy from the reflector gets back to the driven element in phase and thus reinforces propagation in that direction. Recall that the initial energy from the driven element travels a quarter wavelength before reaching the reflector. This is equivalent to 90 electrical degrees of phase shift. An additional 180° of phase shift occurs from the induction of voltage and current into the reflector. The reflector's radiated energy back toward the driven element experiences another 90° of phase shift before reaching the driven element. Thus, a total phase shift of 360° (90° + 180° + 90°) results so that the reflector's energy reaches the driven element in phase.

Yagi–Uda Antenna

The Yagi–Uda antenna consists of a driven element and two or more parasitic elements. It is named after the two Japanese scientists who were instrumental in its development. The version shown in Fig. 14-17(a) has two parasitic elements: a reflector and a director. A *director* is a parasitic element that serves to "direct" electromagnetic energy since it is in the direction of the propagated energy with respect to the driven element. The radiation pattern is shown in Fig. 14-17(b). Notice the two side *lobes* of radiated energy that result. They are generally undesired, as is the small amount of reverse propagation. The difference in gain from the forward to the reverse direction is defined as the *front-to-back ratio* (*F/B* ratio). For example, the pattern in Fig. 14-17(b) has a forward gain of 7 dB and a −1-dB gain (actually, loss, since it is a negative gain) in the reverse direction. Its *F/B* ratio is therefore [7 dB − (−1 dB)], or 8 dB.

This Yagi–Uda antenna provides about 7 dB of power gain with respect to a half-wavelength dipole reference. This is somewhat better than the approximate 3-dB gain of the simple array shown in Fig. 14-16. In practice, the Yagi–Uda antenna often consists of

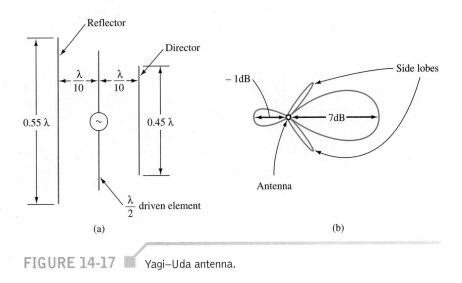

FIGURE 14-17 Yagi–Uda antenna.

one reflector and two or more directors to provide even better gain characteristics. They are often used as HF transmitting antennas and as VHF/UHF television receiving antennas.

The analysis of how the radiation patterns of these antennas result is rather complex and cannot be simply accomplished, as was done for the simple array shown in Fig. 14-16. More often than not, the lengths and spacings of the parasitic elements are the result of experiments rather than theoretical calculations.

Driven Collinear Array

A *driven array* is a multielement antenna in which all of the elements are excited through a transmission line. A four-element collinear array is shown in Fig. 14-18(a). A *collinear array* is any combination of half-wave elements in which all the elements are placed end to end so as to form a straight line. Each element is excited so that their fields are all in phase (additive) for points perpendicular to the array. This is accomplished by the $\lambda/2$ length of transmission line (a $\lambda/4$ twisted pair) between the elements on both sides of the feed point. They are twisted so that the fields created by the line cancel each other to minimize losses.

The radiation pattern for this antenna is provided in Fig. 14-18(b). Energy off the ends is cancelled from the $\lambda/2$ spacing (cancellation) of elements, but reinforcement takes place perpendicular to the antenna. The resulting radiation pattern thus has gain with respect to the standard Hertz antenna radiation pattern shown with dashed lines in Fig. 14-18(b). It has gain at the expense of energy propagated away from the antenna's perpendicular direction. The full three-dimensional pattern for both antennas is obtained by revolving the pattern shown about the antenna axis. This results in the doughnut-shaped pattern for the Hertz antenna and flattened doughnut shape for this collinear array. The array is a more directive antenna (smaller beamwidth). Increased directivity and gain are obtained by adding more collinear elements.

Broadside Array

If a group of half-wave elements is mounted vertically, one over the other as shown in Fig. 14-19, a broadside array is formed. Such an array provides greater directivity in both the vertical and horizontal planes than the collinear array. With the arrangement shown in

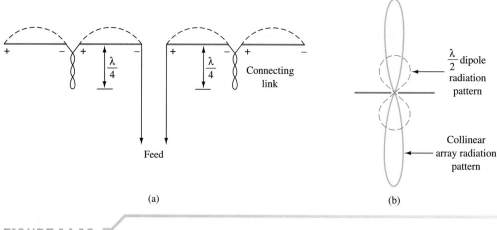

(a)

(b)

FIGURE 14-18 ■ Four-element collinear array.

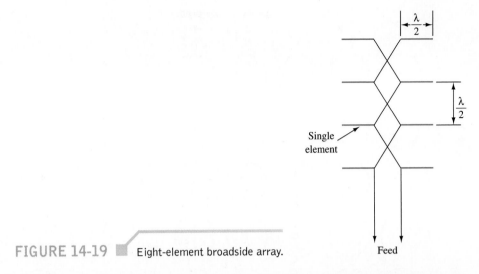

FIGURE 14-19 ▦ Eight-element broadside array.

Fig. 14-19, the separation between each stack is a half wavelength. The signal reversal shown in the connecting wires puts the voltage and current in each element of each stack in phase. The net resulting radiation pattern is a directive pattern in the horizontal plane (as with the collinear array) but also a directive pattern in the vertical plane (in contrast to the collinear array).

Marconi Array

You have probably noticed that standard broadcast AM transmitters usually utilize three or more Marconi antennas lined up in a row with equal spacing between them. The radiation pattern of a single Marconi antenna is omnidirectional in the horizontal plane, which may be undesirable due to interference possibilities with an adjacent channel station or due to geographical population density patterns. For instance, it doesn't make sense for a New York City station to beam half of its energy to the Atlantic Ocean. By properly controlling the phase and power level into each of the towers, virtually any radiation pattern desired can be obtained. Thus, the energy that would have been wasted over the Atlantic Ocean can be redirected to the areas of maximum population density.

This arrangement is called a *phased array* since controlling the phase (and power) to each element results in a wide variety of possible radiation patterns. A station may easily change its pattern at sunrise and sunset because increased skywave coverage at night might interfere with a distant station operating at about the same frequency. To give an idea of the countless radiation patterns possible with a phased array, refer to Fig. 14-20. It shows the radiation for just two $\lambda/4$ vertical antennas with variable spacing and input voltage phase. The patterns are simply the vector sum of the instantaneous field strength from each individual antenna.

14-7 SPECIAL-PURPOSE ANTENNAS

Log-Periodic Antenna

The log-periodic antenna is a special case of a driven array. It was first developed in 1957 and has proven so desirable that its many variations now make up an entire class of anten-

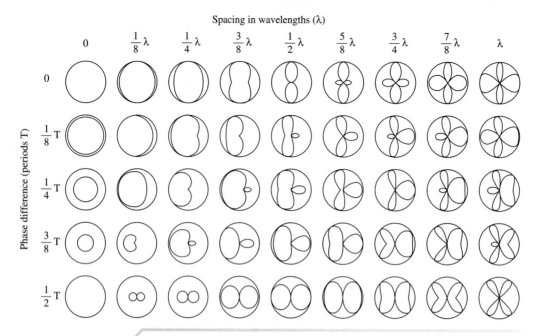

Spacing in wavelengths (λ)

FIGURE 14-20 ▦ Phase-array antenna patterns. (From Henry Jasik, Ed., *Antenna Engineering Handbook,* 1961; courtesy of McGraw-Hill Book Company, New York.)

nas. It provides reasonably good gain over an extremely wide range of frequencies. It is, therefore, useful for multiband transceiver operation and as a TV receiving unit to cover the entire VHF and UHF bands. It can be termed a wide-bandwidth or broad-band antenna. Bandwidth is not to be confused with beamwidth in this situation.

Antenna *bandwidth* is defined with respect to its design frequency, often termed its *center frequency.* If a 100-MHz (center frequency) log-periodic antenna's transmitted or received power is 3 dB down at 50 MHz and 200 MHz, its bandwidth is 200 MHz − 50 MHz, or 150 MHz. This measurement is made in the direction of highest antenna directivity.

The most elementary form of log-periodic antenna is shown in Fig. 14-21(a). It is termed a log-periodic dipole array and derives its name from the fact that its important

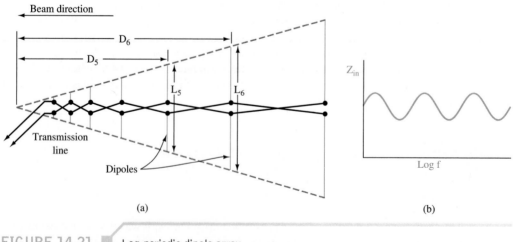

FIGURE 14-21 ▦ Log-periodic dipole array.

characteristics are periodic with respect to the logarithm of frequency. This is true of its impedance, its SWR with a given feed line, and the strength of its radiation pattern. For instance, its input impedance is shown to be nearly constant (but periodic) as a function of the log of frequency in Fig. 14-21(b).

The log-periodic array in Fig. 14-21(a) is seen to consist of a number of dipoles of different lengths and spacings. The dipole lengths and spacings are related by

$$\frac{D_1}{D_2} = \frac{D_2}{D_3} = \frac{D_3}{D_4} = \frac{D_4}{D_5} \cdots = \tau = \frac{l_1}{l_2} = \frac{l_2}{l_3} = \frac{l_3}{l_4} = \frac{l_4}{l_5} \cdots \qquad (14\text{-}4)$$

where τ is called the *design* ratio with a typical value of 0.7. The range of frequencies it is useful over is determined by the frequencies at which the longest and shortest dipoles are a half wavelength.

Loop Antenna

A loop antenna is a single turn of wire whose dimensions are normally much smaller than a wavelength. When this condition exists, the current in it may all be considered in phase. This results in a magnetic field that is everywhere perpendicular to the loop. The resulting radiation pattern is sharply bidirectional, as indicated in Fig. 14-22, and is effective over an extremely wide range of frequencies—those for which its diameter is about $\lambda/16$ or less. The antenna is usually circular, but any shape is effective.

Because of its sharply defined pattern, small size, and broad-band characteristics, the loop antenna's major application is in direction finding (DF) applications. The goal is to determine the direction of some particular radiation. Generally, readings from two different locations will be required due to the antenna's bidirectional pattern. If the two locations are far enough apart, the distance and direction of the radiation source can be calculated using trigonometry. Since the signal falls to zero much more sharply than it peaks, the nulls are used in the DF applications.

While other antennas with directional characteristics could be used in DF, the loop's small size seems to outweigh the gain advantages of larger directive antennas.

Ferrite Loop Antenna

The familiar ferrite loop antenna found in most broadcast AM receivers is an extension of the basic loop antenna just discussed. The effect of using a large number of loops wound about a highly magnetic core (usually ferrite) serves to greatly increase the effective diameter of the loops. This forms a highly efficient receiving antenna, considering its small physical size compared to the hundreds of feet required to obtain a quarter wavelength for the broadcast AM band. The directional characteristics of this antenna are veri-

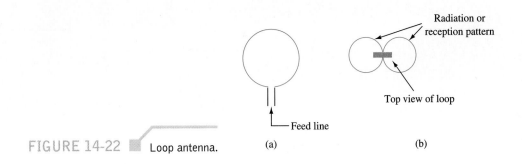

FIGURE 14-22 Loop antenna.   (a)   (b)

fied by the fact that a portable AM receiver can usually be oriented to *null* out reception of a station. You should now be able to determine a line through which that broadcasting station exists when the null is detected.

Folded Dipole Antenna

Recall that the standard half-wavelength dipole (Hertz antenna) has an input impedance of 73 Ω. Recall also that it becomes very inefficient whenever it is not used at the frequency for which its length equals $\lambda/2$ (i.e., it has a narrow bandwidth). The folded dipole antenna shown in Fig. 14-23(a) offers the same radiation pattern as the standard Hertz antenna but has an input impedance of 288 Ω (approximately 4×73 Ω) and offers relatively broadband operation.

A standard Hertz antenna can provide the same broadband characteristics as the folded dipole by incorporating a parallel tank circuit, as shown in Fig. 14-23(b). With the tank circuit resonant at the frequency corresponding to the antenna's $\lambda/2$ length, the tank presents a very high resistance in parallel with the antenna's 73 Ω and has no effect. However, as the frequency goes down, the antenna becomes capacitive, while the tank circuit becomes inductive. The net result is a resistive overall input impedance over a relatively wide frequency range.

The folded dipole is a useful receiving antenna for broadcast FM and for VHF TV. Its input impedance matches well with the 300-Ω input impedance terminals common to these receivers. It can be inexpensively fabricated by using a piece of standard 300-Ω parallel wire transmission line cut to $\lambda/2$ at midband and shorting together the two at each end. Folded dipoles are also invariably used as the driven element in Yagi–Uda antennas. This helps to maintain a reasonably high input impedance since the addition of each director lowers this array's input impedance. It also gives the antenna a broader band of operation.

In applications where a folded dipole with other than a 288-Ω impedance is desired, a larger-diameter wire for one length of the antenna is used. Impedances up to about 600 Ω are possible in this manner.

Slot Antenna

Coupling RF energy into a slot in a large metallic plane can result in radiated energy with a pattern similar to a dipole antenna mounted over a reflecting surface. The length of the slot is typically one-half wavelength. These antennas function at UHF and microwave frequencies with energy coupled into the slot by waveguides or coaxial line feed connected directly across the short dimension of a rectangular slot. These antennas are commonly used in modern aircraft in an array module as shown in Fig. 14-24(a). This 32-element

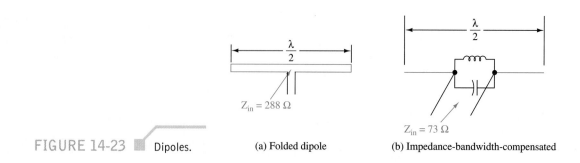

FIGURE 14-23 ▮ Dipoles.   (a) Folded dipole   (b) Impedance-bandwidth-compensated

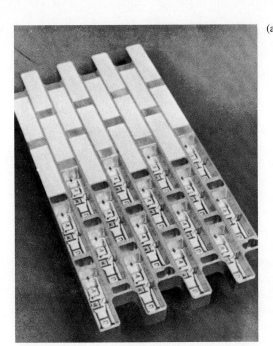

(a)

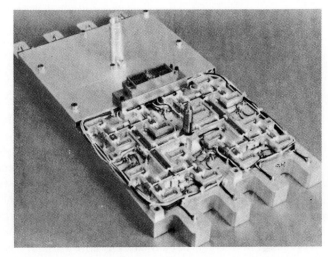

(b)

FIGURE 14-24   Slot antenna array.

(slot) array shows half the slots filled with dielectric material (to provide the required smooth airplane surface) and the others open to show the phase-shifting circuitry used to drive the slots. The rear view in Fig. 14-24(b) shows the coaxial feed connectors used for this antenna array.

The individual drive to each slot is controlled by phase-shifting networks. Proper phasing allows production of a directive radiation pattern that can be swept through a wide angle without physically moving the antenna. This allows a convenient mobile scanning radar system without mechanical complexities. These *phased array* antennas are typically built right into the wings of aircraft, with the dielectric window filling eliminating aerodynamic drag.

TROUBLESHOOTING

Antenna installation is often part of the technician's job. For the antenna to function at peak performance, technicians must follow proper installation techniques. Equipment manufacturers publish guidelines for antenna installation. This section discusses general installation procedures for antenna's and looks at general troubleshooting techniques.

Often antennas are mounted high and are not easily accessible, making them difficult to inspect. The condition of the antenna and the transmission line can be determined, however, by measuring antenna emission and standing waves on the transmission line. In this section you will discover how some simple-to-use basic communication test equipment can check and troubleshoot the antenna systems.

After completing this section you will be able to:

• Identify safety precautions to observe for an antenna installation

• Describe proper antenna grounding

- Describe correct transmission line installation

- Troubleshoot typical antenna problems

- Explain the use of the SWR meter

- Explain how a grid-dip meter is used

- Describe how the SWR meter can help find antenna faults

Installing the Antenna

TV and FM antenna installation practices will be referred to throughout this discussion. Installing TV and FM antennas portrays fairly standard practices that should be followed when doing any kind of communications antenna installation.

Never neglect safety. Standard operating procedure should always make safety first on the list of things to do. Locate power lines, telephone lines, and obstacles that could interfere with the installation or present a hazard to the installers. A tower structure would require a concrete base as the supporting structure. Guy wires need room for proper mounting. Anchor ladders securely. Use safety belts or harnesses whenever climbing towers or other structures. Be aware of building codes and follow installation procedures supplied by the equipment manufacturer. Heed the equipment manufacturer's warnings.

GROUNDING AND LIGHTNING PROTECTION   Antennas on high exposed metal masts are subject to being struck by lightning. Ground the mast by connecting a wire to it and to a ground rod. When the mast is struck by lightning the surge of electricity will be shorted to ground. Some antenna installations require several such ground connections. A typical ground wire size for a TV or FM antenna is 10 AWG. Local ordinances should specify wire size and type for grounding application. A good grounding system will also protect against static charge buildup.

Lightning will follow a second path if a strike occurs. The second path is down the lead-in wire and into the equipment that is connected to it. To protect against this lightning, surge protectors and static discharge devices are placed between the antenna and the receiving equipment. Check the installation manual for recommended lightning surge protectors and static discharge devices.

PLANNING THE INSTALLATION   An antenna installation site must be safe, as previously mentioned. Keep a proper distance from power lines and telephone lines. For good reception, no obstructions should exist between the antenna and the receiving direction. The antenna will generally require a base, a mast, and a good supporting structure. The base may be concrete with anchors for footing, as in a tower structure. The mast may be telescoping poles that are secured to a building or other structure.

If guy wires are used, there must be enough room around the mast to install them. Guy wires usually extend in three equally spaced directions. The wires should intercept the mast at 45° angles for proper support. The guy wires must be well anchored at the ground points where they attach to eyelets. Anchor the eyelets in concrete. This makes a very secure guy wire support system. Tighten the guy wires using turnbuckles.

LEAD-IN WIRES   Proper care should be exercised when running the lead-in line. Improper installation can result in troubles later on. Lead-in wire should be kept as short

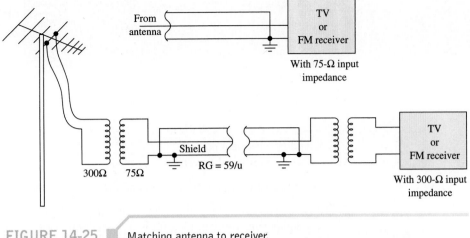

FIGURE 14-25 ■ Matching antenna to receiver.

as possible. Twin-lead 300-Ω antenna wire, often used with TV and FM antennas, must be installed using standoff insulators. This wire should never touch metal. The metal will influence the transmission line's characteristic impedance and cause attenuation and reflections. For areas of high interference, substitute coaxial cable for the twin-lead wire. Coaxial cable does not require insulated standoffs. Taping it to the mast and running it along rain gutters makes for a good installation.

When using coaxial cable instead of twin lead, remember that the antenna impedance must be matched to the transmission line. If the receiver does not permit direct connection of the coaxial cable, an impedance match must be made there also. Coaxial cable normally used with TV and FM antenna installation is RG-59/U. Its characteristic impedance is 75 Ω. Balun transformers match the antenna impedance and receiver input impedance, as illustrated in Fig. 14-25. If the TV or FM receiver has a 75-Ω connection, then an antenna balun is all that is necessary.

Avoid running lead-in wires through windows. Run lines through a special tube that feeds the line into the building via the wall. Some installations may specify using conduit to protect the lead-in line from weathering. Once the line is in the building, distribution outlets can serve to distribute the signal to more than one receiver. If outlets are not used, cut off excess lead in. Do not curl it up behind the receiving equipment.

Typical Troubleshooting Techniques

1. Is the VSWR as low as it should be? Most antennas are designed to operate with a specific type of transmission line, 50-Ω coax being very common. A directional wattmeter can be used to measure VSWR, as shown in Fig. 14-26.

    Sometimes the directional wattmeter takes the form of a directional coupler and a power meter, but the principle is the same. You measure the incident and reflected power and calculate the VSWR with the following formula:

$$\text{VSWR} = \frac{1 + \sqrt{\dfrac{P_r}{P_i}}}{1 - \sqrt{\dfrac{P_r}{P_i}}} \qquad (14\text{-}5)$$

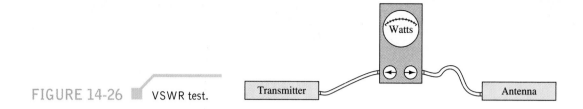

FIGURE 14-26 ▮ VSWR test.

where P_i = incident power
P_r = reflected power

Ideally, there is no reflected power and the VSWR is one.

2. Assume you find a very high VSWR. A common ohmmeter is a very good piece of test equipment. You will look for bad connections, open or shorted tuning elements such as capacitors and inductors.

    Another trick is to sweep the frequency to see if there is a frequency where the VSWR is low. This may lead you to broken elements or show that the design is improper.

    If this antenna is associated with a transmitter, visually inspect all insulators for tracks made by arcs. These arcs may not show up in a low-power test. Attempts to repair such insulators are seldom successful.

3. Has this antenna been subject to severe weather conditions such as ice, wind, and rain? Parts may be broken or full of water.

4. Is the antenna able to handle power applied to it? Antennas such as the Yagi–Uda and log periodic have extremely high voltages at the ends of the elements. Corona will form and sometimes actually melt the ends of the antenna. If this occurs at an AM station, you can hear the audio on the arc. The problem is cured by adding rings or balls to the elements.

5. Installation problems. It is quite often a good idea to question physical dimensions if the antenna is new. People make mistakes during assembly that you can find with a tape measure.

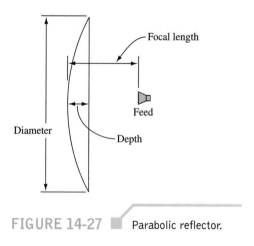

FIGURE 14-27 ▮ Parabolic reflector.

Consider a parabolic reflector used in a TV satellite system. With a tape measure and some string you can find the focal point if you know the following formula:

$$\text{Focal length} = \frac{\text{diameter}^2}{16 \times \text{depth}} \qquad (14\text{-}6)$$

Simply stretch the string across the dish and find the depth D, measure the diameter, and solve for the focal length. Usually the face of the feed should be at the focal point or a very small distance closer to the parabola. Refer to Fig. 14-27. Further information on parabolic antennas is provided in Chapter 16.

Antenna Measurements

Antenna performance predictions are based on free-space operation. Structures located around the antenna can severely affect the performance of the antenna. Building construction, tree growth, towers, and wires can alter the original antenna installation performance. In addition to changes in the area surrounding the antenna, characteristics of the antenna or the transmission line are subject to change over time. Remember, antenna efficiency relies on several factors. Antenna performance relies on electrical length, physical length, and matching the antenna impedance to that of the transmission line. All of these characteristics are subject to change due to storms, weathering, and mishaps.

GRID-DIP METER   The *grid-dip meter* has been used for years to measure radio frequencies. It is a hand-held device that can measure the frequency of tuned circuits and antennas without power being applied to them. Battery-powered and equipped with a tunable oscillator and a scale calibrated in frequency, the meter can measure very accurately the frequency of tuned circuits. It is equipped with plug-in loop coils that serve as probes. The loop couples energy into or out of the grid-dip oscillator circuit. Tuned circuits are checked by bringing the loop coil near them and adjusting the grid-dip oscillator until a dip is seen on the meter (see Fig. 14-28). The scale indicates the tuned circuit's resonant frequency when the dip occurs. A specific frequency is measured by choosing a loop coil for that desired frequency. The grid-dip meter comes with several such loop coils. Sometimes technicians construct their own coil to measure a specific frequency.

The grid-dip meter can measure the resonant frequency of the antenna by connecting it directly to the antenna terminals. The grid-dip can also act as an absorption meter

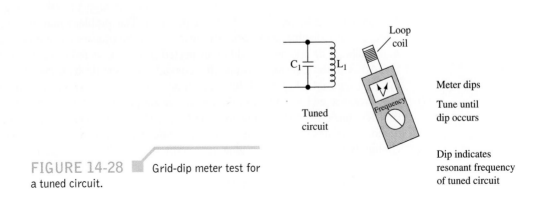

FIGURE 14-28 ■ Grid-dip meter test for a tuned circuit.

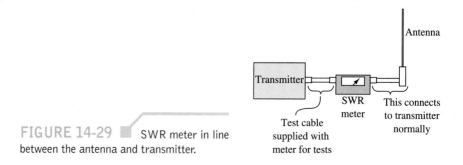

FIGURE 14-29 ■ SWR meter in line between the antenna and transmitter.

that measures radiated antenna energy. By bringing the meter's loop coil close to a radiating antenna and tuning the grid-dip meter for a peak meter indication, the radiating frequency can be determined. Field strength measurements can aid in determining the radiating pattern of the antenna. To do a field strength reading, take a position near the antenna and read the meter after carefully tuning for a peak reading on the scale. After several readings, the antenna radiating pattern can be plotted from the findings.

SWR METER   Another very useful piece of test equipment is the SWR (standing wave ratio) meter. Figure 14-29 shows that the SWR meter is inserted between the transmitter and the antenna. A test cable is attached to the transmitter and to the SWR meter. On the antenna side of the SWR meter connect the transmission line that normally runs from the transmitter to the antenna. An impedance mismatch at the antenna and the transmission line will result in the existence of a voltage standing wave ratio (VSWR) on the transmission line. This test is similar to the one shown in Fig. 14-26.

To measure VSWR with the SWR meter, first calibrate the meter. Follow the instructions supplied with the SWR meter regarding the calibration process. Once the meter is calibrated, switch the meter to the SWR setting to make a reading. A good SWR reading will indicate 1.5 or less. Impedance adjusting of the transmission line or the antenna can be done with the SWR meter in line between them. Changes can be made to the transmission line or the antenna until a minimum SWR reading is obtained. SWR meters can be equipped with an antenna probe to make field strength measurements. These readings are taken at different points around the radiating antenna, and a plot is made to establish the antenna pattern.

TROUBLESHOOTING WITH THE SWR METER   A high VSWR reading, greater than 1.5, on a transmission line indicates a problem. The problem may be from a crimped cable, a crushed cable, an impedance mismatch at the transmitter or antenna, or moisture in the cable. The antenna should be inspected if it is suspected of being faulty. If a problem exists in the transmission line, the coaxial cable should be tested. Use an ohmmeter to check the cable. Figure 14-30(a) illustrates how the cable can be tested. First, measure the resistance end-to-end from the metal outside portions of each connector. Then measure resistance from the connectors' center pin to the other center pin. A small resistance reading indicates a good cable. A high resistance reading indicates an open cable. A cable continuity test can also determine the quality of the cable [see Fig. 14-30(b)]. Short circuit one end of the coaxial cable using a jumper wire while measuring from the other end. A high-resistance reading or open reading indicates an open cable.

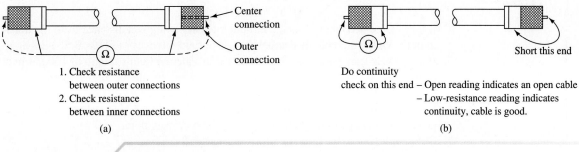

1. Check resistance
   between outer connections
2. Check resistance
   between inner connections

(a)

Center
connection

Outer
connection

Short this end

Do continuity
check on this end – Open reading indicates an open cable
– Low-resistance reading indicates
continuity, cable is good.

(b)

FIGURE 14-30 ▌ Testing coaxial cable.

SUMMARY

In Chapter 14 we examined antenna operation and many of the possible configurations. Be sure to keep in mind that antenna properties apply identically when both transmitting and receiving—the principle of reciprocity. The major topics the student should now understand include:

- the analysis of the Hertz antenna, including its impedance, radiation field, radiation pattern, and gain

- the definition of radiation resistance and related calculations

- the description of antenna feed lines, including resonant and nonresonant

- the operation of impedance matching devices, including the delta match and quarter-wave matching transformer

- the analysis of Marconi antennas, including ground effects, counterpoise effects, radiation pattern, and loading effects

- the effects of parasitic elements, including discussion of the Yagi–Uda antenna, driven collinear array, broadside array, and Marconi array

- the description and operation of various special purpose antennas, including the log-periodic, loop, ferrite loop, folding dipole, and slot antenna's

QUESTIONS AND PROBLEMS

SECTION 1

*1. How should a transmitting antenna be designed if a vertically polarized wave is to be radiated, and how should the receiving antenna be designed for best performance in receiving the ground wave from this transmitting antenna?

*2. A ship radio-telephone transmitter operates on 2738 kHz. At a distant point from the transmitter, the 2738-kHz signal has a measured field of 147 mV/m. The second harmonic field at the same point is measured as 405 μV/m. How much has the harmonic emission been attenuated below the 2738-kHz fundamental? (51.2 dB)

*3. If a field intensity of 25 mV/m develops 2.7 V in a certain antenna, what is its effective height? (108 m)

*4. If the power of a 500-kHz transmitter is increased from 150 W to 300 W, what would be the percentage change in field intensity at a given distance from the transmitter? What would be the decibel change in field intensity? (141%, 3 dB)

*5. If a 500-kHz transmitter of constant power produces a field strength of 100 μV/m at a distance of 100 mi from the transmitter, what would be the theoretical field strength at a distance of 200 mi from the transmitter? (50 μV/m)

*6. If the antenna current at a 500-kHz transmitter is reduced 50%, what would be the percentage change in the field intensity at the receiving point? (50%)

*7. Define *field intensity*. Explain how it is measured.

*8. What is effective radiated power? Given transmitter power output, antenna resistance, antenna transmission line losses, transmitter efficiency, and power gain, show how ERP is calculated.

*9. Define *polarization* as it refers to broadcast antennas.

SECTION 2

10. Explain the development of a Hertz antenna from a quarter-wavelength, open-circuited transmission line.

*11. Draw a diagram showing how current varies along a half-wavelength Hertz antenna.

*12. Explain the voltage and current relationships in a one-wavelength antenna, one-half-wavelength (dipole) antenna, and one-quarter-wavelength *grounded* antenna.

*13. What effect does the magnitude of the voltage and current at a point on a half-wave antenna in *free space* (a dipole) have on the impedance at that point?

*14. Can either of the two fields that emanate from an antenna produce an EMF in a receiving antenna? If so, how?

15. Draw the three-dimensional radiation pattern for the Hertz antenna, and explain how it is developed.

16. Define antenna *beamwidth*.

*17. What is the effective radiated power of a television broadcast station if the output of the transmitter is 1000 W, antenna transmission line loss is 50 W, and the antenna power gain is 3? (2850 W)

18. A $\lambda/2$ dipole is driven with a 5-W signal at 225 MHz. A receiving dipole 100 km away is aligned such that its gain is cut in half. Calculate the received power and voltage into a 73-Ω receiver. (7.57 pW, 23.5 μV)

19. An antenna with a gain of 4.7 dBi is being compared with one having a gain of 2.6 dBd. Which has the greater gain? Express the gain difference in dB and as a simple ratio.

SECTION 3

20. Define *radiation resistance* and explain its significance.

*21. The ammeter connected at the base of a Marconi antenna has a certain reading. If this reading is increased 2.77 times, what is the increase in output power? (7.67)

22. How is the operating power of an AM transmitter determined using antenna resistance and antenna current?

23. Explain what happens to an antenna's radiation resistance as its length is continuously increased.

24. Explain the effect that ground has on an antenna.

25. Calculate the efficiency of an antenna that has a radiation resistance of 73 Ω and an effective dissipation resistance of 5 Ω. What factors could enter into the dissipation resistance? (93.6%)

*26. Explain the following terms with respect to antennas (transmission or reception):
   (a) Field strength.
   (b) Power gain.
   (c) Physical length.
   (d) Electrical length.
   (e) Polarization.
   (f) Diversity reception.
   (g) Corona discharge.

*27. What is the relationship between the electrical and physical length of a Hertz antenna?

*28. What factors determine the resonant frequency of any particular antenna?

*29. If a vertical antenna is 405 ft high and is operated at 1250 kHz, what is its physical height expressed in wavelengths? (5.54 λ)

*30. What must be the height of a vertical radiator one-half wavelength high if the operating frequency is 1100 kHz? (136 m)

SECTION 4

31. What is an antenna feed line? Explain the use of resonant antenna feed lines, including advantages and disadvantages.

32. What is a nonresonant antenna feed line? Explain its advantages and disadvantages.

33. Explain the operation of a delta match. Under what conditions is it a convenient matching system?

*34. Draw a simple schematic diagram of a push-pull, neutralized radio-frequency amplifier stage, coupled to a Marconi-type antenna system.

*35. Show by a diagram how a two-wire radio-frequency transmission line may be connected to feed a Hertz antenna.

*36. Calculate the characteristic impedance of a quarter-wavelength section used to connect a 300-Ω antenna to a 75-Ω line. (150 Ω)

SECTION 5

*37. Which type of antenna has a minimum of directional characteristics in the horizontal plane?

*38. If the resistance and the current at the base of a Marconi antenna are known, what formula could be used to determine the power in the antenna?

*39. What is the difference between a Hertz and a Marconi antenna?

*40. Draw a sketch and discuss the horizontal and vertical radiation patterns of a quarter-wave vertical antenna. Would this also apply to a similar type of receiving antenna?

41. What is an image antenna? Explain its relationship to the Marconi antenna.

*42. What would constitute the ground plane if a quarter-wave grounded (whip) antenna, 1 m in length, were mounted on the metal roof of an automobile? Mounted near the rear bumper of an automobile?

*43. What is the importance of the ground radials associated with standard broadcast antennas? What is likely to be the result of a large number of such radials becoming broken or seriously corroded?

*44. What is the effect on the resonant frequency of connecting an inductor in series with an antenna?

*45. What is the effect on the resonant frequency of adding a capacitor in series with an antenna?

*46. If you desire to operate on a frequency lower than the resonant frequency of an available Marconi antenna, how may this be accomplished?

*47. What will be the effect on the resonant frequency if the physical length of a Hertz antenna is reduced?

*48. Why do some standard broadcast stations use top-loaded antennas?

*49. Explain why a *loading coil* is sometimes associated with an antenna. Under this condition, would absence of the coil mean a capacitive antenna impedance?

Section 6

50. Explain how the directional capabilities of the elementary antenna array shown in Fig. 14-16 are developed.

51. Define the following terms:
(a) Driven elements.
(b) Parasitic elements.
(c) Reflector.
(d) Director.

52. Calculate the ERP from a Yagi–Uda antenna (illustrated in Fig. 14-17) driven with 500 W. (2500 W)

53. Calculate the *F/B* ratio for an antenna with
(a) Forward gain of 7 dB and reverse gain of −3 dB.
(b) Forward gain of 18 dB and reverse gain of 5 dB.

54. Sketch two different Yagi–Uda configurations.

55. Describe the physical configuration of a collinear array. What is the effect of adding more elements to this antenna?

56. Describe the physical configuration of a broadside array. Explain the major advantage they have as compared to collinear arrays.

*57. What is the direction of maximum radiation from two vertical antennas spaced 180° and having equal currents in phase?

*58. How does a directional antenna array at an AM broadcast station reduce radiation in some directions and increase it in other directions?

*59. What factors can cause the directional pattern of an AM station to change?

60. Define *phased array*.

Section 7

61. Describe the major characteristics of a log-periodic antenna. What explains its widespread use? Explain the significance of its shortest and longest elements.

62. Design a log-periodic antenna to cover the complete VHF TV band. (See Chapter 7 for the frequencies involved.) Use a design factor (τ) of 0.7, and provide a scaled sketch of the antenna with all dimensions indicated.

*63. Describe the directional characteristics of the following types of antennas:
   (a) Horizontal Hertz antenna.
   (b) Vertical Hertz antenna.
   (c) Vertical loop antenna.
   (d) Horizontal loop antenna.
   (e) Vertical Marconi antenna.
*64. What is the directional reception pattern of a loop antenna?
65. A loop antenna used for DF purposes detects a null from a signal with the loop rotated 35° CCW from a line of latitude. The antenna is moved 3 mi west along the same line of latitude and detects a null from the same signal source when rotated 45° CW from the line of latitude. On a sketch, show the two points when readings were taken and the exact location of the signal source with respect to the two points.
66. What is a ferrite loop antenna? Explain its application and advantages.
67. What is the radiation resistance of a standard folded dipole? What are its advantages over a standard dipole? Why is it usually used as the driven element for Yagi–Uda antennas instead of the Hertz antenna?
68. Describe the operation of a slot antenna and its application with aircraft in a driven array format.
69. An antenna has a maximum forward gain of 14 dB at its 108-MHz center frequency. Its reverse gain is −8 dB. Its beamwidth is 36° and bandwidth extends from 55 to 185 MHz. Calculate
   (a) Gain at 18° from maximum forward gain. (11 dB)
   (b) Bandwidth. (130 MHz)
   (c) F/B ratio. (22 dB)
   (d) Maximum gain at 185 MHz. (11 dB)
70. Explain the difference between antenna beamwidth and bandwidth.

WAVEGUIDES AND RADAR

OBJECTIVES

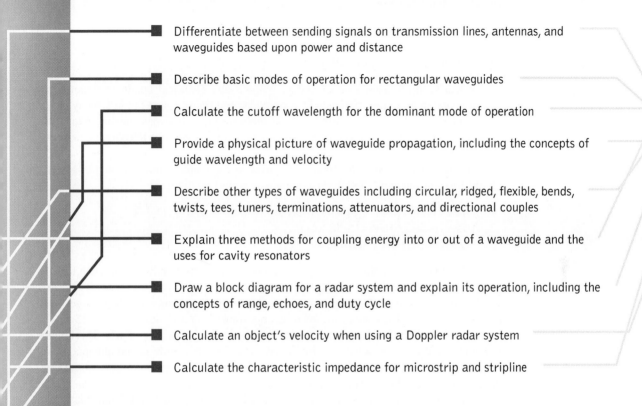

■ Differentiate between sending signals on transmission lines, antennas, and waveguides based upon power and distance

■ Describe basic modes of operation for rectangular waveguides

■ Calculate the cutoff wavelength for the dominant mode of operation

■ Provide a physical picture of waveguide propagation, including the concepts of guide wavelength and velocity

■ Describe other types of waveguides including circular, ridged, flexible, bends, twists, tees, tuners, terminations, attenuators, and directional couples

■ Explain three methods for coupling energy into or out of a waveguide and the uses for cavity resonators

■ Draw a block diagram for a radar system and explain its operation, including the concepts of range, echoes, and duty cycle

■ Calculate an object's velocity when using a Doppler radar system

■ Calculate the characteristic impedance for microstrip and stripline

15-1 COMPARISON OF TRANSMISSION SYSTEMS

The mode of energy transmission chosen for a given application would normally depend on the following factors: (1) initial cost and long-term maintenance, (2) frequency band to be used and its information-carrying capacity, (3) selectivity or privacy offered, (4) reliability and noise characteristics, and (5) power level and efficiency. Naturally, any one mode of energy transmission will have only some of the desirable features. It therefore becomes a matter of sound technical judgment to choose the mode of energy transmission best suited for a particular application.

Transmission lines and antennas are the more commonly known means of high-frequency energy transmission, but waveguides and fiber optics also play an important role. The following examples show that each method of transmission has its proper place. Fiber optics has been left out of this comparison, but this mode of transmission is fully discussed in Chapter 17.

Photo: Watch out for the radar man! (Courtesy of P. David Fisher, Michigan State University.)

It is desired to transmit a 1-GHz signal between two points 30 mi apart. If the received energy in each case were chosen to be 1 nW (10^{-9} W), then for comparison it would be found that for reasonably typical installations, the required *transmitted* power would be on the order of

1. *Transmission lines:* 10^{1500} nW (15,000-dB loss)
2. *Waveguides:* 10^{150} nW (1500-dB loss)
3. *Antennas:* 100 mW (80-dB loss)

Clearly, the transmission of energy without any electrical conductors (antennas) will be found to excel the efficiency of waveguides and transmission lines by many orders of magnitude. As a practical consequence of these results, microwave antenna relay links at about 30-mi intervals are used for cross-country transmission of telephone and television services.

If the transmission path length of the preceding example were shortened by a factor of 100:1 to a distance of 1500 ft, the comparison would become

1. *Transmission lines:* 1 MW (150-dB loss)
2. *Waveguides:* 30 nW (15-dB loss)
3. *Antennas:* 10 μW (40-dB loss)

Quite clearly the waveguide now surpasses either the transmission line or antenna for efficiency of energy transfer.

A comparison of the energy input required versus distance to obtain a received power of 1 nW for these three modes of energy transmission is shown in Fig. 15-1. The frequency is 1 GHz, and the results are expressed on a decibel scale with a 0-dB reference at the required receiver power level of 1 nW. The dashed section of the antenna curve, somewhat beyond 30 mi, indicates that the attenuation becomes severe beyond the line-of-sight distance, which is typically 50 mi.

One final comparison, and then a specific look at waveguides. Transmission of energy down to zero frequency is practical with transmission lines, but waveguides, antennas, and fiber optics inherently have a practical low-frequency limit. In the case of antennas, this limit is about 100 kHz, and for waveguides, it is about 300 MHz. Fiber-optic transmissions take place at the frequency of light, or greater than 10^{14} Hz! Theoreti-

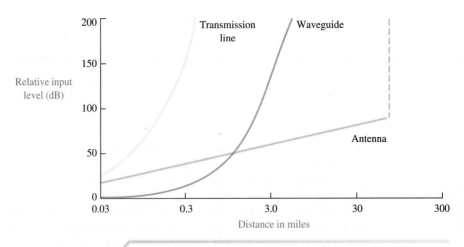

FIGURE 15-1 Input power required versus distance for fixed receiver power.

cally the antennas and waveguides could be made to work at arbitrarily low frequencies, but the physical sizes required would become excessively large. However, with the low gravity and lack of atmosphere on the moon, it may be feasible to have an antenna 10 mi high and 100 mi long for frequencies as low as a few hundred cycles per second. As an indication of the sizes involved, it may be noted that for either waveguides or antennas the important dimension is normally one-half wavelength. Thus, a waveguide for a 300-MHz signal would be about the size of a roadway drainage culvert, and an antenna for 300 MHz would be about $1\frac{1}{2}$ ft long.

15-2 TYPES OF WAVEGUIDES

Any surface separating two media of distinctly different conductivities or permittivities has a guiding effect on electromagnetic waves. For example, a rod of dielectric material, such as polystyrene, can carry a high-frequency wave, somewhat as a glass fiber conducts a beam of light. These phenomena will be further explored in Sec. 15-10 and Chapter 17. The best guiding surface, however, is that between a good dielectric and a good conductor.

In a broad sense, all kinds of transmission lines, including coaxial cables and parallel wires, are waveguides. In practice, however, the term *waveguide* has come to signify a hollow metal tube or pipe used to conduct electromagnetic waves through its interior. They were first extensively used in radar sets during World War II, operating at wavelengths of between 10 and 3 cm.

A waveguide can be almost any shape. The most popular shape is rectangular, but some use of circular and even more exotic shapes is made. We shall mainly study the rectangular waveguide operating in the TE_{10} mode. We shall learn more about this terminology shortly.

Like coaxial lines, waveguides are perfectly shielded—hence, no radiation loss. The attenuation of a hollow pipe is less, and the power capacity is greater, than that of a coaxial line of the same size at the same frequency. Most of the copper loss of a coaxial line occurs in the thin inner conductor; hence, its elimination in a waveguide reduces attenuation and increases the power capacity. It also simplifies the construction and makes the line more rugged.

Waveguide Operation

A rigorous mathematical demonstration of waveguide operation is beyond our intentions. A practical explanation is possible by starting out with a normal two-wire transmission line. You may recall from Chapter 12 that a quarter-wavelength shorted stub looks like an open circuit and in fact is often used as an insulating support for transmission lines. If an infinite number of these supports were added both above and below the two-wire transmission line as shown in Fig. 15-2, you can visualize it turning into a rectangular waveguide. If the shorted stubs were less than a quarter wavelength, operation would be drastically impaired. The same is true of a rectangular waveguide. The *a* dimension of a waveguide, shown in Fig. 15-3, must be at least one-half wavelength at the operating frequency, and the *b* dimension is normally about one-half the *a* dimension.

The wave that is propagated by a waveguide is electromagnetic and, therefore, has electric (*E* field) and magnetic (*H* field) components. The configuration of these two fields determines the mode of operation. If no component of the *E* field is in the direction of propagation, we say that it is the TE mode. TE is the abbreviation for transverse electric. TM is the mode of waveguide operation whereby the magnetic field has no compo-

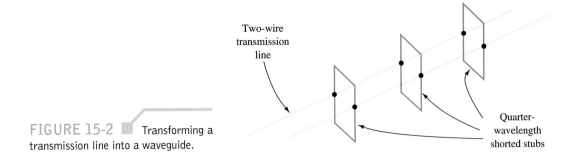

FIGURE 15-2 Transforming a transmission line into a waveguide.

nent in the direction of propagation. Two-number subscripts normally follow the TE or TM designations, and they can be interpreted as follows: For TE modes, the first subscript is the number of one-half-wavelength E-field patterns along the a (longest) dimension, and the second subscript is the number of one-half-wavelength E-field patterns along the b dimension. For TM modes, the number of H fields along the a and b dimensions determine the subscripts. Refer to Fig. 15-4 for further illustration of this process.

The electric field is shown with solid lines and the magnetic field with dashed lines. Notice the end view for the TE_{10} mode in Fig. 15-4. The electric field goes from a minimum at the ends along the a dimension to a maximum at the center. This is equivalent to one-half wavelength of E field along the a dimension, while no component exists along the b dimension. This is, therefore, called the TE_{10} mode of operation. In the TM_{21} mode of operation note that along the a dimension the H field (dashed lines) goes from zero to maximum to zero to maximum to zero. That's two half wavelengths, while along the b dimension one-half wavelength of the H field occurs—thus the TM_{21} designation. Note that in the side views the H fields (dashed lines) are not shown for the sake of simplicity. In these side views the TE modes have no E field in the direction of propagation (right to left or left to right), while in the TM modes the E field (solid lines) does exist in the propagation direction.

Dominant Mode of Operation

The TE_{10} mode of operation is called the dominant mode because it is the most "natural" one for operation. A waveguide is often thought of as a high-pass filter, because only very high frequencies can be propagated. The TE_{10} mode has the lowest cutoff frequency of any of the possible modes of propagation, including both TM and TE types. It is of special interest because there will exist a frequency range between its cutoff frequency (f_c, the lowest frequency that a given waveguide will propagate) and that of the next higher-order mode in which this is the only possible mode of transmission. Thus, if a waveguide is excited within this frequency range, energy propagation must take place in the domi-

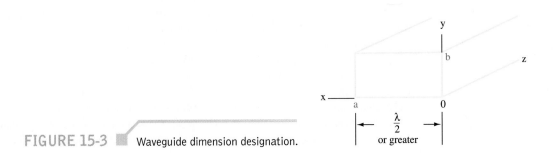

FIGURE 15-3 Waveguide dimension designation.

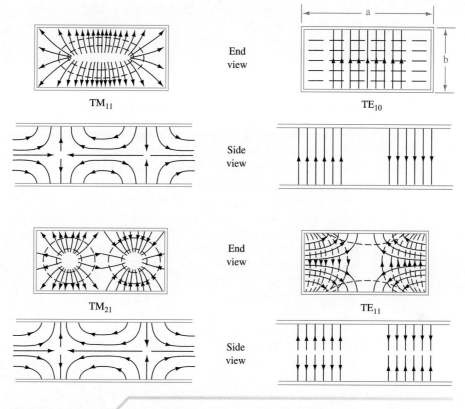

End
view

Side
view

End
view

Side
view

TM$_{11}$

TE$_{10}$

TM$_{21}$

TE$_{11}$

FIGURE 15-4 Examples of modes of operation in rectangular waveguides.

nant mode, regardless of the way in which the guide is excited. Control of the mode of operation is important in any practical transmission system, and thus the TE$_{10}$ mode has a distinct advantage over the other possible modes in a rectangular waveguide. Even more important, however, is the fact that TE$_{10}$ operation allows use of the physically smallest waveguide for a given frequency of operation.

The dimensions for an RG-52/U waveguide are 0.9×0.4 in. This is one of the standard sizes used in the X-band frequency range and is usually called the X-band waveguide. The recommended frequency range for this waveguide size is 8.2 to 12.4 GHz. As a result of this limited range of usefulness, standard sizes of waveguides have been established, each having a specified frequency range. These can be found in most any modern handbook of electronic engineering. The formula for cutoff wavelength is

$$\lambda_{co} = 2a \tag{15-1}$$

for the TE$_{10}$ mode, where a is the long dimension of the waveguide rectangle. Thus, for an RG-52/U waveguide, λ_{co} is 1.8 in, or 4.56 cm. Therefore,

$$f_{co} = \frac{c}{\lambda_{co}} = \frac{3 \times 10^{10} \text{ cm/s}}{4.56 \text{ cm}} = 6.56 \text{ GHz} \tag{15-2}$$

The lowest frequency of propagation (without considerable attenuation) is 6.56 GHz, but the recommended range is 8.2 to 12.4 GHz. The next-higher-order mode is the TE$_{20}$, which has a cutoff frequency of 13.1 GHz. Thus, within the frequency range from 6.56 to

13.1 GHz, only the TE$_{10}$ mode can propagate within the X-band waveguide, in the ordinary sense of the word.

15-3 PHYSICAL PICTURE OF WAVEGUIDE PROPAGATION

For a wave to exist in a waveguide, it must satisfy Maxwell's equations throughout the waveguide. These mathematically complex equations are beyond the scope of this book, but one boundary condition of these equations can be put into plain language: There can be no tangential component of electric field at the walls of the waveguide. This makes sense because the conductor would then *short* out the *E* field. An exact solution for the field existing within a waveguide is a relatively complicated mathematical expression. It is possible, however, to obtain an understanding of many of the properties of waveguide propagation from a simple physical picture of the mechanisms involved. The fields in a typical TE$_{10}$ waveguide can be considered as the resultant fields produced by an ordinary plane electromagnetic wave that travels back and forth between the sides of the guide, as illustrated in Fig. 15-5.

The electric and magnetic component fields of this plane wave are in time phase but are geometrically at right angles to each other and to the direction of propagation. Such a wave travels with the velocity of light and upon encountering the conducting walls of the guide is reflected with a phase reversal of the electric field and with an angle of reflection equal to the angle of incidence. A picture of the wavefronts involved with such propagation for a rectangular waveguide is shown in Fig. 15-6.

When the angle θ (see Fig. 15-5) is such that the successive positive and negative crests traveling in the same direction just fail to overlap inside the guide, it can be shown that the summation of the various waves and their reflections leads to the field distribution of the TE$_{10}$ mode, which travels down the waveguide and represents propagation of energy. The angle that the component waves must have with respect to the waveguide in order to satisfy the conditions for waveguide propagation in a rectangular guide is given by the relation

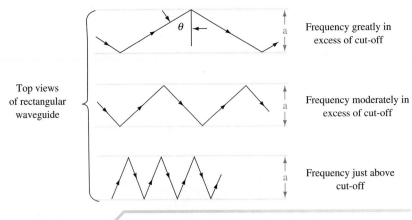

Top views of rectangular waveguide

Frequency greatly in excess of cut-off

Frequency moderately in excess of cut-off

Frequency just above cut-off

FIGURE 15-5 ▦ Paths followed by waves traveling back and forth between the walls of a waveguide.

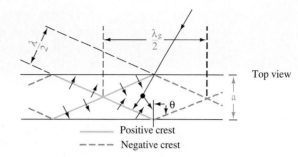

FIGURE 15-6 Wavefront reflection in a waveguide.

------- Positive crest
- - - - Negative crest

$$\cos \theta = \frac{\lambda}{2a} \qquad (15\text{-}3)$$

where a is the width of the waveguide and λ is the wavelength of the wave on the basis of the velocity of light.

Because the component waves that can be considered as building up the actual field in the waveguide all travel at an angle with respect to the axis of the guide, the rate at which energy propagates down the guide is less than the velocity of light. This velocity with which energy propagates is termed group velocity (V_g) and in the case of Fig. 15-6 is given by the relation

$$\frac{V_g}{c} = \sin \theta = \sqrt{1 - \left(\frac{\lambda}{2a}\right)^2} \qquad (15\text{-}4)$$

The guide wavelength (λ_g) is greater than the free-space wavelength (λ). A study of the $\lambda_g/2$ and $\lambda/2$ shown in Fig. 15-6 should help in visualizing this situation. Thus,

$$\frac{\text{wavelength in guide}}{\text{wavelength in free space}} = \frac{\lambda_g}{\lambda} = \frac{1}{\sin \theta} \qquad (15\text{-}5)$$

and therefore

$$\frac{c}{V_g} = \frac{\lambda_g}{\lambda} = \frac{1}{\sqrt{1 - (\lambda/2a)^2}} \qquad (15\text{-}6)$$

In Smith chart solutions of waveguide problems, λ_g should be used for making moves, not the free-space wavelength λ. The velocity with which the wave appears to move past the guide's side wall is termed the phase velocity, V_p. It has a value greater than the speed of light. It is only an "apparent" velocity, however, as it is the velocity with which the wave is changing phase at the side wall. The phase and group velocities V_p and V_g, respectively, are related by the fact that

$$\sqrt{V_p V_g} = \text{velocity of light} \qquad (15\text{-}7)$$

As the wavelength is increased, the component waves must travel more nearly at right angles to the axis of the waveguide, as shown in the bottom portion of Fig. 15-5. This causes the group velocity to be lowered and the phase velocity to be still greater than the velocity of light, until finally one has $\theta = 0°$. The component waves then bounce back

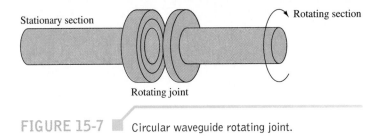

Circular waveguide rotating joint.

and forth across the waveguide at right angles to its axis and do not travel down the guide at all. Under these conditions, the group velocity is zero, the phase velocity becomes infinite, and propagation of energy ceases. This occurs at the frequency that was previously defined as the cutoff frequency, f_{co}. The cutoff frequency for the TE_{10} mode of operation can be determined from the relationship given in Eq. (15-1). It is noted that the waveguide acts as a high-pass filter, with the cutoff frequency determined by the waveguide dimensions. To obtain propagation, the waveguide must have dimensions comparable to a half wavelength, and that limits its practical use to frequencies above 300 MHz.

At frequencies very much greater than cutoff frequency, it is possible for the higher-order modes of transmission to exist in a waveguide. Thus, if the frequency is high enough, propagation of energy can take place down the guide when the system of component waves that are reflected back and forth has the form of the TE_{20} mode. It has a field distribution that is equivalent to two distributions of the dominant TE_{10} mode placed side by side, but each with reversed polarity. This conceptual presentation of waveguide propagation involving a wave suffering successive reflections between the sides of the guide can be applied to all types of waves and to other than rectangular guides. The way in which the concept works out in these other cases is not so simple, however, as for the TE_{10} mode.

15-4 OTHER TYPES OF WAVEGUIDES

Circular

The dominant mode (TE_{10}) for rectangular waveguides is by far the most widely used. The use of other modes or of other shapes is extremely limited. However, the use of a circular waveguide is found in radar applications where it is necessary to have a continuously rotating section such as in Fig. 15-7. Modes in circular waveguide can be rotationally symmetrical, which means that a radar antenna can physically rotate with no electrical disturbance. While a circular waveguide is actually simpler to manufacture than a rectan-

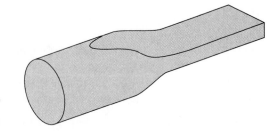

Circular-to-rectangular taper.

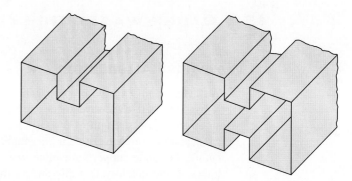

FIGURE 15-9 Ridged waveguides.

gular one, for a given frequency of operation, its cross-sectional area must be more than double that of a rectangular guide. It is, therefore, more expensive and takes up more space than a rectangular guide. Typical radar systems, therefore, consist of a main run with a rectangular waveguide and a circular rotating joint. The transition between rectangular and circular waveguides is accomplished with the circular-to-rectangular taper shown in Fig. 15-8. The transition is accomplished as gradually as possible to minimize reflections.

Ridged

Two types of ridged waveguide are shown in Fig. 15-9. While it is obviously more expensive to manufacture than a standard rectangular waveguide, it does provide one unique advantage. It allows operation at lower frequencies for a given set of outside dimensions, which means that smaller overall external dimensions are made possible. This property is advantageous in applications where space is at a premium, such as space probes and the like. A ridged waveguide has greater attenuation, and this, combined with its higher cost, limits it to special applications.

Flexible

It is sometimes desirable to have a section of waveguide that is flexible, as shown in Fig. 15-10. This configuration is often useful in the laboratory or in applications where continuous flexing occurs. Flexible waveguides consist of spiral-wound ribbons of brass or copper. The outside section is covered with a soft dielectric such as rubber so as to maintain air and watertight conditions. These conditions are desirable so that corrosion of the waveguide's inner surface does not occur.

Corrosion would cause an increase in attenuation through surface current losses and increased reflections. In critical applications, waveguides are filled with inert gas and/or their inside walls are coated with noncorrosive (but expensive) metals such as gold or silver.

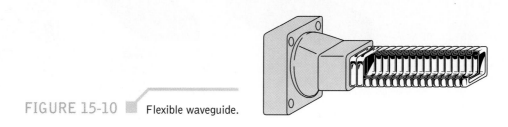

FIGURE 15-10 Flexible waveguide.

Waveguide Attenuation

Waveguides are capable of propagating huge amounts of power. For example, typical X-band (0.9×0.4 in) waveguides can handle 1 million W if operated at 1.5 times f_{co} and an air dielectric strength of 3×10^6 V/m is assumed. At frequencies below cutoff, the attenuation in any waveguide is very rapid, as previously explained. At frequencies above cutoff, the guide supports traveling waves, and they are slightly attenuated because of losses in the conducting walls and in the dielectric that fills the guide. For air-filled guides, the dielectric loss is normally negligible, but if dielectric other than air is used, these losses are often greater than the conductor losses. As the frequency increases, attenuation drops to a broad minimum and then increases slowly with increasing frequency.

The conductor losses are governed in part by the skin effect described in Chapter 12. (At high frequencies the current tends to flow only at the surface of a conductor.) Current that flows in the guide walls is concentrated near the inner surface.

Bends and Twists

It is often necessary to change the physical direction of propagation or the wave's polarization in waveguides.

1. *H bend* [Fig. 15-11(a)]: It is used to change the physical direction of propagation. It derives its name from the fact that the *H* lines are bent in this transition, while the *E* lines remain vertical for the dominant mode.

2. *E bend* [Fig. 15-11(b)]: This section is also used to change the physical direction of propagation. The choice between an *E* or *H* bend is normally governed by mechanical considerations (plumbing considerations if you will) since neither produces large discontinuities if the bends are gradual enough.

3. *Twist* [Fig. 15-11(c)]: A twist section is used to change the plane of polarization of the wave. It can be seen that any desired angular orientation of the wave may be obtained with an appropriate combination of the three types of sections just discussed.

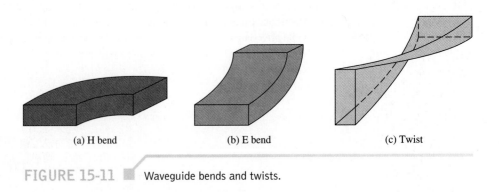

(a) H bend   (b) E bend   (c) Twist

FIGURE 15-11 ■ Waveguide bends and twists.

Rectangular and double-ridge seamless flexible waveguide assemblies, microwave components, waveguide sub-systems, and special purpose waveguide components designed and manufactured to customer specifications. (Courtesy of Continental Microwave & Tool Co., Inc.)

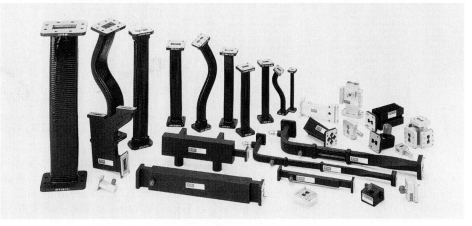

Tees

1. *Shunt tee* [Fig. 15-12(a)]: A shunt tee is so-named because of the side arm shunting the E field for TE modes, which is analogous to voltage in a transmission line. It can be seen that if two input waves at arms A and B are in phase, the portions transmitted into arm C will be in phase and thus will be additive. On the other hand, an input at C results in two equal, in-phase outputs at A and B. Of course, the A and B outputs have half the power (neglecting losses) of the C input.

2. *Series tee* [Fig. 15-12(b)]: If you consider the E field of an input at D, you should be able to visualize that the outputs at A and B will be equal and 180° out of phase, as shown. Once again, the two outputs are equal but are now 180° out of phase. The

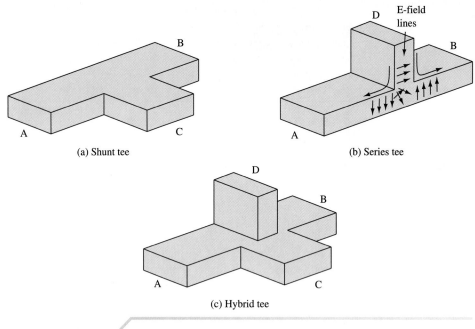

(a) Shunt tee

(b) Series tee

(c) Hybrid tee

FIGURE 15-12 ■ Shunt, series, and hybrid tees.

series tee is often used for impedance matching just as the single-stub tuner is for transmission lines. In that case, arm D contains a sliding piston to effect a short circuit at any desired point.

3. *Hybrid or magic tee* [Fig. 15-12(c)]: This is a combination of the first two tees mentioned and exhibits properties of each. From previous consideration of the shunt and series tees, it can be seen that if two equal signals are fed into arms A and B in phase, there will be cancellation in arm D and reinforcement in arm C. Thus, all the energy will be transmitted to C and none to D. Similarly, if energy is fed into C, it will divide evenly between A and B, and none will be transmitted to D. The hybrid tee has many interesting applications.

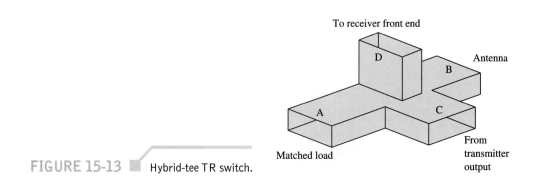

FIGURE 15-13 ■ Hybrid-tee TR switch.

A typical hybrid tee application is illustrated in Fig. 15-13. It is functioning as a transmit/receive switch (TR switch), which allows a single antenna to be used for both transmission and reception. The transmitter's output is fed into arm *C*, where it splits between the *A* and *B* outputs, with virtually no power going to the sensitive receiver at *D*. When the antenna receives a signal at *B*, energy is sent to the receiver at *D* as well as to arms *A* and *C*. The low received power does no damage to the powerful transmitter output. The matched load at *A* is necessary to prevent reflections. Problem 19 at the end of the chapter is used to introduce you to another hybrid tee application.

Tuners

A metallic post inserted in the broad wall of a waveguide provides a lumped reactance at that point. The action is similar to the addition of a shorted stub along a transmission line. When the post extends less than a quarter wavelength it appears capacitive, while exceeding a quarter wavelength makes it appear inductive. Quarter-wavelength insertion causes a series resonance effect whose sharpness (Q) is inversely proportional to the diameter of the post.

The primary usage of posts is in matching a load to a guide so as to minimize the VSWR. The most often used configurations are shown in Fig. 15-14.

1. *Slide-screw tuner* [Fig. 15-14(a)]: The slide-screw tuner consists of a screw or metallic object of some sort protruding vertically into the guide and adjustable both longitudinally and in depth. The effect of the protruding object is to produce shunting reactance across the guide—thus, it is analogous to a single-stub tuner in transmission line theory.

2. *Double-slug tuner* [Fig. 15-14(b)]: This type of tuner involves placing two metallic objects, called slugs, in the waveguide. The necessary two degrees of freedom to effect a match are obtained by making adjustable both the longitudinal position of the slugs and the spacing between them. Thus, it is somewhat analogous to the transmission line double-stub tuner but differs in that the position of the slugs and not the effective shunting reactance is variable.

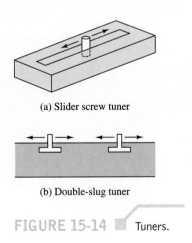

(a) Slider screw tuner

(b) Double-slug tuner

FIGURE 15-14　Tuners.

Since a waveguide is a single conductor, it is not as easy to define its characteristic impedance (Z_0) as it is for a coaxial line. Nevertheless, you can think of the characteristic impedance of a waveguide as being approximately equal to the ratio of the strength of the electric field to the strength of the magnetic field for energy traveling in one direction. This ratio is equivalent to the voltage-to-current ratio in coaxial lines on which there are no standing waves. For an air-filled rectangular waveguide operating in the dominant mode, its characteristic impedance is given by

$$Z_0 = \frac{\mathscr{Z}}{\sqrt{1 - (\lambda/2a)^2}} \tag{15-8}$$

where $\mathscr{Z}$ is the characteristic impedance of free space = 120π = 377 Ω. The guide's characteristic impedance is affected by the frequency of the energy in it since $\lambda = c/f$. Therefore, the guide's impedance is variable and more correctly termed *characteristic wave impedance* rather than just characteristic impedance.

On a waveguide there is no place to connect a fixed resistor to terminate it in its characteristic (wave) impedance as there is on a coaxial cable. But there are a number of special arrangements that accomplish the same result. One consists of filling the end of the waveguide with graphited sand, as shown in Fig. 15-15(a). As the fields enter the sand, currents flow in it. These currents create heat, which dissipates the energy. None of the energy dissipated as heat is reflected back into the guide. Another arrangement [Fig. 15-15(b)] uses a high-resistance rod, which is placed at the center of the E field. The E field (voltage) causes current to flow through the rod. The high resistance of the rod dissipates the energy as an I^2R loss.

Still another method for terminating a waveguide is to use a wedge of resistive material [Fig. 15-15(c)]. The plane of the wedge is placed perpendicular to the magnetic

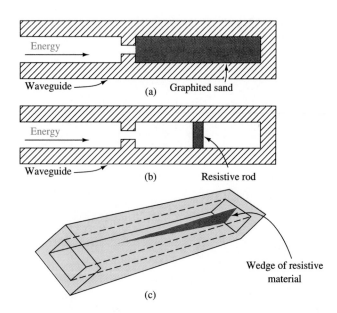

FIGURE 15-15 ■ Termination for minimum reflections.

lines of force. When the H lines cut the wedge, a voltage is induced in it. The current produced by the induced voltage flowing through the high resistance of the wedge creates an I^2R loss. This loss is dissipated in the form of heat. This permits very little energy to reach the closed end to be reflected.

Each of the preceding terminations is designed to match the impedance of the guide in order to ensure a minimum of reflection. On the other hand, there are many instances where it is desirable for all the energy to be reflected from the end of the waveguide. The best way to accomplish this is to simply attach or weld a metal plate at the end of the waveguide.

Variable Attenuators

Variable attenuators find many uses at microwave frequencies. They are used to (1) isolate a source from reflections at its load so as to preclude frequency pulling, (2) adjust the signal level, as in one arm of a microwave bridge circuit, and (3) measure signal levels, as with a calibrated attenuator:

There are two versions of variable attenuators:

1. *Flap attenuator* [Fig. 15-16(a)]: Attenuation is accomplished by insertion of a thin card of resistive material (often referred to as a *vane*) through a slot in the top of the guide. The amount of insertion is variable, and the attenuation can be made approximately linear with insertion by proper shaping of the resistance card. Notice the tapered edges, which minimize unwanted reflections.

2. *Vane attenuator* [Fig. 15-16(b)]: In this type of attenuator the resistance card or vanes move in from the sides, as shown in the figure. It can be seen that the losses (and thus attenuation) will be minimum when the vanes are close to the side walls where E is small and maximum when the vanes are in the center.

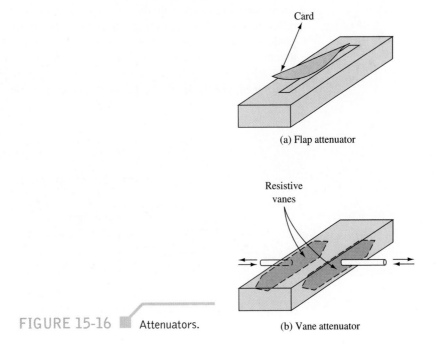

Card

(a) Flap attenuator

Resistive
vanes

(b) Vane attenuator

FIGURE 15-16 ■ Attenuators.

15-7 DIRECTIONAL COUPLER

The two-hole directional coupler consists of two pieces of waveguide with one side common to both guides and two holes in this common side. Its function is analogous to directional couplers used for transmission lines. The sections may be arranged physically either side by side or one over the other. The directional properties of such a device can be seen by looking at the wave paths labeled A, B, C, and D in Fig. 15-17.

Waves A and B follow equal-length paths and thus combine in phase in the secondary guide. If the spacing between holes is $\lambda_g/4$, waves C and D (which are of equal strength) are $\lambda_g/2$ or 180° out of phase and thus cancel. Therefore, if the field within the main guide consists of a superposition of incident and reflected waves, a certain fraction of the wave moving left to right will be coupled out through the secondary guide, and the same fraction of the right-to-left wave will be dissipated in the vane. The wave traveling from right to left causes this energy to be cancelled in the secondary guide's output due to the 180° phase shift caused by the $\lambda_g/2$ path difference of its two equal components through the two coupling holes. This type of coupler is frequency-sensitive, since the spacing between holes must be $\lambda_g/4$ or an odd multiple thereof. The addition of more holes properly spaced can improve both the operable frequency range and directivity. This is known as a multi-hole coupler.

Thus, we see that a directional coupler transfers energy from a primary to an adjacent—otherwise independent—secondary waveguide for energy traveling in the main guide in one direction only. The energy that flows toward the left in the secondary guide is absorbed by the vane in Fig. 15-17, which is a matched load to prevent reflections.

The ratio of P_{out} and the incident power, P_{in}, is known as the *coupling:*

$$\text{coupling (dB)} = 10 \log \frac{P_{in}}{P_{out}} \tag{15-9}$$

We now can understand that a directional coupler can distinguish between the waves traveling in opposite directions. It can be arranged to respond to either incident or reflected waves. By connecting a microwave power meter to the output of the secondary guide a measure of power flow can be made. The coupling is normally less than 1% so that the power meter has negligible loading effect on the operation in the main guide. By physically reversing the directional coupler, a power flow in the opposite direction is determined, and the level of reflections and SWR can be determined.

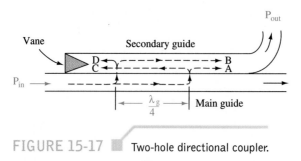

FIGURE 15-17 ▪ Two-hole directional coupler.

15-8 COUPLING WAVEGUIDE ENERGY AND CAVITY RESONATORS

Fundamentally, there are three methods of coupling energy into or out of a waveguide: *probe, loop,* and *aperture.* Probe, or capacitive, coupling is illustrated in Fig. 15-18. Its action is the same as that of a quarter-wave Marconi antenna. When the probe is excited by an RF signal, an electric field is set up [Fig. 15-18(a)]. The probe should be located in the center of the *a* dimension and a quarter wavelength, or odd multiple of a quarter wavelength, from the short-circuited end, as illustrated in Fig. 15-18(b). This is a point of maximum *E* field and, therefore, is a point of maximum coupling between the probe and the field. Usually, the probe is fed with a short length of coaxial cable. The outer conductor is connected to the waveguide wall, and the probe extends into the guide but is insulated from it, as shown in Fig. 15-18(c). The degree of coupling may be varied by varying the length of the probe, removing it from the center of the *E* field, or shielding it.

In a pulse-modulated radar system there are wide sidebands on either side of the carrier frequency. In order that a probe does not discriminate too sharply against frequencies that differ from the carrier frequency, a wideband probe may be used. This type of probe is illustrated in Fig. 15-18(d) for both low- and high-power usage.

Figure 15-19 illustrates loop, or inductive, coupling. The loop is placed at a point of maximum *H* field in the guide. As shown in Fig. 15-19(a), the outer conductor is connected to the guide, and the inner conductor forms a loop inside the guide. The current flow in the loop sets up a magnetic field in the guide. This action is illustrated in Fig. 15-19(b). As shown in Fig. 15-19(c), the loop may be placed in a number of locations. The degree of loop coupling may be varied by rotation of the loop.

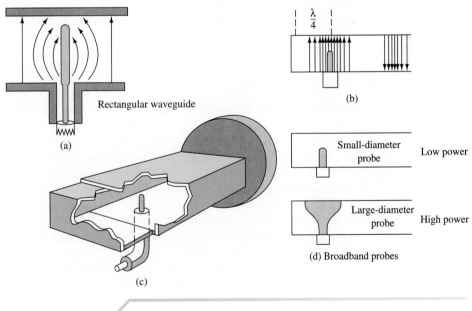

FIGURE 15-18 ▮ Probe, or capacitive, coupling.

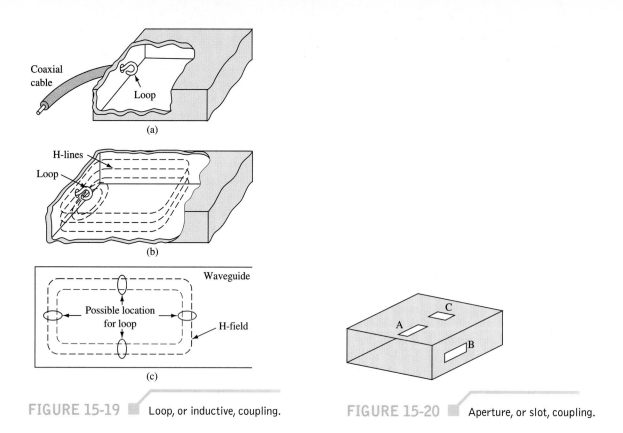

FIGURE 15-19 ▌ Loop, or inductive, coupling.

FIGURE 15-20 ▌ Aperture, or slot, coupling.

The third method of coupling is aperture, or slot, coupling. This type of coupling is shown in Fig. 15-20. Slot *A* is at an area of maximum *E* field and is a form of electric field coupling. Slot *B* is at an area of maximum *H* field and is a form of magnetic field coupling. Slot *C* is at an area of maximum *E* and *H* field and is a form of electromagnetic coupling.

Cavity Resonators

Circuits composed of lumped inductance and capacitance elements may be made to resonate at any frequency from less than 1 Hz to many thousand megahertz. At extremely high frequencies, however, the physical size of the inductors and capacitors becomes extremely small. Also, losses in the circuit become extremely great. Resonant devices of different construction are therefore preferred at extremely high frequencies. In the UHF range, sections of parallel wire or coaxial transmission line are commonly employed in place of lumped constant resonant circuits. In the microwave region, cavity resonators are used. Cavity resonators are metal-walled chambers fitted with devices for admitting and extracting electromagnetic energy. The *Q* of these devices may be much greater than that of conventional *LC* tank circuits.

Although cavity resonators, built for different frequency ranges and applications, have a variety of physical forms, the basic principles of operation are essentially the same for all. Operating principles of cavity resonators are explained in this chapter. These principles are applied in Chapter 16 to the study of important microwave components employing cavity resonators.

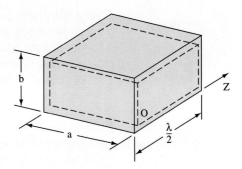

FIGURE 15-21 Rectangular waveguide resonator.

Resonant cavity walls are made of highly conductive material and enclose a good dielectric, usually air. One example of a cavity resonator is the rectangular box shown in Fig. 15-21. It may be thought of as a section of rectangular waveguide closed at both ends by conducting plates. Because the end plates are short circuits for waves traveling in the Z direction, the cavity is analogous to a transmission line section with short circuits at both ends. Resonant modes occur at frequencies for which the distance between end plates is a half wavelength or multiple of a half wavelength.

Cavity modes are designated by the same numbering system that is used with waveguides, except that a third subscript is used to indicate the number of half-wave patterns of the transverse field along the axis of the cavity (perpendicular to the transverse field). The rectangular cavity is only one of many cavity devices useful as high-frequency resonators. By appropriate choice of cavity shape, advantages such as compactness, ease of tuning, simple mode spectrum, and high Q may be secured as required for special applications. Coupling energy to and from the cavity is accomplished just as for standard waveguide, as shown in Fig. 15-19.

Cavity Tuning

The resonant frequency of a cavity may be varied by changing any of three parameters: *cavity volume, cavity inductance,* or *cavity capacitance.* Although the mechanical methods for tuning cavity resonators may vary, they all utilize the electrical principles explained below.

Figure 15-22 illustrates a method of tuning a cylindrical-type cavity by varying its volume. Varying the distance d will result in a new resonant frequency. Increasing dis-

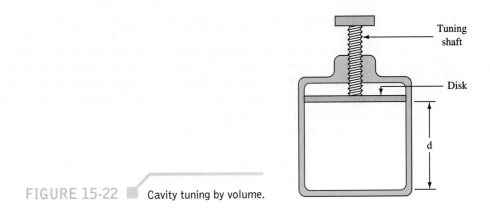

FIGURE 15-22 Cavity tuning by volume.

tance d will lower the resonant frequency, while decreasing d will cause an increase in resonant frequency. The movement of the disk may be calibrated in terms of frequency. A micrometer scale is usually used to indicate the position of the disk, and a calibration chart is used to determine frequency.

A second method for tuning a cavity resonator is to insert a nonferrous metallic screw at a point of maximum H field. This decreases the permeability of the cavity and decreases its effective inductance, which raises its resonant frequency. The farther the screw penetrates into the cavity, the higher is the resonant frequency. A paddle can be used in place of the screw. Turning the paddle to a position more nearly perpendicular to the H field increases resonant frequency.

15-9 RADAR

The first practical use of waveguides occurred with the development of radar during World War II. The high powers and high frequencies involved in these systems are much more efficiently carried by waveguides than by transmission lines. The word *radar* is an acronym formed from the words *ra*dio *d*etection *a*nd *r*anging. Radar is a means of employing radio waves to detect and locate objects such as aircraft, ships, and land masses. Location of an object is accomplished by determining the distance and direction from the radar equipment to the object. The process of locating objects requires, in general, the measurement of three coordinates: range, angle of azimuth (horizontal direction), and angle of elevation.

A radar set consists fundamentally of a transmitter and a receiver. When the transmitted signal strikes an object (target) some of the energy is sent back as a reflected signal. The small-beamwidth transmit/receive antenna collects a portion of the returning energy (called the *echo signal*) and sends it to the receiver. The receiver detects and amplifies the echo signal, which is then used to determine object location.

Military use of radar includes surveillance and tracking of air, sea, land, and space targets from air, sea, land, and space platforms. It is also used for navigation, including aircraft terrain avoidance and terrain following. Many techniques and applications of radar developed for the military are now found in civilian equipment. These include weather observation, geological search techniques, and air traffic control units, to name just a few. All large ships at sea carry one or more radars for collision avoidance and navigation. In space, radars are used for spacecraft rendezvous, docking, and landing, as well as for remote sensing of the earth's environment and planetary exploration.

Radar Waveform and Range Determination

A representative radar pulse (waveform) is shown in Fig. 15-23. The number of these pulses transmitted per second is called the *pulse repetition frequency* (PRF) or *pulse repetition rate* (PRR). The time from the beginning of one pulse to the beginning of the next pulse is called the *pulse repetition time* (PRT). The PRT is the reciprocal of the PRF (PRT = 1/PRF). The duration of the pulse (the time the transmitter is radiating energy) is called the *pulse width* (PW). The time between pulses is called *rest time* or *receiver time.* The pulse width plus the rest time equals the PRT (PW + rest time = PRT). In order for radar to provide an accurate directional picture, a highly directive antenna is necessary. The desired directivity can only be provided by microwave antennas (see Chapter 16), and thus the RF energy shown in Fig. 15-23 is usually in the GHz (microwave) range.

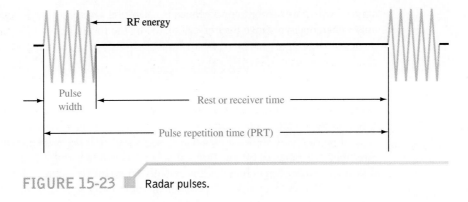

FIGURE 15-23 Radar pulses.

The distance to the target (range) is determined by the time required for the pulse to travel to the target and return. The velocity of electromagnetic energy is 186,000 statute mi/s, or 162,000 nautical mi/s. (A nautical mile is the accepted unit of distance in radar and is equal to 6076 ft.) However, in many instances, measurement accuracy is secondary to convenience, and as a result a unit known as the *radar mile* is commonly used. A radar mile is equal to 2000 yd, or 6000 ft. The small difference between a radar mile and a nautical mile introduces an error of about 1% in range determination.

For purposes of calculating range, the two-way travel of the signal must be taken into account. It can be found that it takes approximately 6.18 μs for electromagnetic energy to travel 1 radar mile. Therefore, the time required for a pulse of energy to travel to a target and return is 12.36 μs/radar mile. This takes into account the two-way travel time. The range, in miles, to a target may be calculated by the formula

$$\text{range} = \frac{\Delta t}{12.36} \qquad (15\text{-}10)$$

where Δt is the time between transmission and reception of the signal in microseconds. However, for shorter ranges and greater accuracy, range is measured in meters.

$$\text{range (meters)} = \frac{c\Delta t}{2} \qquad (15\text{-}11)$$

where c is the speed of light and Δt is in seconds.

Radar System Parameters

Once the pulse of electromagnetic energy is emitted by the radar, a sufficient length of time must elapse to allow any echo signals to return and be detected before the next pulse is transmitted. Therefore, the PRT of the radar is determined by the longest range at which targets are expected. If the PRT were too short (PRF too high), signals from some targets might arrive after the transmission of the next pulse. This could result in ambiguities in measuring range. Echoes that arrive after the transmission of the next pulse are called *second return echoes* (also *second time around* or *multiple time around echoes*). Such an echo would appear to be at a much shorter target range than actually exists and could be misleading if not identified as a second return echo. The range beyond which targets appear as second return echoes is called the *maximum unambiguous range*. Maximum unambiguous range may be calculated by the formula

$$\text{maximum unambiguous range} = \frac{\text{PRT}}{12.2} \tag{15-12}$$

where range is in miles and the PRT is in microseconds. Figure 15-24 illustrates the principles of the second return echo.

Figure 15-24 shows a signal with a PRT of 610 μs, which results in a maximum unambiguous range of 50 mi. Target number 1 is at a range of 20 mi. Its echo signal takes 244 μs to return. Target number 2 is actually 65 mi away, and its echo signal takes 793 μs to return. However, this is 183 μs after the next pulse was transmitted; therefore, target number 2 will appear to be a weak target 15 mi away. Thus, the maximum unambiguous range is the *maximum usable range* and will be referred to from now on as simply *maximum range*. (It is assumed here that the radar has sufficient power and sensitivity to achieve this range.)

If a target is so close to the transmitter that its echo is returned to the receiver before the transmitter is turned off, the reception of the echo will be masked by the transmitted pulse. In addition, almost all radars utilize an electronic device to block the receiver for the duration of the transmitted pulse. However, *double range echoes* are frequently detected when there is a large target close by. Such echoes are produced when the reflected beam is strong enough to make a second trip, as shown in Fig. 15-25. Double range echoes are weaker than the main echo and appear at twice the range.

Minimum range is measured in meters and may be calculated by the formula

$$\text{minimum range} = 150 \text{ PW} \tag{15-13}$$

where range is in meters and pulse width (PW) is in microseconds. Typical pulse widths range from fractions of a microsecond for short-range radars to several microseconds for high-power long-range radars.

A radar transmitter generates RF energy in the form of extremely short pulses with comparatively long intervals of rest time. The useful power of the transmitter is that contained in the radiated pulses and is termed the *peak power* of the system. Because the radar transmitter is resting for a time that is long with respect to the pulse time, the aver-

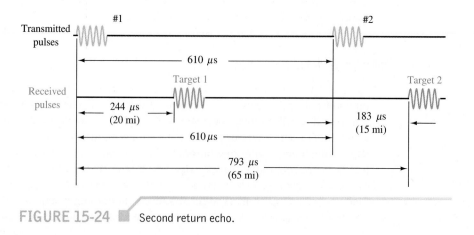

FIGURE 15-24 Second return echo.

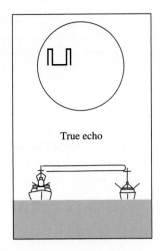

True echo

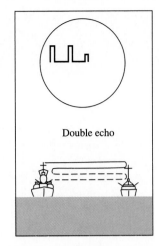

Double echo

FIGURE 15-25 ■ Double range echo.

age power delivered during one cycle of operation is relatively low compared with the peak power available during the pulse time.

The *duty cycle* of radar is

$$\text{duty cycle} = \frac{\text{pulse width}}{\text{pulse repetition time}} \qquad (15\text{-}14)$$

For example, the duty cycle of a radar having a pulse width of 2 μs and a pulse repetition time of 2 ms is

$$\frac{2 \times 10^{-6}}{2 \times 10^{-3}} \text{ or } 0.001$$

Similarly, the ratio between the average power and peak power may be expressed in terms of the duty cycle. In a system with peak power of 200 kW, a PW of 2 μs, and a PRT of 2 ms, a peak power of 200 kW is supplied to the antenna for 2 μs, while for the remaining 1998 μs the transmitter output is zero. Because average power equals peak power times duty cycle, the average power equals $(2 \times 10^5) \times (1 \times 10^{-3})$, or 200 W.

High peak power is desirable in order to produce a strong echo over the maximum range of the equipment. Conversely, low average power enables the transmitter output circuit components to be made smaller and more compact. Thus, it is advantageous to have a low duty cycle. A short pulse width is also advantageous with respect to being able to "see" closely spaced objects.

Basic Radar Block Diagram

A block diagram of a basic radar system is shown in Fig. 15-26. The pulse repetition frequency is controlled by the *timer* (also called *trigger generator* or *synchronizer*) in the modulator block. The pulse-forming circuits in the modulator are triggered by the timer and generate high-voltage pulses of rectangular shape and short duration. These pulses are used as the supply voltage for the transmitter and, in effect, turn it on and off. The modulator, therefore, determines the pulse width of the system. The transmitter generates the high-frequency, high-power RF carrier and determines the carrier frequency. The duplexer is an electronic switch that allows the use of a common antenna for both transmitting and receiving. It prevents the strong transmitted signal from being received by the

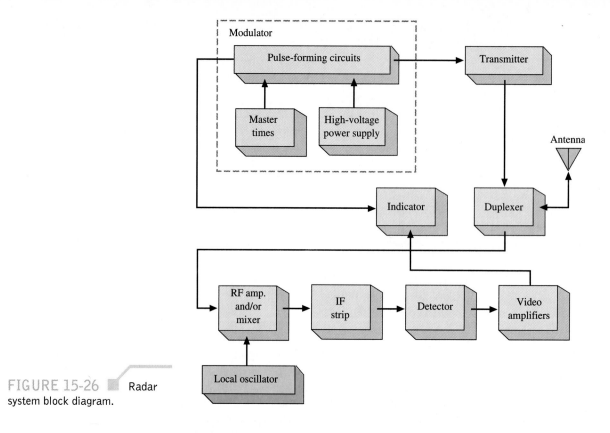

FIGURE 15-26 Radar
system block diagram.

sensitive receiver. The receiver section is basically a conventional superheterodyne receiver. In older radars no RF amplifier is found, due to noise problems with the RF amplifiers of that era.

Doppler Effect

The *Doppler effect* is the phenomenon whereby the frequency of a reflected signal is shifted if there is relative motion between the source and reflecting object. This is the same effect whereby the pitch of a train's whistle is shifted as the train moves toward and then away from the listener. This principle is utilized in many radar systems. For example, moving-target indicator systems (MTI) compare a set of received echoes with those received during the previous sweep and cancel out those whose frequency has remained constant. This removes the *clutter* of stationary targets and permits easier determination of moving targets.

The amount of frequency shift encountered is determined by the relative velocity between transmitter and target. It is predicted by

$$f_d = \frac{2v \cos \theta}{\lambda} \qquad (15\text{-}15)$$

where f_d = frequency change between transmitted and reflected signal
   v = relative velocity between radar and target
   λ = wavelength of transmitted wave
   θ = angle between target direction and radar system

If you have ever received a speeding ticket in a "radar" trap, you now have a better understanding of your downfall.

15-10 MICROINTEGRATED CIRCUIT WAVEGUIDING

The field of communications now makes heavy use of the frequencies from 1 up to 300 GHz—we shall loosely refer to these as *microwave* frequencies. At microwave frequencies, even the shortest circuit connections must be carefully considered due to the extremely small wavelengths involved. In Chapter 16 we shall provide a general study of the microwave field.

The thin-film hybrid and monolithic integrated circuits used at microwave frequencies are called MICs (microwave integrated circuits). Obviously the use of short chunks of coaxial transmission line or waveguides is not practical for the required connections of mass-produced miniature circuits. Instead, either a *stripline* or *microstrip* connection is often used. They are shown in Fig. 15-27. They both lend themselves to mass-produced circuitry and can be thought of as a cross between waveguides and transmission lines with respect to their propagation characteristics.

The stripline consists of two ground planes (conductors) that "sandwich" a smaller conducting strip with constant separation by a dielectric material (printed circuit board). The two types of microstrip shown in Fig. 15-27 consist of either one or two conducting strips separated from a single ground plane by a dielectric. While stripline offers somewhat better performance due to lower radiation losses, the simpler and thus more economical microstrip is the prevalent construction technique.

In either case, the losses exceed those of either waveguides or coaxial transmission lines, but the miniaturization and cost savings far outweigh the loss considerations. This is especially true when the very short connection paths are considered.

As with waveguides and transmission lines, the characteristic impedances of stripline and microstrip are determined by physical dimensions and the type of dielectric. The most often used dielectric is alumina with a relative dielectric constant of 9.6. Proper impedance matching, so as to minimize standing waves, is still an important consideration.

Figure 15-28 provides end views of the three lines. The formulas for calculating Z_0 for them are provided. In the formulas, ln is the natural logarithm and ε is the dielectric constant of the board.

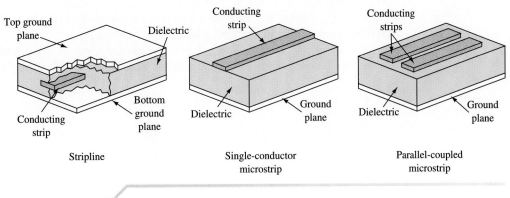

FIGURE 15-27 Stripline and microstrip.

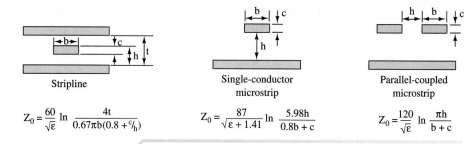

$$Z_0 = \frac{60}{\sqrt{\varepsilon}} \ln \frac{4t}{0.67\pi b(0.8 + {}^c\!/_h)}$$

Stripline

$$Z_0 = \frac{87}{\sqrt{\varepsilon + 1.41}} \ln \frac{5.98h}{0.8b + c}$$

Single-conductor microstrip

$$Z_0 = \frac{120}{\sqrt{\varepsilon}} \ln \frac{\pi h}{b + c}$$

Parallel-coupled microstrip

FIGURE 15-28 ■ Characteristic impedance.

Dielectric Waveguide

A more recent contender for "wiring" of miniature millimeter wavelength circuits is the *dielectric waveguide.* Its operation is dependent on the principle that two dissimilar dielectrics have a guiding effect on electromagnetic waves.

The dielectric waveguide should not be confused with the dielectric-filled waveguide. Both are shown in Fig. 15-29. A regular metallic waveguide is sometimes filled with dielectric, as this decreases the size necessary to allow propagation of a given frequency.

The dielectric waveguide is obviously easy to mass-produce within integrated circuits and offers an advantage over microstrip. At frequencies above 20 to 30 GHz the losses with microstrip become excessive for many system applications as compared to the dielectric waveguide. For example, at 60 GHz microstrip typically attenuates 0.15 dB/cm, while the dielectric waveguide attenuates only 0.06 dB/cm. The figure for a standard rectangular waveguide at 60 GHz is about 0.02 dB/cm but would only be used in systems where cost is not a factor.

Alumina is commonly used as the dielectric material for dielectric waveguides. However, semiconductors such as silicon and gallanium arsenide (GaAs) will undoubtedly be the dielectrics used in the future. This is dictated by the fact that ultimately semiconductor devices will be fabricated directly into the dielectric waveguide.

TROUBLESHOOTING

After completing this section, you should be able to troubleshoot waveguide systems. Waveguide problems are very similar to ordinary transmission line problems. The test equipment may look different, but it is doing the same things.

A word of caution: Waveguide is commonly used to carry large amounts of microwave power. Microwaves are capable of burning skin and damaging eyesight. Never

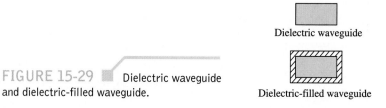

Dielectric waveguide

Dielectric-filled waveguide

FIGURE 15-29 ■ Dielectric waveguide and dielectric-filled waveguide.

work on waveguide runs or antennas connected to a transmitter or radar until you are sure the system is off and cannot be turned on by another person.

After completing this section you should be able to:

- Identify problems caused by joints and flanges

- Detect arcing problems

- Troubleshoot rotary joint failures

- Detect malfunctions by determining VSWR in the guide

Some Common Problems:

1. The joints or flanges between two waveguide sections are the most likely source of a problem. Sometimes, waveguides are pressurized to increase the power rating of the guide and keep water out. Improperly fitted joints can let water in and gas out. Water will raise the VSWR, which can damage most microwave tubes.

   There are two types of flanges: choke and cover. Choke flanges have a groove cut into the face of the flange to keep microwave energy from escaping. There will be a second groove for a gasket. Cover flanges are simply very smooth. Both must be clean and flat. The screws must be the correct size since they help align the two pieces and keep them tightly sealed together.

2. Arcs. Arcs can occur at improperly fitted joints and actually burn holes in the guide. Arcs will occur in the waveguide under high power if some component such as the antenna has failed. You might see evidence of an arc on the broad wall right in the center of the guide.

3. Worn out components. Radar antennas generally have one or more rotary joints. Rotary joints have bearings and sometimes moving contacts. Rotary joints sometimes fail only after the transmitter has had time to heat them sufficiently. By the time the technician has opened the guide and installed test equipment the joint will cool and test well. It is best to have a directional coupler in line while running the transmitter and watch for an increase in reflected power.

   Rotary joints can also be tested on the bench by connecting the joint to a dummy load and measuring VSWR while turning the joint. Still there will be no substitute for the operational test.

   Flexible waveguide is subject to cracking and corrosion. Normally, the loss of a two-foot-long piece of rigid waveguide is so low that the loss of the guide is extremely difficult to measure. A bad piece of flexible guide can usually be detected by connecting a dummy load to it and measuring VSWR while flexing the guide.

Test Equipment:

Figure 15-30 shows how to connect the test equipment for a VSWR test. First connect the power meter to the forward (incident) power coupler and note the reading. Then connect the power meter to the reflected coupler and note the reading.

Reflected power should be very low, and the VSWR should be nearly one. VSWR is given by:

$$\text{VSWR} = \frac{1 + \sqrt{\dfrac{P_r}{P_i}}}{1 - \sqrt{\dfrac{P_r}{P_i}}}$$

(14-5)

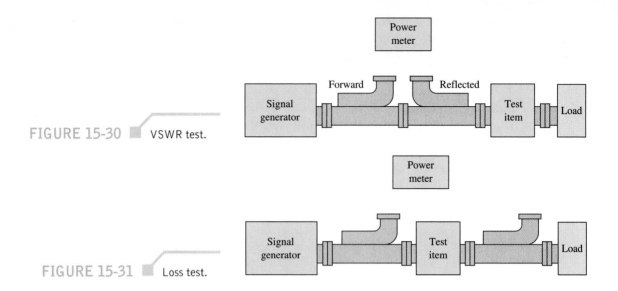

FIGURE 15-30 ▮ VSWR test.

FIGURE 15-31 ▮ Loss test.

where P_i = incident power
P_r = reflected power

The same test equipment is used to measure loss by reversing the reflected coupler and putting the test item between the couplers, as shown in Fig. 15-31. No two couplers are exactly alike, so you must first connect them together without the test item and determine the difference. You should be able to make the loss measurement to within 0.2 dB in this manner.

SUMMARY

In Chapter 15 we studied waveguides and radar. We discovered that waveguides can be derived from transmission line analysis and are capable of handling large amounts of power. The major topics the student should now understand include:

- the comparison of transmission via waveguides, antennas, and transmission lines
- the analysis of waveguide operation and mode designations
- the description and usage of rectangular, circular, ridged, and flexible waveguides
- the application of waveguide bends, twists, tees, and tuners
- the calculation of characteristic impedance for waveguides
- the techniques of waveguide terminations and variable attenuators
- the description and application of directional couplers
- the analysis of coupling waveguide energy using probes, loops, or apertures
- the description and applications for cavity resonators
- the calculation of radar range

- the operational description of radar parameters' maximum range, minimum range, duty cycle, and Doppler effect

- the construction and application of stripline and dielectric waveguides

QUESTIONS AND PROBLEMS

SECTION 1

1. Discuss the relative merits and drawbacks of using antennas, waveguides, and transmission lines as the medium for a communications link.

SECTION 2

2. Broadly speaking, provide a definition of a waveguide. What is normally meant by the term *waveguide?*

3. Explain the basic difference between propagation in a waveguide versus a transmission line.

4. What are the *modes* of operation for a waveguide? Explain the subscript notation for TE and TM modes.

5. What is the *dominant mode* in rectangular waveguides? What property does it have that makes it dominant? Show a sketch of the electric field at the mouth of a rectangular waveguide carrying this mode.

6. A rectangular waveguide is 1 cm by 2 cm. Calculate its cutoff frequency, f_{co}. (7.5 GHz)

7. How does energy propagate down a waveguide? Explain what determines the angle this energy makes with respect to the guide's sidewalls.

SECTION 3

8. Why is the velocity of energy propagation usually significantly less in a waveguide than in free space? Calculate this velocity (V_g) for an X-band waveguide for a 10-GHz signal. Calculate guide wavelength (λ_g) and phase velocity (V_p) for these conditions. (2.26×10^8 m/s, 3.98 cm, 3.98×10^8 m/s)

9. Why are free-space wavelength (λ) and guide wavelength (λ_g) different? Explain the significance of this with respect to Smith chart calculations.

*10. Why are rectangular cross-sectional waveguides generally used in preference to circular cross-sectional waveguides?

11. Calculate λ_g, λ, V_g, and V_p for a 9-GHz signal operating in the dominant mode in a rectangular waveguide 3 × 4.5 cm. The characteristic (wave) impedance is 405 Ω. Using the Smith chart, determine the SWR caused by a horn antenna load of 350 Ω + j100 Ω and the impedance in the guide 4 cm from the antenna load. (3.59 cm, 3.33 cm, 2.79×10^8 m/s, 3.59×10^8 m/s, 1.38, $527 - j81$ Ω)

SECTION 4

12. Why are circular waveguides used much less than rectangular ones? Explain the application of a circular rotating joint.

13. Describe the advantages and disadvantages of a ridged waveguide.

14. Describe the physical construction of a flexible waveguide, and list some of its applications.

15. List some of the causes of waveguide attenuation. Explain their much greater power handling capability as compared to coaxial cable of similar size.
*16. Describe briefly the construction and purpose of a waveguide. What precautions should be taken in the installation and maintenance of a waveguide to ensure proper operation?
17. Why are waveguide bend and twist sections constructed so as to gradually alter the direction of propagated energy?
18. Describe the characteristics of shunt and series T sections. Explain the operation of a hybrid tee when used as a TR switch.
19. Show, with a sketch, how a hybrid tee might be used to feed the first stage of a microwave receiver (the mixer—no RF stage) with the antenna signal and local oscillator signal without any local oscillator radiation off the receiving antenna.
20. Discuss several types of waveguide tuners in terms of function and applications.

SECTION 6

21. Verify the characteristic wave impedance of 405 Ω for the data given in Problem 11.
22. Calculate the characteristic wave impedance for an X-band waveguide operating at 8, 10, and 12 GHz. (663 Ω, 501 Ω, 450 Ω)
23. Explain various ways of terminating a waveguide to minimize and maximize reflections.
24. Describe the action of flap and vane attenuators.

SECTION 7

25. Describe in detail the operation of a directional coupler. Include a sketch with your description. What are some applications for a directional coupler? Define the *coupling* of a directional coupler.
26. Calculate the coupling of a directional coupler that has 70 mW into the main guide and 0.35 mW out the secondary guide. (23 dB)

SECTION 8

27. Explain the basics of capacitively coupling energy into a waveguide.
28. Explain the basics of inductively coupling energy into a waveguide.
29. What is slot coupling? Describe the effect of varying the position of the slot.
*30. Discuss the following with respect to waveguides:
    (a) Relationship between frequency and size.
    (b) Modes of operation.
    (c) Coupling of energy into the waveguide.
    (d) General principles of operation.
31. What is a cavity resonator? In what ways is it similar to an *LC* tank circuit? Dissimilar?
*32. Explain the operation principles of a cavity resonator.
*33. What are waveguides? Cavity resonators?

34. Describe a means whereby a cavity resonator could be used as a waveguide frequency meter.

35. Explain three methods of tuning a cavity resonator.

SECTION 9

*36. Explain briefly the principle of operation for a radar system.

*37. Why are waveguides used in preference to coaxial lines for the transmission of microwave energy in radar installations?

38. With respect to a radar system, explain the following terms:
    (a) Target.   (e) Pulse width.
    (b) Echo.   (f) Rest time.
    (c) Pulse repetition rate.   (g) Range.
    (d) Pulse repetition time.

39. Calculate the range in miles and meters for a target when Δt is found to be 167 μs. (13.5 mi, 25,050 m)

*40. What is the distance in nautical miles to a target if it takes 123 μs for a radar pulse to travel from the radar antenna to the target, back to the antenna, and be displayed on the PPI scope? (10 mi)

41. Explain how multiple targets lead to the term *maximum range* and what is meant by the term. Calculate the maximum unambiguous range for a radar system with PRT equal to 400 μs. (32.8 mi)

42. What are double range echoes? Describe a means of detecting this problem.

43. Why does a radar system have a minimum range? Calculate the minimum range for a system with a pulse width of 0.5 μs.

44. In detail, discuss the various implications of duty cycle for a radar system.

*45. What is the peak power of a radar pulse if the pulse width is 1 μs, the pulse repetition rate is 900, and the average power is 18 W? What is the duty cycle? (20 kW, 0.09%)

46. For the radar block diagram in Fig. 15-26, explain the function of each section.

47. A police radar speed trap functions at a frequency of 1.024 GHz in direct line with your car. The reflected energy from your car is shifted 275 Hz in frequency. Calculate your speed in miles per hour. Are you going to get a ticket? (90 mph, *yes!*)

48. What is the Doppler effect? What are some other possible uses for it other than police speed traps?

SECTION 10

49. Using sketches, explain the physical construction of stripline, single-conductor microstrip, and parallel-coupled microstrip. Discuss their relative merits, and also compare them to transmission lines and waveguides.

50. What is a dielectric waveguide? Discuss its advantages and disadvantages with respect to regular waveguides.

51. Calculate Z_0 for stripline constructed using circuit board with dielectric constant 2.1, $b = 0.1$ in, $c = 0.006$ in, and $h = 0.08$ in. The conductor is spaced equally from the top and bottom ground planes. (50 Ω)

CHAPTER

16

MICROWAVES AND LASERS

OBJECTIVES

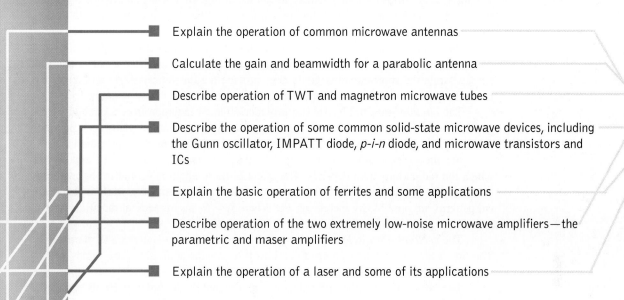

■ Explain the operation of common microwave antennas

■ Calculate the gain and beamwidth for a parabolic antenna

■ Describe operation of TWT and magnetron microwave tubes

■ Describe the operation of some common solid-state microwave devices, including the Gunn oscillator, IMPATT diode, *p-i-n* diode, and microwave transistors and ICs

■ Explain the basic operation of ferrites and some applications

■ Describe operation of the two extremely low-noise microwave amplifiers—the parametric and maser amplifiers

■ Explain the operation of a laser and some of its applications

16-1 MICROWAVE ANTENNAS

The antennas studied in Chapter 14 bear little resemblance to those used for microwave frequencies (>1 GHz). Microwave antennas actually use optical theory more than standard antenna theory. These antennas tend to be highly directive and therefore provide high gain as compared to the reference half-wavelength dipole. The reasons for this include the following:

1. Due to the short wavelengths involved, the physical sizes required are small enough to allow "peculiar" arrangements not practical at lower frequencies.

2. There is little need for omnidirectional patterns since no broadcasting takes place at these frequencies. Microwave communications are generally of a point-to-point nature.

3. Due to increased device noise at microwave frequencies, receivers require the highest possible input signal. Highly directional antennas (and thus high gain) make this possible.

4. Microwave transmitters are very limited in their output power due to the cost and/or availability problems of microwave power devices. This low output power is compensated for by a highly directional antenna system.

Photo: Parabolic antenna for radar system. (Courtesy of Seavy Engineering Associates, Inc.)

Horn Antenna

Open-ended sections of waveguides can be used as radiators of electromagnetic energy. The three basic forms of horn antennas are shown in Fig. 16-1. They all provide a gradual flare to the waveguide so as to allow maximum radiation and thus minimum reflection back into the guide.

Recall that a plain open-circuited waveguide theoretically reflects 100% of the incident energy. In practice, however, the open-circuited guide "launches" a fair amount of energy, while the short-circuited guide does provide the theoretical 100% reflection. By gradually flaring out the open circuit, the goal of total radiation is nearly attained.

The *circular horn* in Fig. 16-1(a) provides efficient radiation from a circular waveguide. The flare angle θ and length l are important to the amount of gain it can provide. Generally, the greater l, the greater the gain, while a flare angle $\theta \simeq 50°$ is optimum.

For the *pyramidal horn* in Fig. 16-1(b) there are two flare angles, θ_1 and θ_2, on which the radiation pattern depends. The effect of horn length is similar to that with the circular horn. Wider horizontal patterns are obtained by increasing θ_2, while wider vertical patterns are possible by increasing θ_1. When $\theta_1 = \theta_2$ a symmetrical radiation pattern is realized.

The *sectoral horn* in Fig. 16-1(c) has the top and bottom walls at a 0° flare angle. The side walls are sometimes hinged (as shown) to provide adjustable flare angles. Maximum radiation occurs for angles between 40° and 60°.

The horns just described provide a maximum gain on the order of 20 dB compared to the half-wavelength dipole reference. While they do not provide the amounts of gain of subsequently described microwave antennas, their simplicity and low cost make them popular for noncritical applications.

Parabolic Antenna

The ability of a paraboloid to focus light rays or sound waves at a point is common knowledge. The same ability is applicable to electromagnetic waves of lower frequency than light as long as the paraboloid's mouth diameter is at least 10 wavelengths. This precludes their use at low radio frequencies but allows use at microwave frequencies.

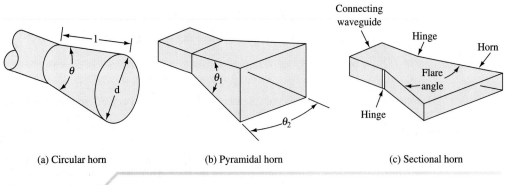

(a) Circular horn (b) Pyramidal horn (c) Sectional horn

FIGURE 16-1 ▮ Horn antennas.

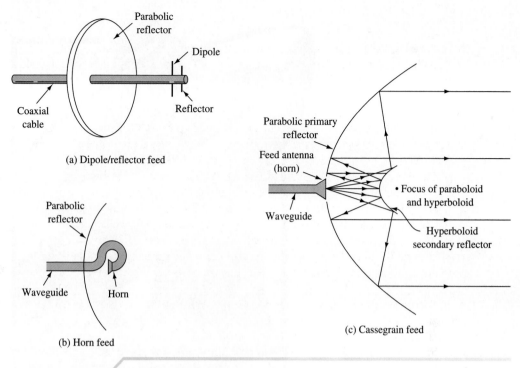

(a) Dipole/reflector feed

(b) Horn feed

(c) Cassegrain feed

FIGURE 16-2 ■ Microwave dish antennas.

There are various methods of feeding the *microwave dish,* as the paraboloid antenna is commonly called. Figure 16-2(a) shows the dish being fed with a simple dipole/reflector combination at the paraboloid's focus. A horn-fed version is shown in Fig. 16-2(b). The *Cassegrain feed* in Fig. 16-2(c) is used to shorten the length of feed mechanism in highly critical applications. It uses a hyperboloid secondary reflector whose focus coincides with the paraboloids. Those transmitted rays obstructed by the hyperboloid are generally such a small percentage as to be negligible.

These dish antennas perform equally well transmitting or receiving, as predicted by antenna reciprocity. They provide huge power gains, with a good approximation provided by the following equation:

$$A_p \simeq 6 \left(\frac{D}{\lambda} \right)^2 \tag{16-1}$$

where A_p = power gain with respect to a half-wavelength dipole
D = mouth diameter of primary reflector
λ = free-space wavelength of carrier frequency

An accurate approximation of the beamwidth in degrees between half-power points is

$$\text{beamwidth} \simeq \frac{70\lambda}{D} \tag{16-2}$$

Parabolic antenna with Cassegrain feed.
(Courtesy of Ortel Corp.)

EXAMPLE 16-1

Calculate the power gain and beamwidth of a microwave dish antenna with a 3-m mouth diameter when used at 10 GHz.

SOLUTION

$$A_p \simeq 6 \left(\frac{D}{\lambda}\right)^2 \tag{16-1}$$

$$\lambda = \frac{c}{f} = \frac{3 \times 10^8 \text{ m/s}}{10 \times 10^9} = 0.03 \text{ m}$$

$$A_p = 6 \times \left(\frac{3 \text{ m}}{0.03 \text{ m}}\right)^2 = 60,000 \quad (47.8 \text{ dB})$$

$$\text{beamwidth} \simeq \frac{70\lambda}{D} \tag{16-2}$$

$$= \frac{70 \times 0.03 \text{ m}}{3 \text{ m}} = 0.7°$$

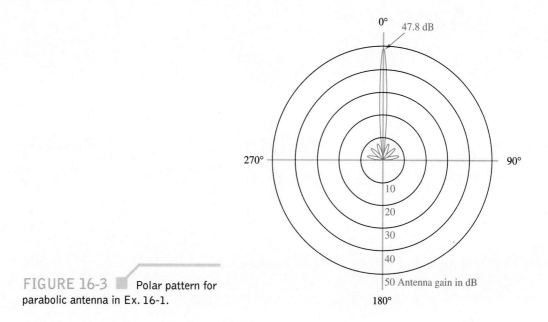

FIGURE 16-3 Polar pattern for parabolic antenna in Ex. 16-1.

Example 16-1 shows the extremely high gain capabilities of these antennas. In this particular case, the dish with a 1-W output is equivalent to a half-wave dipole with 60,000-W output. This power gain is effective, however, only if the receiver is within the 0.7° beamwidth of the dish. Figure 16-3 shows a *polar pattern* for this antenna. It is typical of parabolic antennas and shows the 47.8-dB gain at the 0° reference. Notice the three side lobes on each side of the main one. As you might expect from the antenna's physical construction, there can be no radiated energy from 90° to 270°.

Microwave dish antennas are widely used in satellite communications because of their high gain; they are also used for satellite tracking and radio astronomy. They are also used at 30- to 50-mi intervals (point-to-point line-of-sight conditions) to carry telephone and broadcast TV and other signals throughout the world. Often these antennas have a "cover" over the dish. This is a low-loss dielectric material known as a *radome.* Its purpose may be maintenance of internal pressure or, more simply, environmental protection. The construction of a bird's nest within the dish is undesirable for the bird as well as the antenna user. More detail on these microwave applications was provided in Sec. 15-6.

Lens Antenna

We have all witnessed the effect of focusing the sun's rays into a point using a simple magnifying glass. The effect can also be accomplished with microwave energy, but due to the much higher wavelength, the lens must be large and bulky to be effective. The same effect can be obtained with much less bulk using the principle of *zoning,* shown in Fig. 16-4.

If an antenna launches energy at the focus as shown in Fig. 16-4, its spherical wavefront is converted into a plane (and thus highly directive) wave. The inside section of the lens is made thick at the center and thinner toward the edge in order to permit the lagging portions of the spherical wave to catch up with the faster portions at the center of the

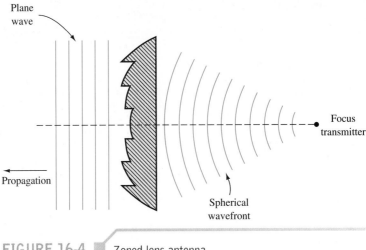

Plane
wave

Focus
transmitter

Propagation

Spherical
wavefront

FIGURE 16-4 ▨ Zoned lens antenna.

wavefront. The other lens sections have the same effect, but they work on the principle that for all sections of the wavefront to be in phase it is not necessary for all paths to be the same. A 360° phase difference (or multiple) will provide correct phasing. Thus, the plane wavefronts are made up of parts of two, three, or more of the spherical wavefronts.

Obviously the thickness of the steps is critical because of its relationship to wavelength. This is, therefore, not a broadband antenna, as is a simple magnifying glass antenna. However, the savings in bulk and expense justify the use of the zoned lens. Keep in mind that these antennas need not be glass as microwaves pass through any dielectric material, though at a reduced velocity compared to free space.

16-2 MICROWAVE TUBES

Yes, tubes still live. In the case of microwave applications, standard triodes or pentodes are not effective due to the interelectrode capacitances and the associated losses. The special-effect tubes presented here do not suffer in that respect and are still in widespread use.

Magnetron

The magnetron is an oscillator unlike any other that has previously been discussed in this text. The magnetron is a self-contained unit. That is, it produces a microwave frequency output within its enclosure without the use of external components such as crystals, inductors, capacitors, etc.

Basically, the magnetron is a diode and has no grid. A magnetic field in the space between the plate (anode) and the cathode serves as a grid. The plate of a magnetron does not have the same physical appearance as the plate of an ordinary electron tube. Since conventional *LC* networks become impractical at microwave frequencies, the plate is fabricated into a cylindrical copper block containing resonant cavities that serve as tuned circuits. The magnetron base differs greatly from the conventional base. It has short,

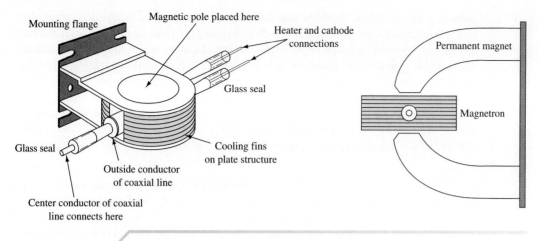

FIGURE 16-5 ▮ Magnetron.

large-diameter leads that are carefully sealed into the tube and shielded, as shown in Fig. 16-5.

The cathode and filament are at the center of the tube. The cathode is supported by the filament leads, which are large and rigid enough to keep the cathode and filament structure fixed in position. The output lead is usually a probe or loop extending into one of the tuned cavities and coupled into a waveguide or coaxial line. The phase structure, as shown in Fig. 16-6, is a solid block of copper. The cylindrical holes around its circumference are resonant cavities. A narrow slot runs from each cavity into the central portion of the tube and divides the inner structure into as many segments as there are cavities. Alternate segments are strapped together to put the cavities in parallel with regard to the output. These cavities control the output frequency. The straps are circular metal bands that are placed across the top of the block at the entrance slots to the cavities. Since the cathode must operate at high power, it must be fairly large and must be able to withstand high

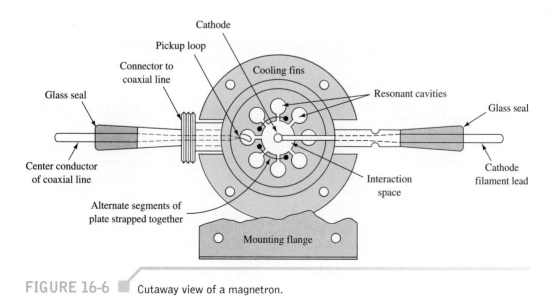

FIGURE 16-6 ▮ Cutaway view of a magnetron.

operating temperatures. It must also have good emission characteristics, particularly under back bombardment, because much of the output power is derived from the large number of electrons emitted when high-velocity electrons return to strike the cathode. The cathode is indirectly heated and is constructed of a high-emission material. The open space between the plate and the cathode is called the *interaction space* because it is in this space that the electric and magnetic fields interact to exert force on the electrons.

The magnetic field is usually provided by a strong permanent magnet mounted around the magnetron so that the magnetic field is parallel with the axis of the cathode. The cathode is mounted in the center of the interaction space.

BASIC MAGNETRON OPERATION   The theory of operation of the magnetron is based on the motion of electrons under the influence of combined electric and magnetic fields. The direction of an electric field is from the positive electrode to the negative electrode. The law governing the motion of an electron in an electric, or *E,* field states that the force exerted by an electric field on an electron is proportional to the strength of the field. Electrons tend to move from a point of negative potential toward a positive potential. In other words, electrons tend to move against the *E* field. When an electron is being accelerated by an *E* field, energy is taken from the field by the electron. The law of motion of an electron in a magnetic, or *H,* field states that the force exerted on an electron in a magnetic field is at right angles to both the field and the path of the electron.

A schematic diagram of a basic magnetron is shown in Fig. 16-7(a). The tube consists of a cylindrical anode with a cathode placed coaxially with it. The tuned circuit (not shown) in which oscillations take place is a cavity physically located in the anode.

When no magnetic field exists, heating the cathode results in a uniform and direct movement in the field from the cathode to the plate, as illustrated in Fig. 16-7(b). However, as the magnetic field surrounding the tube is increased, a single electron is affected, as shown in Fig. 16-8. In Fig. 16-8(a), the magnetic field has been increased to a point where the electron proceeds to the plate in a curve rather than a direct path.

In Fig. 16-8(b), the magnetic field has reached a value great enough to cause the electron to just miss the plate and return to the filament in a circular orbit. This value is the *critical value* of field strength. In Fig. 16-8(c), the value of the field strength has been

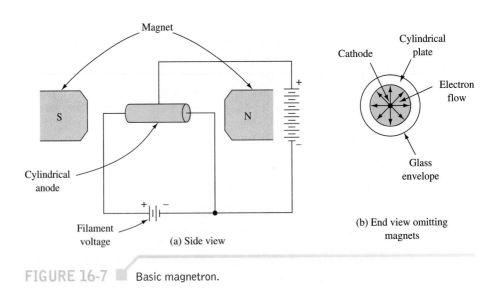

FIGURE 16-7 ▦ Basic magnetron.

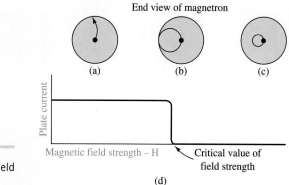

End view of magnetron

(a)   (b)   (c)

Plate current

Magnetic field strength – H

Critical value of
field strength

(d)

FIGURE 16-8 ▦ Effect of magnetic field
on single direct path.

increased to a point beyond the critical value, and the electron is made to travel to the cathode in a circular path of smaller diameter.

Figure 16-8(d) shows how the magnetron plate current varies under the influence of the varying magnetic field. In Fig. 16-8(a), the electron flow reaches the plate so that there is a large amount of plate current flowing. However, when the critical field value is reached, as shown in Fig. 16-8(b), the electrons are deflected away from the plate, and the plate current drops abruptly to a very small value. When the field strength is made still larger [Fig. 16-8(c)] the plate current drops to zero.

When the magnetron is adjusted to the plate current cutoff or critical value and the electrons just fail to reach the plate in their circular motion, the magnetron can produce oscillations at microwave frequency by virtue of the currents induced electrostatically by the moving electrons. This frequency is determined by the size of the cavities. Electrons are accelerated toward the anode by the electric field and bent by the magnetic field so that they travel parallel to the anode. If they pass the anode gap at a time when they are traveling in the same direction as the electric field in the gap, they slow down and thus give up energy (kinetic) to the electric field in the cavity. This reinforces the oscillations and is the basis of operation. A transfer of microwave energy to a load is made possible by connecting an external circuit between the cathode and plate of the magnetron. Magnetrons are widely used as sources of microwave power up to 100 GHz. They can provide up to 25 kW of continuous power at efficiencies up to 80%. Pulsed magnetrons are used in radar applications up to 10,000 kW with low duty cycles. They are also widely used for microwave ovens. They operate at 2.45 GHz with continuous outputs of 400 to 1000 W. Their electromagnetic energy is radiated through the food, which is heated (cooked) from the inside out. These tubes have been highly refined due to this volume application and offer reliability such that they can be expected to outlive the average expected 10- to 15-year life for a home appliance. Since the magnetron is functionally a diode, the only other power supply component required is a transformer for the 2 to 4 kV and the filament voltage.

Traveling Wave Tube

The traveling wave tube (TWT) is a high-gain, low-noise, wide-bandwidth microwave amplifier. TWTs are capable of gains of 40 dB or more, with bandwidths of over an octave. (A bandwidth of one octave is one in which the upper frequency is twice the lower frequency.) TWTs have been designed for frequencies as low as 300 MHz and as high as 150 GHz and continuous outputs to 5 kW. Their wide-bandwidth and low-noise

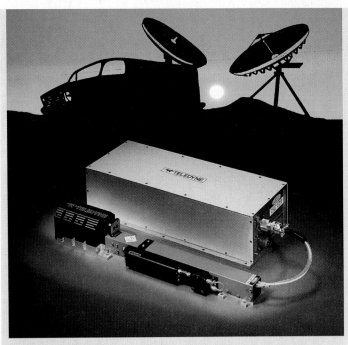

MULTI-BAND TWTA
Pure & Simple

Teledyne Electronic Technologies, Equipment Product Line, a leader in high power microwave amplifiers for almost 30 years, now offers a 325/400 watt multi-band uplink amplifier that provides continuous frequency coverage from 5.875 to 14.500 GHz. Packaged for conduction cooling, the field proven TWT and HVPS give you maximum flexibility at a very competitive price. Single band options available. Rack and hub mount configurations available in the third quarter of 1995.

TWT amplifier system provides up to 400 W at 14.5 GHz. (Courtesy of Teledyne Electronic Technologies. Photo by Jon Ho.)

characteristics make them ideal for use as RF and medium-power amplifiers in microwave and electronic countermeasure equipment. They are widely used as the power output stage in orbiting satellites.

CONSTRUCTION   Figure 16-9 is a pictorial diagram of a traveling wave tube. The electron gun produces a stream of electrons that are focused into a narrow beam by an axial magnetic field. The field is produced by a permanent magnet or electromagnet (not shown) that surrounds the helix portion of the tube. The narrow beam is accelerated, as it passes through the helix, by a high potential on the helix and collector.

OPERATION   The beam in a TWT is continually interacting with a RF electric field propagating along an external circuit surrounding the beam. To obtain amplification, the TWT must propagate a wave whose phase velocity is nearly synchronous with the dc velocity of the electron beam. It is difficult to accelerate the beam to greater than about one-fifth the velocity of light. The forward velocity of the RF field propagating along the helix is slowed to nearly that of the beam due to its travel along the helix. Changing the pitch changes the speed of the RF field.

The electron beam is focused and constrained to flow along the axis of the helix. The longitudinal components of the input signal's RF electric field, along the axis of the

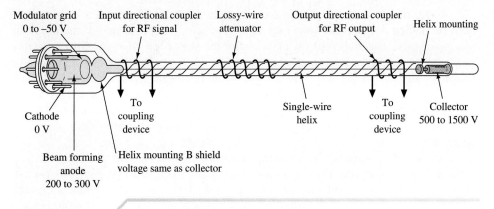

FIGURE 16-9 Pictorial diagram of a traveling wave tube.

helix or slow wave structure, continually interact with the electron beam to provide the gain mechanism of TWTs. This interaction mechanism is pictured in Fig. 16-10. This figure illustrates the RF electric field of the input signal, propagating along the helix, infringing into the region occupied by the electron beam.

Consider first the case where the electron velocity is exactly synchronous with the RF signal passing through the helix. Here, the electrons experience a steady dc electric force that tends to bunch them around position A and debunch them around position B in Fig. 16-10. This action is due to the accelerating and decelerating electric fields. In this case, as many electrons are accelerated as are decelerated; hence, there is no net energy transfer between the beam and the RF electric field. To achieve amplification, the electron beam is adjusted to travel slightly faster (by increasing the anode voltage) than the RF electric field propagating along the helix. The bunching and debunching mechanisms just discussed are still at work, but the bunches now move slightly ahead of the fields on the helix. Under these conditions more electrons are in the decelerating field to the right of A than in the accelerating field to the right of B. Since more electrons are decelerated than are accelerated, the energy balance is no longer maintained. Thus, energy transfers from the beam to the RF field, and the field grows and amplification occurs.

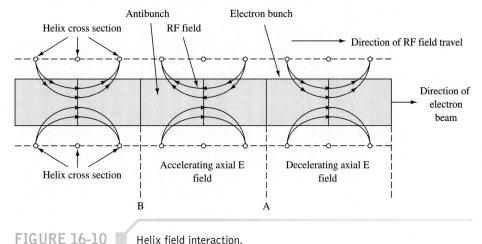

FIGURE 16-10 Helix field interaction.

Fields may propagate in either direction along the helix. This leads to the possibility of oscillation due to reflections back along the helix. This tendency is minimized by placing resistive materials near the input end of the slow wave structure. This resistance may take the form of a lossy wire attenuator (Fig. 16-9) or a graphite coating placed on insulators adjacent to the helix. Such lossy sections completely absorb any backward-traveling wave. The forward wave is also absorbed to a great extent, but the signal is carried past the attenuator by the bunches of electrons. These bunches are not affected by the attenuator and, therefore, reinstitute the signal on the helix, after they have passed the attenuator.

The traveling wave tube has also found application as a microwave mixer. By virtue of its wide bandwidth, the TWT can accommodate the frequencies generated by the heterodyning process (provided, of course, that the frequencies have been chosen to be within the range of the tube). The desired frequency is selected by the use of a filter on the output of the helix. A TWT mixer has the added advantage of providing gain as well as providing mixer action.

A TWT may be modulated by applying the modulating signal to a modulator grid. The modulator grid may be used to turn the electron beam on and off, as in pulsed microwave applications, or to control the density of the beam and its ability to transfer energy to the traveling wave. Thus, the grid may be used to amplitude-modulate the output. The TWT offers wideband performance with high-power outputs up to 150 GHz. TWTs are widely used in wideband communications repeater links. They offer low-noise performance and high-power gains. Their high reliability dictates their use as power amplifiers in communications satellites, where a lifetime in excess of 10 years can be expected.

TWT OSCILLATOR   A forward wave, traveling wave tube may be constructed to serve as a microwave oscillator. Physically, a TWT amplifier and oscillator differ in three major ways. The helix of the oscillator is longer than that of the amplifier, there is no input connection to the oscillator, and the lossy wire attenuator shown in Fig. 16-9 is eliminated. The tube now allows both forward and backward waves and is usually called a *backward-wave oscillator* (BWO). The operating frequency of a BWO is determined by the pitch of the tube's helix. The oscillator frequency may be fine tuned, within limits, by adjusting the operating potentials of the tube.

The electron beam, passing through the helix, induces an electromagnetic field in the helix. Although initially weak, this field will, through the action previously described, cause bunching of succeeding portions of the electron beam. With the proper potentials applied, the bunches of electrons will reinforce the signal on the helix. This, in turn, increases the bunching of succeeding portions of the electron beam. The signal on the helix is sustained and amplified by this positive feedback resulting from the exchange of energy between electron beam and helix.

Klystron

Another common microwave tube is the klystron. It has been widely used in the past and has certain similarities to the TWT. The high-power klystrons are being replaced by either magnetrons or TWTs in new equipment, and solid-state microwave devices are replacing them in low-power applications. The reasons for this are due to the klystron's large size and the complex, costly sources of dc required for operation.

16-3 SOLID-STATE MICROWAVE DEVICES

The advances made in the last decade with microwave solid-state devices have been truly startling. This includes the work with bipolar and field effect transistors as well as a number of special two-terminal devices.

Gunn Oscillator

The Gunn oscillator is a solid-state bulk-effect source of microwave energy. The discovery that microwaves could be generated by applying a steady voltage across a chip of n-type gallium arsenide crystal was made in 1963 by J. B. Gunn. The operation of this device results from the excitation of electrons in the crystal to energy states higher than those they normally occupy.

In a gallium arsenide semiconductor there exist empty electron valence bands, higher than those occupied by electrons. These higher valence bands have the property that electrons occupying them are less mobile under the influence of an electric field than when they are in their normal state at a lower valence band.

To simplify the explanation of this effect, assume that electrons in the higher valence band have essentially no mobility. If an electric field is applied to the gallium arsenide semiconductor, the current that flows will increase with an increase in voltage, provided the voltage is low. However, if the voltage is made high enough, it may be possible to excite electrons from their initial band to the higher band where they become immobile. If the rate at which electrons are removed is high enough, the current will decrease even though the electric field is being increased. Thus it displays the effect of negative resistance.

If a voltage is applied across an unevenly doped n-type gallium arsenide crystal, the crystal will break up into regions with different intensity electric fields across them. In particular, a small domain will form within which the field will be very strong, whereas in the rest of the crystal, outside this domain, the electric field will be weak. The domains formed in the gallium arsenide crystal will not be stationary, since the electric field acting on the electron energy will cause the domain to move across the crystal. The domain will travel across the crystal from one electrode to the other, and as it disappears at the anode a new domain will form near the cathode.

The Gunn oscillator will have a frequency inversely proportional to the time required for a domain to cross the crystal. This time is proportional to the length of the crystal and to some degree to the potential applied. Each domain results in a pulse of current at the output; hence, the output of the Gunn oscillator is a microwave frequency that is determined, for the most part, by the physical length of the chip.

The Gunn oscillator has delivered power outputs of 3 or 4 W at 18 GHz (continuous operation) and up to 1000 W in pulsed operation. The power output capability of this device is limited by the difficulty of removing heat from the small chip. At frequencies above 35 GHz, indium phosphate (InP)-based units are used in place of gallium arsenide (GaAs). These InP devices can deliver continuous powers of 500 mW at 35 GHz and 50 mW at 140 GHz.

The advantages of the Gunn oscillator are its small size, ruggedness, low cost of manufacture, lack of vacuum or filaments, and relatively good efficiency. These advantages open a wide range of application for this device in all phases of microwave operations. This bulk device is the workhorse of the microwave-oscillator field at frequencies above 8 GHz. Below 8 GHz it competes directly with transistor oscillators.

A commercially available Gunn oscillator assembly is shown in Fig. 16-11. It is included within a waveguide section and can be tuned by an included varactor diode over a 4% range at 10 GHz. They are also available in microstrip and coaxial line configurations. Since the Gunn device is a two-terminal solid-state device it is often termed a Gunn diode. They are also identified as transferred electron devices (TEDs) or limited space-charge accumulation devices (LSAs).

The Gunn device can also function as an amplifier. Its negative resistance characteristic allows it to replace the energy consumed by the positive resistance (loss) of either an *LC* tank circuit, shorted transmission line section, or resonant cavity. This *replacement* energy supplied by the negative resistance can also be used for amplification of applied energy. The major problem encountered with amplification using a two-terminal device is the isolation of input and output energy. This is usually accomplished with a device known as a *circulator*. It is similar to the hybrid tee described in the previous chapter in function but differs in physical construction. Its operation is explained in Sec. 16-4.

IMPATT Diode

IMPATT is an acronym for *imp*act ionization *a*valanche *t*ransit *t*ime. The theory of this device was presented in 1958, and the first experimental diode was described in 1965. The basic structure of a silicon *pn*-junction IMPATT diode, from the semiconductor point of view, is identical to that of varactor diodes. The important differences between IMPATT and varactor diodes are in their modes of operation and in thermal design.

Figure 16-12 shows a typical dc current versus voltage (*I–V*) characteristic for a *pn*-junction diode. In the forward-bias direction, the current increases rapidly for voltages above 0.5 V or so. In the reverse direction, a very small current (the *saturation* or *leakage* current) flows until the breakdown voltage, V_b, is reached. Varactor diodes normally operate either forward-biased or reverse-biased with a dc operating point well away from V_b. IMPATT diodes, on the other hand, operate in the avalanche breakdown region, that is, with a dc reverse voltage greater than V_b and substantial reverse current flowing.

Figure 16-13 shows a schematic representation of an IMPATT diode reverse-biased into avalanche breakdown. As in any reverse-biased *pn* junction, a *depletion zone* forms in the *n*-type region of the diode; its width depends on the applied reverse voltage. The depletion zone acts as a nonlinear capacitor if V_{dc} is less than V_b. This property is utilized in varactor diodes. The saturation current, which flows while the reverse voltage is less

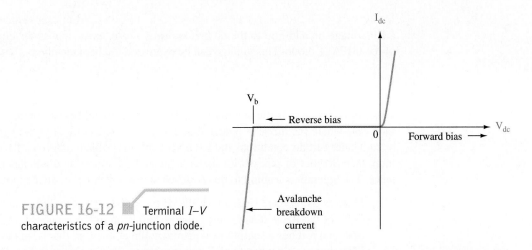

FIGURE 16-12 ■ Terminal *I–V* characteristics of a *pn*-junction diode.

than V_b, is usually on the order of 10 to 100 nA and is depicted in Fig. 16-13 by a small number of electrons flowing to the right from the p^+ region into the *avalanche zone*. When V_{dc} is more negative than V_b, the small number of electrons comprising the saturation current have a very high probability of creating additional electrons and holes in the avalanche zone by the process of avalanche multiplication. The additional electrons are shown in Fig. 16-13 flowing from the avalanche zone into the *drift zone*. In this condition, a large current can flow in the reverse direction with little increase in applied voltage. This is the *avalanche breakdown current,* depicted in Fig. 16-12. The typical dc operating voltage across a diode will be between 70 and 100 V, depending on the diode type, temperature, and the value of the bias current, I_{dc}. Typical avalanche breakdown currents (usually called the *bias current*) range from 20 to 150 mA.

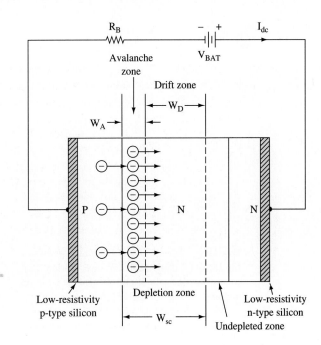

FIGURE 16-13 ■ Schematic representation of reverse-biased *pn*-junction diode.

MICROWAVE PROPERTIES OF THE IMPATT DIODE   Let us assume that somehow a RF voltage, in addition to the dc breakdown voltage, exists across the depletion region of the IMPATT diode. This voltage can be expressed mathematically as

$$V_T(t) = V_b + V_D \sin \omega t \qquad (16\text{-}3)$$

This form of voltage is illustrated in Fig. 16-14(a) and would exist in practice in the common case where the diode is operated in a singly resonant circuit with Q greater than 10 or so. Under certain conditions the RF portion of this voltage induces a RF current that is more than 90° out of phase with the voltage, and therefore the diode has negative resistance. The arguments leading to this conclusion are conveniently divided into two steps:

1. First, as the voltage rises above the dc breakdown voltage during the positive half-cycle of the RF voltage, excess charge builds up in the avalanche region, slowly at first, and reaches a sharply peaked maximum at $\omega t \approx \pi$, that is, in the middle of the RF voltage cycle when the RF voltage is zero. This is shown in Fig. 16-14(b). Thus, the charge generation waveform, in addition to being vary sharply peaked, *lags* the RF voltage by 90°. This behavior arises because of the highly nonlinear nature of the avalanche generation process.

2. The second step in the analysis is to consider the behavior of the generated charge subsequent to $\omega t = \pi$. The direction of the field is such that, referring to Fig. 16-13, the electrons *drift* to the right. The equal number of generated holes move to the left, back into the p^+ contact, and are not considered further in this simple model.

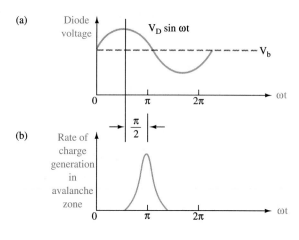

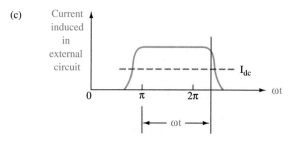

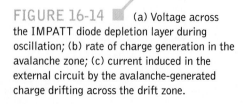

FIGURE 16-14   (a) Voltage across the IMPATT diode depletion layer during oscillation; (b) rate of charge generation in the avalanche zone; (c) current induced in the external circuit by the avalanche-generated charge drifting across the drift zone.

The electrons drift at constant (*saturated*) velocity, v_{sat}, across the drift zone. The time, t, they take to traverse the drift zone is simply the width of the drift zone divided by the constant velocity of the electrons:

$$t = \frac{width}{v_{sat}} \tag{16-4}$$

While the electrons are drifting through the diode, they induce a current in the external circuit, as shown in Fig. 16-14(c). The current is approximately a square wave. By examining Fig. 16-14(a) and (c), it can now be seen that the combined delay of the avalanche process and the finite transit time across the drift zone has caused *positive* current to flow in the external circuit while the diode's RF voltage is going through its *negative* half-cycle. The diode is thus delivering RF energy to the external circuit or, in circuit terms, is exhibiting *negative* resistance. Maximum negative resistance is obtained when

$$\omega t \approx 0.74\pi \tag{16-5}$$

The term ωt is called the *transit angle;* IMPATT diodes are normally designed so that Eq. (16-5) is satisfied at or near the center of the desired operating frequency range.

IMPATT diodes are finding wide application in microwave oscillator and amplification schemes. They can produce 20-W continuous output at 10 GHz and are useful up to 300 GHz. While many two-terminal microwave devices exhibiting negative resistance have been developed in recent years, it appears that the Gunn diode (bulk-effect device) and IMPATT diode (true diode) have the most promising future. Among the other devices that have also been utilized are the following:

1. *TRAPATT diode:* It is similar to the IMPATT diode.
2. *Baritt diode:* It has two junctions separated by a transit time region.
3. *Tunnel diode:* Its use as a microwave power source/amplifier has diminished as Gunn devices and IMPATTs have become readily available.

P-i-n Diode

P-i-n diodes are used as RF and microwave switches whose resistance values are controlled by forward current levels. As shown in Fig. 16-15, a *p-i-n* diode is built from high-resistivity silicon and has an *intrinsic* (very lightly doped) layer sandwiched between a *p*

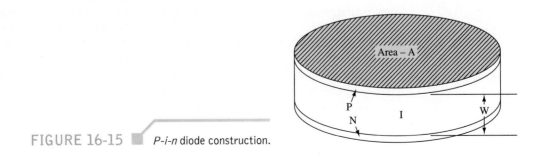

FIGURE 16-15 ■ *P-i-n* diode construction.

| Case style | M1 (inches) | | M2 (inches) | |
Dimension	Min	Max	Min	Max
A	0.08	0.10	0.10	0.12
B	0.12	0.14	0.19	0.21
C	0.01	0.03	0.01	0.03

P-i-n diodes with reverse voltage ratings up to 1000 v. (Courtesy of Loral Microwave-Narda.)

and an *n* layer. When the diode has a forward current, holes and electrons are injected in the *i* region. They do not completely recombine but rather form a stored charge. This stored charge causes the effective resistivity of the *i* region to be much lower than the intrinsic resistivity.

Up to about 100 MHz, this diode acts basically like a conventional rectifier, but at higher frequencies it ceases to rectify. This is due to the stored charge in and the transit time across the intrinsic region. To high frequencies it acts as a variable resistance, easily controlled by the amount of dc forward bias. Resistances of less than an ohm are possible at high-dc forward bias, while a small forward bias may cause it to look like 1 kΩ of resistance to a high-frequency signal.

The *p-i-n* diode is used whenever there is the need to switch microwave energy at frequencies up to 100 GHz, even at high-power levels. A common application is their use as transmit/receive (TR) switches in transceivers operating from 100 MHz and up. As described in Chapter 17, they are also used as photo-detectors in fiber-optic systems.

Microwave Transistor

The three-terminal devices are used primarily to build amplifiers where their inherent input/output isolation permits simpler designs than two-terminal devices. Bipolar transistors are preferred at frequencies below 5 GHz since they provide higher output power and similar noise performance with respect to FETs. Above 5 GHz, bipolars lose power output capability due to inherent high-frequency limitations, and their noise performance is severely degraded. Since bipolar technology is relatively mature, it is not expected that these conditions will be improved to any great extent in the future.

On the other hand, FET technology is still growing. At high frequencies, gallium arsenide (GaAs) FETs offer superior performance over the standard silicon devices. Above 5 GHz, the GaAs FETs offer superior noise performance and output powers compared to the bipolar transistor. The GaAs FET is the most important amplifying device in the 5- to 20-GHz region. Above 20 GHz it is necessary to go to a two-terminal device such as an IMPATT diode or a tube such as the TWT. If high-power output is required (>20 W), the tube device is likely to be used for frequencies down to 2 GHz.

Microwave Integrated Circuits

Microwave monolithic integrated circuits (MMICs) are beginning to make inroads against discrete devices, especially in the lower microwave frequencies of 1–3 GHz. These devices are generally GaAs based and are popular in high-volume applications such as cellular communications.

Figure 16-16 shows the block diagram for a Philips DCS-1800 MMIC. It is a GaAs MESFET device used in cellular communications and includes a power amplifier, transmitter upconversion mixer, LO, and other related components. The LO feeds the upconversion mixer to produce an IF at 400 MHz. The LO is a VCO that is tuned via an external resonator into a frequency ranging from 2110 to 2185 MHz. An external filter is

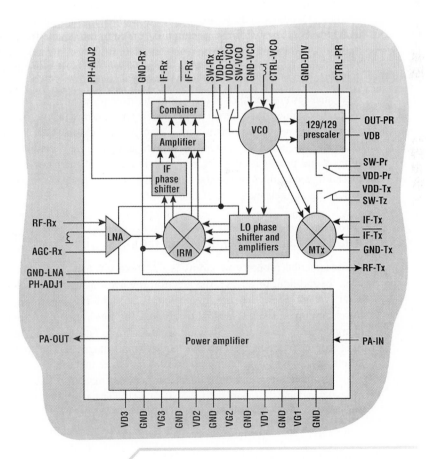

FIGURE 16-16 ■ MMIC used in cellular communications. (Courtesy of *Microwaves and RF.,* March/April 1994)

inserted between the transmitter-mixer (MTx) output (RF-Tx) and the PA input to remove signals generated in the image bandwidth of 2510 to 2585 MHz.

The receiver stage consists of a low-noise amplifier (LNA) and an image-reject mixer (IRM). The RF input (RF-Rx) has a bandwidth from 1805 to 1880 MHz, while the IF output (IF-Rx) is set to 300 MHz, thus resulting in an LO range on the order of 2105 to 2180 MHz. Since the same LO is used for both transmit and receive, it must tune from at least 2105 to 2185 MHz.

The MMIC includes a divide-by-128/129 dual-modulus prescaler. The prescaler converts signals from the local oscillator to an output signal of about 17 MHz, which is compatible with most standard frequency synthesizers. The supply voltages on the receiver and transmitter chains are alternately switched on and off by FETs controlled by external signals.

The power amplifier (PA) is designed to provide output of up to +27 dBm (0.5 W) with a 3.3-V supply. It has an efficiency of more than 30% at 1700 MHz. The input power requirement is 0 dBm. It includes three stages operating Class AB.

16-4 FERRITES

Ferrites are compounds of iron, zinc, manganese, magnesium, cobalt, aluminum, and nickel oxides. They are manufactured by pressing into shape the required mixture of the finely divided metallic oxide powders and then firing the shaped mixture at about 2000°F. The product is a ceramic with high electrical resistance. Ferrites behave as iron alloys at low frequencies, but at high frequencies their high electrical resistance prevents eddy currents, and resonance takes place within the iron atoms themselves. These unusual effects make it possible to use ferrites for special applications in microwave circuits. The most popular ferrite compounds are manganese ($MnFe_2O_3$), zinc ($ZnFe_2O_3$), and yttrium–iron–garnet [$Y_3Fe_2(FO_4)_3$], which is called *yig*. Ferrites are dielectrics, so they support and propagate the electromagnetic energy of microwave signals. When they are placed in waveguides or coaxial circuits they can be transparent, reflective, or absorptive, depending on their magnetic characteristics.

Fundamental Theory of Ferrites

A fundamental property of atoms is that both electrons and protons spin on their own axes. In addition, of course, the electron revolves around the nucleus. An analogy is the solar system, where the earth rotates on its axis as it revolves around the sun. As the electron spins, it creates a magnetic moment, or field, along its spin axis. This spinning charge appears as a current flowing around a loop. The atoms having more electrons spinning in one direction than another act as small magnets. The mutual action of all these atoms explains the magnetic properties of magnetic materials.

If a spinning electron is placed in a static magnetic field, the electron's magnetic moment becomes aligned with the static field. The magnetic moment and its alignment with a dc magnetic field are shown in Fig. 16-17.

Gyroscopic Action

Electrons, because of their spinning motion, behave like very small gyroscopes. When a force is applied to the spin axis of an electron that would cause it to tilt, the electron will

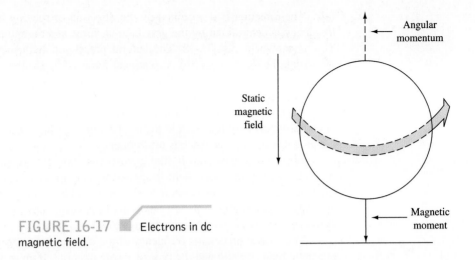

behave like any other gyroscope. It will precess, or wobble. *Precession* is defined as a movement of the axis of rotation at right angles to its original axis. Figure 16-18 shows a gyroscope mounted to a stick so that the stick can pivot freely. Even with the gyroscope spinning, the stick will hang straight down, due to gravity. However, if you try to move the stick from side to side, the gyroscope will force the stick to move around in a circle, or precess.

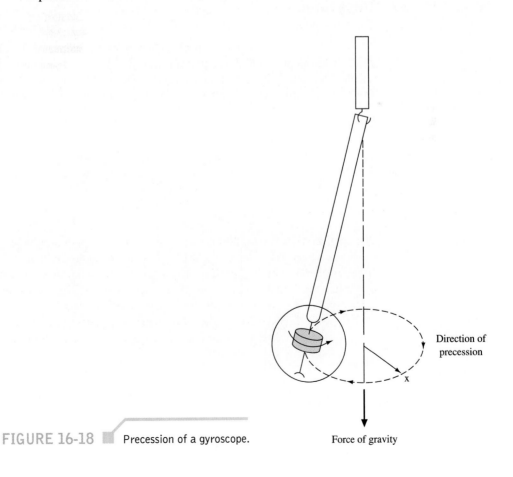

FIGURE 16-18 ■ Precession of a gyroscope.

The direction is determined by the direction of rotation of the gyro rotor, and the frequency is determined by the gravitational force and the momentum of the gyroscope. This is shown in Fig. 16-18. The natural precession frequency could be increased by increasing the force of gravity. A rotational force applied to the stick at the natural precession frequency will displace it from the vertical by a large amount. (The precession path shown in Fig. 16-18 would have a greater diameter.) A rotational force applied at any other frequency will produce a much lower displacement. This is similar to the feedback in an oscillator. With feedback at the right frequency, the amplitude of oscillation is much larger than when the feedback is off frequency.

Electrons behave much like gyroscopes, but gravity has little effect on them. Instead, a steady magnetic field is applied to line up the axes of the spinning electrons. This field causes any precession to die out quickly. Now when an alternating field is applied at right angles to the dc field, the electrons precess, or wobble, just the way the gyroscope and stick did when a sideways force was applied.

The natural precession frequency of the electrons in a ferrite is dependent on the dc magnetic field strength and the type of ferrite material. If an ac field is applied at the natural frequency, the precessional motion builds up. This increases the frictional damping effects because the entire atom is vibrating and the ferrite dissipates as heat the energy extracted from the ac field. The range of natural precession frequencies available with presently used ferrites is from about 30 up to over 200 GHz.

Applications

ATTENUATOR   One application of ferrites is as an attenuator. Figure 16-19 shows a piece of ferrite placed in the center of a waveguide; a steady magnetic field is applied as shown. This arrangement will attenuate frequencies at the resonant frequency of the electrons in the ferrite, whereas other frequencies will be attenuated very slightly. Changing

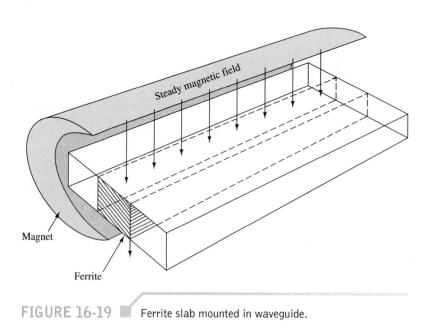

FIGURE 16-19 ■ Ferrite slab mounted in waveguide.

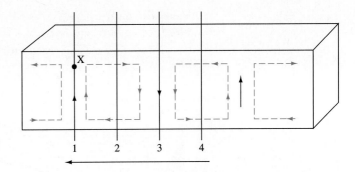

FIGURE 16-20 ■ Effective clockwise
rotation of a magnetic field.

the strength of the dc field produces a change in the frequency that will be attenuated, although this occurs over a limited range.

The dc field is produced by current flowing through a coil wound around the waveguide. The strength of the field, which depends on the current flowing through the coil, determines the frequency of precession. Usually, the ferrite attenuator is in the form of an adjustable vane of ferrite extending into the waveguide. The farther the vane extends into the waveguide, the greater the attenuation, since more of the RF energy must travel through the ferrite.

ISOLATOR   Another application of ferrites is that of an isolator. When used as an isolator, the ferrite allows energy to travel in one direction but absorbs energy traveling in the opposite direction. Figure 16-20 depicts an electromagnetic wave traveling from right to left. It illustrates how the wave, at a point off the center line of the guide, will appear as a rotating magnetic field. At one instant shown in Fig. 16-20, the magnetic field at point *C* is pointed up. When the magnetic field at point 2 reaches point *X*, the magnetic field will be directed to the right. When point 3 reaches *X*, the magnetic field is downward, and when point 4 on the wave arrives at *X*, the field is directed to the left.

Thus, as the wave passes point *X*, the magnetic field appears to rotate in a clockwise direction. At any point off the center of the waveguide, the magnetic field will appear to rotate as the electromagnetic wave passes. This same analysis can be used to show that with a wave traveling from left to right the magnetic field appears to rotate counterclockwise at point *X*.

Now let us place a section of ferrite in the waveguide at *X*. This is shown in Fig. 16-21, which illustrates a simple isolator consisting of a piece of waveguide, a permanent magnet, and a section of ferrite. The ferrite's electron resonant frequency and the microwave frequency are made the same by changing either the magnetic field strength or the microwave frequency. When the frequencies are the same, a wave traveling from left to right in the waveguide produces a rotating force in the direction of the natural precession of the electrons in the ferrite. The amplitude of precession increases, taking power from the electromagnetic wave. This power is dissipated as heat in the ferrite.

A wave that is traveling from right to left in this waveguide acts as a rotating force on the electrons to oppose the natural precession. This will not increase the amplitude of the precession, and energy is not absorbed from the electromagnetic field. About 0.4-dB attenuation takes place in a wave traveling from right to left, but as much as 10-dB attenuation occurs in a wave traveling from left to right.

FARADAY ROTATION   Another effect takes place when microwaves are passed through a piece of ferrite in a magnetic field. The plane of polarization of the wave is rotated if the

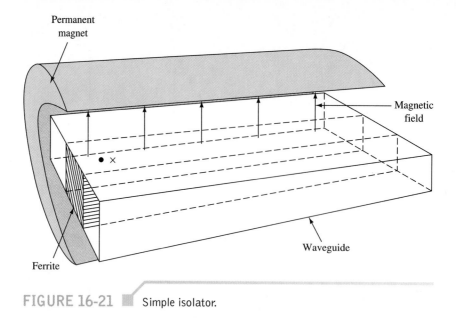

FIGURE 16-21 ■ Simple isolator.

frequency of the microwave is above the resonant frequency of the ferrite electrons. This is known as the *Faraday rotation effect.* When RF energy enters the ferrite material, the magnetic moment of the electron precesses as usual but at a frequency different from the RF. The *H* lines within the ferrite now are the resultant produced by vector addition of the rotating magnetic moment and the RF field. A new RF field, which is rotated from the original RF field, results. The amount of rotation is determined by the dc magnetic field and the length of the ferrite.

Figure 16-22 shows a ferrite rod that is placed lengthwise in the waveguide. The dc magnetic field is set up by a coil. Now assume that a wave that is vertically polarized

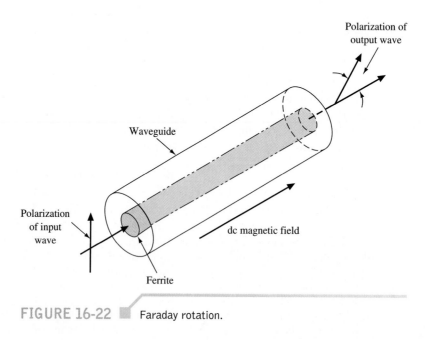

FIGURE 16-22 ■ Faraday rotation.

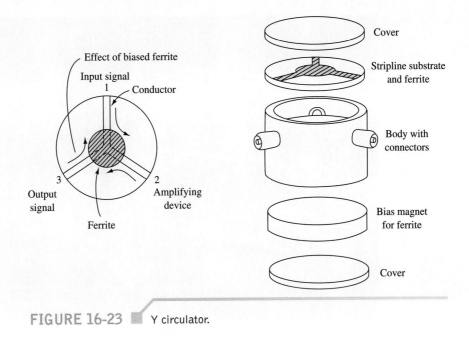

FIGURE 16-23 ■ Y circulator.

enters the left end of the waveguide. As it enters the ferrite section, it will set up limited precession motion of the electrons. The interaction between the magnetic fields of the wave and the precessing electrons rotates the polarization of the wave. With the correct dimensions of the ferrite rod, the wave is polarized at a 45° angle from the original. Different dimensions of the rod and magnetic field strengths produce other shifts in polarization.

CIRCULATOR   As mentioned in Sec. 16-3, two-terminal amplifying devices require a means of isolation between input and output power. A circulator is a ferrite device that is the commonly used solution to that problem. The most popular type of circulator is the Y circulator shown in Fig. 16-23.

Y circulators come in waveguide, coaxial line, or microstrip versions, with the latter shown in Fig. 16-23. With the three lines arranged 120° apart as shown, energy coupled into arm 1 goes only to arm 2, while 2 feeds only 3 and 3 feeds only 1. The ferrite provides the correct rotational shift to provide this operation.

OTHER FERRITE APPLICATIONS   Ferrites find application in many non-microwave applications. They are widely used in portable radio antennas as the core for a winding, IF transformer cores, TV flyback and deflection coil cores, magnetic memory cores in computers, and tape recorder heads. Another application to the communications field is the *ferrite bead*. It is a small bead of ferrite material with a hole through its center such that it can be threaded onto the wires of an electronic circuit. Its effect is to offer virtually no impedance to dc and low frequencies but a relatively high impedance to radio frequencies. They are widely used as inexpensive replacements for radio-frequency chokes (RFC) to obtain effective RF decoupling, shielding, and parasitic suppression without an attendant sacrifice in dc or low-frequency power.

Ferrite applications. (Courtesy of ST Microwave Corp.)

A ferrite bead on a conductor and its inductive effect are shown in Fig. 16-24. As the unwanted high frequency flows through the conductor, it creates a magnetic field around the wire. As the field passes into the ferrite bead, the higher (than air) permeability of the bead causes the local impedance to rise and create the effect of an RFC in that location.

Since ferrite materials are able to attenuate specific microwave frequencies, they are being used for filter applications in place of resonant cavities. The highest-Q ferrite filters are the yig materials. These filters offer small size and electronic (magnetic) tuning advantages over the resonant cavities but are not as high-Q in response. Yig filters are commonly used with the Gunn or IMPATT devices in electronically controlled solid-state microwave sources, as shown in Fig. 16-25. The electromagnet that controls the frequency has been omitted in the figure but must surround the yig sphere shown. The output

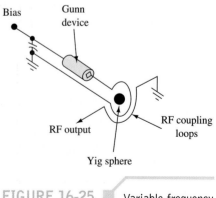

FIGURE 16-25 Variable-frequency microwave signal source.

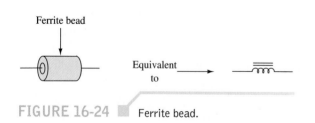

FIGURE 16-24 Ferrite bead.

energy is taken by the RF coupling loop as shown. The simplicity of this variable-frequency microwave source is apparent from the figure.

16-5 LOW-NOISE AMPLIFICATION

Microwave receivers usually deal with a very small input signal. The limiting factor on how small the signal may be is primarily determined by the noise figure of the receiver's first amplifier stage. Several new approaches to microwave amplification offer extremely low-noise characteristics. These include parametric amplifiers and the maser.

Parametric Amplifier

A *parametric amplifier* provides amplification via the variation of a reactance. This reactance is a *parameter* of a tuned circuit—thus the amplifier's name. Consider a *LC* tank circuit that is oscillating at some microwave frequency. If the capacitor's plates are pulled apart at the instant of time that the voltage across them is maximum positive, work has been accomplished. Since the capacitance, *C*, has been decreased and the charge, *q*, must remain the same, the voltage across the capacitor, *V*, must have been increased since $V = q/C$. This is the first step in the amplification provided by a parametric amplifier.

Now the plates are returned to their original separation as the oscillator's signal causes the voltage across the plates to pass through zero. This effort requires no work since now there is no force exerted between the plates. As the voltage across the plates reaches maximum negative, the plates are once again pushed apart, causing the voltage to increase once again. The process is repeated continuously, and amplification has occurred.

The force causing the plates to be pushed apart occurs twice for every cycle of the oscillator's signal and is called the *pump* source. The pump force is thus a signal at twice the oscillator's frequency. It is invariably an ac voltage applied to a varactor diode that is part of the oscillator's tank circuit. The voltage changes the capacitance of the diode at just the right time (as previously described to allow voltage gain). Whereas normally encountered amplifiers provide ac gain with external power obtained from a dc source, the parametric amplifier provides ac gain with external power from an ac source at twice the frequency of the signal being amplified.

A simplified 10-GHz amplifier as just described is shown in Fig. 16-26. Here it is used as the front end (RF stage) of a 10-GHz receiver. As shown, a four-port circulator is required to keep the various signals from interfering with each other. The low-noise characteristic is the result of amplification via a variable reactance, with the noise from resis-

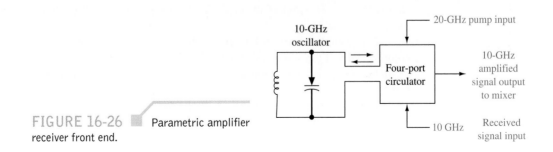

FIGURE 16-26 ■ Parametric amplifier receiver front end.

tance being almost negligible. These amplifiers are capable of noise figures of 0.3 dB, which is an order of magnitude better than that possible with the microwave amplifiers previously discussed. Power gains of 20 dB are realized.

The noise of microwave systems is often given in terms of *noise temperature.* This is sometimes a more convenient method of dealing with noise, as the noise temperature of two devices is directly additive. Keep in mind, however, that noise temperature, like noise resistance, is simply convenient fiction. Noise temperature is related to noise figure (expressed as a ratio and not in decibels) by $T = T_0$ (NF − 1), where T is noise temperature (K) and T_0 is 290 K. See Sec. 1-3 for a discussion of noise figure.

Parametric amplifiers with pump frequencies double the signal frequency are termed the *degenerate mode.* Parametric amplification is also possible (at reduced gain) with pump frequencies other than 2 times the signal frequency. They are termed the *nondegenerate mode.* In this case beating between the two frequencies occurs, and a difference signal called the *idler* signal occurs. This mode of operation allows the parametric amplifier to function directly as a mixer, with its output being the first IF frequency.

The Maser

An even lower-noise microwave amplifier is the *maser.* It stands for *m*icrowave *a*mplification by *s*timulated *e*mission of *r*adiation. It was developed in 1954 by Professor C. H. Townes, who also advanced the theory for the *laser* in 1958. The laser is the optical version of the maser where the "l" stands for light.

It is known that most of the electrons of an atom exist at the lowest energy level when the substance is at a very low temperature (close to absolute zero). If, however, a *quantum,* or bundle of energy, is provided to the atom, the electrons may be raised to a much higher energy level. The applied energy to make this happen is radiation at the frequency of magnetic resonance for the material, as was discussed with ferrites. However, in this case, what is desired is an *emission* of energy, which occurs when the excited electrons return back to the lowest energy level or some intermediate level. The applied frequency is the *pump* signal, while the emitted energy is at some intermediate frequency when the electrons fall back to an intermediate energy level. If an input signal (not the pump signal) is applied at the same frequency as the intermediate frequency, amplification is possible.

A maser amplifier scheme using ruby is shown in Fig. 16-27. Ruby is a crystalline form of silica (Al_2O_3) that has a slight doping of chromium. Its atomic structure has suitably arranged energy levels, and the presence of chromium allows a tuning of usable frequencies of from about 1 up to 6 GHz.

The entire amplifier shown in Fig. 16-27 is enclosed in liquid helium, which maintains it at 4.2 K. This is necessary for the proper electron action within the ruby material, and even the magnetic core is sometimes enclosed so as to take advantage of *superconductivity.* Since the required magnetic field is very high, this allows a reduction in power for its maintenance.

The resonant cavity in Fig. 16-27 should be resonant at both the frequency of the pump signal and the signal being amplified. The pump signal, which is a higher frequency than the signal being amplified, is applied to the cavity via a waveguide. The signal being amplified and the output are connected via a circulator and a coupling probe.

These amplifiers are capable of 25-dB gains, with noise figures as low as 0.2 dB. Needless to say, this is an expensive proposition, and their use is reserved for severe applications such as radio astronomy. Other materials may be used in place of ruby,

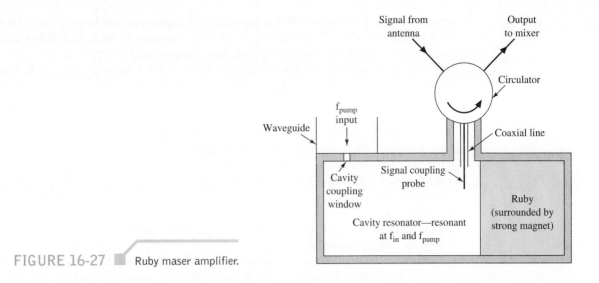

FIGURE 16-27 ▊ Ruby maser amplifier.

including gases such as ammonia. Ammonia was used in the first practical maser but is capable of amplifying only one frequency—about 24 GHz—since no tuning is possible with an external magnetic field. It does find use, however, in *atomic clock* frequency standards. The accuracy of these clocks is about 1 part per million million (1 part per 10^{12}), which means that an error of 1 s every 30,000 years can be expected.

16-6 LASERS

The laser (*light amplification by stimulated emission of radiation*) is similar to the maser except for the frequencies involved. Visible light is electromagnetic radiation at 430 to 730×10^{12} Hz, as opposed to the microwaves we have been dealing with at up to about 0.3×10^{12} Hz (300 GHz). We are concerned with the communication applications of the laser, but you are undoubtedly aware of other uses, such as:

1. Distance measuring equipment
2. Industrial welding
3. Surgical procedures
4. Military applications
5. Producing holograms (three-dimensional photography)
6. Pickup devices in video-disc playback units and compact disc players

As further explained in Chapter 17, the laser is used as the light source for optical-fiber communication systems. The laser can be used to communicate directly by simply transmitting a modulated laser beam through the atmosphere, but too many interferences (fog, dust, rain, and clouds) generally preclude that application. Outer-space communications do not present these limitations. We will look at an application later in this section.

Laser Sources

Many different types of materials have been successfully stimulated to exhibit laser action, including solids, liquids, and gases. The brilliant red, green, blue, and yellow beams seen at laser light shows are produced by gaseous materials. In terms of importance to electronic communications using optical fibers, the semiconductor injection laser is the only useful type. These devices are actually members of the light-emitting diode (LED) family. When the current injected into a diode laser is below a critical value termed the *threshold* (I_{th}), the diode behaves just like an LED and emits a relatively broad spectrum of wavelengths in a wide radiation pattern. As the current is increased to I_{th}, however, the light narrows into a distinct beam and is confined to a very narrow spectrum. *Lasing* action has begun and the device is then functioning as a laser.

The early solid-state lasers were short-lived and could only be operated under short-duty-cycle (pulsed) conditions. The key to the development of continuous-duty, long-lifetime devices was the stripe-geometry injection double-heterojunction (DH) laser. *Heterojunction* refers to a junction of two dissimilar semiconductors, such as gallium arsenide (GaAs) and aluminum gallium arsenide (AlGaAs). This allows the light-emitting *pn*-junction region to be sandwiched between two or more semiconductor layers that confine the generation and emission of light to the junction region. This allows a low threshold current and high efficiency.

The confinement of light between two heterojunctions results when the refractive index of a *pn*-junction material (GaAs) is higher than the semiconductor bordering the *pn* junction (AlGaAs). This causes the heterojunctions to appear as guides to the emitted light just as the clad/core junction does in optical cable (see Chapter 17). The construction of the DH stripe-geometry laser is shown in Fig. 16-28. A *stripe-geometry* device indicates that all but a narrow stripe of one electrode is insulated from the upper surface of one laser. The current flow through the *pn* junction is thereby confined to a thin stripe between the mirror surfaces placed at the two ends. The resulting high-current density in the "stripe" provides very low "lasing" threshold current. These devices can generate several milliwatts of laser light at forward currents of about 100 mA.

The DH lasers are ideally suited for communications through optical fibers (see Sec. 17-5.) They are also finding use in laser printing systems and video/audio disc players. Since these devices operate continuously at room temperature, you may think that one need only connect it to a dc source with an adjustable current-limiting resistance and you're in business. Unfortunately, it is more difficult than that, due to the temperature sensitivity of solid-state lasers. A change of just 1° can halt the lasing action or even destroy the device. Because of this, a DH laser is usually operated in a temperature-controlled oven and/or its forward current is temperature compensated.

Laser Communications

As stated earlier, the laser can be used directly as a communication link. The laser can be modulated to contain a large amount of information. As will be described in Chapter 17, optical fibers are usually used to guide the light to its destination. The direct transmission

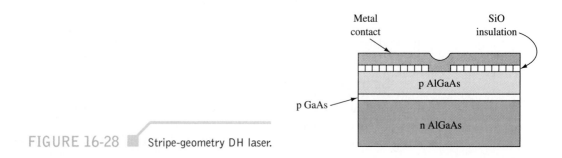

FIGURE 16-28 ■ Stripe-geometry DH laser.

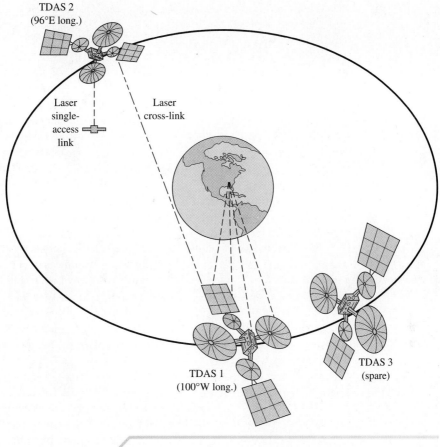

Laser
single-
access
link

Laser
cross-link

TDAS 3
(spare)

TDAS 1
(100°W long.)

FIGURE 16-29   Laser/satellite communication. (Courtesy of *Microwaves and RF,*
April 1985.)

has its problems—it is limited to line of sight, a bird can easily destroy a communication
by flying through the laser beam, and severe attenuation by rainfall or snow cannot be cir-
cumvented. These conditions are not a problem in outer space, however.

A laser communication system is illustrated in Fig. 16-29. This laser provides an
excellent communication link between orbiting communication satellites. The tracking
and data acquisition satellite system shown is used by NASA. The need for a cross-link
between two satellites at up to 2-Gb/s data rate can be handled by a laser link using laser
dishes 1 to 2 ft in diameter. By comparison, a microwave link would require dish antennas 6 to 9 ft in diameter.

Laser Computers

Future computers may be based on optical switches. Such a switch, called a *transphasor,*
is represented in Fig. 16-30. A laser beam is applied to a special crystal made of indium
antimonide. Most of the laser beam bounces off, but some enters the crystal, where it is
trapped, bouncing back and forth. The transphasor is now "off," as shown in Fig. 16-
30(a). At (b) a second, weaker laser is directed at the crystal. It only slightly increases the
light intensity within the crystal but causes the reverberating light waves to start reinforc-

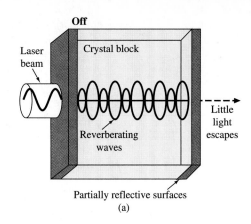

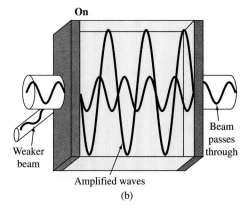

FIGURE 16-30 Transphasor.

ing one another. This causes the laser light suddenly to flash out of the crystal's other side. In effect, a weak beam of photons exerts control over a strong one. This is analogous to the weak base or gate current of a transistor having control over a larger current.

The foremost advantage of optical computers is operating speed. It is expected that speeds 1000 times faster than electronic computers will be attained. The second main advantage comes from the fact that photons have no charge or mass. Unlike electrons, photons have little effect on other nearby photons and can even pass right through each other. This means that multiple beams of light in an optical switch could remain separate, whereas several currents in a single transistor become mixed. This characteristic means that optical computers will lend themselves to parallel processing architecture. Instead of solving problems step by step, parallel machines break apart computational puzzles and solve the many parts or steps all at once, much as the human brain can do. This new technology may also prove useful in fiber-optic systems. The expensive translations from electrons to photons and then (after transmission) back to electrons may be eliminated.

TROUBLESHOOTING

Up to this point we have not considered the power supply's role in electronic communications equipment. The power supply furnishes the voltage and current requirements for electronic circuit operation. Special high-voltage power supplies used in microwave systems and laser systems enable large power handling tube and semiconductor circuits to operate. In this section we are going to look at two popular types of power supply circuits and some faults that can occur in them.

In this section we'll also learn to troubleshoot a traveling wave tube amplifier. Improper operation of a TWTA can nearly always be traced to a power supply. There are a few rules you must know to do the job. Violation of the rules will result in loss of the tube. Please be aware that extreme caution must be observed when dealing with power supplies. The ac line voltage and any high-voltage outputs can be lethal. After completing this section you should be able to

- Identify a switching power supply
- Identify a linear power supply
- Name the cause of high ac ripple on a power supply output
- Name a cause for a blown fuse in primary winding of a power supply transformer
- Identify possible failure modes in a TWT amplifier

Power Supplies

Two popular power supplies are found in electronic equipment today. These are the linear power supply and the switching power supply. Many variations exist of both kinds of supplies. The linear power supply usually has a power transformer that is large and heavy. The linear supply furnishes a constant output voltage to a load. Excessive power is usually wasted in this kind of power supply in the form of heat. Because of this wasted power, efficiency is low. A typical linear power supply is illustrated in Fig. 16-31.

The switching power supply is light and does not use a large bulky power transformer. Instead, a smaller transformer is used. A diagram of the switching power supply is shown in Fig. 16-32. The input ac voltage is rectified, filtered, and applied to the transformer, T_1. The power transistor, Q_1, in the negative supply line is switched on and off at a high frequency rate, 20 to 40 kHz. The load requirement will determine the output of the supply. If the load requirement goes up, the supply furnishes more power. If the load goes down, less power is supplied. Power is not wasted as in the linear power supply.

High efficiency rates are achieved in the switching power supply by feeding a portion of the output voltage back to control an oscillator. The oscillator is part of a pulse-

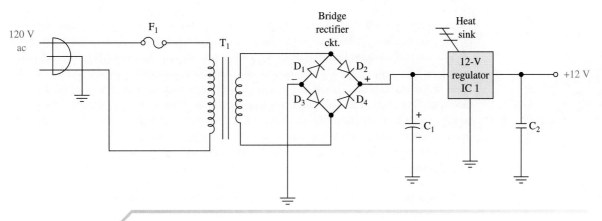

FIGURE 16-31　A linear 12-V regulated power supply using an IC 12-V regulator.

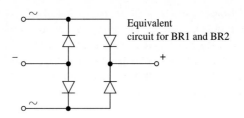

Equivalent circuit for BR1 and BR2

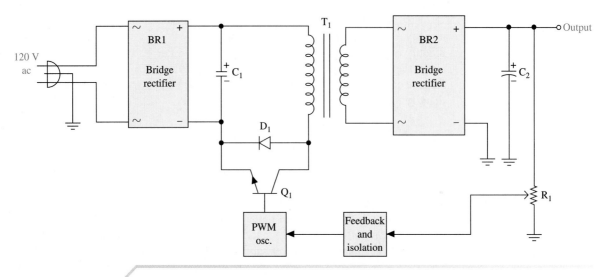

FIGURE 16-32 Simplified diagram of a switching power supply.

width-modulation circuit. A pulse-width-modulated signal is supplied to the base of the power transistor Q_1, which regulates the transistor's duty cycle. An increase of Q_1's duty cycle will increase the output voltage to maintain regulation. A decrease of the duty cycle will decrease the output voltage.

TROUBLESHOOTING THE LINEAR SUPPLY   The linear power supply not only provides operating voltage and current to the electronic circuits connected to it, but must also provide decoupling for them as well. The well-designed power supply will appear as a low impedance to the decoupled frequencies. If the power supply does not provide decoupling, problems like low-frequency oscillation and distortion can occur in the audio circuits. Hum in the audio output is caused by a bad filter capacitor. Capacitor C_1 in Fig. 16-31 is the input filter capacitor. This capacitor reduces the ac ripple in the dc output of the power supply. For a high-ripple problem, replace C_1. The service manual normally specifies the maximum allowable ripple on the dc output voltages. Voltage is regulated by IC1 in Fig. 16-31. If IC1 fails to regulate, a higher than normal voltage will appear at the output. Many power supplies have several regulator circuits operating. Check each specified output voltage against actual measurements to ensure the supply is operating correctly.

The diodes, D_1–D_4, form a full-wave bridge circuit that rectifies the ac from the transformers' secondary winding. If any of the diodes should short, the fuse in the transformers' primary winding will blow. If the power supply blows fuses, suspect a shorted diode. High-wattage resistors in the power supply circuits are subject to changes in value that could change an output voltage. Power transistors are often used in regulator circuits. These are usually mounted on heat sinks and may short out. A shorted power transistor will most certainly blow a fuse and may even burn up any resistor in series with it. When troubleshooting power supplies, look for blown fuses, burned resistors, corroded solder joints, and leaky filter capacitors.

TROUBLESHOOTING THE SWITCHING POWER SUPPLY   Switching power supplies are often more difficult to troubleshoot because of the feedback circuit used to regulate the power supply's output. The feedback circuit is a closed loop system. Breaking the loop is the most effective way to isolate a feedback problem. The closed loop in Fig. 16-32 consists of R_1, the feedback and isolation block, the PWM oscillator, and Q_1. Any one of these can cause the switching power supply to shut down or operate poorly. For example, if all outputs are low or all outputs are high, suspect a feedback loop problem. If the protection diode, D_1, continually blows, the feedback loop is the likely candidate. If only one or two output voltages are low, check the filter capacitors associated with those outputs. The switching power supply can emit EMI radiation (electromagnetic interference) into nearby communications gear or the circuits it is supplying, if not properly filtered. EMI filters protect other circuits from this interference by passing the interference to ground. Bad filters will let the power supply generate noise. Use the oscilloscope to monitor the outputs for noise, ripple, and unusual interference. The switching power supply must not be operated without a load connected to it or damage will most likely occur. As stressed in earlier troubleshooting sections, follow a logical troubleshooting plan when tracking down a power supply fault.

Troubleshooting a Traveling Wave Tube Amplifier

Some initial considerations include:

1. To make construction of the tube RF output circuit simple, the collector and helix are grounded. The cathode will be above ground and have a negative voltage on it (see Fig. 16-33).

2. The helix is delicate and helix current will be small. The collector draws nearly all the current. The helix will be protected with an over current relay that kills the power supplies. Never defeat this relay because the tube can be lost in almost zero time.

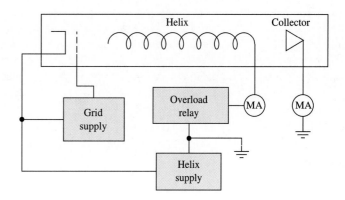

FIGURE 16-33 ■ TWT amplifier dc voltages.

3. Always operate the TWT amplifier into a good dummy load. High VSWR will cause high helix current. The RF input should also be terminated.

TYPICAL DC PROBLEMS

1. Low gain. Gain is a function of helix voltage. Make sure it is correct.

2. Amplifier will not stay on and overload relay trips. Check the overload trip point with an external power supply and milliampere meter as shown in Fig. 16-34. The relay contacts should open when the desired overload current is reached. The resistor may be a potentiometer. You will find the specification for helix current in the TWT manual.

    Check all power supply components with the power off. If you cannot find the problem, you may have to construct a bank of resistors to use in place of the tube to troubleshoot the power supplies while they are operating.

3. Excessive collector current. Make sure the grid supply is OK. Remember, both sides of the grid supply are above ground.

4. You cannot turn the helix supply on. Look for malfunctioning fault relays. TWTs usually have relays that turn the power supplies on in a sequence: heater first, grid next, and helix last.

5. Spurious modulation. Hum or ac ripple on the power supplies will modulate the RF output. The helix supply will be a regulated supply and should have almost no hum. Check this with an oscilloscope with a high-voltage probe.

6. Power output is low. Is too much RF drive being applied? Power output of a TWT increases as drive is increased until saturation is reached. If drive is increased past that point, power output falls off.

TYPICAL RF TROUBLES

1. Poor frequency response or low gain. Most TWTs have a RF bandpass filter in the input to correct the frequency response of the tube. Check the filter for excessive loss at the center frequency and for proper bandpass response. Generally, there should be a loss of several dB at the band edges and very little loss at the band center.

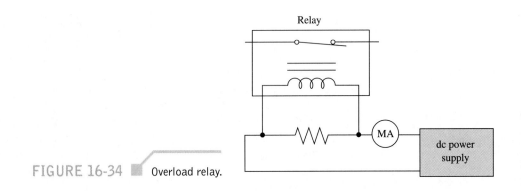

FIGURE 16-34 ▓ Overload relay.

2. Low RF output. At microwave frequencies, coaxial cables and connectors cause a lot of trouble. Assembly procedure is critical for proper operation. Look for loose shields at the connectors and check the cable for loss and VSWR.

Good TWTs will have an isolator in the output circuit. Isolators should show a loss of 1 dB or less in the forward direction and at least a 20-dB loss in the reverse direction.

These components are best checked with a network analyzer and a sweep oscillator. The test methods shown in Chapters 12, 14, and 15 will work too, but they yield less information. Unfortunately, network analyzers are very expensive, and at times you will have to do it the hard way.

SUMMARY

In Chapter 16 we studied microwaves and lasers. We learned that microwaves share many properties with light waves. The major topics the student should now understand include:

- the description and analysis of microwave antennas, including parabolic, horn, and lens varieties

- the calculation of power gain and beamwidth for parabolic antennas

- the description, operation, and application of magnetrons and traveling wave tubes (TWTs)

- the description of solid-state microwave devices, including the Gunn diode, IMPATT diode, *p-i-n* diode, transistors, and monolithic microwave integrated circuits (MMICs)

- the operation and use of ferrites as attenuators, isolators, and filters

- the applications of circulators and ferrite beads

- the description of low-noise amplification techniques using parametric amplifiers

- the description and application of lasers

- the description of laser sources, laser communications, and laser computers

QUESTIONS AND PROBLEMS

SECTION 1

1. Microwave antennas tend to be highly directive and provide high gain. Discuss the reasons for this.
2. What is a horn antenna? Provide sketches of three basic types, and explain their important characteristics.
*3. Describe how a radar beam is formed by a paraboloidal reflector.
4. With sketches, explain three different methods of feeding parabolic antennas.
5. A 160-ft-diameter parabolic antenna is driven by a 10-W transmitter at 4.3 GHz. Calculate its effective radiated power (ERP) and its beamwidth. (29.3 MW, 0.10°)
6. Calculate the minimum acceptable parabolic diameter for a signal at 1.5 GHz. Cal-

culate the resultant antenna gain and beamwidth. If the antenna diameter is now increased by a factor of 10, calculate the gain and beamwidth. (2 m, 600, 7°, 6 × 10^4, 0.7°)

7. A parabolic antenna has a 0.5° beamwidth at 18 GHz. Calculate its gain in dB. (50.7)

8. What is a radome? Explain why its use is often desirable in conjunction with parabolic antennas.

9. Explain the principles of a zoned lens antenna, including the transformation of a spherical wave into a plane wave.

SECTION 2

*10. Explain briefly the principle of operation of the magnetron.

*11. Draw a simple cross-sectional diagram of a magnetron, showing the anode, the cathode, and the direction of electron movement under the influence of a strong magnetic field.

12. Explain how electric and magnetic fields influence electron travel in a magnetron.

13. List and explain the differences between the two classes of magnetron oscillators.

*14. Draw a diagram showing the construction and explain the principles of operation of a traveling wave tube (TWT).

15. Describe four methods of coupling for a TWT.

16. Describe how a TWT can be used as an oscillator.

17. What are some practical applications for a TWT? What are some advantages of this device?

SECTION 3

18. In general, describe some advantages and disadvantages of solid-state microwave devices with respect to tube devices.

19. Describe the principle of operation of the Gunn diode. List some of its applications and its important characteristics.

20. What does IMPATT stand for? Describe the basic operation of the IMPATT diode.

21. Explain the two basic arguments leading to the existence of negative resistance in IMPATT diodes.

22. Compare the use of BJT and FET devices at microwave frequencies. Under what frequency-power output conditions would tubes be used instead?

23. Draw a sketch showing the construction of a p-i-n diode. Describe its operation at low frequencies and at microwave frequencies. List several possible applications for this device.

SECTION 4

24. Describe the composition of a ferrite material. Explain the process of precession as related to ferrite materials.

25. How does ferrite material placed in a waveguide provide attenuation? Explain why this attenuation is effective only at one specific frequency range. How can the frequencies attenuated be varied? Describe a method whereby the amount of attenuation can be varied.

26. What is an isolator? Describe how a ferrite waveguide isolator works.
27. What is a circulator? Draw a sketch of a ferrite Y circulator, and use it to explain the theory of operation.
28. What is a ferrite bead? How is it able to replace the function of a radio frequency choke?

SECTION 5

29. What is a parametric amplifier? In detail, discuss its fundamentals and explain how it differs from a traditional amplifier.
30. Explain the difference between degenerate and nondegenerate mode parametric amplifiers. List some applications for these amplifiers.
31. Calculate the output signal-to-noise ratio for a parametric amplifier with a 0.3-dB noise figure when the input signal-to-noise ratio is 7:1. Calculate the equivalent noise temperature of this amplifier. (6.53, 20.7°)
32. Sketch a ruby maser amplifier and explain its operation. Why is it necessary to maintain the ruby at extremely low temperatures? What side benefits can be obtained from the cooling?
33. What does the acronym *maser* stand for? Why is the ruby maser more useful in communications than the originally developed ammonia maser?

SECTION 6

34. What is a laser? How does it differ from a maser? List some applications of a laser.
35. Describe the construction of a stripe-geometry DH laser. Explain why temperature stabilization is critical to its operation.
36. Explain why direct laser communication schemes are not widely used.
37. Describe the operation of the transphasor. What are the expected advantages of optical computers?

FIBER OPTICS

OBJECTIVES

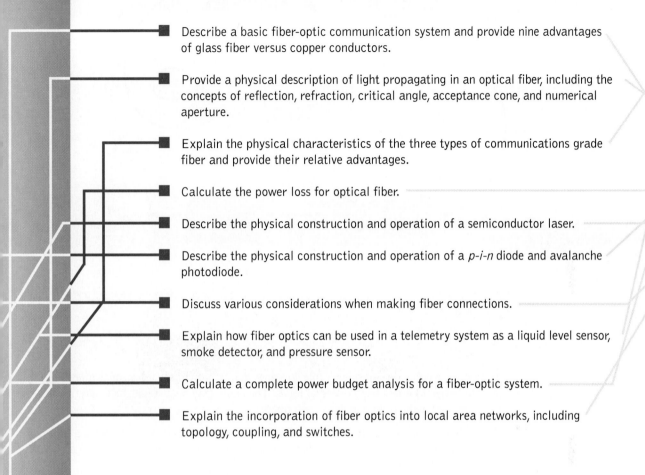

- Describe a basic fiber-optic communication system and provide nine advantages of glass fiber versus copper conductors.

- Provide a physical description of light propagating in an optical fiber, including the concepts of reflection, refraction, critical angle, acceptance cone, and numerical aperture.

- Explain the physical characteristics of the three types of communications grade fiber and provide their relative advantages.

- Calculate the power loss for optical fiber.

- Describe the physical construction and operation of a semiconductor laser.

- Describe the physical construction and operation of a *p-i-n* diode and avalanche photodiode.

- Discuss various considerations when making fiber connections.

- Explain how fiber optics can be used in a telemetry system as a liquid level sensor, smoke detector, and pressure sensor.

- Calculate a complete power budget analysis for a fiber-optic system.

- Explain the incorporation of fiber optics into local area networks, including topology, coupling, and switches.

17-1 INTRODUCTION

Recent advances in the development and manufacture of fiber-optic systems have made them the latest frontier in the field of communications. They are being used for both military and commercial data links and have replaced a lot of copper wire. Their use in telecommunications is extensive. They are also expected to take over much of the long-distance communication traffic now handled by satellite links.

A fiber-optic communications system is surprisingly simple, as shown in Fig. 17-1. It is comprised of the following elements:

Photo: 20-channel optical-fiber system transfers data at 150 Mbits/s per channel. (Courtesy of Motorola. Used by permission.)

655

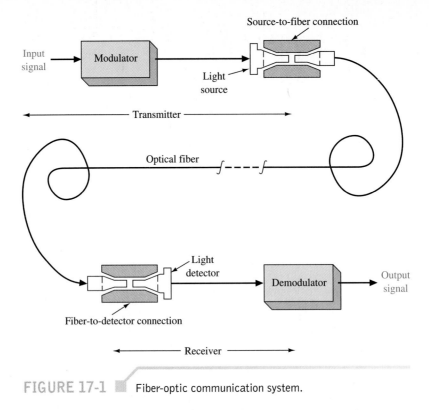

FIGURE 17-1 Fiber-optic communication system.

1. A fiber-optic transmission cable to carry the signal (in the form of a modulated light beam) a few feet or several miles. The cable may be a single, hair-like fiber or a small bundle of hundreds of such fibers.

2. A source of visible or invisible infrared radiation—usually a light-emitting diode (LED) or a solid-state laser—that can be modulated to impress data or an analog signal on the light beam.

3. A photosensitive detector to convert the optical signal back into an electrical signal at the receiver. The most often used detectors are *p-i-n* or avalanche photodiodes.

4. Efficient optical connectors at the light source-to-cable interface and at the cable-to-photodetector interface. These connectors are also critical when splicing of the optical cable is required due to excessive loss that can occur at connections.

5. Standard communications electronics prior to the light source and following the photodetector.

The optical waveguide propagates the light signal in a fashion similar to the standard metallic waveguide described in Chapter 14. The light wave travels down the glass fiber by constant reflection off its side walls. Thus, the trapping of light in a fiber results from the phenomenon of *total internal reflection* (TIR).

The advantages of optical communications links compared to waveguides or copper conductors are enormous. Included are the following:

1. *Substantially lighter weight and smaller size:* The U.S. Navy replaced conventional wiring on the A-7 airplane that transmitted data between a central computer and all

its remote sensors and peripheral avionics with an optical system. In this case, 224 ft of fiber optics weighing 1.52 lb replaced 1900 ft of copper wire weighing 30 lb.

2. *Immunity to electromagnetic interference (EMI):* External electrical noise does not affect energy at the frequency of light.

3. *Virtual elimination of crosstalk:* The light on one glass fiber does not interfere with light on an adjacent fiber. This is analogous to one standard waveguide in close proximity to another. Crosstalk can result from two adjacent copper wires, however.

4. *Lower signal attenuation than other propagation systems:* Typical attenuation figures of a 1-GHz bandwidth signal for optical fibers is 0.03 dB per 100 ft compared to 4.0 dB for both RG-58/U coaxial cable and an X-band waveguide.

5. *Extremely wide system bandwidths:* The intelligence is impressed by varying the light's amplitude. Since the best LEDs have a 5-ns response time, they provide a maximum bandwidth signal of about 100 MHz. Using laser light sources, however, bandwidths of up to 10 GHz are possible with a single glass fiber. The amount of information multiplexed on such a system is indeed staggering. The higher the carrier frequency in a communication system, the greater its potential signal bandwidth. Since fiber-optics systems have carriers at 10^{13} to 10^{14} Hz compared to radio frequencies of 10^6 to 10^9 Hz, signal bandwidths are potentially many times greater.

6. *Lower cost:* Optical-fiber costs are continuing to decline while the cost of copper is increasing. Many systems are now cheaper with fiber, and that trend is accelerating.

7. *Conservation of the earth's resources:* The world's supply of copper (and other good electrical conductors) is limited. The principal ingredient in glass is sand, and it is cheap and in virtually unlimited supply.

8. *Safety:* In many wired systems, the potential hazard of short circuits requires precautionary designs. Additionally, the dielectric nature of optic fibers eliminates the spark hazard.

9. *Corrosion:* Since glass is basically inert, the corrosive effects of certain environments are not a problem.

17-2 THE NATURE OF LIGHT

Before one can understand the propagation of light in a glass fiber, it is necessary to review some basics of light refraction and reflection. The speed of light in free space is 3×10^8 m/s but is reduced in other media. The reduction as light passes into denser material results in refraction of the light. Refraction causes the light wave to be bent, as shown in Fig. 17-2(a). The speed reduction and subsequent refraction is different for each wavelength, as shown in Fig. 17-2(b). The visible light striking the prism causes refraction at both air/glass interfaces and separates the light into its various frequencies (colors) as shown. This same effect produces a rainbow, with water droplets acting as prisms to split the sunlight into the visible spectrum of colors (the various frequencies).

The amount of bend provided by refraction depends on the *refractive index* of the two materials involved. The refractive index, *n,* is the ratio of the speed of light in free space to the speed in a given material. It is slightly variable for different frequencies of light, but for most purposes a single value is accurate enough. The refractive index for free space (a vacuum) is 1.0, while air is 1.0003, water is 1.33, and glass is 1.5.

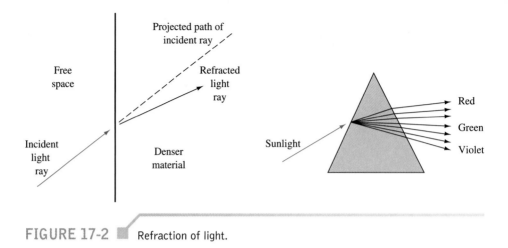

FIGURE 17-2 ■ Refraction of light.

Snell's law predicts the refraction that takes place when light is transmitted between two different materials:

$$n_1 \sin \theta_1 = n_2 \sin \theta_2 \qquad (13\text{-}6)$$

This effect was shown in Fig. 13-4. Figure 17-3 shows the case where an incident ray is at an angle such that the refracted ray goes along the interface, so that θ_2 is 90°. When θ_2 is 90°, the angle θ_1 is at the critical angle (θ_c) and defines the angle at which the incident rays no longer pass through the interface. When θ_1 is equal to or greater than θ_c, all the incident light is reflected and the angle of the incidence equals the angle of reflection, as shown in Fig. 13-4. This characteristic of total reflection for angles greater than the critical angle is of basic importance in fiber-optic communication.

The frequency of visible light ranges from about 4.4×10^{14} Hz for red up to 7×10^{14} Hz for violet.

EXAMPLE 17-1

Calculate the wavelengths of red and violet light.

SOLUTION

For red,

$$\lambda = \frac{c}{f}$$

$$= \frac{3 \times 10^8 \text{ m/s}}{4.4 \times 10^{14} \text{ Hz}} = 6.8 \times 10^{-7} \text{ m}$$

$$= 0.68 \ \mu\text{m or } 0.68 \text{ micron or } 680 \text{ nm}$$

For violet,

$$\lambda = \frac{3 \times 10^8 \text{ m/s}}{7 \times 10^{14}} = 0.43 \text{ micron or } 430 \text{ nm}$$

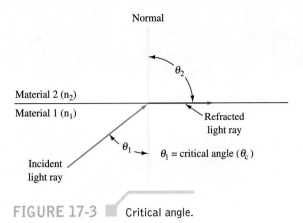

Normal

Material 2 (n$_2$)

Material 1 (n$_1$)

θ_2

Refracted
light ray

θ_1

Incident
light ray

θ_1 = critical angle (θ_c)

FIGURE 17-3 ▨ Critical angle.

The electromagnetic waves just below visible light in frequency are the infrared waves. They are increasingly used in fiber-optic communication schemes. The infrareds of interest to us extend up to wavelengths of about 1600 nm. The wavelengths used in fiber optics range from 600 to 1600 nm. The use of invisible infrared light (above 680 nm) is most common in today's systems.

The amount of energy (U) contained in one photon of light is proportional to frequency.

$$U = hf = \frac{hc}{\lambda} \qquad (17\text{-}1)$$

where h = Planck's constant
$= 6.63 \times 10^{-34}$ J $-$ s

Power is the energy divided by time.

EXAMPLE 17-2

Calculate the energy of one photon for infrared light energy at 1.55 μm.
Calculate the power in a 1-ms pulse of this energy.

SOLUTION

$$U = \frac{hc}{\lambda} \qquad (17\text{-}1)$$

$$= \frac{6.3 \times 10^{-34} \text{ J} - \text{s} \times 3 \times 10^8 \text{ m/s}}{1.55 \times 10^{-6} \text{ m}}$$

$$= 1.22 \times 10^{-19} \text{ J}$$

$$P = \frac{U}{t} = \frac{1.22 \times 10^{-19} \text{ J}}{1 \times 10^{-3} \text{ s}} = 1.22 \times 10^{-16} \text{ W}$$

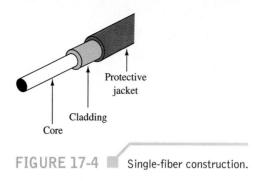

FIGURE 17-4 Single-fiber construction.

Typical construction of an optical fiber is shown in Fig. 17-4. The central core is the portion that carries the transmitted light. In most cases it is glass, but occasionally it is plastic. The cladding is usually glass, but plastic cladding of a glass fiber is not uncommon. In any event, the refraction index for the core and cladding are different. If they are both glass, their manufacturing processes are varied to provide the desired difference. To provide protection, some sort of rubber or plastic jacket may surround the cladding, as shown in Fig. 17-4. This type of fiber is termed the *step-index* variety. "Step index" refers to the abrupt change in refractive index from core to clad. Fibers that include a protective covering are called *cables*. Glass fibers are not rigid as you might expect. A single fiber and its cladding can typically be wound around a pencil without damage. They also have a large tensile strength, greater than equivalently sized steel wire.

Applying the concept of total reflection, propagation of light down the multimode fiber is shown in Fig. 17-5(a). Propagation results from the continuous reflection at the core/clad interface so that the ray "bounces" down the fiber length by the process of total internal reflection (TIR). If we consider point P in Fig. 17-5(a), the critical angle value for θ_3 is, from Snell's law,

$$\theta_c = \theta_3(\min) = \sin^{-1}\frac{n_2}{n_1}$$

Since θ_2 is the complement of θ_3,

$$\theta_2(\max) = \sin^{-1}\frac{\left(n_1^2 - n_2^2\right)^{1/2}}{n_1}$$

Now applying Snell's law at the entrance surface and since $n_{air} \simeq 1$, we obtain

$$\sin \theta_{in}(\max) = n_1 \sin \theta_2(\max)$$

Combining the two preceding equations yields

$$\sin \theta_{in}(\max) = \sqrt{n_1^2 - n_2^2} \tag{17-2}$$

Therefore, $\theta_{in}(\max)$ is the largest angle with the core axis that allows propagation via total internal reflection. Light entering the cable at larger angles will be refracted through the core/clad interface and lost. The value $\sin \theta_{in}(\max)$ is called the *numerical aperture* (NA) and defines the half-angle of the cone of acceptance for propagated light in the fiber. This is shown in Fig. 17-5(b). The preceding analysis might lead you to think

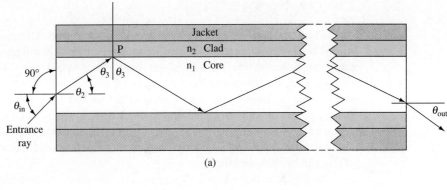

(a)

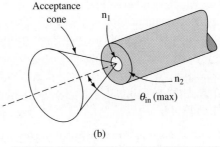

(b)

FIGURE 17-5 (a) Development of numerical aperture; (b) acceptance cone.

that crossing over θ_{in}(max) causes an abrupt end of light propagation. In practice, however, this is not true; thus, fiber manufacturers usually specify NA as the acceptance angle where the output light is no greater than 10 dB down from the peak value. The NA is a basic specification of a fiber provided by the manufacturer that indicates its ability to accept light and shows how much light can be off-axis and still propagated.

EXAMPLE 17-3

An optical fiber and its cladding have refractive indexes of 1.535 and 1.490, respectively. Calculate NA and θ_{in}(max).

SOLUTION

$$\text{NA} = \sin \theta_{in}(\text{max}) = \sqrt{n_1^2 - n_2^2} \qquad (17\text{-}2)$$

$$= \sqrt{(1.535)^2 - (1.49)^2} = 0.369$$

$$\theta_{in}(\text{max}) = \sin^{-1} 0.369$$

$$= 21.7°$$

17-3 OPTICAL FIBERS

Several types of optical fibers are available, with significant differences in their characteristics. The first communication-grade fibers (early 1970s) had light-carrying core diameters about equal to the wavelength of light. They could carry light in just a single waveguide mode. The difficulty of coupling significant light into such a small fiber led to development of fibers with cores of about 20 to 100 μm. These fibers support many waveguide modes and are called *multimode* fibers. The first commercial fiber-optic systems used multimode fibers with light at 0.8- to 0.9-μm wavelengths. A variation of the multimode fiber was subsequently developed, termed graded-index fiber. This afforded greater bandwidth capability.

As the technology became more mature, the single-mode fibers were found to provide lower losses and even higher bandwidth. This has led to their use at 1.3 and 1.55 μm in many telecommunication applications. The new developments have not made old types of fiber obsolete. The application now determines the type used. The following major criteria affect the choice of fiber type:

1. Signal losses
2. Ease of light coupling and interconnection
3. Bandwidth

A fiber showing three different modes (i.e., multimode) of propagation is presented in Fig. 17-6. The lowest-order mode is seen traveling along the axis of the fiber, and the middle-order mode is reflected twice at the interface. The higher-order mode is reflected many times and makes many trips across the fiber. As a result of these variable path lengths, the light entering the fiber takes a variable length of time to reach the detector. This results in a pulse-stretching characteristic, as shown in Fig. 17-6. This effect is termed *pulse dispersion* and limits the maximum rate at which data (pulses of light) can be practically transmitted. You will also note that the output pulse has reduced amplitude as well as increased width. The greater the fiber length, the worse this effect will be. As a result, manufacturers rate their fiber in bandwidth per length, such as 400 MHz/km. That cable can successfully transmit pulses at the rate of 400 MHz for 1 km, 200 MHz for 2 km, and so on. Of course, longer transmission paths are attained by locating repeaters at appropriate locations.

Step-index multimode fibers in common use have core diameters from about 50 to 100 μm. An often used configuration has a 50-μm core and 125-μm cladding. The large core diameter and high NA of these fibers simplifies input coupling and allows the use of relatively inexpensive connectors. Fibers are often specified by the diameters of their core and cladding. For example, the fiber just described would be called 50/125 fiber.

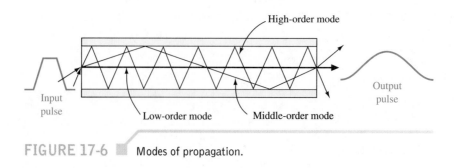

FIGURE 17-6 ▮ Modes of propagation.

An alternative to glass multimode fibers are all-plastic designs. It should be noted that some glass fibers do use plastic cladding. The all-plastic fibers are inexpensive and easy to handle but have very high losses—about 100 dB/km. This limits their use to very short distance systems, such as transmitting signals throughout the interior of an automobile.

A technique used to minimize pulse dispersion effects is to make the core extremely small—on the order of a few micrometers. This type accepts only the lowest-order modes, thereby allowing operation in high-data-rate, long-distance systems. This fiber is quite expensive and requires high-power, highly directional modulated light sources such as a laser. Fiber cables of this variety are called *single-mode* or monomode fibers. Core diameters of only 5 μm are typical.

Graded-Index Fiber

In an effort to overcome the pulse dispersion problem, the *graded-index fiber* was developed. In the manufacturing process for this fiber, the index of refraction is tailored to follow the parabolic profile shown in Fig. 17-7. This results in low-order modes traveling through the constant-density material in the center. High-order modes see lower index of refraction material farther from the core axis, and thus the velocity of propagation increases away from the center. Therefore, all modes, even though they take various paths and travel different distances, tend to traverse the fiber length in about the same amount of time. These cables can therefore handle higher bandwidths and/or provide longer lengths of transmission before pulse dispersion effects destroy intelligibility.

Graded-index multimode fibers with 50-μm diameter cores and 125-μm cladding are used in many telecommunication systems at up to 300 megabits per second over 50-km ranges without repeaters. Graded-index fiber with up to 100-μm core is used in short-distance applications that require easy coupling from the source and high data rates, such as video and high-speed local area networks. The larger core affords better light coupling than the 50-μm core and does not significantly degrade the bandwidth capabilities.

Single-Mode Fibers

The single-mode fiber, by definition, carries light using a single waveguide mode. A single-mode fiber will transmit a single mode for all wavelengths longer than the cutoff wavelength λ_c.

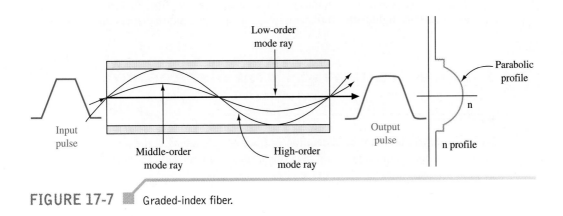

FIGURE 17-7 Graded-index fiber.

$$\lambda_c = \frac{2\pi a n_1 \sqrt{2\Delta}}{2.405} \tag{17-3}$$

where $\Delta = \frac{n_1 - n_2}{n_1}$

$\quad a =$ core radius

At wavelengths shorter than predicted by Eq. (17-3), the fiber supports two or more modes. The core of single-mode fiber must be small, about a few times the cutoff wavelength or several micrometers for operation at the standard 1.3 or 1.55 μm.

EXAMPLE 17-4

Determine the cutoff wavelength for a fiber with a 3-μm-diameter core. The core and cladding indexes of refraction are 1.545 and 1.510, respectively.

SOLUTION

$$\lambda = \frac{2\pi a n_1 \sqrt{2\Delta}}{2.405}$$

$$a = \frac{3 \ \mu m}{2} = 1.5 \ \mu m$$

$$\Delta = \frac{n_1 - n_2}{n_1} = \frac{1.545 - 1.510}{1.545} = 0.023 \tag{17-3}$$

$$\lambda_c = \frac{2\pi \times 1.5 \times 10^{-6} \times 1.545\sqrt{2 \times 0.023}}{2.405}$$

$$= 1.29 \ \mu m$$

Single-mode fibers are widely used in long-haul telecommunications. They permit transmission of about 1 Gb/s and repeater spacing of up to 500 km. These bandwidth and repeater spacing capabilities are constantly being upgraded by new developments.

Figure 17-8 provides a summary of the three types of fiber discussed, including typical dimensions, light paths, refractive index profiles, and pulse dispersion effects.

17-4 FIBER ATTENUATION AND DISPERSION

Attenuation

The single most important consideration in fiber-optic systems is the loss introduced by the fiber. An optical pulse propagating along a fiber is attenuated exponentially. The optical power (P) at a distance (l) from the power transmitted (P_T) is

$$P = P_T \times 10^{-Al/10} \tag{17-4}$$

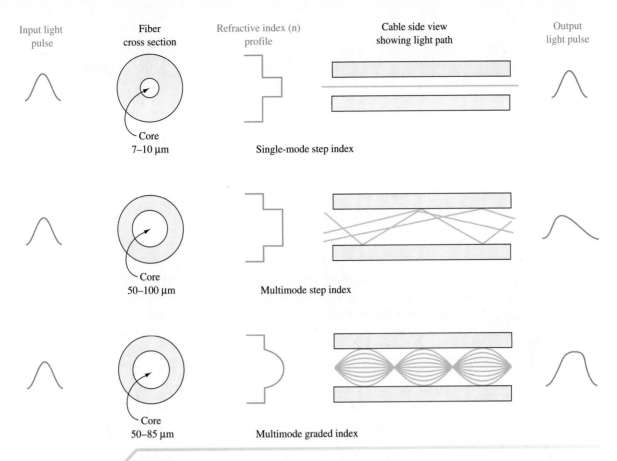

Input light pulse | Fiber cross section | Refractive index (n) profile | Cable side view showing light path | Output light pulse

Core
7–10 μm

Single-mode step index

Core
50–100 μm

Multimode step index

Core
50–85 μm

Multimode graded index

FIGURE 17-8 Types of optical fiber.

where A is the fiber attenuation in dB/km. The lowest attenuation currently available is about 0.1 dB/km for single-mode fiber at 1.55 μm. In general, multimode fibers tend to have higher loss than single mode because of increased scattering from dopants in the fiber core.

EXAMPLE 17-5

Calculate the optical power 50 km from a 0.1-mW source on a single-mode fiber that has 0.25-dB/km loss.

SOLUTION

$$P = P_T \times 10^{-Al/10}$$
$$= 0.1 \times 10^{-3} \times 10^{-(0.25 \times 50)/10} \qquad (17\text{-}4)$$
$$= 0.1 \times 10^{-3} \times 10^{-1.25}$$
$$= 5.62 \ \mu\text{W}$$

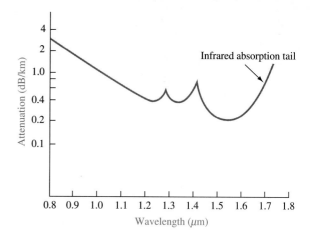

FIGURE 17-9 ▨ Attenuation versus wavelength for typical high-quality fiber.

All transparent materials scatter light because of microscopic density fluctuations (nonuniformities). The most familiar example of this effect is the scattering of sunlight by dust particles in the atmosphere, which yields our "blue" sky. The scattering in a fiber decreases rapidly with increasing wavelength, as shown in Fig. 17-9. It is termed *Rayleigh scattering* and is inversely proportional to the wavelength's fourth power. At longer wavelengths the attenuation increases rapidly due to absorption from the tails of infrared resonances in silica and other fiber constituents.

Notice the attenuation peaks at 1.25 and 1.39 μm in Fig. 17-9. They are caused by minute quantities of water trapped in the glass. These hydroxyl (OH⁻) ions can be minimized in the manufacturing cycle, but it is simpler and less costly to avoid using the wavelengths most affected.

A comparison of the attenuation versus frequency characteristics for various "guiding" media is shown in Fig. 17-10. The very low loss capabilities of glass are ably shown, especially considering that both scales are logarithmic. Glass has the lowest loss and highest bandwidth capability—and also the smallest physical size!

Dispersion

As mentioned previously, dispersion is a particular problem for multimode fiber. The different paths taken by the various propagation modes is the basis of this dispersion. This is called *modal dispersion*. Dispersion is also a factor to consider with single-mode fibers.

Modal dispersion in step-index multimode fibers limits the bandwidth (and bit rate) to about 20 MHz per kilometer of length. The use of graded-index fiber greatly increases this capability to about 1 GHz/km (see Fig. 17-7). Single-mode fiber does not exhibit modal dispersion since only a single mode is transmitted. Two other types of dispersion do limit single-mode fiber bit-rate-distance capability: material and waveguide dispersion. The resultant of these two provides the total dispersion of single-mode fiber.

Material dispersion is caused by the slight variation of refractive index with wavelength for glass. This is the same effect that causes a prism to create the color spectrum. This results in pulse spreading since the light source is not just a single frequency of light. The various Fourier components of a signal thus exhibit slightly different transit times through the fiber.

Waveguide dispersion is caused by a portion of the light energy traveling in the cladding. Typically, 20% of the energy is contained within the cladding. Since the

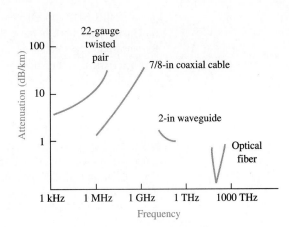

FIGURE 17-10 ▨ Attenuation comparison. (From *IEEE Communications Magazine,* May 1985, © 1985 IEEE, "Introduction to lightwave transmission" by Paul S. Henry.)

cladding has a lower refractive index than the core, velocity variations result that cause pulse dispersion. Waveguide dispersion is wavelength dependent and can have different polarity than material dispersion. They can completely cancel each other out. This occurs near 1.3 μm and is called the *zero-dispersion* wavelength. The combined effects of these two dispersions become significant at the other commonly used wavelength for single-mode fiber, 1.55 μm. Recall, however, that attenuation is significantly lower at 1.55 μm (Fig. 17-9). Current fiber design is taking place to shift the zero-dispersion wavelength up to 1.55 μm. This appears possible by using a complex graded-index fiber.

17-5 LIGHT SOURCES

Two kinds of light sources are used in fiber-optic communication systems: the diode laser (DL) and the high-radiance light-emitting diode (LED). In designing the optimum system, the special qualities of each light source should be considered. Diode lasers and LEDs bring to systems different characteristics:

1. Power levels
2. Temperature sensitivities
3. Response times
4. Lifetimes
5. Characteristics of failure

The diode laser is a preferred source for moderate-band to wideband systems. It offers a fast response time (typically less than 1 ns) and can couple high levels of useful optical power (usually several mW) into an optical fiber with a small core and a small numerical aperture. Recent advances in DL fabrication have resulted in predicted lifetimes of 10^5 to 10^6 hours at room temperature. Earlier DLs were of such limited life as to minimize their use. The DL is usually used as the source for single-mode fiber since LEDs have a low input coupling efficiency.

Some systems operate at a slower bit rate and require more modest levels of fiber-coupled optical power (50 to 250 μW). These applications allow the use of high-radiance LEDs. The LED is cheaper, requires less complex driving circuitry than a DL, and needs no thermal or optical stabilizations. In addition, LEDs have longer operating lives (10^6 to 10^7 h) and fail in a more gradual and predictable fashion than DLs.

Both LEDs and DLs are multilayer devices most frequently fabricated of AlGaAs on GaAs. They both behave electrically as diodes, but their light-emission properties differ substantially. A DL is an optical oscillator; hence it has many typical oscillator characteristics: a threshold of oscillation, a narrow emission bandwidth, a temperature coefficient of threshold and frequency, modulation nonlinearities, and regions of instability.

The light output wavelength spread, or spectrum, of the DL is much narrower than that of LEDs: about 1 nm compared with about 40 nm for an LED. Narrow spectra are advantageous in systems with high bit rates since the dispersion effects of the fiber on pulse width are reduced, and thus pulse degradation over long distances is minimized.

Light is emitted from an LED as a result of the recombining of electrons and holes. Electrically, an LED is a *pn* junction. Under forward bias, minority carriers are injected across the junction. Once across they recombine with majority carriers and give up their energy. The energy given up is about equal to the material's energy gap. This process is radiative for some materials (such as GaAs) but not so for others, such as silicon. LEDs have a distribution of nonradiative sites—usually crystal lattice defects, impurities, and so on. These sites develop over time and explain the finite life/gradual deterioration of light output.

The complete specifications for an LED are provided in Fig. 17-11. The MFOE 3100 series is designed for high-power, medium-response time applications. These LEDs exhibit a typical numerical aperture (NA) of 0.3. They provide a digital data rate of up to 40 M baud and an analog bandwidth of typically 20 MHz.

Figure 17-12 shows the construction of a typical semiconductor laser used in fiber-optic systems. A variation of this laser, the stripe laser, was described in Sec. 16-6. The semiconductor laser uses the properties of the junction between heavily doped layers of *p*- and *n*-type materials. When a large forward bias is applied, a large number of free holes and electrons are created in the immediate vicinity of the junction. When a hole and electron pair collide and recombine, they produce a photon of light. The *pn* junction in Fig. 17-12 is sandwiched between layers of material with different optical and dielectric properties. The material that shields the junction is typically aluminum gallium arsenide, which has a lower index of refraction than gallium arsenide. This difference "traps" the holes and electrons in the junction region and thereby improves light output. When a certain level of current is reached, the population of minority carriers on either side of the junction increases, and photon density becomes so high that they begin to collide with already excited minority carriers. This causes a slight increase in the ionization energy level, which makes the carrier unstable. It thus recombines with a carrier of the opposite type at a slightly higher level than if no collision had occurred. When it does, two equal-energy photons are released.

The carriers that are "stimulated" (remember, laser is an acronym for *l*ight *a*mplification by *s*timulated *e*mission of *r*adiation) as described in the preceding paragraph may reach a density level such that each released photon may trigger several more. This creates an avalanche effect that increases the emission efficiency exponentially with current above the initial emission threshold value. This behavior is usually enhanced by placing mirrored surfaces at each end of the junction zone. These mirrors are parallel, so generated light bounces back and forth several times before escaping. The mirrored surface where light emits is partially transmissive (e.g., partially reflective).

The laser diode functions as an LED until its threshold current is reached. At that point, the light output becomes coherent (spectrally pure), and the output power starts increasing rapidly with increases in forward current. This effect is shown in Fig. 17-13. The typical spectral purity of these lasers yields a line width of about 1 nm versus about

Fiber Optics — MOD Family Infrared LED

MFOE3100
MFOE3101
MFOE3102

. . . designed for fiber optics applications requiring high power and medium-response time. It is spectrally matched to the first window minimum attenuation region of most glass-core fiber optics cables. Motorola's package fits directly into standard fiber optics connector systems. Applications include computer links and industrial controls.

MOD FAMILY FIBER OPTICS INFRARED LED

- Medium Response — Digital Data to 40 Mbaud (NRZ) Typ
- Analog Bandwidth — 20 MHz Typ
- Plastic Package — Small, Rugged and Inexpensive
- Internal Lensing Enhances Coupling Efficiency
- Complements All Motorola Fiber Optics Detectors
- Mates snugly with AMP #228756-1, Amphenol #905-138-5001, OFTI #PCR001 Receptacles
- Low Cost

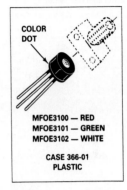

COLOR DOT

MFOE3100 — RED
MFOE3101 — GREEN
MFOE3102 — WHITE

CASE 366-01
PLASTIC

5

MAXIMUM RATINGS

Rating	Symbol	Value	Unit
Reverse Current	I_R	1	mA
Forward Current — Continuous	I_F	60	mA
Total Device Dissipation @ T_A = 25°C Derate above 25°C	P_D	250* 2.63*	mW mW/°C
Operating Temperature Range	T_J	−40 to +100	°C
Storage Temperature Range	T_{stg}	−40 to +100	°C

THERMAL CHARACTERISTICS

Characteristic	Symbol	Max	Unit
Thermal Resistance, Junction to Ambient	θ_{JA}	465 300*	°C/W

ELECTRICAL CHARACTERISTICS (T_A = 25°C)

Characteristic	Symbol	Min	Typ	Max	Unit
Reverse Breakdown Voltage (I_R = 100 μA)	$V_{(BR)R}$	2	8	—	Volts
Forward Voltage (I_F = 50 mA)	V_F	1.5	1.7	2	Volts
Total Capacitance (V_R = 0 V, f = 1 MHz)	C_T	—	70	—	pF
LED Bandwidth, Figure 8 (I_F[DC] = 40 mA, I_F[MOD] = 40 mA p-p)	BWE	—	20	—	MHz

*Installed in compatible metal connector housing.

FIGURE 17-11 ▓ Specifications for MFOE 3100 series LED. (Courtesy of Motorola, Inc.)

MFOE3100, MFOE3101, MFOE3102

OPTICAL CHARACTERISTICS (T$_A$ = 25°C)

Characteristic		Symbol	Min	Typ	Max	Unit
Total Power Output, Figure 2 (I$_F$ = 50 mA, λ ≈ 850 nm)	MFOE3100 MFOE3101 MFOE3102	P$_O$	— — —	850 1650 2200	— — —	μW
Power Launched, Figure 6 (I$_F$ = 50 mA)	MFOE3100 MFOE3101 MFOE3102	P$_L$	10(−20) 50(−13) 80(−11)	— — —	— 100(−10) 160(−8)	μW(dBm)
Numerical Aperture of Output Port (at −10 dB, 250 μm [10 mil] diameter spot), Figure 10		NA	—	0.3	—	—
Wavelength of Peak Emission @ 50 mAdc		λ	—	850	—	nm
Spectral Line Half Width		—	—	50	—	nm
Optical Rise and Fall Times, Figure 7 (I$_F$ = 50 mAdc)		t$_r$	—	19	—	ns
		t$_f$	—	14	—	

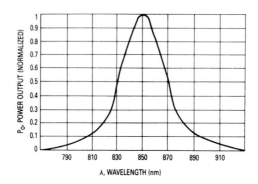

Figure 1. Relative Spectral Output

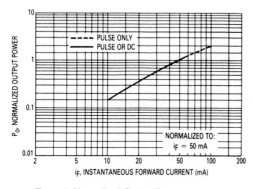

Figure 2. Normalized Output Power versus Forward Current

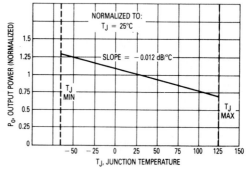

Figure 3. Power Output versus Junction Temperature

FIGURE 17-11 ■ *(Continued)*

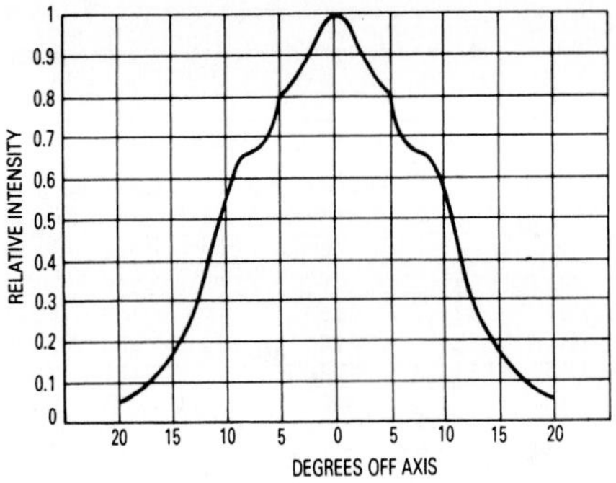

MFOE3100, MFOE3101, MFOE3102

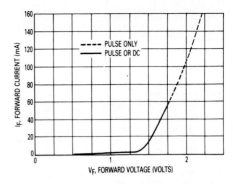

Figure 4. Radial Intensity Distribution

Figure 5. Forward Current versus Forward Voltage

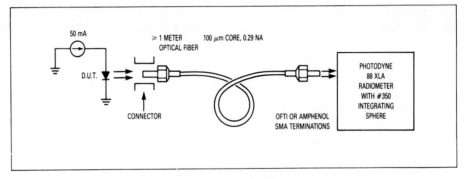

Figure 6. Coupling Efficiency

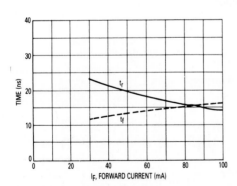

Figure 7. Rise Time and Fall Time versus
Forward Current

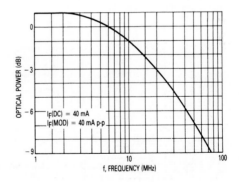

Figure 8. Typical LED Bandwidth

FIGURE 17-11 ■ *(Continued)*

MFOE3100, MFOE3101, MFOE3102

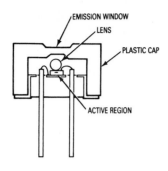

EMISSION WINDOW
LENS
PLASTIC CAP
ACTIVE REGION

CAUTION: Avoid contact with the plastic cap by organic solvents. If contaminated, clean with mild detergent and water.

Figure 9. Package Cross Section

AVERAGE COUPLING EFFICIENCY		
Fiber Core Diameter (μm)	Numerical Aperture	Coupling Efficiency (%)
1000	0.5	67
200	0.4	28
100	0.29	4.5
85	0.26	2.6
62.5	0.28	1.6
50	0.2	0.7

Figure 10. Coupling Efficiency

I_F	R_L
10 mA	330
50 mA	68

NONINVERTING

I_F	R_L
10 mA	330
50 mA	68

INVERTING

Figure 11. TTL Transmitters

Q1, Q2 2N4401
U1 MC3302(1/4)

Figure 12. 1 MHz PIN Receiver

5

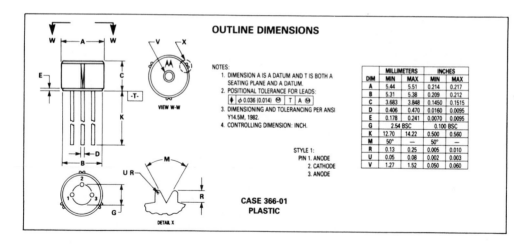

OUTLINE DIMENSIONS

VIEW W-W

NOTES:
1. DIMENSION A IS A DATUM AND T IS BOTH A SEATING PLANE AND A DATUM.
2. POSITIONAL TOLERANCE FOR LEADS:
   [⊕ | φ 0.036 (0.014) Ⓜ | T | A Ⓜ]
3. DIMENSIONING AND TOLERANCING PER ANSI Y14.5M, 1982.
4. CONTROLLING DIMENSION: INCH.

STYLE 1:
PIN 1. ANODE
2. CATHODE
3. ANODE

DETAIL X

**CASE 366-01
PLASTIC**

DIM	MILLIMETERS		INCHES	
	MIN	MAX	MIN	MAX
A	5.44	5.51	0.214	0.217
B	5.31	5.38	0.209	0.212
C	3.683	3.848	0.1450	0.1515
D	0.406	0.470	0.0160	0.0095
E	0.178	0.241	0.0070	0.0095
G	2.54 BSC		0.100 BSC	
K	12.70	14.22	0.500	0.560
M	50°	—	50°	—
R	0.13	0.25	0.005	0.010
U	0.05	0.08	0.002	0.003
V	1.27	1.52	0.050	0.060

FIGURE 17-11 ▦ *(Continued)*

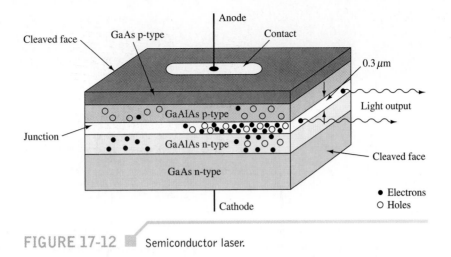

FIGURE 17-12 Semiconductor laser.

40 nm for LED sources. Recall that this is critical to minimize pulse dispersion. The wavelength of light generated is determined by the materials used. The "short-wavelength" lasers at 0.78 to 0.9 μm use gallium arsenide (GaAs) and aluminum gallium arsenide (AlGaAs). "Long-wavelength" (infrared) devices at 1.3 to 1.67 μm are made of layers of indium gallium arsenide phosphide (InGaAsP) and indium phosphide (InP).

17-6 DETECTORS

The devices used to convert the transmitted light back into an electrical signal are a vital link in a fiber-optic system. This important link in a fiber-optic communication system is often overlooked in favor of the light source and fibers. However, simply changing from one photodetector to another can increase the capacity of a system by an order of magnitude. Because of this, current research is accelerating to allow production of improved detectors. For most applications, the detector used is a *p-i-n* diode. Chapter 16 provided detail on *p-i-n* diodes used as microwave switches. The avalanche photodiode is also used in photodetector applications.

Just as a *pn* junction can be used to generate light, it can also be used to detect light. When a *pn* junction is reversed-biased and under dark conditions, very little current flows through it. This is termed the *dark current*. However, when light shines on the

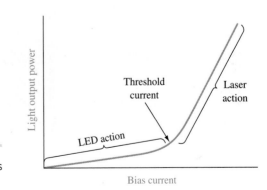

FIGURE 17-13 Light output versus bias current for a laser diode.

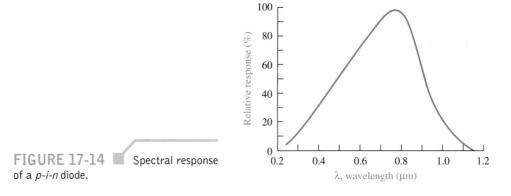

FIGURE 17-14 ▇ Spectral response of a *p-i-n* diode.

device, photon energy is absorbed and hole–electron pairs are created. If the carriers are created in or near the junction depletion region, they are swept across the junction by the electric field. This movement of charge carriers across the junction causes a current flow in the circuitry external to the diode and is proportional to the light power absorbed by the diode.

The important characteristics of light detectors are:

1. *Responsivity:* This is a measure of output current for a given light power launched into the diode. It is given in amperes per watt at a particular wavelength of light.

2. *Dark current:* This is the thermally generated reverse leakage current (under dark conditions) in the diode. In conjunction with the response current as predicted by device responsivity and incident power, it provides an indication of on–off detector output range.

3. *Response speed:* This determines the maximum data-rate capability of the detector.

4. *Spectral response:* This determines the responsivity that is achieved relative to the wavelength at which responsivity is specified. Figure 17-14 provides a spectral response versus light wavelength for a typical *p-i-n* photodiode. The curve shows that its relative response at 900 nm (0.9 *μm*) is about 80% of its peak response at 800 nm.

Figure 17-15 shows the construction of a *p-i-n* diode used as a photodetector. As mentioned previously, light falling on a reverse-biased *pn* junction produces hole–electron

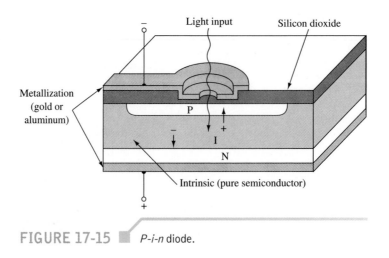

FIGURE 17-15 ▇ *P-i-n* diode.

pairs. The ability of a generated hole–electron pair to contribute to current flow depends on the hole and electron being rapidly separated from each other before they collide and cancel each other out. The reverse-biased diode creates a depletion region at the *pn* junction. The reverse-biased junction can be thought of as a capacitor, with the depletion region as the dielectric. The hole and electron created in the depletion region are rapidly pulled apart by the *p* and *n* materials that act as the capacitor's plates. Widening the depletion region gives more opportunity for hole–electron pairs to form and thus enhances the photodetector operation. The intrinsic (*i*) layer of the *p-i-n* diode in Fig. 17-15 performs that function. The intrinsic layer is a very lightly doped semiconductor material.

The operation of the avalanche photodiode is illustrated in Fig. 17-16. The diode is operated at a reverse voltage near the breakdown of the junction. At that potential, the electrons can be pulled from the atomic structure. With a small amount of additional energy, electrons are dislodged from their orbits, producing free electrons and resulting holes. As shown in Fig. 17-16, a photon of light incident on the junction generates a hole–electron pair in the depletion region. Due to the large electric field, the electron movement is accelerated and they collide with other bound electrons. This creates additional hole–electron pairs that are also accelerated. This produces still more hole–electron pairs, and an avalanche multiplication process (gain) occurs. One electron may produce up to 100 electrons in the avalanche photodiode. The avalanche photodiode is 5 to 7 dB more sensitive than the *p-i-n* diode. This advantage is maintained except when extremely high data rates exceeding 4 Gb/s are experienced. In these cases the better frequency response of the *p-i-n* diode favors its use.

It should be noted that a second role for light detectors in fiber-optic systems exists. Detectors are used to monitor the output of laser diode sources. A detector is placed in proximity to the laser's light output. The generated photocurrent is used in a circuit to maintain the laser's light output constant under varying temperature and bias conditions. This is necessary to keep the laser just above its threshold forward bias current and to enhance its lifetime by not allowing the output to increase to higher levels. Additionally, the receiver does not want to see the varying light levels of a noncompensated laser.

The output current of photodiodes is at a very low level—on the order of 10 nA up to 10 μA. As a result, the noise benefits of fiber optics can be lost at the receiver connection between diode and amplifier. Proper design and shielding can minimize that problem, but an alternative solution is to integrate the first stage of amplification into the same circuit as the photodiode. These integrations are termed *integrated detector preamplifiers* (IDP) and provide outputs that can drive TTL logic circuits directly.

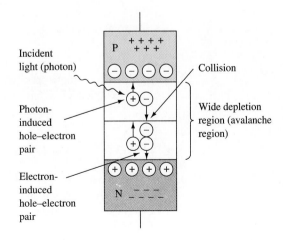

Incident light (photon)

Photon-induced hole–electron pair

Electron-induced hole–electron pair

Collision

Wide depletion region (avalanche region)

FIGURE 17-16 Avalanche photodiode.

MOTOROLA
■ SEMICONDUCTOR ■
TECHNICAL DATA

Fiber Optics — MOD Family
Photo Detector
Diode Output

MFOD3100

**MOD FAMILY
FIBER OPTICS
PHOTO DETECTOR
DIODE OUTPUT**

. . . designed for low cost infrared radiation detection in high frequency Fiber Optics Systems. Motorola's package fits directly into standard fiber optics connectors. Metal connectors provide excellent RFI immunity. Major applications are: CATV, video systems, M68000 microprocessor systems, industrial controls, computer and peripheral equipment, etc.

- Fast Response — 5 ns Max @ 5 Volts
- Analog Bandwidth (−3 dB) Greater Than 100 MHz
- Performance Matched to Motorola Fiber Optics Emitters
- Plastic Package — Small, Rugged and Inexpensive
- Mates snugly with AMP #228756-1, Amphenol #905-138-5001, OFTI #PCR001
  Receptacles
- Low Cost

BLUE COLOR DOT

**CASE 366-01
PLASTIC**

MAXIMUM RATINGS (T_A = 25°C unless otherwise noted)

Rating	Symbol	Value	Unit
Reverse Voltage	V_R	50	Volts
Total Device Dissipation @ T_A = 25°C Derate above 25°C	P_D	50 0.67	mW mW/°C
Operating Temperature Range	T_A	−40 to +100	°C
Storage Temperature Range	T_{stg}	−40 to +100	°C

ELECTRICAL CHARACTERISTICS (T_A = 25°C)

Characteristic	Symbol	Min	Typ	Max	Unit
Dark Current (V_R = 5 V, H ≈ 0, Figure 2)	I_D	—	—	1	nA
Reverse Breakdown Voltage (I_R = 10 μA)	$V_{(BR)R}$	50	—	—	Volts
Total Capacitance (V_R = 5 V, f = 1 MHz)	C_T	—	—	5	pF
Noise Equivalent Power	NEP	—	50	—	fW/$\sqrt{Hz}$

OPTICAL CHARACTERISTICS (T_A = 25°C)

Characteristic	Symbol	Min	Typ	Max	Unit
Responsivity @ 850 nm (V_R = 5 V, P = 10 μW, Figure 3)	R	0.2	0.3	—	μA/μW
Response Time @ 850 nm (V_R = 5 V)	t_r, t_f	—	2	5	ns

FIGURE 17-17 ■ MFOD 3100 photodiode specifications. (Courtesy of Motorola, Inc.)

The complete specifications for the MFOD 3100 photodiode are provided in Fig. 17-17. These devices have a typical responsivity of 0.3 μA/μW and 2-ns response time.

17-7 FIBER CONNECTIONS

Optical fiber is made of ultrapure glass. Optical fiber makes window glass seem opaque by comparison. A window made of this pure glass 1 km thick would be as transparent as a normal pane of glass. It is therefore not surprising that the process of making connections from light source to fiber, fiber to fiber, and fiber to detector becomes critical in a

MFOD3100

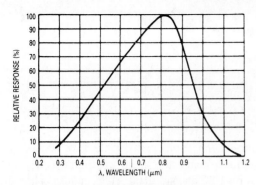

Figure 1. Relative Spectral Response

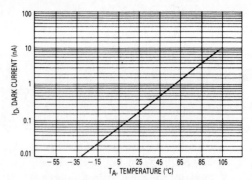

Figure 2. Dark Current versus Temperature

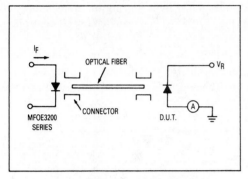

Figure 3. Responsivity Test Configuration

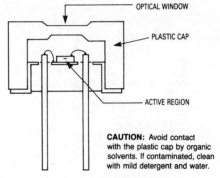

CAUTION: Avoid contact with the plastic cap by organic solvents. If contaminated, clean with mild detergent and water.

Figure 4. Package Cross Section

5

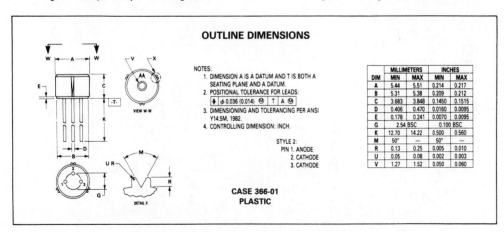

OUTLINE DIMENSIONS

NOTES:
1. DIMENSION A IS A DATUM AND T IS BOTH A SEATING PLANE AND A DATUM.
2. POSITIONAL TOLERANCE FOR LEADS:
   ⊕ ⌀ 0.036 (0.014) Ⓜ T A Ⓢ
3. DIMENSIONING AND TOLERANCING PER ANSI Y14.5M, 1982.
4. CONTROLLING DIMENSION: INCH.

STYLE 2:
PIN 1. ANODE
    2. CATHODE
    3. CATHODE

CASE 366-01
PLASTIC

DIM	MILLIMETERS		INCHES	
	MIN	MAX	MIN	MAX
A	5.44	5.51	0.214	0.217
B	5.31	5.38	0.209	0.212
C	3.683	3.848	0.1450	0.1515
D	0.406	0.470	0.0160	0.0095
E	0.178	0.241	0.0070	0.0095
G	2.54 BSC		0.100 BSC	
K	12.70	14.22	0.500	0.560
M	50°	—	50°	—
R	0.13	0.25	0.005	0.010
U	0.05	0.08	0.002	0.003
V	1.27	1.52	0.050	0.060

FIGURE 17-17 ■ *(Continued)*

system. The low-loss capability of the glass fiber can be severely compromised if these connections are not accomplished in exacting fashion.

Optical fibers are joined either in a permanent splice or with a connector. The connector allows repeated matings and unmatings. Above all, these connections must lose as little light as possible. Low loss depends on correct alignment of the core of one fiber to

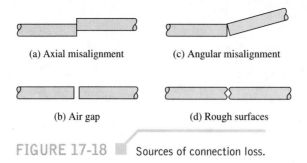

(a) Axial misalignment

(c) Angular misalignment

(b) Air gap

(d) Rough surfaces

FIGURE 17-18 ▮ Sources of connection loss.

another, or to a source or detector. A no-loss connection is impossible. Loss occurs when two fibers are not perfectly aligned within a connector. Axial misalignment typically causes the greatest loss—about 0.5 dB for a 10% displacement. This condition and other loss sources are illustrated in Fig. 17-18. Most connectors leave an air gap, as shown in Fig. 17-18(b). The amount of gap affects loss since light leaving the transmitting fiber spreads conically. Angular misalignment [Fig. 17-18(c)] can usually be well controlled in a connector.

The losses due to rough end surfaces shown in Fig. 17-18(d) can be minimized by polishing (grinding). This is difficult in field installations. Polishing typically takes place after a fiber has been placed in a connector. The alternative to this is the score-and-break

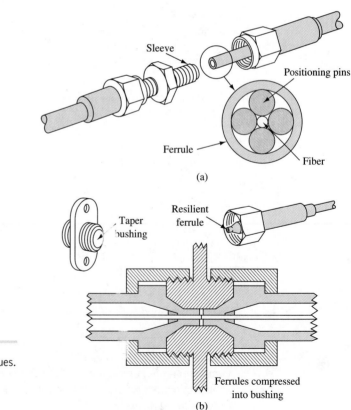

FIGURE 17-19 ▮ Connector techniques.
(Reprinted with permission from *Electronic Design,* June 19, 1986; copyright Hayden Publishing Co., 1986.)

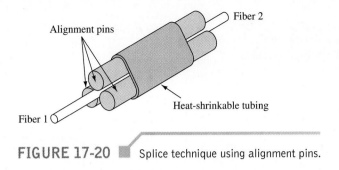

Alignment pins

Fiber 2

Fiber 1

Heat-shrinkable tubing

FIGURE 17-20 ■ Splice technique using alignment pins.

method. With practice, a technician can produce good end finishes by bending the fiber over a finger, applying tension, scribing the fiber with a sharp edge, and breaking the fiber apart. Plastic fiber can be cut with a sharp, hot blade.

The two techniques used in fiber connectors are illustrated in Fig. 17-19. At (a), the excellent alignment properties of the V groove are used. Four round rods circling a fiber create four such grooves and four points of contact to hold the fiber securely in place. The other technique, shown in Fig. 17-19(b), uses a resilient material to enclose the fiber and compensates for variations. The case shown has the fiber within a resilient ferrule. The ferrule is placed in a bushing that tapers inward from both ends. The tapering compresses the ferrule as it is inserted into the bushing. The compression forces the connectors to align in a common axis. Typical connector losses range from 0.5 to 1.5 dB.

A number of techniques are used to splice one fiber to another. The fusion splice involves an electric arc to fuse the two fibers together. This is accomplished after the fibers have been cleaned and cut and then aligned—usually using a microscope. Losses as low as 0.2 dB are attainable using this technique. Another splicing method that is more easily accomplished in the field is the precision pin splice. The fibers to be connected are enclosed by three parallel alignment pins, as shown in Fig. 17-20. An inspection port allows alignment of the fibers to be checked. Once alignment is attained, the heat-shrinkable sleeving shown is heated, causing the pins to squeeze the fibers and thereby maintaining the alignment. A specially formulated epoxy is then applied between the fibers through the port. A final heat-shrinkable tube over the entire assembly completes the splice.

17-8 SYSTEMS

Telemetry

Complete fiber-optic systems are finding widespread use in all types of applications. Figure 17-21 provides a very simple system that can be used in a variety of telemetry applications. Essentially, the transmitter is an astable oscillator using the popular 555 IC. The variable resistance R_1 controls the rate of pulses applied to the LED. R_1 can be any variable-resistance transducer such as a strain gauge for pressure sensing, a thermistor for temperature sensing, or a photoresistor for light sensing. It could also be a carbon microphone so as to transmit audio information. The transmitter is essentially a voltage/frequency (V/F) converter, and thus any of the IC V/F chips (such as the LM331) could be used in this application.

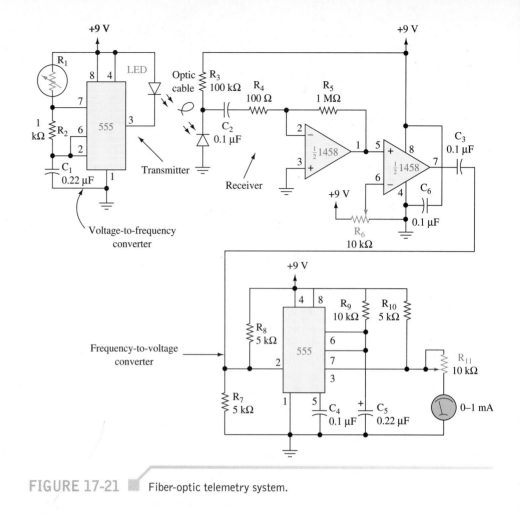

FIGURE 17-21 ■ Fiber-optic telemetry system.

The receiver shown in Fig. 17-21 uses an MC 1458 dual op amp as a preamplifier (first op amp) and as a comparator (second op amp). Once the received pulses are cleaned up by the comparator, they are applied to the 555 IC used as a frequency-to-voltage converter. Adjustment of R_6 controls the threshold level of the comparator, while R_{11} permits calibration of the output meter. The simplicity of this system makes it easily adapted to a variety of applications.

Fiber-Optic Sensors

Interestingly, there is currently much development work taking place to use optical fibers as transducers directly. They generally function by an alteration of light by external stimuli. They can, for example, be used to detect pressure, temperature, magnetism, acidity, and acceleration. They are more rugged and more resistant to corrosion than other sensors and, of course, are compatible with optical-fiber telemetry systems. A ship with an integrated all-optical sensor telemetry system is shown in Fig. 17-22.

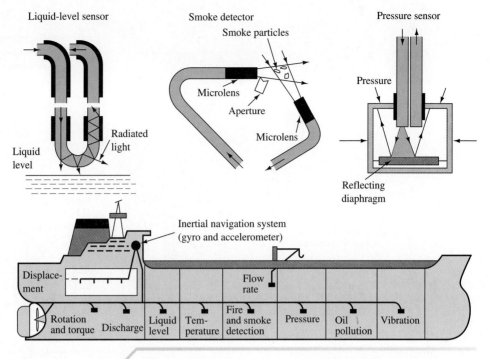

Liquid-level sensor

Smoke detector

Smoke particles

Pressure sensor

Microlens

Aperture

Microlens

Pressure

Radiated light

Liquid level

Reflecting diaphragm

Inertial navigation system
(gyro and accelerometer)

Displace-ment

Flow rate

Rotation and torque   Discharge   Liquid level   Tem-perature   Fire and smoke detection   Pressure   Oil pollution   Vibration

FIGURE 17-22   Optical fibers used as sensors. (From *IEEE Spectrum,* September 1986, © 1986 IEEE, "Optical-fiber sensors challenge the competition" by Thomas G. Giallorenzi, Joseph A. Bucaro, Anthony Dandridge, and James H. Cole.)

The first sensor shown in Fig. 17-22 is a liquid-level sensor. Until the liquid level gets up to the glass fiber (shown in blue), the detected light will remain essentially constant. A rapid change in detected light will occur as the liquid level reaches the fiber, due to the abrupt change from air dielectric "cladding" to that of the liquid. It is then a simple matter to calibrate the received signal with various liquid levels.

The smoke detector shown in Fig. 17-22 will provide essentially zero output under normal conditions. However, when smoke particles are present, an appreciable output will occur due to reflection of the light toward the receiving microlens.

The pressure sensor in Fig. 17-22 has an airtight container that is slightly contorted by changes in the external pressure. This causes a movement in the internally contained reflecting diaphragm that results in a different amount of light to be reflected. This difference is then used to calibrate a meter used to indicate photodetector output.

Communications

A low-cost source detector pair is shown in Fig. 17-23. It is available from GE for less than $2 in production quantities and allows easy application to a variety of relatively high-performance systems. The LED emitter of this low-cost system provides excellent output and is generally designed for short-haul links up to about 1 km without need for repeaters. The threaded cable connector assemblies shown in Fig. 17-23 are becoming a standard for use in fiber-optic systems.

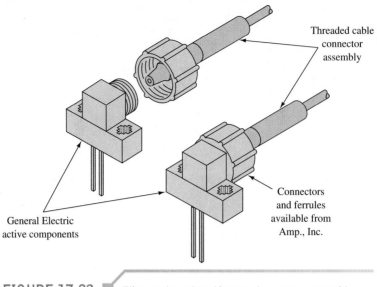

FIGURE 17-23 ▪ Fiber-optic emitter/detector/connector assembly. (Courtesy of General Electric Company.)

Threaded cable connector assembly

General Electric active components

Connectors and ferrules available from Amp., Inc.

A low-cost half-duplex fiber-optic communication system using the light source of Fig. 17-23 is shown in Fig. 17-24. Although full-duplex systems can be implemented in fiber optics, it normally requires two fiber transmitter–receiver sets or a multiplexing system, as described in the next section. The half-duplex system shown allows information to flow in both directions but in only one direction at any given time. The LED emitter light source is not only an emitter, but it is also sensitive to the same light frequency it produces. The system shown in Fig. 17-23 uses the LED as both light transmitter and detector, and this schematic therefore represents one-half of a full system. Biased as a photodiode (reverse-biased or below the positive threshold forward bias of the LED), the GFOE1A1 LED exhibits a sensitivity of about 30 nA per microwatt of irradiation at 940-nM light wavelength.

The light transceiver shown in Fig. 17-24 exhibits a 100-Hz to 50-kHz frequency response and requires no exotic (expensive) components. It has passive receive, transmit priority (voice activated) switching logic, and the frequency-response performance can be upgraded by using higher gain–band-width operational amplifiers. Circuit operation is understood by following the signal through the three portions of the circuit. Both AGC circuits utilize a bilateral analog FET optoisolator's (model H11 F1) variable-resistance characteristic to attenuate the signal or modify a feedback path to provide AGC. In these circuits the peak value of the output signal is compared to the V_{BE}(on) of a Darlington transistor, turning the Darlington on at signal peaks. The two Darlington transistors are the GES 5828 devices shown. Collector current of the Darlington is capacitively filtered and supplies current to the LED of the H11 F1 optoisolator. This lowers the resistance of the analog FET detector that controls signal level.

The voice signal enters via a 4.7-kΩ-H11 F1 attenuator network and is amplified by two BJT amplifier stages (D38S4 and GES 5374 devices). They drive the GFOE1A1 LED with about 50 mA dc carrying 80 mA p-p (max.) voice signal. The vox control logic circuit allows the transmitter output transistor (GES 5374) to function. Otherwise, it is clamped off by the transmit switch transistor and then the system is in the receive mode.

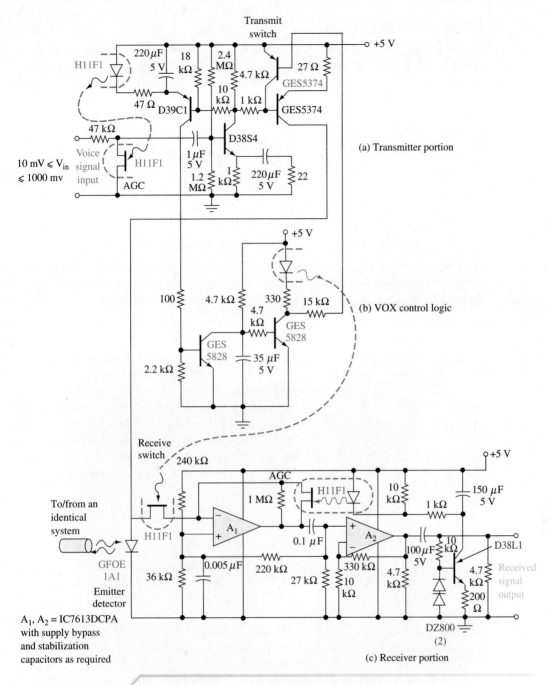

FIGURE 17-24 ■ Half-duplex information link. (Courtesy of General Electric Company.)

In the receive mode the receive switch optoisolator FET exhibits a very low resistance, allowing the LED of the GFOE1A1 to have a low enough forward bias (not conducting) to function as a photodiode and to allow its signal into the A_1, A_2 op-amp stages. This receiver has AGC action provided by another H11 F1 optoisolator. The receiver provides about 2.5 V p-p output signal for light input powers to the GFOE1A1 of about 1 μW to over 200 μW.

Power Budget

The fiber-optic system designer at some point must evaluate all aspects of power to make sure that the system will function properly. That process is called making a "power budget" analysis. The designer must not only make sure that the photodetector receives enough power to produce a usable signal but must also check to make sure that the system is not limited by dispersion effects. As an approximation you can use the following equation to check the dispersion.

$$z \simeq \frac{1}{5B\Delta_t}$$

(17-5)

where z = maximum fiber length (km)

B = maximum bit rate (Mb/s)

Δ_t = dispersion (μs/km)

Once a power budget has been made for a system, it is normal to include a margin of safety to compensate for variations in the source, losses due to bends, and repair splices that may become necessary. Typically, 5–10 dB is used as the safety factor.

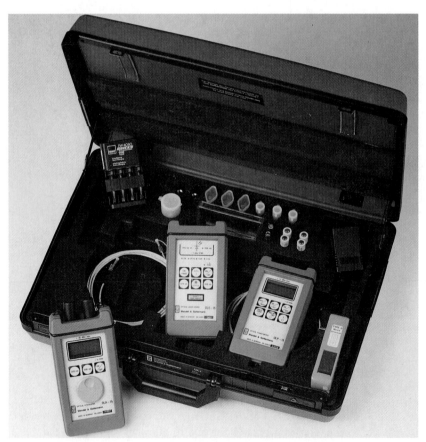

Optical test set for power level measurements. (Courtesy of Wandel & Goltermann Technologies, Inc.)

EXAMPLE 17-6

Make a power budget analysis for a fiber-optic system with the following characteristics:

Losses:

> *LED-to-fiber connection:* 5 dB
> *Three connectors:* 1.5 dB each
> *Six splices:* 0.5 dB each
> *10 km of fiber:* 0.6 dB/km
> *Fiber to detector:* 6 dB

Specifications:

> *LED power output:* 0.1 mW
> *Detector sensitivity:* 0.1 μW
> *Maximum bit rate:* 5 Mb/s
> *Total fiber dispersion:* 4 ns/km

SOLUTION

First sum the total system attenuation:

$$5 \text{ dB} + 3(1.5 \text{ dB}) + 6(0.5 \text{ dB}) + 10(0.6 \text{ dB}) + 6 \text{ dB} = 24.5 \text{ dB}$$

Now, assuming a 5-dB safety factor, we obtain

$$29.5 \text{ dB} = 10 \log \frac{P_t}{P_r}$$

Therefore,

$$\frac{P_t}{P_r} = 891$$

$$P_r = \frac{0.1 \text{ mW}}{891} = 0.112 \ \mu\text{W}$$

Since the detector sensitivity is 0.1 μW, the system should operate satisfactorily without a repeater from the power-loss standpoint. The dispersion effects must now be considered.

$$z \simeq \frac{1}{5B\Delta_t}$$

$$= \frac{1}{5 \times 5 \times 0.004} \qquad (17\text{-}5)$$

$$= 10 \text{ km}$$

Since the system length is equal to the maximum dispersion length (10 km), some testing of the system will be necessary to determine if satisfactory operation without a repeater will be possible.

Until recently, the telecommunications industry has been the major user of fiber-optic systems. The rapid development of fiber-optic technology and resulting cost reductions have allowed other major applications, such as computer communications, local area networks (LANs), CATV, telemetry, and so on.

A typical LAN (see Sec. 11-6) has data rates of 1 to 20 Mb/s and is confined to a span width of about 1 km. These conditions are conveniently met by standard copper wiring techniques. The need for increased capacity is beginning, however, as computer speed and storage capabilities keep moving up. The capacity of the twisted pair or coaxial cable wiring is no longer sufficient for many applications. Newer requirements of 200-Mb/s data rates at up to 10-km span lengths cannot easily and economically be attained with copper.

The cost of installing a modern communications network in a large organization can exceed $1 million. It is imperative that the new systems have enough capacity for present and future needs. Optical-fiber communication systems are the obvious solution for overcoming the bandwidth and attenuation limitations of copper links. Recent developments, primarily in fiber-compatible light sources, have made the fiber systems attractive. A few years ago, optical data rates of 200 Mb/s would have required expensive laser sources. Today, that rate is achieved using LED sources at one-tenth the cost.

When capital cost differences between a fiber system and a coaxial cable system are no longer significant, optical-fiber data networks will usually be chosen. They have orders of magnitude more bandwidth, occupy smaller volume, have much lower signal attenuation (typically, 0.2 dB/km versus 30 dB/km for copper), have no ground loop problems, and neither generate nor show susceptibility to electrical noise. The optical fiber, being much smaller, is more flexible and lighter and eases the installation in already crowded ducts and conduits. Additionally, security is enhanced since it is very difficult to tap an optical fiber without a detectable interruption of service.

Network Topology

As described in Sec. 11-6, the most common LAN configurations are the ring, bus, and star topologies. The ring network has a serial connection between each node. The signals are transmitted node to node and examined and acted upon by each node before being sent to the next one. In the bus topology, all nodes are connected to a common multiple-access transmission medium. This structure has a parallel approach as compared to the ring's serial approach. The star topology generally consists of a controlling node in the center with other active nodes connected to it via a star-shaped network.

Fiber-optic transmission systems can be incorporated in each of the aforementioned topologies. In the bus topology, a passive four-port coupler is used to transmit or receive information on the bus, as shown in Fig. 17-25. Because of the loss characteristics of these couplers, the maximum number of nodes is about 20.

A ring network uses active elements at each node. A failure at one node causes the system to crash unless redundant paths or bypass switches are incorporated at each node. Since fiber loss is so low, the number of nodes in a ring is defined by system restraints, and the distance between nodes can be very large. The star network requires a star coupler (explained later in this section) and can service 64 nodes or more. Even though this system requires more cable than the others, it offers good ability to add additional nodes after the initial system installation. The implementation of optical LANs requires the use of some components that are described in the following paragraphs.

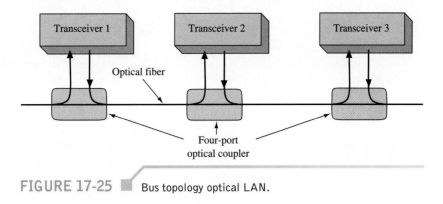

FIGURE 17-25 ▬ Bus topology optical LAN.

Components for Optical LANs

COUPLERS: DEPENDENT AND INVARIANT   When optical-fiber systems are strictly point to point (i.e., the connection of a single transmitter to a single receiver), there is no need for optical couplers. However, transmission of information or data in two directions via a single fiber or linking more than one transmitter and receiver requires some form of optical coupler. Optical couplers come in two basic forms: wavelength-dependent and wavelength-invariant.

There are a number of techniques used to manufacture wavelength-invariant couplers:

1. Three fibers are cut and polished. The input fiber is aligned with the output fibers so that the splitting ratio can be adjusted by changing the relative overlapping of the output fiber's core with the core of the input fiber.

2. A slanted surface that is partially reflective is placed in the propagation path of the main fiber. The reflective surface is typically a semitransparent mirror. One output fiber collects light in the propagation path, and the other output fiber collects the mirror-reflected light.

3. The three fibers are bundled together, heated, twisted, and pulled simultaneously so that the optical modes propagating in one fiber will migrate into another fiber. The desired tap ratio is achieved by controlling the twist and pull operations.

The couplers just described are analogous to the directional couplers described for transmission lines and waveguides in Chapters 12 and 15. In fact, these optical couplers are often called directional couplers. A four-port coupler as shown in the bus topology network of Fig. 17-25 is usually fabricated using the first method just described. There is a high degree of isolation between the two input ports in these couplers. The isolation specification is typically 15 to 40 dB. *Excess loss* is a measure of added losses in addition to the desired splitting ratio. It is caused by inherent coupling losses such as scattering and absorption.

WAVELENGTH-DEPENDENT COUPLERS   These couplers enable more efficient utilization of the already high transmission capabilities of optical fibers. They also allow full-duplex communication using a single fiber. They permit wavelength (frequency)-division multiplexing (WDM) and are therefore usually called WDM couplers. Some

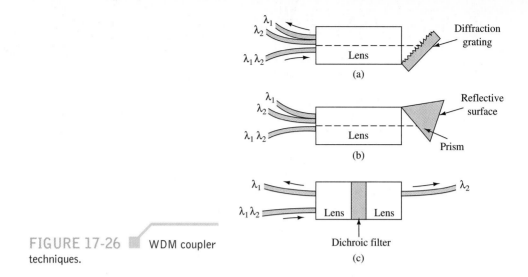

FIGURE 17-26 ░ WDM coupler
techniques.

techniques used to construct them are depicted in Fig. 17-26. They utilize diffraction gratings (a), prisms (b), and dichroic filters (c). Using diffraction gratings or prisms, the various wavelengths of the incoming optical signal are reflected at different angles. They can
thereby be separated and coupled into different output fibers.

A dichroic filter is a coated surface that permits light transmission at some wavelengths and reflection of others. Thus the transmitted wavelength can be coupled into one
fiber and the reflected wavelengths into the other.

Coupler manufacturing techniques for single-mode fiber are now being developed
that require no discrete optical elements. These promise lower cost and performance
advantages. This process is based on gaining access to the field that exists outside the
fiber's core–cladding boundary. This is called the *evanescent field*. The coupler is fabricated by melting or fusing the fibers at high temperature. Precise control of the process
and region of coupling determines both the coupling ratio and this ratio's wavelength
dependence. The WDM couplers available at 1.3 and 1.5 μm offer excess loss as low as
0.05 dB. The wavelength isolation between 1.3- and 1.5-μm signals is about 20 dB. The
commonly required bit error rate (BER) of 10^{-9} corresponds to a 13-dB *S/N*. If the
crosstalk between 1.3 and 1.5 μm is considered as noise, the 20-dB isolation specification
is more than adequate for the 10^{-9} BER.

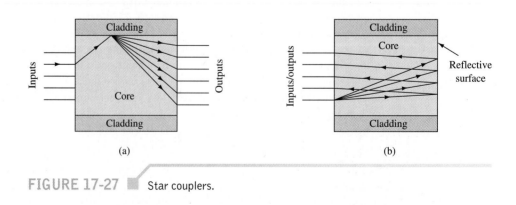

FIGURE 17-27 ░ Star couplers.

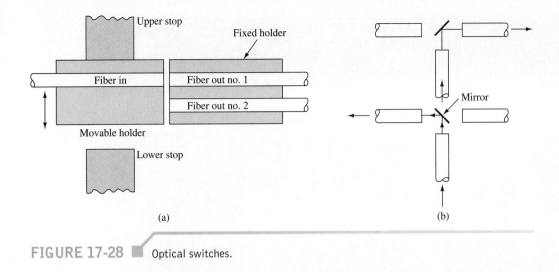

FIGURE 17-28 ■ Optical switches.

STAR COUPLERS   Coupling between multiple nodes of a system as required by star topology systems requires a star coupler. The two basic types of star couplers are shown in Fig. 17-27. The transmissive star at (a) has a designated number of input and output ports. An optical signal at any input port is divided (nearly equally) between all outputs. None of the input ports receive any light. Placing a reflective mirror, as shown at (b), creates a reflective star. In a reflective star, there is a fixed number of ports, and no distinction is made between input and output. An optical input at any port results in nearly equal distribution of light to all other ports of the star. Important specifications for star couplers include port-to-port uniformity (1 dB typical) and excess loss (1.5 dB typical).

SWITCHES   Two types of optical switches are illustrated in Fig. 17-28. They are the moving-fiber and moving-mirror varieties. In the moving-fiber switch a solonoid physically moves fibers to "make" or "break" a light path. The one shown provides light transmission between the input fiber and fiber 1 in one position while transmission between input and fiber 2 occurs in the other position.

The moving-mirror switch in Fig. 17-28(b) contains a mirror that is rotated 90° to cause the transmitted light to be switched between two different destinations. The two critical specifications for optical switches are insertion loss (typically 1.5 dB) and isolation (typically 50 dB).

FDDI

A lack of standards slowed the development of fiber-optic LANs and point-to-point communications. Manufacturers need some sort of standard to design with to reduce costs and to promote compatibility in multiple-vendor component sourcing. The American National Standards Institute (ANSI) has developed the Fiber Distributed Data Interface (FDDI) standard that is now in widespread use.

FDDI utilizes two 100-Mbps token-passing rings. The two independent counterrotating rings are connected to a certain number of nodes (stations) in the network. The primary ring connects only *class A* stations—those offering a high level of protection because of their ability to transfer operation to the secondary ring should the primary ring

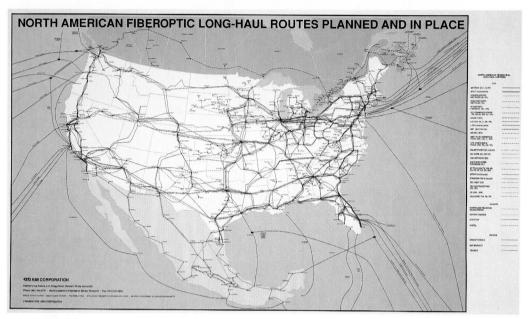

NORTH AMERICAN FIBEROPTIC LONG-HAUL ROUTES PLANNED AND IN PLACE

United States fiber-optic routes. (Courtesy of KMI Corporation.)

fail. The secondary ring reaches all stations and carries data in the opposite direction of the primary ring. The secondary ring can also be used with the primary ring operating so as to allow increased data throughout. The switchover to the secondary path in the event of failure is accomplished by a pivoting spherical mirror within a *dual-bypass* switch. The changeover takes 5 to 10 μs, and a loss of about 1 dB results from the presence of the dual-bypass switch.

Stations on the FDDI ring can be separated by up to 2 km as long as the average distance between nodes is less than 200 m. These limits are imposed to minimize the time it takes a signal to move around the ring. A total of 1000 physical connections and a total fiber path of 200 km are allowed. This allows 500 stations since each represents two physical connections. The type of fiber used is not dictated and is chosen by the user based on the performance required. The fibers used most often are 62.5/125 or 85/125 multimode fiber. LEDs are specified as light sources at 1300 nm.

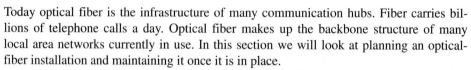

TROUBLESHOOTING

Today optical fiber is the infrastructure of many communication hubs. Fiber carries billions of telephone calls a day. Optical fiber makes up the backbone structure of many local area networks currently in use. In this section we will look at planning an optical-fiber installation and maintaining it once it is in place.

Remember, the diode lasers (DLs) can emit radiation with a far higher energy-density than sunlight, and even though you can't see the radiation, it can easily cause blindness by retinal heating. You should always wear eye protection when working on laser systems. A 1-W or more CW output laser, such as used in medical imaging products, can produce a stunningly high-power density when focused. Even when poorly focused, 1 W across 0.125-in diameter (like the pupil of your eye) means more than 100 times the power density of sunlight! You need to respect the device and follow the rules for working with it. After completing this section you should be able to

- Draw a fiber link showing all components
- Explain the use of the radiometer
- Describe rise time measurement
- Troubleshoot fiber-optic data links

Losses in an Optical-Fiber System

The optical-fiber system in Fig. 17-29 has an emitter, two connectors, the fiber, and the detector. The proper performance of this fiber link depends on the total power losses of the light signal through the link being less than the specified maximum allowable loss. Power is lost in all of the components that make up the system. For example, a connector may have a power loss of 1.5 dB and a splice with 0.5 dB, and the fiber cable itself will also attenuate the light signal. As an example, if a fiber system's maximum allowable losses were 20 dB, and total power losses added up to 17 dB, the system would still have a 3-dB working margin. Of course, this is a small working margin and does not take into account weakening emitters and detectors over a period of time.

Calculating Power Requirements

A power budget should be prepared when installing a fiber-based system. The power budget will specify the maximum losses that can be tolerated in the fiber system. This will help ensure that losses stay within the budgeted power allocation. The launch power is figured first. Launch power is calculated by subtracting the emitter's connector loss from the power being injected into the fiber by the emitter. For example, suppose the emitter produces 150 mW of power, or 20 dB. Then subtract the connector loss of 1.5 dB. Thus, the effective launch power is 18.5 dB. Next, determine the detector's minimum power requirement. Suppose the detector gave us another 10-dB gain. Then the system's maximum losses could not exceed 28.5 dB. Adding up the system losses to include connectors, splices, and fiber cable, the total could not be more than 28.5 dB.

Once the optical fiber system is in place, the *radiometer* would be used to determine the actual power being lost in the system. A calibrated light source injects a known

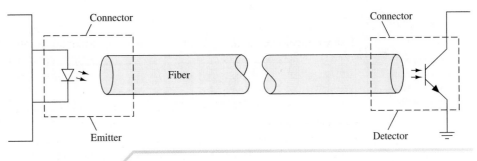

FIGURE 17-29 A fiber link showing emitter, detector, connectors, and fiber cable.

amount of light into the fiber, and the radiometer connected to the other end of the fiber measures the light power reaching it. Periodic checks should be scheduled as preventive maintenance to keep the fiber system in peak performance. Weakening emitters should be replaced before they degrade the system's performance.

Rise Time Measurement

An optical fiber's link performance can be determined by measuring the rise time of injected test pulses. Figure 17-30 shows the test configuration used to inject pulses into the fiber link. The test pulses must have a very fast rise time, typically in pico- or nanoseconds. At the detector end, connect a fast response oscilloscope to the output of the detector and measure the rise time of the injected pulses. Rise time is measured from the 10% point to the 90% point on the positive going edge of the pulse, as illustrated in Fig. 17-30.

The rise time (R_t) measurement must not be greater than the value provided by the following equation:

$$R_t = \frac{0.35}{\text{BW}} \tag{17-6}$$

A 70-ns optical-fiber system will have a bandwidth of 5 MHz based on this equation.

For this system to function properly the measured rise time would have to be less than 70 ns if a 5-MHz bandwidth is needed. Suppose the 5-MHz bandwidth system has the following signal delays: The emitter has a rise time of 1 ns, the detector a rise time of 20 ns, and a delay of 35 ns exists for 1 km of fiber cable. This adds up to a total system delay of 56 ns. This is a functioning system because the 56-ns delay is less than 70 ns. Scheduling periodic checks to keep an eye on the system rise time ensures that the specified system bandwidth is maintained. Weakening emitters and detectors and tight bends in

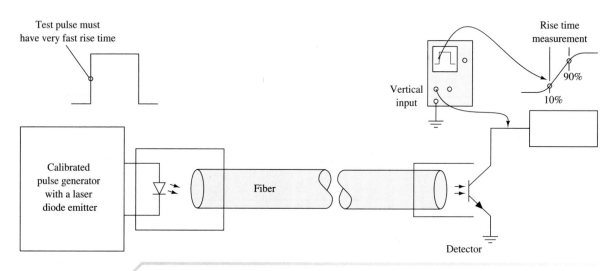

FIGURE 17-30 ■ Rise time test of an optical-fiber link.

the fiber cable will cause an increase in delay time, and the rise time measurement will be greater. This is an excellent means for monitoring a fiber system's performance.

Connector and Cable Problems

Some of the problems associated with fiber-optic links are caused by contact of a foreign substance with the fiber (even the oil from your skin can cause serious trouble). Connectors and splices are potential trouble spots. A back-biased photo diode and an op-amp can be used as a relative signal strength indicator. Looking for the signal while gently flexing cables and connectors can help pinpoint problem areas.

Characteristics of LEDs and DLs

These special purpose diodes are nonetheless diodes and should exhibit the familiar exponential I versus V curve. These diodes do not draw current when forward biased until the voltage reaches about 1.4 V. Some ohmmeters do not put out sufficient voltage to turn a LED on; you may have to use a power supply, a current limiting resistor, and a voltmeter to test the diode.

Reverse voltage ratings are very low compared to ordinary silicon rectifier diodes— as little as 6 V. More voltage may destroy the diode.

A Simple Test Tool

Some systems use visible wavelengths; most use invisible infrared. Another diode of the same type or of similar emission wavelength can be used as a detector to check for output. Use a meter in current mode, not voltage, and compare a good system to the troublesome one.

To increase sensitivity, a simple current-to-voltage converter circuit, made with an op-amp, a feedback resistor, and the detector diode pumping current to the op-amp input, will convert the current from the detector diode to a voltage out of the op-amp. The circuit for this is shown in Fig. 17-31. If signal levels are high, just a resistor across the detector diode is appropriate. Remember to keep the bias voltage small enough so that the voltage developed is well below the maximum reverse voltage allowed for the diode.

If you wish to see the signal modulation, try using an oscilloscope in place of a simple multimeter. A less quantitative check for emitted output can be made using a test card of the type used in TV-repair shops to check for output from infrared remote controls. These cards are coated with a special chemical that in the simultaneous presence of visible and infrared illumination will emit an orange glow.

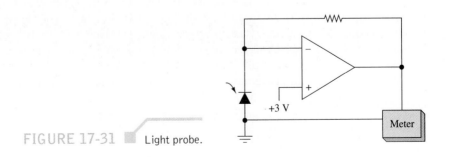

FIGURE 17-31 ■ Light probe.

SUMMARY

In Chapter 17 we introduced the field of fiber optics. We learned that many applications exist in electronic communications for these optical devices. The major topics the student should now understand include:

- the advantages offered by fiber-optic communication

- the analysis and properties of light waves

- the physical and optical characteristics of optical fibers, including multimode, graded index, and single-mode fibers

- the attenuation and dispersion effects in fiber

- the description and operation of the diode laser (DL) and high-radiance light-emitting diode (LED) light sources

- the application of *p-i-n* diodes as light detectors

- the description of common techniques used to connect fibers

- the general applications of fiber-optic systems, including telemetry, sensing applications, and communications

- the power considerations and calculations in fiber-optic systems

- the usage of fiber optics in local area networks (LANs)

- the description of LAN components, including wavelength-dependent and independent couplers, star couplers, and optical switches

QUESTIONS AND PROBLEMS

SECTION 1

1. List the basic elements of a fiber-optic communications system. Explain its possible advantages as compared to a more standard communications system.

SECTION 2

2. Describe the mode of propagation in a fiber-optic communications system. Include the concepts of refraction and critical angle in your description.
3. A fiber cable has the following index of refractions: core, 1.52; and cladding, 1.31. Calculate the numerical aperture for this cable. (0.77)
4. A green LED light source functions at a frequency midway between red and violet. Calculate its frequency and wavelength. (5.7×10^{14} Hz, 526 nm)
5. Calculate the energy for one photon of the light described in Problem 4. Determine the power for a 0.3-ns pulse of this energy. (3.59×10^{-19} J, 1.2 nW)
6. Determine the critical angle beyond which an underwater light source will not shine into the air. (48.7°)

7. An optical fiber has refractive indexes of 1.52 and 1.485 for its core and cladding, respectively. Calculate its NA and θ_{in}(max). (0.324, 18.9°)
8. Describe the significance of NA for a fiber.

SECTION 3

9. Describe the types of fiber-optic cables and provide their relative characteristics.
10. How does a fiber cable adversely affect a short-duration pulse of light during propagation in multimode fibers? This pulse dispersion effect is minimized by use of the graded-index fiber. Explain why this is so.
11. Calculate the NA and cutoff wavelength for single-mode fiber with a core of 2.5 μm and refraction indexes of 1.515 and 1.490 for core and cladding, respectively. (0.274, 1.73 μm)

SECTION 4

12. A system operating at 1.55 μm exhibits a loss of 0.35 dB/km. If 225 μW of light power is fed into the fiber, calculate the received power through a 20-km section. (22.5 pW)
13. A photodetector must receive 125 pW from a fiber to provide satisfactory detection. Calculate the maximum fiber length for the system described in Problem 12. (17.9 km)
14. With reference to Fig. 17-12, explain why fiber offers the largest bandwidth, even though at first glance you might think otherwise.
15. Define *dispersion* and explain its causes and types for both multimode and single-mode fiber.
16. Describe the meaning and cause of the zero-dispersion wavelength near 1.3 μm for single-mode fiber.

SECTION 5

17. Compare the diode laser and LED for use as light sources in optical communication systems.
18. Explain the process of lasing for a semiconductor diode laser. What is varied to produce light at different wavelengths?

SECTION 6

19. List the important characteristics of light detectors.
20. Describe the operation of a *p-i-n* diode used as a photodetector.
21. Describe the operation of an avalanche photodiode. Why is it a more sensitive detector than the *p-i-n* diode?
22. Explain how a photodetector is used to enhance the operation of a semiconductor laser.

23. Why is the connection of glass fibers such an important part of a fiber-optic system? Describe the major sources of loss in a fiber connection.
24. Describe a process whereby a field technician could "break" a glass fiber and end up with two good, smooth surfaces. Repeat this if an all-plastic fiber is being used.
25. Describe the two generally used techniques in glass-fiber connectors.
26. Explain two techniques used to splice glass fibers together.

27. Describe the signal flow for the telemetry system illustrated in Fig. 17-22.
28. How can optical fibers be used as sensors? Devise an optical fiber sensor not described in the text.
29. Explain how the fiber-optic system shown in Fig. 17-25 switches between the transmit mode and the receive mode.
30. A fiber-optic system uses cable with an attenuation of 3.2 dB/km, is 1.8 km long, has one splice with 0.8-dB loss, has a 2-dB loss at both transmitter and receiver due to source/receiver connection, and requires 3 μW of received optical power at the detector. Calculate the required level of optical power from the light source. (34.1 μW)
31. Calculate the maximum length of a dispersion-limited system where the fiber has a 0.0015-μs/km dispersion and a 50-Mb/s bit rate. (2.67 km)
32. Provide a complete power budget analysis for a system with the following losses and specifications:

    Losses:
    Laser-to-fiber connection: 6.5 dB
    Two connections: 1.0 dB each
    Three splices: 0.5 dB each
    20 km of fiber: 0.35 dB/km
    Fiber to detector: 5.5 dB

    Specifications:
    Laser power output: 1.25 mW
    Detector sensitivity: 1.30 μW
    Bit rate: 10 Mb/s
    Total fiber dispersion: 0.35 ns/km

33. Provide a general discussion of the reasons that fiber-optic LANs are being used instead of copper connection schemes.
34. What are some possible disadvantages of fiber-optic LANs compared to copper connection schemes?
35. Sketch fiber-optic LANs in each of the three common LAN topologies.
36. Explain the need for, and operation of, two types of wavelength-invariant couplers.

37. Define the isolation and excess loss specifications for optical couplers.
38. Explain the need for, and operation of, two types of wavelength-dependent optical couplers.
39. Explain the need for, and operation of, the types of star couplers.
40. Describe the operation of three types of optical switches. (Yes, that means you will have to find or design one that is not explained in the text.)
41. What is FDDI? Provide a basic description of an FDDI system.
42. A seven-station FDDI system has spacings of 150 m, 17 m, 270 m, 130 m, 235 m, and 320 m for six of the stations. Determine the maximum spacing for the other station. (278 m)

GLOSSARY

algorithm a plan or set of instructions followed in order to achieve a specific goal

aliasing distortion the distortion that results if Nyquist criteria is not met in a digital communication system using sampling of the information signal; the resulting alias frequency equals the difference between the input intelligence frequency and the sampling frequency

amplitude companding a process of volume compression before transmission and volume expansion after detection

amplitude compandored single sideband (ACSSB) sideband transmission with speech compression in the transmitter and speech expansion in the receiver

amplitude modulation (AM) the process of impressing low-frequency intelligence onto a high-frequency carrier so that the instantaneous changes in the amplitude of the intelligence produce corresponding changes in the amplitude of the high-frequency carrier

angle modulation superimposing the intelligence signal on a high-frequency carrier so that its phase angle or frequency is altered as a function of the amplitude of the intelligence signal

antenna a device that generates and/or collects electromagnetic energy

antenna coupler an impedance matching network in the output stage of an RF amplifier or transmitter that ensures maximum power is transferred to the antenna by matching the input impedance of the antenna to the output impedance of the transmitter

antenna gain a measure of how much more power, in dB, an antenna will radiate in a certain direction with respect to that which would be radiated by a reference antenna, i.e., an isotropic point source or dipole

Armstrong transmitter an FM transmitter that uses a phase modulator to indirectly create FM by first feeding the intelligence

signal through a low-pass filter integrator network

array a group of antennas or antenna elements arranged to provide the desired directional characteristics

ASCII (American Standard Code for Information Interchange) a standardized coding scheme for alphanumeric symbols

aspect ratio in a television picture, the ratio of frame width distance to frame height distance; in the United States it is standardized at 4/3

asynchronous a system in which the transmitter and receiver clocks free-run at approximately the same speed; each transmitted data word contains a start bit and one or more stop bits so that the receiver knows when the actual data occurs

atmospheric noise external noise caused by naturally occurring disturbances in the earth's atmosphere

aural signal the sound or audio portion of a television signal; transmitted using frequency modulation

autodyne mixer another name for self-excited mixer

automatic frequency control (AFC) a negative feedback control system utilized in Crosby FM transmitters to achieve high-frequency stability of the carrier

automatic gain control (AGC) the function in a receiver that allows weak rf signals to be amplified to a high degree and strong rf signals to be amplified to a lesser degree in order to produce a near constant output level

auxiliary AGC an AGC that causes a step reduction in receiver gain at some arbitrarily high level of received signal in order to prevent the receiver from being overloaded

back porch the interval just after the horizontal sync pulse appears on the blanking pulse in a television receiver; this is where an eight cycle sine-wave burst signal appears for the color sync to occur

balanced modulator   a modulator stage that adds intelligence to the carrier in order to produce both sidebands with the carrier eliminated

balanced transmission line   parallel conductors, such as open wire feedline, that carry two equal but opposite phase electrical signals with respect to ground

balanced-ring modulator   a balanced modulator design in which four matched diodes are connected together in a ring configuration

balun   a transformer used to match a balanced transmission line to an unbalanced line

Barkhausen criteria   the two requirements necessary for oscillations to occur; the loop gain must be unity and the loop phase shift must be zero degrees

baseband   the signal is transmitted at its base frequency and no modulation to another frequency range has occurred

Baudot code   a fairly obsolete coding scheme for alphanumeric symbols that is not very powerful but was very popular in the early years of radioteletype

beamwidth   for an antenna, the angular separation between the half-power points on its radiation pattern

Bessel function   a mathematical technique for determining the exact bandwidth of an FM signal

bi-phase code   an encoding format for PCM systems that is very popular for use with optical systems, satellite telemetry links, and magnetic recording systems

binary phase-shift keying (BPSK)   a form of phase shift keying in which the binary "1" and "0" states are represented as no phase shift or phase inversion of the carrier signal

bit   unit of information required to allow proper selection of one out of two equally probable events

bit error rate (BER)   the reciprocal of the error probability, i.e., the average number of bits correct before a bit in error occurs

block check character (BCC)   a more sophisticated method of error detection in high data rate systems in which a block of data is sent, followed by an end of message indicator, then a block check character representing characteristics of the block of data that was sent

BORSCHT   the functions produced on subscriber loop interface circuits in PBX or central office systems; these functions are Battery feeding, Overvoltage protection, Ringing, Supervision, Coding, Hybrid, and Testing

buffer amplifier   an amplifier designed to prevent any amplitude or frequency loading from occurring; it typically has a very high input impedance and low input capacitance in order to remove any amplitude reduction or frequency drifting of the desired signal being amplified

capture effect   an FM receiver phenomena that involves locking onto the stronger of two received signals on the same frequency and suppressing the weaker signal

capture range   the range of applied input frequencies to a PLL that will cause it to lock up

carrier   a radio wave of constant amplitude, frequency, and phase at the frequency of operation for the radio, television, or other type of communication system; this radio wave's amplitude, frequency, or phase is altered by an information signal so that it can carry the information to a distant receiver

Carson's rule   equation for approximating the bandwidth of an FM signal

cathode ray tube (CRT)   a vacuum tube in which the electron can be focused in a small spot on a fluorescent screen at the opposite end of the structure; used in television receivers and oscilloscopes to form the display

cavity resonator   metal-walled chambers in microwave installations fitted with devices for admitting and extracting electromagnetic energy

cell splitting   in cellular telephone, if all traffic in a given cell increases beyond a reasonable capacity, the cell automatically splits into smaller hexagon-shaped coverage areas

cellular telephone   modern mobile telephone system characterized by a network of cell sites, each serving a hexagon-shaped coverage area, and automatic switching of service from one cell site to another as the mobile customer travels from one coverage area to another

central office   a telephone exchange on phone company property that simply switches one telephone line to another so that phone calls can be made

ceramic filter a filter network made from lead zirconate-titanate that exhibits the piezoelectric effect and makes effective filters to convert DSB-SC to SSB in a SSB transmitter

characteristic impedance the input impedance of a transmission line either infinitely long or terminated in a pure resistance exactly equal to its characteristic impedance

characteristic wave impedance characteristic impedance of waveguide that tends to be more variable than that of simple transmission line

charge coupled device (CCD) an integrated circuit imaging unit that is typically used for lower resolution applications such as video conferencing and security cameras

citizen's band a radio service for personal communication or remote control of objects; often used by truck drivers during interstate highway travel

Clapp oscillator a Colpitts oscillator having a third capacitor in series with its inductor in order to produce a more stable output frequency

coaxial cable two conductors, a center conductor and an outer shield, separated by a dielectric at a fixed distance from one another; provides for minimal noise pickup and minimal radiation

code-division multiple-access (CDMA) a satellite communication system in which each earth station uses a different binary sequence to modulate the carrier in a different manner

codec an integrated circuit that contains the A/D encoding circuitry and the D/A decoding circuitry needed in a PCM system

coding the process of transforming messages or signals in accordance with a definite set of rules

collinear array any combination of half-wave elements in which all the elements are excited by a connected transmission line

color burst an eight-cycle sine-wave burst that occurs on the back porch of the horizontal sync pulse in a color television receiver; it is necessary in order to calibrate the receiver's subcarrier generator

Colpitts oscillator a popular LC oscillator, easily recognized by its two capacitors providing the positive feedback path for oscillation

compandor (compress/expand) to provide better noise performance, a variable-gain circuit at the transmitter increases its gain for low-level signals; a complementary circuit in the receiver reverses the process to restore the original information signal

constellation pattern a display used to monitor QAM data signals on an oscilloscope in X:Y mode

continuous wave (CW) an undamped sinusoidal waveform produced by an oscillator in a radio transmitter

convergence in a multibeam cathode-ray tube, a condition in which the beams are adjusted so that they all cross at a specific point; in color television, an alignment procedure used to form the clearest image

converter stage a stage that serves the purpose of converting one frequency into another

corona discharge luminous discharge of energy by an antenna due to ionization of the air around the surface of the conductor

counterpoise the reflecting surface of a vertical antenna if the actual earth ground cannot be used; a flat structure made of wire or screen placed a short distance above ground with at least a quarter-wavelength radius

critical angle the highest angle with respect to a vertical line at which a radio wave of a specified frequency can be propagated and still be returned to the earth from the ionosphere

critical frequency the highest frequency that will be returned to the earth when transmitted vertically under given ionospheric conditions

Crosby transmitter FM transmitters that utilize direct FM modulation with automatic frequency control to correct for carrier drift by comparing it to a reference crystal oscillator

cross modulation a form of distortion that results in mixer stages that are overdriven; characterized by the production of an excessive number of output frequencies by the mixer

crosstalk the unwanted coupling between two electrical conductors by overlapping electric and magnetic fields

crystal filter a crystal network commonly used in SSB transmitters to convert DSB-SC to SSB

crystal oscillator an LC oscillator that uses a piezoelectric crystal in place of the inductor to produce a stable output frequency

cyclic redundancy check (CRC)   a more sophisticated popular method of error detection that involves performing binary division repetitively on each block of data and checking the resulting remainders

damper   a diode in the high voltage oscillator of a television receiver that shorts out unwanted damped oscillations that occur in the yoke-coil's collapsing magnetic fields during the flyback period

Data Communications Equipment (DCE)   indicates the computer peripheral end of a data cable where functions are performed that are not normally performed on the computer end of the cable, such as printing

Data Terminal Equipment (DTE)   indicates the computer terminal end of a data cable where user information is converted into data signals for transmission

dBm   a method of rating power or voltage levels with respect to 1 mW of power

decade   a range of frequencies in which the upper limit is ten times as large as the lower limit

deemphasis   a stage found in an FM receiver that reduces the amplitudes of the high-frequency audio signals back down to their original values to counteract the effect of the transmitter's preemphasis network in order to reduce noise

delay equalizer   a complex LC filter that removes delay distortion from signals on phone lines by providing increased delay to those frequencies least delayed by the phone lines and little delay for all other frequencies

delay line   a length of a transmission line designed to delay a signal from reaching a point by a specific amount of time

delayed AGC   an AGC that does not provide any gain reduction until some arbitrary signal level is attained and therefore has no gain reduction for weak signals

delta modulation (DM)   a digital modulation technique in which the encoder transmits information regarding whether the analog information increases or decreases in amplitude; this information is in digital format and represents the analog information

demodulation   the process of removing the intelligence from the high-frequency carrier in a receiver

diagonal clipping   distortion that occurs in a diode detector if the time-constant of the low-pass filter is set too high

diffraction   the phenomenon whereby waves traveling in straight paths bend around an obstacle

digital communication   the transfer of information from transmitter to receiver by representing it in a digital format before it is transmitted and then converting it back to its original form after it is detected by the receiver

digital signal processing   using programming techniques to process a signal while it is in digital form, either in the transmitter or the receiver

diode detector   the simplest method for detecting an AM signal, consisting of a diode in series with a low-pass filter

diplexer   a filter used in a television transmitter that allows both the video AM signal and the audio FM signal to feed the same antenna

dipole antenna   a straight radiator one-half wavelength long and usually separated at the center by an insulator and fed by a balanced transmission line

direct digital synthesis (DDS)   modern frequency synthesizer designs that offer better repeatability and less drifting but have limited maximum output frequencies and greater complexity and cost

directional coupler   a device that senses how much signal is moving in one direction in a transmission line or waveguide

director   in a Yagi antenna, the parasitic element that effectively directs energy to the driven element

discriminator   an FM detector; a stage in an FM receiver that creates an output dc level that varies as a function of its input frequency

dissipation factor (D)   the inverse of quality factor; often used in rating capacitors

Dolby system   a more advanced noise reduction system used in FM systems in which the preemphasis and deemphasis networks work in a dynamic manner

Doppler effect   a phenomenon whereby the frequency of a reflected signal is shifted if there is a relative motion between the source and reflecting object

double conversion   a superheterodyne receiver design that employs two separate mixers, local oscillators, and intermediate frequencies to avoid image frequency problems

double-sideband suppressed carrier (DSB-SC) the output signal of a balanced modulator; an AM waveform in which both sidebands are present but the carrier has been removed

driven array a multielement antenna in which all the elements are excited through a transmission line

driven element the element of a Yagi antenna that is electrically connected to the transmission line

driver amplifier an amplifier stage that serves to amplify a signal prior to reaching the final amplifier stage in a transmitter

dummy antenna a resistive load used in place of an antenna to test a transmitter under normal loaded conditions without actually radiating the transmitter's output signal

DUT device under test

dynamic range for a receiver, the decibel difference between the largest tolerable receiver input level and its sensitivity

EBCDIC (Extended Binary Coded Decimal Interchange Code) a standardized coding scheme for alphanumeric symbols

electrical noise any undesired voltages or currents that ultimately end up appearing in a receiver's output

electromagnetic interference (EMI) unwanted signals produced by devices that produce excessive electromagnetic radiation, such as oscillators in radios, televisions, and computers

envelope detector another name for a diode detector

equivalent noise resistance used by some manufacturers of microwave devices to represent how noisy a device is by comparing its noise to the resistance value that would produce the same amount of noise

equivalent noise temperature a method of representing how noisy a microwave device actually is

error probability for a digital system, the number of errors per total number of bits received

excess noise noise occurring at frequencies below 1 kHz that varies in amplitude inversely proportional to frequency

exciter the stages necessary in a transmitter to create the modulated signal before subsequent amplification occurs

external noise the noise present in a received radio signal that has been introduced by the transmitting medium

eye pattern using the oscilloscope to display received data bits in a data link; the ideal rectangular edge between the "1" and "0" states appears rounded, forming an eye-shaped display. Undesired phase shifting and jitter can be measured with this technique.

facsimile a system for the transmission of images in which the image is scanned at the transmitter, reconstructed at the receiving station, and duplicated on paper

fading variations in signal strength that may occur at the receiver during the time a signal is being received

ferrite compounds of iron, zinc, manganese, magnesium, cobalt, aluminum, and nickel oxides used in special applications of microwave circuits

ferrite bead a small bead of ferrite material that can be threaded with magnet wire to form a device that offers no impedance to dc and low frequencies, but a high-impedance at RF

fiber-optic detector the device in a fiber optic system that converts the modulated light wave signal back into an electrical information signal

fiber-optic emitter the device in a fiber optic system that converts the information signal into a modulated light wave signal

fiber-optics the use of light wave radiation to send information from a transmitter site to a receiver via fiber optic cable

field frequency the number of times per second that a field of 242.5 horizontal lines are formed on a video image on a television display

filter method the method of creating SSB in a transmitter by first creating DSB-SC and then filtering out the undesired sideband

first detector the mixer stage in a superheterodyne receiver that mixes the received RF signal with a local oscillator signal to form the intermediate frequency signal

flicker another name for excess noise

flyback transformer a transformer used in television receivers to produce the high-voltage needed for the picture tube to produce an image

flywheel effect the repetitive exchange of energy in an LC circuit from the magnetic field of the inductor to the electric field of the capacitor and vice-versa

forward error correcting code (FEC) error checking techniques that allow for subsequent correction at the receiver if an error does occur, rather than allowing the transmitter to retransmit the data

Foster-Seely discriminator an outdated FM discriminator design using two tuned LC networks and two diode detectors to recover the original intelligence in an FM receiver; requires a separate limiter stage but does provide very low distortion

Fourier analysis a method whereby complex repetitious waveforms can be represented by their frequency characteristics

frame frequency the number of times per second that a complete set of 485 horizontal lines are traced in a television receiver

free-running state the undesired unstable operating mode of a PLL (when it is not locked up)

frequency deviation the amount of carrier frequency increase or decrease around its center reference value

frequency division multiplexing simultaneous transmission of two or more signals on one carrier by having each of them on its own separate frequency range

frequency modulation (FM) superimposing the intelligence signal onto a high-frequency carrier so that its frequency departs from its reference value by an amount proportional to the intelligence amplitude

frequency multiplier an amplifier designed so that the output signal's frequency is an integer multiple of the input frequency; accomplished by filtering out a single harmonic of the input signal after it has been greatly distorted to make it rich in harmonics

frequency reuse in cellular telephone, the process of using the same carrier frequency (channel) in different cells that are geographically separated

frequency synthesizer a crystal oscillator capable of generating a large range of output frequencies using only one reference crystal for stability

frequency-division multiple-access (FDMA) satellite communication system in which the satellite contains a wideband receiver/transmitter that includes a number of frequency channels for communciation with many earth stations

frequency-shift keying a form of frequency modulation in which the modulating wave shifts the output between two predetermined frequencies; usually termed the mark and space frequencies

Friiss' formula a method used to determine the total noise produced by amplifier stages that are in cascade

front end the first amplifier stage of a receiver that receives its input signal from the antenna

front porch the interval before the horizontal sync pulse appears on the blanking pulse in a television receiver

front-to-back ratio the difference in antenna gain in dB from the forward direction to the reverse direction

General Purpose Interface Bus (GPIB) another name for the IEEE-488 Interface standard

ghosting a condition that exists if an RF signal takes more than one path to reach the receiver from the transmitter; since one path is usually shorter than the other, one signal reaches the receiver before the other, resulting in a double image or echoing of the information signal

Gray code a numeric code for representing the decimal values from 0 to 9; only one bit changes for each binary increment or decrement

ground wave a radio wave that travels along the earth's surface

Gunn oscillator solid-state bulk-effect source of microwave energy due to the excitation of electrons in the crystal to energy states higher than those they normally occupy

Hamming code a popular forward error checking technique named for R. W. Hamming

handshaking procedures allowing for an orderly interchange of information between a central computer and remote sites

harmonic a sinusoidal wave whose frequency is a multiple of the fundamental frequency

Hartley oscillator a popular LC oscillator, easily recognized by its inductor that is tapped to form positive feedback

Hartley's law the amount of information that can be transmitted in a communication system varies directly proportional to the bandwidth of the channel and the time of transmission

Hertz antenna another name for dipole antenna

heterodyne detector another name for synchronous detector or product detector

high-definition television (HDTV) a new standard being developed that will offer a television picture with the same resolutions as motion picture presentations

high-level modulation in an AM transmitter, when the intelligence is superimposed on the carrier at the last possible point before the antenna

horizontal resolution the number of vertical lines that can be resolved in a television display; in the United States it is set at 428 lines

horizontal retrace time in a television picture, the amount of time it takes to move the electron beam from the right back to the left to start a next line of video

horn antenna microwave antenna consisting of a waveguide that gradually flares out to allow for maximum radiation and minimum reflection back into the guide

iconoscope an early obsolete television camera design

IEEE-488 Interface a standard used in the transmission of parallel data signals from one device to another

image frequency an undesired input frequency in a superheterodyne receiver that will produce the same intermediate frequency as does the desired input signal

IMPATT diode impact ionization avalanche transit time diode; used in the generation of microwave signals

independent sideband (ISB) the transmission of two independent sidebands, each containing different intelligence with the carrier suppressed to a desired level

induction field the radiation that surrounds an antenna but does collapse its field back into the antenna

information theory the branch of learning concerned with optimization of transmitted information

insertion loss the attenuation of a bandpass filter within its specified bandwidth

intelligence the low-frequency information that is modulated onto a high-frequency carrier by a transmitter

intercarrier systems television receivers that process the sound and video signals within the same IF amplifier stages

interlaced scanning the technique of interleaving two fields of 242.5 horizontal lines together to form a video image of 485 horizontal lines in order to use a smaller frame frequency without creating any perceived blinking of the video display

intermodulation distortion the undesired mixing of two signals in a receiver resulting in an output frequency component equal to that of the desired signal

internal noise the noise present in a received radio signal that has been introduced by the receiver circuitry itself

isotropic point source a pinpoint in space that would cause electromagnetic radiation to move equally in all directions in three dimensions; not really possible in practice, but often used as a reference in theory

Johnson noise another name for thermal noise, first thoroughly studied by J. B. Johnson in 1928

klystron an electron tube used at microwave frequencies; it consists of cavity resonators and uses velocity modulation of an electron stream flowing from a heated cathode to a collector anode

laser light amplification by stimulated emission of radiation; low-noise light wave amplifier

limiter a stage found in an FM receiver that removes any amplitude variations of the received signal before it reaches the discriminator stage

local area network (LAN) a network of common users, usually within a mile or two of one another, that share computers and peripheral equipment to complete related tasks

local oscillator (LO) an oscillator used in a superheterodyne receiver to generate a signal to mix with the received RF signal in order to generate a constant intermediate frequency

log periodic antenna a number of dipoles of different lengths and spacing designed to achieve a fairly constant gain and match over a wide range of frequencies

loop antenna an antenna consisting of a single or multi-turn of wire with dimensions much smaller than a wavelength that provides a sharply bidirectional radiation pattern; often used in direction-finding applications

loopback a popular test configuration for a data link in which the receiver takes the received data and sends it back to the transmitter where it is compared with the original data to indicate system performance

low-level modulation   in an AM transmitter, when the intelligence is superimposed on the carrier and then the modulated waveform is amplified before reaching the antenna

lower sideband   a band of frequencies produced in a modulator due to the creation of difference frequencies between the carrier and information signals

magnetron   an electron tube that is surrounded by an electronmagnet that controls the electron flow from cathode to anode; used to generate microwave frequencies in a radio transmitter

man-made noise   external noise produced by man-made devices that is often due to inherant spark producing mechanisms

Manchester code   a popular name for the bi-phase L-code used on Ethernet PCM systems for local area networks

Marconi antenna   another name for vertical antenna

mark   the analog signal representative of a digital high-state; usually a sine wave of a specific frequency

maser   microwave amplification by stimulated emission of radiation; a low-noise microwave amplifier

material dispersion   the spreading of light in fiber optic cable caused by the slight variation of refractive index with wavelength for glass

maximum usable frequency (MUF)   the highest frequency that is returned to the earth from the ionosphere

mechanical filter   a mechanically resonant device that is often used as a sharp band-pass filter to convert DSB to SSB in a SSB transmitter

microcontrollers   microprocessors that are programmed to do a specific task, such as digital signal processing

microstrip   transmission line at microwave frequencies; one or two conductive strips separated from a single ground plane by a dielectric, such as a printed circuit board

microwave   frequencies above 1 GHz having wavelengths between 1 mm and 30 cm; any radio equipment and antennas associated with these frequencies of operation

microwave monolithic integrated circuit (MMIC)   integrated circuits that are used at microwave frequencies for amplification

Miller code   another name for bi-phase codes in PCM systems

modal dispersion   the different paths taken by the various propogation modes in fiber optic cable

modem (modulator/demodulator)   a device that converts digital data to an analog signal that can be transmitted on a telecommunication line and converts the received analog signal to digital data

modulation   the process of putting information onto a high-frequency carrier for transmission

modulation factor   another name for modulation index

modulation index   the measure of the extent to which a carrier voltage is varied by the intelligence; represented as a decimal quantity between 0 and 1 for AM transmitters

multiplexing   the simultaneous transmission of two or more signals onto one carrier

narrowband FM   FM signals that are set up for voice transmissions such as public service communication systems

network   an interconnection of users that allows communication among them

neutralization   a procedure in tuning up a transmitter in which a negative feedback path is introduced in order to counteract the tendancy for a vacuum tube amplifier to self-oscillate due to positive feedback in the tube's interelectrode capacitances

noise figure   a figure of merit describing how noisy a device actually is in decibels

noise floor   on a spectrum analyzer display this is the baseline, usually representing noise of the system under test

noise limiter   a circuit that cuts off all noise pulse peaks that exceed the highest peaks of the desired signal in a receiver

noise ratio   a figure of merit describing how noisy a device actually is; it is not specified in decibels but rather expressed as a ratio quantity having no units

nonlinear device   a device characterized by a non-linear output versus input signal relationship; often used to create new frequencies in radio transmitters and receivers

nonresonant line   a transmission line that is either infinitely long or terminated in a purely resistive load that is equal to its characteristic impedance

NRZ code (nonreturn to zero)   a popular encoding format for PCM systems

Nyquist criteria   the sampling frequency must be at least twice the highest frequency of the input or else distortion that cannot be corrected by the receiver will result

octave a range of frequencies in which the upper limit is double the lower limit

open systems interconnection (OSI) a reference model using standardized procedures developed by the International Organization for Standardization that allows for different types of networks to be interconnected

open-wire cable two conductors spaced a fixed distance apart from each other that connect a transmitter/receiver to an antenna

optimum working frequency (OWF) the frequency that provides for the most consistent communication path via sky waves

oscillator a circuit capable of converting electrical energy from a dc form to an ac form

overmodulation when an excessively large intelligence signal overdrives an AM modulator producing a percent modulation exceeding 100%; produces distortion that cannot be corrected by the receiver

p-i-n diode diodes used as RF and microwave switches that consist of p-type, intrinsic (lightly-doped), and n-type material

packet small segments of data, typically 1000 bits, that may be part of a total message of many packets; the individual packets take unrelated paths within a network to reach the common destination point, where they are reassembled into a complete message

padder capacitor small variable capacitor in series with each ganged tuning capacitor in a superheterodyne receiver to provide for near-perfect tracking at the low end of its tuning range

parabolic antenna microwave antenna consisting of a paraboloid-shaped sheet of metal that reflects received energy to a single point called the focal point, where a small microwave probe picks up the signal and feeds it into the microwave transmission line

parametric amplifier provides low-noise amplification at microwave frequencies via the variation of reactance

parasitic element nondriven element of an antenna

parity the most common method of error detection, in which an extra bit is added to each code representation to give the word either an even number or odd number of 1's

peak envelope power popular method used in rating the output power of a SSB transmitter

percent modulation the measure of extent to which a carrier voltage is varied by the intelligence; represented as a percentage

personal communication network a communication system being developed that will allow for communication from small portable radios on microwave frequencies in a microcellular system similar to cellular telephone

phase method a method of creating SSB in a transmitter without the need for high Q bandpass filters

phase modulation superimposing the intelligence signal onto a high-frequency carrier so that its phase angle departs from its reference value by an amount proportional to the intelligence amplitude

phase noise spurious changes in the phase of a frequency synthesizer's output results at frequencies other than the desired one; a common problem with digital frequency synthesizers

phase-locked loop (PLL) a closed-loop control system that uses negative feedback in order to maintain a constant output frequency

phase-shift keying (PSK) a very efficient, minimal error method of data modulation in which the incoming data causes the phase of the carrier to shift by a predefined amount

phased array a combination of vertical antennas in which there is control of the phase and power of the signal applied at each antenna resulting in a wide variety of possible radiation patterns

piezoelectric effect the property exhibited by crystals that causes a voltage to be generated when they are subject to mechanical stress and, conversely, a mechanical stress to be produced when they are subjected to a voltage

pilot carrier the suppressed carrier in SSB where the carrier is reduced to a lower level but not completely removed

pixel abbreviation for picture element; the smallest resolved area in a given video scanning technique

polarization the direction of the electric field of a given electromagnetic radiated signal

ppm (parts per million) preferred method for rating the frequency stability of crystals

preemphasis a stage found in an FM transmitter that amplifies high-frequency audio signals more than low-frequency audio signals so as to reduce the effect of noise on the received signal

preselector the tuned circuits prior to the mixer in a superheterodyne receiver

private branch exchange (PBX) a telephone exchange on the user's premises that simply switches one telephone line to another so that phone calls can be made

product detector a beat-frequency oscillator, mixer, and low-pass filter stage used in the detection of single-sideband signals in SSB receivers

protocols a complex set of rules designed to force devices sharing a channel to observe orderly communication procedures

pseudorandom a sequence that can be recreated but has the properties of randomness

pulse amplitude modulation (PAM) the process of sampling short pulses of the intelligence signal so that the resulting pulse amplitude is directly proportional to the intelligence signal's amplitude

pulse code modulation (PCM) most common technique for converting an analog signal into a digital word; consists of a sample-hold circuit followed by the actual analog to digital converter circuit

pulse modulation the process of sampling short pulses of the intelligence signal and transmitting them to the receiver with the ability to reconstruct the entire intelligence signal again

pulse-position modulation (PPM) the process of sampling short pulses of the intelligence signal so that the resulting position of the pulses (starting time) is directly proportional to the intelligence signal's amplitude

pulse-width modulation (PWM) the process of sampling short pulses of the intelligence signal so that the resulting pulse-width is directly proportional to the intelligence signal's amplitude

quadrature amplitude modulation (QAM) the most popular method used to achieve high data rates in limited bandwidth channels for data communication links; characterized by two data signals, I and Q, that are 90 degrees out of phase with each other

quadrature detector a popular integrated circuit FM detector that employs two signals that are 90 degrees out of phase with one another in order to recover the original intelligence in an FM receiver

quadrature phase-shift keying (QPSK) a form of phase-shift keying that uses four vectors to represent binary data, resulting in reduced bandwidth requirements for the data transmission channel

quality factor (Q) a measure of the energy stored to the energy lost in an inductor or capacitor

quantization the process of segmenting a sampled signal in a PCM system into different voltage levels with each level corresponding to a different binary number

quarter-wavelength matching transformer a quarter wavelength piece of transmission line of specified line impedance that is used to force a perfect match between a transmission line and its load impedance

quieting the tendency for an FM receiver's audio output signal to reduce in amplitude as the detector responds to an increasing input carrier level

radar (radio detection and ranging) the use of radio waves to detect and locate objects such as aircraft, ships, and land masses

radiation the propagation of energy through space or a material

radiation field the radiation that surrounds an antenna but does not collapse its field back into the antenna

radiation pattern a diagram indicating the intensity of radiation from a transmitter or the response of a receiving antenna as a function of direction

radiation resistance the portion of an antenna's input impedance that is a result of power radiated into space

radio telemetry the process of gathering data on some particular phenomenon without the presence of human monitors and transmitting the data to another site via radio

radio-frequency interference (RFI) undesired radiation from a radio transmitter

raster the illuminated area on the picture of a television receiver when no signal is being received

ratio detector an outdated FM discriminator similar to the Foster-Seely design that uses two tuned LC networks and two diode detectors to recover the original intelligence in an FM receiver; does not require a separate limiter stage but does produce some distortion in its output signal

Rayleigh fading rapid variation in signal strength received by mobile units in urban environments

Rayleigh scattering the scattering of light waves in a fiber that decreases rapidly with increasing wavelength

reactance modulator an amplifier designed so that its input impedance has a reactance that varies as a function of the amplitude of the applied input voltage

reciprocity a property of antennas; refers to an antenna's ability to transfer energy from the atmosphere to its receiver with the same efficiency at which it transfers energy from the transmitter into the atmosphere

reflection for transmission lines, the abrupt reversal in direction of the movement of the electrical voltage and current at its termination due to the impedance mismatch between the line impedance and the load impedance at the termination point; for radiation, the bouncing of radiated energy by any conductive medium, such as metal surfaces or the earth's surface

reflection coefficient the ratio of the reflected voltage to the incident voltage at a termination point in a transmission line

reflector in a Yagi antenna, the parasitic element that effectively reflects energy from the driven element

refraction when electromagnetic waves pass from one density to another, the direction of propagation is altered; i.e., the wave bends

refractive index the ratio of the speed of light in free space to its speed in a given material

relative harmonic distortion an expression specifying the fundamental frequency component of a signal with respect to its largest harmonic in dB

repeater a radio installation consisting of a receiver to pick up a message from one site and a transmitter to transmit the same message on a different frequency to another site

resistor noise another name for thermal noise, due to the fact that it is produced in resistors, especially carbon resistors

resolution the ability to resolve detailed picture elements in a television signal or other imaging systems

resonant transmission line a transmission line terminated with a load that is not matched to its characteristic impedance, causing subsequent reflections

ripple amplitude the variation in attenuation of a sharp bandpass filter within its 6 dB bandwidths

RS-232 a standard of voltage levels, timing, and connector pin assignments for serial data transmission

RZ code (return to zero) an encoding format for PCM systems

S meter signal strength meter that responds to the received signal level; often used as a tuning aid in a receiver

selectivity the extent to which a receiver is capable of differentiating between the desired signal and other undesired signals

self-excited mixer a single stage in a superheterodyne receiver that creates the LO signal and mixes it with the applied RF signal to form the IF signal

sensitivity the minimum input RF signal to a receiver that is required to produce a specified audio signal at its output

shape factor the ratio of the 60 dB and 6 dB bandwidths of a high Q bandpass filter; usually indicative of the steepness of the upper and lower roll-off skirts

shorted-stub matching section a shorted transmission line of specified length that can be used to force a perfect match between a transmission line and its load impedance

shot noise noise introduced by carriers in p-n junctions in all forms of semiconductors

sideband splatter distortion that results in an overmodulated AM transmission creating excessive bandwidths

signal spectrum a method of representing a signal by plotting its amplitude versus frequency characteristics

signal-to-noise ratio a relative measure of desired signal power to noise power at a specific measurement point in a communication system

single sideband (SSB) a form of amplitude modulation whereby the carrier and one sideband are filtered out, leaving the other sideband as the only remaining RF signal

skin effect the tendency for high-frequency electric current to flow mostly near the surface of the conductive material; this is due to self-induction occurring near the center of the conductor that prevents current flow

skip zone the zone between the point where the ground wave is completely dissipated and the point where the first sky wave returns; in this region no sky wave returns to the earth

skipping the alternate refracting and reflecting of a sky wave signal between the ionosphere and the earth's surface

sky wave those radio waves radiated from a transmitting antenna in a direction that produces a large angle with respect to the earth

slope detector a simple FM discriminator that detects FM by first converting FM to AM and then detecting the intelligence by a simple diode detector; usually creates too much distortion to be an acceptable design

slope modulation another name for delta modulation

slope overload a common problem with delta modulators when the analog signal has a high rate of amplitude change; the encoder can fall behind and produce a distorted representation of the analog signal

slot antenna array UHF or microwave antenna, often used on aircraft, that couples RF energy into a slot in a large metallic plane

slotted line a section of waveguide in which a lengthwise slot is cut into its outer conductor and an adjustable probe placed into the slot; often used in determining VSWR, frequency, and load impedances in microwave systems

Smith chart a graph developed by P. H. Smith in 1938 for use as an aid in transmission line, antenna, and waveguide calculations

space the analog signal representative of a digital low state; usually a sine wave of a specific frequency

space noise external noise produced outside of the earth's atmosphere

space wave a radio wave that travels in straight lines between transmitter and receiver, not necessarily close to the ground; they are typically line-of-sight transmissions. If any obstruction exists, the signal is blocked from reaching the receiver.

spectrum analyzer an instrument used to measure the harmonic content of a signal by displaying a plot of amplitude versus frequency

spread spectrum communication systems in which the carrier is periodically shifted about at different nearby frequencies in a randomlike manner determined by a hidden code; the receiver must decode the sequence so that it can follow the transmitter's frequency hops to the various values within the specified bandwidth

spurious frequencies extra frequency components that appear in the spectral display of a signal, signifying distortion in the signal

square-law device a device that exhibits an output versus input signal response resembling a parabola; often used as mixers and detectors in receivers due to its minimum distortion characteristics

squelch a circuit in a receiver that reduces the background noise in the absence of a desired signal; often found in FM receivers

stagger tuning the technique of cascading a number of tuned bandpass filters, each having a slightly offset bandpass frequency, to form a wider flat bandpass with steep high-frequency and low-frequency roll-off skirts

standing wave the fixed voltage and current wave pattern that results in a resonant transmission line due to the vector addition of the incident and reflected traveling waves

static erratic spontaneous electrical noise that may occur in the output signal of a receiver

stereo a radio transmission of two separate signals, left and right, used to create a three-dimensional effect for the receiver's receiving audience

stripline transmission line at microwave frequencies; it consists of two ground planes that sandwich a smaller conducting strip with constant separation by a dielectric material such as a printed circuit board

subsidiary communication authorization (SCA) an additional channel of multiplexed information that is authorized by the FCC for stereo FM radio stations to feed services such as commercial-free programming to select customers

superheterodyne receiver receiver design superior to the simple TRF design due to its extra mixer and local oscillator stages and its ganged tuning characteristics; provides for easier tuning and near-constant selectivity at all frequencies within its tuning range

surface acoustic wave (SAW) device an extremely high Q filter often used in TV and radar applications

surface wave another name for ground wave

surge impedance another name for characteristic impedance

sync separator the circuit in a television receiver that filters the horizontal sync pulses and the vertical sync pulses from the complex video signals

synchronization in television, the precise matching of the movement of the electron beam horizontally and vertically in the recording camera with the electron beam in the television receiver

synchronous a system in which the transmitter and receiver clocks run at exactly the same frequency because the receiver derives its clock signal from the received data stream

synchronous detector a complex method of detecting an AM signal that gives low distortion, fast response, and amplification

tangential method a method of measuring the amplitude of noise on a signal using its oscilloscope display; produces errors of less than 1 dB

TCXO (temperature compensated crystal oscillator) a crystal oscillator that contains circuitry to keep the output frequency constant with respect to changes in temperature

telemetry remote metering; the process of gathering data on some particular phenomenon without the presence of human monitors

television popular communication system that consists of simultaneous transmission of a video picture signal and an audio sound signal

thermal noise internal noise that is due to thermal interaction between free electrons and vibrating ions in a conductor

third order intercept point a receiver figure of merit describing how well it rejects intermodulation distortion due to third order products resulting at the mixer output

time domain reflectometry technique of sending short pulses of electrical energy down a transmission line to determine its characteristics by observing on an oscilloscope for resulting reflections

time-division multiple-access (TDMA) a single satellite to service multiple earth stations simultaneously on the same frequency

time-division multiplexing (TDM) the division of a transmission facility into two or more channels by allotting the common channel into several different information channels, one at a time

tracking range once a PLL is locked up, this is the range of input frequencies that can be applied without having it lose its lock and return to the free-running state

transducer a device that converts energy from one form to another

transit-time noise noise produced in semiconductors due to the fact that when the transit time of the carriers crossing a junction is close to the signal's period, some of the carriers may diffuse back to the source or emitter of the semiconductor

transmission line the conductive connections between communication system elements that carry signal power

transponder an electronic system aboard a satellite that performs the reception, frequency translation, and retransmission of the received radio signals

trapezoidal pattern a measurement technique for checking the purity of an AM modulator by use of the oscilloscope in the X:Y mode

traveling wave tube (TWT) a high gain, low-noise, wide bandwidth microwave amplifier

traveling waves the electrical voltage and current waves that move down a transmission line

trimmer capacitor small variable capacitor in parallel with each ganged tuning capacitor in a superheterodyne receiver that provides for near-perfect tracking at the high end of its tuning range

tropospheric scatter a phenomenon whereby small amounts of radiation are scattered by the troposphere and picked up by receivers at ground level when the transmitted signal is set at a very high power and selected microwave frequencies are used

tuned-radio frequency receiver (TRF) the most elementary receiver design, consisting of RF amplifier stages, a detector, and audio amplifier stages

unbalanced transmission line two conductors, such as coax, that do not carry signals having the same amplitude and opposite phase relationship with respect to ground

Universal Asynchronous Receiver Transmitter (UART) a device that converts parallel computer data into required serial data format

up-conversion   the process of mixing the received RF signal with an LO signal in order to produce an IF signal that is higher in frequency than the original received RF signal

upper sideband   a band of frequencies produced in a modulator due to the creation of sum-frequencies between the carrier and information signals

varactor diode   diode that has a small internal capacitance that varies as a function of its reverse bias voltage

varicap diode   another name for varactor diode

velocity factor   the ratio of the actual velocity of propagation of a signal to the velocity in free space; usually ranging between 0.55 and 0.97

velocity of propagation   the speed at which an electrical signal moves through a conductor

vertical antenna   an antenna that consists of a vertical tower, wire, or rod, usually a quarter wavelength in length, that is fed at the ground and uses the ground as a reflecting surface

vertical resolution   the number of horizontal lines that actually make up a television display; in the United States it is set at 339 horizontal lines

vertical retrace time   in a television picture, the amount of time it takes to move the electron beam from the bottom right corner to the top left corner to start another field of 242 horizontal lines

vestigial sideband   a form of amplitude modulation whereby one of the sidebands is partially filtered out; used in television to reduce the otherwise 8 MHz bandwidth video signal down to a more efficient 6 MHz

video signal   the picture portion of a television signal that contains frequencies as high as 4 MHz and is amplitude modulated onto a carrier

vidicon   a popular television camera design with a much smaller package and lower cost than earlier designs

voltage controlled oscillator (VCO)   an oscillator designed so that its output voltage varies as a function of the amplitude of the applied input voltage

voltage standing wave ratio (VSWR)   the ratio of the maximum voltage to minimum voltage of the standing wave pattern that results on a resonant transmission line

wave propagation   the movement of radio signals through the atmosphere from transmitter to receiver

waveguide   a microwave transmission line consisting of a hollow metal tube or pipe used to conduct electromagnetic waves through its interior

waveguide dispersion   the spreading of light in fiber optic cable caused by a portion of the light energy traveling in the cladding

wavelength   the distance traveled by a wave during one cycle of its periodic variation

wavetrap   a high Q bandstop circuit that attenuates a narrow band of frequencies

white noise   another name for thermal noise, so named because its frequency content is uniform and white light contains light waves of all frequencies in equal concentration

wide area network (WAN)   two or more LANs linked together over a wide geographical area

wideband FM   FM transmitter/receiver systems that are set up for high-fidelity information, such as music, high-speed data, stereo, etc.

Yagi antenna   a popular type of directional antenna consisting of a half-wave dipole as the driven element, one reflector, and several directors

INDEX